D1002271

Myriapoda

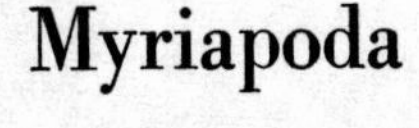

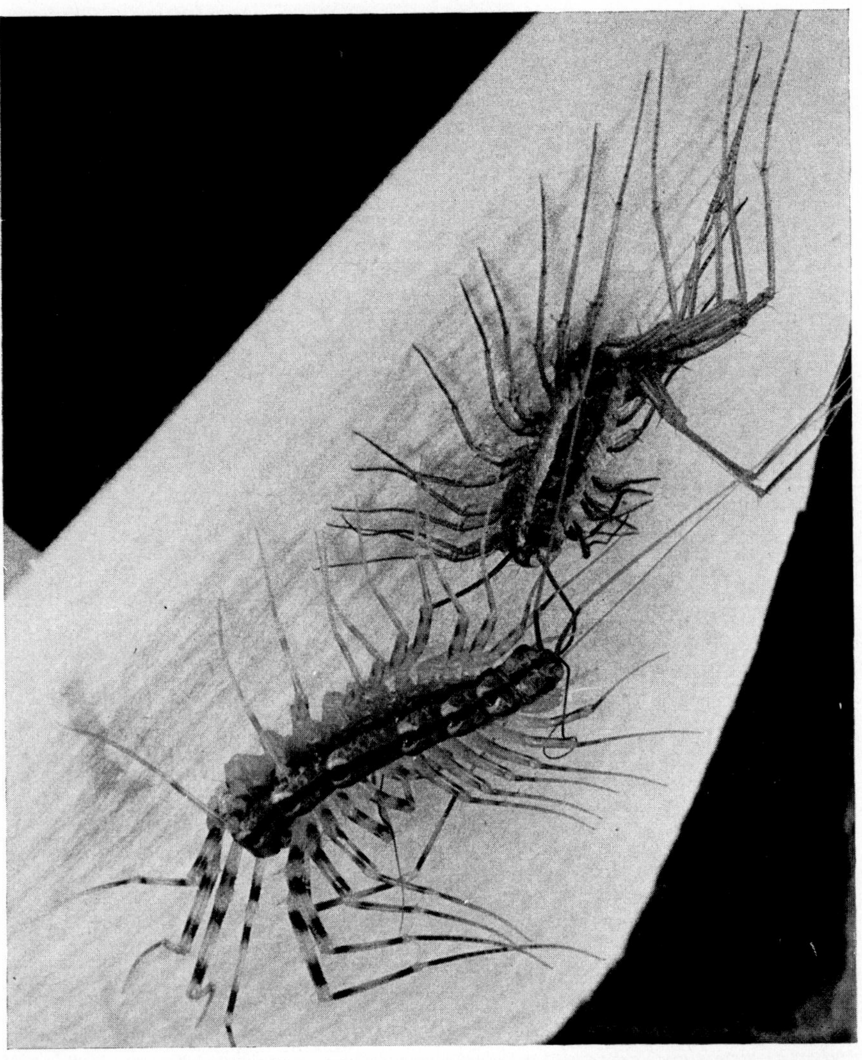

Frontispiece: A female centipede, *Scutigera coleoptrata,* and her cast skin, 53 min after the first appearance of the rupture of the old cuticle at the moulting suture. She is cleaning the long slender hind legs of the right side having already cleaned all the other legs of this side one after the other. (Photograph Dr Wolfgang Dohle).

SYMPOSIA OF THE ZOOLOGICAL SOCIETY OF LONDON
NUMBER 32

Myriapoda

Second International Congress of Myriapodology
(The proceedings of a Symposium organised jointly by the Zoological Society of London and the Centre International de Myriapodologie, Paris, held at the University of Manchester, England, on 5th–12th April, 1972)

Edited by

J. GORDON BLOWER

Department of Zoology
University of Manchester
Manchester, England

Published for

THE ZOOLOGICAL SOCIETY OF LONDON

by

ACADEMIC PRESS

1974

ACADEMIC PRESS INC. (LONDON) LTD
24–28 Oval Road,
London, N.W.1

U.S. Edition published by
ACADEMIC PRESS INC.
111 Fifth Avenue,
New York, New York 10003

Library of Congress Catalog Card Number: 73–7035

ISBN: 0–12–613332–8

PRINTED IN GREAT BRITAIN BY
CLARENDON PRINTERS LTD., BEACONSFIELD

CONTRIBUTORS AND PARTICIPANTS

ANDERSSON, G., *Naturhistoriska Museet, Box 11049, S–40030 Göteborg, Sweden*

*BÄHR, R., *Zoologisches Institut, Westfälischen Wilhelms-Universität, 44 Münster (Westf.), Hindenburgplatz 55, Germany* (p. 383)

*BAKER, A., *Broom's Barn Experimental Station, Higham, Bury St. Edmunds, Suffolk, England* (p. 621)

*BARBER, A. D., *Plymouth College of Further Education, Plymouth PL4 8AA, England* (p. 611)

*BIERNAUX, J., *Chaire de Zoologie et Entomologie Appliquées, Faculté des Sciences Agronomiques de l'Etat, 44 rue de Moulin, 5800 Gembloux, Belgium* (p. 629)

*BLOWER, J. G., *Department of Zoology, University of Manchester, Manchester M13 9PL, England* (pp. 503, 527, 603)

*BOCOCK, K. L., *Soil Ecology Section, Merlewood Research Station, Institute of Terrestrial Ecology, Grange-over-Sands, Lancashire, LA11 6JU, England* (p. 433)

*BRADE-BIRKS, S. G., *The Vicarage, Godmersham by Canterbury, Kent, England* (p. 1)

*BROOKES, C. H., *Department of Chemistry & Biology, John Dalton Faculty of Technology, Manchester Polytechnic, Chester Street, Manchester M1 5GD, England* (p. 485)

*†BÜCHERL, W., *A/c, Instituto Butantan, Caixa Postal 65, São Paulo, SP, Brasil* (p. 99)

*†CAMATINI, M., *Università di Milano, Laboratorio di Zoologia, via Celoria 10, 20133-Milano, Italy* (p. 231)

*CAUSEY, N. B., *College of Arts & Sciences, Zoology & Physiology, Louisiana State University, Baton Rouge, Louisiana 70803, U.S.A.* (p. 23)

*†CHOWDAIAH, B. N., *Department of Zoology, Central College, Bangalore University, Bangalore-1, India* (p. 261)

*†COTELLI, F., *Università di Milano, Laboratorio di Zoologia, via Celoria 10, 20133-Milano, Italy* (p. 231)

*COTTON, M. J., *School of Pharmacy & Biology, Sunderland Polytechnic, Chester Road, Sunderland, Co. Durham, England* (p. 589)

CRABILL, R. E., *Supervisor & Curator, Division of Myriapoda and Arachnida, Smithsonian Institution, Washington, D.C. 20560, U.S.A.*

*CURRY, A., *Department of Zoology, University of Manchester, Manchester M13 9PL, England.* (p. 365)
Present: Public Health Laboratory, Withington Hospital, Manchester M20 8LR, England

*DEMANGE, J-M., *Muséum National d'Histoire Naturelle, Laboratoire de Zoologie (Arthropodes), 61 rue de Buffon, Paris Ve, France* (p. 273)

DESHMUKH, I. K., *Tay Estuary Research Centre, Newport-on-Tay, Fife DD6 8EX, Scotland*
Present: Brathay Hall Field Centre, Ambleside, Westmorland, England

*DOHLE, W., *Freie Universität Berlin, I. Zoologisches Institut, 1 Berlin 33, Königin Luise Strasse 1–3, Germany* (pp. 143, 191)

*EDWARDS, C. A., *Rothamsted Experimental Station, Harpenden, Hertfordshire, England* (pp. 135, 645)

*EASON, E. H., *Bourton Far Hill Farm, Moreton-in-Marsh, Gloucestershire, England* (p. 65)

ENGHOFF, H., *Universitetets Zoologiske Museum, Universitetetsparken 15, DK-2100 København, Denmark*

*FAIRHURST, C. P., *Department of Science, Stockport College of Technology, Stockport SK1 3UK, England* (pp. 575, 611)
Present: Department of Biology, The University, Salford, England.

GABBUTT, P. D., *Department of Zoology, University of Manchester, Manchester M13 9PL, England*

GOUGH, H. J., *I.C.I. Plant Protection Ltd., Jealott's Hill Research Station, Bracknell, Berkshire, England*

*HAACKER, U., *Zoologisches Institut und Zoologisches Museum, 2 Hamburg 13, Von-Melle-Park 10, Germany* (p. 317)

*HEATH, J., *Biological Records Centre, Institute of Terrestrial Ecology, Monks Wood Experimental Station, Abbots Ripton, Huntingdon PE17 2LS, England* (p. 433)

*HERBAUT, C., *Université de Lille, Laboratoire de Biologie Animale, B.P. 36, 59 Villeneuve d'Ascq, France* (p. 237)

HUBERT, M., *Faculté des Sciences, Laboratoire de Zoologie, 35-Rennes-Beaulieu, France*

*HÜTHER, W., *Sammlungen der Abteilung für Biologie, Ruhr-Universität Bochun, 463-Bochum-Querenburg, Germany* (p. 411)

*JEEKEL, C. A. W. *Zoologisch Museum, Universiteit van Amsterdam, Plantage Middenlaan 53, Amsterdam-C* (p. 41)

*†JUBERTHIE, C., *Laboratoire Souterrain, Centre National de la Recherche Scientifique, 09-Moulis (Ariège), France* (p. 199)

*JUBERTHIE-JUPEAU, L., *Laboratoire Souterrain, Centre National de la Recherche Scientifique, 09-Moulis (Ariège), France* (pp. 199, 289)

*†KANAKA, R., *Department of Zoology, Central College, Bangalore University, Bangalore-1, India* (p. 261)
Present: Department of Physiology, St. John's Medical College, Bangalore 560034, India

†KHEIRALLAH, A. M., *Department of Zoology, Faculty of Science, Mohrarrem Bey, Alexandria University, Egypt*

KIME, R. D., *The Royal Grammar School, Guildford, Surrey, England*

*KRAUS, O., *Zoologisches Institut und Zoologisches Museum, Universität Hamburg, 2 Hamburg 13, Papendamm 3, Germany* (p. 13)

*LEWIS, J. G. E., *Dover College, Dover, Kent, England* (p. 423)

MALCOLM, S., *Department of Zoology, The University, Manchester M13 9PL, England*

*MANTON, S. M., *Zoology Department, British Museum (Natural History), Cromwell Road, London, S.W.7* (p. 163)

*MAURIÈS, J-P., *Laboratoire de Zoologie (Arthropodes), Muséum National d'Histoire Naturelle, 61 rue de Buffon, Paris Ve, France* (p. 53)

MEIDELL, B. A., *Zoologisk Museum, University of Bergen, 5000 Bergen Norway*

*MILLER, P. F., *Department of Zoology, University of Manchester, Manchester M13 9PL, England* (pp. 503, 553, 589).
Present: Biological & Chemical Research Institute, Rydalmere, New South Wales, Australia

*†MOUNTFORD, J., *Biometrics Section, Institute of Terrestrial Ecology, 19 Belgrave Square, London SW1X 8PY, England* (p. 433)

*NGUYEN DUY-JACQUEMIN, M., *Muséum National d'Histoire Naturelle Laboratoire de Zoologie (Arthropodes), 61 rue de Buffon, Paris Ve, France* (p. 211)

*PEITSALMI, M., *Kotirannankatu 7, 13200 Hämeenlinna 20, Finland* (p. 471)

*PETIT, G., *Laboratoire de Biologie Animale, Faculté des Sciences, Université de Picardie, 33 rue St. Leu, 80-Amiens, France* (p. 301)

*PETIT, J., *Laboratoire de Biologie Animale, Faculté des Sciences, Université de Picardie, 33 rue St. Leu, 80-Amiens, France* (p. 249)

*PIERRARD, G., *I.R.C.T., B.P. 28, Koutiala, Mali* (p. 629)

*RAJULU, G. S., *Department of Zoology, University of Madras, Triplicane P.O., Madras-5, India* (p. 347)
Present: Department of Zoology, University Extension Centre, P.S.G. College of Technology Campus, Coimbatore-641004, India

*RANTALA, M., *Järvensivuntie 93, 33100 Tampere 10, Finland* (p. 463)

ROUND, J., *The Grammar School, Urmston, Lancashire, England*

*SAHLI, F., *Université de Dijon, Faculté des Sciences de la Vie et de l'Environnement, Laboratoire de Biologie Animale, 21 Dijon, Boulevard Gabriel, France* (p. 217)

*SAITA, A., *Laboratorio di Zoologia, Università di Milano, via Celoria 10, 20133 Milano, Italy* (p. 231)

*SAKWA, W. N., *Department of Zoology, The University, Manchester, England* (p. 329)
Present: Walsall and Staffordshire Technical College, Walsall, Staffs., England

*SCHELLER, U., *Lundsberg, S-68080 Storfors, Sweden* (p. 405)

STANDEN, V., *Department of Zoology, The University, Durham, England*

STRASSER, K., *I-34128-Trieste, via S. Pelagio 16, Italy.*

*TOBIAS, D., *Natur-Museum und Forschungs-Institut Senckenberg, 6 Frankfurt-am-Main, Senckenberganlage 25, W. Germany* (p. 75)

*TURNER, R., *Rothamsted Experimental Station, Harpenden, Hertfordshire, England* (p. 135)

†VEVERS, H. G., *The Zoological Society of London, Regent's Park, London NW1 4RY, England*

WALLWORK, J. A., *Department of Zoology, Westfield College, Kidderpore Avenue, Hampstead, London NW3 7ST, England*

WILLIAMS, R. J., *Design Research Laboratory, Department of Building, University of Manchester Institute of Science & Technology, Manchester, England*

*†WÜRMLI, M. F., *Naturhistorisches Museum Basel, CH-4051 Basel, Augustinergasse 2, Switzerland* (p. 89)

* Contributor.
† Unable to attend.

PRESIDENT, ORGANIZER AND CHAIRMEN

PRESIDENT

THE REV. CANON S. GRAHAM BRADE-BIRKS, *The Vicarage, Godmersham by Canterbury, Kent, England*

ORGANIZER

J. GORDON BLOWER *Department of Zoology, The University, Manchester, England*

CHAIRMEN OF SESSIONS

J. BIERNAUX, *Chaire de Zoologie et Entomologie Appliquées, Faculté des Sciences Agronomiques de l'Etat, 44 rue de Moulin, 5800 Gembloux, Belgium*

N. B. CAUSEY, *College of Arts & Sciences, Zoology & Physiology, Louisiana State University, Baton Rouge, Louisiana 70803, U.S.A.*

J.-M. DEMANGE, *Muséum National d'Histoire Naturelle, Laboratoire de Zoologie (Arthropodes), 61 rue de Buffon, Paris Ve, France.*

W. DOHLE, *Freie Universität Berlin, I. Zoologisches Institut, 1 Berlin 33, Königin Luise Strasse 1-3, Germany*

U. HAACKER, *Zoologisches Institut und Zoologisches Museum, 2 Hamburg 13, Von-Melle-Park 10, Germany*

C. A. W. JEEKEL, *Zoologisch Museum, Universiteit van Amsterdam, Plantage Middenlaan 53, Amsterdam C*

L. JUBERTHIE-JUPEAU, *Laboratoire Souterrain, Centre National de la Recherche Scientifique, 09-Moulis (Ariège), France*

O. KRAUS, *Zoologisches Institut und Zoologisches Museum, Universität Hamburg, 2 Hamburg 13, Papendamm 3, Germany*

J.-P. MAURIÈS, *Musèum National d'Histoire Naturelle, Laboratoire de Zoologie (Arthropodes), 61 rue de Buffon, Paris Ve, France*

G. SUNDARA RAJULU, *Department of Zoology, University of Madras, Triplicane P.O., Madras-5, India*

F. SAHLI, *Université de Dijon, Faculté des Sciences de la Vie et de l'Environment, Laboratoire de Biologie Animale, 21 Dijon, Boulevard Gabriel, France*

C. STRASSER, *I-34128-Trieste, via S. Pelagio 16, Italy*

In memoriam
Ulrich Haacker

1939–1972

Dr Ulrich Haacker was a member of the Second International Congress of Myriapodology; he died only five months after that conference. His colleagues and friends still have a vivid memory of his contribution to the congress and find it difficult to believe that this was his last presentation. Dr Haacker was highly esteemed not only as a scientist but also for his modesty and for his readiness to help whenever he was able to do so. These qualities and his linguistic abilities enabled him to contribute considerably to the harmonious atmosphere and the success of the first two international congresses of myriapodology.

Dr Haacker studied at the universities of Mayence and Bonn and then at Darmstadt, where he received his doctorate in 1967. A year later he went to the University of Hamburg. As a zoologist whose sound knowledge was coupled with an unusually broad scope, he was a stimulating teacher; but at the same time he pursued his research with enthusiasm. His promising plans for the future were curtailed by a malignant disease.

O. Kraus

(A list of Dr. Haacker's publications follows the notice by Professor Magnus published in: *Mitt. hamburg. Zool. Mus. Inst.* **69**: 1–3, 1972).

FOREWORD

The First International Congress of Myriapodology was held in Paris in the spring of 1968 and the *Proceedings* of that congress were published as a special supplement of the *Bulletin du Muséum national d'Histoire naturelle* ((2) *41:* Suppl. no. 2: pp. 1-148, 1970). On this occasion we shared the excellent hospitality of the Natural History Museum in Paris with the Arachnologists who were holding their fourth International Congress and they had a wealth of experience to give to us. One of the fruitful results of this meeting was the setting up of a permanent international secretariat for the documentation of research on the myriapod groups and the provision of a channel of communication between workers throughout the world. Four years later, the myriapodologists, armed with the experience they had gained in Paris, the new cohesion given to them by the Centre International de Myriapodologie and the generous sponsorship of the Zoological Society of London, were able to stage their Second International Congress in Manchester.

It was a great honour for Britain to act as host country. Whilst the foundations of the systematic study of Myriapoda were laid on the continent by Latzel, Attems, Verhoeff, Silvestri, Brandt and Brölemann and by Chamberlin and Cook in the United States, British zoologists had contributed in fair measure. In the last century we had Leach and Newport, and leading us into the new century Pocock, who raised the Chilopoda and Diplopoda to the status of separate classes. The fourth name in this distinguished lineage must surely be that of Brade-Birks, born, as he tells us, three years after the publication in Vienna of Latzel's monumental monograph "which we regard as the beginning of modern work on myriapods". Indeed, it was a great joy to all of us to have the doyen of myriapodologists amongst us and that he should be blessed with good health and vigour in his eighty-fourth year and be able to preside over our deliberations with characteristic enthusiasm.

Manchester had double cause to be proud: first to be chosen as host city for this international meeting and secondly to be the venue for this, the first Symposium of the Zoological Society to be held outside the capital. But the present phase of myriapodology in Britain began here in the north. Brade-Birks himself was a graduate of the University, and he recounts in his Presidential Address how two Lancashire and Cheshire Naturalists on the staff of the Manchester Museum helped to launch him and his wife on their study of the British fauna. Unfortunately Dr Hilda was not able to join us in Manchester but sent her good wishes to the Congress.

The papers gathered here as the *Proceedings* of the Congress range from systematic zoology through physiology, biochemistry and ultrastructure, to ecology and economic status. The volume concludes with details of the Centre International de Myriapodologie (C.I.M.), an appraisal of the first four years and details of future plans. It is something of an event these days to find papers on taxonomy and morphology between the same covers as papers on physiology and ecology. The only things held in common by the contributors were the living animals which formed the subjects of their researches. It is the small size of the groups which makes such an omnibus volume possible. The four classes of myriapods including over 11,000 species together form a smaller assemblage than any other single class of the Arthropoda, that is if we regard the Insecta as a single class; there are twice as many Crustacea and three times more arachnids. The Myriapod groups get an even poorer coverage in the average text book than their variety would seem to merit; they are accorded one sixth of the space afforded to each of the Arachnida and the Crustacea in the latest text. Then again, much of their limited popularity is merely reflected glory, deriving from the conviction of many zoologists that the ancestor of the insects lies somewhere in their midst. Yet the Myriapoda are among the commonest animals on the surface of the earth. In the tropics, millipedes are *the* most conspicuous elements of the ground fauna and, in the temperate zone, they are often the dominant members of the arthropod population of the soil in terms of biomass. Extensive interest in animals on the ground and in the soil dates back no further than the last war. The time is now coming for the Myriapoda to get their just share of attention and it is hoped that this volume might hasten that day.

To some extent this collection of contributions indicates the present state of our knowledge. In some cases the work is still in the descriptive stage, whether it be the simple description of species and their known distributions or the more sophisticated description of fine structure made possible by the advent of the electron microscope. In other examples, the work has clearly left the stage of pure description and has entered the stage of experimentation and quantitative investigation. But perhaps most important are those papers which are examining the very bases of description and classification to find the foundations cracking, or even wholly unsuited to support the superstructure of zoogeographical and evolutionary studies already built upon them.

Myriapodologists could not be accused of working in separate compartments in ignorance of their fellow workers, but their subjects are the very archetypes of compartmentalisation. The number of problems they pose for those interested in the principles of meristic repetition are matched by the many opportunities they provide for their solution. The intriguing variety of methods of developing their segmental pattern, by anamorphosis or epimorphosis or a mixture of the two, besides presenting a challenge to the morphogeneticist also provides relatively simple means for unravelling their life-histories and establishing a firm basis for ecological studies.

If the systematists appear to be so far ahead as to be able to retrace their steps and tidy up the details, the physiologists are advancing more cautiously in a tricky field with difficult experimental material. The details and problems posed by embryonic and post-embryonic growth (both abnormal and normal) are at least clearly stated and defined in several papers presented here and the interesting contributions on the neuro-humoral systems suggest that the physiologists are poised to probe into the very control centres of these processes.

The evolutionary history of the groups elicited much discussion; the more formal parts of these exchanges are included here. The family tree of the Myriapoda appears to have suffered severe coppicing and to have been relegated to the status of a bush with the main branches arising much closer to the roots and beyond our present comprehension. The study of the much later evolutionary processes of niche-diversification represented by the terminal branches is the most recent area of endeavour. As a by-product of this research, the field workers are now able to contrast and specify the harmful effects of millipede population outbreaks, with the possibly beneficial effects of those more 'normal' populations busily degrading the dead vegetation and contributing to soil characteristics. Although zoologists will be the last people to accept the vocationalisation of our seats of learning, it appears that information germane to the real world of today is emerging as the usual and inevitable result of curiosity and disinterested intellectual activity.

The success of the Congress in Manchester was assured by the combined efforts of bodies pledged to sustain this type of activity: the Zoological Society, the permanent secretariat of the C.I.M. and the members of the British Myriapod Group and their ladies. It was predisposed to succeed with such distinguished membership drawn from all parts of the world and with the helpful hospitality of the University, in particular that of my own Department and of Hulme Hall. Ralph Dennell, Beyer Professor of Zoology in the University, was initially responsible for the Manchester commitment to myriapods and at no time other than that leading up to the Congress and beyond it have I valued his support and encouragement more. Many are those who have specifically helped in the publication of this volume. The President, the chairmen of sessions and all the contributors have all proved most helpful in the preparation of material for publication. Dr H. G. Vevers and his staff in the Publications Department of the Zoological Society of London, in particular Miss Unity McDonnell and Mr Leonard Ellis, have contributed greatly towards the accuracy and clarity of the details of each individual paper. My colleagues in the Zoology Department have been most helpful and understanding, in allowing me to monopolize secretarial help; Miss Frances Walker and Mrs Joan Gatehouse, for their part have freely given their secretarial services. Mr Peter Miller gave me considerable help in assembling the verbatim texts of the discussions, in addition to engineering the actual recordings. He did much of the transcribing from the tapes, helped ably through linguistic difficulties by Mr Nick Sakwa. Dr Charles Brookes, Dr Colin Fairhurst and my wife, undertook the preparation of the indexes. To all these ladies and gentlemen I offer my most sincere thanks.

Shortly after the Congress we were saddened by the untimely death of Dr Ulrich Haacker of Hamburg. His elegant and interesting contribution to the symposium and his many incisive comments and questions on other papers will stand as a lasting tribute to him. It is as a gifted scientist and a delightful person that we shall remember him.

J. Gordon Blower

April 1974

CONTENTS

The Group Taxonomy and Geography of the Sphaerotheriida (Diplopoda)

C. A. W. JEEKEL

Intérêt Phylogénique et Biogéographique de quelques Diplopodes récemment décrits du Nord de l'Espagne

J.-PAUL MAURIÈS

On Certain Aspects of the Generic Classification of the Lithobiidae, with Special Reference to Geographical Distribution

E. H. EASON

New Criteria for the Differentiation of Species within the Lithobiidae

D. TOBIAS

Systematic Criteria in the Scutigeromorpha

MARCUS WÜRMLI

Die Scolopendromorpha der Neotropischen Region

WOLFGANG BÜCHERL

Scanning Electron Microscope Studies of Symphyla

R. H. TURNER and C. A. EDWARDS

Étude Ultrastructurale de l'Organe Neurohémal Cérébral de *Spelaeoglomeris doderoi* Silvestri Myriapode Diplopode Cavernicole

C. JUBERTHIE et L. JUBERTHIE-JUPEAU

Les Organes Intracérébraux de *Polyxenus lagurus* et Comparaison avec les Organes Neuraux d'autres Diplopodes

NGUYEN DUY-JACQUEMIN

Sur les Organes Neurohémaux et Endocrines des Myriapodes Diplopodes

FRANÇOIS SAHLI

Spermiogenesis of *Lithobius forficauts* (L.) at Ultrastructural Level

M. CAMATINI, A. SAITA and F. COTELLI

Étude Cytochimique et Origine des Enveloppes Ovocytaires chez *Lithobius forficatus* (L.) (Myriapode, Chilopode)

C. HERBAUT

Contribution à l'Étude de l'Appareil Génital Mâle et de la Spermatogenèse chez *Polydesmus angustus* Latzel, Myriapode Diplopode

JEANNINE PETIT

Studies on the Male Reproductive Pattern in some Indian Diplopoda (Myriapoda)

RANI KANAKA and B. N. CHOWDAIAH

Réflexions sur le Développement de quelques Diplopodes

J.-M. DEMANGE

Action de la Température sur le Développement Embryonnaire de *Glomeris marginata* (Villers)

L. JUBERTHIE-JUPEAU

Sur les Modalités de la Croissance et la Régénération des Antennes de Larves de *Polydesmus angustus* Latzel

GÉRARD PETIT

Patterns of Communication in Courtship and Mating Behaviour of Millipedes (Diplopoda)

ULRICH HAACKER

A Consideration of the Chemical Basis of Food Preference in Millipedes

W. N. SAKWA

A Comparative Study of the Organic Components of the Haemolymph of a Millipede, *Cingalobolus bugnioni*, and a Centipede, *Scutigera longicornis* (Myriapoda)

G. SUNDARA RAJULU

The Spiracle Structure and Resistance to Desiccation of Centipedes

ALAN CURRY

Contribution to the Morphology of Chilopod Eyes

R. R. BÄHR

Pauropoda from Arable Soil in Great Britain

ULF SCHELLER

Zur Bionomie mitteleuropäischer Pauropoden

W. HÜTHER

The Ecology of Centipedes and Millipedes in Northern Nigeria

J. G. E. LEWIS

The Life History of the Millipede *Glomeris marginata* (Villers) in North-West England

J. HEATH, K. L. BOCOCK and M. D. MOUNTFORD

Sex Ratio and Periodomorphosis of *Proteroiulus fuscus* (Am Stein) (Diplopoda, Blaniulidae)

MAIJA RANTALA

Vertical Orientation and Aggregations of *Proteroiulus fuscus* (Am Stein) (Diplopoda, Blaniulidae)

MAIJA PEITSALMI

The Life Cycle of *Proteroiulus fuscus* (Am Stein) and *Isobates varicornis* (Koch) with Notes on the Anamorphosis of Blaniulidae

CHARLES H. BROOKES

The Life-Cycle and Ecology of *Ophyiulus pilosus* (Newport) in Britain

J. GORDON BLOWER and PETER F. MILLER

Food Consumption and Growth in a Laboratory Population of *Ophyiulus pilosus* (Newport)

J. GORDON BLOWER

Competition between *Ophyiulus pilosus* (Newport) and *Iulus scandinavius* Latzel

PETER F. MILLER

The Adaptive Significance of Variations in the Life Cycles of Schizophylline Millipedes

COLIN FAIRHURST

A Population of *Cylindroiulus latestriatus* (Curtis) on Sand Dunes

M. J. COTTON and P. F. MILLER

The Distribution of British Millipedes as known up to 1970

J. GORDON BLOWER

A Habitat and Distribution Recording Scheme for Myriapoda and other Invertebrates

A. D. BARBER and C. P. FAIRHURST

Some Aspects of the Economic Importance of Millipedes

A. N. BAKER

Note à propos des Diplopodes Nuisibles aux Cultures Tempérées et Tropicales

G. PIERRARD et J. BIERNAUX

Some Effects of Insecticides on Myriapod Populations

C. A. EDWARDS

CENTRE INTERNATIONAL DE MYRIAPODOLOGIE

J.-P. MAURIÈS

J.-M. DEMANGE

PROFESSOR OTTO KRAUS

Symp. zool. Soc. Lond. (1974). No. 32, 1–12.

PRESIDENTIAL ADDRESS: RETROSPECT AND PROSPECT IN MYRIAPODOLOGY

THE REV. CANON S. G. BRADE-BIRKS

Fellow of Wye College (University of London), Ashford, Kent, England

It must be very encouraging to the organisers of this gathering to see us assembled here in this great northern university for the Second International Congress of Myriapodology.

One of my first thoughts concerns our indebtedness to the authorities here, to the Zoological Society of London and to the Centre International de Myriapodologie for their support and encouragement in this undertaking.

I should like to make special mention of the wonderful way in which the Zoological Society has sponsored this Congress and of the great interest that Dr Vevers has taken in its preparations. At the same time I should like to pay an especial tribute to Miss Unity McDonnell of the Society's staff; she has been a marvellous and selfless worker behind the scenes and we are deeply indebted to her for all the expert help she has given and for the skill she has shown.

I also thank heartily Miss Frances Walker and Mrs Joan Gatehouse for their invaluable work in Manchester.

It is unnecessary for me to say that without the assistance thus given to us we could not have hoped for the success which we feel sure will now attend this undertaking.

The impetus which our Symposium should give is very important for progress in research in that branch of Zoology to which myriapodologists are devoted.

When in the nineteenth century the English amateur gardener raised a stone and *Lithobius forficatus* rushed out and disappeared among the fallen leaves, not much interest was aroused in the gardener's mind and the incident was soon forgotten. Much greater was the mental impact made upon the garden-lover by the friendly robin that hopped around when digging was in progress and paid such close attention to the turning of the soil.

This contrast of relationship between man and centipede and man and bird probably explains much of the indifference displayed, a hundred years ago, towards the animals in which you and I are interested in the 1970's. And what was true last century had always been true in the past and is largely true today.

Birds and mammals, butterflies, bees and wasps impinge so emphatically upon the economic and aesthetic interests of man that they make a much greater general appeal to people and so receive much fuller attention than you and I can expect for our four classes of Diplopoda, Pauropoda, Chilopoda and Symphyla.

It is natural that this difference of appeal, in which Myriapoda come off so badly, has meant that the English literature of our groups has always been scanty. Especially has there been a lack of guides, in our language, to the study of centipedes, millipedes and their allies. And what has been true with us has had a parallel in many other parts of the world.

The scanty mention of our animals in classical and other early literature has a like explanation and makes it unnecessary for me to say more about these ancient references to myriapods except to suggest that it would be very interesting if someone would now undertake some research into this branch of the subject and give us an historical survey of it.

About three years before I was born Professor Dr Robert Latzel's monumental work, *Die Myriopoden der österreichisch-ungarischen Monarchie*, was completed and I suppose we all regard that as the beginning of modern work on myriapods. Of this study Dr Karl W. Verhoeff has said*:

> R. Latzels Handbuch ist ein so bedeutender Markstein in der Myriapodenkunde, dass man es geradezu als die Grundlage der neueren wissenschaftlichen Arbeit auf diesem Gebiet bezeichnen kann.

or, in English:

> R. Latzel's work is such a distinct milestone in myriapodology that we can, without hesitation, regard it as the foundation of modern scientific work in this field.

In England we had nothing comparable and, in fact, nothing of much use in solving the detailed taxonomic problems which are so serious in the study of Myriapoda.

In 1895, F. G. Heathcote (F. G. Sinclair) was responsible for the section Myriapoda, consisting of some 52 pages, in volume five of *The Cambridge Natural History*. His account included 32 text-figures. Some people may have found this dissertation useful in a general sense but it was hardly adequate.

In the following year (1896) there appeared in *The Royal Natural History*, **6**: 204–213 an apparently little known account of our animals

* Verhoeff, K. W. (1910–1914). *Die Diplopoden Deutschlands*. Leipzig: F. Winter.

from the able pen of R. I. Pocock. And to this piece of work I should like to draw your attention and especially I should like my English-speaking colleagues to look at it, and in so doing, honour one whose labour on the Myriapoda is outstanding among British workers.

It is not because I knew Pocock personally as a friendly, generous zoologist that I especially draw attention to this treatment of our animals, but because I think that it was remarkable that in 1896 he had already given class rank to Chilopoda and Diplopoda and had been able to deal so clearly, concisely and usefully, in less than 10 pages (and they were adequately illustrated pages) with these animals. He wasted no words in giving his account of them and from his own experience of a wide range of species he was able to provide a world survey.

Anyone reading carefully what Pocock here wrote would have as good an idea of the subject as could possibly be given in so short a space. His treatment of comparative anatomy, taxonomy, geographical distribution, habitat and bionomics (including breeding habits and food) was a good guide to these matters and he took the consideration of economic status, the phosphorescence of centipedes and a word about fossil forms in his stride.

Let us take a few sentences from Pocock's work to illustrate the clarity of his style.

First from p. 207:

> The genera of Scolopendridae present a strong family likeness to each other; one of the most remarkable being the African *Alipes*, which has the last three segments of the last pair of legs flattened and leaf-like. The reason of this modification is unknown, but the creature is said to make a noise by knocking and rubbing its legs together.

Then from p. 210:

> In the South African genus *Sphaerotherium* the last pair of legs in the male is furnished with a well developed stridulating apparatus, consisting of a finely ridged plate, which by being rubbed against a set of granules on the inner surface of the last tergal shield, gives rise to an audible sound.

And thirdly and lastly (speaking of millipedes) Pocock remarks, on p. 212:

> Most are slow in their movements, and never trust in speed to escape. When walking the body is kept fully extended, and propelled by the legs, the movements of which resemble a series of waves passing up the body from behind forwards. As already stated, many forms are devoid

> of eyes; but even those possessing well-developed visual organs appear scarcely able to do more than distinguish light from darkness. As they crawl along, every inch of the road is first carefully touched by the antennae, which are tipped with a sensory organ, and the creatures appear to be unaware of the presence of an obstacle until the antennae have actually come into contact with it.

Other concise and useful general pieces of Pocock's work were his articles in the *Encyclopaedia Britannica*.

When, soon after the beginning of the First World War, my wife and I entered upon our joint study of myriapods, faunistic work was beginning to receive a good deal of attention. R. I. Pocock had already given some attention to the distribution of these animals in England but a great deal of his work was concentrated upon exotic forms which gave him the wide view which we have already noted in his writings.

Long before I became acquainted with him, his main interest had been centred upon the Mammalia.

At the Manchester Museum in 1915 were two naturalists with a wide knowledge of the fauna of Lancashire and Cheshire with whom I had become friendly when I was an undergraduate. They were Robert Standen and his son-in-law Wilfred Jackson. The latter, now a nonagenarian, is still deeply interested in a wide spectrum of subjects ranging from conchology to osteology. His contributions to scientific knowledge were eventually recognized adequately when the University of Manchester conferred upon him the honorary degree of Doctor of Science.

Both Standen and Jackson were helpful and encouraging to us in our early studies of Myriapoda and later on Jackson collaborated with us in studying Archipolypoda.

Another noted naturalist, a little our senior, who was a good friend to us in the early days was Richard S. Bagnall, a partner in a firm of engineers in N.E. England. As you know he made notable contributions to the study of the two most difficult of our classes, the Pauropoda and the Symphyla. He was a man of a most generous mind, willing to render every possible aid to his fellow naturalists.

The person who was, above all the rest, our guide and helper in the beginning of our specialization was A. Randell Jackson, M.D., D.Sc., whose pleasant jest was that he posed as a scientist among his medical colleagues and as a medical man among scientists. I suppose that it was through Standen and Wilfred Jackson that we got into touch with him in the first instance. The two Jacksons were not related. Randell Jackson had a medical practice in Chester and Mrs Jackson and he invited me to stay with them for a day or two while he initiated me into the mysteries of the myriapodological craft.

Here again was a naturalist with a generous mind willing and ready to help his young friend. He was friendly, efficient and humorous.

Jackson was essentially an arachnologist and with his tongue in his cheek professed to believe that evil spirits tried to frustrate his study of spiders! But he was able by persistence in his plans to circumvent their designs. I had read every word of Randell Jackson's papers on myriapods, so that I knew them almost off by heart, and when we met, our conversation was of little else than *Lithobius* and *Polyxenus*, of iulids and the Geophilomorpha.

Dr Jackson put us on the track of the best literature for our studies, gave us keys he had perfected for identification and instructed us in the techniques of collecting and laboratory practice, and so we were truly launched in our course of collecting and investigation.

It was not long before we were in frequent correspondence with that amiable and patient French myriapodologist Dr Henry W. Brölemann of Pau.

We also had occasional communication with Professor H. Ribaut of Grenoble.

Brölemann was an especially kind friend to us. He identified material for us, dissected millipedes for us and gave us great encouragement. It was a sorrow to me that we never met this great Frenchman.

Other continental workers with whom we had slight contacts were Carl Graf von Attems and Dr Karl W. Verhoeff.

Prof. F. Silvestri we knew personally. We had a certain amount of correspondence with him and when he came to London to give a course of Special University Lectures he came down and stayed with us one night at my vicarage at Godmersham in Kent. He was a most courteous and friendly person and we greatly enjoyed meeting him.

The scientists at the British Museum (Natural History) have always been willing to help us in our studies and as early as March 1919 we were expressing our thanks* to Mr A. S. Hirst for facilitating our examination of a small part of the valuable collection of specimens under his care at Cromwell Road.

In Northern Ireland at the time we began our studies Nevin H. Foster was a very active field naturalist and we were closely in touch with him because he was interested in the Myriapoda.

Two young and promising myriapodologists who were subalterns in the army in the first war were both killed. They were Colin M. Selbie and Herbert F. Bolton. We had some correspondence with Selbie but

* *Ann. Mag. nat. Hist.* (9) **3**: 253.

never met him. Bolton was an undergraduate at Oxford when he joined the army, but we knew him very well because his home was in Darwen, Lancashire, where we lived. He was a keen collector at Oxford and at Darwen.

THE ECONOMIC STATUS OF MYRIAPODS

When in 1919 I was appointed to a post in an agricultural college in Kent in S.E. England—a school of the University of London—it was natural that I should turn to the study of Chilopoda, Diplopoda and their allies in relation to agriculture and horticulture and so, for several years, this was my main line of research.

I must just mention in passing that I had only been a very short time in Kent before I turned up in woodland at Wye a millipede new to Britain, *Polyzonium germanicum*, which, some time previously, Richard S. Bagnall had felt sure must occur in England.

Broadly speaking, as you must all know, a few millipedes are injurious to crops while centipedes tend to be beneficial to farmers and gardeners as predators upon some pests of plants.

A. Randell Jackson* gave support to the view that centipedes may be beneficial to man when, dealing with *Lithobius forficatus*, he said:

> When out in the garden at night with a lantern hunting for, and destroying slugs, I have several times seen them seized by a *Lithobius* and carried off, their captor shaking them as a terrier shakes rats.

It has however struck me that this likeness to a dog's behaviour when worrying vermin may have been an attempt on the part of the centipede to rid itself of a sticky creature adhering to its mouth-parts.

The mouth-parts of millipedes are not very robust and it is improbable that they are strong enough to penetrate the integument of the underground structures of any but a few crop plants. Nevertheless, the roots of seedlings may be damaged or destroyed. Where damage has already been begun by other pests, millipedes may be present in sufficient numbers to aggravate the injury by invasion through the wounds.

It is to be emphasized that it is the presence of such large numbers of millipedes that makes their depredations serious. I hope that if this subject is not discussed when we come to consider myriapods as pests that someone will take up an enquiry into those favourable conditions for breeding which result in the sudden appearance of sufficiently

* March, 1914 *Lancs & Ches. Nat.*: 456.

serious infestations of injurious species to produce grave losses to farmer or gardener.

In greenhouses, the introduction of overseas species of millipedes has often resulted in serious loss but because of the special conditions, including the local nature of the attack, modern methods of fumigation and other precautions are now very effective. Probably nowadays the principal risk for the grower is failure to recognize, in good time, the nature of a trouble which has not previously been experienced.

MYRIAPODS AND THE SOIL

Many of the animals with which we are concerned are soil dwellers and in this connexion I should like to draw your attention to a matter which, for many years, eluded scientists who were making a study of the soil. Many of them were especially interested in the subject particularly in relation to agriculture and horticulture and a young colleague of mine—the late Basil S. Furneaux—once said something to the effect that the soil had, for far too long, been regarded as dirt by the ignorant and as a mixture of chemical and mechanical particles by the learned.

He might well have added that the learned had generally considered that the soil was a mere 25 cm or less in depth at the surface of the earth and that nothing below that depth need be taken into account in any study of it.

Many of the early soil scientists—pedologists as we term them—were chemists and although they recognized the geological origin of the mineral part of the soil and, indeed, used geological maps to explain some of the differences between soils of different localities, their view of the soil as a whole was largely limited to the mechanical and chemical constitution of the 25 cm which lay nearest to the surface.

But in the latter part of the nineteenth century there was an awakening to a new scientific conception of the nature of the soil.

In Russia a number of investigators were especially concerned to explain the occurrence of belts of soil, many miles in width, which ran roughly east and west over the greater part of Europe. One of these, the Chernozem or Black Soil belt, was of great economic importance because it was the land on which wheat had long been grown with great success. It was eventually recognized that the outstanding characteristics of these rich wheat soils—as well as those of the other belts to the north and south of them—were largely due to climatic factors.

It was in this way that what we may call the *natural history of the soil* became the object of study.

In the twenties of this century, men of science in Great Britain were beginning to look at the soil as a natural object, and one of the ways in which they were able to observe it in this light was to examine vertical sections of soil where they were exposed in such places as quarry faces, sand pits and cuttings made in the construction of roads and railways.

Such sections of the soil are known as *soil profiles*.

Without going into too much detail, I shall be content to draw your attention—by way of illustration—to one example which has a considerable distribution in Kent (in S.E. England) and in other parts of Great Britain.

The horizons of this soil, which is a sandy one, are well developed because the natural drainage is free and the rainwater readily seeps down. The reaction of the soil is acid and this aids the downward removal of organic matter and salts of iron and aluminium to deeper parts of the profile. In deeper parts of the soil, humus, iron and aluminium are deposited. It is obvious that in this way different conditions are provided in the different horizons for the soil fauna.

If faunal sampling is carried out with due regard to the horizons of the soil the sampling is a natural one. Otherwise it will produce misleading results.

There are several points about such a soil which should be taken into account in studying its myriapod fauna.

First, of course, is the nature of the soil surface. Here the sand grains are proportionately numerous when compared with other finer grained mineral constituents and that means that the texture of the soil allows a rapid loss of rainwater falling upon the surface to take place. There is a slight loss of this water by evaporation but much of it passes down into the ground and, passing through the top part of the soil, penetrates to the depths below.

Where there is a good leaf cover, surface evaporation is reduced and a good deal of rainwater is held by the leaves and there is considerable protection for the myriapod fauna inhabiting woodland.

As such soils support trees and are little used as arable land, there is a close relation, at once, between such soils and surface dwelling arthropods.

In such a case there may be a good deal of organic matter in the horizon of the soil just below the leaf cover so that here millipedes will find sustenance as well as in the leaf cover itself. It is obvious that any analysis of population by depth must take this into consideration. And so the top 25 cm of the soil can be expected to give a high population figure.

Below this we may expect some centimetres where there will be little food for our animals and consequently a lower myriapod population.

If we sample in a series of fixed depths we must expect a misleading set of figures but if we take our animals according to the natural divisions, that is to say according to changes in the profile, we shall obtain a natural correlation between soil features and the myriapod population in depth.

Factors in different soils which have a bearing upon the invertebrate population include the availability of calcium carbonate for skeleton building. We have, here, to remember that even in soils in which there is little or no limestone debris, lime may come in, in solution, in underground drainage water from distant calcareous rocks. A good example of this is to be found in some of the wonderfully fertile soils of the artificially drained marshlands of S.E. England, some of them below sea level and protected against marine inundation by sea walls. In such situations sandy soils would be expected to be devoid of calcium carbonate if underground water did not supply the deficiency.

MYRIAPODS AND POLLUTION

We come now to the subject of the relation between Myriapoda and the spread of injurious substances in the atmosphere.

Close to my home in Kent there is a brick boundary wall to a plantation of limes and other trees. It is believed that it was built in or before the first quarter of the nineteenth century. In many places this wall supports considerable patches of lichens (Lichenes).

About 10–15 years ago this wall was the home of the bristly millipede, the Linnean species *Polyxenus lagurus*, the first diplopod to be described in Latzel's monumental work. As you all know it is a tiny creature only 2 mm long. With us in Kent it could be collected in considerable numbers even in winter. I have known it active on Christmas Day.

During the past few months I have sought in vain for this attractive little millipede on this wall. It has completely disappeared.

The brick wall of a churchyard five kilometres away was formerly another habitat of *Polyxenus lagurus*. I have recently looked for it there too; also without success.

What is the reason for this disappearance of *Polyxenus*? Is this the result of atmospheric pollution as distinct from that of ground–water, surface water or soil? Farm tractors pass frequently along the farm road which is bounded by this wall near my home and the churchyard I have mentioned is situated in a small town. I may mention that frogs and toads, formerly plentiful, have, in recent years, disappeared from my home neighbourhood.

LIGHT FROM MYRIAPODS

In the early days of our enthusiasm for the study of Myriapoda we made some enquiries into the subject of luminescence in the Geophilomorpha. In this study we were greatly helped by members of the Dartford Naturalists' Field Club who sent to us in Lancashire specimens they had collected in Kent. This was followed by our discovery of luminous specimens in the north of England.

We found that a number of stimuli resulted in light production by *Geophilus carpophagus*, Leach, by a secretion onto the ventral surface of the body. Examination ventrally under the microscope showed that opaque rounded masses of material, white in reflected light, were often present under the pore-fields and under the surfaces of plates known as 2β and 2γ. These white rounded masses are groups of pyriform, and probably unicellular, glands. Upon stimulation, these glands are discharged onto the ventral plates with the production of light. The luminous material in *Geophilus carpophagus* is a viscous fluid, practically colourless, with a characteristic fruity odour, not unlike that of decaying flowers, drying rapidly in air and strongly acid in reaction. Luciferin (which chemically is an acid) in the presence of the enzyme luciferase is oxidized to form products plus light. The pathway of the reaction has, I am authoritatively informed, been described by Airth, Rhodes & McElroy (1958).*

THE LIGHTER SIDE OF MYRIAPODOLOGY

In introducing the subject of humour in myriapodology with trepidation, I take refuge under the shade of a jingle of unknown authorship:

> A little nonsense, now and then,
> Is relished by the wisest men.

The naturalist is always liable, like the Professor, to be regarded by some sections of the public as a figure of fun. And when his subject of study happens to seem so remote from ordinary life as a millipede or a centipede, the humour of his enthusiasm is considerably heightened.

* Airth, R., Rhodes, W. & McElroy, W. D. (1958). The function of co-enzyme A in luminescence. *Biochim. biophys. Acta* **27**: 519–532.

A good example of this occurred to me when as a young priest in a northern industrial town, I quietly and privately asked the town clerk if there was any objection to my setting traps for millipedes in a public park in the town. He was reassuring, but said that as a matter of form I ought to write him a note which he would bring to the notice of the Parks Committee. I carried out his suggestion and some minor official in the Town Clerk's Office stupidly included the item on the agenda of the Town Council. The result was that there were humorous exchanges at the meeting, questions being asked in fun as to the danger of my traps to visitors to the park. From the local press, a national newspaper picked up the story and published a picture of me setting one of my traps. On my return to my college in Kent after my time in the North, I found notice boards embellished with cuttings of the photograph and general amusement among colleagues and students at what had happened to me.

Even among zoologists themselves, there has been a tendency to make fun and I suppose the first of these to do this was Sir Ray Lankester who in a letter to *Nature* (23rd May 1889) dealing with the gait of centipedes said:

> A Centipede was happy quite
> Until a toad in fun
> Said, "Pray which leg moves after which?"
> This raised her doubts to such a pitch,
> She fell exhausted in a ditch,
> Not knowing how to run.

The true authorship of this oft-quoted verse is not generally known but a slightly different version is credited, in *Cassell's Weekly*, to Mrs Edward Craster in *Pinafore Poems* (1871).

It is well to take all the shafts of humour directed towards us and our animals in the right spirit. It is well if in a sad world we add a little gaiety to life.

PROMOTING INTEREST

One way in which the myriapodologist can promote a more general interest—especially among professional zoologists and amateur naturalists—in the four classes with which he is especially concerned is to pay as much attention as possible to those aspects of myriapodology which have things in common with other zoological studies. For the amateur there are the interests such as life histories, nesting habits, movements, wandering and economic status. For the zoologist there is special interest in our group on account of phylogeny and in such matters as comparative anatomy. It may be along either of these two lines that young scientists may be attracted to specialize in the study of Myriapoda. Established workers in the myriapodological field could render great service by the production of a not too bulky volume to give essential information much on the lines of R. I. Pocock's dissertation in the *Royal Natural History* supplemented by descriptions of individual species and accompanied by keys for identification. Three members of this gathering have already contributed valuable guides to the study, in Britain, of symphyles, centipedes and millipedes*. We need a modern monograph on Pauropoda.

I am sure that we all desire that the present Symposium will not only establish participants in the lines of research they have undertaken but will also give us new enthusiasms to interest others, both professional and amateur, in these four classes which have provided present workers with such interesting fields of study.

All success, then, to our Congress.

FLOREAT SYMPOSIUM!

REFERENCES

* Blower, J. G. (1958). British millipedes (Diplopoda). *Synopsis Br. Fauna* No. 11: 1–74.

Eason, E. H. (1964). *Centipedes of the British Isles*. London and New York: Frederick Warne & Co.

Edwards, C. A. (1959). Keys to the genera of Symphyla. *J. Linn. Soc. (Zool.)* **44**: 164–169.

Edwards, C. A. (1959). A revision of the British Symphyla. *Proc. Zool. Soc. Lond.* **132**: 403–439.

Symp. zool. Soc. Lond. (1974) No. 32, 13–22.

ON THE MORPHOLOGY OF PALAEOZOIC DIPLOPODS

OTTO KRAUS

Zoologisches Institut und Zoologisches Museum,
Universität Hamburg, Hamburg, Germany

SYNOPSIS

The most comprehensive study of palaeozoic millipedes was published by Fritsch in 1899. More recent authors derived their ideas chiefly from his work and copied his figures. There was and still is a more or less general opinion that most of these millipedes of later palaeozoic times were so different that there is no basis for a discussion of potential relationships with recent taxa. On the other hand the present author is convinced that at least some of the better known recent higher taxa can be traced back to late palaeozoic times. To resolve this apparent contradiction a re-examination of the palaeontological objects and of the status of recent forms was carried out.

The main results show that the differences between the late palaeozoic representatives and recent taxa are due to change of habitats. There is much evidence that the fossil forms lived mainly on the surface, especially on plants; this explains the presence of larger eyes in some groups, the occurrence of bifurcated spines, and the special condition of the exoskeleton (free tergites, pleurites, sternites). Perhaps, as the insects evolved, most millipedes were reduced to hypogaeic habitats (under stones, in rotten wood and mould). This change of habitat induced the formation of more rigid exoskeletons (Verhoeff: Einschubzylinder), the reduction of eyes, spines, etc.

But the original condition still seems to be preserved in the special structures of some groups (e.g. Polydesmida, Spirobolida, Spirostreptida), occurring only temporarily when animals are moulting. In this stage their exoskeleton becomes divided into sclerites strikingly similar to those of late palaeozoic representatives. The transitory occurrence of these archaic structures seems due to changed functions: they are still useful when moulting necessitates a flexible, extensible exuvium.

INTRODUCTION AND PRESENT SITUATION

Fossil millipedes are still insufficiently known. Therefore there has been practically no opportunity of using data derived from the fossil record in order to clarify questions of phylogeny. Indeed, with many fossils, even at present, we encounter severe difficulties in establishing a relation with any recent taxon.

There are several reasons for this situation:

(a) The fossil record now available is not very extensive and covers only few formations: most fossil remains originate from Upper Carboniferous deposits and from the Baltic Amber (Oligocene). At all times there must have been little chance of fossilization—perhaps with the exception of the Upper Carboniferous of Central Bohemia.

(b) In many cases the fossils are only moderately or even badly preserved; this precludes detailed statements on their morphology. Fortunately the most comprehensive fossil fauna we know, which was discovered in the so-called "gas-coal" of Central Bohemia, also contains a high proportion of exceptionally well preserved specimens, with microsculpture, ozopores, etc., still visible. But even this seems an insufficient basis for phylogenetic judgements, as the systematics of recent taxa is primarily based on more cryptic details (such as the structure of the gnathochilarium, gonopods, etc.), and not so much on general habitus–characters. Herein lies the main difficulty of determining the systematic position of palaeozoic remains.

(c) Another point of view has already been emphasized by Hoffman (1969: 575): "Virtually the sum of our present knowledge of fossil myriapods stems from the early work of S. H. Scudder during the 1880's and of Anton Fritsch about a decade later". Although additional remains were discovered and described by several later workers, the classic material, especially of the Fritsch collections, was never studied again. Therefore later authors (such as Verhoeff, 1926; Hoffman, 1969) really had to derive their conclusions chiefly from the old original literature.

Of special note is the attempt of Fritsch (1899) to reconstruct the different types of late palaeozoic representatives and to establish relations with recent groups. He remarks (1899: 43): "Im allgemeinen kam ich zu der Ueberzeugung, dass die weite Trennung der fossilen palaeozoischen Myriopoden von den jetzt lebenden allzu gezwungen ist und schon damals Uebergänge zu den recenten vorhanden waren."—On the other hand, Verhoeff (cf. 1926: 330–359) was opposed to Fritsch's ideas, and so he helped to create the impression that many of these palaeozoic representatives were so different in organization from recent ones that discussions on relationships with present taxa might be more or less useless.

But this seems contradicted by arguments indicating a relatively high phylogenetic age of at least some recent taxa of higher rank. The arguments are chiefly derived from zoogeographical patterns. There are several families (e.g. Rhinocricidae, Spirostreptidae and Platyrrhacidae) which occur in different continents, so that Carl (1914) termed this type of discontinuity "transozeanisch". It seems at least imaginable that such distributions might be the results of continental drift. But it must be emphasized that such thoughts—without further arguments—remain pure speculation. But for at least one group with higher categorial rank, the spirostreptomorph diplopods, such additional arguments seem to exist. The relationships of its subtaxa are considered as follows (for detailed reasoning, based on the principles of Hennig's "Phylogenetische Systematik" (1966), see Kraus (1964, 1966).

For the Spirostreptidae it seems possible to estimate the minimum age of this family. Its representatives occur in the Neotropical as well as in the Ethiopian region (Madagassic subregion included). This is a distributional pattern, still preserved, which can be traced back to Middle Triassic times, when both continents had nearly identical saurian faunas (Colbert; Romer (in Mayr, 1952)). In the sense of a *terminus post quem non* it is concluded that the family Spirostreptidae is at least as old as the Middle Triassic (this concerns the family Harpagophoridae as well, for they form the sister-group of Spirostreptidae).

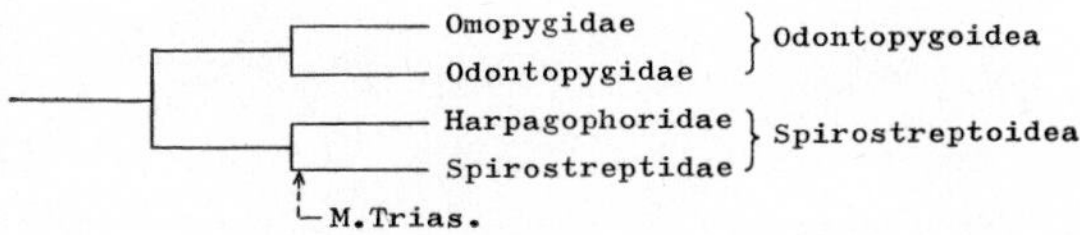

So, the superfamilies Spirostreptoidea and Odontopygoidea must have originated even earlier, perhaps in late palaeozoic times. To some extent this calculation bridges the enormous gap between recent forms and Upper Carboniferous faunas.

In the light of such considerations the original ideas of Fritsch seem more probable, i.e. the concept of a continuous development from Late Carboniferous forms to recent taxa. This necessitates a re-investigation of the classic fossil material. I had the opportunity to see and to work on Fritsch's originals from Central Bohemia, and to include additional specimens from the Národni Museum v. Praze. For comparison I had fossil diplopods from carboniferous deposits in England before me (collections of the British Museum of Natural History).

THE EXOSKELETON OF PALAEOZOIC DIPLOPODA

It is not possible and perhaps not even necessary to report here our results on the structures of fossil diplopods in detail. This will be the object of a later contribution chiefly revising the material from Bohemia. Therefore, the most important characteristics of these palaeozoic representatives are mentioned here in principle.

(a) Larger compound eyes (Fig. 1) evidently occurred more frequently than in present forms. With certainty this was the case not only in the so-called "Archipolypoda" (with the most important genera *Acantherpestes* and *Euphoberia*), but also in the Bohemian forms *Isojulus marginatus*, *I. constans*, *Pleurojulus biornatus*, and in the genus

Glomeropsis; one eye of *G. ovalis* was composed of approximately 1000 single eyes. Perhaps the more frequent occurrence of large eyes may be regarded as a basic feature, as in many other "old" arthropods.

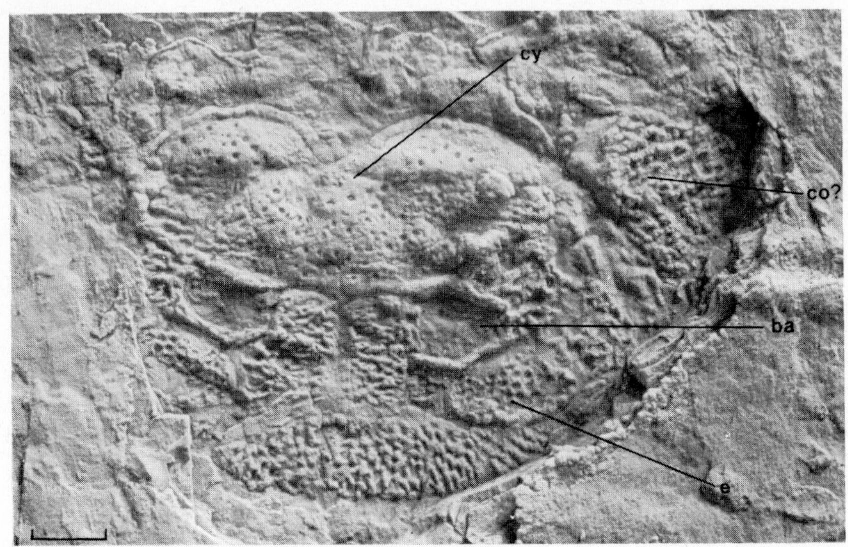

FIG. 1. *Acantherpestes* sp. from Nýřan, Bohemia, Upper Carboniferous. Head from in front (ba: base of antenna; co?: perhaps lateral end of collum; cy: clypeus; e: eye). Scale: 1 mm.

(b) In late palaeozoic faunas from Europe and North America we find a considerable number of large forms (Fig. 3) which, on their tergites, were armed with conspicuous spines (as *Euphoberia, Acantherpestes*). The largest specimen of *Acantherpestes* known to me had a total length of 235 mm. It seems remarkable that these giants must have had such a weak exoskeleton (Fig. 2). This is demonstrated by the Bohemian fossils which in most cases are much more fragmented and disintegrated than the hard parts of other representatives, especially of those with a more julomorph habitus.

The strange and strong spines on the tergites of the metazonites are the best known feature of the "Archipolypoda". They were arranged in (usually four) longitudinal rows, and often bifurcated at the end. In many cases these spines were broken off during the process of fossilization, but their bases on the tergites always remain visible as rounded depressions.

There are no recent diplopods directly comparable with these spined forms. Only Loomis & Hoffman (1962) described two polydesmoid genera also with forked spines. The few species belonging to this group are comparatively small, probably nothing more than a superficial analogy.

Fig. 2. *Acantherpestes vicinus* Fritsch from Nýřan, Bohemia, Upper Carboniferous. Body segments, lateral view: the head of the animal would be at right (bl: base of leg; s: base of spine; st: stigma in the middle of a pleurite; tg: tergite). Scale: 2 mm.

(c) In most recent diplopods the diplosomites appear as homogeneous rings. The two tergites, two pairs of pleurites and two sternites, which form the original components of the exoskeleton, are completely fused to what was called "Einschubzylinder" by Verhoeff. Diplotergites, discrete diplopleurites and discrete sternites are chiefly confined to the

FIG. 3. *Euphoberia ferox* Salter from Coseley (Staffs.), England, Carboniferous. Body segments, dorsal view; the head of the animal would be at left (bd: base of a spine of the paramedian series; fl: forked spine of the lateral series). Scale: 10 mm.

Oniscomorpha and to a part of the Colobognatha (also Pselaphognatha). Apart from the Oniscomorpha one may say that those present forms with free pleurites, etc., belong to small groups containing few species which themselves are rarely dominant in terms of numbers of individuals.

In contrast to this it seems evident that in Late Carboniferous faunas we had considerably more representatives with free pleurites and/or sternites. I was able to confirm that the species of the archipolypod

FIG. 4. *Pleurojulus levis* Fritsch from Nýřan, Bohemia, Upper Carboniferous. Whole animal in lateral position with diplopleurites visible (original in Fritsch, 1899, Pl. 141, Fig. 6). Scale: 10 mm.

genus *Acantherpestes* had discrete sternites (Fig. 2), each of them with paired lateral stigmata; but I could not find any trace of free pleurites, and it seems now that they really did not exist. In the genus *Pleurojulus* we find discrete diplopleurites (Fig. 4), and perhaps the same situation was true in *Anthracojulus*. In the case of *Pleurojulus* Hoffman (1969: 595) thought that the "so-called" 'pleurites' are clearly nothing more than the fractured lower ends of pleurotergites broken when the animal was flattened, not an uncommon occurrence". But I am sure now that these structures cannot be regarded as artefacts.

(d) But we also find other types which seem much more comparable with recent forms. Members of some genera e.g. *Amynilyspes*, *Archiscudderia* and *Glomeropsis* evidently represented the Oniscomorpha (= Pentazonia, = Opisthandria) at this time. Apart from the somewhat strange *Amynilyspes*, which had spines on the tergites, I cannot find characters indicating real differences from recent oniscomorph diplopods Other types might be termed "julomorph" in their appearance. As in recent forms their diplosomites formed continuous rings ("Einschubzylinder"), as in *Xyloiulus*, *Nyranius*, or *Isojulus*. Hoffman (1969), with reservations, grouped them to spirobolids.

BIOLOGICAL INTERPRETATION

It seems possible to interpret the peculiarities of the palaeozoic diplopods mentioned by an analysis of their environment. Representatives of genera such as *Euphoberia* or *Acantherpestes* must have lived on the surface, especially on plants. With their weak exoskeleton and their strong spines these relatively large arthropods must have been unable to burrow in substrates such as rotten wood and mould. Perhaps, as insects evolved, they became more and more restricted to what we may call "underground habitats". At the same time this would explain the tendencies mentioned above (i) towards the reduction of eyes, and (ii) towards more solid diplosomites.

This interpretation is supported by our knowledge of the palaeoecological situation, which in Central Bohemia permitted the fossilization of a very rich fauna of carboniferous diplopods. The remains from the Plzeň (= Pilsen) beds were preserved in sediments deposited in lakes, connected by smaller creeks. The diplopods from the coalfield of Kladno-Rakovnik (= Kladno-Rakonitz) are embedded in sapropelitic sediments; they accumulated at quiet parts of shallow lake basins. Therefore these diplopods must have lived epigaeically on overhanging vegetation and then fallen into the water. I cannot find any proof for Fritsch's idea (mentioned again by Hoffman (1969: 577)) that at least several of the species of "Archipolypoda" lived aquatically or semiaquatically.

But the original composition of diplosomites (tergites, pleurites, etc.) still seems preserved in some groups (e.g. Polydesmida, Spirobolida, Spirostreptida), occurring temporarily when the animals are moulting. In this stage their exoskeleton becomes divided into sclerites similar to those of late palaeozoic representatives. The transitory occurrence of these archaic structures seems due to changed functions: they are still useful when moulting necessitates a flexible, extensible exuvium.

CONCLUSIONS

All arguments favour Fritsch's concept of a continuous development from late palaeozoic diplopods to recent taxa. Striking differences between many of these old forms and present representatives seem due to changed habitats. But despite this better understanding we are still unable to link fossils and recent taxa directly—perhaps this will never be possible.

References

Carl, J. (1914). Die Diplopoden von Columbien, nebst Beiträgen zur Morphologie der Stemmatoiuliden. *Mém. Soc. neuchât. Sci. nat.* **5**: 821–993.

Fritsch (=Frič), A. (1899). *Fauna der Gaskohle und der Kalksteine der Permformation Böhmens.* **4**.

Hennig, W. (1966). *Phylogenetic systematics.* Urbana, Ill.: Univ. Illinois Press.

Hoffman, R. L. (1969). Myriapoda, exclusive of Insecta. In *Treatise on invertebrate paleontology*, R Arthropoda, **4**: 572–606. Moore, R. C. (Ed.) Lawrence, Kansas.

Kraus, O. (1964). Tiergeographische Betrachtungen zur Frage einer einstigen Landverbindung über den Südatlantik. *Natur u. Mus.* **94**: 496–504.

Kraus, O. (1966). Phylogenie, Chorologie und Systematik der Odontopygoideen (Diplopoda, Spirostreptomorpha). *Abh. senckenberg. naturforsch. Ges.* No. 512: 1–143.

Loomis, H. F. & Hoffman, R. L. (1962). A remarkable new family of spined polydesmoid diplopoda, including a species lacking gonopods in the male sex. *Proc. biol. Soc. Wash.* **75**: 145–158.

Mayr, E. (Ed.) (1952). The problem of land connections across the South Atlantic with special reference to the Mesozoic. *Bull. Am. Mus. nat. Hist.* **99**(3): 81–258.

Verhoeff, K. W. (1926–1932). Diplopoda. *Bronn's Kl. Ordn. Tierreichs* **5**(2)(2): 1–2084.

Discussion

FAIRHURST: I wonder if some of the *Amynilyspedida* could be young stadia of other groups? The one you showed (*Archiscudderia*) could perhaps be a very young archipolypodan; perhaps others could be young stadia of iuliforms? There is never any mention of life-history stages in the descriptions, yet the order Amynilyspedida is erected on the possession of 15 segments!

KRAUS: It is very difficult to say anything about life-histories. However, in the Bohemian material in which most of the animals are depressed into two dimensions only, the clear view of surface microsculpture enables close comparison of specimens. If they are identical in surface sculpture *and* have approximately the same number of segments, we can conclude they are the same. But this raises another problem—are the remains the animals themselves or their exuvia? Although our present forms eat their exuvia, perhaps, at these times we speak of, they did not.

DOHLE: Are these glomerid-like fossils really oniscomorphs or are they perhaps polydesmoids which resemble glomerids?

KRAUS: I think that forms such as *Glomeropsis* and *Archiscudderia* really were oniscomorph. *Amynilyspes*, which had spines on the tergites, might perhaps be comparable with the oniscomorph polydesmoids.

DOHLE: What arguments lead you to reject the possibility that these Carboniferous forms were aquatic?

KRAUS: Fritsch considered the so-called Archipolypoda to be aquatic because he occasionally found flattened and paddle-like legs. I am now convinced, having carefully observed their legs under the microscope with light coming from above, that they were perfectly normal diplopod legs with a terminal claw and that the flattened appearance is the result of pressure during fossilization.

DOHLE: Are there double sternites in the anterior segments of the exuvia of recent polydesmoids?

KRAUS: I cannot say—the sternites are partly resorbed during moulting. In exuvia of *Orthomorpha* the posterior segments have paired sternites but in *Polydesmus*, in the same order, there are only very reduced vestiges; so there can be much variation.

BÄHR: There are some of these fossil forms that have eyes with about a thousand ocelli. Do these possibly compound eyes occur in other ancient diplopod groups or are they only found in *Glomeris*-like and *Iulus*-like forms?

KRAUS: They occur in *Pleurojulus, Isojulus*, and *Glomeropsis* and also in some of the Archipolypoda—the large spined forms. Nothing is known of the detailed structure of the eyes of these palaeozoic forms. I know nothing of fossil *Lithobius*. But even in Trilobites, old Chelicerates and in Aglaspida there are large complex eyes. Perhaps this character was lost by many of the early arthropod groups when they took to the cryptic habit.

FAIRHURST: I don't remember any of the *Lithobius* fossils having large eyes.

Symp. zool. Soc. Lond. (1974) No. 32, 23–39.

THE PHYLOGENY OF THE FAMILY PARAIULIDAE (PARAIULOIDEA: BLANIULIDEA: IULIDA: DIPLOPODA)*

NELL B. CAUSEY

Department of Zoology and Physiology, Louisiana State University, Louisiana, U.S.A.

SYNOPSIS

The following combination of characters is unique to the Paraiulidae: the first legs of the male are massive and composed of six articles, and the second are reduced and incorporated into the penial apparatus; the second legs of the female are reduced, and other sexual dimorphism is seen in the gnathochilarium, mandibular cheek, and collum of the male; muscles move the telopodite of the peltogonopods; the gonopods, which typically consist of an unsegmented coxotelopodite and a coxite, have a closed sperm canal and a sperm fovea.

The closest relatives, the Mongoliulidae, are in eastern Asia. From the center of diversity in western North America, the Paraiulidae dispersed to Mexico and Guatemala, the eastern United States and Canada, and to Baja California, Alaska, and Japan. Some 146 species and subspecies are assigned to 26 genera, 11 tribes, and 2 subfamilies.

CLASSIFICATION

Order Iulida (Brandt, 1833)
Suborder Blaniulidea (Koch, 1847)
Superfamily Paraiuloidea (Bollman, 1893)
Family Paraiulidae (Bollman, 1893)

Hoffman (1961) gave four subdivisions of the suborder Blaniulidea family rank and raised the question as to whether one of them, the Zosteractiidae, should have that rank. Mauriès (1970) gave the same four subdivisions superfamily rank. This higher rank is justified by the complexity of the Blaniuloidea (= Nemasomatoidea) and Paraiuloidea and the size of the gaps between each of them and the Paeromopoidea. No recent investigation has been done on the Zosteractoidea, which is known by one species in a few caves in the Mississippi Valley. The Paeromopoidea, composed of one family, occurs in western North America. The Blaniuloidea, all Holarctic and the most widely distributed of the superfamilies, is in need of revision and reassessment of the families. The Paraiuloidea, a less complex group than the Blaniuloidea, is composed of the families Mongoliulidae (Pocock, 1903), in eastern

* This investigation was supported by Grant No. G14486, National Science Foundation.

Asia, and Paraiulidae, all in North and Central America except one species, which is in Japan.

A key to the superfamilies of the Blaniulidea and the families of the Paraiuloidea follows:

1. Peltogonopods consist each of a flattened plate in which coxite and telopodite are fused together; gonopods have no sperm canal or furrow 2
   Peltogonopods consist each of an elongated coxite and a shorter telopodite which is moved by muscles; gonopods may or may not have a sperm canal or furrow.............................. 3
2. Body is large and surface is coarsely furrowed above pores; epigean ....Superfamily Paeromopoidea
   Body is small and unfurrowed; troglobiotic...Superfamily Zosteractoidea
3. First legs of male are minute, consisting of from one to six articles; gonopods have neither a sperm furrow nor a sperm canal...Superfamily Blaniuloidea
   First legs of male are enlarged, consisting of five or six articles; gonopods have either a sperm furrow or a sperm canal...Superfamily Paraiuloidea 4
4. First legs of male consist of five articles; seventh legs are modified; peltogonopods have a flagellum; gonopods have a sperm furrow; sternum of gonopods does not articulate with peltocoxae...Family Mongoliulidae
   First legs of male are massive, consisting of six articles; second legs are reduced in both sexes, and in male are part of penial apparatus; gonopods have a sperm canal and a sperm fovea; sternum of gonopods articulates with lateral surface of peltocoxae...Family Paraiulidae

DIAGNOSIS OF THE FAMILY

Paraiulid millipedes resemble the family Mongoliulidae in that the telopodite of the peltogonopods is moved by muscles and the first legs of the male are enlarged. They differ in the absence of a flagellum on the peltogonopods, the presence of a closed sperm canal and a sperm fovea in the gonopods, a tendency for the gonopods to enlarge and complicate and for the vulvae to enlarge and coalesce, the articulation of the sternum of the gonopods with the peltocoxae, the reduction of the second legs in both sexes, and the greater amount of sexual dimorphism, especially in the male, which has strongly sclerotized and more massive first legs, larger promentum, mandibular cheeks, and collum, and the second legs incorporated into the penial apparatus.

DESCRIPTION

The body consists of from 41 to 74 segments, of which the last two (usually) to five are legless. Moniliform constriction of the body segments is rare. The body length is from 90 to 15 mm and the width from 6 to 0·8 mm. The number of pectinate lamellae on the mandibles, between 6 and 10, is a factor of the body size. The ocular areas are triangular and connected by a wide black band. The body color is brown on a lighter mottled pattern (Fig. 1). The body surface is seldom shining but is smooth above the pores except in *Ptyoiulus*, which is faintly grooved. The caudal spine is varied.

After the final moult the sexual dimorphism is conspicuous; it is evident to a lesser extent in the three stadia preceding the last (Hefner, 1929). In the female, the second legs, vulvae, and ends of the neighbouring pleurites are modified. The range in the size of the second legs is

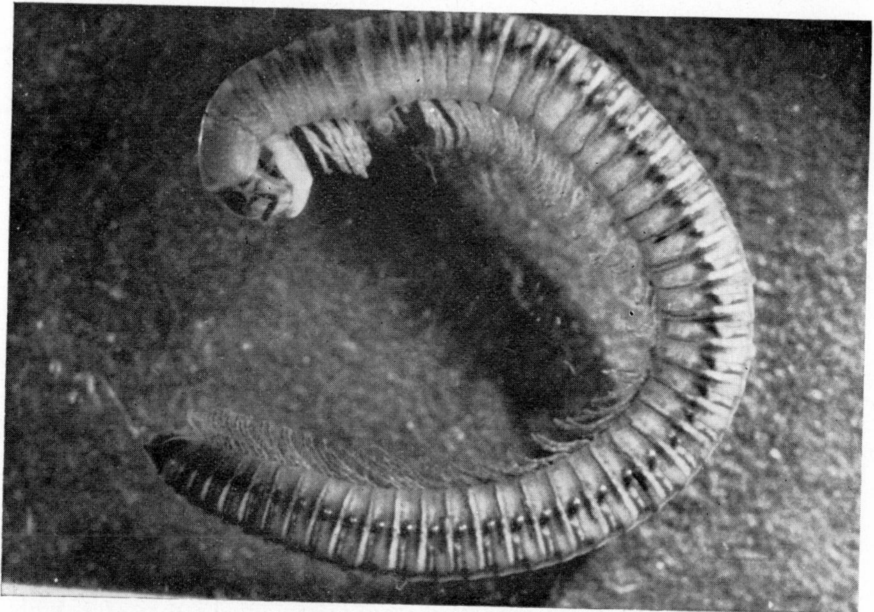

Fig. 1. Male of the tribe Aniulini showing the collum, massive first legs, peltogonopod, keel of the sternum behind the gonopods, and typical body proportions and color pattern. The salient gonopods are shielded from view by the peltocoxites. Body width 2 mm.

great; in *Ptyoiulus* only a pigmented trace remains, while in *Mulaikiulus* almost no reduction has occurred. The majority of females are between these extremes and have minute, articulated second legs attached either to or near the anterior surface of the vulval apparatus. There is also a great range in the modification of the two vulvae. In their simplest form, they are small, ovoid, discrete, and composed of an operculum and a bursa. In the majority of females, however, the opercula form a single enlarged vertical plate, a synoperculum, and the two bursae are attached to its caudal surface (Fig. 3). The form, size, and amount of coalescence and sclerotization of the vulval apparatus is varied. This organ reaches its greatest development in the Aniulini and the Paraiulini, where a mature female is easily recognized by the swollen appearance of segments 2 and 3. In the Bollmaniulina there is a vertical plate but no vulval apparatus. The vertical plate is formed from the enlarged and flattened coxosternum of the second legs, and the small opercula are in the primitive position with respect to the bursae (Fig. 2a).

In the subfamily Ptyoiulinae, where the peltogonopods have taken over the role of sperm transfer, there is no semblance of a vertical plate; and the vulvae—the smallest in the family—lack the typical bursa.

Sexual dimorphism is extensive in the male. The promentum is enlarged and the surrounding parts of the gnathochilarium are correspondingly reduced. The gula is usually enlarged. The mandibular cheeks, enlarged and of varied form, tend to assume a characteristic shape in each tribe. The collum is enlarged, with the shape tending to be characteristic for each tribe. The claspers, or modified first legs, appear as conical segmented structures in stadium VIII of *Aniulus bollmani*, the only species for which the number of stadia is known. After the final moult to stadium XI they are composed of six thickened, heavy articles and are long enough to extend well beyond the mouth. The penial apparatus, which is unique to the paraiulids, varies little. It is composed of the bilobed penis, the minute telopodites of the second legs, and the broad coxosternum of those legs. Occasionally the caudal spine is sexually dimorphic. The seventh legs are never modified. In some of the Bollmaniulina the immatures have coxal lobes, but adult males never have them. The two pairs of legs of segment 7 appear as minute rudiments of gonopods (rather than as minute mounds) in stadium VIII in *A. bollmani* and are completely differentiated by stadium XI (Hefner, 1929). The ventral margins of the seventh pleurites, the sternum of the tenth legs, the penial apparatus, and, in some genera, the sixth pleurites are modified during the last moult.

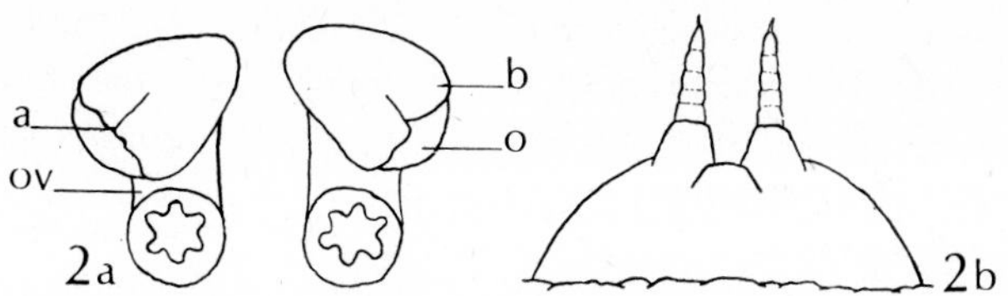

FIG. 2. Generalized scheme of the vulvae and vertical plate of the Bollmaniulini, caudal view. 2a, vulvae; 2b, vertical plate. a, aperture; b, bursa; o, operculum; ov, oviduct.

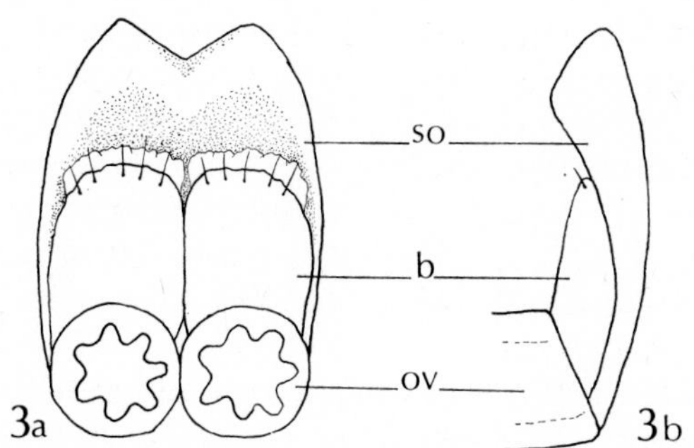

FIG. 3. Vulval apparatus of *Sophiulus tivius*. 3a, caudal view; 3b, lateral view; so, synoperculum (vertical plate).

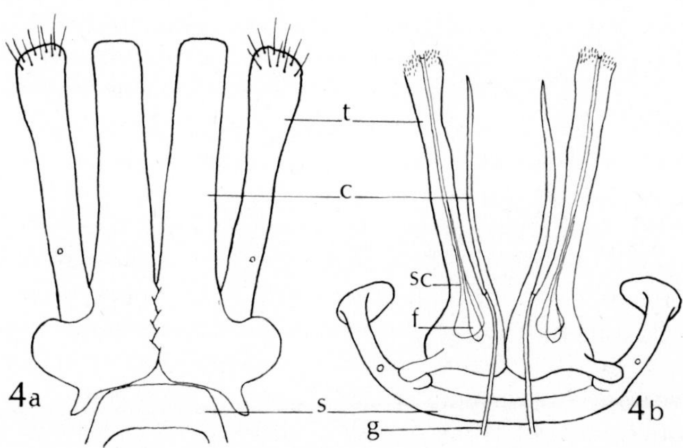

FIG. 4. Generalized scheme of the anterior and posterior gonopods of the Bollmaniulini. 4a, peltogonopods and sternum, cephalic view; 4b, gonopods and sternum, caudal view. c, coxite; f, sperm canal fibril; g, accessory gland canal; s, sternum; sc, sperm canal; t, telopodite.

The peltogonopods (Fig. 4a) consist each of a large coxa, an elongated coxite, and a lateral telopodite which is moved by muscles. The coxites assume many forms in their rôle of protecting the gonopods behind them; there is a characteristic form for each tribe, with variations in each genus and lesser variations in each species. The unsegmented, clavate, apically setose telopodites tend to be of more uniform shape than the coxites; the shape characteristic of each tribe varies little. The sternum of the peltogonopods is small and semicircular; the large peltocoxae meet in the midline behind the sternum.

The shape, position, and amount of movement of the peltogonopods vary within each tribe. The position of the peltocoxites varies from the horizontal, when they form a flat shield level with the body surface, to the vertical, when they serve as lateral shields for the exposed gonopods; a nearly horizontal position, with a podlike excavation in the caudal surface, is common in the most primitive members of the family. In the Paraiulini the principal shield is the peltotelopodites rather than the peltocoxites. There is relatively little modification of the muscles and apodemes of the peltogonopods as compared with those of the gonopods.

The gonopods (Fig. 4b) consist each of a piece, usually unsegmented, divisible into a coxal region, a coxite, and an elongated telopodite. A sperm canal opens at the apex of the telopodite on a setose area and connects to the sperm fovea in the base; an accessory gland canal opens on either the coxite or the coxa. The rodlike sternum is U-shaped, with the ends bent forward and articulated with the lateral surface of the peltocoxae.

The numerous modifications of the gonopods provided evidence from which phylogeny is inferred. Those which characterize the tribes are correlated with marked differences in the movement of the gonopods during the transfer of sperm. Most of the modifications of the external parts of the gonopods are changes to the primitive pieces rather than additions. In each of the large tribes there is a trend toward enlargement and/or elaboration of the gonopods. Fusion, shortening, and reduction have occurred, also. Other changes are the loss of the setose area, the lengthening and multiplication of the fibrils in the sperm canal, the loss of the fibrils, the addition of the large setae, the multiple branching of the telopodite and the coxite, and changes in the position and size of the sperm fovea and the opening of the accessory gland canal.

The movements of the gonopods tend to be the same throughout a tribe. The modifications of the muscles and apodemes associated with these movements are less variable than the external changes to the gonopods and are a good source of data for inferring phylogeny at the tribal level. In tribes with large, heavy gonopods, as the Aniulini and

Paraiulini, the modifications are especially remarkable. The sternum is also modified, the most notable change being the loss of the middle region. This adaptation has been acquired two, or possibly three, times independently in the Disjuncta (Fig. 5).

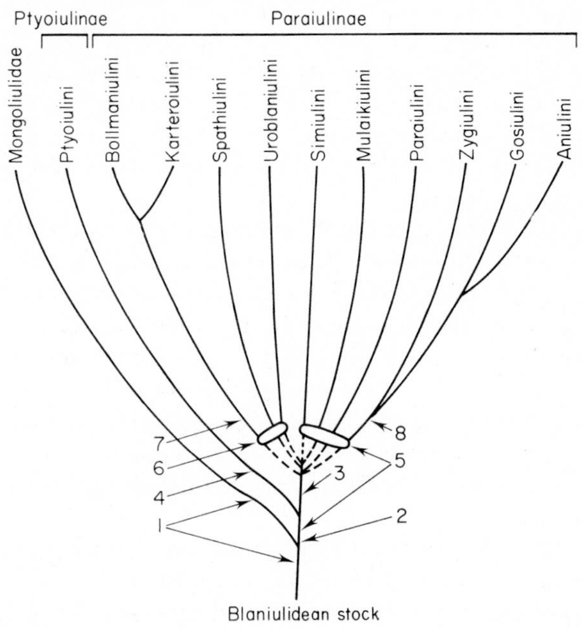

FIG. 5. Phylogenetic tree of the superfamily Paraiuloidea. 1, the gonopods have a sperm furrow; 2, the gonopods have a sperm canal; 3, the gonopods function in sperm transfer; 4, the peltogonopods have assumed the principal rôle in sperm transfer; 5, the gonopods are disjunct; 6, the gonopods are nondisjunct; 7, the supertribe Bollmaniulina; 8, the supertribe Aniulina.

SUBFAMILIES AND TRIBES

Basis for the determination of suprageneric taxa

A paraiulid tribe is an easily determined taxon. Each is assumed to be monophyletic, the members share a general similarity in the genital and somatic adaptations and, except for geographic relics, occupy a continuous range.

The genitalia of both sexes of the tribes Bollmaniulini and Kartero-iulini are similar, making the supertribe Bollmaniulina easily recognizable. In the Aniulini and Gosiulini of the supertribe Aniulina, the gonopods have a general similarity and the sternum is divided, but the vulval apparatus has not evolved at a similar rate in each tribe. The Zygiulini, the most primitive tribe of the Aniulina, diverged from the line before the sternum divided. The Paraiulini is the only tribe of the Disjuncta with fibrils in the sperm canal. Some of the characteristics which the Paraiulini share with the Aniulini, including the division of the sternum, must be attributed to parallel evolution.

The Spathiulini and Uroblaniulini are similar in having flanges on the caudal surface of the peltocoxites and vulval apparatus with similar proportions. The absence of sperm canal fibrils in the Spathiulini may be correlated with the small body size.

The distribution of sperm canal fibrils is erratic and of limited help in determining the history of the family. There are many long fibrils in the Uroblaniulini, two or three short ones in the Paraiulini, and one short one in the Bollmaniulini, Karteroiulini, and Ptyoiulini. They are absent in the small bodied tribes, the Mulaikiulini, Simiulini, Spathiulini, and Zygiulini. They are also absent in the Gosiulini and Aniulini, which are of moderate size. It can be inferred that primitive members of the family had one fibril, since all of the Ptyoiulinae and some of the Paraiulinae have one. In subsequent evolution of the Paraiulinae, the fibril was either retained or lost. If retained, it either remained single or underwent multiplication and lengthening.

It is possible to infer a general relationship of the tribes from the muscles and apodemes involved in the movement of the gonopods. On this basis the tribes of the Paraiulinae fall into two groups, and in each group there are both primitive and derived tribes. In the more primitive group, the Disjuncta, the gonopods are moved separately and there are no podlike modifications on the telocoxites except in the primitive Simiulini. In the females of the most primitive tribes of the Disjuncta, the vertical plate is much shorter than in the Nondisjuncta. The disjunct tribes are Simiulini, Mulaikiulini, Zygiulini, Gosiulini, Aniulini, and Paraiulini. In the remaining tribes of the Paraiulinae, the Nondisjuncta, the gonopods are connected at the base or to the sternum in various ways and are moved in unison, and the vulvae either remain in the primitive condition or become united in the vulval apparatus. The variations in the base of the gonopods suggest that the disjunct condition was acquired independently at least three times. The nondisjunct tribes are Spathiulini, Bollmaniulini, Karteroiulini, and Uroblaniulini.

The subfamily Ptyoiulinae is characterized by unusual somatic and genital modifications, the most notable being those concerned with the functional change in the rôle of the peltocoxites in transferring sperm. The males resemble the Bollmaniulini in the simple structure of the gonopods and in that there is one fibril in the sperm canal; they differ in that the gonopods are disjunct. The similarities suggest that the Ptyoiulinae diverged either from the Bollmaniulini or near them and that disjunction was acquired later. If so, they are younger than some of the tribes of the subfamily Paraiulinae. An alternate view (see Fig. 5), which is supported by the remarkable somatic and genital differences between the members of the two subfamilies, is that the divergence occurred soon after the family evolved.

A synopsis of the subfamilies and tribes of the Paraiulidae

Subfamily. Paraiulinae, **new rank.** Body is asetose except at hind end. Peltocoxites lack a calyx; are of numerous shapes. Gonopods are of numerous shapes and sizes; either with or without sperm canal fibrils. Vulvae may be discrete and with the opening guarded by a small operculum, but usually they form a vulval apparatus, in which the opercula form the vertical plate; 2nd legs are varied in size. 40–74 body segments. 10 tribes.

Tribe Aniulini, **new.** Peltocoxites are usually erect and leaflike. Gonopods are elongated, heavily sclerotized, composed of a telopodite and a coxite; sternum is divided. Vulval apparatus is large, thickened, sclerotized; telopodites of 2nd legs are minute. 48–61 segments. 7 gen., 56 spp. and/or subspp. Atlantic Coastal Plain west to Rocky Mountains; rare in southern Canada and northeastern Mexico.

Tribe Bollmaniulini, **new.** Peltocoxites are elongated, podlike. Gonopods are coalesced at base, elongated, sclerotized, composed of a telopodite and coxite; one short sperm canal fibril. Vulvae are discrete; flattened coxosternum of 2nd legs forms a vertical plate. 44–60 segments. 3 gen., 17 spp. and/or subspp. Pacific Coast from Baja California to British Columbia and east to Montana and Alberta.

Tribe Gosiulini, **new.** Peltocoxites are subcylindrical. Gonopods are ventrally directed, elongated, telopodite is divided, coxite is simple; lobes of seventh pleurites encircle gonopods. Vulval apparatus is small, unsclerotized; coxosternum of 2nd legs forms a flap on anterior surface of vulval apparatus. 46–57 segments. 1 gen., 4 spp. Arizona and Colorado southeast to northeastern Mexico.

Tribe Karteroiulini, **new.** Peltocoxites are elongated, podlike. Gonopods are sclerotized, connected at base by a sclerite or by a muscle; telopodite and coxite of gonopods are coalesced; one short sperm canal fibril. Vulvae are discrete; flattened coxosternum of 2nd legs forms a vertical plate. Body surface shines. 41–56 segments. 2 gen., 2 spp. Japan and Pacific Coast from northern California to southcentral Alaska.

Tribe Mulaikiulini, **new.** Peltocoxites are elongated, podlike. Gonopods are unsclerotized; telopodite is divided, coxite is simple; sternum is divided. Vulval apparatus is thin, unsclerotized, vulval plate is wider than its length. 2nd legs are almost as large as 3rd. Body surface shines. 66–74 segments. 1 gen. 4 spp. Southcentral California.

Tribe Paraiulini, **new.** Peltocoxites are wide at base and narrowed distad; peltotelopodites are broad and flattened, covering gonopods. Gonopods are heavily sclerotized, complicated by division of telopodite and coxite; 2 or 3 short sperm canal fibrils; sternum is divided. Vulval apparatus is large, thickened, sclerotized, openings are on ectal surfaces; 2nd legs are minute rudiments. 40–55 segments. 3 gen., 38 spp. and/or subspp. Tamaulipas south to western Guatemala.

Tribe Simiulini, **new.** Peltocoxites are elongated, podlike. Gonopods are weak, coxa is a distinct segment, telopodite is divided, coxite is simple. Female is unknown. 56–66 segments. 1 gen., 2 spp. Central California.

Tribe Spathiulini, **new.** Peltocoxites are elongated, with flanges on caudal surface. Gonopods are elongated, coalesced at base, telopodite is divided, coxite is simple. Vulval apparatus is slightly thickened, unsclerotized, vulval plate is longer than its width; 2nd legs are minute rudiments. 44–53 segments. 2 gen., 4 spp. Central and south-central California.

Tribe Uroblaniulini, **new.** Peltocoxites are elongated, with flanges on caudal surface. Gonopods are elongated, moderately sclerotized, complicated by lobes, a large column, heavy setae, and numerous fibrils in the sperm canal. A wide apodeme is added to base of gonopods. Vulval apparatus is thickened, unsclerotized, longer than its width; 2nd legs are minute rudiments. Caudal spine is usually hooked. 43–61 segments. 4 gen., 17 spp. Discontinuous distribution: Puget Sound east to Idaho and Georgia north to near Hudson's Bay.

Tribe Zygiulini, **new.** Peltocoxites are elongated, quadrate. Gonopods consist of a curved, sclerotized telopodite and an unsclerotized coxite. Pleurites of segment 7 and peltocoxites cover gonopods. Vulval apparatus is thin, weakly sclerotized, wider than its length; 2nd legs are minute rudiments. 45–50 segments. 1 sp. New Mexico.

Subfamily Ptyoiulinae, **new.** Body is weakly caniculate above pores and setose; peltocoxites have a calyx on the apex, are podlike, function as sperm channel; gonopods are disjunct, have one short sperm canal fibril; vulvae are small, ovoid, discrete, and aperture is guarded by a yolk rather than an operculum; 2nd legs and their sternum are degenerate. 55–70 segments. 1 tribe.

Tribe Ptyoiulini, **new.** With the characteristics of the subfamily. 1 gen., 3 spp. and/or subspp. Centered in the Appalachian Mountains, extending north into southern Ontario, west to Arkansas, and south to Florida.

Note: The tribes Aniulini, Gosiulini and Zygiulini constitute the supertribe Aniulina and the tribes Bollmaniulini and Karteroiulini the supertribe Bollmaniulina. No other supertribes have been designated because the relationship of the remaining tribes is uncertain.

A diagrammatic report of the relationship of the tribes and some of the higher taxa, based on the evidence summarized in this section, is presented in Fig. 5.

ORIGIN AND DISTRIBUTION

Paraiulids range from 61°N Lat. in southcentral Alaska and 50°N Lat. in Ontario south to 15°N Lat. in southwestern Guatemala. The largest populations are in temperate deciduous forests or in once forested grasslands where the annual precipitation is more than 30 in. They occur where the annual precipitation is under 20 in. only as small isolated populations. No cave adapted populations are known.

The nearest relatives of the Paraiulidae, the Mongoliulidae, are in northeastern China, Korea, and Japan. The essential characters which separate them tend to be of a more primitive grade in the Mongoliulidae than in the Paraiulidae. Pocock (1903) defined both families adequately, but Japanese workers have ignored his work and assigned many species of the Mongoliulidae to the Paraiulidae in the erroneous belief that all species with enlarged first legs in the male are confamilial.

All known species of the Paraiulidae except one are in North and Central America. The exception, *Karteroiulus niger*, is in the Japanese islands of Honshu and Shikoku. The combination of derived and generalized characters in this species negates the probability that it is very near the parent stem of the family; nevertheless, it may be a relic. If so, the paraiulids moved from a center of origin in Japan or elsewhere in eastern Asia eastward across the Bering bridge. The nearest relative of *K. niger*, *Litiulus alaskanus*, ranges along the Pacific Coast from southcentral Alaska to northern California. The next nearest relatives are the more generalized members of the tribe Bollmaniulini, which are abundant at the southern end of the range of *L. alaskanus*. This distribution suggests that either *K. niger* or a direct antecedent was a westward migrant across the Bering bridge and that the center of origin of the family must be sought for in North America rather than in eastern Asia. Such an occurrence of a single migrant representative of a large taxon is not unusual (Ross, 1962). *L. alaskanus* is now in a favorable position for becoming a migrant across the Bering bridge, should it become available. There may be other migrants or relics in eastern Asia. Except for Japan, the diplopod fauna of eastern Asia is scarcely known. This area is tantalizing to students of the Paraiuloidea, for collecting there will surely yield more taxa.

Evidence from distribution and morphology points toward California as the center from which early dispersion occurred. The most primitive species are in California, while the most highly derived are towards the limit of the family's range. Californian species are in four tribes, of which the Simiulini, Mulaikiulini, and Spathiulini occur only in that state and are small bodied, scarce, and little known. The larger, better known Bollmaniulini range beyond California. It is not possible to designate any one of these tribes as the one nearest the parent tribe, for each has derived as well as primitive characters. Bollmaniulini is the only tribe of the four with the vulvae in the primitive, discrete condition and a fibril in the sperm canal; its gonopods are coalesced. In the Simiulini the gonopods are disjunct and the coxae are divided from the telopodites, but like the other small bodied tribes, there are no sperm canal fibrils.

Dispersion occurred when conditions across the continent were favorable. The members of a community moved in unison, thus tending to keep monophyletic taxa together. There is little evidence of the dispersal routes, for great areas in the middle of the continent are so arid the intervening populations have not survived. From the present records it can be inferred that there were two dispersion centers, one in the Puget Sound area and the other in central California (Fig. 6). The Bollmaniulini moved south into Baja California, north into British Columbia, and spilled out through the Rocky Mountains as far as Montana. They gave rise to the Karteroiulini, which moved on up the Pacific coast to southern Alaska and eventually to Japan. The Uroblaniulini, the only tribe with members on both sides of the continent, probably moved eastward from the northern dispersal center. The Ptyoiulini may have taken the northern route also. Their greater degree of modification suggests greater age, and hence they may have dispersed before the younger tribes of the Paraiulinae. Dispersion from the southern center probably took a more southern route and included the tribes Zygiulini, Gosiulini, Aniulini, and Paraiulini or their antecedents. The Aniulini underwent extensive speciation in the Southern States and radiated north, west to the Rocky Mountains, and south to northeastern Mexico. The route of the Paraiulini is difficult to reconstruct; severe climatic conditions have destroyed the intervening populations in northern Mexico.

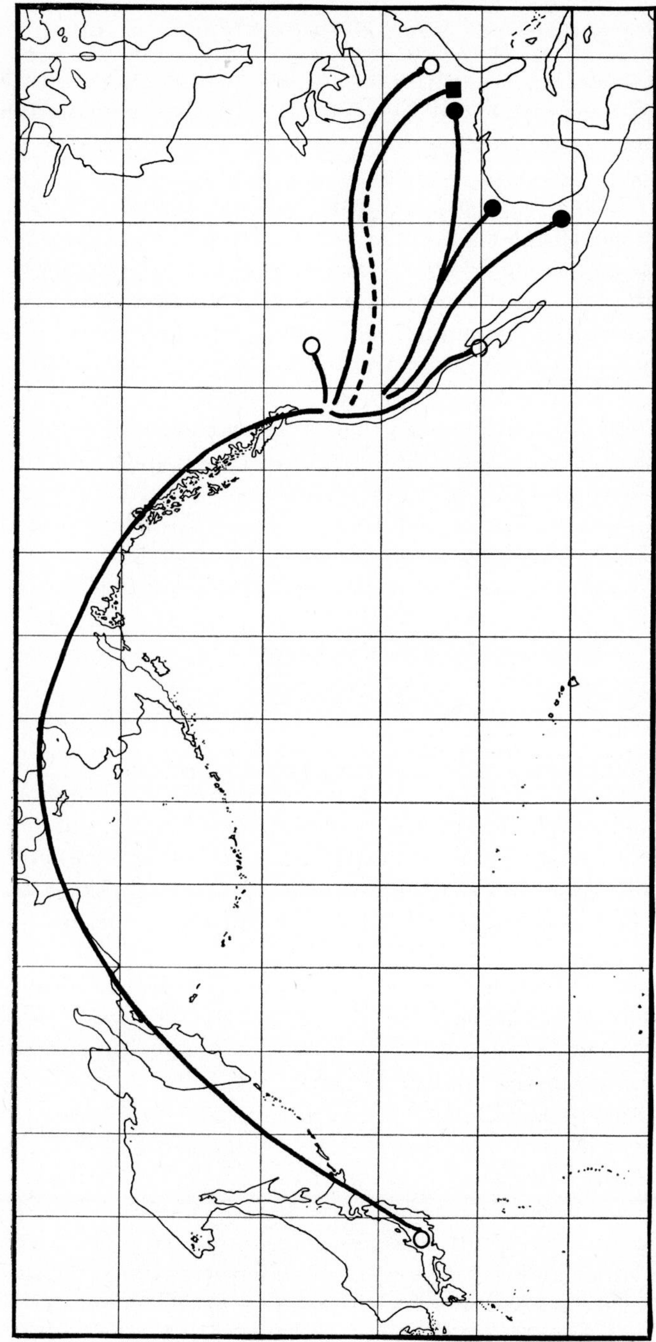

FIG. 6. Main dispersal routes of the Paraiulidae. Subfamily Ptyoiulinae, square symbol; subfamily Paraiulinae, circles; group Disjuncta, solid circles; group Nondisjuncta, open circles.

Generalizations derived from the distribution of the family are:

1. The ranges of the most primitive members are in western North America and the most highly evolved are in eastern North America, Central America and Japan.

2. The range overlap is greatest and populations are largest in ecologically favorable areas, i.e., well watered temperate hardwood forests, as in the Puget Sound area and the Appalachian Mountains.

3. Unless interrupted by unfavorable climatic or geological conditions, the range of a tribe is continuous, suggesting that the members of a tribe constitute a monophyletic group derived from the tribe's founders, who underwent further differentiation during and after dispersion.

Under present conditions the range of most tribes is receding. Range recession has resulted in the isolation of numerous populations, of which many have become the centers of rapid speciation. This is especially apparent in parts of Mexico and Texas. Range expansion at the tribal level is probably occurring in southeastern Canada.

The only known fossil species (Miner, 1926), from the Oligocene of Florissant, Colorado, is some 15 million years old. It cannot be placed with certainty in any of the present tribes.

DISCUSSION AND SUMMARY

Data for inferring the phylogeny of the Paraiulidae are mainly from functional and morphological characters of the genitalia, but some are from somatic characters and distribution. Somatic characters are important but not numerous. Distribution is useful because the low vagility of diplopods tends to keep taxa within a common descent together.

Inasmuch as form and function of genitalia evolve simultaneously, they cannot be separated in inferring phylogeny. The modifications of the external parts of the genitalia are mainly changes to parts already present rather than the addition of new parts. The principal changes are in the nature of enlargement, strengthening, and elaboration. Fusion, reduction, and loss are less common. The same sorts of changes have occurred to the structures concerned with the functioning of the genitalia: the muscles, apodemes, and sterna. These parts are more stable throughout the family than are the external parts of the genitalia and at the subgeneric level they present fewer changes; hence, they provide reliable data for inferring the relation of the higher taxa of the family.

The early paraiulids had the following combination of characters: 60 to 70 body segments and a width of 1·5 to 2·5 mm; the male had an enlarged promentum, mandibular cheeks, and collum, a penial apparatus incorporating the second legs, disjunct gonopods, a closed sperm canal and a sperm canal fibril; the female's second legs were almost as large as the third, and the small, discrete vulvae consisted each of a bursa and a small operculum.

Very early there developed, possibly independently at least three times, methods of moving the gonopods in unison. The members with this character are the Nondisjuncta. In the more primitive Disjuncta, the gonopods are moved independently. Both groups underwent further differentiation and the descendent tribes dispersed to the limits of the family's range. The most primitive members of each group are now in California, where the original differentiation is assumed to have occurred.

The tribe is a logical taxon to work with in studying the early phylogeny of the family. The monophyly of each one is reflected in its distribution within a continuous area and in the common peculiarities of the functional morphology of the genitalia of its members. Gaps between tribes are obvious and unequal; less obvious are the relationships of most of them.

References

Bollman, C. H. (1893). The Myriapoda of North America. *Bull. U.S. natn. Mus.* No. 46: 1–120.

Brandt, J. F. (1833). Tentaminum quorundam monographicorum Insecta Myriapoda Chilognatha Latreilli spectantium. *Byull. mosk. Obshch. Ispȳt. Prir.* **6**: 194–209.

Hefner, R. A. (1929). Studies of parajulid diplopods. 1. The development of the external sexual structures of *Parajulus impressus* Say. *J. Morph. Physiol.* **48**: 153–163.

Hoffman, R. L. (1961). A new genus and subfamily of the diplopod family Nemasomatidae from the Pacific Northwest. *Proc. ent. Soc. Wash.* **63**: 58–64.

Koch, C. L. (1847). System der Myriapoden. In *Kritische Revision der Insectenfauna Deutschlands* **3**. Herrich-Schaeffer, G. A. W. (Ed.).

Mauriès, J.-P. (1970). Un nouveau Blaniulide cavernicole du pays Basque Français. Eléments d'une nouvelle classification des diplopodes Iulides (Myriapoda). *Annls Spéléol.* **25**: 711–719.

Miner, R. W. (1926). A fossil myriapod of the genus *Parajulus* from Florissant, Colorado. *Am. Mus. Novit.* No. 219: 1–5.

Pocock, R. I. (1903). Remarks upon the morphology and systematics of certain chilognathous diplopods. *Ann. Mag. nat. Hist.* (7) **12**: 515–532.

Ross, H. H. (1962). *A synthesis of evolutionary theory.* Englewood Cliffs, N.J.: Prentice Hall.

Discussion

KRAUS: Are the paraiulids which extend into Central and South America, as far as Guatemala, restricted to higher altitudes?

CAUSEY: Yes, they are mainly on the Mexican plateau. In Mexico they extend down into the temperate zone near the coast but not into the low *tierra caliente* along the coast. They are rare in northern Mexico which is very arid. The largest populations are on the plateau, following the mountains down into Guatemala. In western Guatemala they occur only in the mountains.

KRAUS: So they are perhaps younger invaders?

CAUSEY: Yes, I am sure they are, but their invasion route is now quite arid and those from the north could surely not cross that way now.

HAACKER: Do you have any ideas or observations on the function of the very peculiar vulval apparatus and the male appendages?

CAUSEY: No. I have found them mating but have been concerned mainly with taxonomy. Their function must be quite different in the various super-tribes. Brölemann had a paper on the vulval apparatus of paraiulids but he dealt entirely with the animals which I put into one tribe, the Aniulini. There is a good picture in his paper of the vulval apparatus but the vulvae in the other tribes are very different.

HAACKER: It would be very interesting because these structures seem to be very peculiar among Diplopoda—especially in the females.

STRASSER: It would be interesting to know something about the Asian populations since their systematic status seems obscure.

CAUSEY: Yes. Pocock described some and the Japanese have described 20 or so species, perhaps more. There are two families, one characterized by the reduction of the seventh legs in the male, the other with normal seventh legs. I have not studied the females extensively in the former, the family Mongoliulidae. It appears an interesting area to explore—they occur in Japan, Korea and Mongolia. I do not know how far they extend.

STRASSER: Is there any hypothesis about the connexion between western and eastern groups?

CAUSEY: I think the superfamily evolved in Asia and that either immediate precursors or the early members of the family moved into North America by the Alaskan land bridge and then one species returned that way.

KRAUS: But the pattern of distributions in North and Central America which you showed us seems so complicated. Might there have been more than one invasion from Asia to North America?

CAUSEY: I am willing to entertain the idea—but I think the family underwent the diversification described here in western North America, because only one species has been found in Japan, and while it belongs to a primitive tribe, it is a highly evolved member of that tribe. It must have moved there from North America. Probably much evidence of the movements has been lost due to changes in climate and the death of many populations *en route*.

Symp. zool. Soc. Lond. (1974) No. 32, 41–52.

THE GROUP TAXONOMY AND GEOGRAPHY OF THE SPHAEROTHERIIDA (DIPLOPODA)

C. A. W. JEEKEL

Institute of Taxonomic Zoology (Zoological Museum), University of Amsterdam, The Netherlands

SYNOPSIS

The diplopod order Sphaerotheriida (the so-called "giant pill-millipedes") occupies a discontinuous geographical area which includes South Africa, Madagascar, the entire Oriental region, the eastern part of Australia, and New Zealand. A formerly adopted classification of the group was based primarily on the structure of the antennae, and must be discarded as artificial.

Following a study of material and literature a new classification, based primarily on female sexual characters, is proposed. It is shown that the taxonomic subdivision of the order largely coincides with the geographical isolation of the various groups: the species from South Africa, from Madagascar, from India and Ceylon, from Australia and New Zealand, and from the area between the Himalayas and the Celebes, each forming taxonomically well-defined categories.

It is suggested that the present geographical distribution of the various sphaerotheriid families, subfamilies and tribes may be traced back to the Mesozoic period.

The following new suprageneric taxa are proposed: Cyliosomatini nov. trib., Arthrosphaerinae nov. subfam., Arthrosphaerini nov. trib., Zoosphaeriini nov. trib., and Castanotheriinae nov. subfam.

INTRODUCTION

The primary division of the chilognath diplopods concerns the distinction between Pentazonia or Opisthandria on the one hand, and the Proterandria on the other. Whereas the latter group contains the vast majority of recent diplopods, the first is composed of three well-separated groups with a comparatively small number of genera and species.

These three groups, which I regard as having each the taxonomic status of an order, are the Glomerida, the Sphaerotheriida, and the Glomeridesmida. In most classifications Glomerida and Sphaerotheriida are regarded as more closely related to each other than to the Glomeridesmida, and are brought into one order for that reason. However, the differences between Glomerida and Sphaerotheriida seem to be of such importance that to combine them in one order is just too hypothetical to have any practical value at all. From the morphological differences between the three orders and the uniformity within each of them it is evident that Glomerida, Sphaerotheriida, and Glomeridesmida must be very old groups, probably dating from the Palaeozoic.

The Glomeridesmida is a small order consisting of a few genera with a small number of species occurring in the tropical parts of the Oriental and Neotropical regions.

The Glomerida and Sphaerotheriida are larger, having some 200 to 250 described species each. Both have a discontinuous geographical distribution, their range being largely complementary. The Glomerida belong essentially to the northern hemisphere. They are represented in North America by a small number of species, and are furthermore well differentiated in the western and eastern areas of the Palaearctic region, as well as in the Oriental region.

The Sphaerotheriida, on the other hand, are not found in the Americas or in the Palaearctic region. They occur in South Africa, in Madagascar, in the whole Oriental region south-eastward to Java, the Celebes and the Moluccan island of Halmahera, in Australia along the east coast from Cape York down to Tasmania, and in the two main islands of New Zealand.

PREVIOUS CLASSIFICATIONS AND THEIR GEOGRAPHICAL IMPLICATIONS

Former attempts to subdivide the Sphaerotheriida into families or lower suprageneric categories have been abortive.

After a study of some Australian material of the group, and a comparison of the Australian species with a species from Madagascar and data from the literature, Brölemann (1913: 79) set up the family Sphaeropoeidae for *Sphaeropoeus* Brandt, which tentatively included also the genera *Arthrosphaera* Pocock and *Zephronia* Gray, leaving the South African and Malagassy species of *Sphaerotherium* Brandt and the genus *Cyliosoma* Pocock from Australia in the family Sphaerotheriidae. The two families were distinguished mainly on account of some differences in the anatomy of the head.

Sphaerotheriidae and Sphaeropoeidae were accepted as two distinct families by Attems (1914: 140, 142) who, however, characterized them exclusively by the structure of the antennae, more specifically by the number of visible antennomeres: seven in the Sphaerotheriidae, six in the Sphaeropoeidae. Later, in 1926, Attems (1926: 119) dropped this division, and once again recognized only one single family. As was pointed out by Verhoeff (1929: 1378) quite correctly, the number of visible antennomeres is closely connected with the number of sensory cones at the end of the terminal antennomere, of which there may be four or many. These features probably developed polyphyletically and are hardly suitable for differentiating suprageneric categories.

As usual, the result of the artificial subdivision of the Sphaerotheriida into two families was that the geographical distribution of the one largely overlapped that of the other. Roughly, the range of the Sphaerotheriidae in the sense of Attems, 1914, included South Africa, Madagascar, Australia, and New Zealand, whereas the Sphaeropoeidae were dominant in the Oriental region.

MAIN CHARACTERS

The descriptions of the sphaerotheriid species which have been published over the years are of very unequal value and in most cases insufficient for recognition of the species. An extensive revision of the type material of all described species is obligatory if any progress towards a modern taxonomy of the group is to be made at all.

In the course of a preliminary study of material of Sphaerotheriida from each of the main distributional areas, and a perusal of the pertinent literature, it was first noted, that within certain geographical areas there exists a remarkable stability in certain features of the vulvae of the females. In the species of the Indian peninsula and Ceylon, generally referred to *Arthrosphaera* Pocock, the vulvae are characterized by the separation of the opercula by the distal parts of the medial sclerites of the bursae. Exactly the same condition was found in the Sphaerotheriida of Madagascar, with the only apparent difference that the operculum in these species is subreniform, i.e. constricted in the middle (Fig. 1a). The structure of the vulvae in the Indian and Malagassy species is clearly different from that in the sphaerotheriids in the remaining distributional areas. In all these the opercula of the opposite coxae are side by side and never separated by parts of the medial bursal sclerite.

A closer examination of species from South Africa, Australia and New Zealand, and South-East Asia, showed that in spite of a superficial similarity in the basic structure of the vulvae, the latter showed another important difference. Apparently in all South-East Asian species the distal part of the bursa is applied to the basal part of the operculum. The operculum is not in the slightest way embraced by bursal membranes (Fig. 1c). In the South African and Australian species the bursa embraces the operculum at its medial and lateral edges (Fig. 1b). This condition, which can be established by close observation, is also found in the Indian and Malagassy species.

It is evident that these particular features of the vulvae of the females are of the greatest importance for the classification of the Sphaerotheriida because of their stable nature. Together with the already known characters of the male telopods and the undoubtedly existing

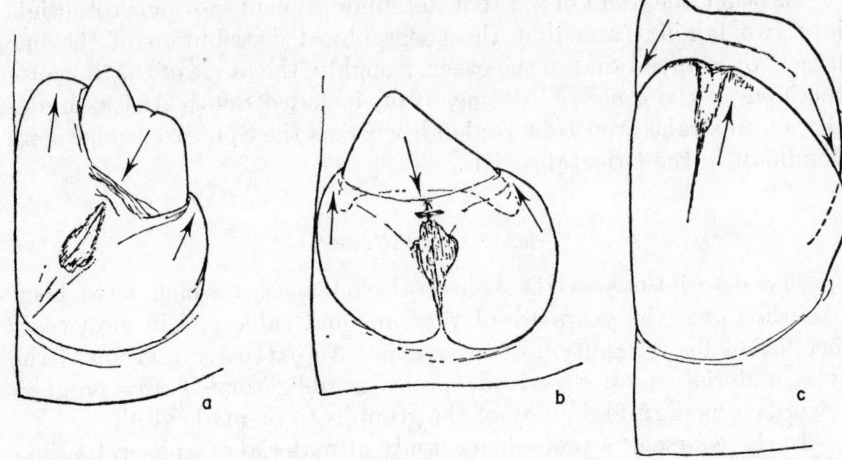

FIG. 1. (a) Left vulva, posterior aspect, of a Malagassy sphaerotheriid. (b) The same of an Australian sphaerotheriid. (c) The same of a Malayan sphaerotheriid. The black arrows indicate the important structural details.

but not yet fully known characters of the non-sexual morphology they will enable us to come to a sound classification of the order.

Another interesting female character indicative of the close relationship between the Malagassy and Indian Sphaerotheriida concerns the stridulatory apparatus. This structure has been recorded already by de Saussure & Zehntner (1902: 41, pl. 15, Figs 11, 11a) for *Sphaerotherium actaeon* White from Madagascar. It is, unfortunately, unknown whether or not it occurs in all Malagassy sphaerotheriids. The apparatus consists of three or four oblique carinae on each side of the hypoproct, and a knob near the condylus of the prefemur of the last pair of legs. Apparently sound is produced by quick rhythmical movements of the knobs across the carinae.

A basically similar structure apparently not previously noticed is found in the Indian and Ceylon species. Here, however, there is only one carina on each side of the hypoproct. It corresponds with a whole series of knobs all along the distal margin of the prefemur of the last pair of legs.

As far as is known, a female stridulatory apparatus does not occur elsewhere in the Sphaerotheriida.

NEW CLASSIFICATION

With the aid of the new characters we come to the following classification.

Family Sphaerotheriidae Brandt, 1833

Basis of operculum of vulva embraced by bursa.

Subfamily Sphaerotheriinae Brandt, 1833

Opercula of opposite vulvae not separated by the protruding medial pieces of the bursae. Females without stridulatory apparatus.

Tribe Sphaerotheriini Brandt, 1833. Posterior telopods with a stridulatory apparatus consisting of several strongly chitinized ridges.

Genera: *Sphaerotherium* Brandt, 1833, *Kylindotherium* Attems, 1926. Moreover the following generic names have been based on species belonging to, or probably belonging to the same tribe: *Bournellum* de Saussure & Zehntner, 1902, *Eubournellum* Attems, 1908, *Natalobelum* Verhoeff, 1924, *Neobournellum* Attems, 1908, *Oligaspis* Wood, 1865, and *Tetraconosoma* Verhoeff, 1924.

Tribe Cyliosomatini **nov. trib.** Posterior telopods without stridulatory apparatus.

Genera: *Cyliosoma* Pocock, 1895, *Cyliosomella* Verhoeff, 1924, *Epicyliosoma* Silvestri, 1917, *Paracyliosoma* Verhoeff, 1928, *Procyliosoma* Silvestri, 1917, and *Syncyliosoma* Silvestri, 1917. A revision of this group is necessary to evaluate the taxonomic status of most of these generic names.

Subfamily Arthrosphaerinae **nov. subfam.**

Opercula of opposite vulvae separated medially by the protruding medial pieces of the bursae. Females with a hypoproct stridulatory apparatus.

Tribe Arthrosphaerini **nov. trib.** Prefemur of anterior telopods without a stridulatory apparatus composed of one or more chitinous carinae. Operculum of vulva not constricted in the middle.

Genera: *Arthrosphaera* Pocock, 1895, *Odontosternum* Attems, 1943, *Trivandrobelum* Verhoeff, 1937, and *Trochosoma* Chamberlin, 1921, most of which remain to be re-examined.

Tribe Zoosphaeriini **nov. trib.** Prefemur of anterior telopods with a stridulatory apparatus composed of one or more chitinous carinae. Operculum of vulva constricted in the middle, subreniform.

Genera: *Zoosphaerium* Pocock, 1895, *Globotherium* Brölemann, 1922, *Heligmasoma* Chamberlin, 1921, and *Sphaeromimus* de Saussure & Zehntner, 1902. The latter two genera need a re-examination.

Family Sphaeropoeidae Brölemann, 1913

Basis of operculum of vulva not embraced by the bursa; the distal part of the bursal sclerites applied to the basal part of the operculum.

Subfamily Sphaeropoeinae Brölemann, 1913

Movable digit of the posterior telopods composed of a separate tibia and tarsus.

Genera: *Bothrobelum* Verhoeff, 1924, *Chinosphaera* Attems, 1935, *Indosphaera* Attems, 1935, *Kophosphaera* Attems, 1935, *Leptotelopus* Silvestri, 1897 (synonym: *Lophozephronia* Attems, 1936, new synonym) *Prionobelum* Verhoeff, 1924, *Sphaerobelum* Verhoeff, 1924, *Sphaeropoeus* Brandt, 1833 (synonym: *Leptoprotopus* Silvestri, 1897, *Lissosphaera* Attems, 1943, *Pantitherium* Attems, 1932, *Tonkinobelum* Verhoeff, 1924, new synonyms).

Subfamily Castanotheriinae **nov. subfam.**

Movable digit of posterior telopods composed of a single podomere, the tibiotarsus.

Genera: *Borneopoeus* Verhoeff, 1924, *Castanotherium* Pocock, 1895, *Castanotheroides* Chamberlin, 1921, *Luzonosphaera* Wang, 1951, *Pulusphaera* Attems, 1935, and *Rajasphaera* Attems, 1935.

The above classification is a preliminary outline which is based on the present very insufficient knowledge of the Sphaerotheriida. The reference to various categories of certain genera may be somewhat arbitrary and based on indirect evidence rather than on morphological facts. Certainly a more thorough study of the group will result in more extended diagnoses of the suprageneric categories.

Regarding the status of certain names the following remarks must be made. The family name Zephroniidae Gray, 1843, will probably replace Sphaeropoeidae eventually. The type genus, *Zephronia* Gray, 1832, which is based on a virtually unknown species *Zephronia ovalis* Gray, 1832, may belong either to the subfamily Sphaeropoeinae or to the Castanotheriinae. Most of the species referred to *Zephronia* by Attems and others belong to the Sphaeropoeinae.

The name Oligaspidinae Bollman, 1893, based on *Oligaspis* Wood, 1865, is likely to be a synonym of Sphaerotheriini, *Oligaspis* probably being synonymous with *Sphaerotherium* Brandt.

The genus *Cryxus* Leach, 1814, based on *Julus ovalis* Linnaeus, 1758, cannot be referred to any of the above categories because of the enigmatic status of the type-species.

It is important to note that, whereas the family Sphaerotheriidae is little differentiated on the genus level, this differentiation in the Sphaeropoeidae is much more conspicuous. One gets the impression that the differentiation within the latter family is of a geologically more recent period.

GEOGRAPHY OF THE SPHAEROTHERIIDA

The new concept of the two families, Sphaerotheriidae and Sphaeropoeidae, coincides geographically with two separate distributional areas. The Sphaerotheriidae are representing the order in South Africa, Madagascar, India and Ceylon, and Australia and New Zealand, whereas the Sphaeropoeidae occupy the remaining part of the ordinal range, i.e. the Oriental region from Assam to Celebes and Halmahera, and from the Philippines to Java. The two families are spatially separated by the valley of the Ganges river and by the eastern part of the Indo-Australian archipelago.

The family Sphaerotheriidae falls naturally into four tribes which are as well-defined morphologically as they are widely separated geographically. The Sphaerotheriini represent the order in South Africa, the Zoosphaeriini in Madagascar, the Arthrosphaerini in the Indian Peninsula and Ceylon, and the Cyliosomatini in Australia and New Zealand. At the time it is not yet clear whether the Sphaerotheriini are related most closely to the Zoosphaeriini or to the Cyliosomatini. It is, however, quite evident that the Zoosphaeriini and the Arthrosphaerini are more closely related to each other than to any other sphaerotheriid group; the subfamily Arthrosphaerinae in which they are combined suggests a strong tie between the faunas of Madagascar and India and Ceylon. The wide geographical gap between these two tribes contrasts markedly with the much narrower space separating the Zoosphaeriini from the Sphaerotheriini or the Arthrosphaerini from the Sphaeropoeinae. It is also interesting to note the close relationship between Australian and New Zealand sphaerotheriids notwithstanding the distance between them.

The family Sphaeropoeidae occurs in a coherent area in South-East Asia, but, contrary to what we have seen in the Sphaerotheriidae, the two well-defined subfamilies occupy partially overlapping areas. The Sphaeropoeinae range from Assam to Borneo and Sumatra, whereas the Castanotheriinae are confined to the Philippines, the northern Moluccas, and the larger Sunda islands. The two subfamilies meet in the Sunda area, and Borneo and Sumatra seem to be the only islands where representatives of both subfamilies have been recorded with certainty. As far as is known at present the Castanotheriinae have never been reported from the South-East Asian mainland.

DISTRIBUTIONAL HISTORY

The distinct and uniform morphology of the Sphaerotheriida and its wide discontinuous range seem to indicate that the order is very old, and may have existed as a separate taxonomic unity since the Palaeozoic.

Each of the four main geographical areas occupied by the family Sphaerotheriidae has its own systematic category. This suggests that we are dealing here with the remnants of a once continuous range divided by geographical barriers into four isolated populations. These then developed their own characters, deviating from the original type only in morphological details (Fig. 2).

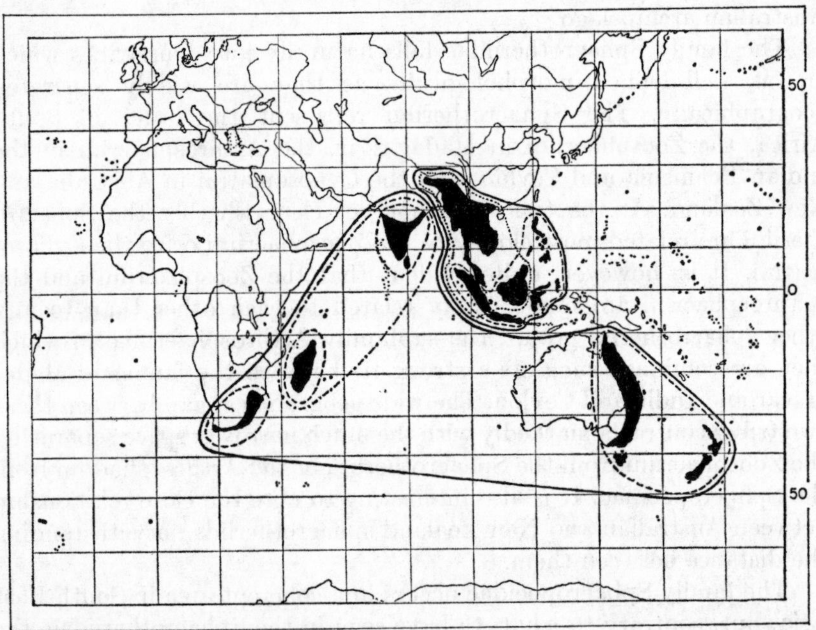

Fig. 2. Map showing the world-distribution of the order Sphaerotheriida in black. The solid lines indicate the areas of the Sphaeropoeidae (South-East Asia) and the Sphaerotheriidae (remaining distribution). Within the range of the Sphaeropoeidae the areas of the subfamilies Sphaeropoeinae and Castanotheriinae are indicated by a dotted and an interrupted line, respectively. Within the range of the Sphaerotheriidae the areas of the various tribes are indicated by interrupted lines. The distribution of the Arthrosphaerinae is particularly indicated by – · – – · in order to emphasize the close relationship between the two composing tribes.

The areas in which the four tribes of the Sphaerotheriidae occur are old continental shields or parts of such shields, which were not disturbed by submergence or orogenesis for a very long time, at least not since the Palaeozoic or the Mesozoic. The present distribution of the sphaerotheriid tribes appears to be largely autochthonous.

The lack of important geological activity for a long time may also account for the slight morphological differentiation within each of the sphaerotheriid tribes. Probably the limited morphological variation originated mainly in climatic changes rather than in orogenetic disturbances.

By contrast an important part of the present range of the Sphaeropoeidae includes tertiary mountain chains. Furthermore the two subfamilies Sphaeropoeinae and Castanotheriinae have a partly overlapping distributional area. Although these two subfamilies might also be separated by a geographical barrier, orogenetic activites probably resulted in a secondary overlap of primarily separated areas. Possibly the same geological disturbances during the Tertiary resulted in the morphological differentiation within the Sphaeropoeinae and the Castanotheriinae, which is much more conspicuous than in the family Sphaerotheriidae.

The taxonomic and geographical hiatus between the families Sphaerotheriidae and Sphaeropoeidae possibly reflects the long separation of the two groups by the Tethys depression.

It is difficult to find the palaeogeographical maps necessary to trace the time of origin of the sphaerotheriid subfamilies and tribes. Possibly the two maps (Fig. 3a & b) illustrating the land-sea distribution in the Upper Jurassic and the Upper Cretaceous, respectively, after Furon (1959: pl. VI, pl. VIII) give an indication of the period when these groups evolved.

In the first map the coherence of the African, Malagassy, and Indian areas is still indicated. The range of the Cyliosomatini was apparently divided at an earlier period. Even from this map it is clear that the distribution of the family Sphaerotheriidae cannot be explained without reference to continental drift. In fact the distribution of this family still seems to include the remnants of the old southern continent Gondwanaland. Remarkably enough no exchange of faunal elements took place after the Indian peninsula became connected with the northern continent, which is contrary to the condition described in the family Paradoxosomatidae (Jeekel, 1968: 135).

As far as is known there seems to be no clear evidence in geology that Madagascar and India were connected at one time. The close relationship between the Arthrosphaerini and the Zoosphaeriini points

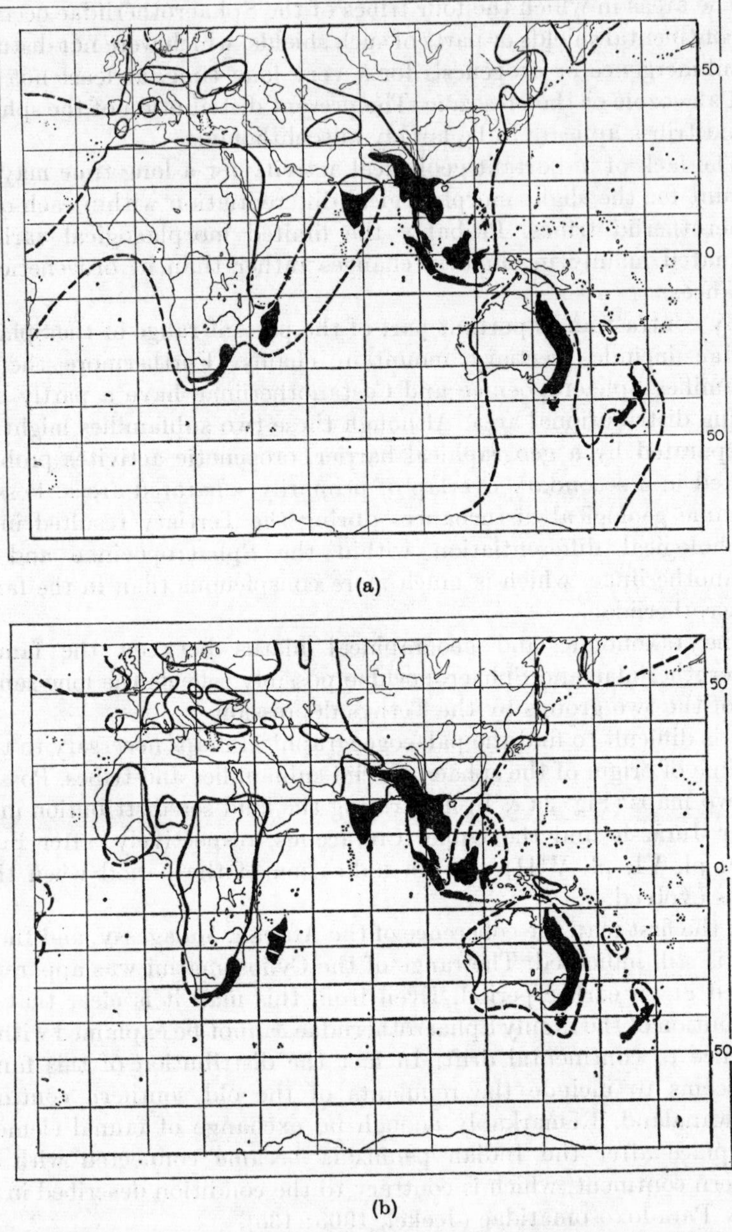

FIG. 3. Two maps showing the world-distribution of the Sphaerotheriida in black against the land-sea distribution in the Upper Jurassic (a) and Upper Cretaceous (b) (after Furon, 1959) as evidence for a possible Mesozoic origin of the current distributional pattern of the order.

to such a connexion. One might even say that the connexion between Madagascar and India existed more recently than that between Madagascar and Africa.

The current range of the Sphaeropoeidae largely existed already in the Upper Jurassic, the separation of the areas of the Sphaerotheriidae and Sphaeropoeidae became a fact probably in the late Palaeozoic or in the early Mesozoic.

It is interesting to note the division into two areas of the sphaeropoeid distributional range in the Upper Cretaceous. Possibly the eastern area gave rise to the Castanotheriinae, and the western to the Sphaeropoeinae.

References

Attems, C. (1914). Die Indo-Australischen Myriopoden. *Arch. Naturgesch.* **80A** (4): 1–398.

Attems, C. (1926). Myriopoda. *Handb. Zool.* **4**: 1–402.

Brölemann, H. W. (1913). The Myriapoda in the Australian Museum. Part ii.—Diplopoda. *Rec. Aust. Mus.* **10**: 77–158.

Furon, R. (1959). *La Paléogéographie.* Paris: Payot.

Jeekel, C. A. W. (1968). *On the classification and geographical distribution of the family Paradoxosomatidae (Diplopoda, Polydesmida).* Amsterdam: privately published.

Saussure, H. de, & Zehntner, L. (1897/1902). Myriapodes de Madagascar. In *Histoire physique, naturelle et politique de Madagascar* **27** (53): i-viii, 1–356, Grandidier, A. (ed.) Paris.

Verhoeff, K. W. (1928–1932). Klasse Diplopoda. *Bronn's Kl. Ordn. Tierreichs* **5** (II) (2): i–vi, 1073–2084.

Discussion

Haacker: I agree with your doubts about the utility of head characters, especially antennal characters. I have collected South African species and have the impression that they vary even within one population and so are not useful for classification. I think the telopods are very useful and also the female characters you describe and the stridulatory organs. I was surprised by the close relationship you find between the South African and the New Zealand species.

Jeekel: It is only a hypothesis. The African and New Zealand species appear to me to be closer to each other than to the Madagascan and Indian species. Brölemann had the same idea and created the family Sphaerotheriidae, in the strict sense, because he found such close similarity between the Australian *Cyliosoma* and the African and Madagascan species of *Sphaerotherium.*

HAACKER: *Cyliosoma* has no stridulatory organs?

JEEKEL: No.

HAACKER: So you consider the male stridulatory organs of *Sphaerotherium* and, for instance, *Zoosphaerium*, to be convergent?

JEEKEL: Yes, I think I must. Perhaps *Sphaerotherium* may be intermediate between *Cyliosoma* and *Zoosphaerium* from Madasgascar—but I am not sure. It is the overall idea of the close relation of Australian and South African species which is important.

CAUSEY: I have been curious about this stridulation. Can it really be heard?

JEEKEL: I think Dr. Haacker can answer.

HAACKER: I have a tape recording with me; you will hear it when I present my paper.

JEEKEL: But only the sound of the South African males; not the females—which would be most interesting to hear, I think.

STRASSER: Have these new female characters profoundly altered the previous classification based on male characters?

JEEKEL: Yes, the male telopods were only used for characterizing the original genera—many of which I now regard as tribes. The telopods give very few clues of inter-relationships between tribes but the vulvae give evidence of two distinct groups; because of their constancy and stability I think the female characters are very useful for determining higher categories—but we must look for other supporting characters.

STRASSER: Is there any hope of finding other characters? Haacker told me he was working on this question.

HAACKER: I think it is very difficult to find other morphological characters. Schubart used the structure of the margins of the tergites but I am not very convinced by the results.

JEEKEL: Material from all the regions should be taken into consideration. Only the South African sphaerotheriids are reasonably worked out. The Madagascan fauna, for instance, should be re-investigated, most of it is incompletely known.

Symp. zool. Soc. Lond. (1974) No. 32, 53–63.

INTÉRÊT PHYLOGÉNIQUE ET BIOGÉOGRAPHIQUE DE QUELQUES DIPLOPODES RÉCEMMENT DÉCRITS DU NORD DE L'ESPAGNE

J.-PAUL MAURIÈS

Muséum national d'Histoire naturelle, Laboratoire de Zoologie-Arthropodes, Paris, France

SYNOPSIS

Attention is drawn to:

1. *Aragosoma barbieri* Mauriès (1971a) from the Central Pyrenees. It is very close, geographically and morphologically, to *Pyreneosoma* (Craspedosomoidea). Because of its ecological requirements (nivicole, troglophile) it has a very limited area of distribution. Its primitive gonopods confirm that it is a relict species surviving from an early cold period.

2. *Cantabrodesmus lorioli* Mauriès (1971c). This is a cavernicole from the Cantabrique mountains, a polydesmoid showing affinity with the neotropical Platyrrhacidae which, by contrast, is evidence of a period when this region experienced a tropical climate.

3. *Mesoiulus* spp. of caves in the Cantabrique mountains occupy the same places which *Blaniulus* and *Typhloblaniulus* occupy in the Pyrenees and the Cevennes and which *Typhloiulus* and *Trogloiulus* occupy in Italy and the Balkans.

INTRODUCTION

La présente note a pour but essentiel d'attirer l'attention sur quelques diplopodes récemment décrits du Nord de l'Espagne, afin de montrer une fois de plus l'intérêt que présentent, à des degrés divers, et pour diverses raisons, soit écologiques, soit biogéographiques, soit phylogéniques, les diplopodes d'une manière générale.

La faune diplopodologique du Nord de l'Espagne, du moins celle qui peuple le versant sud des Pyrénées et plus à l'ouest les Monts Cantabriques, commence à être connue, notamment à la suite d'une série de récoltes effectuées au cours de ces 20 dernières années et dont les résultats ont été publiés récemment (Mauriès, 1971a,b,c, 1972). Malgré les lacunes encore nombreuses, on connaît actuellement une soixantaine d'espèces, ce qui est sans doute loin du chiffre réel, puisqu'on en compte 114 sur le versant nord-pyrénéen! Sur ces 60 espèces, le tiers environ a été récolté dans des grottes, contre moins de 1/6 ème pour le versant français; cette différence ne s'explique pas seulement par le fait qu'un climat plus sec et plus chaud favorise la colonisation des grottes, mais aussi par le fait que la faune cavernicole, du moins en ce qui concerne les myriapodes, a été plus prospectée que la faune épigée du côté espagnol.

La liste des espèces recensées au Nord de l'Espagne, si elle reste encore relativement faible, nous permet déjà de pouvoir effectuer quelques comparaisons avec la faune nord-pyrénéenne. Cependant, tel n'est pas notre propos ici. Nous dirons simplement qu'en ce qui concerne la faune épigée, la coupure n'est pas aussi nette que ce que l'on aurait pu le supposer, surtout aux extrémités de la chaine; on remarque par exemple, du côté occidental, une véritable faune de type atlantique répartie sur les deux versants des Pyrénées (*Polydesmus coriaceus*, alias *atlanticus*—*Blaniulus dollfusi*—*Loboglomeris*—*Protoglomeris*—etc.) Nous verrons que cette continuité est bien loin d'être la règle chez les cavernicoles.

Nous examinerons ici trois espèces (ou groupes d'espèces) appartenant à trois genres et trois ordres différents qui ont particulièrement attiré notre attention du fait de leur singularité par rapport à l'ensemble de la faune diplopodologique de la région pyrénéo-cantabrique. Nous verrons qu'elles suscitent des commentaires fort différents.

ARAGOSOMA BARBIERI MAURIÈS (1971a)

Ce craspedosomide, de grande taille (3 cm de long sur 2·8 mm de large) est une belle espèce découverte dans les grottes d'altitude de la province d'Huesca, par Barbier et Dumont en 1962 et 1963. Ce diplopode rappelle, par sa taille, sa coloration et sa morphologie externe, de manière frappante, les *Pyreneosoma* Mauriès (1959), qui sont d'ailleurs, comme le montre la carte (Fig. 1), géographiquement très voisins. Les deux genres se ressemblent aussi sur le plan écologique, puisque *Pyreneosoma* se trouve dans les cirques et vallons froids et argileux des Pyrénées centrales françaises entre 2000 et 2600 m d'altitude, tandis qu'*Aragosoma* peuple les grottes froides d'altitude, vers 1500–2000 m, comme la grotte dite des "Tesserefts du Collerada", où la neige persiste même pendant la belle saison.

Par ailleurs, la parenté de ces deux genres est évidente si on compare leurs gonopodes (voir Mauriès, 1959, 1971a) qui se signalent, dans l'un comme dans l'autre des deux genres, par leur grande simplicité de structure, ce qui est peu fréquent chez les craspedosomides.

Tout, aussi bien leur isolement géographique que leur écologie si particulière et la structure primitive de leurs gonopodes, semble plaider en faveur du caractère relictuel des *Aragosoma* et *Pyreneosoma*. Ces diplopodes, cryophiles aux exigences écologiques strictes, apparaissent comme des vestiges de périodes froides réfugiés au voisinage des

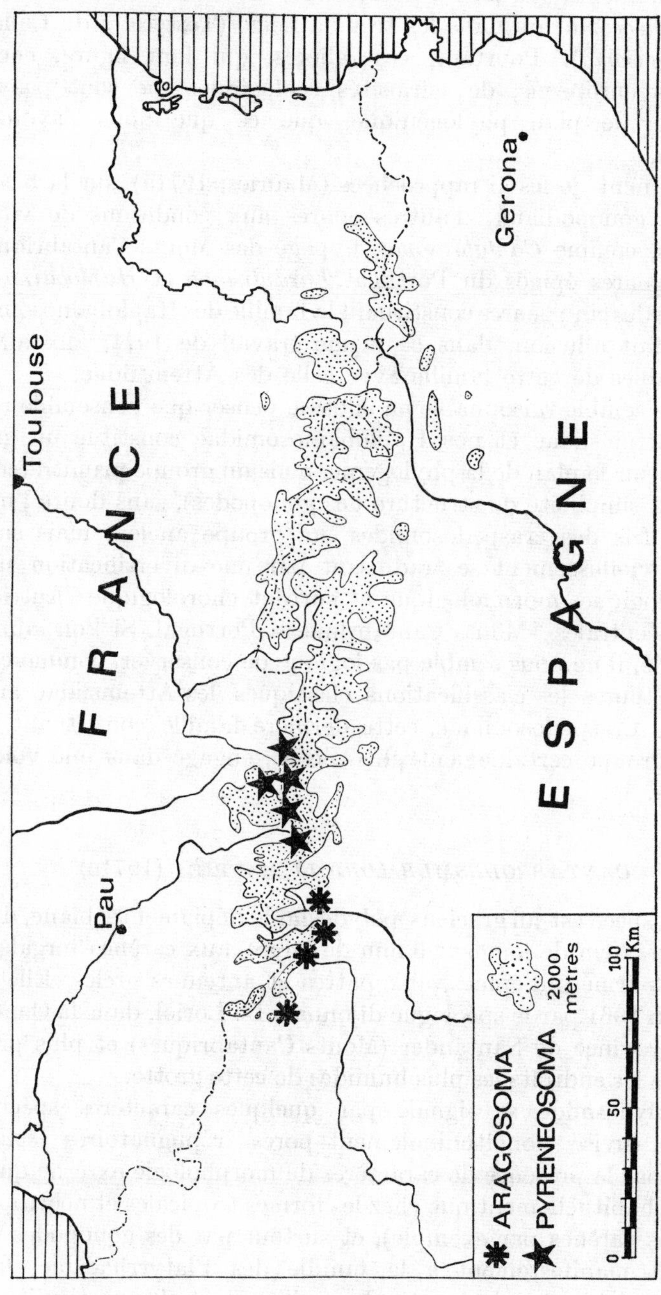

FIG. 1. Localisation des genres *Aragosoma* et *Pyreneosoma* dans la chaîne pyrénéenne.

névés permanents (Cirques d'Estarranhe, de Troumouse, de Gavarnie pour *Pyreneosoma*) ou dans les trous à neige (Tesserefts du Collerada pour *Aragosoma*). Pourtant, ces relictes, qui font figure, chez les diplopodes européens, de curiosités écologiques, ne sont pas aussi isolées, sur le plan phylogénique, que ce que nous l'avons cru longtemps.

Récemment, je les ai rapprochées (Mauriès, 1971a), sur la base des caractères gonopodiaux, d'autres genres aux conditions de vie très différentes, comme *Cantabrosoma*, hypogé des Monts Cantabriques et les deux genres épigés du Portugal *Turdulisoma* et *Haplobainosoma*; l'ensemble des cinq genres constituant la famille des Haplobainosomidae; j'ai déjà fait allusion, dans ce même travail de 1971, aux affinités incontestables de cette famille avec celle des Attemsiidae.

Il nous semble raisonnable en effet de penser que l'ensemble réalisé par ces Attemsiidae et nos Haplobainosomidae constitue un groupe homogène sur le plan de la phylogénie, mais un groupe primitif (comme l'atteste la simplicité de structure des gonopodes), sans doute l'un des plus primitifs des craspedosomides, un groupe ancien, mais surtout vieilli, ce vieillissement se traduisant par une diversification sur les plans écologique, morphologique et surtout chorologique: Autriche—Pyrénées centrales—Monts Cantabriques—Portugal. Si l'on admet ce qui précède, il ne nous semble pas logique de conserver, comme celà se voit dans toutes les classifications classiques, les Attemsiidae au voisinage des Craspedosomidae, cette dernière famille constituant à nos yeux un groupe certainement plus récent, engagé dans une voie très spécialisée.

CANTABRODESMUS LORIOLI MAURIÈS (1971c)

Cette espèce est un gracieux polydesmide dépigmenté, blanc, d'assez belle taille (2 cm de long sur 5 mm de large), aux carènes tergales très longues et armées d'épines, aux pattes et antennes grêles. Elle a été récoltée en 1961 par le spéologue dijonnais De Loriol, dans la Cueva del Molino, province de Santander (Monts Cantabriques) et plus précisément dans les endroits les plus humides de cette grotte.

Ce polydesmide se signale par quelques caractères spécifiques (mentum divisé longitudinalement-pores répugnatoires *sous* les carènes), par la présence de caractères de morphologie externe que l'on ne trouve habituellement que chez les formes tropicales et néotropicales (forme des carènes par exemple), et surtout par des gonopodes qui le rattachent manifestement à la famille des Platyrrhacidae, famille propre aux régions tropicales et indo-pacifiques, et plus particulièrement à un sous-genre de *Platyrrhacus* Koch: *Tirodesmus* Cook, d'Amérique Centrale.

Ce n'est pas la première fois que l'on découvre en Europe un polydesmide cavernicole aux affinités tropicales (voir Tabacaru, 1970, qui cite les genres *Devillea*, *Trichopolydesmus*, *Verhoeffodesmus*, *Galliocookia*, *Bacillidesmus*, *Serradium* et *Eroonsoma*); mais c'est sans doute le premier qui appartient à la famille des Platyrrhacidae.

Cette découverte doit être mise en parallèle avec celle de l'isopode terrestre *Cantabroniscus* décrit par Vandel en 1965, et dont le plus proche parent est le genre mexicain *Mexiconiscus*. La découverte de De Loriol vient donc fournir un nouvel exemple d'affinités incontestables et de relations sans doute très anciennes entre certains éléments de la faune européenne et la faune de l'Amérique centrale.

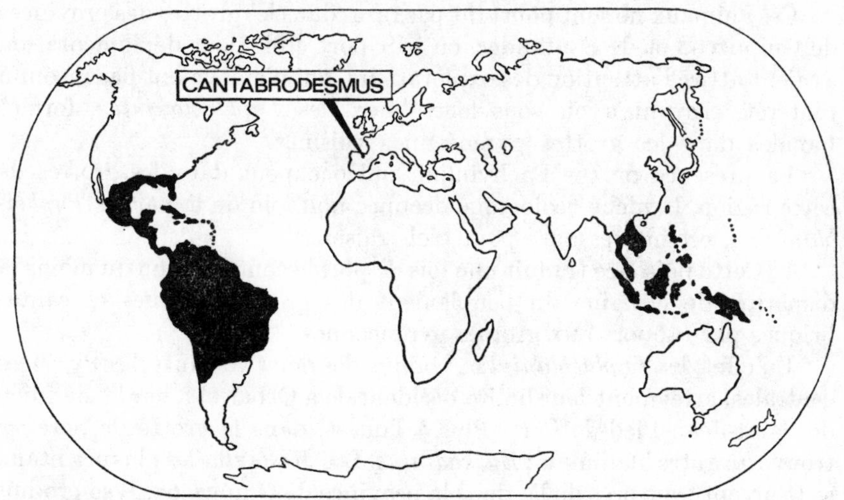

Fig. 2. Répartition mondiale des Platyrrhacidae (d'après Attems, 1926; Fage, 1937, modifié); on notera l'isolement géographique de *Cantabrodesmus*.

D'autre part, le fait que *Cantabrodesmus* appartienne aux Platyrrhacidae n'est pas dépourvu d'intérêt si l'on se souvient que cette famille a déjà retenu l'attention d'auteurs tels que Fage (1937), et surtout Jeannel (1942). En effet, cette famille est complètement absente du continent africain (carte, Fig. 2); Jeannel la prend comme exemple de témoin d'une faune très ancienne, dite Inabrésienne, qui se serait scindée en trois îlots, (indo-pacifique, néotropical, africain), dont seuls auraient subsisté les deux premiers sous l'effet de variations climatiques. Cette hypothèse est assez séduisante, mais on ne peut l'appliquer au cas qui nous intéresse ici que si on en élargit quelque peu le cadre, en admettant que les ancêtres des Platyrrhacidae ont peuplé non seulement l'Amérique, l'Afrique et la zone indo-pacifique, mais aussi une partie de l'Europe; les éléments de cette dernière région auraient, comme ceux d'Afrique, et sans doute pour des raisons similaires, disparu; *Cantabrodesmus* représenterait un élément qui ne devrait sa survie qu'au refuge dans le domaine souterrain.

LE GENRE *MESOIULUS* BERLESE

Les *Mesoiulus* des Monts Cantabriques et du Pays Basque espagnol, dont nous connaissons aujourd'hui cinq espèces (Mauriès, 1971b), n'ont été signalés pour le première fois qu'en 1939 par Verhoeff qui, croyant avoir affaire à un nouveau genre, les avait nommé *Baskoiulus*.

Ces animaux ne sont pourtant pas rares dans les grottes des provinces de Guipuzcoa et de Santander, où leur port grêle, leur dépigmentation avaient attiré l'attention des explorateurs, qui n'en avaient pas reconnu l'intérêt, croyant avoir sous leurs yeux des *Typhloblaniulus*, formes banales dans les grottes pyrénéennes voisines.

La présence de ces Pachyiulini, qui occupent dans les grottes de cette région, la niche écologique occupée non loin de là par les *Typhloblaniulus*, est intéressante pour trois raisons:

1. Cette présence traduit une fois de plus le contraste, ou du moins la disparité, de l'origine du peuplement des grottes basques et cantabriques par rapport aux grottes pyrénéennes.

En effet, les *Typhloblaniulus*, connus des deux versants des Pyrénées centrales, atteignent leur limite occidentale à Orbaiceta, sur le méridien de Saint-Jean-Pied-de-Port. Plus à l'ouest, dans la grotte de Sare, se trouve un autre blaniulide: *Euzkadiulus*. Les *Mesoiulus* les plus orientaux se trouvent tout près de là, dans la province de Guipuzcoa. Il se produit dans cette zone du Massif de la Rhune, des changements de caractère de la faune cavernicole, soit qu'un groupe existe d'un côté et pas de l'autre (cas des *Spelaeoglomeris*), soit qu'un groupe est remplacé par un autre (cas de *Mesoiulus* qui remplace *Typhloblaniulus*, des Haplobainosomidae qui remplacent les Anthogonidae), soit enfin qu'un groupe devienne cavernicole (cas des *Trachysphaera* alias *Gervaisia* et du blaniulide *Blaniulus dollfusi*).

2. Nos *Mesoiulus* occupent une place particulière par rapport aux Pachyiulini.

Rappellons d'abord que le genre *Mesoiulus*, constitué de formes aveugles et dépigmentées, comprend une quinzaine d'espèces le plus souvent terricoles, disséminées des Monts Cantabriques à la Turquie d'Europe, et qu'à l'exception de celles qui nous occupent ici, deux seulement ont été récoltées dans des grottes (*M. scossirolii* Manfredi, de Toscane et *M. kosswigi* Verh., de la région d'Istanbul).

Notons ensuite qu'il existe d'autres Pachyiulini qui peuplent le domaine souterrain: il s'agit des *Apfelbeckiella*, dont plusieurs espèces sont cavernicoles en Bulgarie et en Dobroudja roumaine.

L'originalité des *Mesoiulus* cantabriques, par rapport aux autres *Mesoiulus* et aux *Apfelbeckiella*, c'est qu'ils sont fixés dans une région située loin de l'aire d'extension des Pachyiulini en général; ce groupe n'est, en effet, pas représenté dans la péninsule ibérique ni dans le sud de la France. Si l'on se souvient que l'aire de prédilection des Pachyiulini se trouve dans la partie orientale du bassin méditerranéen, avec des avancées dans les péninsules balkaniques et italiennes ainsi que dans le Maghreb et jusqu'aux Canaries, si l'on note leur grande diversité morphologique et notamment les tendances régressives qui se manifestent chez les formes endogées et troglophiles, on est amené à penser que ces iulides constituent un groupement ancien, qui a du avoir une répartition plus vaste et surtout plus homogène.

3. Il est également intéressant d'observer la place occupée par les *Mesoiulus* et les *Apfelbeckiella* cavernicoles par rapport aux deux autres grands groupes d'Iulida cavernicoles d'Europe: les Blaniulini et les Typhloiulini.

La carte (Fig. 3) montre:

—qu'il n'y a pas de chevauchement d'aire de répartition entre les trois groupes, pas plus entre les *Mesoiulus* et *Typhloblaniulus*, qu'entre les *Apfelbeckiella* et les *Typhloiulus*, pas plus d'ailleurs qu'entre les *Typhloblaniulus* et les *Typhloiulus;* ce qui tendrait à montrer que ces trois groupes s'excluéraient mutuellement et occuperaient des niches écologiques comparables.

—les Blaniulini peuplent les grottes de la partie la plus occidentale de l'Europe (Sardaigne, Pyrénées, Cévennes), tandis que les Typhloiulini occupent les grottes italiennes et balkaniques. Ceci a déjà été montré par Tabacaru, 1970, qui pense avec raison que les premiers ont évolué sur la tyrrhénide, tandis que les seconds ont évolué sur l'égéide.

Quant aux Pachyiulini, ils apparaissent, en dehors de quelques stations disséminées, comme repoussés ou concentrés, ou mieux comme ne persistant surtout qu'aux deux extrémités de l'aire considérée: *Mesoiulus* cantabriques à l'ouest, *Apfelbeckiella* à l'est. Outre le fait que cette répartition plaide en faveur de leur origine mésogéenne, antérieure aux transgressions lutétiennes, donc plus ancienne, elle semble montrer que les Pachyiulini ont cédé la place à des formes plus récentes et plus dynamiques, comme les Blaniulini et les Typhloiulini.

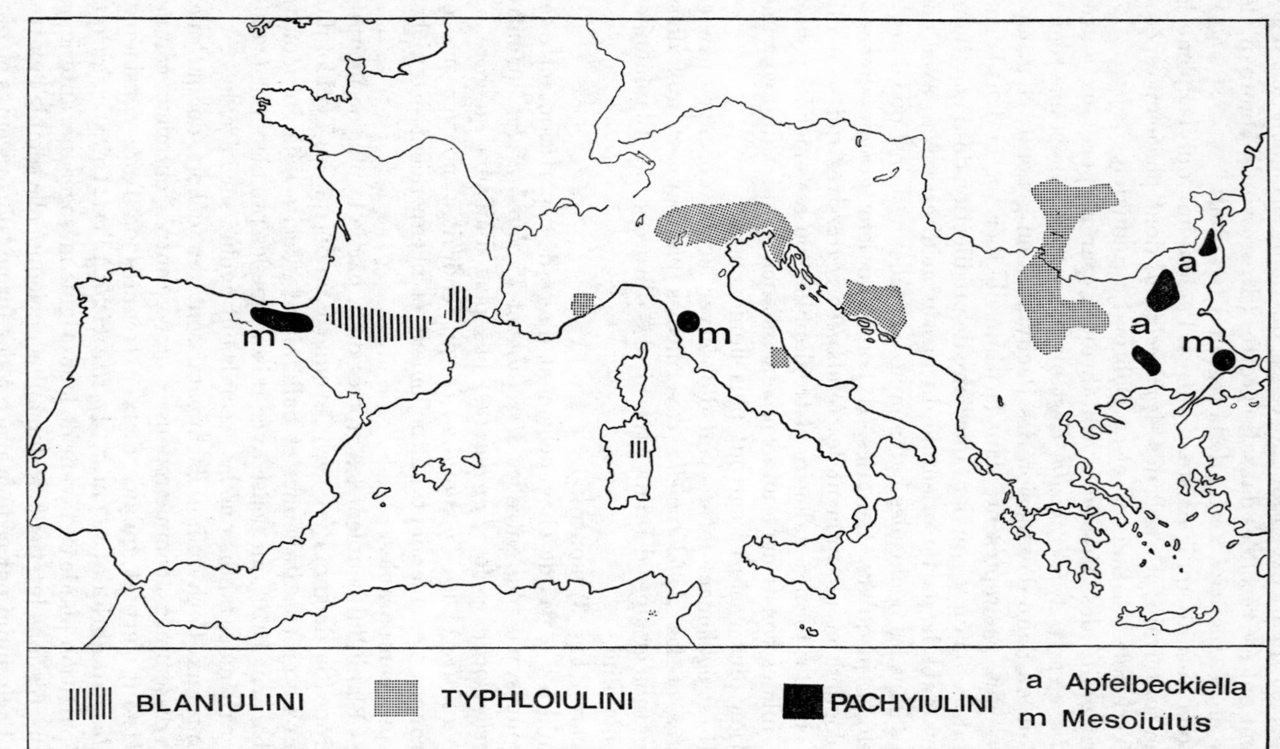

FIG. 3. Répartition des trois principales lignées cavernicoles d' Iulida européens; noter la dispersion des Pachyiulini par rapport aux deux "blocs" des Typhloiulini et des Blaniulini (d'après Strasser, 1962; Tabacaru, 1970, schématisé).

CONCLUSIONS

Ces quelques exemples n'ont pas d'autre prétention que d'attirer l'attention:

1. Sur l'intérêt des formes cavernicoles en tant que témoins historiques des déplacements et successions de faunes qui ont régi le peuplement actuel.

2. Sur l'intérêt particulier des diplopodes, qui du fait de leurs grandes exigences écologiques, et de leurs faibles possibilités de déplacement, ne peuvent être que d'excellents indicateurs écologiques et biogéographiques.

Mais ils montrent aussi, qu'il est assez souvent difficile, lorsque le besoin s'en fait sentir, de déterminer avec précision les affinités phylogéniques de tel ou tel diplopode. Cela tient, pour une grande part, aux classifications classiques, qui non seulement sont anciennes, mais restent souvent boiteuses parce qu'elles sont basées de façon non méthodique à la fois sur des caractères de morphologie externe et des caractères sexuels, souvent choisis plus en fonction de leur commodité immédiate d'utilisation que de leur signification phylogénique. Il importe donc, pour que les diplopodes puissent fournir à l'écologie et la biogéographie tous les éléments que ces sciences sont en droit d'attendre d'un groupe zoologique dont les quelques exemples donnés plus haut suffisent à montrer l'intérêt, qu'une systématique moderne, essentiellement basée sur l'utilisation des caractères gonopodiaux, traduise plus fidèlement la phylogénie du groupe. C'est ce que nombre de nos collègues s'efforcent de faire actuellement. La tâche est immense, mais elle est nécessaire si nous voulons qu'une meilleure connaissance de la phylogénie retentisse favorablement sur le développement de la biologie, de l'écologie et de la biogéographie des diplopodes.

REFERENCES

Attems, C. (1926). Myriapoda. *Handb. Zool.* **4**.

Fage, L. (1937). Les Myriapodes. In *Encyclopédie Française* **5** (B): 5.74 (10 & 11). Paris: Larousse.

Jeannel, R. (1942). *La genèse des Faunes terrestres.* Paris: P.U.F.

Mauriès, J.-P. (1959). *Pyreneosoma,* genre nouveau de Craspedosomide des Hautes-Pyrénées. *Bull. Soc. Hist. nat. Toulouse* **94**: 203–208.

Mauriès, J.-P. (1971a). Diplopodes épigés et cavernicoles des Pyrénées espagnoles et des Monts Cantabriques. I–III. Introduction. *Bull. Soc. Hist. nat. Toulouse* **106**: 401–422.

Mauriès, J.-P. (1971b). IV–V. Blaniulides et Iulides. *Bull. Soc. Hist. nat. Toulouse* **107**: 103–116.

Mauriès, J.-P. (1971c). VI. Polydesmides. *Bull. Soc. Hist. nat. Toulouse* **107**: 117–124.

Mauriès, J.-P. (1972). VII. Glomérides. Essai de classification des Glomeroidea. *Bull. Soc. Hist. nat. Toulouse* **107**: 423–436.

Strasser, C. (1962). Die Typhloiulini (Diplopoda Symphyognatha). *Atti Mus. civ. Stor. nat. Trieste* **23**: 1–77.

Tabacaru, I. (1970). Sur la répartition des diplopodes cavernicoles européens. *Livre Centenaire E. G. Racovitza, 1868–1968*: 421–444.

Vandel, A. (1965). Sur l'existence d'Oniscoides très primitifs menant une vie aquatique et sur le polyphylétisme des Isopodes terrestres. *Annls Spéléol.* **21**: 643–650.

Verhoeff, K. W. (1939). Eine neue cavernicole Juliden-Gattung. *Mitt. Höhl. u. Karstforsch.* **1938**: 11–14.

Discussion

STRASSER: Have you compared *Aragosoma* with the genus *Fuentea* of Brölemann? You said that this genus could belong to the Attemsiidae but I doubt this since the larva is not known and one of the main characters of Attemsiidae is that the parasoma and lateral keels are placed lower than in other Craspedosomoidea. Secondly as regards the affinity of *Cantabrodesmus* with the Platyrrhacidae which you show to be distributed only in north America and oriental Asia. Would it not be more plausible to suggest a journey from Africa into northern Spain since we already know of tropical African elements there, such as *Macellolophus*. Could *Cantabrodesmus* be related to this?

MAURIÈS: *Macellolophus* could not be related to *Cantabrodesmus*.

STRASSER: Regarding *Mesoiulus*, I have found that *Mesoiulus scossirolii* Manfredi, is not a *Mesoiulus*, it belongs to *Elbaiulus*. You have also mentioned Tabacaru's idea that the Typhloiulidae come from the Aegean. I believe that this family is polyphyletic and that iulids have lost their eyes and transformed themselves into *Typhloiulus* not once, but two, three or four times. This would account for the curious distribution of the Typhloiulidae. A *Typhloiulus* sp. occurs in the Salzburg area; it could not have crossed the Alps, or even gone round them. These difficulties disappear if the Typhloiulidae are polyphyletic.

KRAUS: It was always difficult to explain the distribution of the Platyrrhacidae in the neotropical region and in south east Asia but not in Australia and New Zealand. If *Cantabrodesmus* really belongs to this family we may suppose that platyrrhacids originally occupied the continents of the northern hemisphere but only survived in south east Asia and south America which they invaded from the north.

CAUSEY: I am confused about the family you are talking about. The Platyrrhacidae do not occur in the southern United States. They occur in parts of

Mexico, not all of Mexico by any means, but not in the southern United States.

JEEKEL: I think the map is wrong. Brazil should be excluded except the Eastern provinces. The eastern border of the range is about half way up the north coast—then northward up to Costa Rica. I think your map includes the Euryuridae which do occur in the United States and in South and Central America; this family is closely related to the Platyrrhacidae. However, this does not alter the argument put forward by Professor Kraus.

MAURIÈS: Indeed, Attems' map is wrong, but only in the details. I am only concerned to show the peculiar position of *Cantabrodesmus* in relation to other Platyrrhacidae. Details of the distribution of the family in America, or elsewhere, I prefer to leave to the specialists working on this group.

SAHLI: Is it certain that there are no platyrrhacids in Australia and southern Africa?

MAURIÈS: Fage (1937)* gives a map but I do not know his sources. Africa has been quite well-studied and yet none have been found.

* See list of References p. 61.

Symp. zool. Soc. Lond. (1974) No. 32, 65–73.

ON CERTAIN ASPECTS OF THE GENERIC CLASSIFICATION OF THE LITHOBIIDAE, WITH SPECIAL REFERENCE TO GEOGRAPHICAL DISTRIBUTION

E. H. EASON

Bourton Far Hill Farm, Moreton-in-Marsh, England

SYNOPSIS

The Lithobiidae have not made their due contribution to zoogeographical knowledge because of the confused state of their generic classification. Of the two systematists who have done most work on this family, R. V. Chamberlin was a splitter and C. G. Attems a lumper. Chamberlin's and Attems' systems are discussed and it is suggested that neither has produced a satisfactory classification. Examples are given of the misleading zoogeographical conclusions which could be drawn from the present inadequate classification and the more valid conclusions which might be drawn from its revision.

INTRODUCTION

The generic classification of the Lithobiidae is, at present, in disarray owing to a general lack of co-ordination between the systems used on either side of the Atlantic, failure on the part of European workers to accept one another's views, and the undoubted fact that we are dealing with a very difficult group of animals from the point of view of the systematist. Briefly, we have Chamberlin's classification of the American and Australasian species together with a few of those from Europe and Asia with which he dealt, Attems' final classification of the Lithobiidae of the world, and the classification of European species used by most modern European authors which consists of acceptance of a few of the more distinctive genera, leaving the majority of species, which show considerable diversity, in either *Lithobius* Leach or *Monotarsobius* Verhoeff.

Many of Chamberlin's earlier genera seem to be valid but latterly he became a confirmed "splitter" and named a large number of genera, many of them monotypic, with little justification. Attems, on the other hand, named a few genera, some very weakly defined, when he first approached the problem (Attems, 1926) but finally became a "lumper", lumping many of Chamberlin's valid genera with each other or with those from Europe and producing some very unnatural groups (Attems, 1938). Bearing in mind the purpose of a generic classification as being a nice compromise between a convenient system with each genus definable by means of easily seen characters suitable for use in a key, and a system reflecting what seem to be phyletic relationships, neither Chamberlin's nor Attems' systems can be accepted as a whole. Ideally, the genera should be co-ordinate not only with each other but also with those recognized in other parts of the animal kingdom. Although this ideal is unattainable it is worth bearing in mind when deciding on the rival merits of lumping and splitting. A revised classification of the whole family is obviously required but will be a vast undertaking, quite beyond the scope of this short paper, and I shall only give a few examples of the more striking defects in the present systems and my attempts to correct them.

Throughout this paper I shall use the term "genus" for any species-group which has received a name. Many were originally named as subgenera of *Lithobius*, or have been subjected to promotion, demotion or synonymy by different authors.

ATTEMS' COMPOSITE GENERA

Attems (1938) redefined a number of genera in senses quite different from those intended by their original authors. The three likely to cause most confusion, one of which he had erected himself and all of which he reduced to subgenera of *Lithobius*, are briefly described:

Monotarsobius *Verhoeff*

Although limiting this genus to those species with only 20 antennal articles, Attems took no account of the degree of fusion of the tarsal articulations of the first to thirteenth legs or of the shape of the short tergites. By excluding those species with conspicuous modification of the male fourteenth or fifteenth legs he actually excluded *L. curtipes* C. L. Koch, the species designated by Verhoeff (1905) as generotype, as well as other European species such as *L. turkestanicus* Attems which Verhoeff would have included in *Monotarsobius*. Attems proposed six of Chamberlin's North American genera as synonyms.

Pokabius *Chamberlin*

Attems included in this North American genus all those species with only 20 antennal articles and with some dorsal modification of the male fourteenth or fifteenth legs. He proposed seven genera, all North American or Asiatic, as synonyms one of which (*Arenobius* Chamberlin) was included by Chamberlin in the family Gosibiidae. Although he made no special mention of *L. curtipes* he would, by definition, have included it in *Pokabius*.

Alokobius Attems

Even the genus as originally described (Attems, 1926), which included only those species with some modification of the male fourteenth or fifteenth tibiae and numerous antennal articles, is unnatural as few of the species placed in it by Attems (1927) have many features in common. Attems' final enlargement of this genus to include species with every type of secondary sexual modification of the posterior legs (Attems, 1938) brought together some even more diverse species. He proposed nine of Chamberlin's genera as synonyms, four of which Chamberlin included in the Gosibiidae. The secondary sexual modifications of *Alokobius* will be discussed further (p. 68).

WALESOBIUS CHAMBERLIN, 1920

The genus *Walesobius* was erected by Chamberlin on the basis of Pocock's (1891) description of *Lithobius sydneyensis* from Sydney, Australia, a species also recorded from Auckland, New Zealand (Archey, 1937). Another species, *Walesobius excrescens*, was described by Attems (1928) from specimens of both sexes from Cape Town, South Africa and is probably identical with *sydneyensis* (Eason, 1973). We thus have a genus with one or possibly two species apparently distributed in southern Australia, New Zealand and South Africa which seems to support Brinck's (1960) observations on the relationship between the South African and Australasian fauna. Add to this the fact that *Chilebius*, which was erected by Chamberlin (1955) to receive *C. coquimbo* Chamberlin from the Chilean coast of South America and *C. platensis* (Gervais, *sensu* Silvestri) from the same region and also from the mouth of the River Plate, is indistinguishable from *Walesobius* (as far as one can tell from Chamberlin's description) and we might have genus with a most unusual and significant distribution. As Brinck pointed out, faunistic connexions at generic level are frequent between South Africa and Australasia and also between South Africa and temperate South America, and connexions between all three regions, although rare, do occur. But *L. sydneyensis* Pocock is none other than *L. araïchensis* Brölemann, a species fairly widely distributed in Morocco and southern Spain (Eason, 1973), and a natural distribution including the whole south temperate zone and also the western Mediterranean seems altogether unlikely. The probability is that the species spread along the trade-routes from the Iberian peninsula or north Africa to Cape Town,

Sydney and Auckland, all seaports. Whether *C. coquimbo*, which is almost certainly identical with *C. platensis* (*sensu* Silvestri), belongs to *sydneyensis* is uncertain but it may well do as all the South American records for these forms are from seaports or coastal localities. It is, of course, possible that *sydneyensis* and *coquimbo* are no more than congeneric, indicating a faunistic connexion between the western Mediterranean and southern South America, but this seems unlikely. A further possibility is that their resemblance is due to convergent evolution and this can only be decided by examining the South American specimens.

MALE SECONDARY SEXUAL MODIFICATIONS AS GENERIC CHARACTERS

The generic status of *Walesobius* depends on the wart-like outgrowth on the male fifteenth femur and Attems (1938) made it a synonym of *Alokobius*. But secondary sexual modifications of the posterior legs take many forms and not only is there no homology between, for example, the shallow dorsal sulci on the fourteenth and fifteenth tibiae of *Lithobius dentatus* C. L. Koch and the tuft of setae on the fourteenth tibia of *L. muticus* C. L. Koch (both of which Attems (1927) included in *Alokobius*), but very similar structures occurring on the same article of the same leg are not necessarily homologous as they have, I believe, been acquired independently during the course of evolution in distantly related species and should not, by themselves, be given very much weight as generic characters. *L. calcaratus* C. L. Koch has a very similar structure to that found in *L. sydneyensis* on the same article of the same leg at much the same site, but its other features suggest quite different affinities. Modern European authors have not seen fit to place *L. araïchensis* (= *sydneyensis*) in a separate genus and I think *Walesobius* should be disregarded. *Alokobius* is, of course, altogether unnatural—perhaps the most unnatural group that has ever been suggested. On the other hand some of Chamberlin's North American genera such as *Nampabius* (which Attems made a synonym of *Pokabius*) are characterized by this kind of modification and seem to be valid, but their species have other features in common.

INDIGENOUS SPECIES OF LITHOBIIDAE IN THE SOUTHERN TEMPERATE ZONE

If the species of *Walesobius* both belong to an introduced species of *Lithobius* the question arises as to whether there are any indigenous species of Lithobiidae at all in the southern temperate zone where the Lithobiomorpha are well-represented by the Henicopidae. The only species recorded from New Zealand other than *L. sydneyensis* is *L. argus* Newport from Wellington and this has been shown to be based on introduced specimens of *L. forficatus* Linn. (Eason, 1972). The only species recorded from Australia other than *L. sydneyensis* is *Australobius scabrior* Chamberlin based on a single specimen from Queensland (Chamberlin, 1920) and this bears such a close resemblance to species of *Australobius* found in the East Indies, with one of which it may be identical, that introduction is a possibility. The only species recorded from South Africa other than *Walesobius excrescens* (= *sydneyensis*) is *L. peregrinus* Latzel and although this was found in an inland locality (Attems, 1928) *peregrinus* is notorious for the way in which it has spread, presumably by artificial means, far beyond its natural European range (Eason, 1973). There are a few recorded species from the tropical parts of South America (Turk, 1955) which probably belong to the group of species referred by Chamberlin to the Gosibiidae, distributed chiefly in the southern United States, Mexico and Central America, but the few records from the temperate region might all refer to *L. sydneyensis*; and the more southerly parts of Chile and the Argentine have not been surveyed for centipedes as far as I know. It certainly seems that the natural range of the family is not so far south as was previously supposed.

SOWUBIUS CHAMBERLIN, 1912

The genus *Sowubius* was erected by Chamberlin on the basis of Pocock's (1895) description of *Lithobius stolli* from Guatemala. Chamberlin (1922) placed *Sowubius* in his family Gosibiidae so it could hardly be further removed from any known European species in the current classification. But *L. stolli* is none other than *L. castaneus* Newport, a common Mediterranean species (Eason, 1973). Its presence in Guatemala is, I suspect, due to artificial introduction. *L. castaneus* is the generotype of *Euporodontus* Verhoeff (1942) which is almost indistinguishable from *Kosswigibius* Chamberlin (1952) with a closely related Turkish species as generotype. Verhoeff (1942) and Chamberlin (1952) stress the characteristic structure of the prosternum as a generic character in *Euporodontus* and *Kosswigibius* respectively whereas Chamberlin (1912) was only concerned to differentiate *Sowubius* from other Central American genera and defined it in a very negative way, making no mention of the prosternum. But all three terms refer to the same zoological entity and one of them has stood until now as an indigenous Central American genus. Whether the genus should be recognized as valid is rather doubtful and I am inclined to agree with Matic (1970) that it should be disregarded.

MESOBIUS CHAMBERLIN, 1951

Chamberlin must have attributed Brölemann's figure of the male fifteenth leg of *L. calcaratus* (Brölemann, 1930: Fig. 443) to *L. castaneus* owing to his misreading of Brölemann's legend to the figure, because he described the genus *Mesobius* with the prosternum of *castaneus* and the fifteenth leg of *calcaratus*, naming *L. castaneus* Newport as generotype (Chamberlin, 1952). If this genus were, indeed, based on a non-existent animal it would have to be rejected but in an earlier publication Chamberlin (1951) clearly designated *Mesobius danianus* from Florida as the generotype and gave a fair description of what appears to be *L. calcaratus*, mentioning *L. castaneus* as a constituent species only. The earlier description was probably based on examination of introduced specimens of *L. calcaratus* whereas the later description seems to have been based, not on any specimen, but on a misunderstanding of Brölemann's figures of *L. calcaratus* and *L. castaneus*: these figures and their corresponding legends are certainly arranged in a confusing way and it is easy to see how this mistake arose. However, the earlier description and designation are authentic and Matic (1970) was mistaken when he quite understandably equated *Mesobius* with *Euporodontus*.

L. calcaratus ought, I believe, to be placed in a separate genus along with two or three European species and should *M. danianus* prove to be identical with *L. calcaratus*, *Mesobius* will be its correct name.

AMPHIATLANTIC GENERA

Throughout this paper I have assumed, when two forms are discovered in widely separate regions and are found to be not only generically but also specifically and subspecifically identical, that their distribution is not natural but due to artificial introduction. I realize that I could be wrong in making this assumption: the Chilopoda are an ancient group and species have probably evolved slowly: as far as Europe and America are concerned some authorities believe in a land-bridge as recently as the Pliocene (Wegener, 1929). But I find the record of *L. castaneus* from Guatemala and of *L. calcaratus* (assuming this species to be identical with *Mesobius danianus*) from Florida unconvincing evidence of a natural faunal exchange across the North Atlantic. But when such forms appear to be congeneric although specifically distinct the presumption is that their distribution is natural and that when they occur on either side of the Atlantic we have evidence of such faunal exchange and are dealing with Amphiatlantic genera. By Amphiatlantic one means genera occurring in both Europe and North America but not central or eastern Asia, which excludes such Holarctic genera as *Paobius* Chamberlin which are more likely to owe their present distribution to exchange across the Bering bridge rather than across the Atlantic. The term Amphiatlantic is also usually used, as it is here, to exclude those genera (if any) occurring throughout the tropical regions of both Old and New Worlds and also those with an essentially Circumpolar distribution (see also Lindroth, 1957).

The three Amphiatlantic genera of Lithobiidae which, I suggest, can be defined at present are:

1. *Lithobius* Leach, *sensu* Chamberlin, 1925. In addition to *L. forficatus* which may be introduced and one or two species which may belong to the introduced *L. peregrinus*, there are a few North American species of this genus which seem to be indigenous. Although *Lithobius* in Chamberlin's sense, which excludes the majority of European species at present included in *Lithobius*, may be too restricted there is no doubt that these American species are congeneric with many of those indigenous in Europe.

2. *Eulithobius* Stuxberg, *sensu* Crabill, 1958. There are three recorded indigenous North American species of this genus which seem to be congeneric not only with *L. validus* Meinert (= *L. punctulatus* C. L. Koch of some authors) but also with a few other European species which at present remain in *Lithobius* but which constitute a well-defined species-group (Prunesco, 1966) and are certainly indigenous in Europe.

3. *Sigibius* Chamberlin. There are five recorded American species of this genus, four from North America and one from Honduras, which are certainly congeneric with quite a large group of European species at present included in *Monotarsobius*. These species, which have 25 or more antennal articles, should, I think, all be transferred to *Sigibius*, leaving only those with the antennal articles restricted to 20 in *Monotarsobius*.

When more is known of the Asiatic fauna it is, of course, possible that all these genera will prove to be Holarctic and not Amphiatlantic. On the other hand, when there is more co-ordination between the American and European classification more of these Amphiatlantic genera may be recognized.

CONCLUSION

Conclusions as to the zoogeographical significance of the distribution of a family in which the generic classification is so unsatisfactory and the species in many parts of the world so incompletely known can at present be only very speculative. Attems' genera *Alokobius*, *Monotarsobius* and *Pokabius*, even if they were natural groups, are so large that their distribution gives little more information than does that of the family as a whole. Many of Chamberlin's genera are too small to give any more information than that given by the distribution of individual species. There are probably a number of genera which are identical and differ from one another in name only, but whose synonymy has not been recognized, and which give a false picture of generic distribution. But a revised classification, a compromise between the lumping of Attems and the splitting of Chamberlin, and more attention to the possibility of artificial introduction, should enable the Lithobiidae to make their due contribution to zoogeographical knowledge.

References

Archey, G. (1937). Revision of the Chilopoda of New Zealand. *Rec. Auckland Inst. Mus.* **2**: 43–100.

Attems, C. G. (1926). Myriapoda. *Handb. Zool., Berl.* **4**: 1–402.

Attems, C. G. (1927). Myriopoden aus dem nördlichen und östlichen Spanien, gesammelt von Dr F. Haas in den Jahren 1914–1919. *Abh. senckenb. naturforsch. Ges.* **39**: 233–290.

Attems, C. G. (1928). The Myriopoda of South Africa. *Ann. S. Afr. Mus.* **26**: 1–431.

Attems, C. G. (1938). Die von Dr C. Dawydoff in franzosich Indochina gesammelten Myriopoden. *Mem. Mus. natn. Hist. nat. Paris* (N.S.) **6**: 187–353.

Brinck, P. (1960). The relations between the South African fauna and the terrestrial and limnic animal life in the southern cold temperate zone. *Proc. R. Soc.* (B) **152**: 568–571.

Brölemann, H. W. (1930). Myriapodes. Chilopodes. *Faune Fr.* **25**: 1–405.

Chamberlin, R. V. (1912). New genera of North American Lithobiidae. *Can. Ent.* **44**: 173–178.

Chamberlin, R. V. (1920). The Myriapoda of the Australian region. *Bull. Mus. comp. Zool. Harv.* **64**: 1–269.

Chamberlin, R. V. (1922). The centipeds of Central America. *Proc. U.S. natn. Mus.* **60** (7): 1–17.

Chamberlin, R. V. (1925). The genera *Lithobius*, *Neolithobius*, *Gonibius* and *Zinapolys* in America north of Mexico. *Bull. Mus. comp. Zool. Harv.* **57**: 439–504.

Chamberlin, R. V. (1951). On five new American Lithobiid centipeds. *Gt Basin Nat.* **11**: 115–118.

Chamberlin, R. V. (1952). On the Chilopoda of Turkey. *İstanb. Üniv. FenFak. Mecm.* (B) **17**: 183–258.

Chamberlin, R. V. (1955). The Chilopoda of the Lund University and California Academy of Science Expeditions. *Acta Univ. lund.* N.F. (2) **51** (5): 1–61.

Crabill, R. E. (1958). A new *Eulithobius*, with a key to the known American species (Chilopoda: Lithobiidae). *J. Wash. Acad. Sci.* **48**: 260–262.

Eason, E. H. (1972). The type specimens and identity of the species described in the genus *Lithobius* by George Newport in 1844, 1845 and 1849 (Chilopoda, Lithobiomorpha) *Bull. Br. Mus. nat. Hist.* (Zool.) **21**: 297–311.

Eason, E. H. (1973). The type specimens and identity of the species described in the genus *Lithobius* by R. I. Pocock from 1890 to 1901 (Chilopoda, Lithobiomorpha). *Bull. Br. Mus. nat. Hist.* (Zool.) **25**: 41–83.

Lindroth, C. H. (1957). *The faunal connections between Europe and North America.* Stockholm: Almqvist & Wiksell.

Matic, Z. (1970). Contributo alla conoscenza dei chilopodi di Turchia. *Fragm. ent.* **7**: 5–13.

Pocock, R. I. (1891). Descriptions of some new species of Chilopoda. *Ann. Mag. nat. Hist.* (6) **8**: 152–164.

Pocock, R. I. (1895). Chilopoda and Diplopoda. *Biologia cent.-am.* **14**: 1–217.

Prunesco, C. (1966). Groupe des espèces *Lithobius punctulatus* C. Koch—*Lithobius matici* nom. nov. (Chilopoda) en Europe. *Acta zool. cracov.* **11**: 51–62.

Turk, F. A. (1955). The chilopods of Peru with descriptions of new species and some zoogeographical notes on the Peruvian chilopod fauna. *Proc. zool. soc. Lond.* **125**: 469–504.

Verhoeff, K. W. (1905). Über die Entwicklungsstufen der Steinlaüfer, Lithobiiden, und Beiträge zur Kenntnis der Chilopoden. *Zool. Jb.* (Supplement) **8**: 195–298.

Verhoeff, K. W. (1942). Zur Kenntnis mediterraner Chilopoden besonders der Insel Ischia. *Z. Morph. Ökol. Tiere* **38**: 483–525.

Wegener, A. (1929). *Die Entsthehung der Kontinente und Ozeane.* 4. Aufl. Brunswick: Vieweg & Sohn.

Discussion

Causey: What do you mean by a genus?

Eason: Any group of species resembling one another so closely as to suggest a relatively recent common ancestor.

Jeekel: Is *Lithobius hageni* described by Lawrence from Tristan da Cunha the same as *Lithobius araïchensis* Brölemann?

Eason: No, it is identical with *Lithobius melanops* Newport.

Symp. zool. Soc. Lond. (1974) No. 32, 75–87.

NEW CRITERIA FOR THE DIFFERENTIATION OF SPECIES WITHIN THE LITHOBIIDAE

D. TOBIAS

Natur-Museum und Forschungs-Institut Senckenberg, Frankfurt-am-Main, West Germany

SYNOPSIS

Studies of extensive material of lithobiid centipedes from the Pyrenees have led to new conceptions regarding the taxonomic value of characters and the definition of species within the family Lithobiidae.

Statistical analyses of population samples taken throughout the various regions of distribution are essential prerequisites for taxonomic evaluations in this group, and the observation of living animals is also of great importance.

By these means, groupings can be discerned which differ from each other in significant combinations of characters. A method of indirect reasoning is presented for deciding the question of species delimitation (in the sense of biological species).

In view of the information gained from these new procedures, it seems necessary to make a complete reassessment of the existing classification of the Lithobiidae, in order to obtain an insight into the true relationships within the family.

INTRODUCTION

Since the time of Newport, numerous specialists have dealt with the taxonomy of Lithobiidae and nearly 1400 species have now been described. An attempt to refer one individual to a particular "species" frequently fails, however, because it usually possesses only a few of the given characters of that group. The obvious reason for this is that hardly any of the authors investigated the variability of the characters used to separate the species.

Thus, the taxonomic categories of the Lithobiidae were for the most part laid down according to a rigid typological concept, and this often resulted in the description of individuals or even individual aberrations.

Nowadays, animals with special structures, such as the secondary sexual characters of the males, can be assigned to an appropriate species with no great difficulty. Other Lithobiidae, however, belong to a diffuse assemblage within which all characters are more or less variable (Fig. 1) and delineation of separate forms is much more difficult. This paper seeks to discover if further species within this group can be identified.

To begin with, I examined the validity of the characters used up to the present, using statistical methods.

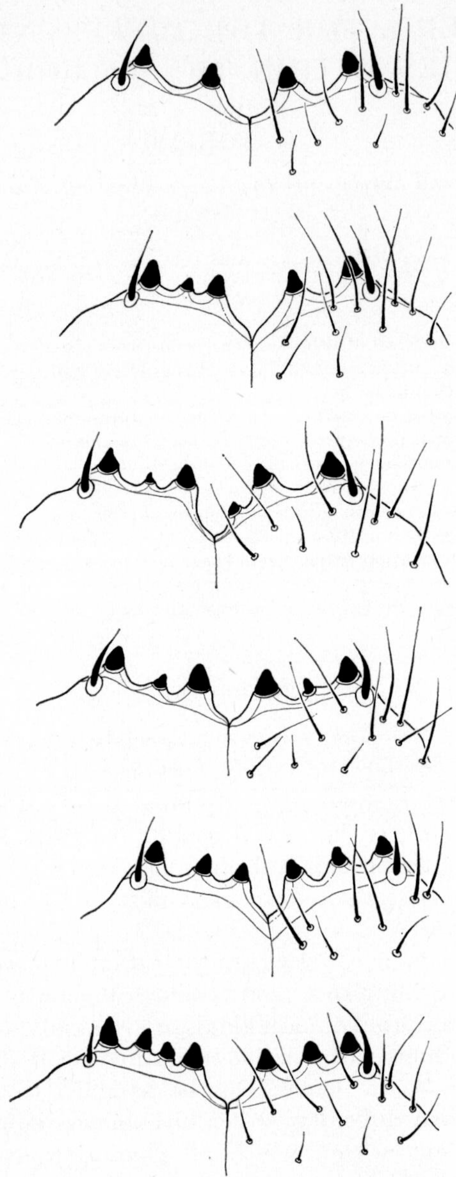

FIG. 1. Forcipular teeth of a series of adult Lithobiidae from a narrowly circumscribed habitat. Variation of number of teeth of 2+2 to 3+4.

THE STATISTICAL SEPARATION OF FOUR GROUPS OF LITHOBIIDS FROM MOUNT CANIGOU

Specimens suitable for this investigation were collected on a series of trips to Mount Canigou in the French East Pyrenees. This mountain was chosen because its unusually high population of Lithobiidae provided a sufficiently large number of specimens for statistical analysis. Also, Mount Canigou rises steeply from a height of 800 m to 2785 m from the valley of the Têt. Thus a comparison of a number of different biotopes was possible—from the Garigue at the foot of the mountain through a wooded area and on through areas with dwarf scrub-heather to the treeless upper region. These trips were undertaken at different times of the year.

The lithobiid fauna in this area conformed to the typical picture already described. Out of 657 adult animals 13 (nearly 2%) differed significantly from the others. These 13 individuals fell naturally into four groups. They were assigned to species but were omitted from further investigation, mainly because their small number does not permit an accurate statistical judgement to be made. These four groups, as understood in the sense of Brölemann's (1932) diagnoses are as follows:

Lithobius castaneus Newport 1844	1♂	3♀
Lithobius calcaratus C. L. Koch 1844	2♂	
Lithobius pilicornis doriae (Pocock 1890)	3♂	3♀
Lithobius microps Meinert 1868		1♀

The remaining 644 animals could also be divided up into four groups according to three classical structural characters. To begin with, these groups will not be assigned to species but to forms A, B, C and D:

Form A = without accessory apical claw on the 15th pair of legs,
Form B = more than 3 + 3 forcipular teeth,
Form C = with VaC on the 15th pair of legs,
Form D = the remainder, not possessing characters A, B and C.

Thus, with these three characters, four forms can be identified. Some animals cannot be definitely assigned to forms B, C and D using these three characteristics alone, but additional characters such as female gonopods, the number of coxal pores and the number of spines on the 14th pair of legs show a clear trimodal distribution and allow a classification to be made.

When the values of these simple characters were compared it became evident that in nearly all cases the averages were significantly different, but their ranges overlapped considerably (Fig. 2).

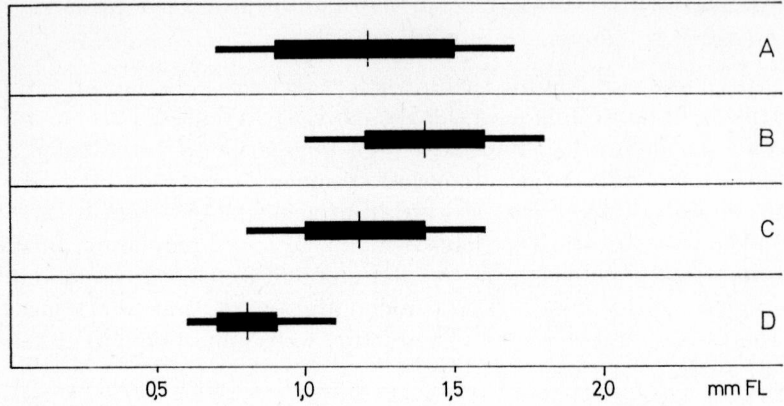

FIG. 2. Variation in the lengths of femur 15 in forms A, B, C and D. Each histogram represents 100 % of the animals of one form. The thicker part shows the range of variation in this character for 70 % of these animals; the two thinner parts each show the range of variation for 15 % of these animals. The vertical mark gives the position of the arithmetical mean. mmFL—length of femur in millimetres; where left and right femora differed in length, the mean of the two was used.

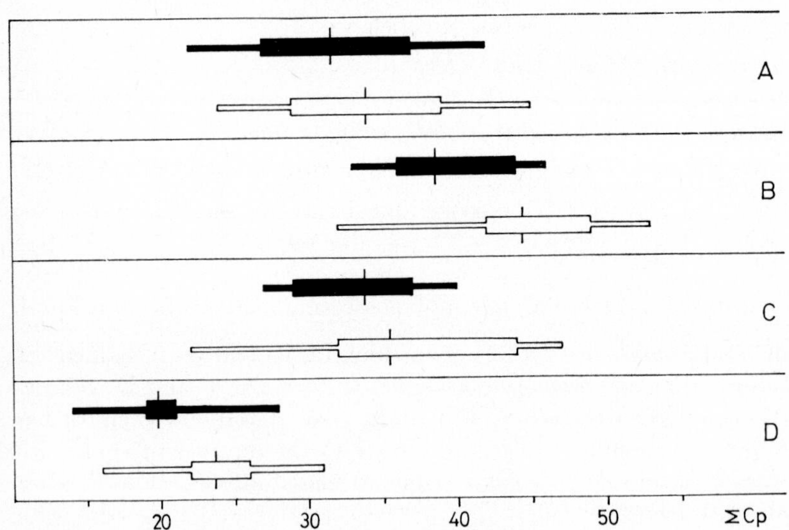

FIG. 3. Variation of the total number of the coxal pores (Σ Cp) on the last four pairs of legs in forms A, B, C and D.

Open histograms: female. Closed histograms: male. Further explanation as in Fig. 2.

This particular example shows that the identification of a species according to the characters of one individual is not admissible. Only the knowledge of the scattering of the characters allows a statement to be made in this report. In the case of the Mount Canigou specimens, the analysis of characters reconfirmed the original separation of the population into four forms. These can be differentiated not only according to the originally mentioned main characters, but also by an extensive catalogue of other characters.

But we clearly see that the taxonomic value of the single characters has to be judged differently from form to form. In Fig. 3 the ranges of variation of the character "sum of the coxal pores" are shown for the four forms (divided up into males and females). This character separates B from D; that is to say the ranges do not overlap. Forms A and C cannot be distinguished by this character as we are unable even to differentiate the averages. For the separation of the other forms, it is of accessory value, because with larger populations the averages can be differentiated from each other.

Basically, a character is of taxonomic value only when, in association with other characters, it can be used to differentiate one form from another.

THE CHANGE OF CHARACTERS IN SUCCESSIVE ADULT STADIA

What are the reasons for this considerable variation in the size or number of the characters? After thorough statistical evaluation, it became evident that most of the characters correlate positively with the body length of the animals. With the exception of only a few cases, this correlation was statistically significant ($p = 0{\cdot}001$); thus, long animals have more spines, teeth and coxal pores than shorter ones!

Verhoeff (1925) observed that an adult female lithobiid, having laid its eggs, moulted repeatedly. This observation suggested that forms B, C and D could be adults of different ages.

In order to examine this theory, 300 Lithobiidae from Mt. Canigou were kept alive, anaesthetized at intervals of several months, and tested for possible changes of characters.

Almost 70% altered their characters within six months! For instance, the number of coxal pores increased by an average of two in 29 out of 74 animals of form D: in two cases, the increase was by six pores, and with one female of form B by eight pores. It was also noticed that the number of articles of the undamaged antennae usually increased by one or two. With the regeneration of amputated antennae, a sharp increase followed; eventually the antennae attained the original

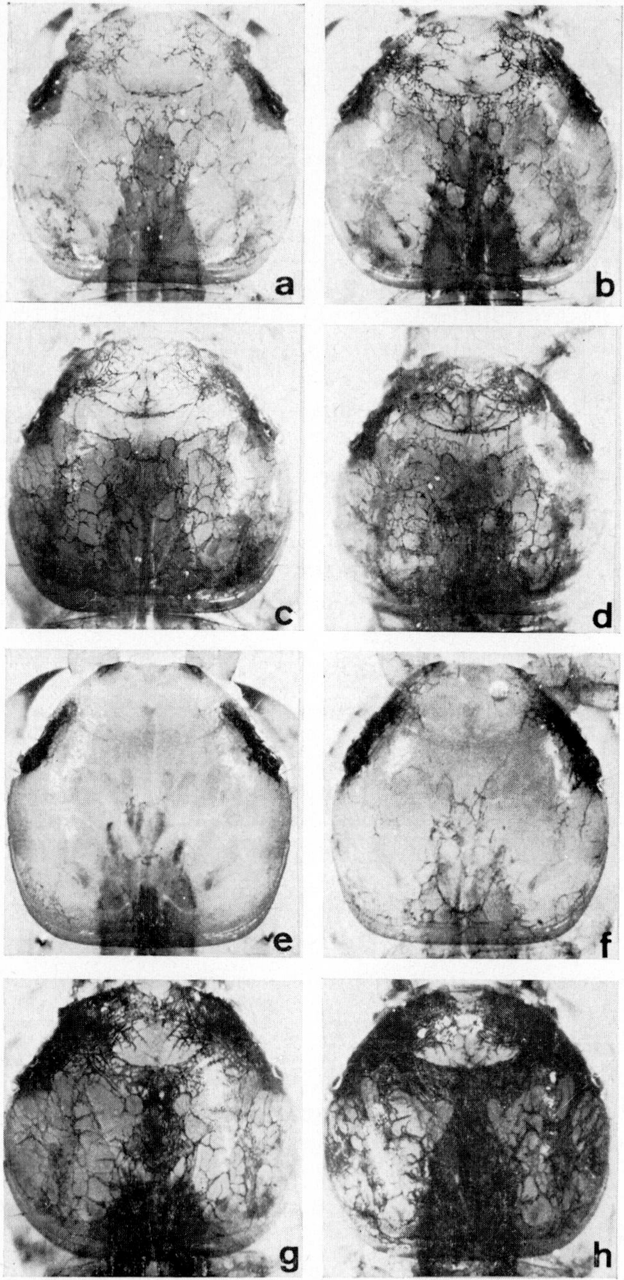

FIG. 4. Cephalic plate pattern of the forms A (a + b), B (c + d), C (e + f) and D (g + h). Left column: most frequent expression; right column: extreme variations.

number of articles. One antenna of a female of form B regenerated 24 articles within eight months after two moults. The number of spines on the 14th and 15th pair of legs increased only slightly, the number of forcipular teeth, the number of spines on the female gonopods, and the form of the claws also remained quite stable.

Although it was established that a change of characters does occur during the life of an adult lithobiid, no individual exceeded the range established for the form to which it was assigned, even after 22 months of observation. Animals of form D remained in form D although they grew older and larger. They never developed more than 3 + 3 teeth or more than a maximum of 32 spines on the 14th pair of legs; they never became B-animals. The initial classification into four forms, therefore, proved again to be valid.

NEW CHARACTERS OBSERVED IN LIVING ANIMALS

Investigations of the anaesthetized Lithobiidae drew attention to a pattern on the cephalic plate which was not present in preserved specimens. This took the form of a reticular structure which appeared over the whole body, but formed a clearly distinguishable pattern only on the cephalic plate. It was a pigmentary deposit in the reticular connective tissue. These patterns were present in the living Mount Canigou animals in four different variations (Fig. 4):

1. dark band over the suture of the head (D),
2. light band over the suture of the head (B),
3. weak reticulation over the suture of the head (A) and,
4. hardly any noticeable pigmentation on the cephalic plate (C).

Classification of the animals according to these hitherto unknown characters corresponded exactly to the original classification according to classical characters—once more verifying the authenticity of the initial categories. Here there were no intermediates between forms B, C and D, and only a small number between A and D animals.

This new character is already obvious at the end of the anamorphic stage and allows a young animal to be referred to its appropriate form. Without this character this would not be possible until after the moult to the mature adult (Fig. 5).

The individual expression of the pattern then remains unchanged in adult animals in spite of further moults. It is unmistakable, like a fingerprint, and each individual can be differentiated from all others of its species at any time by means of a photograph of its cephalic plate. This character disappears immediately the animal dies. In spite of many

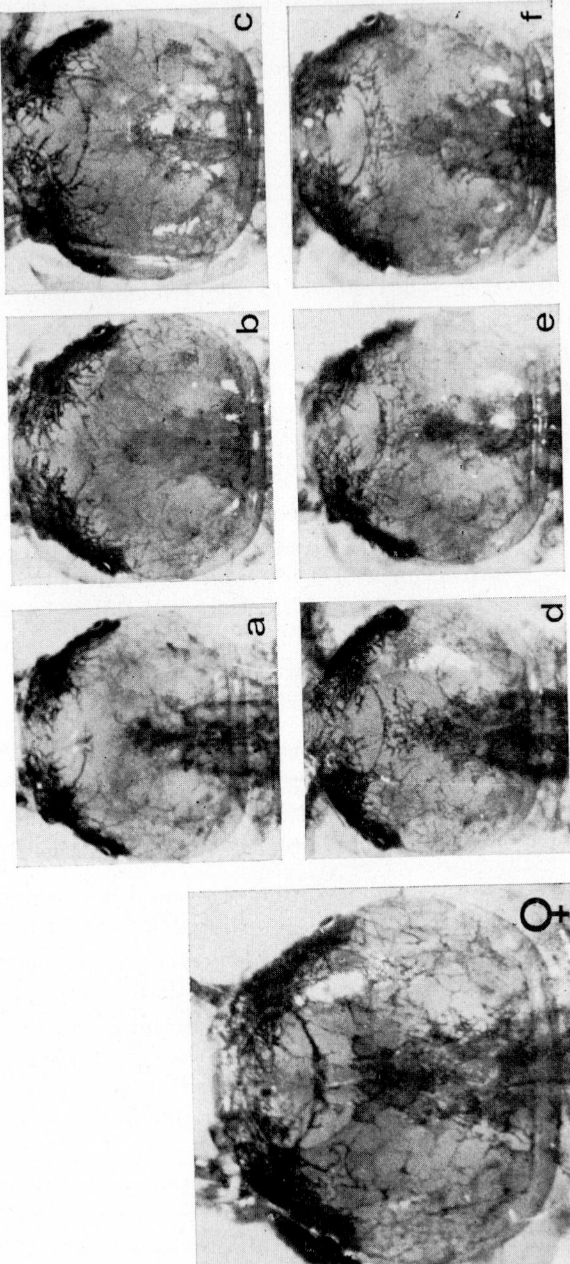

Fig. 5. Cephalic plate patterns of a female of the form D with young animals, a–f.

attempts, a method of preserving these taxonomically valuable structures has not been found.

ALTITUDINAL AND ECOLOGICAL DISTRIBUTION OF THE FOUR FORMS

Yet another classification into the four forms can be made according to the preferred biotope! The forms A, B and C are clearly restricted to different altitudes (Fig. 6).

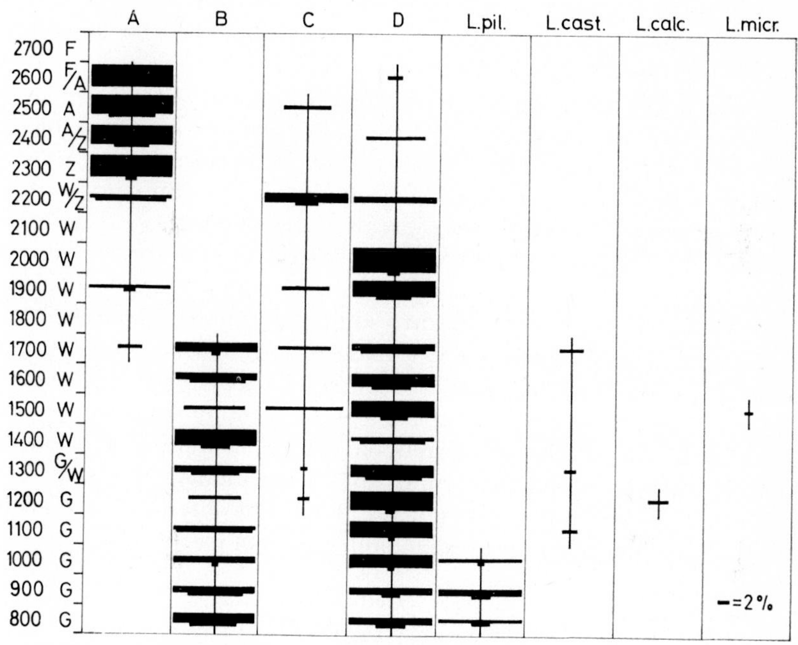

FIG. 6. Distribution of the forms A to D and the other four species (see p. 77) in the particular altitudes of Mount Canigou (number of all individuals found per altitude differences of 100 m = 100%). Each rectangle is built up from 2% sections (see key), with a portion added below to make up the exact percentage. The vertical line through each rectangle represents the altitude range. G = Garigue, W = Wood, Z = Dwarf scrub-heather, A = grassland, F = Stony region of mountain summit.

Form A can be found exclusively at an altitude of over 1700 m, with the greatest abundance in the treeless region. The six specimens found in the wooded zone were gathered from a small meadow; on the border of the wood and the dwarf scrub-heather area they obviously preferred the treeless zone, too. In this border area, animals of forms C and D were found almost exclusively in the wood. On the edge of the forest, on two occasions specimens of forms A and D were found under one rock.

Form B extends from 800 m (in the Garigue) to 1700 m within the wooded zone, and is most frequent in the lowest part of the densely wooded area. Animals of forms B and D were often found there together under the same rock.

Form C was found mainly in the wooded region at an altitude of 2200 m but this not very common species is scattered over an area from 1200–2550 m.

Only form D occurs at all altitudes, that is from 800 m almost to the summit at 2650 m, although it distinctly prefers the wooded regions. The few that were found above the wooded zones came from rocks under small groups of trees and bushes. Examinations of the Mount Canigou population revealed that the altitude has an effect on the characters within the forms A, B and D. With form D, a negative correlation between altitude and head-length was apparent. Specimens from lower altitudes were generally larger than those from higher zones. Since all characters correlate with the size of the animal, different combinations appear in animals from different altitudes. These, however, are not significantly different from one another.

Forms A and C, which prefer the higher regions of Mount Canigou, displayed a positive correlation between locality and head length. For example, the size of the specimens of form A from the wooded zone and all the characters correlated with size, were well below the average which had been determined for all the animals in group A.

Thus the four forms investigated differed not only in their morphological characters but also in their preferred biotope.

CONCLUSIONS AND DISCUSSION

It is evident that form groups can be established according to the main characters and that these forms can also be distinguished by additional characters, although only with the aid of statistics. From my observations of living lithobiids, the supposition that they might be different age stages was invalidated. The very stable character of the pattern of the cephalic plate confirmed the separation of the four forms. There remains only the question: to which taxonomic categories do they belong—are they variations, sub-species or species?

We must eliminate the possibility of their being sub-species, i.e. geographical races in the sense of Rensch (1959), since the four forms are sympatric, and often even syntopic under the same rock. It is therefore also not possible to regard them as ecophenotypes of a single species. Furthermore the distinguishing characters of young animals reared under the same conditions correspond to those of their respective mothers.

Other groups in the infra-sub-specific range as categorized by Mayr (1963) with the expression "non-inherited variation" could not be considered in this examination. Only within form A was a seasonal variation of generations apparent.

Biological proof by means of cross-breeding could not be achieved in the space of time available for the work, although such results as are available indicate that the forms in question must be biospecies.

1. Of the 900 living and conserved Lithobiidae, only three hybrids were found, indicating the existence of reproductive limits between the forms.
2. The characters of the immature animals correspond to those of their mothers.
3. In spite of repeated moults as adults, none of the animals changed its characters to such an extent as to negate such significant differences as have been established between the four forms.
4. In most cases the character averages were significantly different between the forms.
5. Living specimens can be assigned with certainty to one of the four forms according to the cephalic plate patterns.
6. Environmental modification differs from form to form.

If, as Herre (1964) states, the biospecies can be classified by their morphogenic performance—i.e. the morphospecies—then the four forms presented here are certainly four species. If they are assigned to species already described, the oldest names for these taxa would be:

A = *Lithobius mononyx* Latzel 1888
B = *Lithobius piceus* L. Koch 1862
C = *Lithobius melanops* Newport 1845
D = *Lithobius tricuspis* Meinert 1872.

Here these four species have been categorized exclusively from the specimens obtained from Mount Canigou. But are these representative of all populations from different habitats? Do the spectra of characters presented here give a true picture of these widespread species? Unfortunately they do not!

A comparison with Spanish animals of these species quickly showed that the range of variation of the Mount Canigou specimens by no means included the complete spectrum of characters within one form.

For example, animals of the species *Lithobius mononyx* (A) and *Lithobius melanops* (C) are smaller in Spain than on Mount Canigou. They are found at altitudes lower than 1000 m; but as the body length of these species correlates positively with the altitude, the catalogue of all characters correlated with size changes, too.

The limits of a species cannot be determined by investigation of a population from a single area only, however extensive. The true spectra of variations can only be established with certainty by taking a suitable number of random samples from the complete range of distribution.

With the results presented here, we are obliged to revise essential points in the prevailing taxonomic conceptions about Lithobiidae. First of all, it is no longer admissible to describe and to define one species on the basis of data gathered from one individual. Furthermore, it is impossible to make a generally valid statement about the taxonomic value of single characters for the entire family of Lithobiidae. The value of a character must be determined anew for each taxon! For each taxonomic work on Lithobiidae there are a number of necessary prerequisites:

1. A large number of specimens should be collected so that statistical analysis can be applied.
2. The animals should be taken from different biotopes because of the environmental influence on the development of the characters.
3. Random samples should be taken from the entire range of distribution to give a complete picture of the species.
4. Material should be collected in more than one year. A population can consist of different generations which have had different developmental histories according to the year of their birth, and consequently different degrees of environmental modifications of certain characters (Tobias, 1969: 42–43).
5. Living animals should be observed. Firstly to investigate the cephalic plate pattern which is also characteristic of forms other than those under discussion; secondly to observe how the characters change with each moult, as this is the only way in which statements about the stability of individual features can be confirmed.

In this way, forms can be identified which differ significantly in the combination of their characters. From these, one can ascertain the validity of the species without resorting to cross-breeding experiments.

The prevalent state of the taxonomy of Lithobiidae does not correspond to these demands and therefore has to be revised. The 1400 different species of Lithobiidae mentioned at the beginning certainly do represent the variability of the characters. This means that all previous descriptions of species must be re-examined and, in most cases, revised.

References

Brölemann, H. W. (1932). Éléments d'une faune des Myriapodes de France. *Faune Fr.* **25**: 1–405.

Herre, W. (1964). Zur Problematik der innerartlichen Ausformung bei Tieren. *Zool. Anz.* **172**: 403–425.

Mayr, E. (1963). *Animal species and evolution.* Cambridge, Mass.: Harvard Univ. Press.

Rensch, B. (1959). *Evolution above the species level.* London: Methuen.

Tobias, D. (1969). Grundsätzliche Studien zur Art-Systematik der Lithobiidae. *Abh. senckenb. naturforsch. Ges.* No. 523: 1–51.

Verhoeff, K. W. (1925). Chilopoda. *Bronn's Kl. Ordn. Tierreichs* **5** (2): 1–725.

Discussion

EASON: Has Mrs Tobias discovered any method of fixing the characteristic pattern she finds on the head of live specimens of *Lithobius* species, so that it can be seen in preserved material?

TOBIAS: No, not so far.

CRABILL: I believe I have found such a method, but I cannot give details of it at this time until I can be certain through further tests that it is really effective. I intend to publish on the whole matter.

KRAUS: You presented very good photographs of the characteristic pattern on the head of live specimens. It would be of interest to know some details of how you immobilized these lively animals satisfactorily.

TOBIAS: I found a moment's immersion in soda-water (ice-cooled) immobilized the animals satisfactorily.

BRADE-BIRKS: I used two glass plates, of correct weight, hinged in the form of a book, to restrict the movement of live animals for photography.

Symp. zool. Soc. Lond. (1974) No. 32, 89–98.

SYSTEMATIC CRITERIA IN THE SCUTIGEROMORPHA

MARCUS WÜRMLI

Natural History Museum, Basel, Switzerland

SYNOPSIS

The present situation in the taxonomy of the Scutigeromorpha is reviewed. The delineation of old taxa is criticized and guidelines for the description of new taxa are given. A firmer base for assessing the importance of given characters is recommended; in the author's view, this can only be obtained by describing the range of variation in the complete ontogenetic and metameric series.

INTRODUCTION

This preliminary review of the present state of scutigeromorph systematics* is based on the material in several European museums (e.g. Vienna, Hamburg, Munich, Basle), and on an analysis of the literature. Since the characters used are weighted *a posteriori* the approach used is that of the traditional, "alpha"-taxonomy (Sokal & Sneath, 1963: 10).

The taxonomy of the Scutigeromorpha is less advanced than that of the Lithobiomorpha. There are several reasons for this; disregarding the fact that certain species are naturally very rare, we have little material because of the extreme agility of these animals. Scutigeromorpha have never been popular among collectors. Moreover captured specimens are always more or less damaged. Certain characters seem to be very variable, but we have little quantitative measure of this in the absence of adequate series of specimens. It is a particular disadvantage that we are not acquainted with details of the postembryonic development (apart from certain species of *Thereuonema* Verhoeff, 1904, *Thereuopoda* Verhoeff, 1904, and also *Podothereua* Verhoeff, 1905, *Tachythereua* Verhoeff, 1905 (Verhoeff 1905b,c; Murakami 1956a,b, 1959a,b). The best known species in this respect of which we have an abundance of material is our common house centipede *Scutigera coleoptrata* (Linné, 1758) (cf. Verhoeff 1904a, 1905b, 1937a). At present, this has to serve as a model for all Scutigeromorpha. However, notwithstanding the apparently great homogeneity of the group, the use of *S. coleoptrata* to exemplify the whole order might lead to severely

* My use of morphological terms and names of taxa does not imply any judgement regarding their comparative morphology or their systematic validity.

misinterpreting other genera. There are also purely systematic difficulties. There is great variability, especially of metameric structures (see Verhoeff, 1904a; Chamberlin, 1920 (*Scutigera coleoptrata* (Linné, 1758)); Lignau, 1929 (*Thereuonema turkestana* Verhoeff, 1905); Kraus, 1954 (*Scutigera linceci* (Wood, 1867)); Murakami 1956a (*Thereuonema hilgendorfi* Verhoeff, 1905); Murakami, 1959b (*Thereuopoda ferox* (Verhoeff, 1936)), and there are a great number of possible characters. These factors, together with the relative homogeneity of the whole group and the lack of invariable and unequivocal diagnostic structures such as the gonopods of Diplopoda, have resulted in some degree of taxonomic confusion. In regard to our knowledge of evaluating the characters (except the female gonopods), I agree with the remarks of Tobias (1969: 5): "It is impossible to make a generally valid statement about the taxonomic significance of any particular character throughout the Lithobiidae; this must be evaluated separately for each different taxon", and I think they may be equally true of the Scutigeromorpha.

We can distinguish two main periods in the development of scutigeromorph taxonomy. The first extends from Linné to around the beginning of this century, at which time the initial contributions of Verhoeff were published. Only few descriptions of this period are clear enough for certain identification of the species in question (e.g. Meinert, 1886a,b; Haase, 1887; de Saussure & Zehntner, 1902). Since Verhoeff and later workers, including Chamberlin, have not usually been able to check the old types we cannot be certain whether their new species had been described previously; nor can we be sure, in all instances, of the exact identity of the species attributed to the older authors. Thus nomenclatural difficulties are added to the taxonomic ones.

The generic classification which Verhoeff (1904b, 1905a,b, 1925, etc.) founded seems to me still partly valid but needs revision because of his use of characters which appear to me to be excessively variable. Chamberlin also described many genera (*Gonethina*, 1918, *Gonethella* 1918—both forgotten because they never appeared in the Zoological Record—*Diplacrophor* 1922, *Gomphor* 1940, *Thereulla* 1955, *Phanothereua* 1958). But apart from three new and possibly questionable genera (*Brasilophora* Bücherl, 1939; *Brasiloscutigera* Bücherl, 1939; *Thereuoquima* Bücherl, 1949) nothing really basic has been added since Verhoeff's time.

In addition to assessing and criticizing existing work, the present paper offers some practical suggestions for the future description and redescription of new and existing taxa.

THE SYSTEMATIC IMPORTANCE OF THE SINGLE CHARACTERS

General features

Size

In my experience, many mature scutigeromorph centipedes measure at least about 20 mm, and therefore an older pseudomaturus of about 16 mm will probably not differ fundamentally from a maturus (of about 20 mm). Thus, although smaller specimens can be described, they should not be named as new, because even the generic characters, like the exoskeletal prominences of the tergal plates, may not yet be well developed. The new genera described by Chamberlin—*Gomphor* 1940, *Gonethina* 1918, *Phanothereua* 1958, *Thereulla* 1955—seem to me to be unnecessary, although I cannot be certain of this until I have examined the types.

It is possible that important characters of scutigeromorphs alter during the post-maturational moults as happens in the Lithobiidae (Tobias, 1969, and this Symposium, pp. 75–87) but this possibility has not yet been entertained.

Pigmentation

Although Verhoeff denied a more than incidental importance of pigmentation I think (cf. Kraus, 1957) that it can be important in certain genera. In many groups consisting predominantly of species with a good visual faculty the pigmentation appears to play an important role (e.g. Lycosidae, Salticidae). We are led to believe from the studies of Tobias (1969) that in certain European species of *Lithobius* the pattern of pigmentation of the cephalic plate is peculiar to all stages (and even individuals) of a species. I have started similar investigations on the importance of distribution—not of colour quality (Murakami, 1959b)—of pigment in Scutigeromorpha.

Unfortunately it is quite impossible to determine the pigmentation of most long-preserved specimens. According to the method of fixation and conservation the pigment can be partly or completely lost. In some cases every fine detail remains unaltered for more than a century.

Thus descriptions should contain notes about the gross and fine distribution of pigment on the head, tergal plates (cf. Miyoshi, 1939, pl. III) and legs (anuli on prefemur, femur and tibia).

Head

The macrosculpture (grooves, depressions, swelling and the general shape) and the distribution of exoskeletal outgrowths (cf. *Tergal plates*, p. 92) must be studied, even when their importance is not yet established.

The shape (form and relative length of the protrusions) of the cross-striped cephalic sutures is of great importance for generic and specific classification (Ribaut, 1923; Verhoeff, 1937b: *Thereuopoda* Verhoeff, 1904 and *Parascutigera* Verhoeff, 1904). According to my own observations, these characters are not important in *Scutigera* Lamarck, 1801.

Antennae

The ratio of length/width of the single articles and the relative length of the whole antenna are characters which separate some families. The morphologically homogeneous antennae carry many hairs and bristles of various shapes (e.g. hooked or not; cf. *Tergal plates*, p. 92) and lengths. Their arrangement relative to each other, and their pattern and density per article seem to be typical for genera, but not for single species. I cannot decide in what groups and to what extent the number of articles of the first and second flagellum (also duploflagellum) is important in taxonomy. But I am inclined to believe that it would be very slight. The number of articles of such highly metameric organs as the tarsi also seem too variable both within and between stadia for them to be taxonomically useful.

Mouth parts

There may or may not be labral and epipharyngeal differences. Sometimes the differences between single species (of the same genus) are greater than between genera (as defined by Verhoeff).

The teeth and pectinate lamellae of the mandibles bear no systematically relevant feature. The same is valid for the first maxillae. On the contrary, the shape and length of the second maxillae and especially the arrangement of their exoskeletal prominences are very important. It is always useful to note the relative length and the form (microsculpture, extremity broadened or not) of the long spine-bristles.

Prehensors

Important characters for generic classification are the shape, length and width of the single articles, especially the coxa and the prominences of the inner margin of the coxa.

Tergal plates (including stomatotergites)

One should not neglect to draw and describe the contours of the tergal plates (cf. Lawrence, 1960). Indeed we know that the whole structure of the tergal plates (especially of the posterior margin) can alter in the course of the postembryonic development (cf. Kraus, 1957).

The alveolar microsculpture is apparently not a systematic criterion (but cf. *Scutigerina* Silvestri, 1903 and *Madagascophora* Verhoeff, 1936).

The exoskeletal prominences are some of the best features for distinguishing the genera. The following catalogue of forms (partly see Attems, 1926) does not claim to be complete; on the contrary one may often find special forms and all transitions between the single types:

Simple, rigid, immovable, non-innervated prominences

1. Hairs: short and slender.
2. Spinules, spinulae (= German "Haardörnchen"): short, sturdy hairs (Fig. 1).
3. Spiculae (= German "Haarspitzen"): spike-shaped long hairs (Fig. 2).
4. Short spiculae.
5. Spines, spinae: very often associated with bristles (Fig. 3).

Movable, innervated prominences

1. Bristles, setae (Fig. 3).
2. Spine-bristles, spinosetae: enlarged, thickened bristles. The microsculpture of the spine-bristle (especially of the second maxilla) (Fig. 4) can be a criterion for the distinction of species.

It is necessary to describe the shape, length and width (both in absolute units) of the prominences. The various dimensions effectively show specific differences (cf. Ribaut, 1923). The associations between the various forms of outgrowth are also significant. Figures 5 and 6 show two possible associations between a bristle and two spines. In Fig. 6 the two spines have fused into one double-toothed structure. One should always note the ratio of size of such partners.

The quantitative distribution of the single types of prominences is also of the greatest importance. One should note the number per unit area or, preferably, per entire tergal plate. A heterogeneous distribution on one tergum or between different tergites is worth noting. The differences must be described in detail; it is not sufficient to describe the situation of the sixth and seventh tergal plates only because, according to Verhoeff's "Dornenregel", they carry more prominences. As a general rule the whole range of variation of metameric organs should be described. I emphasize once again that the tergite armour can vary during postembryonic development.

The saw of spines and bristles lying on the margin of the stomatotergites often shows specific differences e.g. number per side, alternation of tall and shorter spines and bristles, direction of prominences.

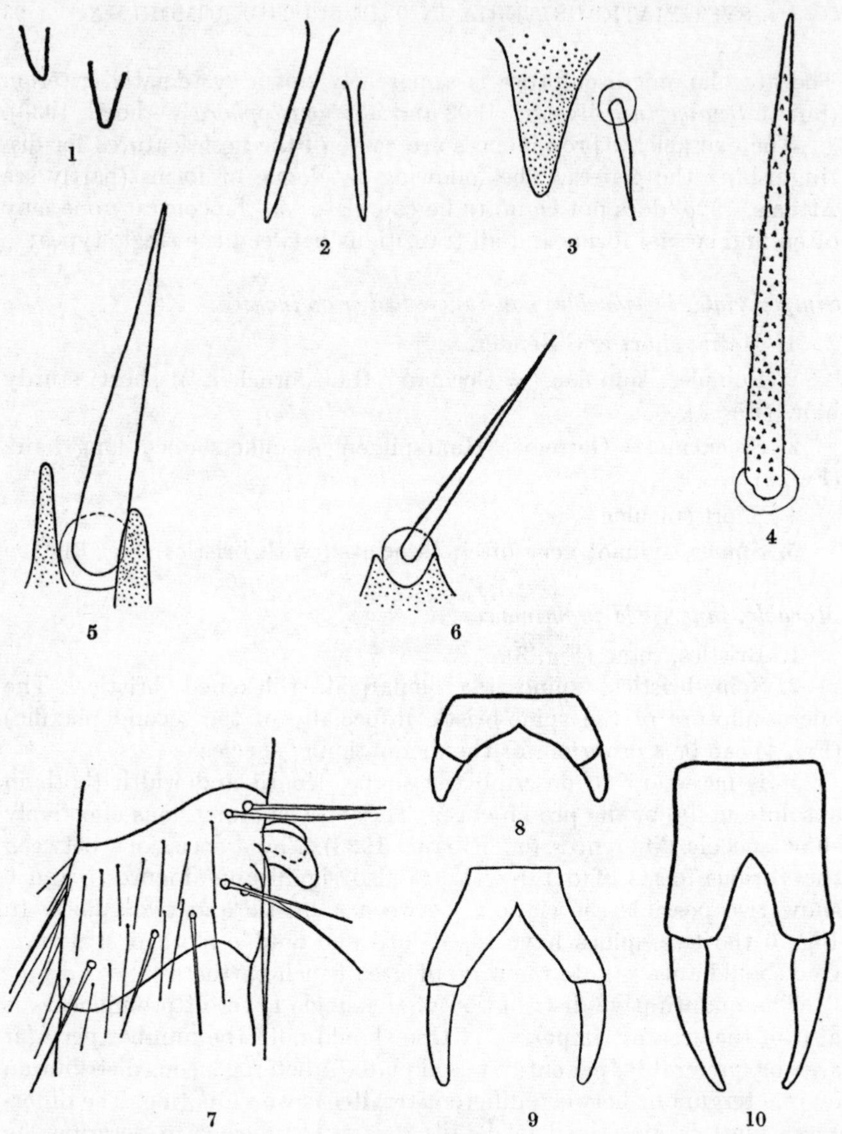

FIGS 1–10

FIG. 1. Spinulae from tergal plate of *Pselliodes* sp. n. FIG. 2. Spicule from tergal plate of *Thereuonema* sp. n. FIG. 3. Association between spine and bristle from tergal plate of *Thereuonema* sp. n. FIG. 4. Spine-bristle from second maxilla of *Allothereua* sp. n. FIG. 5. Association between one bristle and two spines from tergal plate of *Allothereua* sp. n. FIG. 6. Association between one bristle and two spines from tergal plate of *Thereuonema* sp. n. FIG. 7. Tarsale sinuatum of second tarsus, eighth leg of gen. n. sp. n. *Note:* FIGS 1–3, 5–7 nearly same scale. FIGS 8–10. Schematic development of hypothetical gonopods (with reference to Murakami, 1959b: Figs 4, 5, 7). All same scale. Bristles and hairs omitted. FIG. 8. Agenitalis. FIG. 9. Younger pseudomatura. FIG. 10. Matura.

Sternites

The sternites show features similar to the tergites. But we know little about the systematic importance of the features, such as the shape of the posterior margins (sinuous or not), the exoskeletal prominences, the microsculpture and the longitudinal ridge.

Legs

The number of articles of the tarsi is very distinctive for the single stadia of postembryonic development (cf. Verhoeff, 1904a). But the number in mature specimens is of unknown but probably little importance. As in the antennae, the disposition, density, shape and length of the exoskeletal prominences are very variable; probably some of these features will prove to be distinctive. In general one can find the following four types (Fig. 7):

1. Hairs (moulting hairs, "Häutungshaare" sec. Verhoeff, 1904a).
2. Bristles (often various forms in the same animal).
3. "Resilient sole-hairs" ("federnde Sohlenhaare", Crines appressi subpedales, Verhoeff, 1904a): hairs thickened at the base, pressed against the under surface of the article and extending to the middle of the under surface of the next one.
4. Tarsal papillae ("Tarsalzapfen"), often very specific in form.

The presence of one or two thick bristles lying at the end of the first tarsus is of considerable generic importance, as is an indication of where they first appear in the series.

The shortest way to describe the distribution of the above characters and the proportion of Tarsalia asinuata—Tarsalia sinuata of the second tarsus (Verhoeff, 1904a: 211) is to use a table like those proposed for the Lithobiomorpha.

The prefemur, femur and tibia frequently bear on their keels (if present, cf. *Scutigerina* Silvestri, 1903), faces and extremities very long spine-like bristles. Their form and distribution should be described. The ratio of the length of legs to the length of the trunk may be important in defining the higher categories and may have ecological importance. The number of articles is highly variable even between the two sides of the same specimen; new species perhaps should not be based on this character alone.

Gonopods

The tripartite (pro-, mes- and metarthron) female gonopods are, as Verhoeff eventually recognized, *the best criteria for species segregation.* In generic classification they play a rôle in only a few cases. A clearly visible suture separating the proarthron and mesarthron is a primitive feature. The author (Würmli, in press) proposed to use the following indices for describing the gonopods:

A/B = greatest length/greatest width
$C+D/E$ = length of proarthron + mesarthron/length of metarthron
C/D = length of proarthron/length of mesarthron
F/G = width of one mesarthron/width of the sinus between the two mesarthra, both measured at the level of the point of inflexion of the inner margin
$\frac{H-I}{C+D}$ = a measure for the divergence of the outer margins of the pro- and mesarthron (H = width at the base of proarthron, I = width of gonopods, measured at the points of insertion of the metarthra
α = angle of the sinus at the medial distal end of the proarthra.

In the course of my investigations it appeared that the characters of form not covered by these indices, e.g. the shape and contour of the margins of pro-, mes- and metarthron (serration of inner margin of metarthron and the curvature of the top of metarthron), are much more important. Moreover, there is no correlation between the single indices. Here too, the distribution of hairs and spines compared with the naked areas can be used as a criterion.

The gonopods are very simple organs in Scutigeromorpha. They show little variation of form and are very similar in various genera.

In judging the gonopods one must take into account the artefacts. If the alcohol has been too strong the gonopods can be folded keel-shaped; in that case the metarthra overlap. After maceration the gonopods can become completely flat; then the mesarthra and metarthra gape strongly. Praematuri and younger pseudomaturi can be recognized by their similarly strongly gaping gonopods which can be understood from their development (cf. Kraus, 1957; Murakami 1956a,b) (Figs 8–10). The maturi of *Tachythereua maroccana* Verhoeff, 1905 and of *Prionopodella* Verhoeff, 1924 are exceptional in possessing wide gonopods.

From my own observations on some species of *Scutigera* Lamarck, 1801 and *Thereuonema* Verhoeff 1904, the inner and outer genital styles of the male do not appear to provide characters other than those at family level.

CONCLUSIONS

For promoting the taxonomy of the Scutigeromorpha we need investigations into variability, postembryonic development and allometry, ecology, behaviour and genetics. For descriptive taxonomy it is urgently necessary to find and redescribe the older types (especially those of Chamberlin). The search for new systematic criteria that are valid for all stadia must continue (possibly pigmentation?). Perhaps the generic classification may need revision.

We do not yet know the importance of single characters and thus we are forced to describe all the somatic features mentioned on pp. 91–96. Ribaut (1923) and Lawrence (1960) may serve as models. It does not seem advisable to erect new species on the basis of female gonopods alone. Verhoeff, supposedly, was able to recognize most species correctly without the knowledge of the female gonopods. Indeed, in *Scutigera* Lamarck, he recognized too many.

The whole range of variability along the axis of metamerism should be described as Bücherl (1939) has done. It is insufficient to describe a limited number of the segmental organs.

It is not advisable to erect new species from isolated males (even mature ones) nor from younger female pseudomaturae. The situation is more critical within new genera because the characters needed for their definition are generally not linked with sexual ones. It is also inadvisable to erect new species on differences in the female gonopods, unless they are manifest in individuals of the same developmental stage.

Acknowledgement

I wish very much to thank my dear friend Chr. Stücklin, D.D., who has revised my stylistic expression.

References

Attems, C. (1926). Myriopoda. *Handb. Zool.* **4** (1): 1–402.

Bücherl, W. (1939). Os quilopodos de Brasil. *Mems Inst. Butantan* **13**: 49–362.

Chamberlin, R. V. (1920). The myriopod fauna of the Bermuda Islands, with notes on variation in *Scutigera*. *Ann. ent. Soc. Am.* **13**: 271–302.

Haase, E. (1887). Die Indisch-Australischen Myriopoden. I. Chilopoden. *Abh. Ber. K. zool. anthrop.-ethn. Mus. Dresden* **1886/87** (5): 1–118.

Kraus, O. (1954). Myriapoden aus El Salvador. *Senckenberg. biol.* **35**: 293–349.

Kraus, O. (1957). Myriapoden aus Peru. VI. Chilopoda. *Senckenberg. biol.* **38**: 359–404.

Lawrence, R. F. (1960). Myriapodes Chilopodes. *Faune Madagascar* **12**: 1–121.

Lignau, N. (1929). Neue Myriapoden aus Zentralasien. *Zool. Anz.* **85**: 204–218.

Meinert, F. (1886a). Myriapoda Musaei Hauniensis. III. Chilopoda. *Vidensk. Meddr dansk. naturh. Foren.* **1884–86**: 100–150.

Meinert, F. (1886b). Myriapoda Musaei Cantabrigensis, Mass. Part I. Chilopoda. *Proc. Am. phil. Soc.* **23** (122): 161–233.

Miyoshi, S. (1939). Chilopoda of Jehol. *Rep. scient. Exped. Manchoukou* Sect. V, Div. I, Part IV, art. 14: 20–29.

Murakami, Y. (1956a). The developmental stadia of *Thereuonema hilgendorfi* Verhoeff (Chilopoda, Scutigeromorpha). *Zool. Mag., Tokyo* **65**: 37–41.

Murakami, Y. (1956b). The life history of *Thereuonema hilgendorfi* Verhoeff (Chilopoda, Scutigeridae). *Zool. Mag., Tokyo* **65**: 42–46.

Murakami, Y. (1959a). Postembryonic development of the common Myriapoda of Japan. I. The anamorphic development of the leg-bearing segments of Scutigeridae (Chilopoda) and a new aspect on the problem of its tergite. *Zool. Mag., Tokyo* **68**: 193–199.

Murakami, Y. (1959b). Postembryonic development of the common Myriapoda of Japan. II. *Thereuopoda ferox* Verhoeff (Chilopoda, Scutigeridae). *Zool. Mag., Tokyo* **68**: 324–329.

Ribaut, H. (1923). Chilopodes de la Nouvelle-Calédonie et des Îles Loyalty. *Nova Caledonia.* (A. Zoologie) **2** (L. 1): 1–79.

de Saussure, H. & Zehntner, L. (1902). Myriapodes de Madagascar. In *Histoire physique, naturelle et politique de Madagascar* **27**: 1–356. Grandidier, A. (ed.). Paris: Imprimerie Nationale.

Sokal, R. R. & Sneath, P. H. A. (1963). *Principles of numerical taxonomy.* San Francisco & London: Freeman & Co.

Tobias, Dagmar (1969). Grundsätzliche Studien zur Artsystematik der Lithobiidae (Chilopoda: Lithobiomorpha). *Abh. senckenb. naturforsch. Ges.* No. 523: 1–51.

Verhoeff, K. W. (1904a). Mittheilungen über die Gliedmassen der Gattung *Scutigera. Sber. Ges. naturf. Freunde Berl.* **1904**: 198–236.

Verhoeff, K. W. (1904b). Ueber Gattungen der Spinnenasseln. *Sber. Ges. naturf. Freunde Berl.* **1904**: 243–285.

Verhoeff, K. W. (1905a). Ueber Scutigeriden. 5. Aufsatz. *Zool. Anz.* **29**: 73–119.

Verhoeff, K. W. (1905b). Zur Morphologie, Systematik und Hemianamorphose der Scutigeriden. *Sber. Ges. naturf. Freunde Berl.* **1905**: 9–60, 2 pl.

Verhoeff, K. W. (1905c). Ueber Scutigeriden. 6. Aufsatz, Variabilität und *Thereuonema-Arten.* Tarsen mit sprungweiser Abänderung. *Zool. Anz.* **29**: 353–371.

Verhoeff, K. W. (1925). Results of Dr E. Mjöberg's Swedish Scientific Expeditions to Australia 1910–1913. 39. Chilopoda. *Ark. Zool.* **17A** (3): 1–62.

Verhoeff, K. W. (1937a). Zur Biologie der *Scutigera coleoptrata* und über die jüngeren Larvenstadien. *Z. wiss. Zool.* **150**: 262–282.

Verhoeff, K. W. (1937b). Chilopoden aus Malacca, nach den Objecten des Raffles Museum in Singapore. *Bull. Raffles Mus.* No. 13: 198–270.

Würmli, M. (in press). Zur Systematik der Scutigeriden Europas und Kleinasiens. *Ann. naturh. Mus. Wien.*

Symp. zool. Soc. Lond. (1974) No. 32, 99–133.

DIE SCOLOPENDROMORPHA DER NEOTROPISCHEN REGION

WOLFGANG BÜCHERL

z.Z. Secção de Artropodos Peçonhentos, Instituto Butantan, Caixa Postal 65, São Paulo, SP, Brasil

SYNOPSIS

All the scolopendromorphs are poisonous animals, killing their prey with venom (Bücherl, 1946b). From Linné (1758) until today, 27 authors published 78 taxonomic papers on the neotropical Scolopendromorpha; Pocock, Brölemann, and Verhoeff four each, Kraus three, Bücherl 13, Chamberlin 17, and the others, one, two or three.

Only references subsequent to the complete bibliography published by Attems (1930) are given. References after this date, in the text, give only the Journal and the pertinent page and figure numbers; titles and complete page numbers are given in the list of references.

The number of known species has increased markedly, but little can be concluded about the geographical distribution, the frequency and ecological properties of the scolopendromorphs in the neotropical region. Nearly 70% of the species are known only from one or from a few specimens. A few genera have been described from only one species and one specimen, e.g. *Scolopendropsis bahiensis*, and nobody can find another specimen in the type-locality. *Arthrorhabdus spinifer, Paracryptops inexspectus, Mimops occidentalis, Kartops guianae*, etc. are extremely rare.

The following genera: *Trachycormocephalus, Campylostigmus, Asanada* (Fig. 12 a, b), *Pseudocryptops, Digitipes, Alipes* (Fig. 13 a, b) *Ethmostigmus, Alluropus, Arrhabdotus, Theatops, Plutonium, Kethops, Tidops* and the subgenus *Otostigmus*, are not represented in the neotropical region.

A list of neotropical genera, together with the number of species and subspecies they contain, is appended.

SCOLOPENDRIDAE

Scolopendrinae

Scolopendrini

Arthrorhabdus Pocock, 1891.

1. *Arthrorhabdus spinifer* (Kraepelin) 1903—Holotypus von Belém, Staat von Pará, äquatoriales Brasilien.

Cormocephalus Newport, 1845.

2. *Cormocephalus amazonae* (Chamberlin) 1914—(Fig. 17)—Holotypus von Manaus, Staat von Amazonas, äquatoriales Brasilien.

3. *Cormocephalus andinus* (Kraepelin) 1903:

a. *Cormocephalus andinus andinus* (Kraepelin) 1903—(Figs 3, 18: a, b, c, d)—*Perustigmus alticolus* + *rapax* Verhoeff 1941 (Fig. 16: a, b);

Cupipes andinus Chamberlin, 1944; *Cormocephalus andinus* Bücherl, 1943b; Kraus, 1954; Turk, 1955; Kraus, 1957. *Cormocephalus andinus andinus* Bücherl, 1972—Holotypus wahrscheinlich von Sorate, Bolivien. *Verbreitung* in Peru: Santa Ana, Cusco, Olmos, Cordillera Azul, Colcabamba, Tapacocha, Huanuco, Zorate, Carpis, Carhuamayo, Paucartambo, Tarma, Huayuncape, Arequipa, Sivia, Sodondo-Fluss (*Perustigmus*), Amancais, Cerro de Pasco. Von 670 (Tingo Maria) bis 3800 m (Cerro de Pasco); feuchte Waldungen; hohe Nebelgebiete; oft unter Steinen; auf sandigen Steppenböden.

b. *Cormocephalus andinus rubrifrons* Bücherl, 1950a—*Mems Inst. Butantan* **22**: 177, Figs 7, 8 (Fig. 19: a, b, c), Turk, 1955; *Cormocephalus* (*C.*) *andinus* Kraus, 1957—Nur 6 Basalglieder der Antennen kahl; erst vom 15. Tergit nach rückwärts sind die Seitenränder vollständig; 2–20. Sternit mit vorderen Paramedianfurchen und ein vorderes ovales Grübchen; Kopf und 1. Rumpfsegment braun, der Rest olivenfarbig—Holotypus aus Huanuco, 1900 m; MHNL No. 10 025, Weibchen. *Verbreitung:* Huanuco, Abancay, Sahuayaco, Huallaga, Cajacudro, Ayacucho; vegetationsarme Gegenden bevorzugend (Abancay); oft fast arides Klima.

4. *Cormocephalus anechinus* Chamberlin, 1957—*Gr. Basin Nat.* **17**: 30—Holotypus 48 km, Paratypus 37 km östlich von Carhuamayo, Provinz Junin, Peru.

5. *Cormocephalus bonaerius* Attems, 1928—(Fig. 20)—Holotypus von der Insel Bonaire, vor der Küste Venezuelas.

6. *Cormocephalus brasiliensis* Humbert & Saussure, 1870—Der Holotypus soll von Manaus, Amazonas, stammen.

7. *Cormocephalus annectans* (Chamberlin) 1941—*Cupipes annectans* Chamberlin, 1941; *Cormocephalus andinus* Kraus, 1957; *Cormocephalus annectans* Bücherl, 1972—Holotypus von Pongo de Manseriche, Loreto, Marañon-Fluss, 170 m; Bassler legit, 1924; im AMNH, No. 9510.

8. *Cormocephalus carolus* Chamberlin, 1955—Holotypus und ein zweites Exemplar von den Galapagos-Inseln.

9. *Cormocephalus guildlingi* Newport, 1845—*Cupipes guildlingi* Chamberlin, 1944—Holotypus von Skt. Vincent. Weitere Funde von Haiti unter Steinen und aus der Grotte Diquini, im Westen von Port au Prince.

10. *Cormocephalus impressus* Porat, 1876.

a. *Cormocephalus impressus impressus* Porat, 1876; Bücherl, 1943; *C.* (*C.*) *impressus* Kraus, 1957; *Verbreitung:* Mexico, Haiti, Skt. Barthelemy, Equador, Peru: Cerro Carambajoij, Cajamarca, 3000 m; Ainín, 3300 m; El Infernillo, 3370 m; Chuquibamba, 3500 m; Bergsteppe; Exemplare im SMFI.

b. *Cormocephalus impresses birabeni* Bücherl, 1953—*Mems Inst. Butantan* **25**: 109, Figs 15–18. (Figs 2, 21: a, b)—Holotypus im La Plata Museum, Argentinien; Biraben legit bei San Pedro, Prov. Salta, West-argentinien.
c. *Cormocephalus impressus glabrus* Bücherl, 1950a—*Mems Inst. Butantan* **22**: 173, Fig. 22; Bücherl, 1953; Turk, 1955; *C.* (*C.*) *impressus* Kraus, 1957—Holotypus im MHNL, No. 10 128; San Mateo, am Rimac-Flusse, 3000 m, Peru; Weyrauch legit, 1948; *Verbreitung* Peru: San Mateo; Andahuaylas, 3100 m; Tarma, 3300 m; Huamachuco 2300 m; Cajamarca; Canta; Cerro Chamis. Von *C. i. impressus* vor allem durch das meist völlige Fehlen von Längs-oder Querfurchen am Coxosternum und hellen Längsstreifen in der Mitte der Tergite gut zu unterscheiden.
d. *Cormocephalus impressus neglectus* (Chamberlin) 1914—*Cupipes neglectus* Chamberlin, 1914; *Cormocephalus* (*C.*) *impressus neglectus* Bücherl, 1939; 1950a; *Cormocephalus* (*C.*) *impressus* Kraus, 1957—Holotypus aus Rondonia, an den Ufern des Madeira-Flusses, West-brasilien. Kopfplatte so lang wie breit; nur das 1. Grundglied der Antennen völlig kahl, die drei folgenden schon spärlich beborstet; letztes Sternit mit Längsdepression.
e. *Cormocephalus impressus pernuanus* Bücherl, 1953—*Mems Inst. Butantan* **25**: 118, Figs 3–5; Turk, 1955; *Cormocephalus impressus* Kraus, 1957. Truncus und Beine grün; Kopfplatte mit zwei Längsfurchen nur in der hinteren Hälfte; 4 Grundglieder der Antennen kahl; das 5. ventral 6. und 7. ventral und dorsal beborstet; nur das 21. Tergit seitlich berandet; Coxopleuren, ausser 1 winzigen Enddörnchen, keine Seitendörnchen; 21. Praefemur ohne Ventraldornen.—Holotypus vom Cerro de Pasco, San Rafael, 3800 m; Weyrauch legit, 1946, Peru; IB No. 683. *Verbreitung:* Peru: San Rafael u. Abhänge von Cerro de Pasco.
f. *Cormocephalus impressus unimarginatus* Bücherl, 1941—*Mems Inst. Butantan* **15**: 123, Figs 4–6—Holotypus im NMR, No. 141, aus Veadeiros, im Staate Goias, Brasilien. Kopfschild breiter als lang; Antennen den Hinterrand des 1. Tergites kaum erreichend; schon die Grundglieder deutlich, wenn auch spärlich beborstet.

11. *Cormocephalus lineatus* Newport, 1845—Holotypus von Skt. Vincent.

12. *Cormocephalus mediosulcatus* Attems, 1928—(Fig. 23)—Bücherl, 1939—Holotypus vom Distrikt von Caldeirão, Marajo-Insel, Nordost—Brasilien.

13. *Cormocephalus mundus* Chamberlin, 1955—*Acta Univ. lund.* N.F.(2) **51**; 46. Kraus, 1957—Holotypus 40 min östlich von Abancay, Peru.

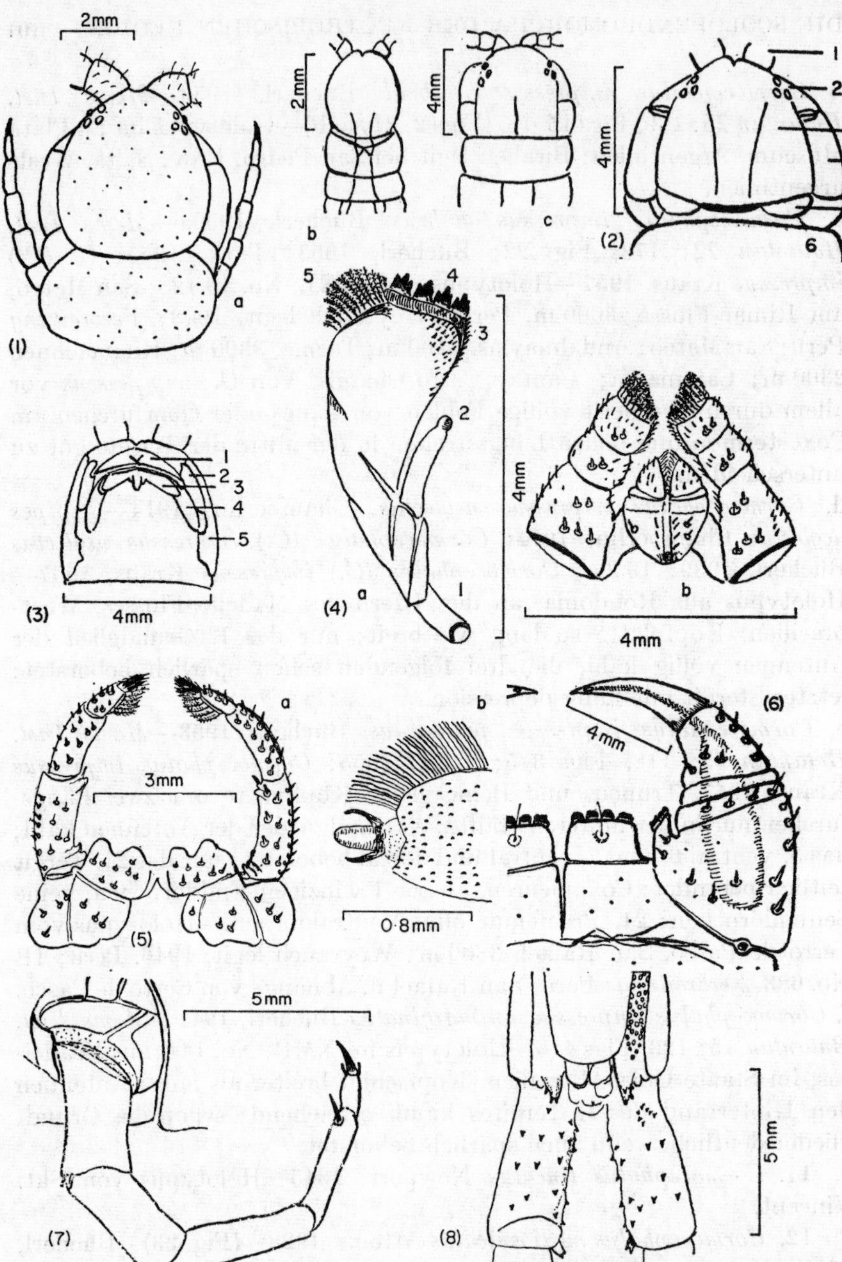

FIGS 1–8. 1. (a) *Scolopendra amazonica;* (b) *Newportia fuhrmanni;* (c) *Rhoda thayeri.* 2. *Cormocephalus impressus birabeni.* 3. Unterseite des Kopfschildes von *Cormocephalus andinus.* 4. *Scolopendra viridicornis nigra:* (a) Mandibel; (b) 1. Maxillenparr. 5. *Scolopendra spinipriva* (a) 2. Maxillenpaar; (b) Endabschnitt desselben. 6. *Scolopendra spinipriva:* Coxosternum. 7. *Scolopendra viridicornis:* Tarsal-und Klauensporne. 8. *Scolopendra amazonica:* 21. Segment-ventral.

14. *Cormocephalus tingonus* Chamberlin, 1957—*Gr. Basin Nat.* **17**: 31–32. Holotypus von Ross u. Schlinger 1954 gefangen im Monsontal bei Tingo Maria, Peru. Keine Ventraldornen am Praefemur der Analbeine.

15. *Cormocephalus ungulatus* (Meinert) 1886—*Cupipes ungulatus* Meinert, 1886; *Cormocephalus* (*C.*) *ungulatus* Attems, 1930, Chamberlin, 1957—Holotypus von Grande Anse, Haiti. *Verbreitung:* Grande Anse; Equador: Pungo; Peru: Yurso; Lachai bei Lima.

16. *Cormocephalus venezuelanus* (Brölemann) 1898—Venezuela, Ohne Angabe des Fundortes.

Hemiscolopendra Kraepelin, 1903.

17. *Hemiscolopendra chilensis* (Gervais) 1847, Chamberlin, 1955—Holotypus wahrscheinlich südlich von Concepcion, Chile. (Figs 14a: 2; 24a). *Verbreitung:* Chile: Concepcion; Valparaiso; Aconcagua; Copiapó; Talcahuano; Villa Rica; Juncal; Coquimbo; Guayacan, am 30. S. Breitengrad; San Carlos. Argentinien: 5 min nördlich von Dean Funes bei Cordoba; Pampa, Prov. Cordoba.

18. *Hemiscolopendra galapagosa* Chamberlin, 1955—*Acta Univ. lund.* N.F. (2) **51**: 34 Holotypus aus Galapagos, Crocker Exped. legit, 1932.

19. *Hemiscolopendra laevigata* Porat 1876—(Fig. 24 b)—*Cormocephalus* (*Hemiscolopndra*) *laevigata* Bücherl, 1941. *Verbreitung:* Argentinien: Buenos Aires; Uruguay: Montevideo, Maldonado; Chile: Santiago; Kolumbien; Cayenne in Franz. Guayana.

20. *Hemiscolopendra michaelseni* (Attems) 1903—(Fig. 24c)—*Cormocephalus* (*Hemiscolopendra*) *michaelseni* Bücherl, 1941. *Verbreitung:* Chile: Valparaiso, Viña del Mar, Quilpué, Coquimbo.

21. *Hemiscolopendra perdita* Chamberlin, 1955—*Acta Univ. lund.* N.F.(2) **51**: 34. Holotypus von der Halbinsel Coquimbo bei Forte. *Verbreitung:* Chile: Coquimbo, Aconcagua. Argentinien: Humahuaca, 2900 m; Cerro San Xavier; Tucuman.

22. *Hemiscolopendra platei* (Attems) 1903—(Fig. 24d)—*Cormocephalus* (*Hemiscolopendra*) *platei* Bücherl, 1941; Chamberlin, 1955; Kraus, 1957. *Verbreitung:* Chile: Valparaiso (Holotypus); Coquimbo; Quilpué; Viña del Mar; Las Villas. Argentinien: Chubut; Las Plumas. Peru: Mejica; Mollendo; der Südküste entlang.

Rhoda Meinert 1886.

23. *Rhoda calcarata* (Pocock) 1891.

a. *Rhoda calcarata calcarata* (Pocock) 1891—Holotypus wahrscheinlich aus Salvador, Bahia; ein zweites Exemplar aus Recife; Pernambuco, Brasilien.

b. *Rhoda calcarata carvalhoi* Bücherl, 1941, *Mems Inst. Butantan* **15**: 126–127—(Figs 14a: 3; 25: a, b)—Holotypus—Barra do Tapirapé-Fluss, Mato Grosso, Brasilien; IB No. 291; 3 Paratypen; Carvalho legit 13/5/45. *Verbreitung:* Zentralbrasilien: Tapirapé; Mato Verde; São Domingos; Chavantina; Aragarças; Abaetetuba, bei Belém; Pará.

24. *Rhoda thayeri* Meinert, 1886, Bücherl, 1939; 1941—(Fig. 1: c)—Holotypus von Santarém, Staat Pará, Brasilien.

Scolopendra Linné, 1758—(Figs 14a: 1; 15).

25. *Scolopendra alternans* Leach, 1815; Bücherl, 1941; Chamberlin, 1944. *Verbreitung:* Cuba, Portorico, Skt. Thomas, Skt. Croix, Guadelupe, Antigua, Montserrat; Zidadelle von Cristophe am Capo von Haiti. U.S.A.: Miami (FM); Venezuela (MNR).

26. *Scolopendra amazonica* (Bücherl) 1946a—*Scolopendra morsitans amazonica* Bücherl, 1946a. *Mems Inst. Butantan* **19**: 135–137, Figs 1, 2, 3; Bücherl, 1953; *Scolopendra morsitans* Jangi, 1955; *S. amazonica* Lewis, 1968; 1969. (Figs 1: a, 8)—Holotypus, Männchen, IB No. 242 mit 2 Paratypen: Manaus, Amazonas, Brasilien. *Verbreitung:* Manaus, Uaupés, etwas südlich vom Aequator, an der Grenze mit Kolumbien; Belém, im Staate Pará, Sioli legit 1944; unterscheidet sich von *S. morsitans morsitans:* Kopf, 1. und 21. Segment mit den Analbeinen braunrot; Tergite gelbbraun mit dunkelgrünen Querbändern vor dem Hinterrande. 20 Antennenglieder, 6 kahl; Paramedianfurchen des 2. Tergites nur im 1. Drittel; Tergite 3–20, neben den kompleten Paramedianfurchen, noch ein kurze Mittelfurche in der 2. Körperhälfte; nur 20. und 21. Tergit mit Seitenrändern, manchmal einige vordere scheinberandet; 21 Tergit mit vollständiger Medianfurche; Coxosternum der Kieferfüsse furchenlos; Coxopleurenfortsatz mit 1 Enddorn; sonst wie *m. morsitans.*

27. *Scolopendra angulata* Newport 1844.

a. *Scolopendra angulata angulata* Newport, 1844—Bolivien, Equador, Venezuela, Trinidad, Skt. Thomas, Skt. Vincent, Granada.

b. *Scolopendra angulata explorans* (Chamberlin) 1914—Holotypus am Mamoré-Flusse, Rondonia, Brasilien.

c. *Scolopendra angulata moojeni* Bücherl, 1941, *Mems Inst. Butantan* **15**: 119, Figs 1, 2, 3—(Fig. 26)—Holotypus MNR No. 3, Tapirapé-Fluss, Mato Grosso, Brasilien; Paratypen MNR Nos. 4, 6, 7. 8, 9, 10, 19; *Verbreitung:* Brasilien: Mato Grosso: Tapirapé-Fluss; Santa Isabel; Ilha do Bananal; Mato Verde; São Domingos; Araguaia-Fluss; Ceará: Fortaleza, in feuchtem Walde; Roraima, im Territorium von Rio Branco; parque vom Xingú-Flusse, Amazonas. Unterscheidet sich von *a. angulata* und *a. explorans* durch das Fehlen des Furchendreieckes am Coxosternum der Kieferfüsse, aber statt dessen ein Furchenkreuz; letztes Tergit mit Längsdepression; Coxopleurenfortsatz ohne Dornen.

28. *Scolopendra armata* Kraepelin, 1903.

a. *Scolopendra armata armata* Kraepelin, 1903—(Fig. 27)—Holotypus aus Venezuela.

b. *Scolopendra armata amancalis* Bücherl, 1943b. *Mems Inst. Butantan* **17**: 19–26, Figs 1–3; Kraus, 1957—Holotypus IB No. 361; Amancais, bei Lima, Peru, im sandigen Terrain, unter Steinen, 300 m; Paratypus, Quebrada Verde, bei Lima. Nur ganz kurze hintere Furchen auf der Kopfplatte; nur 2 Grundglieder der Antennen kahl; mit durchgehender Querfurche auf dem Coxosternum; Praefemur aller Beine mit 3 dorso-apicalen Dörnchen; am 20. ein Eckdorn mit 4 Spitzen und 1 auf der Mittelfläche; 21. mit 17–19 Dornen, Eckdorn mit 6 Spitzen.

29. *Scolopendra arthrorhabdoides* Ribaut, 1914—(Fig. 28)—Holotypus von Gadua, Kolumbien.

30. *Scolopendra crudelis* C. L. Koch, 1847—Holotypus von Skt. Barthelemy, Kleine, Antillen.

31. *Scolopendra galapagoensis* Bollmann, 1890; Chamberlin, 1944—Holotypus von den Galapagos-Inseln. Unter Steinen, im Grase.

32. *Scolopendra gigantea* Linné, 1758.

a. *Scolopendra gigantea gigantea* Linné, 1758; Bücherl, 1939, 1941; Chamberlin, 1955. *Verbreitung:* Chile, Kolumbien, Venezuela, Trinidad, Jamaica, Honduras. MNR No. 86 und 140, ohne Fundort.

b. *Scolopendra gigantea weyrauchi* Bücherl, 1950a—*Mems Inst. Butantan* **22**: 174, Figs 1–4—(Fig. 29)—Kraus, 1955, 1957; Chamberlin, 1957—Holotypus, Weibchen, MHNL No. 10 035, Weyrauch legit 1947 bei Pucara, auf einer Steppe, 900 m, bei Jaen, Norden von Peru. *Verbreitung:* Peru: Cerro Campana; Tambo Tingo; Mirador; Lives; Quebrada Verde bei Lima; San Bartolomé; La Libertad; Caja Bamba; Cajacay; Pariacoto; Campana. Nur 5–6 Grundglieder der Antennen kahl auf der dorsalen und 3–4 ventralen Seite; vom 2. bis 18. Praefemur mit 3 dorso-apicalen Dörnchen; 19. mit 4; 20. mit 4–5 und ebensovielen an einem mesalen apicalen Eckdorn; Coxopleurenanhang mit 9–12 Dörnchen an der Spitze, seitlich 1 (manchmal 3–4), am Hinterrande in Tergitnähe 1–2; Praefemur der Analbeine mit 24–28 Dornen, Eckdorn mit 6–8 Dornspitzen.

33. *Scolopendra hermosa* Chamberlin, 1941, *Bull. Am. Mus. nat. Hist.* **78**: 500, Fig. 225; Turk, 1955—(Fig. 30)—Holotypus vom Ufer des Pisqui-Flusses, im AMNH No. 9511. *Verbreitung:* Pisqui-Fluss, Bombo-Fluss, Pampa Hermosa am Ucayali; colinas Contayo, am Tapiche Fluss; Huallaga.

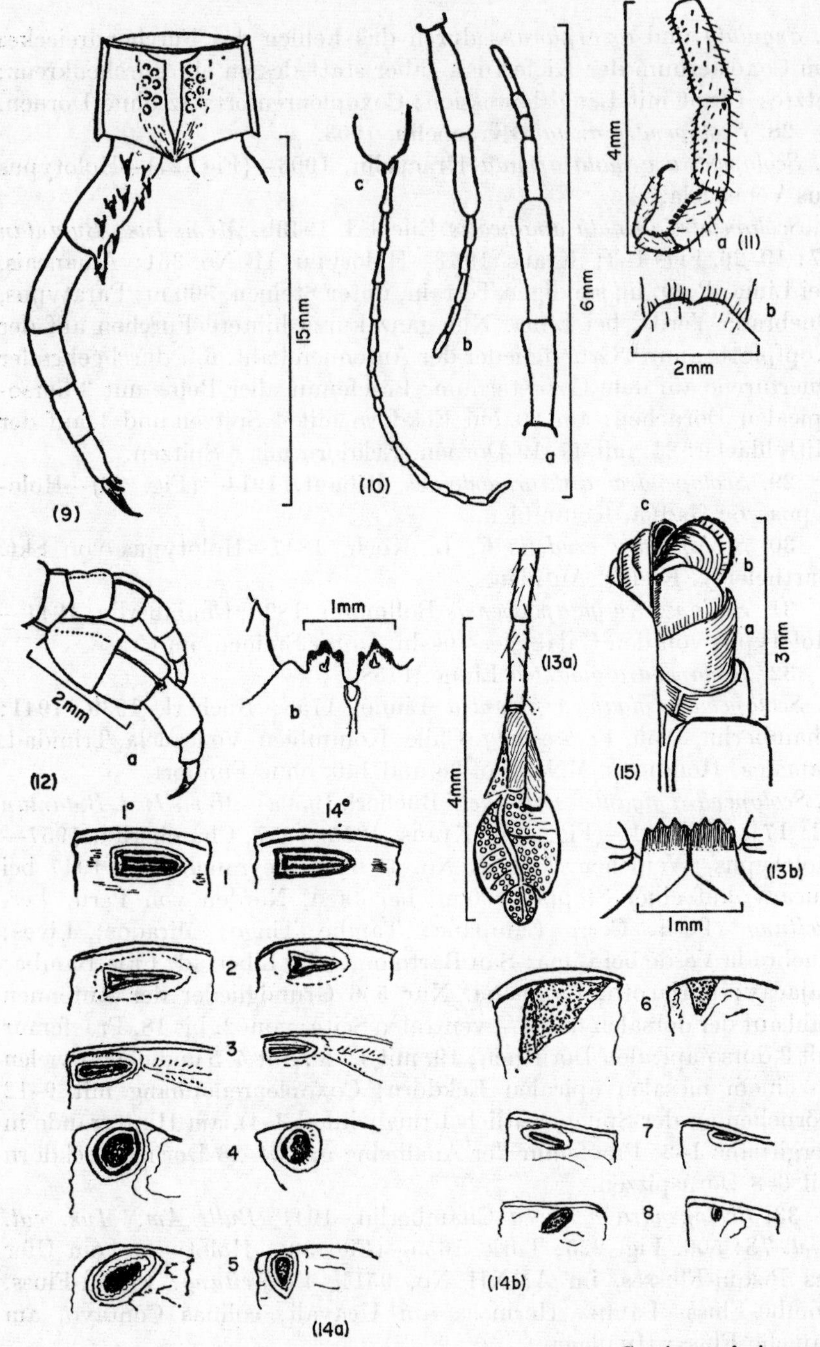

Caption on facing page

34. *Scolopendra morsitans* Linné, 1758.

a. *Scolopendra morsitans morsitans* Linné, 1758—IB No. 241, Stadt São Paulo, Staat São Paulo, Lane legit 1944; IB No. 986, Salvador, Bahia; Peru: Yanjui, Huallaga-Fluss, 350 m, Weyrauch legit 1947, SMFI No. 2964; Hacienda Sta. Helena, am Chusgon-Flusse, nördlich von Huamachuco, unter Steinen, im Busche, 1550 m, SMFI No. 3059 Weyrauch legit 1955.

35. *Scolopendra pomacea* C. L. Koch, 1847.

a. *Scolopendra pomacea minuscula* Bücherl, 1946a, *Mems Inst. Butantan* **19**: 135–136, Figs 13–14—Holotypus IB No. 461, Veadeiros, Staat Goias, Zentralbrasilien, Blaser legit 1941.

36. *Scolopendra subspinipes* Leach, 1815.

a. *Scolopendra subspinipes fulgurans* Bücherl, 1946a, *Mems Inst. Butantan* **19**: 135–142, Fig. 12—Holotypus von Rubião Junior, Staat São Paulo, Brasilien, IB No. 183; *Verbreitung:* Brasil, Staat São Paulo: Rubião Junior; Socorro; Avaré; verschiedene Vororte der Hauptstadt São Paulo; Staat Paraná: Ponta Grossa; Staat Guanabara: Rio de Janeiro; Stadtteil Sta. Tereza, Tijuca und Alto da Bôa Vista. Unterscheidet sich von *S. s. subspinipes* durch den matten jedoch sehr auffallenden Glanz des ganzen Körpers; Antennen meist 18-selten 19-gliederig, 6 kahl; Zahnplatten und Zähne ausserordentlich klein, 5 + 5, der innere etwas grösser und deutlich isoliert; Medialzahn am Telopodit der Kieferfüsse fast rudimentär, mit 4 winzigen Zacken, 3 davon auf einem gemeinsamen Sockel; Längsfurchen auf dem 2.–6. Sternite nur in der vorderen Hälfte.

37. *Scolopendra sumichrasti* Saussure 1860—Mexico: Vera Cruz; San Luis de Potosi; Ciudad Vallas; Guatemala; Honduras: Cueva de Los Sabinos.

38. *Scolopendra viridicornis* Newport, 1844 (Fig. 7)—

a. *Scolopendra viridicornis viridicornis* Newport, 1844—Die häufigste,

---

Figs 9–15. 9. *Scolopendra viridicornis nigra:* Letzter Tergit und Präfemur. 10. 1. und 2. Endbeintarsus von *Newportia monticola:* (a) Die Form *cuzcona;* (b) Die Form *perucola;* (c) Die Form *weyrauchi* (frei nach Chamberlin, 1955). 11. *Cryptops debilis* Chamberlin: (a) letztes Bein; (b) Vorderrandwulst des Coxosternums (vom Autor frei nach Chamberlin, 1955). *Asanada sokotrana attemsi:* (a) letztes Bein; (b) Zahnplatten (n. Attems). 13. *Alipes congoensis* (frei nach Attems, 1930): (a) letztes Bein (b) Zahnplatten (zum Vergleich mit Fig. 40). 14. Aussehen und Lage der Stigmenöffnungen an 1. und am 14. Körpersegmente: (a) 1—von *Scolopendra;* 2—*Hemiscolopendra;* 3—*Rhoda calcarata carvalhoi;* 4—*Otostigmus* (*Parotostigmus*)*;* 5—*Rhysida;* (b) 6—*Dinocryptops miersi;* *Trigonocryptops debilis;* 8—*Scolopocryptops ferrugineus soucupi.* 15. *Scolopendra* sp.: die normalerweise unsichtbaren, weil eingestülpten Prägenital- Genital-und Analsegmente (das Sichtbarmachen derselben ist wichtig, um die Geschlechter zu unterscheiden).

bis 21-und mehr cm grosse Scolopendra-Art Brasiliens. Uber 200 Exemplare in den Sammlungen von IB, MNR, Depto. Zoologia, Alto do Ipiranga, Hauptstadt São Paulo. *Verbreitung* Brasilien, Staat Mato Grosso: Terrenos; Três Lagoas; Campo Grande; Agachi; am Garça-Flusse; am Itacy-See; Caceres; Chibarro; Staat São Paulo: Hauptstadt São Paulo in den Stadtteilen Santana, Butantan, am Pinheiros-Flusse, Cerqueira Cesar, Tremembé, Vila Mariana, Barra Funda, Campos Elíseos, Bras; Alto da Serra, in Richtung Santos, 25 km von S.Paulo; Mogi das Cruzes; Rubião Junior; Santo André; Cotia, Suzano; Ouro Branco; Guarani; Mogí Mirim; Osasco; Araçatuba; Rincão; Rancharia; Corumbataí; São Manoel; Pederneiras; Sorocaba; Ubirama; São Miguel Paulista; Praia Grande; São Vicente; Porto Ferreira; Jaboticabal; Santos; Apiaí; Campinas; Rio Verde; Capão Bonito; Bauru; Avaré; Staat Guanabara: Rio de Janeiro, in den Stadtteilen Santa Tereza, Tijuca, Alto da Boa Vista; Staat Rio de Janeiro: Itaipú; Staat Espirito Santo: Vitoria; Staat Bahia: São Felix; Staat Sergipe: Aracajú; Staat Ceará, Fortaleza, Staat Maranhão: São Luis; Staat Amazonas: Manaus; Staat Rio Grande do Sul: Passo Fundo; Erechim; Bento Gonçalves; Staat Paraná: Palmeira; Piraí; Rio Negro; Staat Santa Catarina; Blumenau; Lagôa do Norte; Perdizes; Morangaba; Staat Pará: Belém. Paraguay: Assunsión; Capiata. Argentinien: Palmar. Kolumbien: Medellin; Cuba: Havana.

b. *Scolopendra viridicornis nigra* Bücherl, 1939, *Mems Inst. Butantan* **13**: 237–238, Fig. 54; Bücherl, 1946a—(Figs 4a, b; 9)—Holotypus, Weibchen, Três Pontas, Paraná, Brasil. *Verbreitung:* Paraná: Três Pontas; Sta. Catarina: Perdizes; Rio Grande do Sul: Santo Angelo; Minas Gerais: Belo Horizonte; Amapá: Serra do Navio, Macapá; Mato Grosso: Terrenos. Hintere Hälfte der Tergite fast schwarz; Kopf, 1. und 21. Segment rostfarben; Paramedianfurchen auf den Tergiten fast unsichtbar, etwas ausgeprägter auf den vorderen Segmenten.

39. *Scolopendra spinipriva* (Bücherl) 1946.

Scolopendra viridicornis spinipriva Bücherl, 1946c, *Mems Inst. Butantan* **19**: 146–147, Fig. 11—Figs 5a, b; 6)—Holotypus, Weibchen, IB No. 228, von Santo Angelo, bei Santo André, zirka 18 km von der Stadt São Paulo in Richtung Santos, 800 m; am 12/2/1944 gefangen. Nur 2 Grundglieder der Antennen kahl, das 3.schon im distalen Drittel der Innenseite beborstet; 1. Tergit mit Cervical-und 2 schwachen keine W-Zeichnung bildenden Längsfurchen; 1. bis 16. Praefemur keine dorso-apicalen Dörnchen; 17.–18. mit 0 bis 1, 19. mit 1, selten mit 2, 20. mit 2 dorso-apicalen und 1 medialen Dörnchen; Coxopleurenfortsatz mit 2 Enddornen und 1 sub-apicalen.

40. *Scolopendra viridis* Say 1821.

a. *Scolopendra viridis viridis* Say, 1821, Bücherl, 1939, 1941; Chamberlin, 1944; Kraus, 1954—Verbreitung in der Neotropischen Region: Guatemala, Honduras, Nicaragua, Costa Rica, San Salvador.

Scolopendropsis Brandt, 1841.

41. *Scolopendropsis bahiensis* Brandt, 1841—Salvador, Bahia, Brasil. Chamberlin (1914) identifizierte sie mit *Rhoda thayeri;* auch nach der Wiederbeschreibung von Attems (1930) ist man versucht, anzunehmen, dass das einzige Exemplar nur 21 und nicht 23 Beinpaare besitze.

Otostigminae

Otostigmini

Otostigmus Porat, 1876 (Figs 14a: 4; 31)

Subgenus: *Parotostigmus* Pocock, 1896

Synonymien: *Dactylotergitius*, *Ecuadopleurus*, *Androtostigmus* Verhoeff, 1937, *Zool. Jb.* (Syst.) **70**: 1–16; *Coxopleurotostigmus* Bücherl, 1939, *Mems Inst. Butantan* **13**: 256–259.

42. *Otostigmus* (*P.*) *amazonae* Chamberlin, 1914—Holotypus von Manaus. *Verbreitung:* Brasilien: Amazonas: Manaus; Pará: Belém: IB No. 137; 741/4; 835/1; 897/1: A. R. Hoge und H.Sioli legerunt 1952 und 1958. Amapá: Serra do Navio, IB No. 921, bei Perimirim, 20 m; No. 864; A. R. Hoge legit Mai 1962. Mato Grosso: Mündung des Tapirapé- Flusses, in faulendem Holze, MNR No. 72, 73. 74, 77, 80, 108 (72 mit zirka 30 pulli), 73 mit 45 Eiern; Mato Verde, IB No. 570. Peru: Payas Mayo; Atocongo; Lachay; Amancais; Chincheros; Urubamba-Fluss bei Cusco; Huamachuco; Matucana; Pariacoto; Asia; Cajacay, 3300 m; Ainín; Sisicaya; San Juan; Cerro dos Caracoles; Matucana, 2600 m; San Bartolomé. Die peruanischen Gebirgsformen scheinen zur Gruppe *muticus* sensu Karsch 1884 zu gehören.

43. *Otostigmus* (*P.*) *brunneus* Chamberlin, 1921—Holotypus von Dunoon, Brit. Guayana.

44. *Otostigmus bürgeri* Attems, 1903.

a. *Otostigmus* (*P.*) *bürgeri bürgeri* Attems, 1903—Holotypus von Villa Vicencio, Kolumbien. *Verbreitung:* Kolumbien; Peru: Tarma, 3300 m, MHNL No. 12 009, Weyrauch legit.

b. *Otostigmus* (*P.*) *bürgeri monsonus* Chamberlin, 1957, *Gr. Basin Nat.* **19**: 30–41—Holotypus vom Monsontal, bei Tingo Maria, Peru; Ross und Schlinger legerunt 1954.

45. *Otostigmus* (*P.*) *calcanus* Chamberlin, 1944, *Publs. Field Mus. Nat. hist.* (*zool. Ser.*) **28**: 175. Holotypus und 2 Paratypen aus Urco,

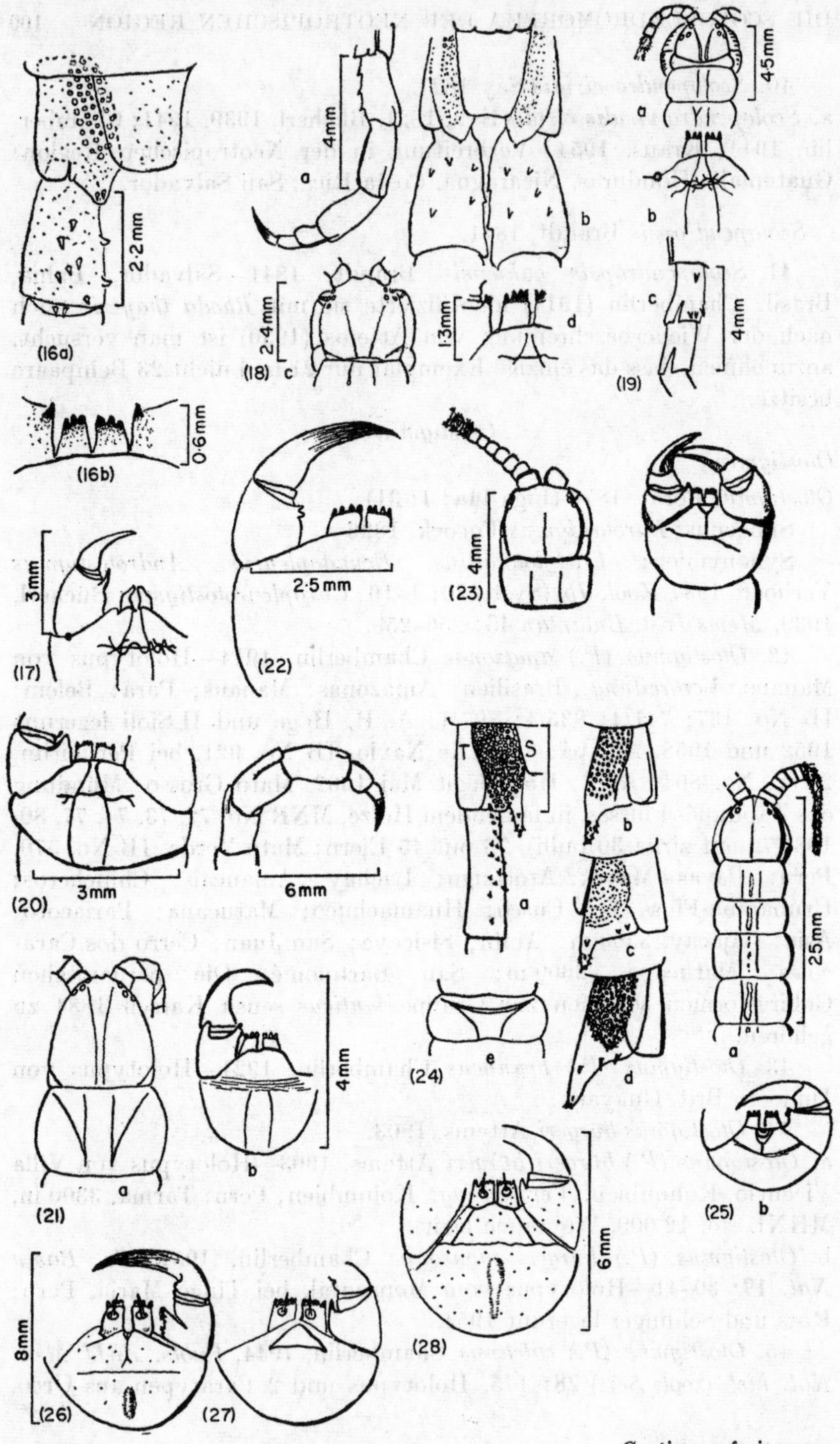

Caption on facing page

bei Calca, Provinz Cusco, Peru; Schmidt leg. 1939; im FM.-*Verbreitung:* Peru: Calca, Tarmatambo, Pichita Caluga.

46. *Otostigmus (P.) caraibicus* Kraepelin, 1903—Holotypus von der Insel Skt. Thomas.

47. *Otostigmus (P.) casus* Chamberlin, 1914—Holotypus von Madeira-Mamoré, Mato Grosso, Brasilien.

48. *Otostigmus caudatus* Brölemann, 1902.

a. *Otostigmus (P.) caudatus caudatus* Brölemann, 1902—(Fig. 32)—Holotypus von Belém, einem Stadtteil der Hauptstadt São Paulo. *Verbreitung:* Brasilien: Staat São Paulo: zirka 20 Exemplare aus verschiedenen Stadtteilen; Santo Angelo; Vicente Carvalho: Itaquera; São Sebastião; Rio Claro; Campinas; Cocaia; São Miguel Paulista; Limeira; Atibaia; Porto Ferreira. Paraná: Londrina; Ponta Grossa; Palmeira; Porto Amazonas; Canavieiras; Rio Negro. Santa Catarina: Lagôa. Rio Grande do Sul: Uruguaiana. Mato Grosso: Terrenos; Rio de Janeiro: Volta Redonda, Terezópolis.

b. *Otostigmus (P.) caudatus hogei* nomen novum für *Otostigmus (P.) caudatus insularis* Bücherl, 1949, *Mems Inst. Butantan* **21**: 1–8, Figs 1, 2—ein nomen praeoccupatum (Fig. 46: a, b)—Holotypus, Männchen, IB No. 404: Paratypen, 7 Männchen, 4 Weibchen, von Queimada Pequena-Insel, zirka 40 km südlich von Santos, im Atlantik. *Verbreitung:* Inseln Queimada Pequena, Queimada Grande, Alcatrazes, alle in der Nähe von Santos.

Unterscheidet sich von *O. caudatus caudatus* durch die 2 kräftigen, bis über die Mitte, reichenden Sternitfurchen; der keulenförmige Anhang am letzten Tergit der Männchen ist fast $1\frac{1}{2}$ mal so lang als der Tergit. Zirka 100 Exemplare von den drei Inseln, in den Jahren, 1947, 1948 und 1950, von A. R. Hoge gefangen.

49. *Otostigmus (P.) cavalcantii* Bücherl, 1939b, *Revta Biol. Hyg.* **10**: 54–64, Figs 1–3—(Fig. 33: a, b)—Holotypus IB No. 48, Männchen

---

FIGS 16–28. 16. *Cormocephalus andinus:* Die Form *alticolus:* (a) Coxopleuren und Endbeinpräfemur (b) Zahnplatten (frei nach Verhoeff, 1941). 17. *Cormocephalus amazonae.* 18. *Cormocephalus andinus*—(a) letztes Bein; (b) letztes Präfemurpaar-ventral; (c) Basalplatten und Längsfurchen auf der Kopfplatte und dem 1. Tergite; (d) Coxosternum. 19. *Cormocephalus andinus rubrifrons:* (a) 1. Tergit; (b) Coxosternum mit Längsfurchen und Querfurchennetz; (c) letzter Präfemur und Femur. 20. *Cormocephalus bonaerius:* Coxosternum und letztes Bein (nach Attems). 21. *Cormocephalus impressus birabeni:* (a) Längsfurchen auf dem Kopfe und 1. Tergite; (b) Furchen auf dem Coxosternum. 22. *Cormocephalus impressus glabrus.* 23. *Cormocephalus mediosulcatus:* Kopf, dorsal und ventral. 24. *Hemiscolopendra* (a) *chilensis;* (b) *laevigata;* (c) *michaelseni;* (d) *platei;* (e) *punctiventris:* 1. Tergit. 25. *Rhoda calcarata carvalhoi:* (a) Kopf und die 4 ersten Rumpfsegmente; (b) Coxosternum. 26. *Scolopendra angulata moojeni:* Coxosternum. 27. *Scolopendra armata armata:* Coxosternum (frei nach Attems). 28. *Scolopendra arthrorhabdoides:* Coxosternum (nach Attems).

und Alotypus, Weibchen, von einem Garten in Perdizes, Capital von São Paulo. *Verbreitung:* Stadt São Paulo; Araçatuba; São Carlos; Mogi das Cruzes; Alto da Serra; Morro Agudo; Pirassununga; Emas; Rio Claro.

Gehört zur Gruppe *caudatus;* davon durch das Vorhandensein eines Coxopleurenfortsatzes der Männchen leicht zu unterscheiden.

50. *Otostigmus* (*P.*) *perdicensis* (Bücherl) 1943a. *Otostigmus* (*P.*) *cavalcantii perdicensis* Bücherl, 1943, *Mems Inst. Butantan* **16**: 85–89, Figs 1–2—Holotypus, Männchen, IB No. 296; Paratypen, IB, im ganzen 19 Exemplare, Männchen und Weibchen, alle von Perdizes, Santa Catarina. Gehört zur Gruppe *caudatus-cavalcantii*, aber nur mit 2 Tarsalspornen auf den Beinen 1–7 oder höchstens 9, bei *cavalcantii* auf den Beinen 1–16.

51. *Otostigmus* (*P.*) *clavifer* Chamberlin, 1921—(Fig. 34)—Holotypus von Dunoon, Brit. Guayana.

52. *Otostigmus* (*P.*) *cooperi* Chamberlin, 1942a, *Pan Pacif. Ent.* **18**: 125–126. Holotypus und 1 Paratypus aus der Grotte Chilibrillo, bei Buenos Aires, Panama; K. W. Cooper legit.

53. *Otostigmus* (*P.*) *demelloi* Verhoeff, 1937, *Zool. Jb.* (Syst.) **70**: 1–16—Holotypus von Brasilien, Minas Gerais, ohne Fundortangabe.

54. *Otostigmus* (*P.*) *dentifusus* Bücherl, 1946e, *Mems Inst. Butantan* **19**: 135, Figs 15, 16—(Fig. 45a, b)—Holotypus, Männchen und 7 Paratypen, IB No. 477, von Itanhaem, an der Atlantikküste, südlich von Santos; in einem Krabbenloch, neben dem Brackwasser, von J. Kretz am 4/4/1940 gefangen.

55. *Otostigmus* (*P.*) *diringshofeni* Bücherl, 1969, *Beitr. neotrop. Fauna* **6**: 90–91, Figs 1–2—(Fig. 39)—Holotypus, Männchen, IB No. 987, aus der Sammlung des Herrn von Diringshofen. Eingefangen in São Paulo de Olivença, Westamazonas, Brasilien, nahe der peruanischen Grenze. Der Gruppe *silvestrii* nahe, aber besonders durch den 20. Sternit recht verschieden.

56. *Otostigmus* (*P.*) *dolosus* (Attems) 1928—Paraguay, ohne Fundortangaben.

57. *Otostigmus* (*P.*) *ethonyx* Chamberlin, 1955, *Acta Univ. lund. N.F.* (2), **51** (5): 41–42.—Holotypus 40 min östlich von Abancay, Peru.

58. *Otostigmus* (*P.*) *fossulatus* Attems, 1928—(Fig. 35)—Holotypus ZMUH, H. Schmidt legit 1926 bei Finca La Caja, 8 km von San José, zwischen den Flüssen Virilla und Torres, Costa Rica.

59. *Otostigmus* (*P.*) *goeldii* Brölemann, 1898—Holotypus aus Belém, Pará, äquatoriales Brasilien.

60. *Otostigmus* (*P.*) *inermis* Porat, 1876—*Verbreitung:* Argentinien: Quilmes, IB No. 619/3, A. Barrios legit, 1950; Palmas, Prov.

Jujuy, M. Biraben legit Mai 1947; Campo Gallo, Prov. Santiago del Estero. Brasilien: Paraná: Volta Grande, R. Hertel legit 1945. Kolumbien: Santa Marta, H. N. Rowland legit 1902.

61. *Otostigmus* (*P.*) *carbonelli* (Bücherl) 1959—*Otostigmus* (*P.*) *inermis carbonelli* Bücherl, 1959a, *Mems Inst. Butantan* **29**: 233–241—Holotypus IZUC, Carbonell legit, von Cerro Delgado Chalbaud, Venezuela. Unterscheidet sich von *inermis:* durch die braune Farbe des Kopfes und der vorderen Tergite; die 4–5 hinteren Tergite haben keine erhabenen Längskiele, jedoch Längsreihen feiner Dörnchen; erster und letzter Tergit glatt; Sternite ganz ohne Längsfurchen; Coxosternum der Kieferfüsse mit vorderer schwacher Medianfurche.

62. *Otostigmus* (*P.*) *insignis* Kraepelin 1903—(Fig. 36: a, b)—Holotypus von Loja, Ecuador.

63. *Otostigmus* (*P.*) *kretzi* Bücherl, 1939b, *Revta Biol. Hyg.* **10**: 59, Figs 1–2 (Fig. 41)—Holotypus IB No. 38, Weibchen, J. Kretz legit, Igarapava, São Paulo, Brasil, unter Baumwurzeln. *Verbreitung:* Brasil, São Paulo: Igarapava; Nova Roma; Analandia; Baguassú.

64. *Otostigmus* (*P.*) *langei* Bücherl, 1946a, *Mems Inst. Butantan* **19**: 1, Figs 1–2—(Fig. 43: a, b)—Holotypus im Museu Paranaense, Curtiba, R. Lange legit 1/7/1945 in Bariguí, Staat Paraná.

65. *Otostigmus* (*P.*) *latipes* Bücherl, 1954, *Mems Inst. Butantan* **26**: 1–6, Figs 1–2, Fotos 1–2—(Fig. 40)—Holotypus in Canada balsam, IB, H. Sick legit 1950, am Flusse das Mortes, Mato Grosso; 2 Paratypen IB No. 811 und 817.

66. *Otostigmus* (*P.*) *lavanus* Chamberlin, 1957, *Gr. Basin Nat.* **17**: 30–41—(Fig. 37: a, b)—Holotypus, Männchen, von Ross und Schlinger 6–12 min südwestlich von Banos, Ecuador, gefangen.

67. *Otostigmus* (*P.*) *leior* Chamberlin, 1955, *Acta Univ. lund. N.F.* (2) **51** (5): 42—Holotypus 20 min südlich von Cusco, Peru, gefangen.

68. *Otostigmus* (*P.*) *limbatus* Meinert, 1886—*Verbreitung:* Brasilien: Santa Catarina: Lagoa; Ilha do Bom Abrigo; Parana: Curitiba; Porto do Cisne; Mato Grosso: Terrenos; São Paulo: die Inseln Alcatrazes; Queimada Pequena und Queimada Grande; Alto da Serra; Argentinien: Entre Rios.

69. *Otostigmus* (*P.*) *diminutus* (Bücherl) 1943—*Otostigmus* (*P.*) *limbatus diminutus* Bücherl, 1943b, *Mems Inst. Butantan* **17**: 19–26—Holotypus IB No. 408, von Santo André, São Paulo, Brasil. 3 Paratypen—Unterscheidet sich leicht von *limbatus:* die 2 Sternitlängsfurchen werden immer kürzer, je weiter die Sternite von vorne entfernt sind; am 19. gehen sie nur mehr bis zur Hälfte, am 20. Sternite fehlen sie ganz; die 2 Längseindrücke in der Mitte sind sehr schwach, oft kaum bemerkbar; Beine 1–3 mit 2 Tarsalspornen; Kopf, 1. und letzter

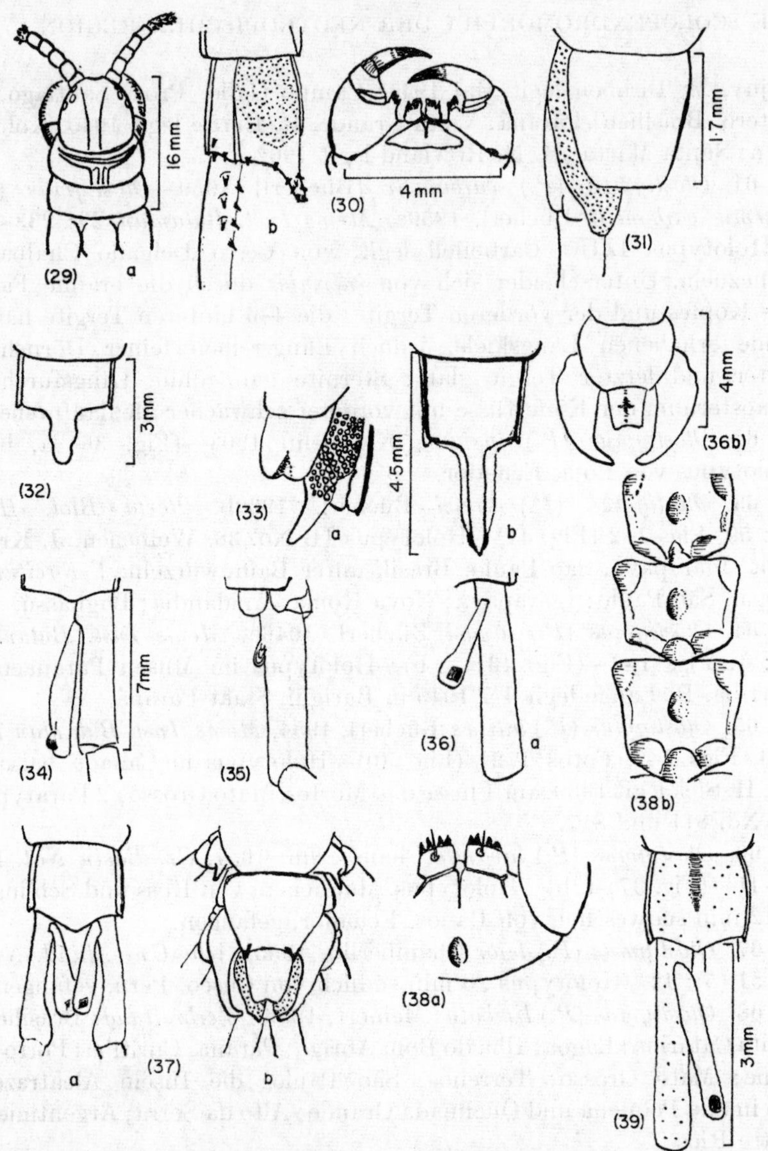

Figs 29–39. 29. ***Scolopendra gigantea weyrauchi:*** **(a) Kopf und die ersten drei Tergite; (b) Letztes Segment und Präfemur-seitlich. 30.** ***Scolopendra hermosa:*** **Coxosternum (nach Chamberlin). 31.** ***Otostigmus*** **(*P.*)** ***amazonae:*** **letztes Sternit und Coxopleuren. 32.** ***Otostigmus*** **(*P.*)** ***c. caudatus:*** **letztes Tergit mit fingerförmigen Fortsatz. 33.** ***Otostigmus*** **(*P.*)** ***c. cavalcantii:*** **(a) Coxopleurenfortsatz; (b) letztes Tergit mit Fortsatz. 34.** ***Otostigmus*** **(*P.*)** ***clavifer:*** **Anhang am letzten Präfemur des Männchens. 35.** ***Otostigmus*** **(*P.*)** ***fossulatus:*** **Anhang am letzten Präfemur des Männchens (nach Attems). 36.** ***Otostigmus*** **(*P.*)** ***insignis*** **(a) Anhang am letzten Präfemur; (b) Fortsätze am vorletzten Sternite. 37.** ***Otostigmus*** **(*P.*)** ***lavanus:*** **(a) Anhänge am letzten Präfemurpaar; (b) Ventralseite der 3 letzten Sternite des Männchens (frei nach Chamberlin). 38.** ***Otostigmus*** **(*P.*)** ***longipes:*** **(a) Coxosternum; (b) Grübchen auf den Sterniten. 39.** ***Otostigmus*** **(*P.*)** ***diringshofeni:*** **Anhang am letzten Präfemur des Männchens.**

Tergit bräunlich, der Rest blau; nur 2 Grundglieder der Antennen kahl; die 4 Kieferfusszähne stehen auf einem gemeinsamen Grundblock, nur ihre Spitzen sind frei; Coxosternum mit vorderer mittlerer Längsfurche; Tergite 2, 3 und 4 schon mit kurzen vorderen Paramedianfurchen, bei 3 und 4 auch schon hintere.

70. *Otostigmus* (*P.*) *longipes* Bücherl 1939b, *Revta Biol. Hyg.* **10**: 54–64, Figs 1–3—(Fig. 38: a, b)—Holotypus, Männchen, IB No. 33; T. Delarmo legit unter Orchideen, in Lobo, Staat São Paulo, Brasilien. *Verbreitung:* Lobo und Chavantina, Mato Grosso, H. Sick legit 4/8/49.

71. *Otostigmus* (*P.*) *longistigma* Bücherl 1939, *Bolm biol. Clube zool. Bras.* (N.S.) **4**: 444–447—Holotypus, Weibchen, IB No. 58, von Campo Limpo, bei Jundiaí, Staat São Paulo, Brasilien; in den seitlichen Hohlräumen eines Termitenbaues, am 24/1/1947 eingefangen.

72. *Otostigmus* (*P.*) *mesethus* Chamberlin 1957, *Gr. Basin Nat.* **17**: 30–41—(Fig. 47: a, b)—Holotypus und 1 Paratypus, 6 bis 12 min südwestlich von Banos, am Abhange von Tungurahua, in Ecuador, von Ross und Schlinger eingefangen.

73. *Otostigmus* (*P.*) *muticus* Karsch, 1884—Typus von Peru, ohne Fundort. *Verbreitung:* Peru: Asia, 120 km von Lima; San Bartolomé; Quebrada Verde; Oxapampa; Lima.

74. *Otostigmus* (*P.*) *occidentalis* Meinert, 1886—patria?

75. *Otostigmus* (*P.*) *parvior* Chamberlin, 1957, *Gr. Basin Nat.* **17**: 30–41—Holotypus von Ecuador, 20 min südöstlich von Ambabo. (Fig. 48).

76. *Otostigmus* (*P.*) *pococki* (Kraepelin) 1903—Holotypus von Haut Carsevenne, Guayana (Brit.). *Verbreitung:* Brit. Guayana; Venezuela: Rancho Grande, 11 Exemplare; Peru: Divisoria; Acomayo; Huanuco; Tingo Maria, Huacapistana; Brasil: Insel Skt. Ana im Amazonasdelta; bei Belém, Pará (Fig. 49).

77. *Otostigmus* (*P.*) *exspectus* (Bücherl) 1959—*Otostigmus* (*P.*) *pococki exspectus* Bücherl, 1959a,—*Mems Inst. Butantan* **29**: 233–241—Holotypus aus dem Oberen Orenoco, Venezuela, 1100 m, Weibchen, IZUC; bei Cerro Delgado Chalbaud, Carbonell legit 1951. Coxosternum mit vorderer Längsfursche; nur 21. Tergit berandet, einige vordere scheinberandet; Sternite 6–19 mit 2 nur vorderen Längsfurchen.

78. *Otostigmus* (*P.*) *pradoi* Bücherl, 1939a—*Bolm biol. Clube zool. Bras.* (*N.S.*) **4**: 444–447, Figs 1–3. (Fig. 42: a, b)—Holotypus, Männchen, IB No. 52, 1 Paratyp, Weibchen, von Campinha und Lobo, Staat São Paula, Brasilien.

79. *Otostigmus* (*P.*) *rex* Chamberlin 1914—Holotypus vom Mamoré-Flusse, im Westen von Mato Grosso, Brasilien. *Verbreitung:* Westbrasilien, Madeira-Mamoré; Peru: Huamachuco, Celendin, Tingo Maria.

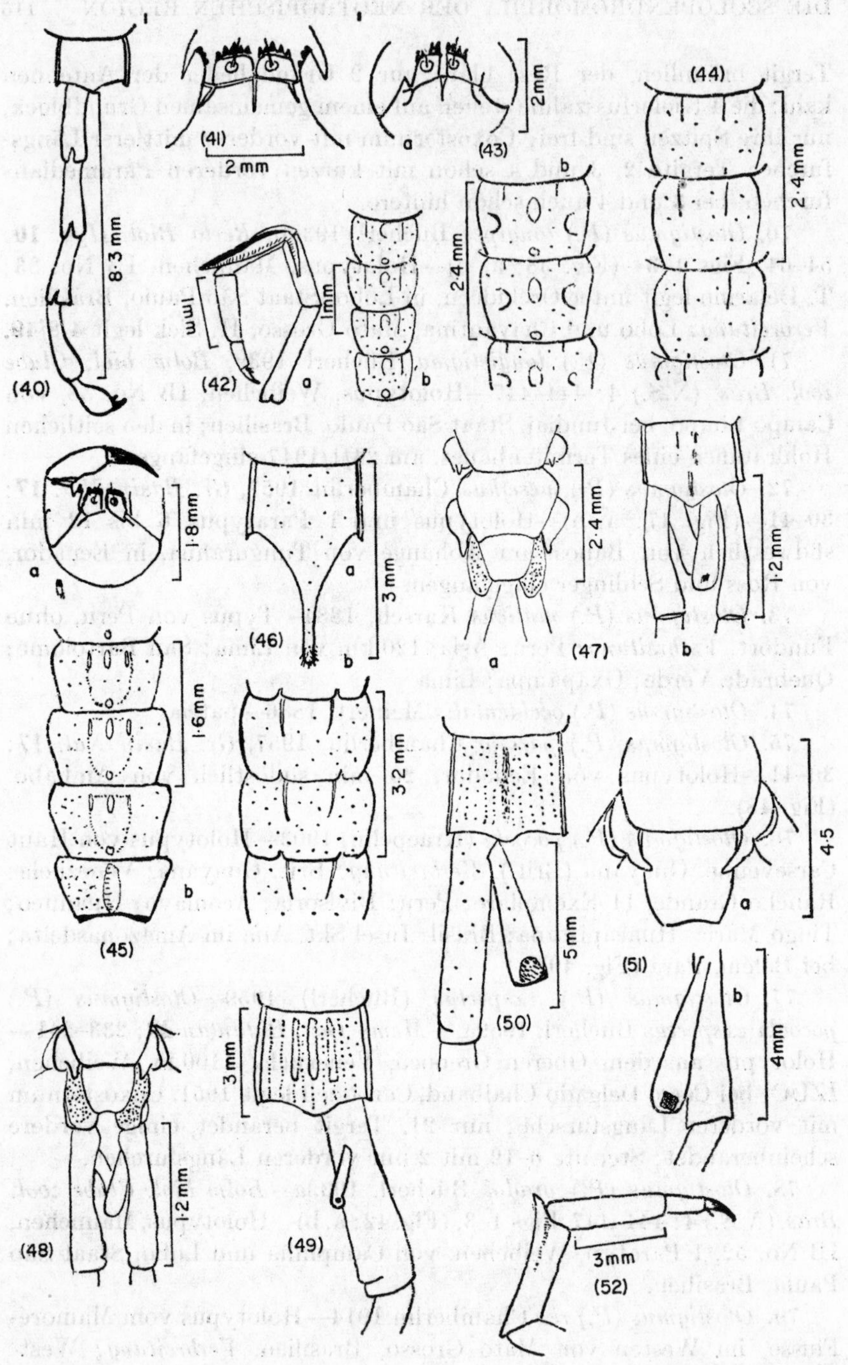

Caption on facing page

80. *Otostigmus* (*P.*) *samacus* Chamberlin, 1944 *Publs Field Mus. nat. Hist.* (*Zool. Ser.*) **28** : 175—Holotypus, FM, Männchen, von Samac, Alta Verapaz, Guatemala; D. Clark legit.

81. *Otostigmus* (*P.*) *scabricauda* (Humbert & Saussure) 1870—(Fig. 50)—Holotypus von Rio de Janeiro. *Verbreitung:* Peru: Pisqui-Fluss, zwischen Cajamarca und Celendin, 2300 m; Llama, zwischen Chiclayo und Cutervo; Carpish, zwischen Huanuco und Tingo Maria; Pichita Caluga. Ecuador: Corazon; Kolumbien: Popayan; Brasil: Rio de Janeiro (Stadt) (locus typicus); Staat Rio de Janeiro: Nova Iguassú; Staat São Paulo: die Inseln Queimada Grande und Pequena; Alcatrazes; Ubatuba; São Sebastião; Praia Grande; Pedro de Toledo; Ribeirão Pires; Santo André; Juquiá; Ana Dias; Igarapava; Corumbataí; Lobo; Vargem Grande; Bueno de Andrade; Itapecerica; Pindamonhangaba; Cocaia; Capão Redondo; Alto da Serra; die Stadt São Paulo; Rio Grande; Butantan; Vila Mirim; Mogí das Cruzes. Paraná: Piarí; Londrina; Palmeira; União da Vitoria; Ilha do Mel (Insel); Goiás: Cana Brava; Mato Grosso: Terrenos; Agachi; Santa Catarina: Joinville; Amazonas: Manaus; Jacarecanga; Amapá: Serra do Navio; Rondonia: am 54. km der Fernstrasse BR2;

82. *Otostigmus* (*P.*) *silvestrii* Kraepelin 1903.

a. *Otostigmus* (*P.*) *silvestrii silvestrii* Kraepelin 1903—(Fig. 51 a, b)—Holotypus von Pifo, Ecuador. *Verbreitung:* Ecuador: Pifo; Brasil: Westamazonas: Tabatinga, IB No. 893, von Diringshofen legit, 1960.

b. *Otostigmus* (*P.*) *silvestrii intermedius* Kraepelin, 1903—Ecuador, ohne Fundort.

83. *Otostigmus* (*P.*) *spiculifer* Pocock, 1893—Typus von Skt. Vincent.

84. *Otostigmus* (*P.*) *sternosulcatus* Bücherl, 1946a, *Mems Inst. Butantan* **19**: 1–9, Figs 3–4. (Fig. 44). Holotypus Weibchen und 2 Paratypen, No. 16, Museu Paranaense, in Curitiba, von R. Lange im Orte Rio d'Areia und Contenda, Staat Paraná, Südbrasilien, gefangen.

FIGS 40–52. 40. *Otostigmus* (*P.*) *latipes:* letztes Bein des Männchens. 41. *Otostigmus* (*P.*) *kretzi:* Coxosternum. 42. *Otostigmus* (*P.*) *pradoi:* (a) Tibialapophyse am letzten Beine des Männchens; (b) Sternitfurchen und Medianeindrücke. 43. *Otostigmus* (*P.*) *langei:* (a) Coxosternum; (b) Furchen, Median- und Seiteneindrücke auf den Sterniten. 44. *Otostigmus* (*P.*) *sternosulcatus:* Längsfurchen der Sternite. 45. *Otostigmus* (*P.*) *dentifusus:* (a) Coxosternum; (b) die 4 letzten Sternite. 46. *Otostigmus* (*P.*) *caudatus hogei:* (a) letztes Tergit des Männchens; (b) vordere Längsfurchen der Sternite. 47. *Otostigmus* (*P.*) *mesethus:* (a) Sternitanhänge an den 3 letzten Segmenten; (b) Anhänge am letzten Präfemurpaar des Männchens. 48. *Otostigmus* (*P.*) *parvior:* Anhänge am letzten Präfemurpaare des Männchens. 49. *Otostigmus* (*P.*) *pococki:* Letztes Tergit und Anhang am Präfemur des Männchens (nach Attems). 50. *Otostigmus* (*P.*) *scabricauda:* letztes Tergit und Präfemuralanhang des Männchens. 51. *Otostigmus* (*P.*) *silvestrii silvestrii:* (a) Anhange am 20. Sternit; (b) Präfemuralanhang des Männchens. 52. *Otostigmus* (*P.*) *tibialis:* Tibialprocess am letzten Beine des Männchens.

85. *Otostigmus* (*P.*) *suitus* Chamberlin, 1914—Holotypus von Madeira-Mamoré, West Mato Grosso, Brasilien.

86. *Otostigmus* (*P.*) *sulcatus* Meinert, 1886—Typus von Montevideo, Uruguay.

87. *Otostigmus* (*P.*) *therezopolis* Chamberlin, 1944, *Publs Field Mus. nat. Hist.* (*Zool. Ser.*) **28**: 183—Holotypus im FM, 2 Paratypen; 5 Meilen nördlich von Terezópolis, im Staate von Rio de Janeiro, von K. P. Schmidt 1926 gefangen.

88. *Otostigmus* (*P.*) *tibialis* Brölemann 1902—Holotypus von Piquete, Staat São Paulo. *Verbreitung:* Staat São Paulo: Piquete, die Inseln Queimada Grande, Pequena, Alcatrazes, São Sebastião, Alto da Serra; die Stadt São Paulo; Tietê; Jardinópolis: Itapira; Jundiaí; Ouro Branco; Santo Amaro; Banhado; Rio Grande do Sul: Passo Fundo; Santa Catarina: Lagoa; Insel Bom Abrigo, Insel Castilho; Paraná: Campininha; Mato Grosso: Terrenos; Minas Gerais: Poços de Caldas; Rio de Janeiro: Im Staatspark von Terezópolis. Argentinien: Prov. von Jujuy: Yala; Prov. von Misiones: San Xavier (Fig. 52).

89. *Otostigmus* (*P.*) *tidius* Chamberlin, 1914—Holotypus von Manaus, Amazonas.

90. *Otostigmus* (*P.*) *vulcanus* Chamberlin, 1955, *Acta Univ. lund.* N.F. (2) **51**(5): 34—Holotypus aus der Prov. Jujuy, 3 min im Süden des Vulkans, 2000 m.

Rhysida Wood, 1862—(Fig. 14a: 5).

91. *Rhysida brasiliensis* Kraepelin, 1903—Typus vom Staate Minas Gerais, Brasilien. *Verbreitung:* Staat São Paulo: Die Stadt (Stadtteil Limão); Campinas; Helvetia; Barretos; São Carlos; Jaboticabal; Moreira Cesar; Pirassununga; Staat Rio de Janeiro: Volta Redonda, Terezópolis, Nova Friburgo; Staat Guanabara: Rio de Janeiro; Staat Espírito Santo: Vitoria; Staat Minas Gerais: Passagem.

92. *Rhysida rubra* (Bücherl) 1939—*Rhysida brasiliensis rubra* Bücherl, 1939, *Mems Inst. Butantan* **13**: 279–280—(Fig. 53)—Holotypus, IB No. 92, Paratypus IB 134, P. Schleich legit 8/9/1944 in Terrenos, Mato Grosso, Brasilien. *Verbreitung:* Terrenos; Corumbatai; Palmeira; Chavantina. Hellbraun bis rötlich, mit hellem Längsband auf den Tergiten und in der Mitte ein schmaler dunkler Längsstreifen; 21 Antennenglieder, 3 Grundglieder kahl; kleine Zahnplatten, jede mit 5–6–7 winzigen Zähnen; Sternite mit 3 vorderen und 3 hinteren Grübchen, ausser den 2 kurzen Längsfurchen; 16.–20. Beinpaar nur mit 1 Tarsalsporn; 1. Beinpaar mit 1 Femural-und 1 Tibialsporn, 2. Paar mit 1 Tibialsporn.

93. *Rhysida celeris* (Humbert & Saussure) 1870.

a. *Rhysida celeris celeris* (Humbert & Saussure) 1870—*Verbreitung* (in der neotropischen Region): Antillen, Venezuela, Bolivien, Argentinien, Brasil: Staat São Paulo: Piratininga, Ouro Branco; Bahia: Salvador; Goias: Rio Verde; Mato Grosso: Tapirape; Amazonas: Manaus, National-Reservat am Xingu-Flusse; Jacaré, am Tapajóz-Flusse; Pará: Aurá; Amapá: Serra do Navio; Belém; Santarém.

b. *Rhysida celeris andina* Bücherl, 1953, *Mems Inst. Butantan* **25**: 109—(Fig. 54)—Holotypus IB No. 657, 1 Paratypus MHNL No. 12 008, von Weyrauch am 1/4/1942 bei Chandemayo, am Colorado-Flusse, 200 m, in Peru, gefangen. *Verbreitung:* Peru: Chandemayo; San Ramon, Agauytia-Fluss; San Luis, am Paucartambo, Purus: mehrere Exemplare im SMFI No. 2959, 2960, 2961, 2962, 2963. Unterscheidet sich von *celeris celeris:* der ganze Körper einfarbig gelblich; Zahnplatten mit 5–6 Zähnen, ohne Basalfurchen, jede mit einem Grübchen, mit Borste; Tergite glatt, ohne Runzeln, vom 16.–bis 19. mit. Scheinrändern; Coxopleurenfortsatz mit 1–2 Dornchen; letztes Bein so lang wie die 5 letzten Körpersegmente.

94. *Rhysida longipes* (Newport) 1845—(Fig. 55)—*Verbreitung* in Brasilien: Rio de Janeiro: Barra Mansa; Guanabara: Rio (Santa Tereza); São Paulo: Santos.

95. *Rhysida nuda* (Newport) 1845.

a. *Rhysida nuda nuda* (Newport) 1845—(Fig. 56: a, b)—*Verbreitung* in der Neotropischen Region: Brasil: Rio Grande do Sul: Rosario; Porto Alegre; Tupacertã; Passo Fundo; São Paulo: Santos; Santo Angelo; die Hauptstadt; Iguape; Rio de Janeiro: Terezópolis; Roraima: São Marcos; Argentinien: Prov. Corrientes: Manantiales; Bolivien: San Ignacio; Peru: Bombo-Fluss; Orellana; El Salvador: Depto. La Libertad: Santa Tecla; Finca del Paraiso; San Salvador; Depto. San Miguel: Laguna de Jacotal.

b. *Rhysida nuda immarginata* (Porat) 1876—Altweltlich; in der Neotropischen Region.: Guatemala, Venezuela, nach Attems (1930).

96. *Rhysida riograndensis* (Bücherl) 1939—*Rhysida nuda riograndensis* Bücherl, 1939, *Mems Inst. Butantan* **13**: 277–278—Holotypus IB No. 80, Paratypus IB No. 54, Gerson legit 18/5/1936 bei Rosario im Staate Rio Grande do Sul, Brasilien. *Verbreitung:* Rio Grande do Sul: Rosario; São Leopoldo; Paraná: Cascavel; Palmeira; Rio Negro; Araucaria—Tergite oliv grün, Kopf und die 2 letzten Segmente braun; 4+4 Zahne, die 2 inneren verwachsen, mit geteilten Spitzen; die ersten 3–5 Beinpaare mit 1 Tibialsporn.

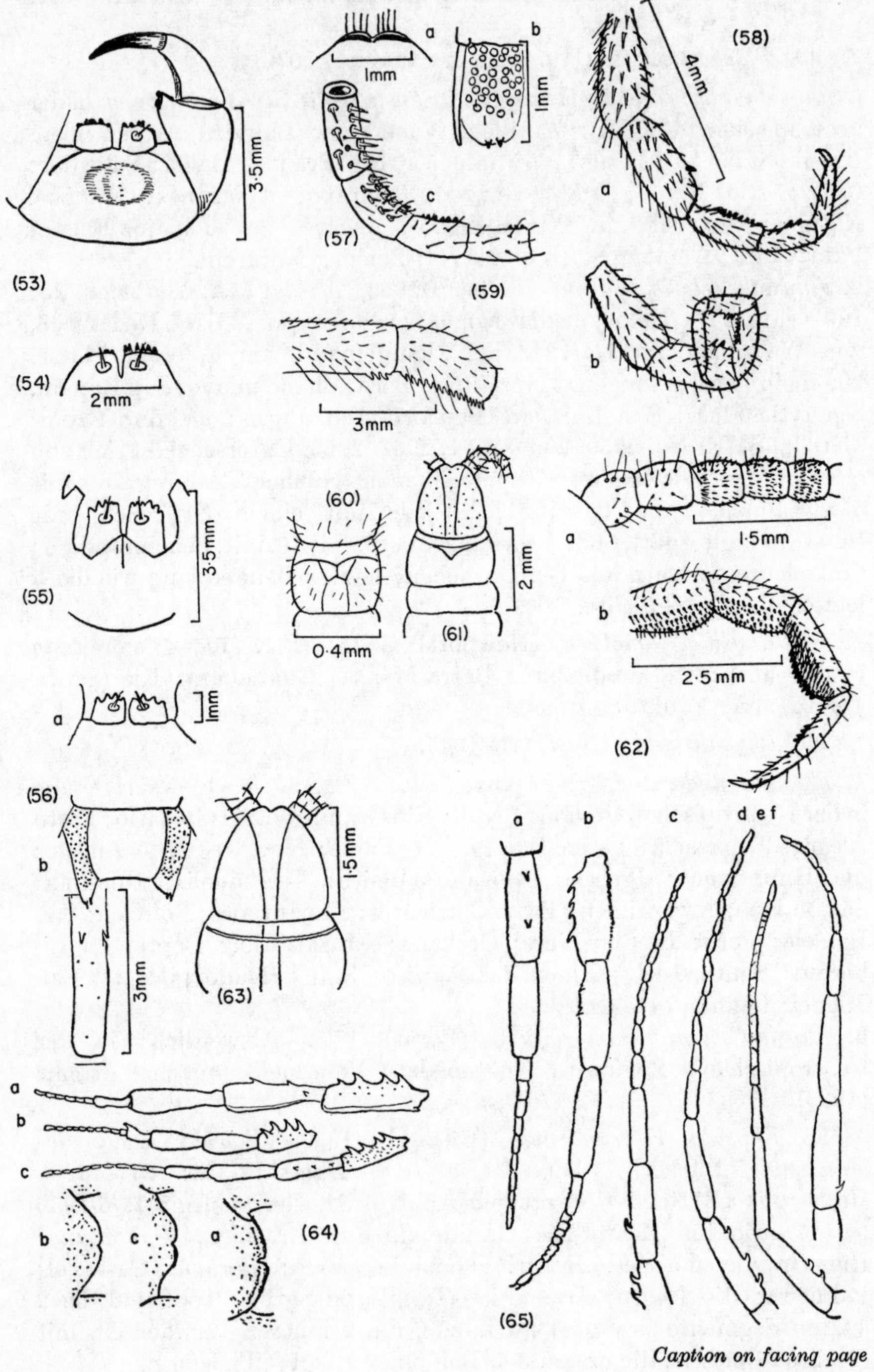

Caption on facing page

CRYPTOPIDAE

Cryptopinae

Chromatanops Pocock, 1893.

97. *Cryptops* (*Chromatanops*) *bivittatus* Pocock, 1893—*Verbreitung* (nach Attems 1930): Skt. Vincent; Costa Rica; Kolumbien: Sierra de Sta. Martha.

Cryptops Leach, 1815.

98. *Cryptops annectus* Chamberlin, 1947, *Ent. News* **58**: 146–147—Holotypus von El Valle: Panamá; N. L. H. Krauss legit.

99. *Cryptops annexus* Chamberlin, 1962, *Univ. Utah biol. Ser.* **12** (4): 12—Holotypus von Chiloe, Depto. Chepu, Chile; M. W. Holdgate legit 1958 (Fig. 58: a).

100. *Cryptops argentinus* Bücherl, 1953, *Mems Inst. Butantan* **25**: 125, Figs 19–22—Holotypus IB No. 770, Paratypen IB 771 und 757, von Manantiales, Prov. von Corrientes, Westargentinien; Biraben legit 1949 (Fig. 57: a, b, c).

101. *Cryptops armatus* Silvestri, 1899—Typus von Santiago, Chile.

102. *Cryptops beebei* Chamberlin, 1924—Typus von den Galapagen Inseln.

103. *Cryptops calinus* Chamberlin, 1957, *Gr. Basin Nat.* **17**: 30–41—Holotypus 11 min westlich von Cali, Kolumbien; Ross und Schlinger legerunt, 1955.

104. *Cryptops crassipes* Silvestri, 1895—Holotypus von Resistencia, Argentinien.

105. *Cryptops debilis* Chamberlin, 1955, *Acta Univ. lund.* N.F. (2) **51** (5): 34–63. Holotypus von Llanguihue, Chile (Fig. 11: a, b).

106. *Cryptops detectus* Silvestri, 1899—Holotypus von Temuco, Chile.

107. *Cryptops dubiotarsalis* Bücherl, 1946a, *Mems Inst. Butantan* **19**: 1–9, Figs 6–7. Holotypus, Männchen, No. 10 im Museu Paranaense,

---

FIGS 53–65. 53. *Rhysida brasiliensis rubra:* Coxosternum mit eiförmiger Vertiefung. 54. *Rhysida celeris andina:* Zahnplatten. 55. *Rhysida longipes:* Zahnplatten. 56. *Rhysida nuda nuda:* (a) Zahnplatten; (b) Coxopleuren und Endbeinpräfemur. 57. *Cryptops argentinus:* (a) Doppelbogiger Vorderrand des Coxosternums; (b) Porenfeld an den Coxopleuren; (c) letztes Bein- seitlich. 58. *Crytops:* (a) *annexus;* (b) *frater* (frei nach Chamberlin). 59. *Crytops triserratus* (nach Attems). 60. *Crytops heathi:* 1. Tergit. 61. *Cryptops galatheae:* Kopf und erste Tergite (nach Attems). 62. *Cryptops dubiotarsalis:* (a) Antennenglieder; (b) letztes Bein. 63. *Crytops schubarti:* Kopf und 1. Tergit. 64. *Newportia:* Vorderrand des Coxosternums und letztes Bein von *monticola* mit Variationen, die (a) *koepkei;* (b) *peruviana;* (c) *occidentalis* genannt wurden (nach O. Kraus). 65. *Newportia:* letztes Bein von (a) *atopa;* (b) *albana;* (c) *caldes;* (d) *schlingeri;* (e) *rossi;* (f) *ecuadorana;* (frei nach Chamberlin, 1957).

Curitiba; Paratypen: 1 Männchen. 2 Weibchen, 1 iuvensis; R. Hertel legit von Volta Grande, Staat von Paraná, Brasilien. *Verbreitung:* Brasilien, Staat von Paraná: Volta Grande (Typenort). Vila Velha; Curitiba; Guaraqueçaba (Fig. 62: a, b).

108. *Cryptops frater* Chamberlin, 1962, *Univ. Utah biol. Ser.* **12** (4): 1–23—Holotypus von Wellington, bei Puerto Eden, Chile; M. W. Holdgate legit 1958 (Fig. 58: b).

109. *Cryptops furciferens* Chamberlin, 1921—Holotypus von Demerara, Brit. Guayana.

110. *Cryptops galatheae* Meinert, 1886—(Fig. 61)—*Verbreitung:* Argentinien: Prov. Cordoba: Cabana; Uruguay: Montevideo; Brasil: Sta. Catarina: Insel Castilho; Florianópolis; Paraná: Curitiba; Hafen "De Cima"; Canavieiros; Staat São Paulo: Insel Queimada Grande; Pirassununga; Vassununga; am Fall des Flusses Emas; Porto Ferreira; Mogí Guassú; Sete Lagôas; Santa Rita Passa Quatro; Descalvado: 19 Exemplare in der Sammlung IB.

111. *Cryptops heathi* Chamberlin, 1914—Holotypus von Independencia, Staat Paraíba, Brasilien (Fig. 60).

112. *Cryptops manni* Chamberlin, 1915—Typus von Haiti.

113. *Cryptops melanifer* Chamberlin, 1955, *Acta Univ. lund.* N.F. (2), **51** (5) 34–53. Typus aus Kolumbien, zwischen Orchideen gefangen. 3 Paratypen.

114. *Cryptops micrus* Chamberlin, 1922—Holotypus von Trece Aguas, Guatemala.

115. *Cryptops monilis* Gervais 1849—Typus von Valdivia, Chile. Nach Chamberlin (1955) ist die Art zwischen Valdivia und Valparaiso häufig.

116. *Cryptops nahuelbuta* Chamberlin, 1955, *Acta Univ. lund.* N.F. (2), **51** (5) 34–53—Holotypus von Nahuelbuta, 1200 m, 7 Paratypen, Ross und Michelbacher legerunt, 1950.

117. *Cryptops navigans* Chamberlin, 1913—Galapagos Inselgruppe (Clipperton Insel).

118. *Cryptops nivicomes* Verhoeff, 1938, *Zool. Jb.* (Syst.) **71**: 367–368—Holotypus, Weibchen, von Fierro Carrera, über Santiago, 2350 m, an der Schneegrenze.

119. *Cryptops patagonicus* Meinert, 1886—Puerto Bueno, Patagonien.

120. *Cryptops positus* Chamberlin, 1940—*Psyche* **47**: 66—Holotypus und 2 Paratypen in der Quarentäne in New Orleans zwischen Importen aus Honduras und Nicaragua gefangen.

121. *Cryptops pugnans* Chamberlin, 1922—Typus von Progreso, Honduras.

122. *Cryptops rossi* Chamberlin, 1955—*Acta Univ. lund.* N.F. (2), **51** (5): 34–61—Holotypus von Buenaventura, Kolumbien. 1 Paratypus.

123. *Cryptops schubarti* Bücherl, 1953, *Mems Inst. Butantan* **25**: 109–151, Figs 12–14—(Fig. 63)—Holotypus IB No. 711, Schubart legit 14/2/1942 in feuchter Erde, von losem Laub bedeckt. *Verbreitung:* Brasilien, Staat São Paulo: Porto Ferreira (Typenort); Jaboticabal; Insel "do macaco"; Santa Rita Passa Quatro; Baguassú.

124. *Cryptops triserratus* Attems, 1903—(Fig. 59)—Typus aus Valdivia, Chile.

125. *Cryptops vector* Chamberlin, 1931—*Pan Pacif. Ent.* **7**: 189–190—Typus in der Quarentäne in New Orleans zwischen Importen aus Honduras gefangen.

126. *Cryptops venezuelae* Chamberlin, 1939—Typus in der Quarentäne Station in Washington zwischen Importen aus Venezuela gefangen.

127. *Cryptops watsingus* Chamberlin, 1939—*Pan Pacif. Ent.* **15**: 63–65—Typus in der Quarentäne in New Orleans zwischen Pflanzenimporten aus Guatemala gefangen.

Trigonocryptops Verhoeff, 1906.

128. *Trigonocryptops debilis* Bücherl, 1950a, *Mems Inst. Butantan* **22**: 173–186, Figs 10–12—(Fig. 14b: 7)—Holotypus MHNL No. 10 044; Weyrauch legit 4000 m bei Abancay, unter Steppenvegetation. *Verbreitung:* Peru: Abancay; San Luis de Suharo, 500 m (SMFI 3067/3); Atiquippa, SMFI 3024/7; Capa, SMFI 3025; Chala, an der Pacifikküste, SMF 3067/3; Koepcke legit 1956.

129. *Trigonocryptops iheringi* Brölemann, 1902—Holotypus von Alto da Serra, zirka 30 km von der Stadt São Paulo, Richtung Santos, zirka 900 m, Waldgegend. *Verbreitung:* Brasilien: Staat São Paulo: Stadt São Paulo: zirka 20 Exemplare aus verschiedenen Stadtteilen; São Roque; Eleuterio; São Miguel; Varzea; Tietê; Itapira; Itararé; Pirapora; Pirassununga; Emas; Corumbatai; Atibaia; Mato Grosso: Terrenos; Pontal; Campo Grande; Rio de Janeiro: Nova Iguassú; Paraná; Rio Negro; Caiacanga; Morretes; Santa Catarina; Insel Bom Abrigo; Perdizes; Guanabara: Alto da Boa Vista; Amazonas: Manaus.

130. *Trigonocryptops triangulifer* Verhoeff, 1937, *Zool. Jb.* (Syst.) **70**: 1–16. Typus aus Minas Gerais, Brasilien, ohne Fundort.

Paracryptops Pocock, 1891.

131. *Paracryptops inexspectus* Chamberlin, 1914—Dunoon, Brit. Guayana.

Mimops Kraepelin, 1903.

132. *Mimops occidentalis* Chamberlin, 1914—Typus von Rio de Janeiro, Staat Guanabara.

Dinocryptopinae

Dinocryptops Crabill, 1953, *Ent. News* **64**: 96.

133. *Dinocryptops miersi* (Newport) 1845—(Fig. 14b: 6)—*Verbreitung:* USA: Georgia, Virginia, etc. . . . ; America Central; Venezuela, Guayanen; Brasilien: Rio Grande do Sul: Uruguaiana; Passo Fundo; Paraná: Palmeira; Caiacanga; Uraí; Mato Grosso: Terrenos; Campo Grande; Caceres; São Paulo: Lôbo; Corumbataí; Sarandy; Bueno de Andrade; Ubirama; Piracicaba; Sorocaba; Embú-Guassú; Cubatão; Alto da Serra; Capital; Argentinien: Caraguatay; Peru: Tingo Maria; 4 min von Otusco; Ecuador: Quito; Alausi; Tixan: zusammen mehr als 200 Exemplare.

134. *Dinocryptops puruensis* (Bücherl) 1941—*Scolopocryptops miersii puruensis* Bücherl, 1941, *Mems Inst. Butantan* **15**: 129–130, Figs 9–12—Holotypus No. 27 im MNR und 3 Paratypen, No. 24 und 49 MNR und IB 293, vom Ufer des Sees Mapixi, am Flusse Purús, Staat Amazonas, Brasilien. *Verbreitung:* Lago Mapixi; Staat Mato Grosso: Xavantina, IB No. 633/7, Sick legit, 1949; Goiás: Rio Verde; Pará: Belém, Ledoux legit 11/6/1951; A. R. Hoge legit 4 Exemplare, 17/9/52; Amapá: Serra do Navio; Venezuela: Randal, am Orinoco, Carbonell legit, 4/6/1951—Unterscheidet sich von *miersi:* Coxosternum der Kieferfüsse mit transversal Furche; Tergite mit kurzer Mittelfurche; alle Tergite scheinberandet aber nur in der vorderen Hälfte; der Coxopleurenanhang ist viel länger als die Länge des letzten Sternites; Basalglieder der Antennen fast kahl.

Newportia Gervais 1847.

Subgenus *Newportia* Gervais, 1847.

135. *Newportia* (*N.*) *atopa* Chamberlin, 1957, *Gr. Basin Nat.* **17**: 37, Fig. 8—(Fig. 65: a)—Holotypus 36 min südlich von Alausi, Ecuador.

136. *Newportia* (*N.*) *balzani* Silvestri, 1895—(Fig. 66)—Typus von den Ufern des Flusses Apa, Paraguay.

137. *Newportia bicegoi* Kraepelin, 1903.

a. *Newportia* (*N.*) *bicegoi bicegoi* Kraepelin, 1903—(Fig. 67: a, b)—Holotypus von Manaus.

b. *Newportia* (*N.*) *bicegoi collaris* Kraepelin, 1903—Typus von Bas Carsevenne, Franz. Guayana. *Verbreitung:* Franz. Guayana; Venezuela: Horquatas Minas und Zulia, Carbonell legit 1950.

138. *Newportia* (*N.*) *bollmani* Attems, 1930—Typus von Cuba.

139. *Newportia* (*N.*) *caldes* Chamberlin, 1957, *Gr. Basin Nat.* **17**: 37–38, Fig. 10—(Fig. 65: c)—Holotypus 2 min westlich von Calarca, Caldes, Kolumbien.

140. *Newportia* (*N.*) *cubana* Chamberlin, 1915—Holotypus von Cuba, Panamo. *Verbreitung:* Cuba: Panamo. Guantamano.

141. *Newportia* (*N.*) *dentata* Pocock, 1890—Typus von Chimborasso, Ecuador.

142. *Newportia* (*N.*) *diagramma* Chamberlin, 1921—Typus von Dunoon, Brit. Guayana. *Verbreitung:* Brit. Guayana, Venezuela: Zulia; Humana Selva; Rancho Grande; El Junquito, Marcuzzi legit, 2/7/1950.

143. *Newportia* (*N.*) *aureana* (Bücherl) 1941—*Newportia diagramma aureana* Bücherl, 1941, *Mems Inst. Butantan* **15**: 139–140, Figs 13–15—(Fig. 68: a, b)—Holotypus im MNR No. 144, A. de Carvalho legit am Fluss Aurá, Staat Pará, Brasilien. *Verbreitung:* Aurá; Aragarças, am Araguaia Flusse.

144. *Newportia* (*N.*) *ecuadorana* Chamberlin, 1957, *Gr. Basin Nat.* **17**: 37, Fig. 7—(Fig. 65: f)—Holotypus 2 min westlich von Banos, Tungurahua, Ecuador.

145. *Newportia* (*N.*) *fuhrmanni* Ribaut, 1914.

a. *Newportia* (*N.*) *fuhrmanni fuhrmanni* Ribaut, 1914—Typus vom Cafetal Camelia, bei Angelopolis in den Zentral Cordilleren, Kolumbien —(Fig. 1:b).

b. *Newportia* (*N.*)*fuhrmanni ignorata* (Kraus) 1955, *Newportia ignorata* Kraus, 1955, *Senckenberg. biol.* **36**: 173–200, Figs 8–11—(Fig. 71: d)—Holotypus SMFI No. 2447, Zorate, am San Bartolomé Flusse, Peru, Koepcke legit 1954. Noch 5 weitere Exemplare von Zorate.

146. *Newportia* (*N.*) *heteropoda* Chamberlin, 1918—Cuba.

147. *Newportia* (*N.*) *longitarsis* (Newport) 1845.

a. *Newportia* (*N.*) *longitarsis longitarsis* (Newport) 1845—(Fig. 71: a)—*Verbreitung:* Venezuela: El Junquito; Peru: Huanuco; Tingo Maria; Brasil: Mato Grosso: am oberen Cuminá-Flusse; São Paulo: Corumbataí; Tres Pontes; Pirassununga.

b. *Newportia* (*N.*) *longitarsis sylvae* Chamberlin, 1914—Typus vom Mamoré-Flusse, Mato Grosso (Rondonia).

c. *Newportia* (*N.*) *longitarsis stechowi* Verhoeff, 1938, *Zool. Anz.* **123**: 123–130. Typus von Maracay, bei Caracas, Venezuela.

d. *Newportia* (*N.*) *longitarsis tropicalis* Bücherl, 1959a, *Mems Inst. Butantan* **29**: 236—Holotypus von Zulia, Kunana Selva, Venezuela.

148. *Newportia* (*N.*) *monticola* Pocock, 1890—(Figs 71: b, 72). Synonymien: *N. cuzcona* Chamberlin 1955, *Acta Univ. lund.* N.F. (2)

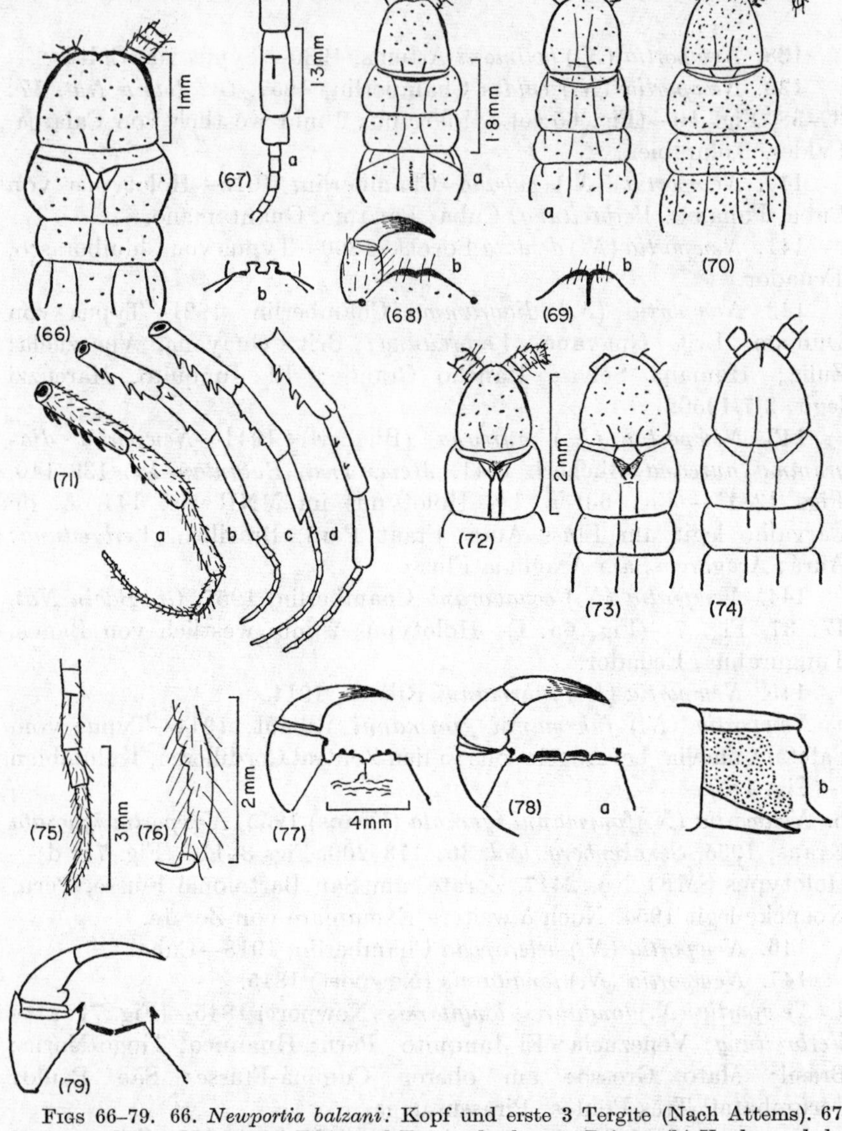

FIGS 66–79. 66. *Newportia balzani:* Kopf und erste 3 Tergite (Nach Attems). 67. *Newportia bicegoi bicegoi:* (a) Tibia und Tarsen des letzten Beines; (b) Vorderrand des Coxosternums. 68. *Newportia aureana:* (a) Kopf und erste Tergite; (b) Vorderrand des Coxosternums. 69. *Newportia ernsti fossulata:* (a) Kopf und vordere Tergite; (b) Vorderrand des Coxosternums. 70. *Newportia maxima.* 71. *Newportia:* letztes Bein von (a) *longitarsis longitarsis;* (b) *monticola;* (c) Die Form *weyrauchi;* (d) *fuhrmanni ignorata.* 72. *Newportia monticola:* Kopf und 1. Segment. 73. *Newportia spinipes* (Mexico) (nach Attems). 74. *Newportia divergens:* Kopf und erste Segmente. 75. *Newportia paraensis:* 1. und 2. Tarsus des letzten Beines. 76. *Newportia amazonica:* 2 Tarsus mit Kralle am letzten Beine. 77. *Scolopocryptops denticulatus:* Vorderrand des Coxosternums. 78. *Scolopocryptops f. soucupi:* (a) Vorderrand des Coxosternums; (b) Coxopleuren. 79. *Scolopocryptops spinulifer:* Coxosternum.

51 (5): 39, Fig. 25—(Fig. 10: a); *N. koepckei* Kraus 1954, *Senckenbergiana* **34**: 320, Pr. 3, Figs 30–31—(Fig. 64: a)—*N. occidentalis* Kraus 1954, *Senckenbergiana* **34**: Figs 33–34:—(Fig. 64: c)—*N. perucola* Chamberlin 1955 *Acta Univ. lund.* N.F. (2) **51** (5): 39, Fig. 26—(Fig. 10, b); *N. peruviana* Kraus 1954, *Senckenbergiana* **34**: 319, Pr. 3, Fig. 29—(Fig. 64: b); *N. weyrauchi* Chamberlin 1955, *Acta Univ. lund.* N.F. (2) **51** (5): 40, Fig. 27—(Figs 10: c, 71: c). *Verbreitung:* Kolumbien: Caldes; Cali; Ecuador: Zurucuchu—See; Alausi; Quito; Pichilinque; Banos; Peru: Chaquil; Lupin; Cusco; Hacienda Llaquen; Hacienda Montesco; El Tunel; Andahuaylas; Asia.

149. *Newportia* (*N.*) *oligopla* Chamberlin, 1945, *Ent. News* **56**: 171–174—Holotypus von Skt. Augustine, Trinidad.

150. *Newportia* (*N.*) *pusilla* Pocock, 1893—*Verbreitung:* Kolumbien: Zentral Cordilleren; Camelia und Oriental Cordilleren: Bogotá; Paramo Cruz Verde; Tambo; Viota; Venezuela: Rancho Grande.

151. *Newportia* (*N.*) *rogersi* Pocock, 1896—Vulkan Irazu, Costa Rica.

152. *Newportia rossi* Chamberlin 1957, *Gr. Basin Nat.* **17**: 40, Fig. 12—(Fig. 65: e)—Typus 6 min westlich von Cali, Kolumbien.

153. *Newportia* (*N.*) *schlingeri* Chamberlin, 1957, *Gr. Basin Nat.* **17**: 40–41, Fig. 11—(Fig. 65: d)—Typus 98 min östlich von Olmos, Peru.

154. *Newportia* (*N.*) *simoni* Brölemann, 1898—Typus von La Guayra, Venezuela.

Subgenus *Scolopendrides* Humbert & Saussure 1896.

155. *Newportia* (*S.*) *albana* Chamberlin, 1957, *Gr. Basin Nat.* **17**: 30—41—(Fig. 65: b)—Holotypus und 1 Paratypus 2 min westlich von Alban, Condin Amarca, Kolumbien.

156. *Newportia* (*S.*) *brevipes* Pocock, 1891. Typus von Demerara, Brit. Guayana. Ein zweites Exemplar vom Aurá-Flusse, Staat Pará; im MNR No. 147; Moojen legit.

157. *Newportia* (*S.*) *divergens* Chamberlin, 1922—(Fig. 74)—Typus von Joyabay, Guatemala—Längsfurchen des 1. Tergites mit W-Formation, vorne und fast unmerklicher Vertiefung. Vergleiche dieselbe Ausprägung bei der mexikanischen Art, *N. spinipes* Pocock, 1896 (Fig. 73).

158. *Newportia* (*S.*) *ernsti* Pocock, 1891.

a. *Newportia* (*S.*) *ernsti ernsti* Pocock, 1891—*Verbreitung:* Skt. Vincent; Venezuela: Caracas.

b. *Newportia* (*S.*) *ernsti fossulata* Bücherl, 1941, *Mems Inst. Butantan* **15**: 140, Figs 16–17—(Fig. 69: a, b)—Holotypus No. 146, MNR; Moojen legit, Aurá, Staat Pará, Brasilien. Ein zweites Exemplar aus Chavantina, Mato Grosso, an den Ufern des oberen Tapajoz; in faulem

Holze; Dr. Sick legit 1948—21. Praefemur, Femur, Tibia von gleicher Länge, 1. Tarsus halb so lang wie Tibia, 2. Tarsus abrupt dünner als 1. Tarsus; die Transversalfurche auf der Kopfplatte nur zwischen den beiden Längsfurchen sichtbar; Coxosternum mit einer kurzen Längsfurche; Tergit 2.–4. schon mit abgekürzten Seitenfurchen; Sternite der hinteren Körperhälfte mit 1 Mittel-und 2 Seitenlängsfurchen und noch 1 vorderen und 1 hinteren Querfurche; Analpraefemur mit 5, Femur mit 0–2 Dornen.

159. *Newportia* (*S.*) *lasia* Chamberlin, 1921—Typus von Dunoon, Brit. Guayana.

160. *Newportia* (*S.*) *maxima* Bücherl, 1941. *Mems Inst. Butantan* **15**: 143–144, Figs 18–20—(Fig. 70)—Holotypus und 1 Paratypus im MNR, No. 148, zirka 35 km südlich des Fangortes von *ernsti fossulata*, am Aurá-Flusse. Ein Exemplar, IB No. 572, von der Bananal—Insel, am Araguaia-Flusse; Hoge legit 29/9/48.

161. *Newportia* (*S.*) *mimetica* Chamberlin, 1922—Typus von Lombardia, Honduras.

162. *Newportia* (*S.*) *paraensis* Chamberlin, 1914—(Fig. 75)—Typus von Belém, Pará, Brasilien.

163. *Newportia* (*S.*) *stolli* Pocock, 1896—Typus von Quezaltenango, Guatemala.

Subgenus *Newportides* Chamberlin, 1921.

164. *Newportia* (*N.*) *amazonica* Brölemann, 1904—Fig. 76)— Typus von Manaus, Amazonas, Brasil.

165. *Newportia* (*N.*) *sulana* Chamberlin, 1922—Forma ignota.

166. *Newportia* (*N.*) *unguifer* Chamberlin, 1921—Typus von Labba Creek, bei Dunoon, Brit. Guayana.

Scolopocryptopinae

Kartops Archey, 1923.

167. *Kartops guianae* Archey, 1923—Typus von Kartabo, District Bartica, Brit. Guayana.

Scolopocryptops Newport, 1845

168. *Scolopocryptops denticulatus* (Bücherl) 1946—*Otocryptops denticulatus* Bücherl, 1946a, *Mems Inst. Butantan* **19**: 7–8, Fig. 5—(Fig. 77)—Holotypus, Weibchen, No. 7 im Museu Paranaense, Curitiba; Paratypen, Männchen und Weibchen; alle von Bariguí, Staat Paraná, Brasilien; R. Lange legit 1943 und 1945.

169. *Scolopocryptops ferrugineus* (Linné) 1767.

a. *Scolopocryptops ferrugineus ferrugineus* (Linné) 1767—Synonymien: *S.f. inversus* (Chamberlin) 1921; *Scolopocryptops ferrugineus genuinus*

Verhoeff, 1941; *Scolopocryptops ferrugineus parcespinosus* (Kraepelin) 1903; *Scolopocryptops miersii peruanus* Verhoeff, 1941. Uber diese Synonymien siehe Kraus, 1957, *Senckenberg. biol.* **38**: 380–392. *Verbreitung:* Afrika (Camerun), Mexico, Antillen, Guatemala: Chichivac; Vulkan Tajumulco; El Salvador: Depto. Sant' Anna; serra von Matapan; Depto. San Vicente; Ecuador: Alausi; Peru: Palco, Sivia; Carhuamayo; Otusco; Tingo Maria; Taulis; Pajonal; La Florida; Llaguén; Ijicucho; Borgapampa, Jaen, Atocongo; Bolivia: San Ignacio; Venezuela; Rancho Grande; Brasil: Amazonas: Bôca Copeá; Staat Maranhão: Fazenda Canaã; Mato Grosso: Campo Grande; Jacaré, am obren Tapajoz; Minas Gerais: Vicente Carvalhaes; Staat Goias: Uberaba; São Paulo: Ouro Fino; Pirassununga; Emas; Baruerí; Três Pontes; Corumbataí; Conchas; Caraguatatuba; São Pedro; Bueno de Andrade; Ribeirão Bonito; Serra Negra; Morro Agudo; Itanhaém; Jaboticabal; Francisco Morato; Jundiaí; Atibaia; die Stadt São Paulo; Staat Rio de Janeiro: Volta Redonda; Itatiaia; Paraná; Londrina; Campo Mourão; Pôrto Amazonas; Iraí; Cruz Machado; Palmeira; Santa Catarina: Ilha do Bom Abrigo; Rio Grande do Sul: Passo Fundo; Porto Alegre; Pederneira; Tupaceretã.

b. *Scolopocryptops ferrugineus guacharensis* Manfredi, 1957, *Boln Soc. venez. Cienc. nat.* **18**: 178, Figs 1–2; Holotypus von Venezuela, Staat Monagas, Grotte Guacharo, in 700 m Tiefe, im Salon "Humboldt", in absoluter Finsternis, Temperatur 10°C, 100% Feuchtigkeit.

c. *Scolopocryptos ferrugineus macrodon* (Kraepelin) 1903—Brasilien, Staat Paraná, ohne Fundort.

d. *Scolopocryptops ferrugineus riveti* (Brölemann) 1919—Typus von Narihuana, Ecuador.

e. *Scolopocryptops ferrugineus soucupi* (Bücherl) 1943—*Otocryptops f. soucupi* Bücherl, 1943B, *Mems Inst. Butantan* **17**: 24, Figs 1–3 (Figs 14b; 8; 78: a, b). Holotypus, Weibchen, IB No. 349 und 3 Paratypen, Männchen und Weibchen, von Amancais, bei Lima, Peru; Soucup legit 1941. *Verbreitung:* Peru: Amancais; Sivia. Atocongo; Infernilo; Matucana; Rimac-Fluss; San Mateo; Zorate; Tapacocha; Palca; Acobamba; Tarmatambo; Pichita Caluga; San Ramon; Oxapampa; Campanillaya; Mantaro-Fluss; Colcabamba; Recuay; Huancayo; Celendin; Cajamarca; Canta; Neuquem; Tarma; Cerro de Pasco; Andahuaylas.

170. *Scolopocryptops melanostomus* Newport, 1845—*Verbreitung:* Alte Welt, Zentral-und Südamerika; Venezuela: Ayapaina; Rancho Grande; Zulia; Franz. Guayana: Saint Claire; Peru: Iquitos; Porto Rico: Villalba; Torro Negro; Gilla de Guilarte; Haiti: Kenskoff;

Guatemala: Escobar; Brasilien: Terezópolis; Petropolis im Staat Rio de Janeiro.

171. *Scolopocryptops spinulifer* (Bücherl) 1949—*Otocryptops spinulifer* Bücherl, 1949, *Mems Inst. Butantan* **21**: 5–6, Figs 3–4; (Fig. 79). Holotypus und 2 Paratypen, IB No. 31; A. R. Hoge legit auf der Insel Queimada Grande, bei Santos, Brasilien. Weitere Exemplare IB No. 31/3; 112; 158/3; 405/2, Männchen und Weibchen, von demselben Fundorte.

ZOO-GEOGRAPHISCHE BETRACHTUNGEN

Endemisch in der Neotropischen Region sind *Hemiscolopendra* (mit nur einer nearktischen Art als Ausnahme), *Rhoda, Scolopendropsis,* fast alle *Parotostigmus*-Arten, *Chromatanops, Kartops* und die grosste Zahl der *Newportia* Arten.

Die bisher am häufigsten aufgefundenen Arten sind: *Scolopendra viridicornis, Cormocephalus andinus, C. impressus, Parotostigmus scabricauda, P. tibialis, Cryptops monilis, Trigonocryptops iheringi, Dinocryptops miersi, Newportia monticola* und *Scolopocryptops ferrugineus. Hemiscolopendra* und *Cryptops* wiegen in Chile und im vorandinischen West-Argentinien bei weiten vor, *Cormocephalus* und *Newportia* in Peru und die nördlichen andinischen Länder. Von der wirklichen Arten- und Individuenzahl der Scolopendromorphen im Neotropischen Grossraum kann man sich, an Hand der bisherigen Ausbeuten und Fundplätzen, kaum ein korrektes Bild machen. Es ist überraschend, dass keine neuen Gattungen mehr aufgefunden wurden.

ZUSAMMENFASSUNG

In Tabelle I sind Gattungen, Arten und Unterarten der Neotropischen Scolopendromorphen bis Attems, 1930 (in Klammern) und von diesem Datum bis 1971 (soweit mir die Literatur zugänglich war) verzeichnet.

DANKSAGUNG

Ich danke besonders dem Conselho Nacional de Pesquisas, von Rio de Janeiro, für das Stipendium, das mir diese Arbeit ermöglichte, der Direktion des Institutes Butantan, des Museu Nacional in Rio de Janeiro und des Museu Paranaense von Curitiba, die mir das Scolopendromorphen Material zur Verfügung stellten. Ich danke Herrn. Prof. O. Kraus, der mir Gelegenheit gab, das Material im Senckenberg Museum und Forschungsinstitut einzusehen. Besonders bin ich auch

TABELLE I

Zahl der Gattungen, Arten und Unterarten

Scolopendridae		Gattungen	Artenzahl		Unterarten	
	Scolopendrinae	*Arthrorhabdus*	(1)	1		
		Cormocephalus	(10)	15	(2)	8
		Hemiscolopendra	(5)	6		
		Rhoda	(2)	2		2
		Scolopendra	(13)	15		11
		Scolopendropsis	(1)	1 ?		
	Otostigminae	*Otostigmus (P.)*	(22)	49	(2)	6
		Rhysida	(4)	6		4
Cryptopidae						
	Cryptopinae	*Cryptops (CH.)*	(1)	1		
		Cryptops (C.)	(14)	30		
		Trigonocryptops	(1)	3		
		Paracryptops	(1)	1		
		Mimops	(1)	1		
	Dinocryptopinae	*Dinocryptops*	(1)	2		
		Newportia	(23)	25	(4)	4
		Tidops	(1)	1		
	Scolopocryptopinae	*Kartops*	(1)	1		
		Scolopocryptops	(3)	4	(5)	5
2	5	17	(105)	164	(13)	40

den Herren Marcuzzi, Carbonell, Biraben, P. G. Aguilar, W. Weyrauch, H. Sick, H. Sioli, O. Schubart und allen anderen, welche mir Scolopendromorphen zugesandt haben, verpflichtet. Herrn. Dr. J. Gordon Blower und dem Publications Department von der The Zoological Society of London bin ich ebenfalls zu grossem Dank verpflichtet.

LITERATURVERZEICHNIS

Attems, C. (1930). Scolopendromorpha. *Tierreich* **54**.

Bücherl, W. (1939). Os Quilópodos do Brasil. *Mems Inst. Butantan* **13**: 49–362.

Bücherl, W. (1939a). Dois novos quilopodos do subgênero *Parotostigmus* da coleção do Instituto Butantan. *Bolm biol. Clube zool. Bras.* N.S.**4**: 444–447.

Bücherl, W. (1939b). Tres Escolopendrídeos novos na coleção miriapodológica do Instituto Butantan. *Revta Biol. Hyg.* **10**: 54–65.

Bücherl, W. (1941). Quilópodos novos da coleção miriapodológica do Museu Nacional do Rio de Janeiro. *Mems Inst. Butantan* **45**: 119–158.

Bücherl, W. (1943a). Descrição de uma nova subespécie do gênero *Otostigmus* Porat, subgênero *Coxopleurotostigmus* Bücherl. *Mems Inst. Butantan* **16**: 85–91.

Bücherl, W. (1943b). Quilópodos do Peru I. *Mems Inst. Butantan* **17**: 19–26.

Bücherl, W. (1946a). Quilópodos do Museu Paranaense de Curitiba. *Mems Inst. Butantan* **19**: 1–10.

Bücherl, W. (1946b). Ação do veneno dos Escolopendromorfos do Brasil sôbre alguns animais de laboratório. *Mems Inst. Butantan* **19**: 181–197.

Bücherl, W. (1946c). Novidades sistematicas na ordem Scolopendromorpha. *Mems Inst. Butantan* **19**: 135–158.

Bücherl, W. (1949). Quilópodos das Ilhas da Queimada Grande e Pequena. *Mems Inst. Butantan* **21**: 1–8.

Bücherl, W. (1950a). Quilópodos do Peru II. *Mems Inst. Butantan* **22**: 173–186.

Bücherl, W. (1950b). Quilópodos da Venezuela I. *Mems Inst. Butantan* **22**: 187–198.

Bücherl, W. (1953). Quilópodos, aranhas e escorpiões, enviados ao Instituto Butantan para determinação. *Mems Inst. Butantan* **25**: 109–151.

Bücherl, W. (1954). *Otostigmus* (*Parotostigmus*) *latipes* n.sp. *Mems Inst. Butantan* **26**: 1–6.

Bücherl, W. (1959a). Chilopoden von Venezuela II. *Mems Inst. Butantan* **29**: 233–241.

Bücherl, W. (1959b). Kritische Untersuchungen der *Newportia*-Arten. *Beitr. neotrop. Fauna* **1**: 229–242.

Bücherl, W. (1969). *Otostigmus diringshofeni* n. sp. *Beitr. neotrop. Fauna* **6**: 90–91.

Bücherl, W. (1972). Sistematica e ecologia dos Escolopendromorfos do Peru. *Revta peru. Ent. agric. Lima* **14**: 160–169.

Chamberlin, R. V. (1931). On three new Chilopods. *Pan Pacif. Ent.* **7**: 189–190.

Chamberlin, R. V. (1939). Four new Centipeds of the genus *Cryptops*. *Pan Pacif. Ent.* **15**: 63–65.

Chamberlin, R. V. (1940). On some Chilopods from Barro Colorado Island. *Psyche* **47**: 66–74.

Chamberlin, R. V. (1941). On a collection of Millipedes and Centipedes from North-Eastern Peru. *Bull. Am. Mus. nat. Hist.* **78**: 473–535.

Chamberlin, R. V. (1942a). Two new Centipeds from the Chilibrillo Caves, Panama. *Pan Pacif. Ent.* **18**: 125–126.

Chamberlin, R. V. (1942b). On ten new centipedes from Mexico and Venezuela. *Proc. biol. Soc. Wash.* **55**: 17–24.

Chamberlin, R. V. (1944). Chilopods in the collection of Field Museum of Natural History. *Publs Field Mus. nat. Hist.* (*Zool. Ser.*) **28**: 175–216.

Chamberlin, R. V. (1945). Two new Centipeds from Trinidad. *Ent. News* **56**: 171–174.

Chamberlin, R. V. (1947). A few Chilopods taken in Panama by N. L. H. Krauss. *Ent. News* **58**: 146–149.

Chamberlin, R. V. (1952). The Centipeds of South Bimini, Bahamas Islands, British West Indies. *Am. Mus. Novit.* No. 1576: 1–8.

Chamberlin, R. V. (1955). The Chilopoda of the Lund University and California Academy of Sciences Expeditions. *Acta Univ. lund.* N.F. (2) **51**(5): 1–61.

Chamberlin, R. V. (1957). Scolopendrid chilopods of the Northern Andes region taken on the California Academy South America Expedition of 1954–1955. *Gr. Basin Nat.* **17**: 30–41.

Chamberlin, R. V. (1962). Chilopods secured by the Royal Society Expedition to Southern Chile in 1958/59. *Univ. Utah biol. Ser.* **12** (4): 1–23.

Crabill, R. E. (1953). Concerning a new genus *Dinocryptops* and the nomenclatural status of *Otocryptops* and *Scolopocryptops*. *Ent. News* **64**: 96.

Jangi, B. S. (1955). Some aspects of the morphology of the Centipede *Scolopendra morsitans* Linn. *Ann. Mag. nat. Hist.* (12) **8**: 597–607.

Kraus, O. (1954). Myriapoden aus Peru I. *Senckenbergiana* **34**: 311–323.

Kraus, O. (1955). Myriapoden aus Peru III. *Senckenberg. biol.* **36**: 173–200.

Kraus, O. (1957). Myriapoden aus Peru VI. *Senckenberg. biol.* **38**: 359–404.

Lewis, J. G. E. (1968). Individual variation in a population of the centipede *Scolopendra amazonica* from Nigeria and its implications for methods of taxonomic discrimination in the Scolopendridae. *J. Linn. Soc.* (Zool.) **47**: 315–326.

Lewis, J. G. E. (1969). The variation of the centipede *Scolopendra amazonica* in Africa. *Zool. J. Linn. Soc. Lond.* **48**: 49–57.

Manfredi, P. (1957). Nuovo Scolopendride cavernicolo americano. *Boln Soc. venez. Cienc. nat.* **18**: 175–180.

Porter, C. E. (1911). *Introducion al estudio de los miriapodos.* Santiago, Chile: Imprensa Universitaria, Bandera 130.

Turk, F. A. (1955). The Chilopods of Peru with descriptions of new species and some zoogeographical notes on the Peruvian Chilopod fauna. *Proc. zool. Soc. Lond.* **125**: 469–504.

Verheoff, K. W. (1937). Über einige Chilopoden aus Australien und Brasilien. *Zool. Jb.* (Syst.) **70**: 1–16.

Verhoeff, K. W. (1938). Chilopoden–Studien, zur Kenntnis der Epimorphen. *Zool. Jb.* (Syst.) **71**: 339–388.

Verhoeff, K. W. (1938a). Über einige Chilopoden des zoologischen Museums in München. *Zool. Anz.* **123**: 123–130.

Verhoeff, K. W. (1951). Chilopoden und Diplopoden. In *Beiträge zur Fauna Perus* **2**: 5–68. Titschack, E. (ed.) Jena: Fischer.

Abkürzungen

FM — Field Museum.
HMW — Hof Museum, Wien.
IB — Instituto Butantan, São Paulo.
IZUC — Instituto de Zoologia, Universidade de Caracas.
MHNL — Museu de Historia Natural, Universidade San Marcos, Lima.
MNR — Museu Nacional, Rio de Janeiro.
SMFI — Senckenberg Museum und Forschungs Institut, Frankfurt.
ZMUH — Zoologisches Museum der Universität Hamburg.
AMNH— American Museum Natural History, New York.

Symp. zool. Soc. Lond. (1974) No. 32, 135–142.

SCANNING ELECTRON MICROSCOPE STUDIES OF SYMPHYLA

Rothamsted Experimental Station, Harpenden, Hertfordshire, England

R. H. TURNER and C. A. EDWARDS

SYNOPSIS

Taxonomic features of common representatives of the Scolopendrellidae, Bagnall (*Symphylella vulgaris* Hansen) and the Scutigerellidae, Bagnall (*Scutigerella immaculata* Newport) were studied using the scanning electron microscope. The mouthparts, scuta, legs and coxal sacs, calicles and cerci of both genera were compared.

INTRODUCTION

The taxonomy of the Symphyla has been little studied. Two families, the Scutigerellidae, Bagnall, and the Scolopendrellidae, Bagnall, and 15 genera are well-defined (Edwards, 1959), but many of the morphological characters used to differentiate them remain only partially understood. One reason is that the main taxonomic characters are difficult to see with a light microscope, because good mounts are difficult to make, there is little depth of focus at high magnifications and resolution cannot be better than 0·5 μm.

The scanning electron microscope (SEM) provides much better resolution (10–25 nm), great depth of focus at all magnifications and three-dimensional manoeuvrability of the specimen, so that many characters can be studied on the same specimen.

We compared the taxonomic morphological structures of common representatives of the Scolopendrellidae (*Symphylella vulgaris* Hansen) and the Scutigerellidae (*Scutigerella immaculata* Newport).

We used both fresh specimens and others fixed in 70% alcohol and 5% glycerol. Living animals were anaesthetized with ether and mounted on 1 cm dia. specimen stubs with double-sided adhesive tape ("Twin-stick"). Preserved animals were mounted similarly after arranging the legs and antennae. All specimens were coated with gold, approximately 30 nm thick, in an Edwards 12E6 coating unit and examined in a Cambridge "Stereoscan" Mark IIA SEM at accelerating voltages between 3 kV and 30 kV.

The morphological characters studied are illustrated in Figs 1–24.

The antennae often collapsed in the high vacuum during coating and in the microscope, so we shall not discuss antennal characters, hoping that more refined techniques such as freeze-drying or critical point drying will overcome this difficulty.

SCUTA

The scuta differ in shape and number between genera. Increased scutal numbers seem to be due to subdivision of the basic scutal pattern, as shown in *Scutigerella* (Fig. 1). However, relation between the position of each pair of legs and a scutum with a curved (Fig. 3) or pointed (Fig. 4) posterior margin remains a constant feature (Ribaut, 1931; Edwards & Belfield, 1967). Figures 3 & 4 compare the first five scuta of *Scutigerella* and *Symphylella* and clearly show that the first scuta are reduced and immediately behind the head. These structures are rarely seen in preparations for light microscopes. The consistent difference is that the scuta of *Scutigerella* are scaly (Fig. 3), but those of *Symphylella* have a granular appearance (Fig. 4). The granules appear to be clumps of tiny setae (Fig. 5) which are even more pronounced on the skin on the sides of the body (Figs 4 & 6).

Figure 2 shows a character commonly used to separate *Scutigerella* and *Symphylella*, namely a cavity under the last scutum, between the cerci, of *Scutigerella*. The cavity is completely lacking in *Symphylella*.

MOUTHPARTS

Figures 7–12 illustrate the mouthparts and front of the head. Structural differences were expected because *Scutigerella* is often phytophagous and a crop pest, whereas *Symphylella* usually eats fungi and other small organisms. The most distinct difference is the much larger dorsal pad at the front of the head of *Symphylella* (Fig. 7) than of *Scutigerella* (Fig. 8). The pad in *Symphylella* bears grooves corresponding to setae and rather broad, pointed appendages, but *Scutigerella* has no grooves or setae and the appendages are narrower and finger-like. Both genera have well developed teeth on the first maxillae and three lobes at the anterior end of both second maxillae (labium) (Fig. 12); these appear to bear sense organs. The vestigial palps on the first maxillae are clearly visible (Fig. 10). The form of the mandibles could not be distinguished in intact specimens as they are obscured by the maxillae. The setae on the second maxillae are sculptured with longitudinal grooves (Fig. 11).

The most obvious difference is the smooth scale-like patterns on the cuticle of all parts of the mouthparts of *Scutigerella* (Fig. 10) compared with a densely granular appearance in *Symphylella* (Fig. 9).

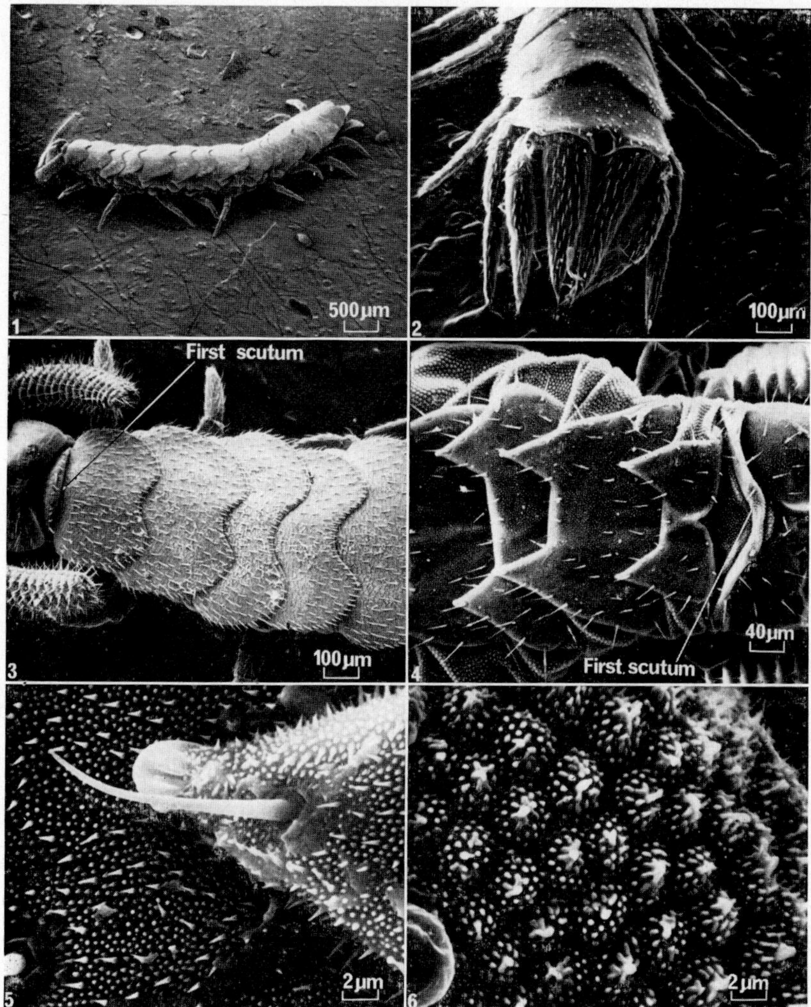

Figs 1–6. 1. *Scutigerella immaculata* Newport showing the basic scutal pattern. 2. The posterior of *Scutigerella* showing the characteristic cavity. 3. First five scuta of *Scutigerella*. 4. First five scuta of *Symphylella vulgaris* Hansen. 5. Detail of the tip of a scutum of *Symphylella*. 6. Detail of the side of the body of *Symphylella*.

LEGS, STYLI AND COXAL SACS

As Figs 13–18 show, one of the principal differences between the genera is again the scaly nature of parts of the legs of *Scutigerella* (Figs 13 & 15) compared with the densely pilose legs of *Symphylella*

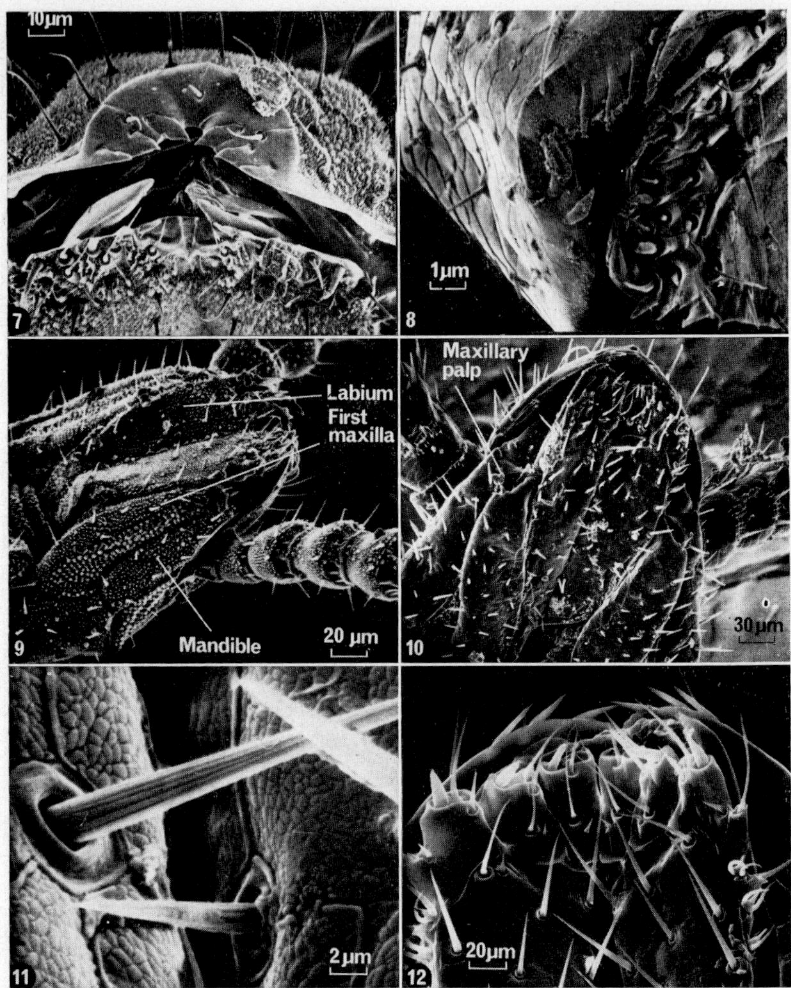

Figs 7–12. 7. Head of *Symphylella* showing a large dorsal pad. The teeth of the second maxillae are also visible. 8. Reduced dorsal pad on the head of *Scutigerella*. 9. Ventral view of the mouthparts of *Symphylella*. 10. Ventral view of the mouthparts of *Scutigerella*. Note the vestigial maxillary palps. 11. Detail of the setae on the second maxillae of *Scutigerella*. 12. Second maxillae of *Scutigerella* showing three lobes at the anterior of each, these lobes possibly bear sense organs.

(Figs 14, 16 & 18). *Scutigerella* has very distinct styli (Figs 13, 15 & 17) whereas *Symphylella* has none. The mouths of the coxal sacs of *Scutigerella* are smooth (Fig. 17) whereas those of *Symphylella* are pilose. Figure 14 shows a coxal sac of *Symphylella* with its aperture wide open.

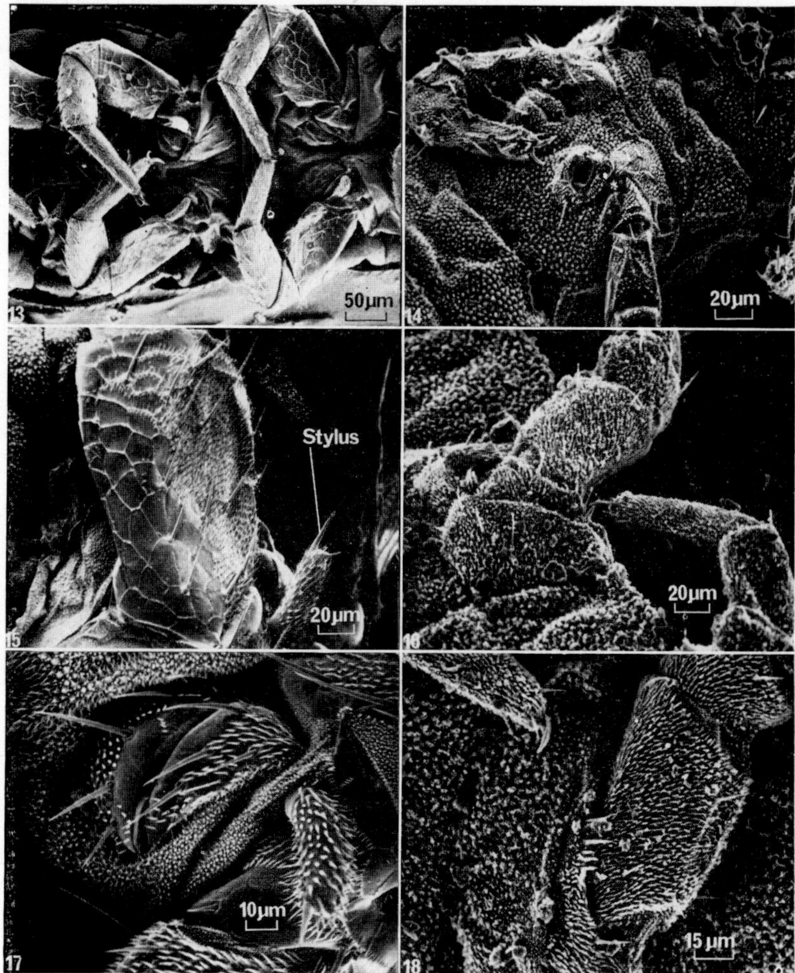

FIGS 13–18. 13. Ventral surface of *Scutigerella*. 14. Ventral view of *Symphylella* showing a coxal sac with an open aperture. 15. Detail of a leg (prefemur) and stylus of *Scutigerella*. 16. Detail of a leg of *Symphylella*. 17. Detail of a coxal sac and stylus of *Scutigerella*. 18. Detail of a leg of *Symphylella*.

This may support the hypothesis of Williams (1907) and Verhoeff (1933) that the coxal sacs have a respiratory function.

POST-ANTENNAL ORGAN

Michelbacher (1938) said the post-antennal organ of *Scutigerella immaculata* was a sac at the base of the antenna with a single opening to the exterior. We can distinguish openings only in specimens of *Scutigerella* and there appear to be five openings rather than one (Figs 19 & 20).

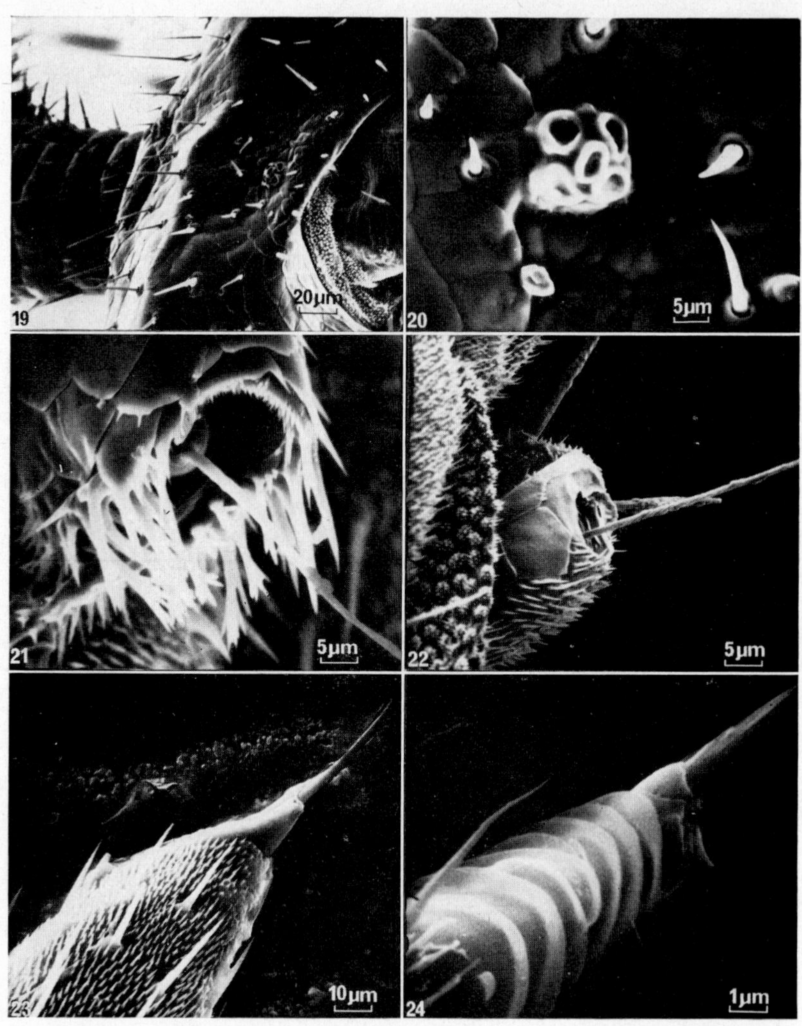

FIGS 19–24. 19. Head of *Scutigerella* showing the post-antennal organ. 20. Detail of the post-antennal organ shown in Fig. 19. 21. Sense calicle of *Scutigerella* showing the bulb at the base of the sensory seta and the branched setae around the cup-like cavity. 22. Sensory calicle of *Symphylella* showing the longitudinal, spiral sculpturing and the absence of branched setae around the cavity. 23. Cercus of *Scutigerella* with a smooth pre-terminal area. 24. Pre-terminal area of the cercus of *Symphylella* showing a plate-like structure and the stout terminal seta.

CALICLES

On either side of the posterior end of the body are long, sensory hairs with swollen bases that emerge from cup-like cavities (trichobothria). These have generally been assumed to be tactile (Hansen, 1903; Michelbacher, 1938). They differ considerably between the two genera; the cavity of *Scutigerella* is surrounded by a dense ring of intricately branched setae or projections from the overlapping plates or scales that surround it (Fig. 21), but that of *Symphylella* has only a few plates and no branched setae around the cavity (Fig. 22). The long hairs have swollen bases in both genera; no sculpturing can be distinguished in *Scutigerella* but there is a spiral, longitudinal groove in the long hair of *Symphylella*.

CERCI

These are the posteriorly directed last pair of appendages. They are covered in large and small setae and terminate in a long, stout, articulated seta. The structure of the tips of the cerci differs greatly in the two genera; in *Scutigerella* the pre-terminal area is quite smooth (Fig. 23) whereas in *Symphylella* it consists of a number of overlapping plates or scales that appear as rings under light microscopes (Fig. 24).

Both the calicles and cerci differ from other characters; whereas most parts of the body of *Scutigerella* bear scales and *Symphylella* is pilose, the opposite is true of the calicles and cerci.

References

Edwards, C. A. (1959). Keys to the genera of *Symphyla*. *J. Linn. Soc. Lond.* **44**: 164–169.

Edwards, C. A. & Belfield, W. (1967). A new genus and species of Symphyla, *Neosymphyla ghanensis*, with comments on Symphyla segmentation. *Revue Ecol. Biol. Sol* **4**: 517–521.

Hansen, H. G. (1903). The genera and species of the order Symphyla. *Q. Jl microsc. Sci.* **47**: 1–101.

Michelbacher, A. E. (1938). The biology of the garden centipede, *Scutigerella immaculata*. *J. Agric. Sci., Calif. Agric. exp. Stn* **11**: 55–148.

Ribaut, H. (1931). Observations sur l'organisation des Symphyles. *Bull. Soc. Hist. nat. Toulouse* **62**: 443–465.

Verhoeff, K. W. (1933). Symphyla and Pauropoda. *Bronn's Kl. Ordn. Tierreichs* Band **5**: Abt. 2: Buch 3: Lief 1: 1–120.

Williams, S. R. (1907). Habits and structure of *Scutigerella immaculata* Newport. *Proc. Boston Soc. nat. Hist.* **33**: 461–485.

Discussion

Demange: (Trans. Haacker) Have you found structures which may be of use in systematics; do you think this technique of scanning electron microscopy might be used to detect characters which have not been found by normal microscopy?

Turner: Yes. Details of the surface not easily visible in the light microscope may be quickly identified in the scanning electron microscope although probably by the time you have obtained details of the body surface you can identify the specimen by more conventional methods. The SEM helps to provide better structural details, e.g. the number of setae on the legs, and the method is simple and rapid—only 15 min are required to prepare a living specimen—this is probably the greatest advantage of the method.

Curry: Would freeze-drying methods stop the antennae from collapsing?

Turner: We have not yet looked at freeze-dried specimens. I have looked at some which were critical-point-dried; this is probably better than freeze-drying but the specimens were still collapsed; symphylans have very soft antennae.

Curry: Surely these things still contain water and will tend to distort under high vacuum?

Turner: Most specimens were distorted especially when viewed at low magnifications but this distortion is not so important at higher magnifications.

Dohle: I did not understand the structure of the post antennal organ. It looked as though there was no hole.

Turner: Yes, it appears so under the SEM. Whether the structure was the post antennal organ or not we are not quite sure. I certainly did not see a hole in any of the specimens examined.

Symp. zool. Soc. Lond. (1974) No. 32, 143–161.

THE SEGMENTATION OF THE GERM BAND OF DIPLOPODA COMPARED WITH OTHER CLASSES OF ARTHROPODS

WOLFGANG DOHLE

WE Zoologie I, Freie Universität Berlin, Berlin, German Federal Republic

SYNOPSIS

The author's results on the segmentation of the germ band of different diplopod species are reported and compared with other investigations. It is suggested that the segmentation of the germ band follows a general plan. The question of homonomous segmentation and the possible factors determining it is discussed in the light of experiments performed on other arthropods especially insects.

INTRODUCTION

The segmentation of the Diplopoda seems to be a rather puzzling question. In a previous paper on the embryology of *Glomeris marginata*, *Orthomorpha gracilis*, *Polydesmus complanatus*, and *Polyxenus lagurus* (Dohle, 1964) I expressed the view that all Diplopoda originally have the same segmentation of the germ band. All former work, on embryonic as well as on adult morphology, was full of contradictions. More recently there appeared some papers (Demange, 1967; Bodine, 1970) that offered completely different interpretations to my own. I therefore reinvestigated my material and preparations and began a new investigation on *Ommatoiulus* (= *Schizophyllum*) *sabulosus aimatopodus* (Risso).

GERM BAND FORMATION

The cleavage of the egg of Diplopoda leads to a blastoderm that completely surrounds the yolk. Energids have, as in most myriapods and insects, remained in the interior as yolk-cells. Eventually the blastoderm cells on the future ventral side become more crowded, the cells showing a columnar shape in sagittal sections. This thickened area of the blastoderm will become the ectodermal germ band. The posterior end of the future germ band is marked by a plug of cells that pushes into the yolk and which can be compared with the cumulus primitivus of *Scolopendra* and spiders.

The mesoderm migrates into the yolk from the plug, perhaps also from intersegmental spaces as Pflugfelder (1932: 671) described. The first indication of segmentation is a thickening of ectodermal cells and an arrangement of the mesoderm into transverse bands. These bands hollow-out to form coelomic sacs. When the segmentation is clearly visible in whole-mounts, there are already the vestiges of the following parts of the embryo (Fig. 1): head lobes (kl); a clypeal rudiment (cl) in

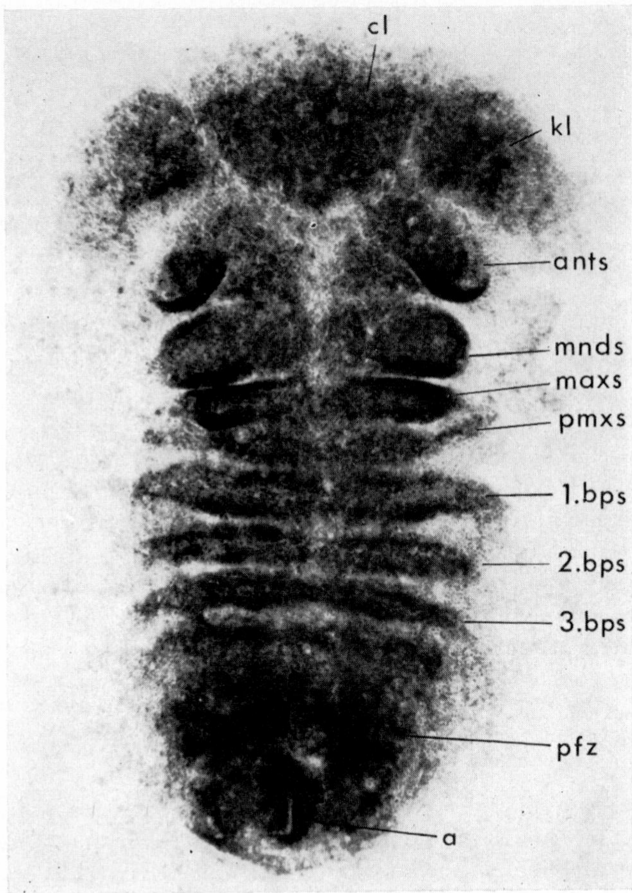

FIG. 1. *Glomeris marginata*. Early germ band.

the anterior median part; paired antennal vestiges (ants) with ganglia not connected in the median line, but separated by the posterior part of the clypeal cell mass, which will give rise to the stomodaeal invagination; the premandibular (= intercalary) segment in the posteromedian space behind the antennae; mandibular (mnds), maxillary (maxs), and post-maxillary (pmxs) segments; three trunk segments (1 bps–3 bps); an undifferentiated proliferation zone (pfz); at the posterior end the anus (a) as the aperture of the proctodaeal invagination.

APPENDAGE BUD FORMATION

I will not give the details of the differentiation of the first segmental rudiments, but will describe a later stage of *Glomeris* in which the first

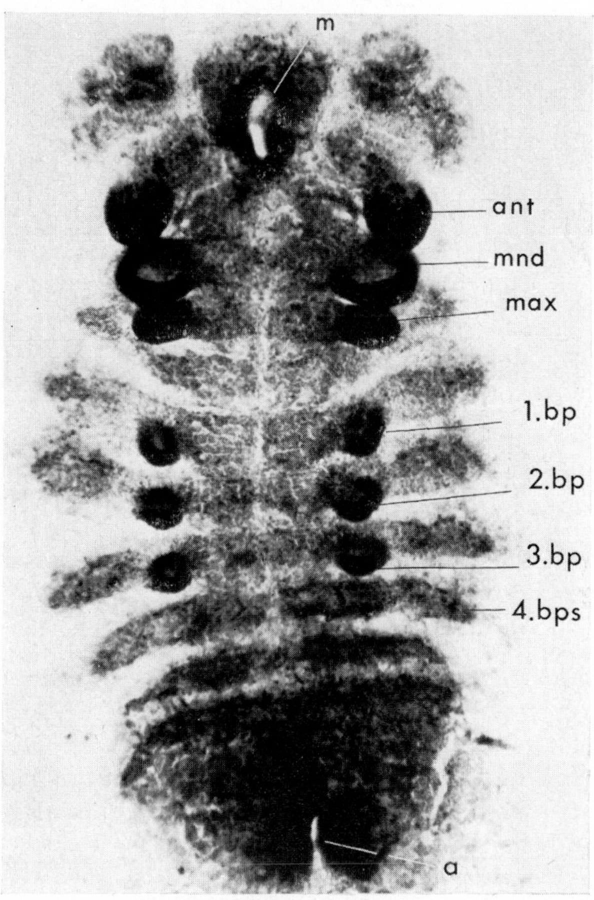

FIG. 2. *Glomeris marginata.* Germ band with appendage buds.

appendage buds are formed (Fig. 2). We see: the stomodaeal invagination (m) in a separated clypeal field; head lobes divided into two parts, each with a ganglion groove; antennae (ant) with their ganglia; a premandibular segment behind the stomodaeum; mandibles (mnd); first maxillae (max); a post-maxillary segment without appendages, but with well defined lateral ectoderm and mesoderm, thus resembling the following trunk segments; three trunk segments with appendage buds (1 bp–3 bp); a fourth trunk segment (4 bps); a fifth trunk segment recently separated from the proliferation zone; the anus (a).

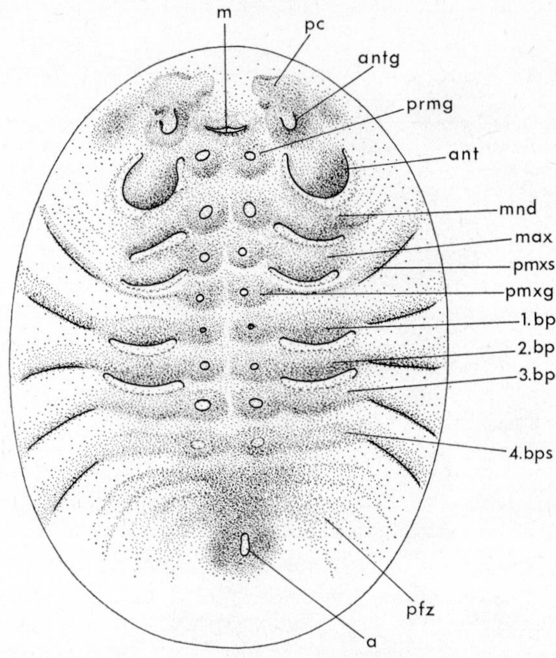

Fig. 3. *Ommatoiulus* (= *Schizophyllum*) *sabulosus aimatopodus*. Germ band with appendage buds.

Corresponding stages of *Ommatoiulus* (Figs 3, 4) show the same principal characteristics: antennae, mandibles, and maxillae having been formed; a post-maxillary segment without appendages; first to third leg pair buds; a fourth leg pair segment being formed. In contrast to *Glomeris* the ganglia are more concentrated and with ganglion grooves, and the lateral parts of the segments have already surrounded the yolk on the dorsal side. The post-maxillary segment (pmxs) marks the posterior border of the future head.

Further differentiation of segmental structures and separation of new segments from the proliferation zone in *Glomeris* leads to a stage

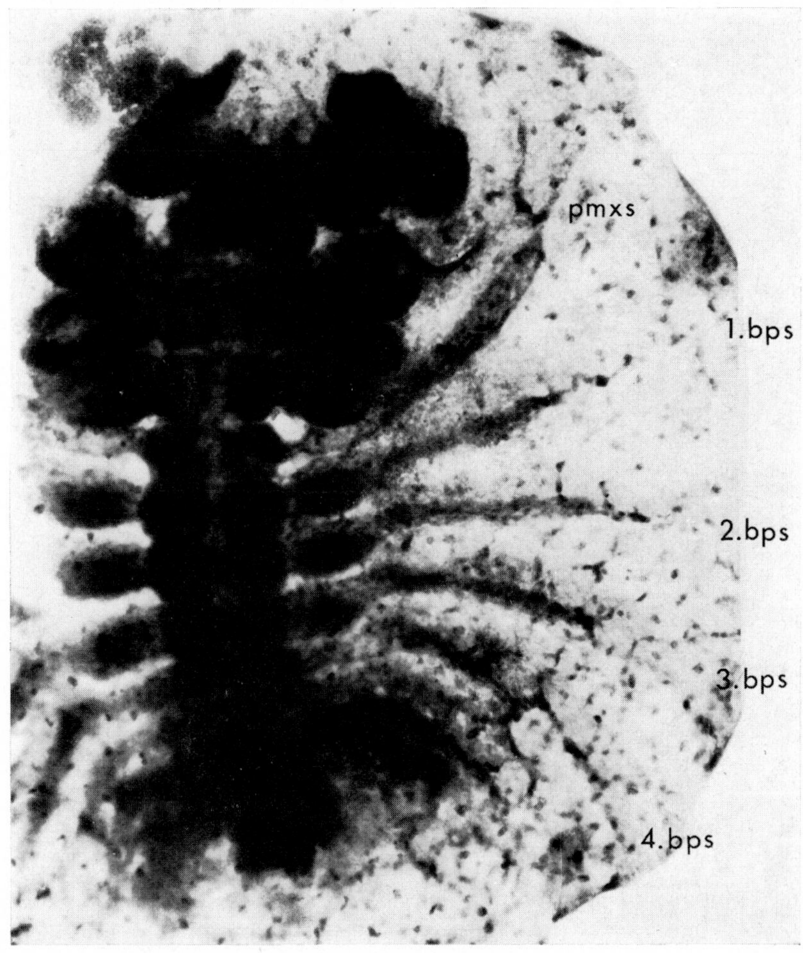

FIG. 4. *Ommatoiulus* (= *Schizophyllum*) *sabulosus aimatopodus*. Approximately the same stage as Fig. 3.

shown in Fig. 5. Anteriorly there is the protocerebrum with lateral optic lobes; the labrum overlapping the stomodaeal opening; the antennae and their ganglia; and behind the mouth are the ganglia of the premandibular segment (never has a trace of appendages been found on

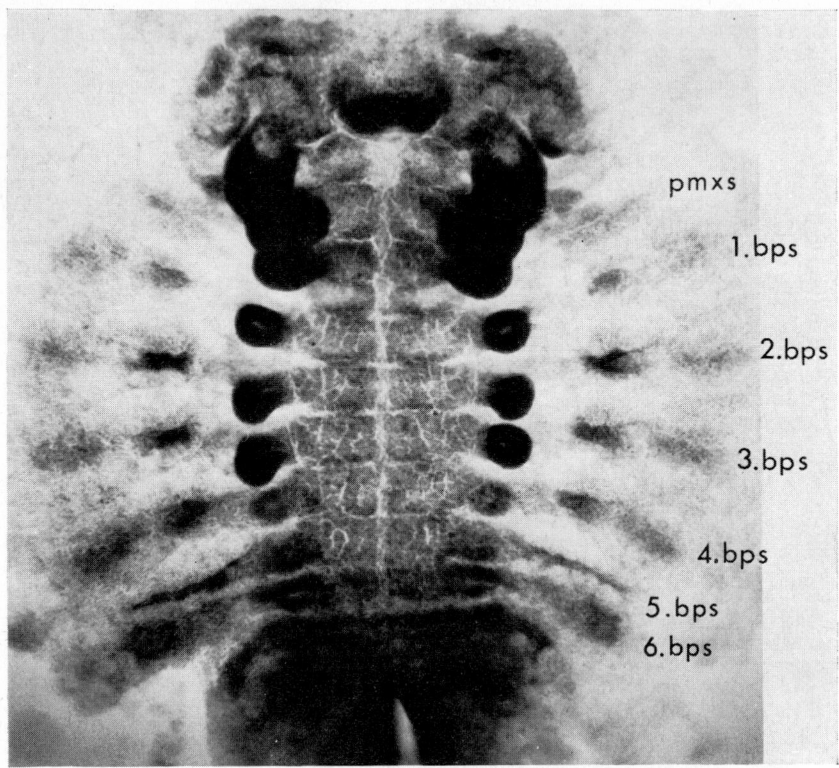

Fig. 5. *Glomeris marginata*. Embryo just at the beginning of ventral flexure.

this segment in diplopods). The mandibles and maxillae can be seen but the post-maxillary segment is partly obscured in the median line because blastokinesis is beginning and a ventral furrow is forming; however, the lateral parts of this segment are clearly visible (pmxs).

In comparison with the stage described before (Fig. 2) the lateral parts of the trunk segments (1 bps–4 bps) have migrated towards the dorsal side. The tips of the first three leg pairs are longer, a fourth leg pair is being formed; the lateral parts of the 5th and 6th trunk segment lie very close together (5 bps–6 bps). Eventually they will unite to form one tergite, the first so-called "double segment" (tergite V).

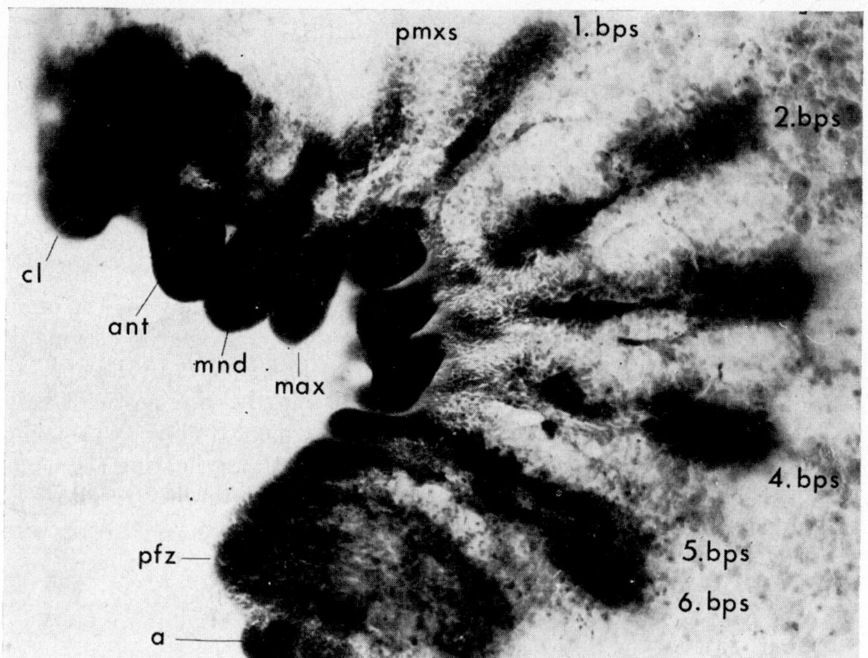

FIG. 6. *Glomeris marginata*. Embryo during ventral flexure.

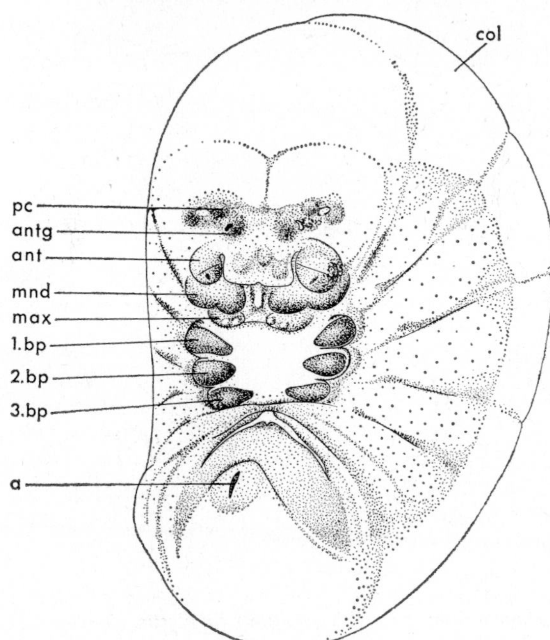

FIG. 7. *Ommatoiulus* (= *Schizophyllum*) *sabulosus aimatopodus*. Pupoid separated from the embryonic membrane.

FORMATION OF TERGITES

Figure 6 shows a slightly older stage during ventral flexure. The connexions between the three parts of a segment, the ganglia, appendages and lateral lobes, are still detectable without doubt. They are ascertained by serial sections (Dohle, 1964). The lateral part of the post-maxillary segment (pmxs) marks the hind border of the head; the lateral part of the first leg bearing segment (1 bps) will become the hind border of tergite I or the collum; the second leg bearing segment (2 bps) will form tergite II, which is known in *Glomeris* by its special size and function as "Brustschild"; the third leg bearing segment will form tergite III; the fourth leg bearing segment (4 bps) will form tergite IV; the lateral parts of the fifth and sixth leg bearing segment (5 bps + 6 bps) will form only one pleurite and one tergite (tergite V). The same holds for the following segments that are separated from the

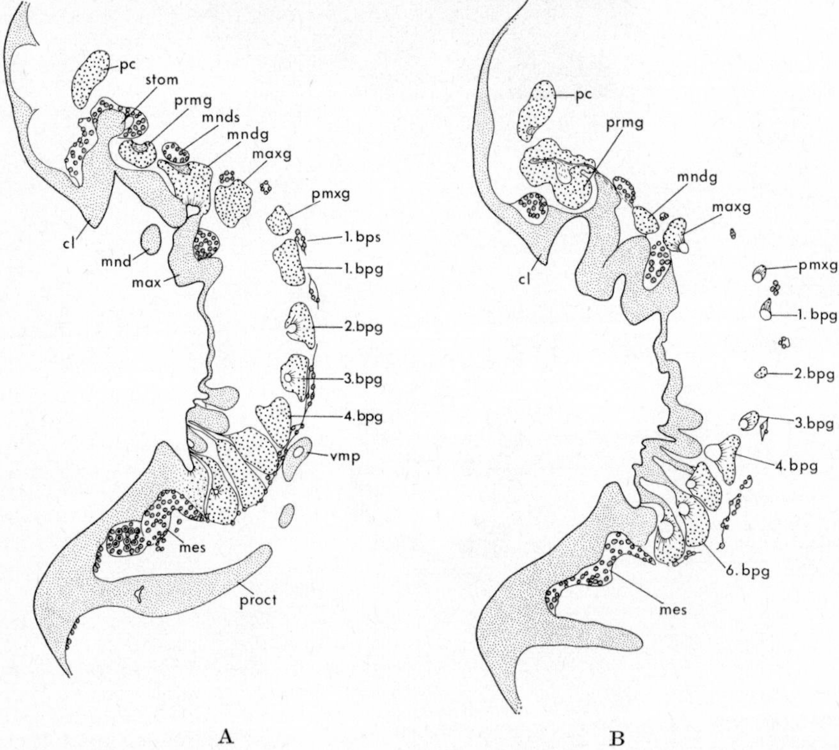

FIG. 8 A–D. *Ommatoiulus* (= *Schizophyllum*) *sabulosus aimatopodus*. Pupoid. Four sections from a series of sagittal sections showing the connexion of the segmental mesodermal somites in the median (A–B), the appendicular (C) and the lateral part (D). Ectoderm finely stippled, ganglia roughly stippled, mesoderm cellular.

proliferation zone (pfz); the seventh and eighth leg bearing segments will unite to form tergite VI, the eighth and ninth will form tergite VII, and so on.

These relations: the first leg bearing segment forming the hind border of the collum, the second to fourth leg bearing segments forming the hind border of one tergite each and the fifth and sixth leg bearing segments uniting, in their lateral parts, to form only one pleurite and one tergite, apply also to species which have the tergite, pleurites and two sternites enclosed in a complete ring. I have verified these relations for *Orthomorpha gracilis*, *Polydesmus complanatus* and recently for *Ommatoiulus sabulosus* by whole-mounts (Fig. 7) and serial sections (Fig. 8).

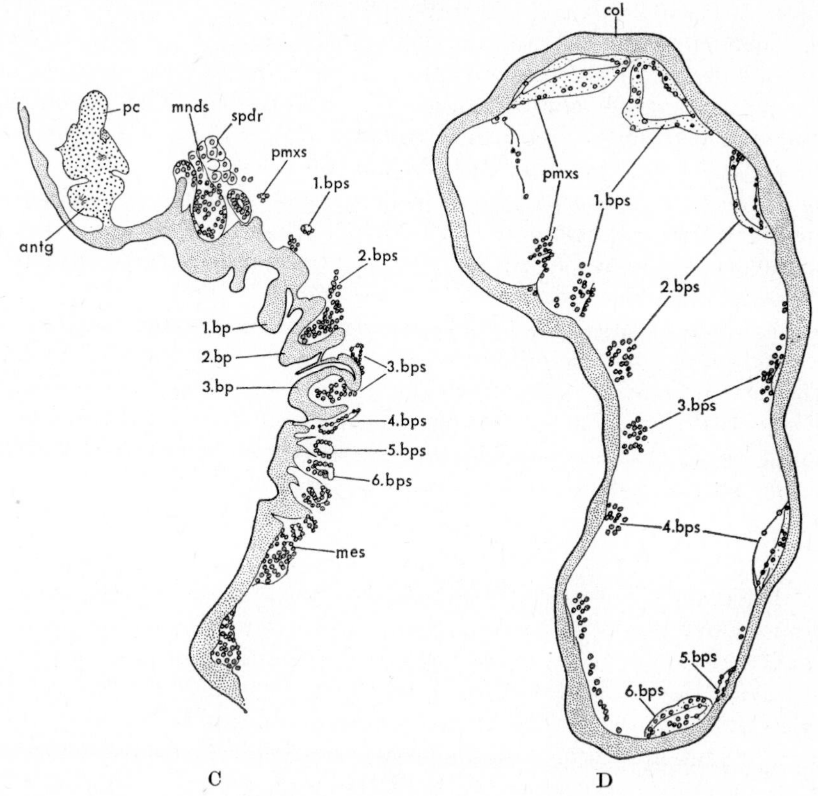

DISCUSSION

Many investigators do not agree with these conclusions, and this deserves further comment.

There are mainly three questions around which differences of opinion have gathered.

1. Is the gnathochilarium composed of one or of two pairs of appendages?
2. Does the collum belong to the first leg bearing segment or to the post-maxillary segment?
3. Are the first trunk segments simple or double segments?

Gnathochilarium

Nearly all embryologists agree that there is but one pair of maxillae and that appendage buds on the post-maxillary segment are lacking (Metschnikoff, 1874; Lignau, 1911a,b; Heymons, 1897; Silvestri, 1898, 1903, 1933, 1950; Dohle, 1964; Bodine, 1970).

There are only two exceptions. Pflugfelder (1932: Fig. 9) reported an appendage pair on the post-maxillary segment in *Platyrrhacus amauros*. I suppose that he figured a stage before appendage bud formation. Embryos with only three trunk segments usually have not developed appendage buds. In other Polydesmoidae (*Polydesmus abchasius*, *P. complanatus*, *Orthomorpha gracilis*) the ventral furrow is formed before bud formation. Judging from Pflugfelder's figure, there is not yet a distinction between the appendicular and the lateral parts of the trunk segments. The observations of Robinson (1907), who claimed to have seen a pair of appendage buds between mandibles and maxillae in *Archispirostreptus*, have already been corrected by Silvestri (1933). Therefore, it may be accepted that Diplopoda have—as the Pauropoda (Tiegs, 1947)—but one pair of maxillae and that the gnathochilarium is composed of the maxillae and their sternal part. The sternum of the legless post-maxillary segment forms the gula.

Collum

The majority of embryologists were of the opinion that the collum, i.e. the first tergite of the diplopods, which has a characteristic triangular shape and never forms a ring, is the tergite of the post-maxillary segment (Heymons, 1897; Silvestri, 1898, 1903, 1950; Pflugfelder, 1932; Bodine, 1970). The truth is rather difficult to establish. The segments are first formed on the ventral side, the tergites being formed by migration of the lateral parts of the segments to the dorsal side. Therefore one must investigate this migration step by step. As it is not yet possible to observe the migration directly in living embryos, a very full series of stages is needed, and, in addition to external observation of fixed and stained embryos, one must reconstruct the connexion of the mesodermal parts of a segment very carefully. I cannot understand, therefore, how it seemed possible for Bodine (1970), for instance, to make her statements without histological work. Lignau (1911a,b), in an exhaustive investigation of *Polydesmus*, first proved that the collum belongs to the first leg pair. All embryological work that refutes this statement, seems to me not very convincing; nor does refutation based on morphological work, for there is no competent investigation of the innervation of the integument. In embryos of *Glomeris*, *Orthomorpha* and *Ommatoiulus* (Fig. 8) the connexion of the mesoderm of the first leg bearing segment with the mesoderm underlying the hind border of the collum has been reconstructed by serial sections.

But the whole question is not so simple. First of all one does not know whether it is really the lateral ectoderm that migrates to the dorsal side, or whether formerly extra-embryonic ectoderm is induced to form definite dorsal ectoderm. Has the mesoderm a segment-forming influence on the ectoderm or vice-versa, or are the two independent of one another?

What we can observe is that the furrow which delimits the head from the collum dorsally runs through the thickened lateral ectoderm of the post-maxillary segment. Most of the post-maxillary lateral mesoderm inserts on the anterior part of the collum (Fig. 8D). The collum is thus a complex structure. Its anterior border is built-up by post-maxillary ectoderm, its posterior border by ectoderm of the first leg bearing segment. Often a forward shifting of anterior legs with regard to other segmental structures has been assumed. This assumption does not at all agree with the facts. In the course of ventral flexure and formation of the head a marked concentration of the segments takes place (Fig. 9A–D). Especially the lateral parts of the post-maxillary segments and of the first leg bearing segment are drawn forwards, as well as the median part of the mesoderm (Fig. 8, 1 bps). Therefore, the legs insert rather on the posterior ridge of the segment in later stages (Fig. 6, Fig. 9D).

"Diplosegments"

The first leg bearing segment forms the hind part of the collum, the second leg bearing segment forms the hind part of tergite II, the third leg bearing segment forms tergite III, the fourth forms tergite IV, and the fifth plus the sixth leg bearing segments unite in their lateral ectodermal parts to form tergite V, i.e. the first double tergite.

Considering the architecture of the body of adult ring-forming diplopods, this observation seems to be most improbable. But I have

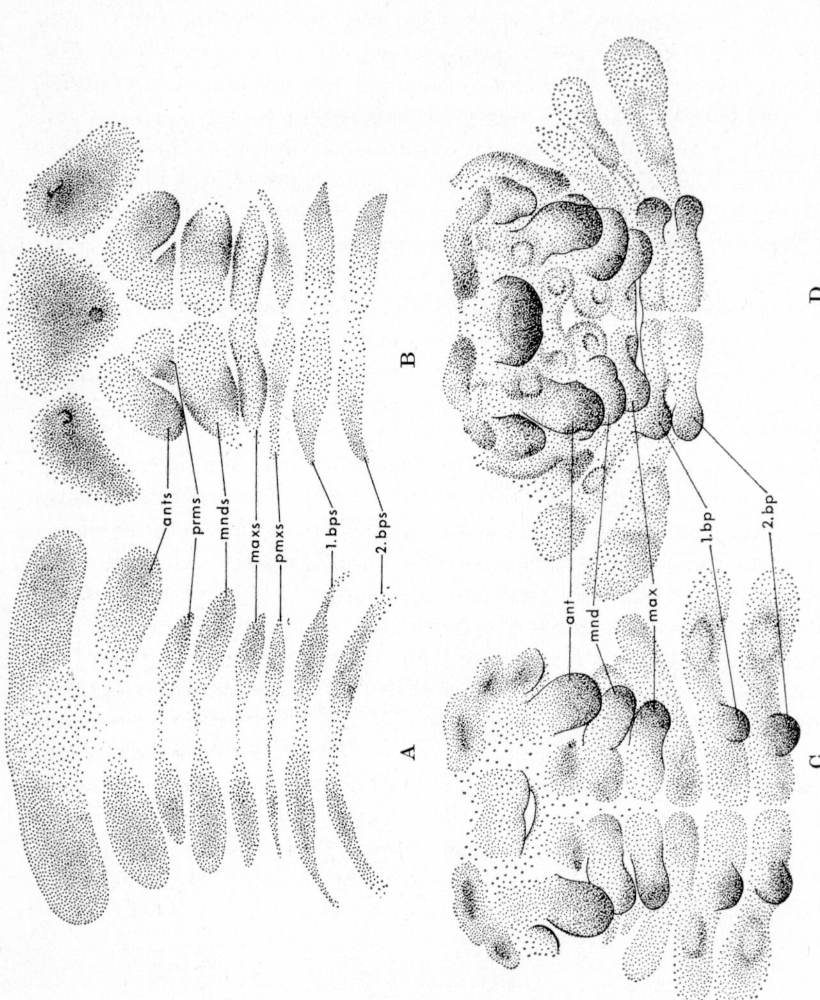

FIG. 9 A–D. *Glomeris marginata.* Anterior part of four stages, showing the contraction of the segments of the head region and the forward shifting of the lateral parts of the postmaxillary and first leg bearing segment in relation to their median parts.

verified this observation in *Orthomorpha* and *Ommatoiulus*, two ring-forming species of the orders Polydesmoidea and Iuliformia. We must take into account that through formation of firm rings which include the tergite, the pleurites and two sternites, there is always one "false" sternite included in a ring. Tergite IV, having been formed mainly by the 4th trunk segment, includes the third leg pair, tergite V having been formed by the 5th and 6th trunk segments, includes the fourth and the fifth leg pairs, and so on.

The question, which segments are double segments and which are simple segments, proves to be irrelevant. In the trunk there are segments which possess all the criteria of their segmental nature whether they contribute to the formation of tergites individually or in pairs. Each of the first four segments form the hind part of one tergite while pairs of the succeeding segments, 5th and 6th, 7th and 8th, 9th and 10th and so on, unite to form the hind parts of tergites V, VI and VII.

Segmentation in general

At this stage one must ask: what is a segment and what is segmentation?

One may define segmentation in purely morphological terms as the repetition of structures along a main axis. It does not in principle matter where one draws the limit, but one may for convenience draw the limit between two segments in the intersegmental membrane. This homonomy on the ventral side is sometimes even maintained in arthropods that have an accentuated dorsal heteronomy. The reason may be, as in diplopods, that the segmentation is first laid down on the ventral side of the embryo and that, while the dorsal parts are formed, other influences may be superimposed on this primary homonomous segmentation.

It is true that the segmentation even in the germ band is never absolutely homonomous. This may be due to other presegmental determining factors. But the segmental nature of many structures, e.g. of the head, could never have been established without the observation of segmentation of the embryo.

One may remember that the main criteria for the segmental nature of arthropod structures formulated by Tiegs and accepted by many authors, are clearly embryological criteria: intersegmental furrows, mesodermal somites, ganglionic centres with ventral grooves (=ventral organs), and appendage buds. The intersegmental furrows often disappear, the borders of the adult sclerites not matching the furrow. The mesodermal somites differentiate into musculature that inserts on more than one tergite. The metameric origin of the nerve cord is sometimes only faintly visible in adults; the nerves of one ganglion may innervate muscles of other segments. Appendage buds are in some cases, e.g. on the premandibular segment in the symphylan *Hanseniella* (Tiegs, 1941) or in the collembolan *Orchesella* (Bruckmoser, 1965), completely reduced in the course of development. This means that it is necessary to take the embryological segmentation as a basis and to describe the alterations and the possible destruction of primary homonomous segmentation in later phases of development. It is not sufficient to take a supernumerary sclerite or an additional muscle in the adult as an indication of a reduced segment.

But segmentation is not a problem of geometry, but of developmental physiology. There have been many experimental attempts to discover the reasons for the repetition of structures at certain intervals. In insects, transplantation experiments have led to the conclusion that there must be repeated gradients of some determining substance (Locke, 1967; Lawrence, 1970; Stumpf, 1966). This means that a certain level of concentration of this substance is responsible for the development of a certain structure. A "neutral" point is reached in the intersegments.

Where is the segmentation–determining gradient localized? In contrast to what one would suppose, it seems to be localized in the ectoderm rather than in the mesoderm. In embryos of *Chrysopa* and *Leptinotarsa*, ectoderm differentiates into segmental structures without underlying mesoderm; the mesoderm does not differentiate when the corresponding part of ectoderm has been destroyed by thermocautery (Bock, 1942; Seidel, Bock & Krause, 1940; Haget, 1953).

Is then segmentation a preformed pattern localized in the cortex or in the blastoderm?

If it were so, reversions of polarity of the germband, as were experimentally produced by pushing pole material to the anterior end in *Euscelis* (Sander, 1961), by centrifugation in *Chironomus* (Yajima, 1960) or by U.V. light in *Smittia* (Kalthoff & Sander, 1968), would be impossible. Schnetter (1936) produced dwarf honey bee embryos by ligation of the egg, Küthe (1966) revealed a field pattern migrating on the ventral side without any obvious migration of cells. By this, cells with a common genealogical origin may come under the influence of different segmental morphogenetic units. It is well known that in insects most of the mesoderm of the neck region comes from the somite of the prothorax (Scholl, 1969). But according to the analysis of the teratological defects in *Tachycines* by Wada (1966) the neck musculature is only differentiated under the influence of a labial morphogenetic unit.

But the segmentation may not even be determined by the mode of germ band formation. In higher Crustacea it is possible to follow the exact lineage of each cell produced by the teloblasts up to the differentiation of ganglia, and appendage buds (Dohle, 1972). The intersegmental furrows do not match the genealogical limits between the descendants of one cell row. On the contrary, the genealogical limit runs transversely through an appendage bud.

I think that we have to keep all these questions in mind when performing an investigation on differentiation of segmental structures. When describing the alterations of structure we do not see the responsible factors. Therefore, the concept of descriptive and comparative embryology is mainly a concept of integrity and self-differentiation of rudiments. If there is induction, substitution, histolysis and regeneration in the course of metamorphosis, the embryologist is always somewhat baffled, and severe objections have been put forward against the homology of structures that do not have exactly the same development. However what is important for differentiation is the genetic programme which leads to the visible structures at the periphery. Cell material may only be competent or even indifferent for a certain differentiation, far from being determined irreversibly. But in normal development this cell material always becomes the same adult structure. This may be independent of the strict cell-lineage. This means that a structure connected with another even when forming the same "organ" has not necessarily the same origin. I cannot say much about the ontogenetic origin without knowing it. Adult morphology is not a good basis for drawing conclusions on the primary embryonic segmentation.

CONCLUSION

Finally, I would like to point out some possible investigations that might help to reconcile the opposite opinions on the segmentation of the Diplopoda.

A careful examination of the differentiation of the muscles from the mesodermal somite is highly desirable.

It would be most helpful to label small areas of the lateral ectoderm of the segments by irradiation in order to determine their fate on the dorsal side.

Morphological work on adults should no longer be done without investigating the innervation of the muscles and of the integument; the exact origin of nerves must be determined by sections.

The formation of rings in a polydesmoid or iuliform should be carefully examined, on the basis of external morphology as well as by serial sections.

These investigations would at least help to define the areas in the adult that are occupied by cells descendent from the embryonic area that we call a segment.

ACKNOWLEDGMENTS

I wish to thank Mr J. G. Blower and Dr C. H. Brookes for correcting my English manuscript, and Mr J. G. Blower and Dr P. D. Gabbutt for helpful suggestions.

References

Bock, E. (1942). Wechselbeziehungen zwischen den Keimblättern bei der Organbildung von *Chrysopa perla* (L.). *Wilhelm Roux' Archiv EntwMech* **141**: 159–247.

Bodine, M. W. (1970). The segmental origin of the appendages of the head and anterior body segments of a Spiroboloid milliped, *Narceus annularis*. *J. Morph.* **132**: 47–68.

Bruckmoser, P. (1965). Embryologische Untersuchungen über den Kopfbau der Collembole *Orchesella villosa* L. *Zool. Jb.* (Anat.) **82**: 299–364.

Demange, J.-M. (1967). Recherches sur la segmentation du tronc des Chilopodes et des Diplopodes Chilognathes. *Mém. Mus. nat. Hist. nat.* (A) **44**: 1–188.

Dohle, W. (1964). Die Embryonalentwicklung von *Glomeris marginata* (Villers) im Vergleich zur Entwicklung anderer Diplopoden. *Zool. Jb.* (Anat.) **81**: 241–310.

Dohle, W. (1972). Über die Bildung und Differenzierung des postnauplialen Keimstreifs von *Leptochelia spec.* (Crustacea, Tanaidacea). *Zool. Jb.* (Anat.) **89**: 503–566.

Haget, A. (1953). Analyse expérimentale des facteurs de la morphogénèse embryonnaire chez le Coléoptère *Leptinotarsa*. *Bull. biol. Fr. Belg.* **87**: 123–217.

Heymons, R. (1897). Mittheilungen über die Segmentirung und den Körperbau der Myriopoden. *Sber. preuss. Akad. Wiss.* **1897**: 915–923.

Kalthoff, K. & Sander, K. (1968). Der Entwicklungsgang der Missbildung "Doppelabdomen" im partiell UV-bestrahlten Ei von *Smittia parthenogenetica* (Dipt., Chiromidae). *Wilhelm Roux' Archiv EntwMech* **161**: 129–146.

Küthe, H.-W. (1966). Das Differenzierungszentrum als selbstregulierendes Faktorensystem für den Aufbau der Keimanlage im Ei von *Dermestes frischi* (Coleoptera). *Wilhelm Roux' Archiv EntwMech* **157**: 212–302.

Lawrence, P. A. (1970). Polarity and patterns in the postembryonic development of insects. *Adv. Insect Physiol.* **7**: 197–266.

Lignau, N. (1911a). Über die Entwicklung des *Polydesmus abchasius* Attems. *Zool Anz.* **37**: 144–153.

Lignau, N. (1911b). Die Embryonalentwicklung des *Polydesmus abchasius* Attems. Ein Beitrag zur Morphologie der Diplopoden. *Zap. novoross. Obshch. Estest.* **38**: 1–249.

Locke, M. (1967). The development of patterns in the integument of insects. *Adv. Morphogen.* **6**: 33–88.

Metschnikoff, E. (1874). Embryologie der doppeltfüssigen Myriapoden (Chilognatha). *Z. wiss. Zool.* **24**: 253–283.

Pflugfelder, O. (1932). Über den Mechanismus der Segmentbildung bei der Embryonalentwicklung und Anamorphose von *Platyrrhacus amauros* Attems. *Z. wiss. Zool.* **140**: 650–723.

Robinson, M. (1907). On the segmentation of the head of Diplopoda. *Q. Jl microsc. Sci.* **51**: 607–624.

Sander, K. (1961). Umkehr der Keimstreifpolarität in Eifragmenten von *Euscelis* (Cicadina). *Experientia* **17**: 179–180.

Schnetter, M. (1936). Die Entwicklung von Zwerglarven in geschnürten Bieneneiern. *Zool. Anz.* Suppl. **9**: 82–88.

Scholl, G. (1969). Die Embryonalentwicklung des Kopfes und Prothorax von *Carausius morosus* Br. (Insecta, Phasmida). *Z. Morph. Tiere* **65**: 1–142.

Seidel, F., Bock, E. & Krause, G. (1940). Die Organisation des Insekteneies (Reaktionsablauf, Induktionsvorgänge, Eitypen). *Naturwissenschaften* **28**: 433–446.

Silvestri, F. (1898). Sulla morfologia dei Diplopodi III. Sviluppo del *Pachyiulus communis* (Savi). *Atti R. Accad. Lincei* (5) **7**: 178—180.

Silvestri, F. (1903). *Acari, Myriopoda et Scorpiones hucusque in Italia reperta.* Classis Diplopoda, **1** *Anatome.* Portici.

Silvestri, F. (1933). Sulle appendici del capo degli "Japygidae" (Thysanura entotropha) e rispettivo confronte con quelle dei Chilopodi, dei Diplopodi e dei Crostacei. *Int. Congr. Ent.* **5**: 329–343.

Silvestri, F. (1950). Segmentazione del corpo dei Colobognati (Diplopodi). *Int. Congr. Ent.* **8**: 571–576.

Stumpf, H. F. (1966). Über gefälleabhängige Bildungen des Insektensegmentes. *J. Insect Physiol.* **12**: 601–617.

Tiegs, O. W. (1941). The embryology and affinities of the Symphyla, based on a study of *Hanseniella agilis*. *Q. Jl microsc. Sci.* **82**: 1–225.

Tiegs, O. W. (1947). The development and affinities of the Pauropoda, based on a study of *Pauropus silvaticus*. *Q. Jl microsc. Sci.* **88**: 165–267.

Wada, Sh. (1966). Analyse der Kopf-Hals-Region von *Tachycines* (Saltatoria) in morphogenetische Einheiten II. Mitteilung: Experimentell-teratologische Befunde am Kopfskelett mit Berücksichtigung des zentralen Nervensystems. *Zool. Jb.* (Anat.) **83**: 235–326.

Yajima, H. (1960). Studies on embryonic determination of the Harlequin-fly, *Chironomus dorsalis* I. Effects of centrifugation and of its combination with constriction and puncturing. *J. Embryol. exp. Morph.* **8**: 198–215.

Abbreviations

a — anus
ant — antennae
antg — antennal ganglia
ants — antennal segment or somite
bp — leg pair
bpg — leg pair ganglia
bps — leg bearing segment or somite
cl — clypeus
col — collum (1st tergite)
kl — head lobes
m — mouth
max — maxillae (1st maxillae)

maxg — maxillary ganglia
maxs — maxillary segment or somite
mes — undifferentiated mesoderm
mnd — mandibles
mndg — mandibular ganglia
mnds — mandibular segment or somite
pmxg — post-maxillary ganglia
pmxs — post-maxillary segment or somite
prmg — premandibular ganglia
prms — premandibular segment or somite
proct — proctodaeum
stom — stomodaeum
vmp — Malpighian tubules

Discussion

MANTON: I would like to say a word, if I may, in appreciation of Dr Dohle's work. The beautiful slides he has shown to us, his sections and his comments on them all go to show that his work is extremely sound. As he has said, the basic segmentation of these animals is laid down in the embryo and that is where one must first seek the vital information. I think his work is very fine indeed and I would like to express that view as an embryologist.

DEMANGE: I would like to ask you a question more morphological than embryological. If I have understood you well the first ambulatory appendages are attached to the collum?

DOHLE: Yes.

DEMANGE: But is it not generally accepted that the gula is the sternite of the post-maxillary segment and the collum is the tergite of this segment?

DOHLE: The gula represents part, but not the whole, of the sclerotization of the sternal region of the post-maxillary segment. The lateral parts of this segment, which migrate dorsally, form the posterior border of the head and also the anterior border of the collum. This means that the neck is intrasegmental. The intersegmental limit between the post-maxillary segment and the segment of the first pair of appendages lies within the collum somewhat behind its anterior border. One cannot determine this limit more precisely. The lateral parts of the segments are masses of cells, of ectoderm and of mesoderm cells beneath. In preparations these masses of cells are more deeply coloured than the furrows between them, but these furrows are also composed of cells; the intersegmental furrow is not a cell-less space.

DEMANGE: Between embryologists and morphologists working on adults or subadults there is apparently a certain opposition of ideas. I think it is apparent that the morphologists cannot dispense with embryology and the embryologists cannot do without a knowledge of the morphology of adults. One can say, with Professor Denis, that one sees more things in an adult than in an embryo. It is evident that masses of cells in an embryo are difficult to understand. I think that there are between us false problems. If, nevertheless, the first segments of the body are really double as a morphologist would like to think, it is probably the result of evolution, evolution sufficient to suppress one of the segments. It is evident that a segment which has almost disappeared during the course of evolution is more difficult to discern in an embryo and the question remains open. I have raised these points of interest and I should like you to add to your exposition and the documents you have brought.

DOHLE: What I think is that there are perhaps two forms of segmentation. There is an embryonic segmentation and a secondary segmentation which is superimposed on it. In the Crustacea there is a distinct "segmentation" suggested by rows of cells which are proliferated by teloblasts. But the genealogical limits between the descendants of these rows of cells run transversely across the rudiments of ganglia and appendages and do not match the intersegmental furrows. These intersegmental furrows nevertheless, by analogy with transplantation experiments in insects, may be regarded as "neutral" points, as real limits between the segments of an adult. Because of this possibility of different forms of segmentation I think it is really necessary to describe in detail the differentiation of the somites, of the anterior as well as of the posterior somites. There is, for the anterior segments, only one pair of somites—this is very clear.

DEMANGE: Have you examined the formation of the nervous system in relation to ectodermal invaginations?

DOHLE: Yes, up to the formation of fibrils, and up to the formation of nerves.

DEMANGE: And you have seen only one pair of invaginations per neuromere?

DOHLE: One pair only.

SAHLI: How many neuromeres did you finally find in the head?

DOHLE: The protocerebrum is divided into two parts, one of which is more medial and the other more lateral. Then there are the antennal, premandibular, mandibular, maxillary and post-maxillary ganglia; that is all . . . there is no supplementary neuromere belonging, for example, to a superlingual segment as postulated by Chaudonneret (1950).*

* Chaudonneret, J. (1950). La morphologie céphalique de *Thermobia domestica* (Packard) (Insecte Aptérygote Thysanoure). *Annls Sci. nat.* (*Zool.*) (11) **12**: 145–302.

Symp. zool. Soc. Lond. (1974) No. 32, 163–190.

SEGMENTATION IN SYMPHYLA, CHILOPODA AND PAUROPODA IN RELATION TO PHYLOGENY

S. M. MANTON

Research Fellow, Queen Mary College, London, and Zoology Department, British Museum (Natural History) London, England.

SYNOPSIS

The fundamental segmentation of the major parts of the trunk of the various classes and orders of Myriapoda has been decisively shown by embryological data. Series of mesodermal somites are the determinants of segmentation, and this is followed by the formation of ganglia, segmental limbs and basic segmental patterns of musculature. Dorsally there are divergencies in the adult states which call for explanation. These are: (1) the extra tergites on at least segments 4, 6 and 8 in the Symphyla; (2) the well formed intercalary tergites and sternites in the Geophilomorpha; (3) the poorly defined intercalary tergites in the Scolopendromorpha; (4) heteronomy in tergite length in the Scolopendromorpha, Lithobiomorpha and Scutigeromorpha, the degree of heteronomy increasing in the scutigeromorph direction, and with long tergites over legs 7 and 8; (5) the absence of alternate tergites in the Pauropoda; (6) certain incompletely and erroneously described features of the muscular systems.

All these six phenomena are readily understandable as a result of a thorough examination of comparative morphology and comparative movements and capabilities of the various animals, together with a recognition of the habits of evolutionary significance in the various taxa, analysis of their locomotory mechanisms and other mechanisms of essential movements.

INTRODUCTION

In the absence of conclusive fossil records of the evolution of arthropods, the two most important sources of evidence concerning the past history of the phylum Uniramia, comprising the Onychophora, Myriapoda, and Hexapoda (Manton, 1972), are the studies of embryonic development and of functional morphology on a broad comparative basis. The latter is the newer approach and during the last 25 years spectacular advances have been made in our understanding of arthropodan relationships and evolution, although this was not the primary object of the work when it was originally undertaken.

The appreciation of habits, which fit an animal to no particular habitat, but lead by associated morphological changes to better living in the same or in a variety of environments, is of paramount importance. The study of vertebrate palaeontology shows that habits may persist unchanged for many millions of years and such habit persistence has also been important in the evolution of arthropods. It is sometimes easy to apprehend which, of the many things an animal does, has been of greatest survival value, but this is not always so and many of the most fundamental habits have hitherto passed unrecognized.

It is probable that the early terrestrial ancestors of the Uniramia were little sclerotized and walked upon unjointed lobopodial limbs (Manton, 1972). A differentiation of habits at this early stage has led to the separation of the Onychophora and the four myriapod and five hexapod classes (Manton, 1972). These animals all needed some protection, such as obtained under stones, in decaying vegetation or in the surface layers of the soil. They all needed food, of vegetable or animal origin, and they needed mates, so that moving about was equally essential. The first and most important habit divergencies may be summarized by considering the reactions of such primitive animals to obstructions in their path.

1. A tendency to squeeze through narrow spaces, by deforming the body without pushing, would lead to the present-day structure and proficiencies of the Onychophora (Figs 22–24). All their essential trunk features today, besides being entirely arthropodan, are associated with this habit which is of great survival value because the animals can reach commodious spaces where predators large enough to harm them cannot follow (Manton, 1958b, 1972).

2. A tendency to seek a way through narrow channels by simply turning and flexing the body, without distortion or pushing, leads to the characteristics of the Symphyla (Figs 19, 20).

3. A head-on bulldozer-like shoving into and through the substratum, by the motive force of the legs (Fig. 10) has led to the trunk features of the Diplopoda.

4. Running round an obstruction, also seeking shelter in crevices by using the ability to flatten the body somewhat (Figs 12, 14), has led to the faster running Chilopoda; the Scolopendromorpha and Anamorpha perfecting their running ability in different ways (Figs 11, 12, 17, 18), while the Geophilomorpha have exploited the earthworm-like technique of burrowing (Figs 14–16), a thrust being exerted against the substratum from the body surface, the force coming from the trunk musculature and not from that of the limbs.

5. The hexapod classes have avoided all structural features associated with habits 1–4 (Manton, 1972, 1974).

The trunk segmentation of arthropods is established early in embryonic development, and it is the series of mesodermal somites which evokes the more superficial segmental structures, the limbs, the ganglia and visible ectodermal features. We have sound and reliable information concerning the laying down of the series of trunk segments in all arthropodan classes (Anderson, 1973), but we must look to other disciplines for the meaning of the many divergencies which exist in the serial repetition of certain, mainly dorsal, features which appear in later developmental stages. Consideration below will be limited to the trunk region of Myriapoda, excluding the more anterior and posterior segments.

Diplopoda

A brief consideration of the diplopod trunk is a convenient preliminary to the conclusions reached concerning the trunk regions of Symphyla, Chilopoda and Pauropoda. Most of the trunk of diplopods is formed by the fusion together of two successive segments to form diplosegments. Each of these is established by two pairs of mesodermal somites and their derivatives, and contains two pairs of ganglia, limbs, tracheal pouch apodemes, heart, ostia etc. It has been shown (Manton, 1954) that burrowing by the motive force of the legs has been the prime factor associated with diplopod trunk evolution. Greatest head-on shoving, or pushing with the anterodorsal surface, is obtained by many legs pushing against the ground simultaneously. In slow patterned gaits the forward stroke of the leg is rapid and the backstroke slow so that a large proportion of each cycle of leg movement is spent on the backstroke and most of the legs are pushing at once, a small proportion performing the forward swing. Metachronal waves of limb movement travel over the body from before backwards. The slowest patterns of gait, in which the backstroke is of the longest duration, can only be performed when many legs are present and each metachronal wave is long (Fig. 10 shows a diplopod with about 22 propulsive legs in each wave). In surface running there are about 5 propulsive legs in each metachronal wave. The more leg pairs a diplopod possesses the greater will be the pushing force and the stronger (and slower) will be the practicable gait patterns (see, however, Manton, 1954, 1974). But a trunk that is very long is usually unwieldy and unsuited to most arthropods' needs. If the diplopod trunk was formed of single segments it would usually be very long. The fusion together of segments in pairs to form diplosegments enables the body to be much shorter, with every other joint providing sufficient mobility, rigidity and incompressibility, so that the motive force from the legs is transmitted unimpaired to the head end. The construction of inter-diplosegmental joints shows micro-engineering efficiency and the hard cuticle, protective anterior pushing shield (collum or the equivalent), keels when present, etc., are all essential for the various burrowing techniques (Manton, 1961).

Diplosegments are usually cylindrical and remarkably short and wide; and many species are at the limit of practicable shortness, if suitable joint construction is to be maintained. The reasons for the evolution of diplosegments are quite clear; they assist in the provision of a maximal burrowing force and keep the body from becoming too long for the usual habits. No such fusion of trunk segments takes place in any other class of myriapods and indeed is not required.

SERIAL REPETITION OF ORGANS ON THE TRUNK REGIONS OF SYMPHYLA, CHILOPODA AND PAUROPODA

In these classes the establishment and growth of the embryonic germ band, with its segmental series of mesodermal somites, nerve ganglia, limb buds, etc., first proceeds in a simple manner, differentiating from before backwards (Heymons, 1901; Tiegs 1940, 1945, 1947). In later embryonic or in larval stages divergencies appear in the serial repetition of certain structures, mainly dorsal in position. These divergencies comprise:

1. The formation of extra tergites so that they total more than the number of paired legs arising from the same segments.
2. The differentiation of the anterior ends of the tergites of a transverse line of lesser sclerotization which delimits an anterior piece, the intercalary tergite, possessing various degrees of flexibility on the main tergite behind.
3. Mobile parts of the sternal and ventro-lateral cuticle, partially corresponding with the dorsal features, provide extra hinge lines in these regions.
4. A heteronomy in tergite length is present in many Chilopoda, so that the tergites are alternately long and short along the body; with two successive long tergites over leg pairs 7 and 8 in the Scolopendromorpha and Lithobiomorpha; and fusion together of these successive long tergites in the Scutigeromorpha. The short tergites are almost eliminated in the Scutigeromorpha and alternate tergites are absent in the Pauropoda.
5. Heteronomy in musculature.

There have been various unsatisfactory attempts to explain the occurrence of the above peculiarities and always without any understanding of the functions of the structural features under consideration. Not one of these peculiarities is explicable on theories of complete or partial elimination of segments locally along the trunk e.g. Demange (1967). Faulty descriptive morphology is no evidence of derivation from diplosegments. The excellent studies of embryonic and larval development show no such postulated changes. The functional usefulness of the divergencies of structure have never been elucidated by other workers and neither have any workable hypotheses been put forward to account for the evolution of such features on a functional basis. Without some functional advantages the features in question could hardly have been evolved. A study of habits, proficiencies, and facilitating morphology of the myriapod and hexapod classes shows how and why the above structures have evolved. These features and the associated changes in the basic musculature, are all derived from the serial embryonic or larval metameric segmentation in which each segment carries one leg pair. Ample functional reasons for the evolution of these structures are provided by the elucidation of their present day uses.

Brief reference will be made below to muscular systems. For the first time a comparative and functional classification of uniramian muscle systems has been arrived at as the result of a functional analysis of muscles, skeleton, movements and locomotory mechanisms throughout the Uniramia (Manton, 1974: Appendix—gives a summary and full references to figures and descriptions). Five categories of muscles exist, each with recognizable functions and positions in the body. The exact functions show variations according to different lengths and dispositions of the muscles within each category. Moreover, where the functions of the category of muscles are not required by certain taxa this category is absent.

1. The superficial muscles, mainly oblique and dorso-ventral, correspond with the circular and oblique layers of the Onychophora and are mainly concerned with the maintenance of alignment in all taxa and with the promotion of trunk flexibility in epimorphic Chilopoda and Symphyla and enrolment in Diplopoda.

2. The dorsal, lateral and sternal longitudinals correspond with the same in the Onychophora. These muscles provide cohesion for the whole body, and, where the lateral component is present, deal with the control of hydrostatic pressures and muscle forces used in jumping (Collembola) and burrowing (Geophilomorpha) and general purposes in the Onychophora.

3. The deep oblique and

4. the deep dorso-ventral muscles form antagonistic pairs of systems, concerned with the maintenance of trunk rigidity, necessary during fast running with few legs in contact with the ground at one moment (Chilopoda, Pauropoda), and with sudden increases in hydrostatic pressure used for jumping (Collembola).

5. The extrinsic leg muscles, which provide (i) the promotor–remotor swing of the leg and also (ii) the rocking of the dorsal face of the leg forwards during the backstroke, and vice versa, in many classes.

The data concerning the divergent segmental structures listed above which have emerged from present knowledge of habits, functions, locomotory mechanisms, gaits and musculature are consistent and harmonious. No satisfactory conclusions can be based upon inaccurate and incomplete descriptions of musculature. A brief account is given below of the conclusions reached for each class under consideration; but reference should be made to the fuller accounts where all relevant data and figures are to be found (Manton, 1950, 1974).

SYMPHYLA

The excellent developmental studies of *Hanseniella* by Tiegs (1940, 1945) leave no doubt about the formation of one extra tergite on segments 4, 6 and 8 and on further segments in other genera, these sclerites being a duplication of the normal single dorsal sclerite on just these segments. Ravoux (1962) has confirmed the basic segmentation of the symphylan trunk. It remains to show here why these extra tergites were evolved and what purposes they serve. The symphylan habits of survival value concern their ability to seek a passage through the soil, etc., by flexing and turning (Fig. 20) into narrow tortuous existing channels without pushing and without actual body deformation (Manton, 1966), such as is seen in Onychophora (Manton, 1958b).

Extra tergites of Symphyla

The positions of the extra tergites on segments 4, 6 and 8 of *Scutigerella* are shown in lateral view in Fig. 1, the body slightly flexed dorso-ventrally in (a) and markedly so in (b); a tracing of a photograph of *Symphylella* crawling round the edge of a thin leaf (c) shows how the extra dorsal hinge lines, formed by the extra tergites, are used. They provide a contribution of 75° to the total 180° flexure of the trunk. Remarkable intersegmental flexibility also exists in the horizontal plane (Fig. 2b, c). The postero–lateral tergal lobes slide over one another on the concavity of the bend and on the convexity they give protection to the dorso–lateral surfaces which would otherwise be exposed by such great flexures. Extra tergites are also present in the chilopod *Craterostigmus*, where they contribute to the extreme secondary flexibility of this animal (see below).

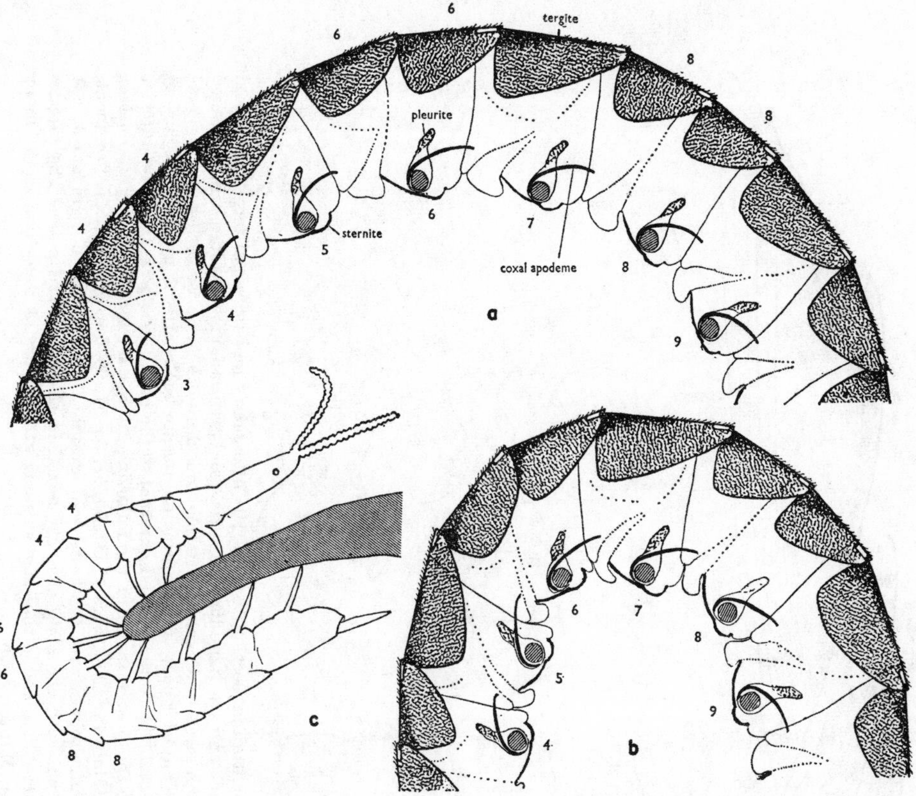

FIG. 1. (a–b) *Scutigerella immaculata*. Diagrams of the left side of the body to show tergites (mottled), pleuron (white), coxal apodemes (black) and sternites (black). In (a) the body is slightly and in (b) markedly dorso-ventrally flexed.
(c) *Symphylella* sp. larval stage crawling round the edge of a leaf, from a photograph. After Manton, 1966.

Intercalary tergites of Symphyla

In addition to the extra tergites, all tergites of Symphyla show transverse intercalary tergites differentiated by transverse hinge lines (Fig. 2a). These delicate intercalary pieces are plainly revealed in sections stained with Mallory's triple stain, although they are not easy to see by other means. Intercalary pieces are very mobile and can even fold back over the tergite behind, as shown by intercalary tergite 3 and anterior intercalary tergite 4 on Fig. 2a, posterior intercalary tergite 4 being in the normal position. The presence of the intercalary tergites aids the general flexibility of the body. They occur wherever marked trunk flexibility is required, e.g. in some Chilopoda and in the japygid

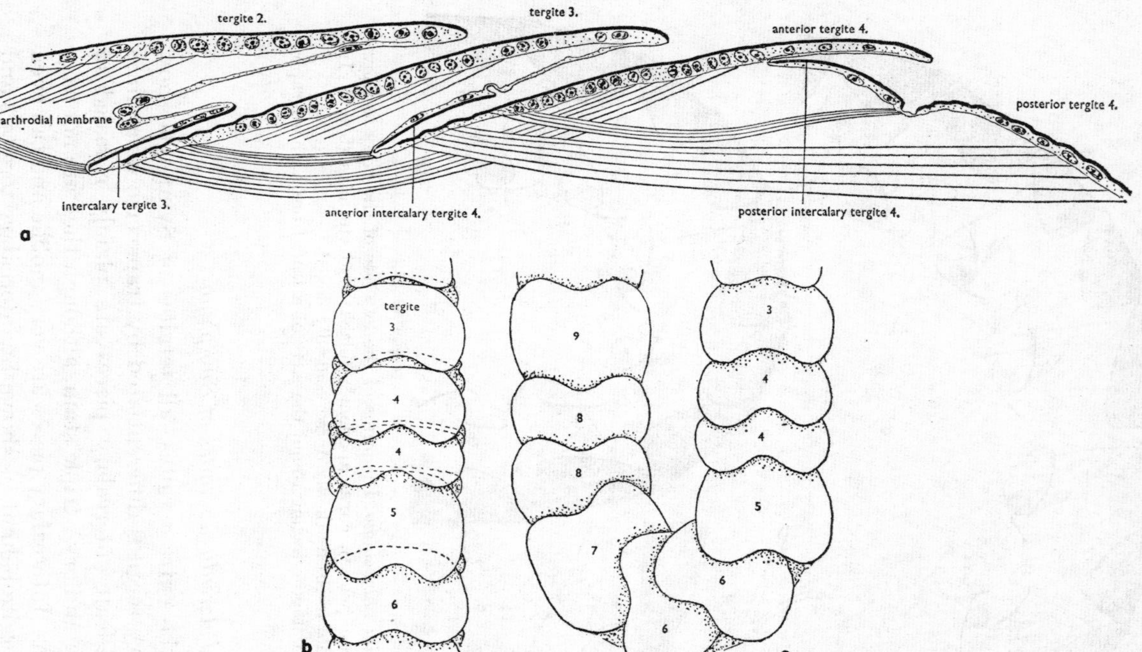

FIG. 2. *Scutigerella immaculata*. (a) Longitudinal section to show the intercalary and principal tergites, marked sclerotization being indicated by heavy black and the more flexible cuticle and arthrodial membranes being bounded by thin lines. The lateral and the submedian series of dorsal longitudinal muscles both comprise the two sectors here shown: the larger muscles extend from the anterior part of each principal tergite backwards to much of the surface of the following principal tergite, and the smaller sectors, situated external to the larger ones, pass from the anterior end of each principal tergite backwards to the posterior end of the following intercalary tergite. There are no long dorsal longitudinal muscles (cf. Chilopoda, Manton, 1965: Figs 35–37, 50, 51, 55–57, 60–62). (b) Tergites of segments 3 to 6, drawn from a photograph of a living animal with the body held straight. (c) Tergites of segments 3 to 9, drawn from a photograph of a living animal with the body strongly flexed in the horizontal plane. The tergites are drawn as if their margins are constant; it is possible that slight rolling at the margins takes place as in many Chilopoda (Manton, 1965: Fig. 6a–d). After Manton 1966.

and proturan thorax. In the chilopod *Craterostigmus*, using marked antero-posterior movements of the head for its peculiar feeding mechanism, the intercalary tergite on the poison claw segment can similarly fold back over the main tergite behind (Manton, 1965: Fig. 83a). There is reason to suppose that neither the Diplura nor the Protura are ancestral to any other type of hexapod (Manton, 1972). They share with the Symphyla and Chilopoda the separation of tergites from sternites by a wide pleuron which bears separate sclerites in various numbers and arrangements (Manton, 1972: Fig. 9). The differently contrived dorsal and ventral intercalary features are associated with the need of flexibility in all named classes. They can only be partial retentions of primitive features on which are superimposed structures used for specialized and particular purposes.

Musculature and other features of Symphyla

The marked trunk flexibility of Symphyla is aided by the massive superficial pleural muscles. These muscles were first described by Tiegs (1940, 1945) and are shown in Manton, 1966, Fig. 4a, marked muscle *pct*. Equally striking is the entire absence of the rigidity promoting deep dorso-ventral and deep oblique muscles, together with long sectors of the dorsal longitudinal system (Manton, 1965, 1966, 1974, and see below).

The Symphyla are faced with having to use a locomotory mechanism within the limitations imposed by the trunk flexibility, a higher priority than normal fast running by the employment of fast patterned gaits. This they do in a remarkable manner. They are fast running on the soil surface, ceaselessly sensing the ground with very rapid antennal movements (Fig. 19), and gaining safety from pursuing arachnids of small size, which cannot turn with ease, by changing direction frequently and by turning through very acute angles. Their trunk flexibility makes it impossible for them to use gaits with fast patterns which require few legs to be in contact with the ground at one moment (Manton, 1952: Fig. 3, 1965: Fig. 28). Instead they use slow or moderate patterns of gait in which stability is gained by the many legs on the ground simultaneously and speed is gained by rapid stepping (short pace duration). The speed of Symphyla, alone among the Uniramia, depends on reducing the duration of the forward stroke. Leg morphology and leg muscles show many peculiarities which assist a rapid forward swing (Manton, 1966).

Conclusions concerning Symphyla

Thus all the functional and morphological evidence concerning the symphylan trunk shows peculiarities of structure which harmoniously

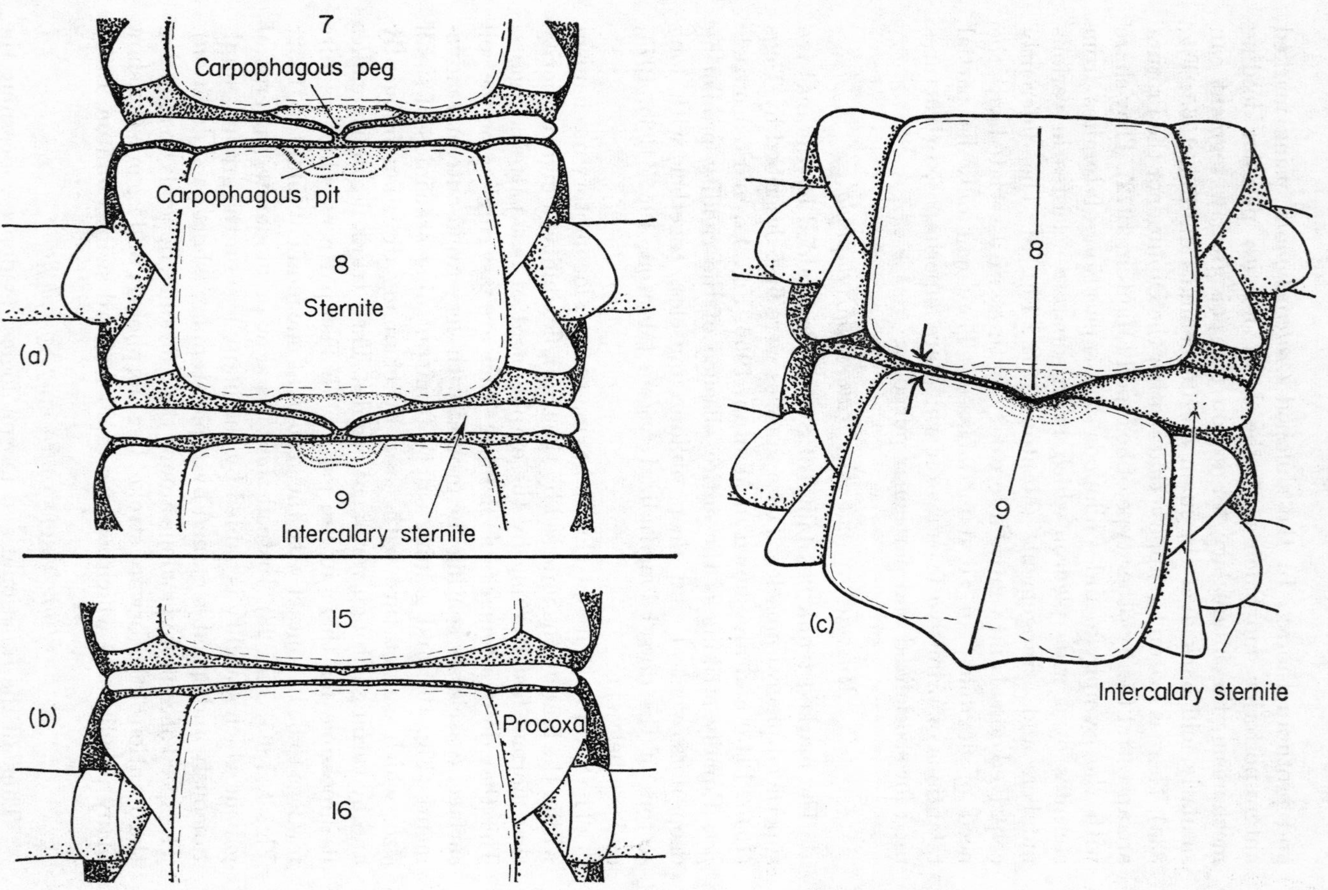

Caption on facing page

fit together with the way the whole works and with the needs of the animals. Symphylan evolution has been bound up with the habits named above, and along with habit perfection have evolved the facilitating skeleto-muscular systems. All are based upon the simple segment series marked by the trunk limbs and revealed by developmental studies. No case can be made for regarding the symphylan trunk as fundamentally diplosegmental (Demange, 1967) or for partial elimination or fusion, of segments along the trunk.

CHILOPODA

In the Chilopoda also the study of development shows that the series of metameric trunk segments each carry one pair of legs, mesodermal somites and their derivatives, ganglia, etc. But the needs of these animals are more complex than those of Symphyla. Flexibility requirements will first be considered, followed by those of rigidity of the trunk.

Intercalary tergites, intercalary sternites, body shaping and relevant musculature in Geophilomorpha

Intercalary sclerites are best developed in chilopods showing greatest trunk flexibility, the Geophilomorpha (Figs 3, 4, 14–16), and ill-defined or absent in the Scolopendromorpha and Anamorpha which use fast patterns of gait needing greatest trunk rigidity. All Chilopoda can traverse narrow crevices by dorso-ventral flattening; the slits in the cards shown in Figs 12 and 14 can all be traversed by the animals (see also Fig. 21). The Geophilomorpha employ also the earthworm-like technique for better burrowing; a group of segments becomes short and thick (Fig. 16) and delivers a heave against the soil from the body surface. Segments add themselves to the posterior part of the thickening and pay out from it in front, so that the anterior elongating part of the body progresses further into a widened crevice. The surface armour

---

FIG. 3. Ventral view of a potash preparation of the cuticle of *Geophilus carpophagus* to show the general organization of the ventral sclerites in the Epimorpha and the carpophagous structures peculiar to certain Geophilomorpha. In (a) and (b) the body is a little longer than the resting length in order to show the limits of the intercalary sternites.

(a) Segments 7 to 9, from the middle of the region bearing the carpophagous process from the posterior part of each sternite fitting into the carpophagous pit of the sternite behind. (b) Segments 15 and 16 where the intercalary sternites are just united across the middle line. (c) Segments 8 to 9 in their natural positions when flexed laterally as far as can be done without inrolling of the sternite edges at the arrows, some 12°. By inrolling, flexures of 25° to 30° are possible. After Manton, 1966.

remains remarkably intact throughout these shape changes (Manton, 1952: pl. 31; 1965: pl. 3). Each intercalary tergite slides over the tergite behind; buckling of both can be effected by the dorsal musculature; and sternites and pleurites change their shape enormously by marginal sinking into the body or the reverse. These furling and unfurling movements are made possible by cones of greater sclerotization

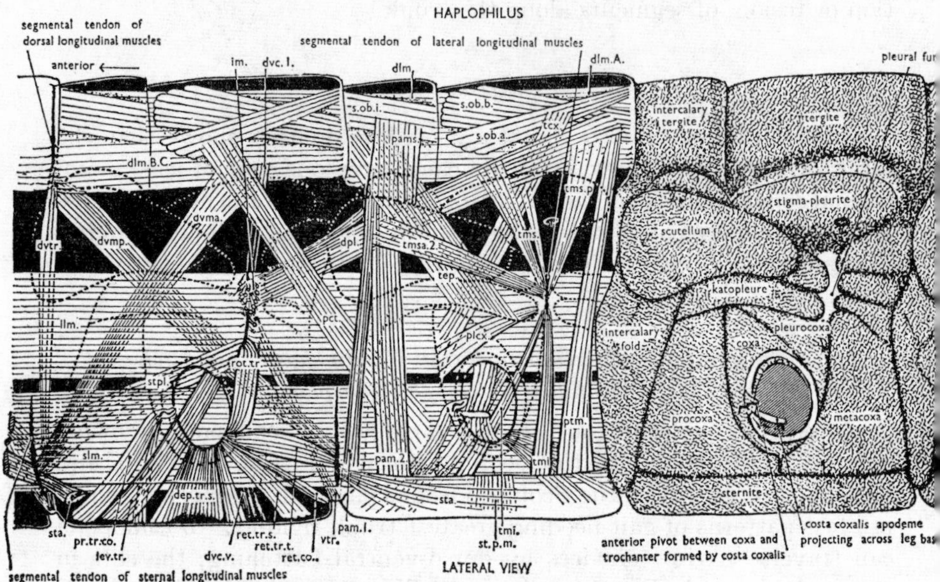

FIG. 4. Lateral view of three segments of *Haplophilus subterraneus* (Geophilomorpha), with legs cut off at the trochanter, to show right: the superficial sclerites (mottled), pleural furrow (white) and costa coxalis apodeme projecting across the leg base internally; middle: the superficial musculature; and left: the deeper musculature and extrinsic leg muscles viewed from the body surface. The principal muscles only are shown, there are many smaller pleural muscles. The unlabelled muscles from the costa coxalis apodeme are *lev.tr.co.* above and *dep.tr.* below. The superficial pleural muscle *pam.1* from the sternal intercalary fold inserts dorsally on the intercalary tergite; *pam.2* from the sternite inserts largely on the scutellum and partly on the intercalary tergite. Muscle *pams.* arises from the scutellum. After Manton, 1965.

being set in flexible endocuticle in the marginal zones (Blower, 1951), thus providing great strength and flexibility (Manton, 1965: Figs 21, 23). Shape changes of the body are also mediated by different movements on the two sides when the body flexes horizontally; intercalary sternites open out or fold away as shown (Fig. 3). Most of the burrowing is done by the anterior third of the body where the legs, used to pull the body forwards and hold it during a burrowing heave, are much stouter than they are posteriorly. Anteriorly also are the carpophagous pits and pegs. These structures are at maximal development about leg-segments 7–8 in *Geophilus carpophagus*. When burrowing commences at an angle to the general surface of the substratum, the anterior segments must flex dorso-ventrally as well as execute burrowing heaves of thickening. The carpophagous structures interlock as the inter-segmental folds of cuticle deepen and the pegs and pits prevent undue infolding and damage to internal organs (see further Manton, 1965: 282). Geophilomorpha with no carpophagous structures may be expected to burrow at right angles to a surface less easily.

Geophilomorph musculature shows correlations with function. The dorsal, lateral and sternal longitudinals are very stout and supply the main force giving the burrowing heave against the soil, aided by the deep oblique muscles *dvmp*. The deep dorso-ventral muscles *dvtr*. act as antagonists to the longitudinals and deep oblique muscles in restoring the elongated shape. All the components of the dorsal longitudinals are short, stout muscles crossing only one joint. There is an abundance of superficial muscles (Fig. 4). (Details of all muscles and many other associated features are given in Manton, 1965, 1974: Appendix.)

The Geophilomorpha show no trace of heteronomy in any serial structures and have no need of this feature. There is no evidence to show that they ever possessed diplosegments.

Facilitations of fast running in Chilopoda by cuticular and muscular features

Faster running is practised by the Scolopendromorpha, using fast patterns of gait with very short pace durations. Such a rapidity of stepping is not seen in the Geophilomorpha. The massive longitudinal muscles of these animals maintain trunk rigidity when using fast patterns of gait with few legs in contact with the ground and slow stepping. The Anamorpha, with longer legs, use a different type of fast gait (Manton, 1952, 1965), and also rapid stepping. The Scolopendromorpha possess considerable flexibility of trunk, used in fitting themselves under cover, but they need facultative rigidity during fast running, as do the Anamorpha. The Cryptopidae, with the most flexible trunks, cannot employ the fast patterns of gait used by the Scolopendridae. The gait patterns of the former are moderately fast and speed is gained by remarkably rapid stepping.

The simplest use of intercalary tergites and intersternite infoldings during fast running is shown by the Cryptopidae (Manton, 1965: Fig. 16). Good dorso-ventral bending could not take place without a hinge between intercalary and principal tergites. The Scolopendridae use faster gait patterns with as few as 3 out of 40 legs propulsive at one moment and greater trunk rigidity is required. Well hinged intercalary tergites would be unsuitable and sufficient mobility is provided by the incomplete hinge lines situated anteriorly on the tergites (Manton, 1965: Figs 35–37). The suggestion that the large tendinous junction uniting the tergite with muscles *dom.* and *dvmp.* (see the above figures) represents the vanishing suture between intercalary and principal tergite is invalid since this tendon can lie right across the hinge line (Manton, 1965: Fig. 14). Tendons in the Uniramia anchor muscles marginally or onto the faces of sclerites, or even onto flexible cuticular membrane near to sclerites, as is mechanically suitable. The Anamorpha with longer legs have an even greater need to control unwanted flexibility and intercalary tergites are here absent and there are no sclerotized intercalary sternites (Manton, 1965: Figs 13, 15). Ridges, marked on Manton, 1965: Fig. 15 provide essential rigidity to the anterior parts of the tergites, which must endure greater tensions of extrinsic muscles from the longer legs and other muscles than in the Epimorpha.

Tergite heteronomy in Scolopendromorpha and Anamorpha

The control or limitation of unwanted yawing during fast running is essential to all fleet chilopods. Widely separated single, or groups of propulsive legs, always moving in opposite phase, tend to throw the body into horizontal undulations and these are controlled in large measure by the different degrees of tergite heteronomy and the associated changes in the segmental musculature. The movement of the head and antennae in sensing a path wide enough for the legs to follow also tends to start trunk undulations (Manton, 1965: Fig. 28).

Cinematography shows that the most stable region of the body lies at legs 7–8 in both Scolopendromorpha and Anamorpha; see Fig. 5 abc (7) showing the position of the mid-posterior margins of this segment as the animal ran from bottom to top. Segments posterior to this level show increasing undulations.

A recording of the flexibility which actually exists along the body can be made from good quality photographs of animals flexed in the horizontal plane, resting or otherwise. In Fig. 6 the middle of such a bend on the trunk is marked by an arrow. In the geophilomorph centipede and in the diplopod the line joining the points forms an even curve. In *Lithobius* the small dots show the degrees of flexure at the anterior end of the short tergites and the large dots show the much smaller flexures at the anterior end of the long tergites. There is greater tergite heteronomy in *Lithobius* than in the Scolopendromorpha

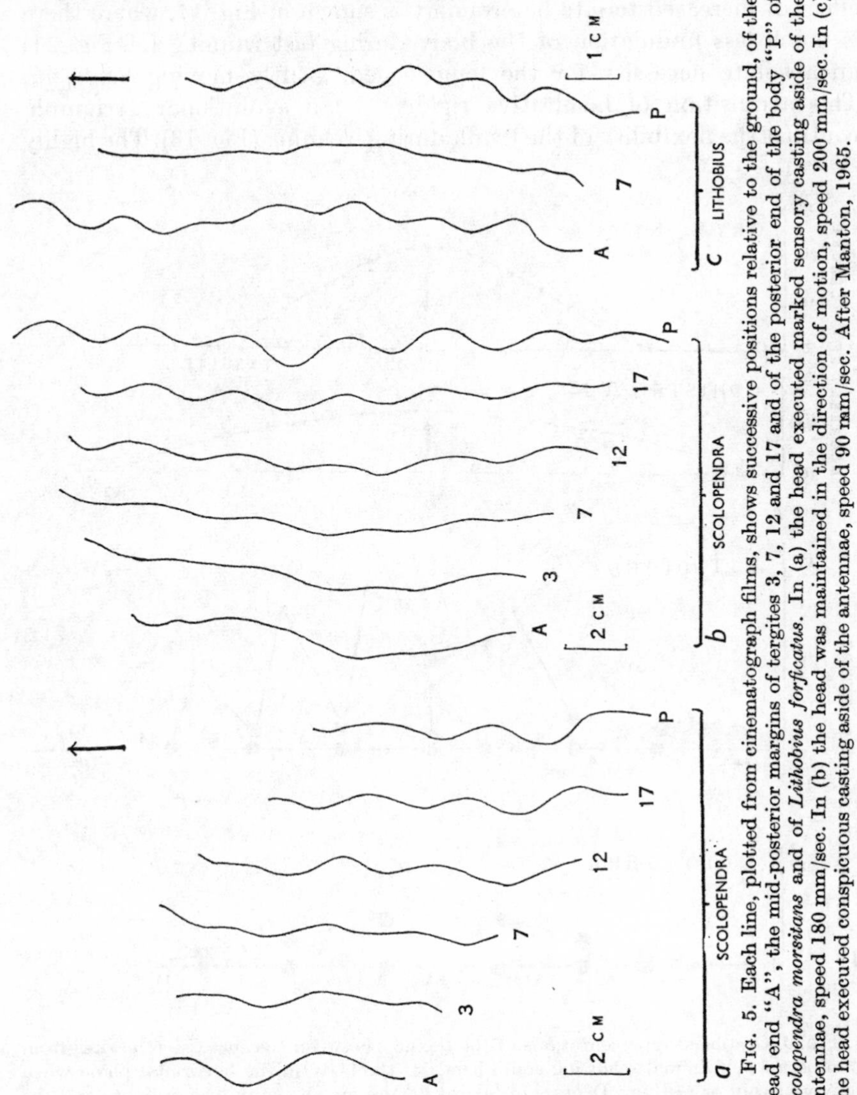

Fig. 5. Each line, plotted from cinematograph films, shows successive positions relative to the ground, of the head end "A", the mid-posterior margins of tergites 3, 7, 12 and 17 and of the posterior end of the body "P" of *Scolopendra morsitans* and of *Lithobius forficatus*. In (a) the head executed marked sensory casting aside of the antennae, speed 180 mm/sec. In (b) the head was maintained in the direction of motion, speed 200 mm/sec. In (c) the head executed conspicuous casting aside of the antennae, speed 90 mm/sec. After Manton, 1965.

and plottings for the latter show less rigidity at the anterior end of the long tergites, and it is least in *Cryptops* where heteronomy is very slight. In *Scutigera*, Fig. 6, the flexibility at the anterior end of the short tergites has been eliminated by their small size and position under the edge of the long tergites. The general rigidity has been increased by fusion of the two large central tergites which now cover legs 6–9. The effect of increased tergite heteronomy is shown in Fig. 17, where there is much less undulation of the body during fast running (cf Fig. 11) an absolute necessity for the long legged, swiftly moving *Scutigera*. This acquisition of facultative rigidity is an evolutionary triumph, contrast the flexibility of the trunk during cleaning (Fig. 18). The highly

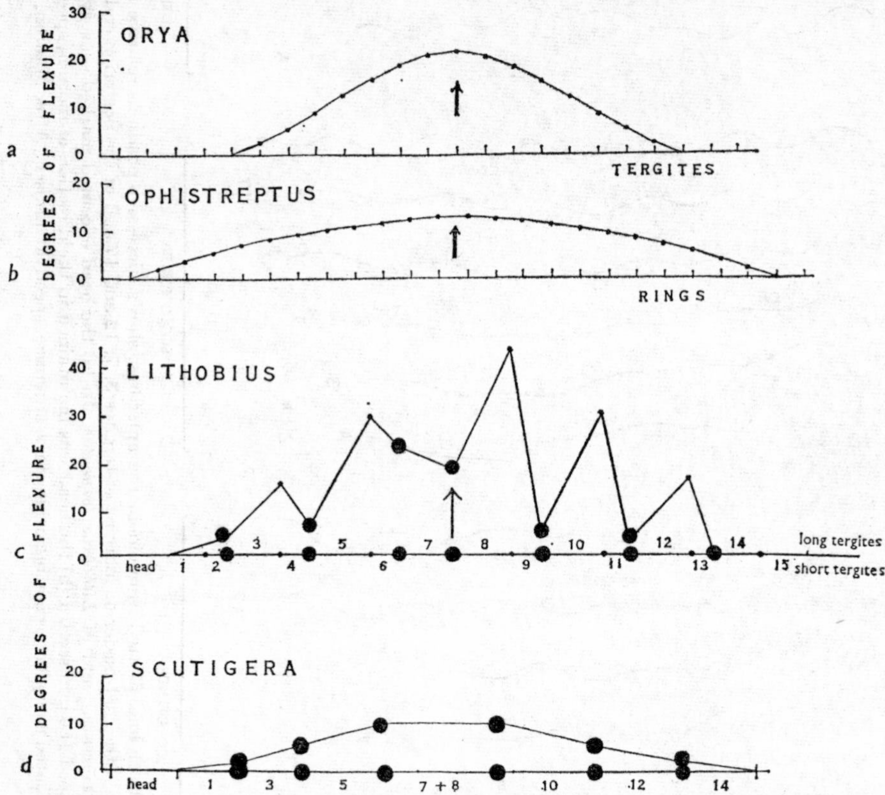

Fig. 6. Graphical representations of the flexures between the successive tergites from photographs of animals showing acute bends of the body in the horizontal plane when moving freely or resting. Degrees of lateral flexure are shown on the ordinate and the lengths of successive tergites on the abscissa. The heavy dots on (c) and (d) show the points of flexure at the anterior ends of the long tergites, and the vertical arrows mark the middle of each bend of the body.

(a) The geophilomorph centipede *Orya barbarica*, middle part of the body, in a "hairpin" bend, 16 segments effecting a total flexure of about 180°. (b) The iuliform millipede *Ophistreptus guineensis*, middle part of the body, in a "hairpin" bend, 23 diplosegments effecting a total flexure of about 180°. (c) *Lithobius forficatus* in a "hairpin" bend of about 190° by segments 1–13. (d) *Scutigera coleoptrata* in as acute a bend as is possible to the animal, about 35° by segments 1–14. After Manton, 1965.

advanced vision; mode of feeding; specialized poison claws (they are not primitively simple); and the most wonderfully contrived and enormous mandibles used for cutting up hard, fast moving prey (in contrast to the Epimorpha, where strong poison claws can tear into prey, the weaker mandibles only dealing with the soft part); all demonstrate an advanced and not a primitive group of chilopods (Manton, 1965). It is only dorsally that there is an approach to segment reduction and this is only external. There is no change in the serial repetition of leg-bearing segments which make up the body.

Musculature and heteronomy

The basis of the control of yawing associated with tergite heteronomy is not far to seek. Every segment possesses a certain complement of muscles all along the body. As noted above, the segments at the head end and at the posterior extremity of the body are not being considered here because their peculiarities do not illuminate the composition of the major part of the trunk. The serial segmental complements of muscles are uniform in number of main muscles, but sometimes one muscle may possess more than one sector (e.g. *dpl.* and *dpll.*, *dvmpa.* and *dvmpb.*, Manton, 1965: Figs 51, 56) and this is a common phenomenon associated with function. A deep oblique muscle, with much force to generate, but no complexity of movement to create, may be unusually large, as in Collembola (Manton, 1972: Fig. 21a, muscle *ob.*, abdominal segment 3), and its functions in the thysanuran abdomen dictate subdivision into a number of sectors with varied dorsal insertions in association with more complex movements (Manton, 1972: Figs 38d, 39, muscles 24, 25A, 25B, 26 which correspond with *ob.* of Collembola and *dvma.* of Chilopoda). The presence or absence of more than one sector to a muscle is correlated with functional needs and provides no evidence of partial elimination of segments along the trunk.

Three features of the musculature of the fast running Chilopoda contribute towards this accomplishment by providing facultative rigidity controlling the tendency to yaw as well as rigidity preventing sagging of the body on to the substratum between widely spaced propulsive legs.

1. Well developed deep oblique muscles *dvmp.*, *dvma.* and their antagonistic deep dorso-ventral muscles, *dvtr.*, etc.; these categories are entirely absent in Symphyla where flexibility of trunk is the priority.

2. The simple short sectors of the dorsal longitudinal muscles crossing one joint in Geophilomorpha are represented by a complex system composed partly of short muscles, *dlm.*, *dom.*, and more bulky

long sectors, muscles *dlm.A.*, *dlm.B.*, *dlm.C.*, are present crossing more than one intersegment (Manton, 1965: Figs 50, 51, 55, 56, 57, 60–62).

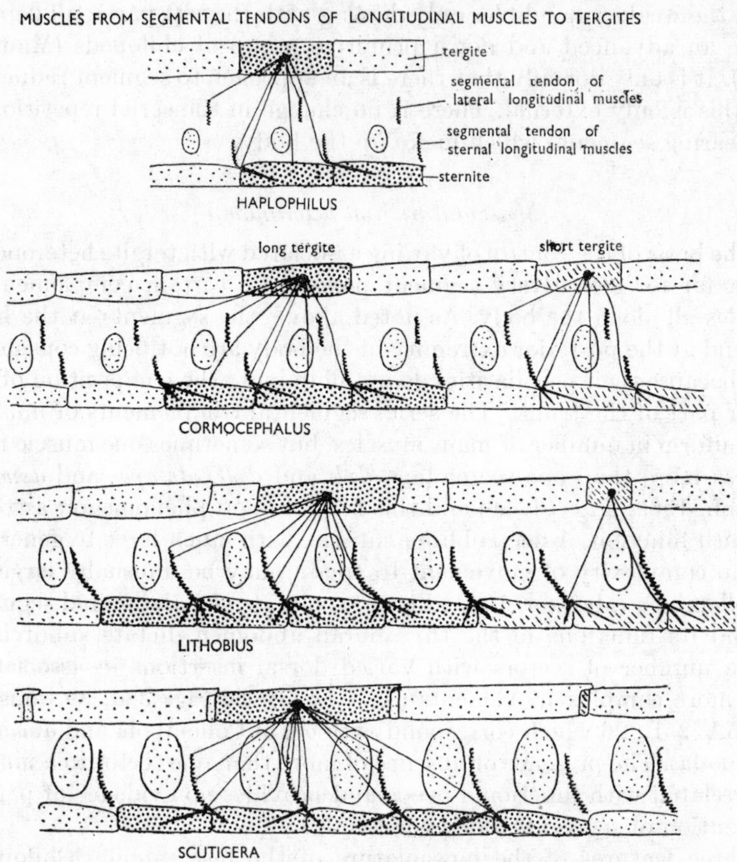

FIG. 7. Diagrams showing the linkage of the tergites to the sternites by muscles via the segmental tendons of the longitudinal muscles (zig-zag lines) and the coxal apodemes (not shown) in examples from the four chilopodan orders. One tergite of the geophilomorph *Haplophilus subterraneus* and one long tergite of each of the other animals, together with their associated sternites, are marked by heavy stippling. One short tergite of *Cormocephalus calcaratus*, *Lithobius forficatus* and *Scutigera coleoptrata* are marked by interrupted hatching and their corresponding muscular connections shown. Each line represents one muscle. From left to right the muscles attached to one stippled tergite are:- *Haplophilus* (6 muscles), *dpl.*, *dvma.*, *dvtr.*, *dvc.*, *llm.t.*, *dvmp.*, *Cormocephalus* (7 muscles), *dpll.*, *dpl.*, *dvc.2.*, *dvma.*, *dvtr.*, *dvc.*, *dvmp.*, *Lithobius* (8 muscles), *dpll.* (two sectors) from preceding segment, *dpl.* from own segment, *llm.t.p.*, *dvtr.*, *dvma.*, *dvmp.*, *llm.t.a.*, *llm.t.p.*, and *Scutigera* (12 muscles), *dpl.*, *dvma.*, *dvtr.s.*, *llm.t.a.*, *llm.t.p.*, *dpl.*, *dvma.*, *dvtr.*, *dvmp.*, *llm.t.a.*, *llm.t.p.*, *dvmp.* For details see Figs 41, 42, 49, 60, 67 in Manton, 1965.

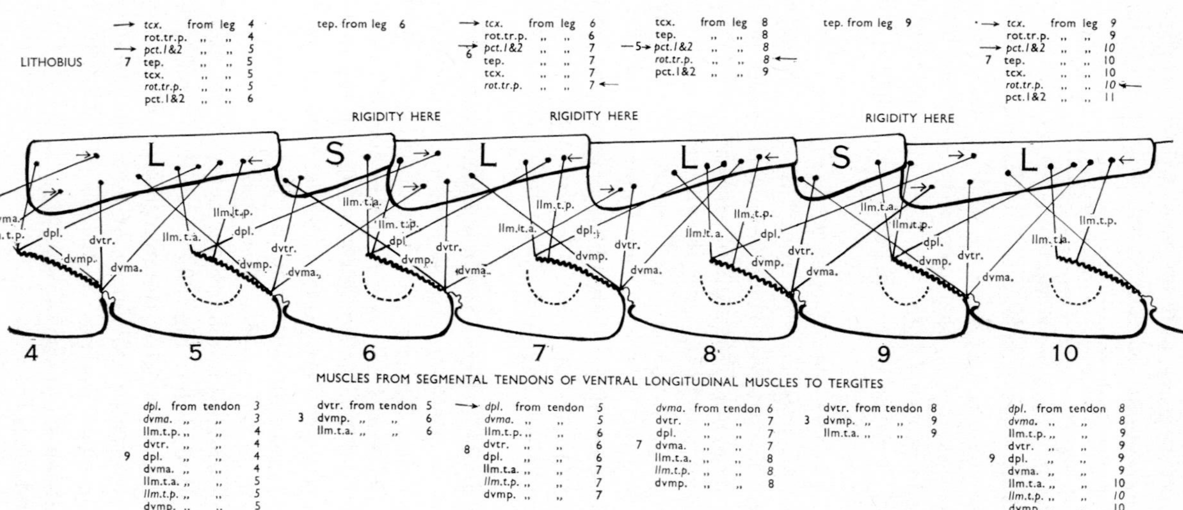

FIG. 8. Diagram showing the disposition of the principal dorso-ventral and oblique muscles on tergites 5 to 10 of *Lithobius forficatus* in order to depict the comparison between the muscle insertions of a typical consecutive long and short tergite, such as those of leg-segments 5 and 6, and the insertions present on the consecutive long tergites on leg-segments 7 and 8. The muscles arise from segmental tendons (zig-zag lines) of the ventral longitudinal musculature (fused tendons of the sternal and lateral longitudinal muscles). Below each sternite is listed the dorso-ventral and oblique muscles inserting on the corresponding tergite, and above is listed the extrinsic coxal muscles and those from the katopleure and procoxa which insert upon each tergite. The arrows and italics indicate muscles whose dorsal insertions have migrated, and the arrows indicate the direction from which they have come. A simplification in the diagram is explained in Manton, 1965, concerning muscles *pct.1* and *pct.2*, here shown as one muscle, and muscle *dpl.* in the S-segments (bearing short tergites), which divides into three sectors, only one being shown here. The dotted arcs show the positions of the proximal rims of the trochanter in each segment. For further description see Manton, 1965.

3. A shifting of the dorsal insertions of many trunk and extrinsic limb muscles off the short tergites and on to the long tergites, in progressive measure as heteronomy in tergite length increases. It has been shown (Manton, 1961 §19) how diplopod muscles can change their origin or insertions and occasionally both. The shifting of the dorsal insertions of chilopod muscles is shown diagrammatically in Fig. 7. In the Geophilomorpha (no heteronomy in tergite lengths) every tergite is linked with the segmental tendons and sternites while in the lowermost figure each long tergite is linked with five successive sternites instead of with two. The extrinsic muscles from the legs also migrate progressively onto the long tergites (Manton, 1965, Fig. 33). Overall rigidity is thus well provided for.

The extra rigidity at legs 7–8 is neatly contrived (Fig. 8). These two long tergites receive muscles respectively from the short tergites in front and from behind so that between them they carry more muscles than do any two normal segments. This zone of greatest rigidity lies at a workable distance behind the head with its antennal movements, giving a maximal damping out of undulations. The process is completed by the fusion of these tergites in the Scutigeromorpha and by the absence of a long posterior part of the body in the Anamorpha, so that the tergites of legs 7–8 become central. Segment numbers are controlled by other factors, such as even loading on the legs during fast gaits, etc. (Manton, 1952, 1965).

An attempt has been made in Fig. 9 to show how some of the long and short sectors of the dorsal longitudinal muscles cooperate with the deep oblique muscles *dvmp*. Tension on a muscle is shown in black and by stippling. In Fig. 9a *dlm.A* and *dlm.B.*, extending between two long tergites and *dvmp.5*, cannot do other than flex the body dorsoventrally, aided by tension on the lateral and sternal longitudinals. In Fig. 9b the tension on the black muscles cannot do other than straighten the body, here drawn in the flexed position. Corresponding diagrams for *Scolopendra viridicornis* possessing much greater heteronomy in tergite lengths, are given in Manton, 1965: Fig. 39, where it is seen how the greater heteronomy and the further differentiation of the dorsal musculature provide even greater efficiency for both the movements shown. This is not a complete analysis of the coordination between muscle systems, but just an example of the inevitable effect of the muscle changes associated with heteronomy.

Craterostigmidae (*Chilopoda*)

The work recorded on living and preserved material of the Australasian *Craterostigmus* (Manton, 1965: §§8 (xiii), 9) shows it to be an

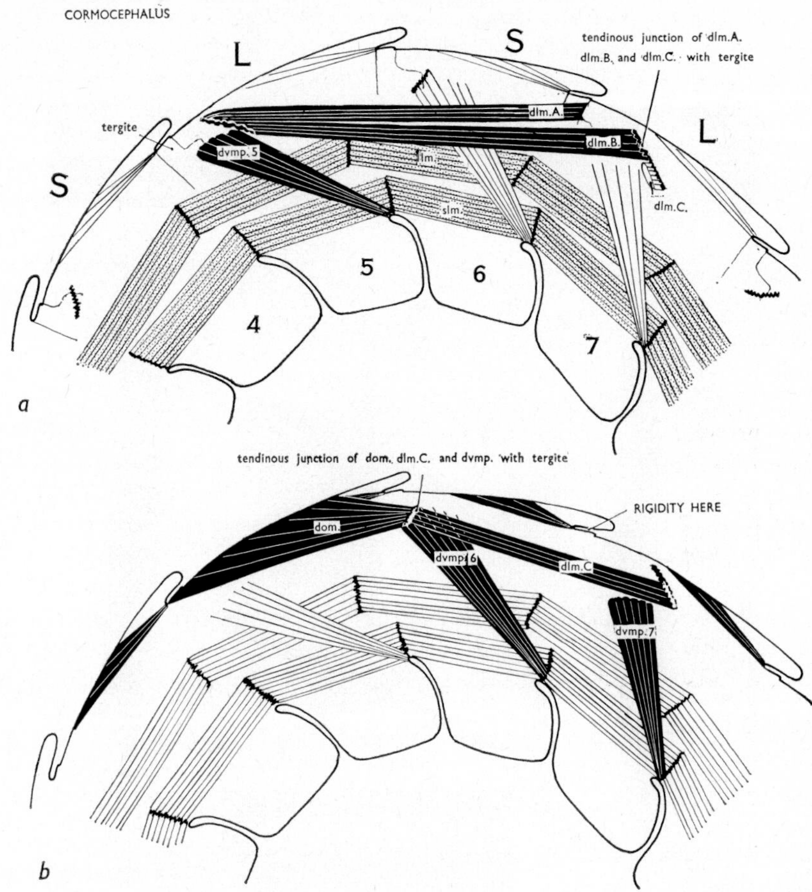

FIG. 9. Diagram of part of the trunk of *Cormocephalus* showing dorsal muscles *dlm.*, *dlm.A.*, *dlm.B.*, *dlm.C.*, deep dorso-ventral muscles *dvmp.*, lateral and sternal longitudinal muscles *llm.* and *slm.*, viewed from the sagittal plane, the long and short tergites being marked L and S. Strong tension on a muscle is indicated in black or by stippling. Muscle tensions in (a) tend to flex the body and in (b) tend to straighten the body. For further description see text. After Manton, 1965.

aberrant scolopendromorph, not a lithobiomorph, in which habit reversal has taken place. The once well developed heteronomy in tergites has given place to secondary trunk flexibility for quite different habits of life. Every long tergite has divided into two (Manton, 1965: Figs 74, 75). Two transverse hinge lines lie across each sternite. Paired intercalary sternites lie between the sternites in a ventro-lateral position and intercalary tergites are very well jointed from the anterior ends of the original long tergites, but not from the posterior products of partition

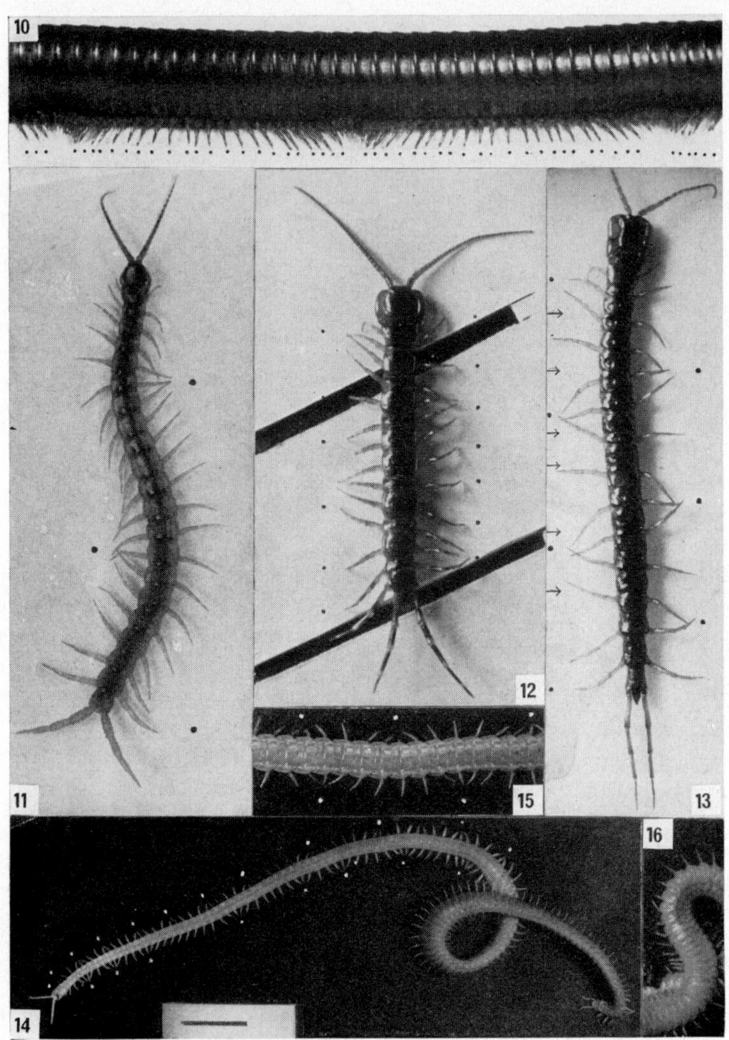

Figs 10–16

See caption on page 186

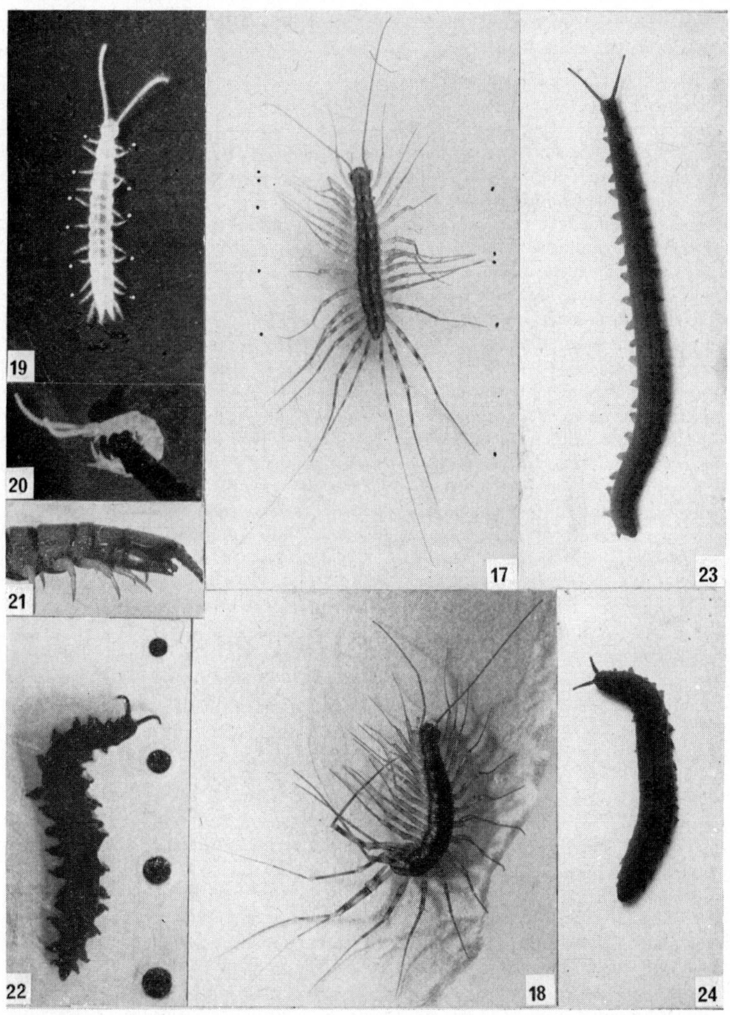

FIGS 17–24

See caption on page 186

(cf. Symphyla, Fig. 2, p. 170). The overlaps of the tergal series and infolding of the sternal structures (Manton, 1965: Figs 77, 78), together with modifications of the scolopendromorph series of long dorsal muscles and a development of extra tendons of insertion on the tergites,

---

FIGS 10–24. Gaits and associated features of trunk morphology in some myriapods and Onychophora. (Black or white dots level with the legs indicate that their tips are on the ground during the propulsive stroke.) (After Manton, 1958b.) Fig. 10. A iuliform millipede performing a slow gait giving a strong push at the head end where the collum transmits the thrust. Each of the two metachronal waves in the photo (about half the length of the body) shows 20–22 legs propelling and 8–10 legs recovering. Fig. 11. *Scolopendra cingulata* 110 mm, showing its fastest gait; the recovery stroke is much slower than the propulsive (8·5 : 1·5) thus fewer legs are on the ground (5–7, 18–20 on right, 12–14 on left) than off, at any one moment. These points of support are staggered on the two sides (black dots) and conspicuous yawing or horizontal undulation results. This is partly damped by a small amount of heteronomy of tergites (more obvious in *Lithobius* and *Scutigera*, Figs 12, 17) Note that the propulsive legs *converge* whereas in *Lithobius* and *Scutigera* (Figs 12, 17) they *diverge*. Fig. 12. *Lithobius forficatus* 25 mm long running slowly. Although the pleuron is flexible, there is not the same degree of dorso-ventral flattening as in *Himantarium* (cf. Fig. 14). The upper slit in the card is freely passable, the lower one, with difficulty. Tergite heteronomy is more marked than in *Scolopendra* (cf. Fig. 11); the two adjacent long tergites 7 and 8 occupy a central position between head and tail which is consequently the most stable part of the body. Although yawing does occur at faster speeds than shown here, it is more under control than in *Scolopendra* as a result of the more pronounced heteronomy of tergites and muscles. The propulsive legs *diverge* as in *Scutigera* (Fig. 17). Fig. 13. *Craterostigmus tasmanianus* 46 mm long. Control of lateral undulation has been sacrificed in secondary acquisition of lateral flexibility due to sub-division of the long tergites at the points indicated by the arrows. Fig. 14. *Himantarium* sp., a geophilomorph centipede 222 mm long when extended. A soil crevice is widened by local thickening and shortening of the body, the thrust being supplied by the body muscles and not by the limbs. The anterior half is extended and the posterior half is longitudinally contracted. The slit in the piece of white card shows the smallest gap through which the animal can pass by dorso-ventral flattening. Fig. 15. A section of *Himantarium* elongated; the principal and intercalary tergites can be seen. Fig. 16. The same section of *Himantarium* contracted; each principal tergite completely overlaps the intercalary and is itself shorter because it is convexly arched. Fig. 17. *Scutigera coleoptrata* 22 mm. The fleetest of all centipedes; running very fast (*c.* 500 mm/sec) with very little lateral undulation apparent, countered by the much more exaggerated heteronomy of tergites. Only the longer tergites 1, 3, 5, 7 and 8 (fused), 10, 12 and 14 visible—they cover the very much reduced intervening tergites. Note that the long propulsive legs diverge as in *Lithobius* (Fig. 12). Fig. 18. *S. coleoptrata* resting and cleaning its legs. Left leg 12 is being passed between the mouthparts; the horizontal flexure necessary for this operation shows that the control of horizontal undulation during running (Fig. 17) is facultative and not due to permanent rigidity of the trunk. Fig. 19. *Scutigerella immaculata* (Symphyla), 4 mm long. The flexibility of the trunk is enhanced by the divided tergites on segments 4, 6 and 8, which allow for the acute flexure seen in Fig. 20. Fig. 20. *S. immaculata* running round a plant fibre. Fig. 21. *Scolopendra cingulata*, side view of the head to show the great flattening of the head capsule; the head, mouthparts and poison claws are together less deep than the body, so enabling the manipulation of prey in very shallow crevices. Fig. 22. *Peripatus novae-zealandiae* (Onychophora). All the holes in the card are passable. It walked rapidly through the largest hole but took 20 min to deform itself, one leg at a time, in passing through the smallest hole, 2·5 mm in diam. Fig. 23. *Peripatopsis moseleyi*, extended (60 mm) and walking fast. Fig. 24. *P. moseleyi*, contracted and resting.

all provide extreme trunk flexibility. As might be expected, the animal is incapable of using fast patterns of gait; an attempt was made momentarily under brilliant light for cinematography, but resulted in hopelessly uncontrolled yawing, crossing of propulsive legs, etc. Only slow patterns of gait and slow movements are now compatible with the extreme modifications of the trunk providing secondary flexibility (see below).

PAUROPODA

These minute animals form a fitting conclusion to the analysis presented above. Again Tiegs (1947) has demonstrated by his fine embryological studies how the trunk of *Pauropus sylvaticus* is established by a series of mesodermal somites, ganglia, etc., corresponding with the adult series of legs. Tergites develop on every other segment and spread anteriorly and posteriorly, but do not quite touch one another. These animals normally live in draught-free environments inside decaying logs, etc. They move about rapidly for their size. They use fast patterned gaits with few legs in contact with the ground at one moment. They show great rigidity of body and no flexibility. It comes as no surprise to find that the flexibility promoting superficial oblique muscles are entirely absent (cf. Symphyla) and that the rigidity promoting deep dorso-ventral and deep oblique muscles are well developed as in Chilopoda, Collembola, Thysanura, etc. The absence of alternate tergites represents a more extreme form of heteronomy than is reached by the Chilopoda and only the long, stability promoting dorsal longitudinal muscles are present, extending between the tergites of alternate segments. There is no special rigidity by two long central tergites on successive segments, there being no casting sideways of the head in running and the legs being short. The whole organization of sclerites and muscles forms an end term to the known modifications suiting the use of fast gait patterns (Manton, 1966).

CONCLUSIONS

Thus the evidence provided by embryological studies and accurate descriptive morphology, combined with a study of habits and accomplishments of the animals concerned, shows decisively that the major part of the trunk in the Diplopoda alone is diplosegmental and that the trunk legs of Symphyla, Chilopoda and Pauropoda reflect the basic simple serial segmentation of this part of the body.

There is no acceptable evidence in support of the contention by Demange (1963) that in the Chilopoda "la zone comprise entre les 7°

et 8° segments est une région perturbée"; neither is there any acceptable evidence for the statement (Demange, 1967) that "Les blocs bisegmentaires des Chilopodes constituent donc des unités indissociables, donc la musculature est caractéristique de chaque groupe dimétamérique. La 'diplopodie' des Chilopodes ne se limite donc pas seulement aux plaques tergales mais à l'ensemble des deux métamères. Elle est plus profonde, chez les Chilopodes, qu'on ne peut le soupçonner *a priori*." The work which invalidates all these statements is presented at this Symposium as a resumé. It was published in full (Manton, 1954, 1956, 1958ab, 1961, 1965, 1966), yet it is entirely ignored by Demange (1967), apart from inaccurate inclusions in his list of references.

The evidence presented in outline above also demonstrates how and why the main divergencies from serial metamery have turned up, together with the guiding forces which have been at work in directing the evolution of the four myriapod classes along quite different lines. Evolutionary progress in all has been based upon different habits which have persisted for very long periods of time; and there are many examples, such as the Craterostigmidae amongst the Chilopoda, in which secondary, more recent, changes in habits have been associated with changed morphology. The Diplopoda also show similar examples in *Polyxenus* and the Lysiopetaloidea (Manton, 1956, 1958a), where secondary and more recent alterations in habits are associated with changes in the typical diplopodan organization.

It is to be hoped that personal ideas about the existence of diplosegments throughout the Myriapoda, which do not agree with modern knowledge, will not find their way into new text-books and so mislead large numbers of persons.

Acknowledgment

Permission from the Linnean Society of London's Zoological Journal to reproduce the figures in this paper is gratefully acknowledged.

References

Anderson, D. T. (1973). *Embryology and phylogeny in annelids and arthropods.* Oxford: Pergamon Press.

Blower, J. G. (1951). A comparative study of chilopod and diplopod cuticle. *Q. Jl microsc. Sci.* **92**: 141–161.

Demange, J.-M. (1963). La segmentation dorsale des Myriapodes Chilopodes au niveau de la zone des 7° et 8° segments. *C.r. hebd. Séanc. Acad. Sci., Paris* **257**: 514–517.

Demange, J. M. (1967). Recherches sur la segmentation du tronc des Chilopodes et des Diplopodes Chilognathes (Myriapodes). *Mém. Mus. natn. Hist. Nat. Paris.* (A) **44**: 1–188.

Heymons, R. (1901). Die Entwicklungsgeschichte der Scolopender. *Zoologica, Stutt.* **13**: 1–244.

Manton, S. M. (1950). The evolution of arthropodan locomotory mechanisms. Part 1. The locomotion of *Peripatus. J. Linn. Soc.* (Zool.) **41**: 529–570.

Manton, S. M. (1952). Part 3. The locomotion of the Chilopoda and Pauropoda. *J. Linn. Soc.* (Zool.) **42**: 118–166.

Manton, S. M. (1954). The evolution of arthropodan locomotory mechanisms. Part 4. The structure, habits and evolution of the Diplopoda. *J. Linn. Soc.* (Zool.) **42**: 299–368.

Manton, S. M. (1956). The evolution of arthropodan locomotory mechanisms. Part 5. The structure, habits and evolution of the Pselaphognatha (Diplopoda). *J. Linn. Soc.* (Zool.) **43**: 153–187.

Manton, S. M. (1958a). The evolution of arthropodan locomotory mechanisms. Part 6. Habits and evolution of the Lysiopetaloidea (Diplopoda), some principles of leg design in Diplopoda and Chilopoda, and limb structure of Diplopoda. *J. Linn. Soc.* (Zool.) **43**: 487–556.

Manton, S. M. (1958b). Habits of life and evolution of body design in Arthropoda. *J. Linn. Soc.* (Zool.) **44**: 58–72.

Manton, S. M. (1961). The evolution of arthropodan locomotory mechanisms. Part 7. Functional requirements and body design in Colobognatha (Diplopoda), together with a comparative account of diplopod burrowing techniques, trunk musculature and segmentation. *J. Linn. Soc.* (Zool.) **44**: 383–462.

Manton, S. M. (1965). The evolution of arthropodan locomotory mechanisms. Part 8. Functional requirements and body design in Chilopoda, together with a comparative account of their skeleto-muscular systems and an Appendix on a comparison between burrowing forces of annelids and chilopods and its bearing upon the evolution of the arthropodan haemocoel. *J. Linn. Soc.* (Zool.) **46**: 251–483.

Manton, S. M. (1966). The evolution of arthropodan locomotory mechanisms. Part 9. Functional requirements and body design in Symphyla and Pauropoda and the relationships between Myriapoda and pterygote Insects. *J. Linn. Soc.* (Zool.) **46**: 103–141.

Manton, S. M. (1972). The evolution of arthropodan locomotory mechanisms. Part 10. Locomotion, habits, morphology and evolution of the hexapod classes. *J. Linn. Soc.* (Zool.) **51**: 203–400.

Manton, S. M. (1974). The evolution of arthropodan locomotory mechanisms. Part 11. Habits and evolution of the Uniramia (Onychophora, Myriapoda, Hexapoda) and a survey of arachnid locomotory mechanisms. *J. Linn. Soc.* (Zool.). **53**: 257–375.

Ravoux, P. (1962). Étude sur la segmentation des Symphyles (Fondée sur la morphologie définitive et la postembryogenèse). Suivie de considérations sur la segmentation des autres myriapodes. *Annls Sci. nat.* (12) **4**: 141–472.

Tiegs, O. W. (1940). The embryology and affinities of the Symphyla, based on a study of *Hanseniella agilis. Q. Jl microsc. Sci.* **82**: 1–225.

Tiegs, O. W. (1945). The post-embryonic development of *Hanseniella agilis* (Symphyla). *Q. Jl microsc. Sci.* **85**: 191–329.

Tiegs, O. W. (1947). The development and affinities of the Pauropoda, based on a study of *Pauropus sylvaticus*. *Q. Jl microsc. Sci.* **88**: 165–336.

Discussion

BRADE-BIRKS: I should like to say how much impressed we have always been by the wonderful patience of Dr. Manton in her work. I am not qualified to say anything about the specific subject of the lecture but it has been a revelation tonight, to hear her speak and to see her slides.

DOHLE: You said that you did not start with the intention of drawing phylogenetic conclusions, but afterwards, having done your work, you draw them. You gave some examples, for instance, your Fig. 7, and I think one should read from above to below what you believe to be the specializations in phylogenesis. I think that every species is a mixture of specialized and more generalized or, say, "original" characters. You base your judgements principally on the skeleto-muscular systems. Could you say anything about the other organ systems? Do they point in the same direction?

MANTON: I would like to answer at great length but must be brief. Take the series in Chilopoda ending with the Scutigeromorpha. From the limited amount of data I have given to you today the Scutigeromorpha represent the most advanced members of the Chilopoda. I came to the same conclusion after studying the morphology of the head and the mechanisms of the mandibles and poison claws. The same conclusion is also supported by our knowledge of their visual acuity and by the presence of a respiratory pigment in their blood associated with their very rapid movements (*Scutigera* is the fleetest chilopod I have studied). I do not know of any feature of their anatomy, provided it is correctly interpreted and understood, which is not harmonious with this conclusion. I should have stressed the harmonious nature of all the features that I have mentioned today and a great many that I have not mentioned. All contribute to better living by facilitating certain habits rather than adapting the animals to specific habitats. I am aware that in the literature you will find the notion that the Scutigeromorpha are the most primitive of centipedes, but this cannot be substantiated by logical arguments if these are based on the whole anatomy and the functional relations of the various parts.

DOHLE: Yes, I was thinking of the work of Fahlander (1938)* for instance, who took quite the contrary view to yourself.

MANTON: Yes, but with far less evidence, I think.

* Fahlander, K. (1938). Beiträge zur Anatomie und Systematischen Einteilung der Chilopoden. *Zool. Bidr. Upps.* **17**: 1–148.

Symp. zool. Soc. Lond. (1974) No. 32, 191–198.

THE ORIGIN AND INTER-RELATIONS OF THE MYRIAPOD GROUPS

A summary of a free discussion prepared by the Chairman:

WOLFGANG DOHLE

The Free University, Berlin, Germany

As a basis for the discussion the relationship diagrams of Tiegs (1947), Dohle (1965) and Manton (1970) were presented (Figs 1–3).

MANTON gave a preview of her ideas on the classification of what she now calls the Phylum Uniramia (Onychophora—Myriapoda—Hexapoda)*, as follows:

It is probable that the Monognatha, Dignatha and Trignatha represent grades of organization, reached independently, and that they do not constitute taxonomic categories. There is no valid reason for supposing that Dignatha came from ancestral Monognatha. It is probable that Chilopoda and Symphyla have had dignathan ancestors, but this is not so for the Hexapoda, their trignathy having been directly acquired.

The basis of these conclusions lies in the comparative functional morphology of the head. All Crustacea and Chelicerata are primarily aquatic and deal with their food by gnathobases. The Chelicerata possess an anterior series of gnathobases reduced to one pair in most Arachnida. A fundamentally different gnathobase is seen in the mandible of Crustacea, working in quite a different manner from that of the Chelicerata. The Uniramia contrast in using the tip of a whole limb for manipulating food. The onychophoran jaw works by backward and forward slicing, the movement being near to the promotor–remotor movement of a walking limb (Manton, 1937, 1950); the terminal claws are enlarged to form cutting blades. The myriapodan mandible is quite different. It is jointed and bites by adduction in the transverse plane. Abduction by extrinsic mandibular muscles is an impossibility; the movement is effected by swinging anterior tentorial apodemes pushing the mandibles apart. The hexapod mandible is different again. It is unjointed and its primitive movement is not in the transverse plane, as in myriapods, but a rolling movement, derived from the promotor–remotor swing of a walking limb, which produces a grinding action of the molar lobes. Many changes have led to the evolution of

* This work has been published (Manton, 1972).

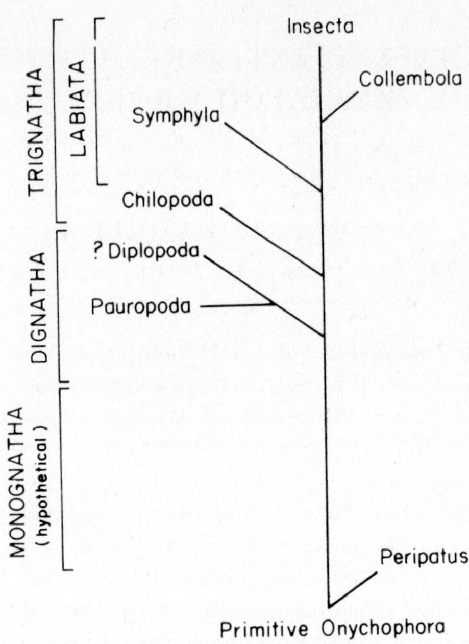
Insecta
Collembola
Symphyla
Chilopoda
? Diplopoda
Pauropoda
Peripatus
Primitive Onychophora
TRIGNATHA
LABIATA
DIGNATHA
MONOGNATHA
(hypothetical)

FIG 1

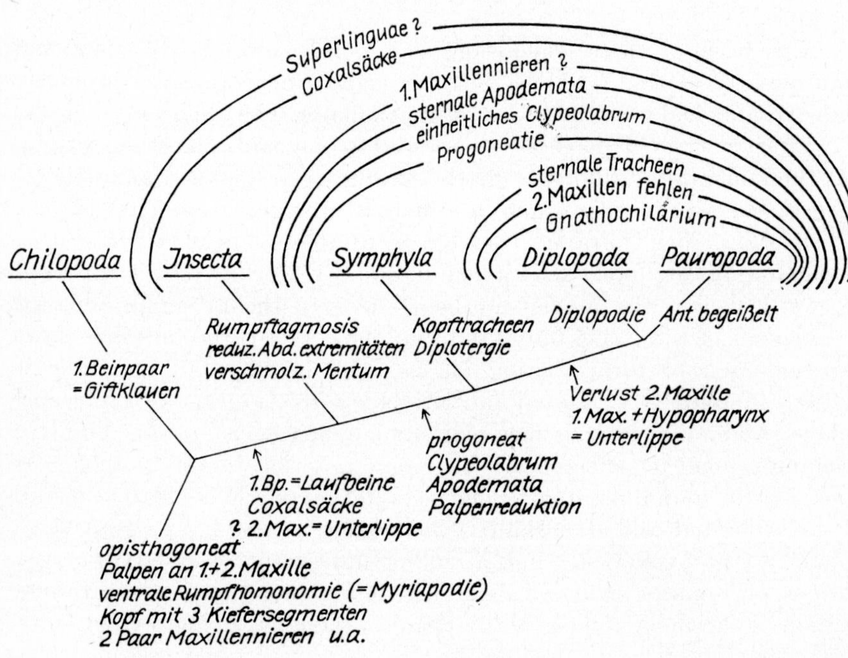
Superlinguae ?
Coxalsäcke
1.Maxillennieren ?
sternale Apodemata
einheitliches Clypeolabrum
Progoneatie
sternale Tracheen
2.Maxillen fehlen
Gnathochilarium
Chilopoda
Jnsecta
Symphyla
Diplopoda
Pauropoda
Rumpftagmosis
reduz. Abd. extremitäten
verschmolz. Mentum
Kopftracheen
Diplotergie
Diplopodie
Ant. begeißelt
1. Beinpaar
= Giftklauen
Verlust 2.Maxille
1. Max. + Hypopharynx
= Unterlippe
progoneat
Clypeolabrum
Apodemata
Palpenreduktion
1. Bp. = Laufbeine
Coxalsäcke
? 2. Max. = Unterlippe
opisthogoneat
Palpen an 1.+2. Maxille
ventrale Rumpfhomonomie (= Myriapodie)
Kopf mit 3 Kiefersegmenten
2 Paar Maxillennieren u.a.

FIG. 2

Captions on facing page

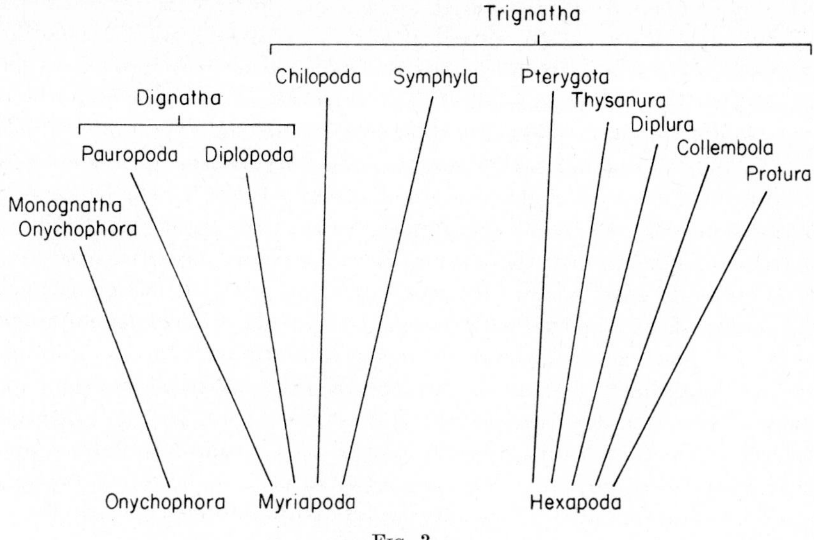

FIG. 3

FIGS 1–3. Relationship diagrams: Fig. 1: Tiegs, 1947; Fig. 2: Dohle, 1965, Verwandtschaftsschema der Antennaten auf Grund von wahrscheinlich synapomorphen Homologien aufgestellt und dann als Stammbaum gelesen; Fig. 3: Manton, 1970, the proximity of the roughly vertical lines indicates phylogenetic affinity. The grades of organization, Monognatha, Dignatha and Trignatha, do not indicate relationship.

secondary transverse biting in hexapods, just as transverse biting has been evolved independently a number of times within the Crustacea from their entirely different basic type of mandible. The hexapod mandibles need stable and fixed anterior tentorial apodemes and a head capsule braced also by the presence of a posterior pair of tentorial apodemes.

In the Onychophora the jaws function in association with entognathy and a dilatable circular lip, the pharynx being suctorial. In the Myriapoda and Hexapoda, the mandibles cannot function alone, whether they are entognathous or not (entognathy has been independently acquired many times in most groups of arthropods). In the Myriapoda the jaws and swinging tentorial apodemes, supported by the head capsule, depend on the limbs of the second maxilla segment in various ways. The hexapod mandible depends on the first maxilla and the second maxilla as well as the two pairs of tentorial apodemes for its normal functioning.

Thus the Myriapoda possess a functional and highly integrated unit comprising— a particular type of mandible, a swinging anterior tentorial apodeme and a maxilla, all with appropriate muscle connections and movements. It is possible to regard all the myriapod classes as basically dignathan, with varied posterior additions to the head in some classes, in which the additions lack both structural and functional uniformity. The Pauropoda show no additions to the basic dignathan head. The Diplopoda possess a post-gnathochilarial segment without limbs which, embryologically, contributes both to the posterior part of the head and to the collum segment (Dohle, 1964). In the Chilopoda a second maxilla segment, with limbs, forms part of the head behind the basic dignathan unit; these limbs, as mouth parts, are little modified. The Symphyla also possess a second maxilla segment which is very little modified as compared with the trunk segments in early development but does become part of the head at the same time as the mandibular and first maxillary segments. But this is no typical labial segment, such as occurs in the trignathan Hexapoda; there are no posterior tentorial apodemes and any resemblance to a labium is partial and convergent (Manton, 1964). The trignathy of Chilopoda and Symphyla is superimposed upon basic myriapodan dignathy. The diplopod condition has not advanced so far as in the Symphyla; only part of a post-gnathochilarial segment has been incorporated into the head. The Chilopoda have progressed even beyond a trignathan state, since the poison claw segment functionally forms part of the head. Instead of singling out the Pauropoda and Diplopoda as two divergent branches of a single dignathan stem, I prefer to regard the whole of the Myriapoda as basically dignathan, with various additional head features of a diverse nature.

The trignathan heads of the hexapod classes contrast with those of the Myriapoda. Hexapod trignathy is based upon a unique functional unit comprising—a particular type of mandible, requiring fixed apodemal support, a first maxilla, labium (second maxillae) and two pairs of fixed tentorial apodemes. This more elaborate unit cannot have arisen by fundamental alterations of the basic dignathan unit and the addition of a second maxilla segment with posterior tentorial apodemes. The well integrated hexapod trignathy has been very flexible in its evolutionary adaptations, but cannot be regarded as having evolved through a dignathan state as has the trignathy of Chilopoda and Symphyla. Hexapod heads have probably been trignathan as far back as their earliest differentiation. Thus the basic functional dignathan unit of all Myriapoda and the trignathan unit of Hexapoda are entirely distinct, the one could not have arisen from the other but both could be descended from early arthropods showing different patterns of cephalization (Fig. 4 level c.). Similarly a monognathan jaw unit, such as seen in the Onychophora, cannot have been a precursor of the dignathan or trignathan units because the differences between them are too great.

It is considered that the evidence to date does not justify either a very close alignment or a fundamental dichotomy between the Pauropoda and Diplopoda alone, terming them the Dignatha in a taxonomic sense. The morphological differences between the trunks of Pauropoda and Diplopoda are great, but they are explicable in terms of locomotory habits. It is much more realistic to regard all Myriapoda as basically dignathan, with various additions of an unconformable nature, to the hinder part of the head. The trignathy of the myriapods (Chilopoda and Symphyla) is quite distinct from that of the hexapods, and the latter cannot be derived from the former, or from any dignathan state, without proposing functionally impossible ancestral stages.

DEMANGE supported the idea of grouping all Myriapoda together. An additional argument for this arrangement would be that all chilopods and diplopods have *en principe* diplosegments, as Duboscq had suggested in 1900 and as Demange had concluded in 1967 ("tout Chilopode est fondamentalement Diplopode"), on the basis of the comparative anatomy of the skeleto-muscular system.

MANTON replied that Tiegs (1941, 1947) had decisively shown that the Symphyla and the Pauropoda have simple segments. She also affirmed that there is no evidence for diplosegments in the Chilopoda.

DOHLE pointed out that there is not only a difference of detail between the schemes of Tiegs (1947) and Manton (1970), but that there is also a principal difference. Tiegs drew connections between the different groups and thereby decided that in his opinion these groups have a single stem, i.e. a single common ancestor. Manton draws the lines without connecting them at their base. Therefore, in her classification the question remains open as to whether the undoubted similarities in structure and development between some of the groups indicate common ancestry or whether they have to be regarded as mere convergencies.

MANTON said that she chose this kind of presentation "to avoid all pitfalls". She said that she is convinced that there are many examples of convergence in arthropodan evolution, for instance, sclerotization, mandible formation, entognathy. While admitting that the different groups or classes that she assembles under the name Myriapoda and Hexapoda may have had a common stem, she regarded the concept of dichotomy in evolution as too much of an over-simplification.

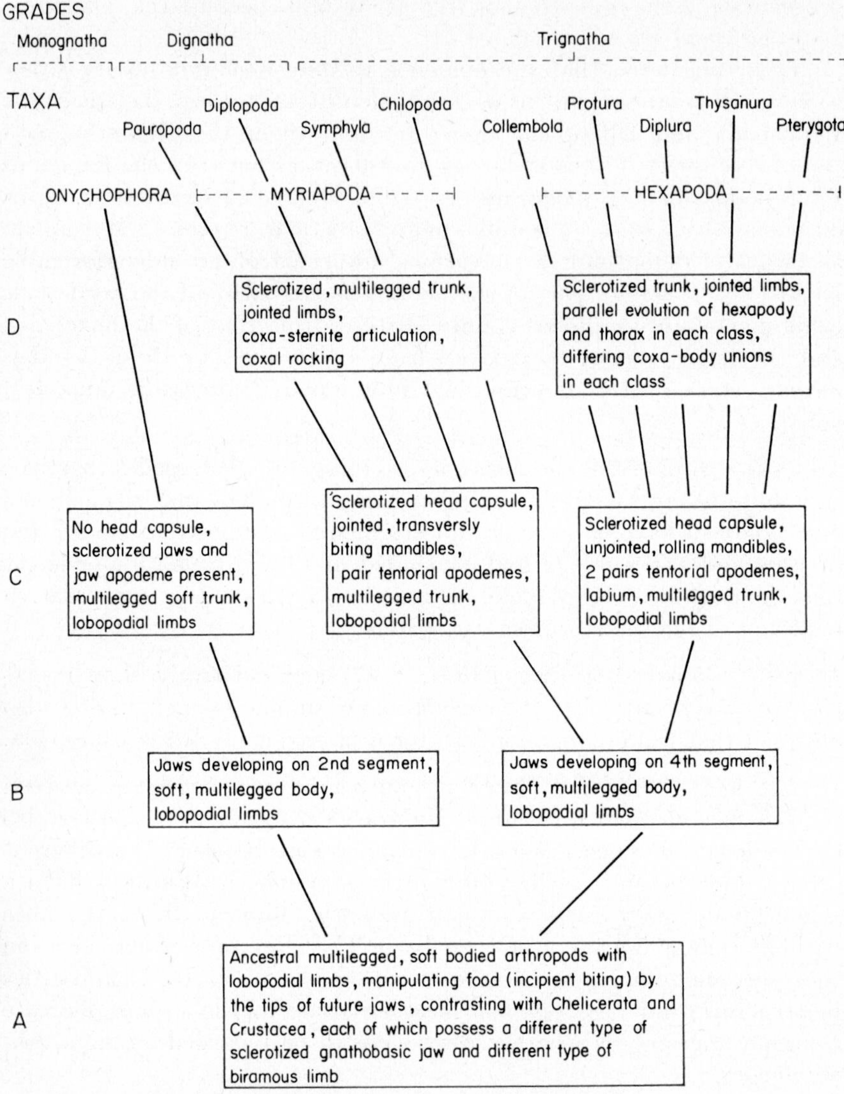

FIG. 4. Relationship diagram of the Phylum Uniramia. This diagram was presented to Congress as a preview of Manton (1972: Fig. 40).

MEIDELL said that at the Museum in Bergen the students are taught a system that in principle resembles that given by Tiegs. He said that his working-group had discussed the theories of Hennig (1966) on phylogenetic systematics and that they had come to the conclusion that Hennig's method is—in contrast to the opinion of many systematists—more applicable to the species and genus level than to the higher levels of class and order. Meidell criticized the diagram of Manton (1970*) in which all lines emerge isolated from each other. He said that both Diplopoda and Pauropoda lack appendages on the post-maxillary segment and that this lack has to be regarded as a derived character held in common by these two classes. If this interpretation is admitted, then, according to Hennig's principle of synapomorphy, they must have a common ancestor and they are to be regarded as a taxon. This means that the lines for Diplopoda and Pauropoda should be connected.

Enghoff on the basis of Meidell's arguments, also thought it desirable to connect Diplopoda and Pauropoda. He interpreted Manton's summary as pointing to a common stem for the Myriapoda. He said that perhaps Dr Manton did not regard her diagram as an evolutionary tree.

Dohle called attention to the possibility that all the characters enumerated in Dr Manton's new diagram in favour of a unification of the myriapod groups may be original characters ('symplesiomorphies' in the terminology of Hennig) i.e. characters of the common ancestor of Myriapoda and insects. If so, these characters would not provide justification for proposing a close phylogenetic relationship between the myriapod classes.

Kraus asked Dr Manton whether the jointed mandibles with transversely biting movement of Myriapoda and the unjointed mandibles with rolling movement of Hexapoda could not be interpreted as "Specialisationskreuzungen"; on the one hand the jointed mandibles of the Myriapoda may be an original feature whilst their transverse movement is specialized and derived, and on the other hand the unjointed mandibles of the Hexapoda may be a derived character although they retain the original feature of a rolling movement similar to the leg movement.

Manton denied that possibility. She does not regard the movement of the myriapod mandible as derived. She said that in some chilopods the first pair of walking legs has a function and movement similar to the gnathal appendages in holding the prey.

The discussion was concluded with the remark that this discussion as well as that of the afternoon should be accepted as a challenge. Every systematist and taxonomist should pay regard to the theoretical

* Reproduced here as Fig. 3.

background of systematics. He should always state precisely to what sort of system he makes his contribution, whether he is trying to reconstruct the phylogenetic evolution of the species by the methods of phylogenetic systematics, whether he wants to give a summary of all the similarities between the species that have been quantified in a typological system, or whether he merely wishes to give a symbolized key for determination.

References

Demange, J. -M. (1967). Recherches sur la segmentation du tronc des Chilopodes et des Diplopodes Chilognathes (Myriapodes). *Mém. Mus. natn. Hist. nat. Paris* (A) **44**: 1–188.

Dohle, W. (1964). Die Embryonalentwicklung von *Glomeris marginata* (Villers) im Vergleich zur Entwicklung anderer Diplopoden. *Zool. Jb.* (Anat.) **81**: 241–310.

Dohle, W. (1965). Über die Stellung der Diplopoden im System. *Zool. Anz.* (Suppl.) **28**: 597–606.

Duboscq, O. (1900). Le développement de la Scolopendre d'après Heymons. *Archs Zool. exp. gén.* (Notes et revue). **12**: XXVI–XXXII.

Hennig, W. (1966). *Phylogenetic systematics.* Urbana–Chicago–London: University of Illinois Press.

Manton, S. M. (1937). Studies on the Onychophora. II. The feeding, digestion, excretion and food storage of *Peripatopsis*. *Phil. Trans. R. Soc.* (B) **227**: 411–464.

Manton, S. M. (1950). The evolution of arthropodan locomotory mechanisms. Part I. The locomotion of *Peripatus*. *J. Linn. Soc.* (Zool.) **41**: 529–570.

Manton, S. M. (1964). Mandibular mechanisms and the evolution of arthropods. *Phil. Trans. R. Soc.* (B) **247**: 1–183.

Manton, S. M. (1970). Arthropods: Introduction. In *Chemical zoology* **5**: 1–34. Florkin, M. and Scheer, B. T. (eds.). New York and London: Academic Press.

Manton, S. M. (1972). The evolution of arthropodan locomotory mechanisms. Part 10. Locomotory habits, morphology and evolution of the hexapod classes. *J. Linn. Soc. (Zool.)* **51**: 203–400.

Tiegs, O. W. (1941). The embryology and affinities of the Symphyla, based on a study of *Hanseniella agilis*. *Q. Jl microsc. Sci.* **82**: 1–225.

Tiegs, O. W. (1947). The development and affinities of the Pauropoda, based on a study of *Pauropus silvaticus*. *Q. Jl microsc. Sci.* **88**: 165–267, 275–336.

Symp. zool. Soc. Lond. (1974) No. 32, 199–210.

ÉTUDE ULTRASTRUCTURALE DE L'ORGANE NEUROHÉMAL CÉRÉBRAL DE *SPELAEOGLOMERIS DODEROI* SILVESTRI MYRIAPODE DIPLOPODE CAVERNICOLE

C. JUBERTHIE and L. JUBERTHIE-JUPEAU

Laboratoire souterrain du C. N. R. S., Moulis, France.

SYNOPSIS

A study using both light and electron microscopy reveals that the cerebral neurohaemal organs of *Spelaeoglomeris doderoi* have no true secretory cells but accumulate the neurosecretory product, most of which is liberated into the haemolymph in the region of the organ, a smaller part passing by the para-oesophageal nerves to the para-oesophageal bodies. The neurohaemal organ contains four types of axons distinguished by their diameters and by the size and electron density of the inclusions of secretion. The axon-types are compared with those recently described in *Craspedosoma.*

The cerebral nerve which supplies the organ contains the same types of axons, at least three of which probably correspond to three different neurosecretory products and not stages in the elaboration of a single product within the organ. The absence of true secretory cells is probably a primitive condition as also described in other millipedes and in the corpora cardiaca of collemboles. The corpora cardiaca of higher insects differ in possessing secretory cells.

INTRODUCTION

Nos connaissances sur les centres nerveux des diplopodes oniscomorphes sont toutes à rapporter à Saint-Remy (1890). Cet auteur a réalisé en effet l'étude architectonique du cerveau chez cinq espèces de *Glomeris,* qui toutes ont montré une structure identique.

Un complexe endocrinien du même type que celui qui a été décrit chez de nombreux arthropodes a été mis en évidence chez plusieurs diplopodes dont *Glomeris marginata* par Gabe (1954) qui a d'une part localisé les cellules neurosécrétrices protocérébrales, et d'autre part découvert un organe d'accumulation du produit de neurosécrétion. Par la suite, l'existence de cet organe d'accumulation a été reconnue chez plusieurs autres oniscomorphes: *Loboglomeris pyrenaica* et *L. rugifera, Glomeris intermedia* et *Glomeridella kervillei* (Juberthie-Jupeau, 1967a). Ce travail s'est accompagné de la mise en évidence chez toutes ces espèces et également chez *Glomeris marginata* de corps paraeosophagiens pairs, situés sous le cerveau, au voisinage du collier périoesophagien et en relation chacun avec la "glande cérébrale" correspondante par l'intermédiaire d'un nerf qui traverse l'organe; tout le long de ce nerf les flaques de neurosécrétion ont été observées. Sahli (1966) fait une mise au point de ces organes déjà decrits chez d'autres diplopodes.

Les oniscomorphes inféodés au milieu souterrain n'ont fait l'objet jusqu'à ce jour d'aucune investigation dans le domaine de l'endocrinologie. Ces animaux présentent comme la plupart de ceux qui se rencontrent dans ce milieu une extrême lenteur de développement et une fécondité très faible (Juberthie-Jupeau, 1967b). Le but de ce travail est donc l'étude de l'un des éléments du complexe endocrinien du diplopode oniscomorphe troglobie *Spelaeoglomeris doderoi* Silvestri.

Cette étude porte sur la glande cérébrale ou mieux, organe neurohémal cérébral; elle est faite en microscopie photonique et électronique, et prélude à une étude histophysiologique de différents organes.

Matériel et techniques

Les animaux étudiés, des adultes, proviennent de la grotte du Bédat, Bagnères de Bigorre (Hautes-Pyrénées).

Pour la microscopie photonique ils ont été fixés aux liquides de Carnoy, de Bouin, de Bouin-Duboscq et de Halmi. Les animaux ont été débités en coupes sériées de 7 μm d'épaisseur et colorés au bleu alcian-hémalun-phloxine, à la fuchsine-paraldéhyde, à l'hématoxyline chromique phloxine et au vert de méthyle-pyronine.

Pour la microscopie électronique, le système nerveux cérébral a été fixé après dissection dans le Ringer ou dans le fixateur; les fixations ont été faites soit à l'acide osmique à 2% dans le tampon Millonig II pendant 1 h à 1–2°C, soit au glutaraldéhyde à 3% dans le tampon Millonig II, 30 min à 1 h à 1–2°C suivi d'une seconde fixation à l'acide osmique à 2% dans le tampon Millonig II durant 30 min à 1–2°C. Les pièces ont été déshydratées à l'éthanol, passées à l'oxyde de propylène puis dans un bain d'oxyde de propylène et d'Epon à parties égales et incluses à l'Epon. Les blocs ont été débités en coupes sériées au microtome Reichert avec un couteau en diamant, montées sur grilles sans support, et contrastées à l'acétate d'uranyle puis au citrate de plomb. Elles ont été observées au Microscope Sopelem Micro 75 du Laboratoire sous une tension de 50 kv.

ÉTUDE ANATOMIQUE ET STRUCTURALE

Il existe une paire d'organes neurohémaux cérébraux chez *Spelaeoglomeris doderoi*. Chacun d'eux est situé latéralement sous le cerveau et sous le nerf de l'organe de Tömösvary près du tégument de la capsule céphalique. Vu de dessus il se situe entre la partie latéro-externe de la base de l'antenne et l'organe de Tömösvary correspondant. Notons que chez cette espèce dépourvue d'yeux le nerf de l'organe de Tömösvary quitte seul les parties tout-à-fait latérales du protocérébron.

Chaque organe neurohémal est ovoïde, allongé, son grand axe étant grossièrement parallèle au nerf de l'organe de Tömösvary; il mesure environ 50 à 70 μm dans son grand axe et 20 à 30 μm dans son petit axe perpendiculaire au premier. Il est appendu par son extrémité interne à un nerf très court qui quitte le cerveau face ventrale et antérieure à la limite du deuto- et du protocérébron (Fig. 1).

Chaque organe est enveloppé par une fine enveloppe en continuité avec l'épinèvre du nerf qui le pénètre. Cette enveloppe s'épaissit par endroits et possède à ce niveau quelques cellules qui semblent de type glial. A l'intérieur de cette enveloppe l'organe se présente comme une masse compacte mais hétérogène qui se caractérise par l'absence totale de cellules sécrétrices propres. Cette absence s'est révélée de façon constante chez tous les animaux observés et quelle que soit la technique mise en oeuvre. Les méthodes de détection des produits de neurosécrétion mettent en évidence de nombreux grains et flaques plus ou moins volumineux de produit neurosécrété. Ces flaques et ces grains se colorent en effet en violet après l'emploi de la fuchsine paraldéhyde, en bleu après l'emploi du bleu alcian et en bleu noir après la mise en oeuvre de la technique de Gomori. A côté de ces éléments il en existe d'autres plus nombreux qui se teintent en rose par la phloxine.

Notons que des fibres nerveuses et neurosécrétrices quittent l'organe; en particulier, il en part un nerf qui longeant le cerveau face ventrale

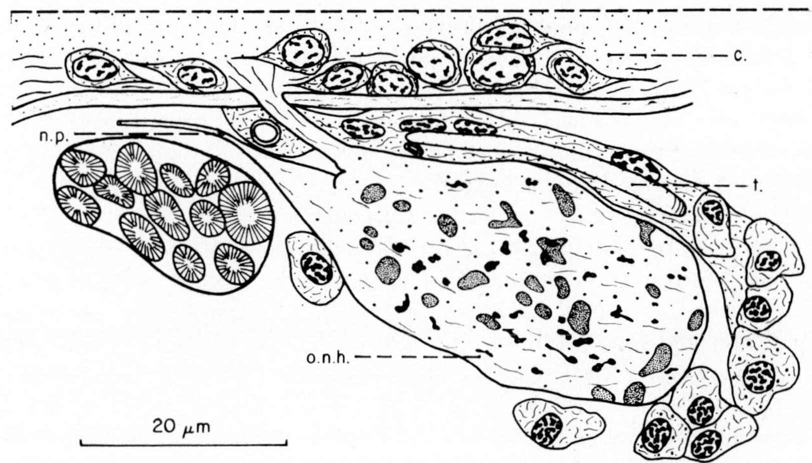

FIG. 1. Organe neurohémal cérébral (o.n.h.). *c.* = cerveau; *n. p.* = nerf du corps para-oesophagien; *t.* = trachée. Halmi, Bleu alcian-hémalun-phloxine.

gagne un corps para-oesophagien qui semble identique à celui des autres oniscomorphes.

Notons enfin que l'organe est entouré d'un tissu assez lâche et discontinu au sein duquel se reconnaissent des cellules sanguines et des trachées.

ULTRASTRUCTURE DE L'ORGANE NEUROHEMAL CEREBRAL

Enveloppe

L'organe neurohémal baigne en totalité dans l'hémolymphe. Il est limité par une lamelle neurale très mince, de 200 à 400 Å d'épaisseur, tandis que celle-ci, autour du cerveau, mesure de 4500 à 5500 Å environ. Elle est formée d'une substance amorphe et elle est dépourvue, comme autour du cerveau, de fibres de collagène. Sur une partie de l'organe, la lamelle neurale est la seule enveloppe et la seule barrière entre les axones neurosécréteurs périphériques et le sinus sanguin. Ailleurs, un périneurium, formé de quelques prolongements gliaux, s'interpose entre les axones et la lamelle neurale notamment aux points d'arrivée et de départ des nerfs; au point d'arrivée du nerf cérébral il est bien développé et se compose de prolongements et de noyaux de cellules gliales; il s'insinue également dans certaines digitations qui pénètrent dans l'organe; quelques trachéoles accompagnent ces prolongements internes.

Types d'axones neurosécréteurs

L'organe neurohémal est presque exclusivement formé d'axones et de terminaisons axoniques contenant de la neurosécrétion (Fig. 2); il ne renferme que de très rares cellules gliales et aucune cellule sécrétrice propre. Sur une coupe transversale on peut dénombrer de 400 à 500 axones renfermant des grains de neurosécrétion, de plusieurs types, imbriqués les uns dans les autres.

Les différences entre les types d'axones sont plus tranchées après fixation au tétroxyde d'osmium qu'après double fixation au glutaraldéhyde et tétroxyde d'osmium. En revanche la membrane des grains élémentaires est mieux conservée après double fixation, et les microtubules ne sont bien préservés qu'après ce type de fixation. Le diamètre des grains élémentaires reste du même ordre de grandeur dans les deux cas.

Un premier groupe d'axones, les plus nombreux, renferme des grains élémentaires de neurosécrétion, denses aux électrons, que l'on peut rapporter au type A par leur taille, de 900 à 2200 Å et par leur densité aux électrons; ils rappellent les grains de neurosécrétion renfermant des hormones polypeptidiques. Un second groupe d'axones que l'on peut rapporter au type B renferme des grains de faible densité aux électrons et de petite taille 800 à 1150 Å (Fig. 2).

Les axones de type A peuvent être divisés en trois sous-groupes (A1, A2, A3) en se fondant sur la taille, le nombre et la densité des grains qu'ils renferment. Les images sur lesquelles repose cette classification des axones rappellent celles qui ont été obtenues chez le diplopode *Craspedosoma* (Seifert, 1971), dans la glande du sinus des crustacés (Bunt & Ashby, 1967), et en partie dans les corpora cardiaca des insectes (Cazal, Joly & Porte, 1971).

Type A1

Environ 30% des axones appartiennent à ce type (Fig. 3). Ce sont eux qui atteignent les plus grands diamètres, la majorité d'entre eux mesurent de 2 à 3,5 μm en section transversale; quelques-uns cependant n'excèdent pas 1 μm à 1,5 μm de diamètre et correspondent probablement à des ramifications latérales ou terminales des premiers. Ils renferment un nombre élevé de grains élémentaires de neurosécrétion très denses aux électrons parmi lesquels se trouvent les plus gros de l'organe neurohémal, avec 2500 Å de diamètre; leur taille est éminemment variable et va de 1000 à 2500 Å. Ces grains denses sont répartis au sein de grains gris, de faible densité aux électrons, (ceci est surtout net après fixation au tétroxyde d'osmium seul), dont le bord est indistinct et qui ne sont que des coupes tangentielles plus minces de granules denses car ainsi que le montrent des coupes sériées successives, à l'emplacement d'un grain gris sur une coupe on trouve un dense sur la suivante. Les axones renferment quelques mitochondries ovoïdes ou allongées et à crêtes peu nombreuses, ainsi que de très rares microtubules au niveau des dilations axonales où s'accumulent les grains.

Ils renferment un petit nombre de vésicules synaptoides claires, à membrane nette, de 300 à 550 Å de diamètre. Elles sont éparses entre les granules de neurosécrétion ou près de la membrane plasmique; parfois elles sont groupées près de l'axolemme en petits bouquets de quatre à dix unités en général. On n'observe pas de groupement important de ces vésicules.

Ces axones présentent également entre les grains quelques vésicules de pinocytose de 900 à 1100 Å de diamètre environ.

Type A2 (*Fig. 5*)

Les axones de ce type se différencient des précédents par un diamètre moyen plus petit et par des grains denses aux électrons de taille plus

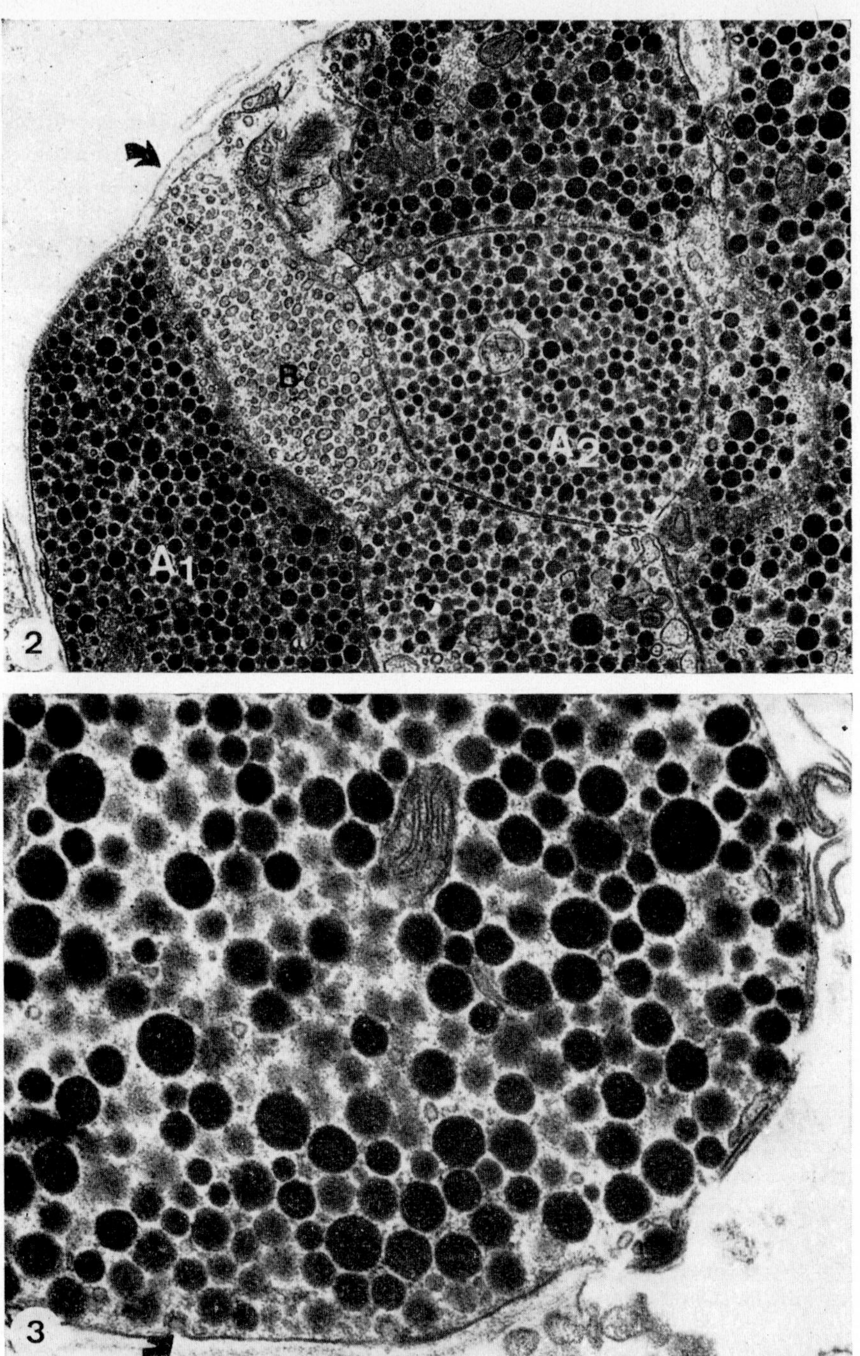

FIG. 2. Organe neurohémal cérébral. Vue générale montrant trois types d'axones neurosécréteurs A_1, A_2 et B et la mince lamelle neurale périphérique (flèche). Tétroxyde d'osmium: × 14 500.

FIG. 3. Axone de type A_1. Formation de vésicules de pinocytose (flèche) sous la lamelle neurale périphérique: × 36 000.

faible. Environ 30% des axones appartiennent à ce type. Leur diamètre varie de 1 à 2 μm et la taille de leurs grains élémentaires de 900 Å à 1200 Å. L'espace entre les grains denses est occupé par des grains de faible densité aux électrons, qui ne sont que des grains denses coupés tangentiellement. Les autres éléments figurés sont identiques à ceux des axones de type A1.

Type A3 (*Fig. 6*)

Les axones de ce type sont très peu nombreux, à peine une dizaine dans l'ensemble de l'organe. Ils sont voisins du type A1 par le diamètre élevé des axones, la taille et la forte densité aux électrons des granules élémentaires. Il s'en distingue au premier chef par le nombre peu élevé des grains, ce qui donne à l'axoplasme un aspect clair caractéristique. Le diamètre des grains s'échelonne entre 900 et 2200 Å.

Les vésicules synaptoides y sont également éparses, leur nombre et leur taille semblent du même ordre que dans les axones A1; les éléments de réticulum endoplasmique lisse sont plus nombreux et plus dilatés; des corps multilamellaires sont parfois présents.

Type B (Fig. 7)

Près de 40% des axones de l'organe appartiennent à ce type mais ce sont des fibres de diamètre relativement faible s'échelonnant entre 1 et 2 μm, de sorte qu'en volume leur importance est moindre.

Après fixation au tétroxyde d'osmium, les axones de type B, contrairement aux axones de type A, renferment presque exclusivement des grains de faible densité électronique, grains "gris" de certains auteurs, limités par une membrane très nette. Ces grains élémentaires, très nombreux, mesurent de 800 Å à 1150 Å; rarement sphériques, leur aspect rappelle souvent celui d'une sphère déformée de façon irrégulière, parfois leur membrane semble éclatée. A un grossissement de 150 000, leur contenu n'apparaît pas homogène mais comme composé soit de grains contigus laissant entre eux des vides soit de microvésicules sans qu'il soit possible de trancher. Ils rappellent ainsi le contenu des grains décrits par Cazal (1971) chez *Locusta migratoria*.

Après double fixation les grains sont mieux conservés et se différencient un peu moins nettement des grains des fibres de type A2.

Quelques-uns de ces axones renferment de rares grains denses de même diamètre que les grains "gris"; le plus souvent ils présentent une membrane nette décollée du grain central sphérique et dense et représentent un sous-type.

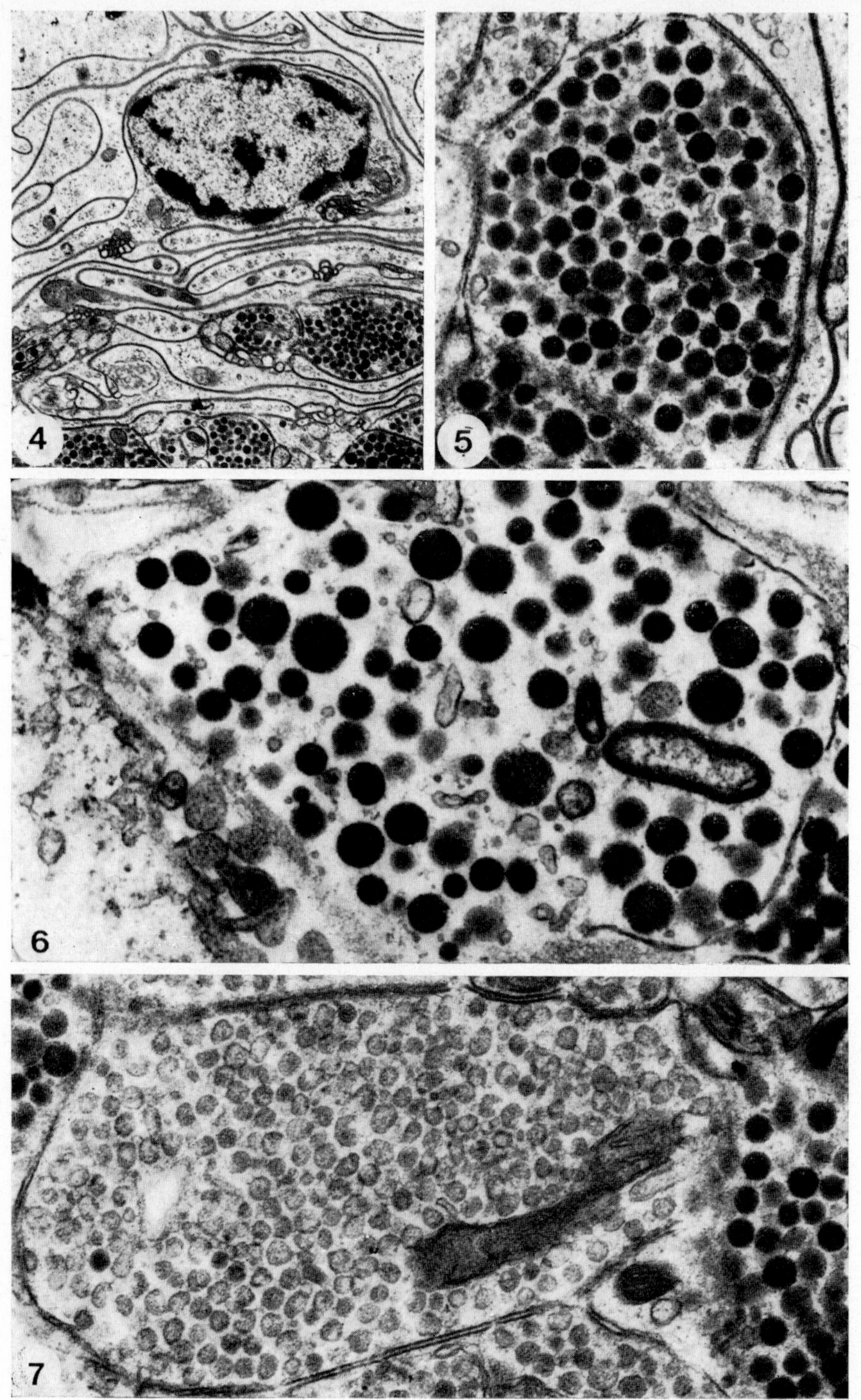

Captions on opposite page

Les mitochondries sont longues ou ovoides, à matrice dense, à crêtes peu nombreuses, les vésicules synaptoïdes éparses et en petit nombre.

Cellules gliales (Fig. 4)

L'organe neurohémal renferme, entre certains paquets d'axones, des prolongements gliaux, qui ont deux origines. Les uns proviennent des très rares cellules gliales que renferme l'organe, les autres de cellules gliales externes périneurales dont les prolongements pénètrent dans l'organe, leur invagination, lorsqu'elle est importante, étant accompagnée et limitée par la lame neurale.

Les noyaux gliaux, globuleux, mesurent de 3 à 5 μm; le cytoplasme renferme quelques mitochondries courtes plus rarement allongées à crêtes peu nombreuses, des ribosomes épars, de nombreux éléments courts du réticulum endoplasmique, la plupart lisses, plusieurs appareils de Golgi apparemment sans activité, et souvent des corps denses. Les prolongements gliaux s'unissent entre eux par des tight junctions particulièrement longues à la périphérie.

Pinocytose

Les axones périphériques séparés de l'hémolymphe par la mince lamelle neurale, aussi bien que les axones du centre de l'organe contigus à d'autres axones ou à des prolongements gliaux, présentent un grand nombre de formations en "omega" sur leur membrane plasmique. Elles sont particulièrement développées dans les axones de type A1 (Fig. 3); elles donnent des vésicules de pinocytose de 1000 Å de diamètre environ, au sein de l'axoplasme.

ULTRASTRUCTURE DU NERF CÉRÉBRAL DANS SON TRAJET EXTERNE

Le nerf cérébral par lequel arrivent à l'organe neurohémal les axones des cellules neurosécrétrices du cerveau est en première approximation cylindrique et se compose:

1. d'une lamelle neurale de faible épaisseur (450 à 500 Å),
2. d'une couche périneurale bien développée, formée de cellules gliales, dont les prolongements laissent entre eux des espaces vides.

Fig. 4. Cellule et noyau glial de l'enveloppe périneurale du nerf cérébral près de son entrée dans la glande: × 6000.

Fig. 5. Axone de type A_2: × 36 000.

Fig. 6. Axone de type A_3: × 36 000.

Fig. 7. Axone de type B: × 36 000.

3. d'une centaine d'axones neurosécréteurs de trois types différents (A1, A2, B), qui ont les mêmes caractéristiques que ceux décrits dans l'organe neurohémal. Ces axones sont un peu moins groupés que dans l'organe, certains paquets étant en effet, isolés dans le périneurium.

Le type A3 n'est représenté que par une seule fibre; il est peu probable que la dizaine de fibres de ce type que renferme la glande provienne de la division de cette fibre unique. Dans l'organe leur origine est soit différente ou bien elle représente un stade de décharge avancé des fibres de type A1.

CONCLUSIONS

1. De l'ultrastructure de "l'organe neurohémal cérébral" de *Spelaeoglomeris* l'on peut conclure qu'une partie du produit de neurosécrétion cérébral se stocke à son niveau et y est libérée dans l'hémolymphe, et que cette formation est dépourvue de cellules sécrétrices propres.

Chez *Spelaeoglomeris*, l'organe neurohémal cérébral sans sécrétion intrinsèque constitue un dispositif que nous considérons comme primitif et qui a été précédemment décrit en microscopie photonique chez les symphyles sous le nom de glandes céphaliques (Juberthie-Jupeau, 1961, 1963), chez les diplopodes iulides sous le nom d'organe de Gabe (Sahli, 1966), chez *Jonespeltis splendidus* (Prabhu, 1961), chez *Polyxenus* (Nguyen Duy-Jacquemin, 1971) et récemment retrouvé en microscopie électronique chez le diplopode nématophore *Craspedosoma* (Seifert, 1971). Les diplopodes étudiés semblent donc homogènes à ce point de vue, et la microscopie électronique confirme les interpretations des auteurs cités ci-dessus, sur des préparations observées en microscopie photonique. De plus, l'on sait maintenant que ce dispositif simple caractérise le groupe inférieur d'insectes collemboles, et que des cellules sécrétrices intrinsèques ne s'y adjoignent que chez les aptérygotes les plus évolués et les insectes supérieurs pour former les corpora cardiaca (Juberthie & Cassagnau, 1971).

2. Les auteurs tendent de plus en plus à distinguer plusieurs types d'axones neurosécréteurs en fonction de la taille, de la structure, et de la densité aux électrons des granules élémentaires. Il est pour l'instant impossible de raccorder avec certitude un type de granule donné à une hormone et certains auteurs pensent qu'un même grain pourrait renfermer plus d'une hormone. En effet, le nombre de types d'axones reconnus dans la glande du sinus (Bunt & Ashby, 1967) et dans les corpora cardiaca de *Locusta migratoria* (Cassier & Fain-Maurel, 1970) est inférieur à celui des hormones connues.

Le fait que le nerf cérébral qui se rend à l'organe neurohémal renferme les mêmes types d'axones que cet organe, tend à prouver qu'au moins trois types A1, A2, et B, correspondent bien à trois types de produit de neurosécrétion et ne représentent pas des stades de transformation d'un seul produit dans l'organe neurohémal.

3. On peut homologuer avec vraisemblance le type A1 de *Spelaeoglomeris* au type T3 de *Craspedosoma* et A2 au T2. Chez les deux espèces, quelques rares axones à axoplasme clair, à grains peu nombreux, sont observés, mais la forte différence de taille des grains ne permet pas de les homologuer; s'ils traduisent des phénomènes de réplétion, ils seraient à rapprocher chez *Spelaeoglomeris* du type A1 et chez *Craspedosoma* du type T2. Les autres types ne peuvent être homologués.

BIBLIOGRAPHIE

Bunt, A. H. & Ashby, E. (1967). Ultrastructure of the sinus gland of the crayfish, *Procambarus clarkii*. *Gen. comp. Endocr.* **9**: 334–342.

Cassier, P. & Fain-Maurel, M. A. (1970). Contribution à l'étude infrastructurale du système neurosécréteur rétrocérébral chez *Locusta migratoria migratorioïdes* (R. et F.) I. Les corpora cardiaca. *Z. Zellforsch. mikrosk. Anat.* **111**: 471–482.

Cazal, M. (1971). *Les corpora cardiaca chez* **Locusta migratoria** *L. et leurs fonctions.* Thèse, Montpellier: 1–192.

Cazal, M., Joly, L. & Porte, A. (1971). Etude ultrastructurale des corpora cardiaca et de quelques formations annexes chez *Locusta migratoria* L. *Z. Zellforsch. mikrosk. Anat.* **114**: 61–72.

Gabe, M. (1954). Emplacement et connexions des cellules neurosécrétrices chez quelques Diplopodes. *C. r. hebd. Séanc. Acad. Sci. Paris* **238**: 828–830.

Juberthie, C. & Cassagnau, P. (1971). L'évolution du système neurosécréteur chez les Insectes; l'importance des Collemboles et des autres Aptérygotes. *Rêvue Ecol. Biol. Sol* **8**: 59–80.

Juberthie-Jupeau, L. (1961). Données sur la neurosécrétion protocérébrale et mise en évidence de glandes céphaliques chez *Scutigerella pagesi* Jupeau (Myriapode Symphyle). *C. r. hebd. Séanc. Acad. Sci. Paris* **253**: 3081–3083.

Juberthie-Jupeau, L. (1963). Recherches sur la reproduction et la mue chez les Symphyles. *Archs Zool. exp. gén.* **102**: 1–172.

Juberthie-Jupeau, L. (1967a). Données sur le système endocrinien de quelques Diplopodes Oniscomorphes (Myriapodes). *C. r. hebd. Séanc. Acad. Sci. Paris* **265**: 1527–1529.

Juberthie-Jupeau, L. (1967b). Etude du biotope et du développement d'un Diplopode cavernicole, *Spelaeoglomeris doderoi* Silvestri. *Spelunca Mem.* **5**: 273–276.

Nguyen Duy-Jacquemin, M. (1971). Mise en évidence de glandes cérébrales chez le Diplopode Pénicillate *Polyxenus lagurus*. *C. r. hebd. Séanc. Acad. Sci. Paris* **272**: 1984–1986.

Prabhu, V. K. K. (1961). The structure of the cerebral glands and connective bodies of *Jonespeltis splendidus* Verhoeff (Myriapoda, Diplopoda). *Z. Zellforsch. mikrosk. Anat.* **54**: 717–733.

Sahli, F. (1966). *Contribution à l'étude de la périodomorphose et du système neurosécréteur des Diplopodes Iulides*. Thèse, Dijon: 5–228.

Saint-Remy, G. (1890). *Contribution à l'étude du cerveau chez les Arthropodes Trachéates*. Thèse, Paris.

Seifert, G. (1971). Ein bisher umbekanntes Neurohämalorgan von *Craspedosoma rawlinsii* Leach (Diplopoda, Nematophora). *Z. Morph. Tiere* **70**: 128–140.

Discussion

Strasser: You said that cavernicoles have a longer development and lay fewer eggs than epigeic forms. I think this would be interesting to pursue, and at what better place than Moulis!

Juberthie-Jupeau: It is a fact which has been stated in a general way from the biological point of view. Other studies have also shown that the metabolism of cavernicolous species is slowed down compared with that of epigeic animals. There is in fact a large literature on the biology of cavernicoles but up to now this has not dealt with these modifications and the related endocrine phenomena. We have begun to do this but we do not know which organs are implicated and in what manner.

Symp. zool. Soc. Lond. (1974) No. 32, 211–216.

LES ORGANES INTRACÉRÉBRAUX DE *POLYXENUS LAGURUS* ET COMPARAISON AVEC LES ORGANES NEURAUX D'AUTRES DIPLOPODES

NGUYEN DUY-JACQUEMIN

Muséum national d'Histoire naturelle, Laboratoire de Zoologie Arthropodes, Paris, France

SYNOPSIS

The intracerebral organs of *Polyxenus lagurus* are subspherical bodies symmetrically placed on each side of the medial groove, in the dorso-medial part of the protocerebrum. The nuclei of these cells look similar to those of the cerebral cortex and have a reduced cytoplasm. These cells gather around a mass of material, which appears to be colloidal in nature. The organs are present in both larva and adult. They can be compared with the "neural organs" described in iulids and glomerids. A study of their embryology will establish their origin.

INTRODUCTION

Sous le neurilemme protocérébral de certains diplopodes un organe pair a été décrit sous des noms divers: ilôts intraglobulaires (Sahli, 1966), glandes neurales (Junqua, 1966), organes neuraux (Juberthie-Jupeau, 1967). Ce dernier terme est adopté actuellement. Ces formations sont à rapprocher des organes ventraux des pycnogonides et des péripates, des organes neuraux des opilions et des glandes neurales des solifuges, qui dérivent d'invaginations ectodermiques embryonnaires: les organes ventraux. Chez le diplopode: *Platyrrhacus amauros* (Attems)* et les chilopodes, tels les scolopendres, ils sont présents chez l'embryon, disparaissent chez les larves; les adultes en sont donc dépourvus.

Au cours de nos recherches sur la neurosécrétion de *Polyxenus lagurus*, nous avons découvert des formations analogues dans le cerveau de cette espèce. Leur étude fait l'objet de la présente note.

DESCRIPTION DE LA PAIRE D'ORGANES INTRACÉRÉBRAUX DE *POLYXENUS LAGURUS*

Ce sont deux formations subsphériques, occupant une position

* Depuis les travaux de Pflugfelder (1932) sur cette espèce nous savons que les organes disparaissent très tôt au cours du développement postembryonnaire (2ème larve). Heathcote (1888) admet chez "*Iulus terrestris* Leach" la disparition de ces formations peu après l'éclosion; ayant rencontré celles-ci chez les adultes de *Tachypodoiulus albipes* (C.L.K.) et *Schizophyllum sabulosum* (L.) il paraît nécessaire de vérifier ses assertions.

symétrique par rapport au sillon sagittal, dans la zone médio-dorsale du protocérébron (Fig. 1). Elles comprennent une cavité centrale entourée de noyaux, sauf en profondeur où la cavité est en contact direct avec le neuropile. Ces noyaux appartiennent à des cellules à cytoplasme réduit, à membrane cellulaire peu visible et qui se distinguent mal des cellules du cortex cérébral. Dans certaines préparations les noyaux sont légèrement aplatis contre la sphère centrale; parfois ils sont isolés de ceux de l'écorce cérébrale par suite de la rétraction due à la fixation (Fig. 1). Cela prouve l'autonomie de l'organe. Ce dernier est séparé du neurilemme recouvrant le lobe frontal protocérébral par deux ou trois couches de noyaux.

Sous l'action des colorants, le contenu de la cavité prend une teinte à peine plus foncée que celle du neuropile. Il s'en distingue par un aspect colloïdal et une structure légèrement radiée pouvant présenter quelques lacunes (Fig. 1). Le diamètre de la cavité, mesuré sur une coupe horizontale, est d'environ 9 μm chez les adultes, alors que, sur une coupe sagittale, il atteint seulement 7 μm.

La paire d'organes intracérébraux est présente à tous les stades larvaires; le plus grand diamètre de la cavité centrale est d'environ 4 μm chez la larve du premier stade; il croît progressivement au cours du développement postembryonnaire.

COMPARAISON AVEC LES ORGANES NEURAUX D'AUTRES DIPLOPODES

Localisation

Les organes neuraux ont été décrits par Sahli chez sept espèces de Iulida dont en particulier *Tachypodoiulus albipes* (C.L.K.) et *Schizophyllum sabulosum* (L.) en 1966, et par Juberthie-Jupeau chez les Glomerida en 1967; ils ont été signalés par Junqua (1966) chez *Pachybolus laminatus* (Cook), *Peridontopyge junquai* Demange et *Schizophyllum rutilans* (C.L.K.). Ils sont localisés dans le protocérébron de tous ces diplopodes au contact des globuli*. Chaque organe se trouve sur la face antéro-interne, ou latéro-externe, ou médiane du globulus des divers glomérides observés. Chez *Tachypodoiulus* et *Schizophyllum* le globulus 1 est situé sur la face antéro-dorsale du lobe frontal et l'organe neural est appliqué contre sa face latéro-interne. Chez *Polyxenus*, la formation intracérébrale a une position assez voisine, proche du sillon médian séparant les deux lobes frontaux (Fig. 1), mais moins antérieure. La localisation de

* Chez les pénicillates les globuli ne sont pas identifiables au microscope optique, les différences de taille et de structure entre les noyaux de l'écorce cérébrale étant faibles.

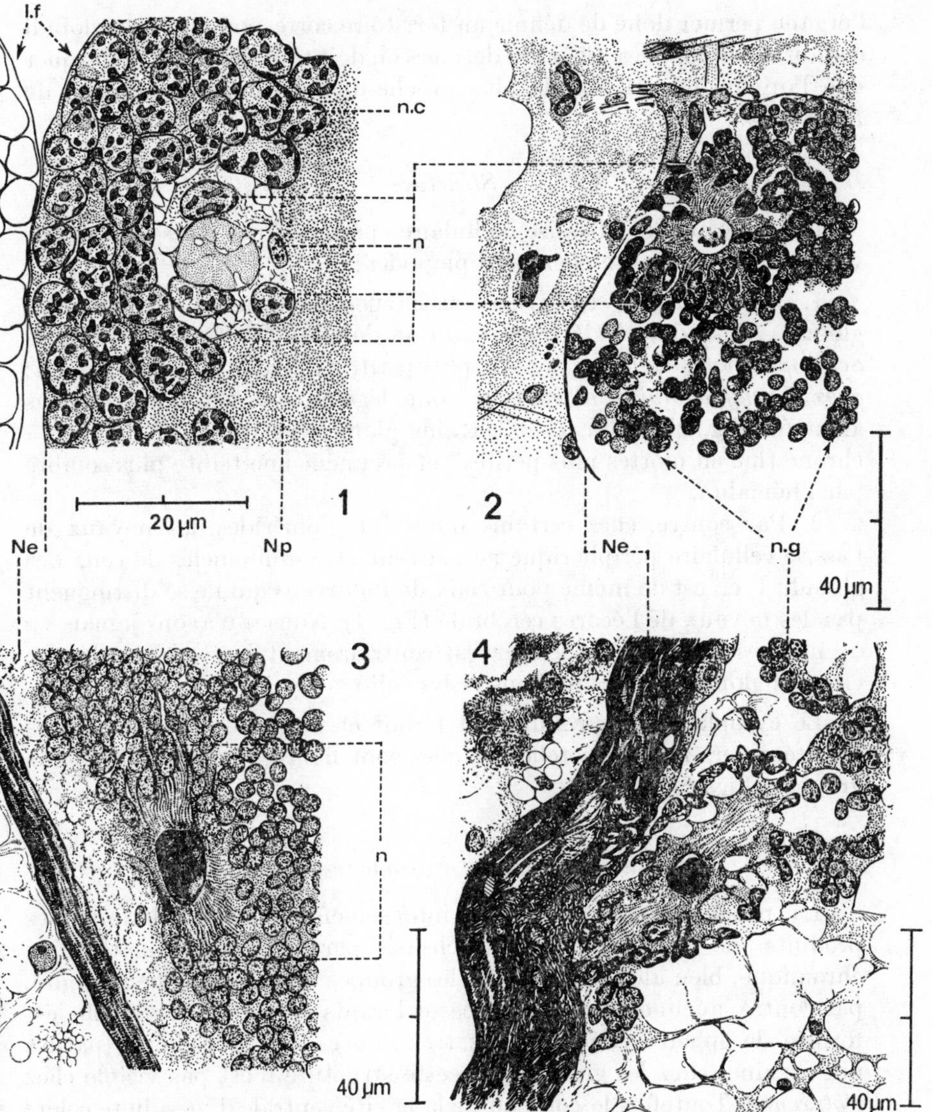

Fig. 1. Coupe frontale du cerveau de *Polyxenus lagurus* montrant l'organe intracérébral droit.

Fig. 2. Coupe transversale du cerveau de *Tachypodoiulus albipes* (C.L.K.) passant par l'organe neural (préparation de J.-P. Mauriès).

Fig. 3. Coupe parasagittale d'un Odontopygidae gen. sp. passant par l'organe neural (préparation de J.-M. Demange et C. Junqua).

Fig. 4. Coupe parasagittale de *Pachybolus* sp. passant par l'organe neural (préparation de J.-M. Demange et C. Junqua).

l'organe permet donc de définir un territoire correspondant aux globuli des autres diplopodes. Chez ces derniers on doit tout de même remarquer que l'organe neural semble plus proche du neurilemme que celui de *Polyxenus*.

Structure

Les noyaux de l'assise cellulaire périphérique présentent une variabilité importante chez les diplopodes:

1. Dans certains cas ils se distinguent nettement de ceux des globuli. Ainsi, d'après la représentation de Sahli (1966), les noyaux de *Schizophyllum sabulosum* sont de plus petite taille que ceux des globuli; chez *Tachypodoiulus albipes* ils sont légèrement plus gros et plus allongés (Fig. 2); enfin, chez certains glomérides ils sont ovoïdes, "à chromatine en mottes plus petites" et prennent une teinte plus sombre par l'hémalun.

2. Par contre, chez certains iulides et glomérides, les noyaux de l'assise cellulaire périphérique ne peuvent être differenciés de ceux des globuli. Il en est de même pour ceux de *Polyxenus* qui ne se distinguent pas des noyaux de l'écorce cérébrale (Fig. 1). Nous n'y avons jamais vu de mitoses (même chez les larves) contrairement à ce qu'on observe chez les glomérides, les opilions et les solifuges.

Le cytoplasme des cellules est réduit et ne présente pas de stries orientées vers le centre, telles qu'elles sont figurées pour les différents groupes (Figs 2, 3 et 4).

Affinités tinctoriales du contenu des organes

La masse centrale se colore intensément par les colorants des produits de neurosécrétion (fuchsine paraldéhyde, hématoxyline chromique, bleu alcian) dans tous les groupes étudiés. Elle ne présente, par contre, aucune affinité pour ces colorants chez *Polyxenus*. Elle est formée de sphères concentriques très minces à la périphérie (parfois une dizaine) chez les glomérides; cette structure n'est pas visible chez *Polyxenus*. Toutefois le contenu de la cavité centrale d'un adulte coloré à la fuchsine paraldéhyde-azan, présente un cercle bleu périphérique très mince entourant une substance colloïdale radiée, colorée en jaune; on peut voir là un début de différenciation du produit sécrété.

Dans certains cas chez *Schizophyllum sabulosum* et *Tachypodoiulus albipes* la cavité centrale est presque entièrement vide avec seulement quelques gouttelettes (Fig. 2); cela n'a pas été observé chez nos exemplaires de *Polyxenus*.

CONCLUSION

Les organes intracérébraux de *Polyxenus lagurus* présentent donc peu de différenciation, comme d'ailleurs l'ensemble de l'écorce cérébrale. Par leur position et leur aspect général ils peuvent néanmoins être homologués aux organes neuraux des autres diplopodes. Seule l'étude du développement embryonnaire déterminera leur origine, c'est-à-dire montrera si, comme chez les autres diplopodes, les pycnogonides, les péripates, les solifuges et les opilions, ils résultent de la persistance chez l'adulte d'une paire d'organes ventraux d'origine ectodermique.

Remerciements

Je remercie M. A. Badonnel et M. J.-M. Demange pour leurs précieux conseils et M. M. Gaillard pour l'iconographie.

References

Heathcote, F. G. (1888). The post-embryonic development of *Iulus terrestris*. *Phil. Trans. R. Soc.* **179**: 157–179.

Juberthie-Jupeau, L. (1967). Existence d'organes neuraux intracérébraux chez les Glomeridia (Diplopodes) épigés et cavernicoles. *C. r. hebd. Séanc. Acad. Sci., Paris* **264**: 89–92.

Junqua, C. (1966). Recherches biologiques et histophysiologiques sur un solifuge saharien *Othoes saharae* Panouse. *Mém. Mus. natn. Hist. nat., Paris* (N.S.) **43A**: 1–124.

Pflugfelder, O. (1932). Über den Mechanismus der Segmentbildung bei der Embryonalentwicklung und Anamorphose von *Platyrrhacus amauros* Attems. *Z. wiss. Zool.* **140**: 650–723.

Sahli, F. (1966). *Contribution à l'étude de la périodomorphose et du système neuro-sécréteur des Diplopodes Iulides.* Thèse Doct. Sciences, Dijon.

Abréviations

l.f: lobes frontaux protocérébraux; n: noyaux des cellules périphériques de l'organe neural; n.c: noyaux du cortex; Ne: neurilemme; n.g: noyaux du globulus; Np: neuropile.

Discussion

Dohle: Have you any experiments or at least any information as to the physiological effect of these organs?

Nguyen Duy-Jacquemin: No. Unfortunately not.

DEMANGE: I apologise for intervening in a subject which is not my speciality. In reply to Dr Dohle, I think that Junqua performed some experiments on Solifugidae, without success. He had tried ablation of these organs to see if there were any effects on the course of the moult. He obtained no relevant results, but died before he was able to conclude his experiments.

DOHLE: Junqua?

DEMANGE: Yes, Claude Junqua, who presented a thesis on the biology of Solifugidae in 1966*.

SAHLI: Also in reply to Dr Dohle, there are no experiments apart from those which Demange mentioned, from the physiological viewpoint. In iulids, I searched for variations in relation to the different phases of the cycle, at the time of egg laying, at the moment of coupling, at the moment of moulting, before moulting, during the moult and after moulting. But the neurosecretory products stocked in Gabe's organ show no variations related to the physiological state of the animal. With light microscopy at least, it appears that we cannot derive information from histophysiology.

* See list of references, p. 215.

Symp. zool. Soc. Lond. (1974) No. 32, 217–230.

SUR LES ORGANES NEUROHÉMAUX ET ENDOCRINES DES MYRIAPODES DIPLOPODES

FRANÇOIS SAHLI

Laboratoire de Biologie Animale et Générale, Faculté des Sciences de la Vie et de l'Environnement, Université de Dijon, France

SYNOPSIS

Our knowledge of the organs of Gabe ("cerebral glands"), the para-oesophageal bodies and the connective bodies of diplopods is reviewed. Notwithstanding the differences of opinion as to whether true secretory cells occur in Gabe's organ and the inclusion within the organ of both acidophil and basiphil products, they are here considered as simple organs of accumulation. They are not homologous with the corpora cardiaca of pterygote insects but are the histophysiological equivalents of the neurohaemal organs of the Collembola, Diplura and certain Thysanura and Symphyla.

Various arguments support the hypothesis that the para-oesophageal bodies of diplopods are equivalent to the corpora cardiaca of apterygote insects.

The connective bodies are, like Gabe's organs, cephalic neurohaemal organs, but they differ from Gabe's organs since after Gabe's trichrome staining (paraldehyde-fuchsin-picroindigocarmine) or Gomori's chrome hematoxylin-phloxine they do not contain an acidophil component. They are connected with a tract of secretion running from the brain along the ventral nerve cord. A comparable tract exists in certain annelids and insects but in these there are no individual organs of accumulation histophysiologically equivalent to the connective bodies of diplopods.

The connective bodies are perhaps vestiges of a primitive metameric arrangement of paired neurohaemal organs. They are the analogues and not the homologues of the perisympathetic organs of pterygote insects which also accumulate an azocarminophil secretion.

ORGANES DE GABE ("GLANDES CÉRÉBRALES")

Ces organes sont actuellement connus chez bon nombre de diplopodes (Tableau I). Leur emplacement est variable.

Chez les iulides ils occupent une position céphalique postéro-ventrale, de part et d'autre de la masse nerveuse sous-oesophagienne.

Chez les pénicillates leur emplacement rappelle celui des iulides. La formation décrite par Nguyen Duy Jacquemin (1971b) chez *Polyxenus* sous le nom de "glande cérébrale" est effectivement un organe de Gabe, comme nous avons pu nous en convaincre après examen des préparations de cet auteur. Quant à la "glande cérébrale" décrite par Seifert & El-Hifnawi (1971, 1972), il s'agit d'une formation autre que l'organe de Gabe. Ce dernier a été signalé chez *Polyxenus* par les deux mêmes auteurs (1972) sous le nom de "Neurohämalorgan am Nervus protocerebralis". (Donc: glande cérébrale de Nguyen Duy-Jacquemin = Neurohämalorgan am Nervus protocerebralis de Seifert & El-Hifnawi =

Tableau I

Diplopodes dont les organes de Gabe ont fait l'objet de recherches histologiques

Espèces	Références
ONISCOMORPHA Plesiocerata	
Glomeris marginata (Villers)	Gabe, 1954; Juberthie-Jupeau, 1967
Glomeris intermedia Latzel	Juberthie-Jupeau, 1967
Glomeris rugifera Verhoeff	Juberthie-Jupeau, 1967
Loboglomeris pyrenaica (Latzel)	Juberthie-Jupeau, 1967
Glomeridella kervillei (Latzel)	Juberthie-Jupeau, 1967
NEMATOPHORA Ascospermophora	
Chordeumidae	
Chordeuma silvestre Latzel	Gabe, 1954
Craspedosomidae	
Craspedosoma simile (Verhoeff)	Palm, 1955; Seifert, 1971*
Craspedosoma alemannicum Verhoeff	Sahli, 1966
PROTEROSPERMOPHORA Polydesmoidea	
Polydesmidae	
Polydesmus angustus Latzel	Gabe, 1954†; Sahli, 1966
Polydesmus helveticus Latzel	Gabe, 1954
Polydesmus testaceus C. L. Koch	Glaser, 1958; Gersch, 1958
Strongylosomidae	
Orthomorpha gracilis (C. L. Koch)	Sahli, 1962, 1966
Pratinidae	
Jonespeltis splendidus Verhoeff	Prabhu, 1959, 1961, 1962, 1964
OPISTHOSPERMOPHORA Symphyognatha	
Blaniulidae	
Blaniulus guttulatus (Bosc)	Gabe, 1954
Typhloblaniulus troglobius Brol.	Sahli, 1958, 1961, 1966
Iulidae	
Tachypodoiulus albipes (C. L. Koch)‡	Gabe 1954, Sahli 1958, 1966
Schizophyllum sabulosum (Linné)	Sahli, 1961, 1966
Cylindroiulus londinensis (Leach)§	Gabe, 1954.
Cylindroiulus teutonicus (Pocock)	Sahli, 1961, 1966
Cylindroiulus silvarum (Meinert)	Sahli, 1961, 1966
Cylindroiulus nitidus (Verhoeff)	Sahli, 1966
Iulus albolineatus Leach	Gabe, 1954
Iulus scandinavius Latzel	Sahli, 1961, 1966
Leptoiulus simplex glacialis (Verh.)	Sahli, 1961, 1966
PSELAPHOGNATHA SCHIZOCEPHALA	
Polyxenidae	
Polyxenus lagurus (Linné)	Nguyen Duy-Jacquemin, 1971 a & b; Seifert & El-Hifnawi, 1972‖

* *C. rawlinsii* (Leach) in Seifert; en synonymie avec *C. simile* (Verh.).

† *P. complanatus* in Gabe.

‡ *T. albipes* (C. L. Koch) = *T. niger* (Leach).

§ *C. londinensis* in Gabe = ? *C. teutonicus* (Pocock).

‖ les formations décrites par Seifert & El-Hifnawi (1971) ne sont pas des organes de Gabe.

organe de Gabe *nobis*; *non*: glande cérébrale de Seifert & El-Hifnawi 1971).

Chez les Oniscomorpha, les Nematophora et les Proterospermophora, les organes de Gabe sont beaucoup moins éloignés du cerveau que chez les iulides: en position latérale, ils se situent plus ou moins à son voisinage.

Ces organes sont innervés par un nerf protocérébral qui, chez les iulides, tire, au moins en partie, son origine des cellules neurosécrétrices des globuli I. Ils comportent des produits de sécrétion, des noyaux, des trachées et des terminaisons de fibres nerveuses.

Après coloration à la fuchsine paraldéhyde-picroindigocarmin de Gabe ou à l'hématoxyline chromique-phloxine de Gomori, on distingue typiquement *deux sortes de produits*: l'un retenant la fuchsine paraldéhyde ou l'hématoxyline chromique, l'autre le picroindigocarmin du Gabe ou la phloxine du Gomori. En outre, il existe bien souvent des flaques présentant une gamme de teintes intermédiaires, avec une prédominance marquée pour l'un des composants (p. ex. la fuchsine paraldéhyde ou le picroindigocarmin du Gabe). Une même flaque peut être, soit uniformément, soit diversement teintée.

Le produit retenant la fuchsine paraldéhyde ou l'hématoxyline chromique est indubitablement d'origine protocérébrale: les organes de Gabe sont des centres d'accumulation d'un produit de neurosécrétion protocérébral.

Comment expliquer la pluralité des produits de sécrétion dans l'organe de Gabe des diplopodes? Est-elle liée à une activité sécrétrice propre?

A l'exception de Gabe (1954) et de Seifert & El-Hifnawi (1972), les auteurs n'ont pas pu déceler dans les glandes cérébrales des diplopodes des cellules pouvant être indubitablement considérées comme sécrétrices. Chez les iulides, il y a lieu de penser que les organes reçoivent un produit venant des corps para-oesophagiens. Mais même chez les diplopodes dépourvus de corps para-oesophagiens (comme les polydesmes et apparemment les craspédosomides), il existe, dans les organes de Gabe, à la fois un matériel se colorant franchement à la fuchsine paraldéhyde ou à l'hématoxyline chromique et un matériel gardant complètement ou partiellement son acidophilie après oxydation permanganique.

D'une part, dans certains cas, les mêmes produits de sécrétion s'observent indubitablement dans l'organe et dans son nerf afférent. D'autre part, on trouve les mêmes éléments nucléaires dans l'organe et dans son nerf afférent. Ces deux points semblent aller à l'encontre de la conception d'une activité sécrétrice propre. L'hypothèse de transformations chimiques du produit de neurosécrétion protocérébral au cours de son cheminement n'est pas à exclure; elle permet d'expliquer la présence d'un produit acidophile, indépendamment de toute activité sécrétrice propre de l'organe, autrement dit, sans que l'on soit obligé de faire intervenir une activité sécrétrice propre.

Les observations en microscopie électronique plaident en faveur de l'absence de cellules sécrétrices dans l'organe de Gabe de *Craspedosoma* (Seifert, 1971) et de *Polyxenus* (Seifert & El-Hifnawi, 1972). Chose étonnante, celui du *Schizophyllum sabulosum* contiendrait, par contre, quelques éléments sécréteurs "Drüsenparenchymzellen"; Seifert & El-Hifnawi, 1972).

Jusqu'à plus ample informé et à la lumière des travaux de Juberthie & Cassagnau (1971), nous pensons néanmoins que les organes de Gabe des diplopodes peuvent être considérés comme de simples organes neurohémaux, histophysiologiquement équivalents, non pas des corpora cardiaca, mais des organes neurohémaux des collemboles, des campodés, de certains machilides (Fig. 1) et des symphyles (c.f. Juberthie-Jupeau, 1963).

Nous rejoignons les vues de Prabhu (1961) et de Seifert (1971), qui, respectivement chez *Jonespeltis* et chez *Craspedosoma*, considèrent les "glandes cérébrales" des diplopodes comme de simples organes d'accumulation. Gabe (1967) considère également les "glandes cérébrales" comme des organes neurohémaux; mais il les compare à ceux des arachnides, des chilopodes et aux corpora cardiaca des insectes aptérygotes et des ephéméroptères: il les rattache au type d'organe neurohémal (type II) caractérisé par la présence de fibres neurosécrétoires, de la névrolgie et de cellules capables d'une sécrétion propre.

D'après nos observations et celles de divers auteurs, il convient plutôt de rattacher les "glandes cérébrales" des diplopodes au type I de Gabe, c'est-à-dire au type qui comporte les simples organes d'accumulation.

Sous réserve de précisions nouvelles que pourra nous apporter la microscopie électronique, il semble qu'on puisse considérer les organes de Gabe comme des organes neurohémaux plus primitifs que ceux des chilopodes (Gabe, 1952; Scheffel, 1965; Joly, 1966a,b); ils occupent chez les diplopodes une position plus latérale que les organes neurohémaux de certains insectes (Cazal, 1948; Juberthie & Cassagnau, 1971), où la disposition primitive serait celle d'une relation de voisinage étroit avec la paroi de l'aorte (Fig. 1 *A*, *B*, *C*). Il se peut néanmoins que l'organe neurohémal des diplopodes ne soit pas l'homologue *morphologique* (*sensu* des morphologistes Snodgrass, Denis, Chaudonneret, etc.) de celui des insectes aptérygotes.

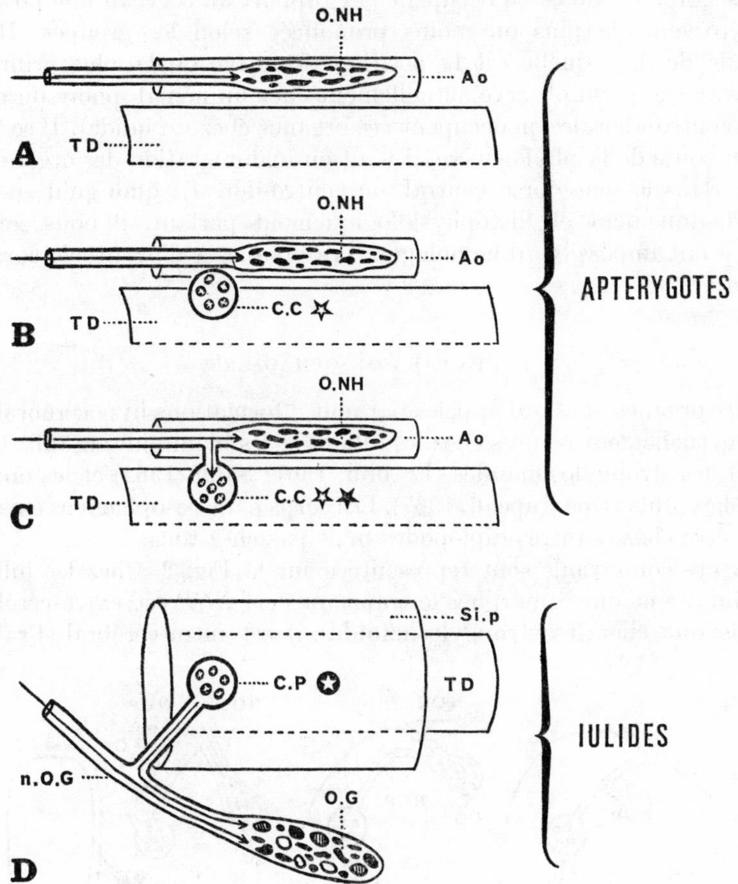

FIG. 1. Organes neurohémaux (*O.NH*), *corpora cardiaca* (*C.C*) et corps paraoesophagiens (*C.P*).

A, B, C: Insectes aptérygotes, d'après Juberthie & Cassagnau (1971) (3 types seulement ont été représentés). **D**: Diplopodes iulides, d'après Sahli (1966). **A**: *Campodea, Machilis, Dilta* et *Lepismachilis*; des organes neurohémaux simples, sans corpora cardiaca. **B**: *Ctenolepisma*; des organes neurohémaux simples; des corpora cardiaca ne recevant aucun produit de neurosécrétion cérébral, mais renfermant un produit de sécrétion fortement acidophile (Gabe, 1967), d'origine intrinsèque, indiqué par une étoile blanche. **C**: *Thermobia domestica*: le produit de neurosécrétion cérébral se déverse en partie dans les organes neurohémaux, en partie dans les corpora cardiaca, comme l'indiquent les flèches; ces derniers contiennent 2 produits: l'un d'origine intrinsèque (étoile blanche), l'autre d'origine extrinsèque cérébral (étoile noire). **D**: Iulidae: l'organe neurohémal (= organe de Gabe, *O.G*) reçoit un produit de neurosécrétion cérébral et, selon l'hypothèse la plus vraisemblable, un produit du corps paraoesophagien (étoile blanche entourée d'un cercle noir). *Ao*: aorte; *n.O.G*: nerf de l'organe de Gabe; *Si.p*: sinus péri-oesophagien; *TD*: tube digestif.

Les organes de Gabe occupent par rapport au cerveau une position postéro-ventrale plus ou moins prononcée selon les groupes. Il est difficile de dire quelle est la position primitive ou la plus primitive (p. ex. celle qu'on observe actuellement chez un nématophore ou celle, plus ventro-aborale, qu'occupent ces organes chez un iulide). Il se peut qu'au cours de la phylogénèse, il y ait eu une migration des organes de Gabe dans le sens dorso-ventral (ou ventro-dorsal): quoi qu'il en soit, histologiquement et histophysiologiquement parlant, il nous semble tout à fait impossible d'homologuer ces organes à d'éventuels corpora allata.

CORPS PARA-OESOPHAGIENS

Ces organes, d'abord appelés par nous "formations hypocérébrales", sont actuellement connus chez les iulides et les blaniulides (Sahli, 1961, 1966), les strongylosomoïdes (Prabhu, 1961; Sahli, 1962) et les oniscomorphes (Juberthie-Jupeau, 1967). Les corps para-oesophagiens existent peut-être chez d'autres diplopodes mais pas chez tous.

Leurs connexions sont représentées sur la Fig. 2. Chez les iulides, blaniulides et oniscomorphes le trajet du nerf *NCP* est extra-cérébral, tandis que chez les strongylosomoïdes, il est intra-cérébral (Prabhu,

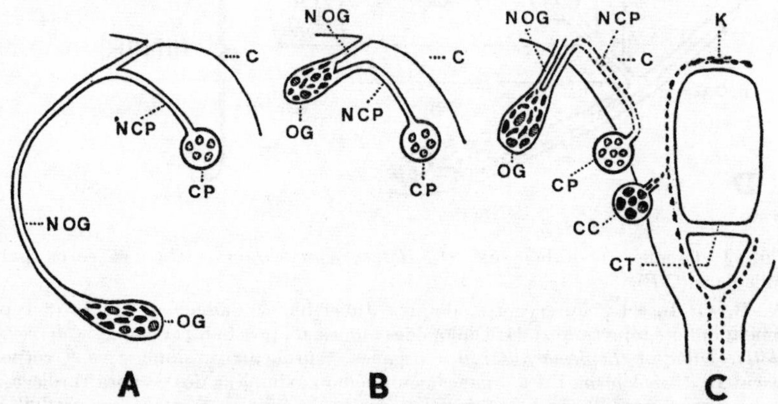

FIG. 2. Organes de Gabe et corps para-oesophagiens des diplopodes. **A**: chez les Opisthospermophora Symphyognatha (Iulidae et Blaniulidae), d'après Sahli (1966). **B**: chez les Oniscomorpha Plesiocerata, d'après Juberthie-Jupeau (1967). **C**: chez les Proterospermophora Strongylosomidae et Pratinidae, d'après Prabhu (1961) et Sahli (1966). Les corps connectifs ne sont connus que chez *Orthomorpha* et *Jonespeltis* (**C**); ils sont absents en **A** et vraisemblablement en **B**. *C*: cerveau; *C C*: corps connectifs; *C P*: corps para-oesophagiens; *C T*: commissure tritocérébrale; *K*: commissure de la voie de neurosécrétion TR 4; *NCP*: nerf du corps para-oesophagien; *NOG*: nerf de l'organe de Gabe; *O G*: organe de Gabe.

1961). Chez les opisthospermophores le nerf *N OG* est particulièrement long et le nerf *N CP* ne part pas directement de l'organe de Gabe. Avec les données que nous possédons il nous est difficile de dire quelle est la condition primitive et de préciser le sens dans lequel l'évolution s'est faite.

Si l'on s'en réfère aux travaux de Cassagnau & Juberthie (1967a,b) et de Juberthie & Cassagnau (1971) sur les aptérygotes, il ressort que, primitivement, il n'y a pas de corpora cardiaca: il existe seulement un organe neurohémal (Fig. 1*A*). Quand tous deux sont présents, organe neurohémal et corpus cardiacum peuvent être distincts chez certains aptérygotes (Fig. 1*B*, 1*C*). Les fibres qui parviennent aux corpora cardiaca peuvent parfois, comme chez *Ctenolepisma*, (Fig. 1*B*), ne transporter aucun produit de neurosécrétion protocérébral: le stockage de ce produit se fait alors uniquement (et donc entièrement) dans l'organe neurohémal.

S'il est possible d'établir des comparaisons ou des homologies valables entre les formations endocrines des insectes et celles des diplopodes—arthropodes assez différents les uns des autres à bien des égards—on peut penser que ce sont les corps para-oesophagiens qui constituent éventuellement l'équivalent des corpora cardiaca des insectes aptérygotes (Fig. 1). Leur position le long du sinus péri-oesophagien, leur connexion avec le système "cellules neurosécrétrices des globuli I—organes de Gabe (organes neurohémaux)" et la présence de grosses cellules ayant une activité sécrétrice propre sont des arguments qui peuvent plaider en faveur de cette hypothèse.

Il convient néanmoins d'insister sur une différence non négligeable entre corps para-oesophagiens et corpora cardiaca. Ces derniers reçoivent—généralement (*cf.* les restrictions émises ci-dessus)—un produit de neurosécrétion protocérébral; par contre le produit de leur activité sécrétrice propre n'est pas déversé dans l'organe neurohémal, lorsque ce dernier existe.

Chez les diplopodes, au contraire, d'après nos observations (Sahli, 1966) et selon l'hypothèse la plus vraisemblable, d'une part les corps para-oesophagiens ne reçoivent pas de neurosécrétat protocérébral, d'autre part ils déversent—au moins en partie—leur produit de sécrétion dans l'organe neurohémal pair (organe de Gabe). Ce dernier serait donc, chez les diplopodes possédant des corps para-oesophagiens, un centre d'accumulation à la fois du produit de neurosécrétion protocérébral et d'une partie au moins du produit de sécrétion des corps para-oesophagiens. L'hypothèse selon laquelle le produit de neurosécrétion protocérébral se déverserait, en partie dans les organes de Gabe, en partie dans les corps para-oesophagiens, pour séduisante qu'elle soit, ne nous paraît pas conforme aux faits, du moins chez les iulides.

CORPS CONNECTIFS

Le terme est dû à Prabhu (1961). Ces formations ne sont connues que chez certains Proterospermophora: le pratinide *Jonespeltis* et le strongylosomoïde *Orthomorpha*, où ils ont été décrits pour la première fois en 1961 respectivement par Prabhu et par Sahli. Les iulides, les blaniulides et les polydesmides ne possèdent pas de corps connectifs.

Ce sont de petits organes pairs, insérés sur chaque branche du collier périoesophagien, un peu au-dessus de la commissure tritocérébrale. Le produit de neurosécrétion cheminant le long de la voie cerveau-collier-masse nerveuse sous-oesophagienne-chaîne nerveuse ventrale (= voie "*TR 4*"; Sahli, 1966) se déverse en partie dans les corps connectifs (Fig. 2*C* et Fig. 3*F*). Ceux-ci, organes d'accumulation d'un produit de neurosécrétion cérébral, ne possèdent aucune activité sécrétrice propre. Ce sont donc de simples *organes neurohémaux* dont les flaques se colorent intensément à la fuchsine paraldéhyde (ou à l'hématoxyline chromique); à l'inverse des organes de Gabe, ils ne contiennent aucun produit colorable au picroindigocarmin. Après coloration a l'azan, ils se teintent généralement dans les tons rouges; mais parfois ils se colorent à l'orange G et au bleu d'aniline.

Orthomorpha et *Jonespeltis* possèdent donc deux organes neurohémaux céphaliques pairs distincts: l'organe de Gabe et le corps connectif. Puisque chez certains diplopodes, il existe deux organes neurohémaux céphaliques distincts, on ne saurait, sous risque de confusion ou d'imprécision, remplacer le terme d'organe de Gabe ou celui de glande cérébrale par le terme d'organe neurohémal. Il serait souhaitable que les auteurs puissent s'accorder sur la terminologie. Personnellement nous proposons le terme "organes de Gabe"—qui ne préjuge de rien—de préférence à "glandes cérébrales", puisque, chez les diplopodes, l'activité glandulaire de ces formations est loin d'être universellement admise. Admettant le bien-fondé de notre argumentation, Seifert & El-Hifnawi (1972) viennent d'ailleurs d'adopter notre terminologie.

Il existe chez certains insectes un cheminement cerveau-collier périoesophagien—ganglion sous-oesophagien—chaîne nerveuse ventrale; parfois on n'observe qu'un ou plusieurs cheminements dans la chaîne nerveuse ventrale. Dans l'état de nos connaissances, il y a lieu de distinguer chez ces arthropodes plusieurs types de cheminements, d'une part selon les affinités tinctoriales, d'autre part selon la voie empruntée.

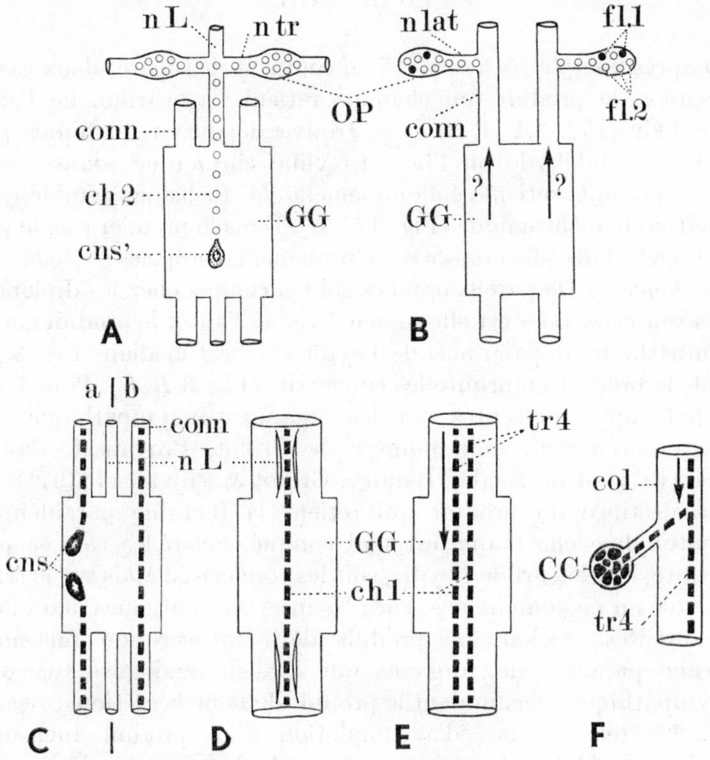

FIG. 3. Organes neurohémaux et voies de cheminement de produits de neurosécrétion dans la chaîne nerveuse ventrale et dans le collier péri-oesophagien de certains insectes et diplopodes.

Pour le produit de neurosécrétion retenant l'azocarmin, la voie de cheminement (*ch 2*) est conventionnellement représentée par des cercles (haut de la figure: **A** et **B**); pour le produit retenant la fuchsine-paraldéhyde ou l'hématoxyline chromique les voies (*ch 1*) sont représentées par de petits rectangles noirs (bas de la figure: **C, D, E, F**). **A**: concerne les ganglions abdominaux des phasmides (d'après Raabe, 1965). **B**: les ganglions thoraciques d'un hyménoptère symphyte (d'après Provansal, 1971); les flaques *fl 1* sont peu abondantes, les flaques *fl 2* sont nombreuses. **C**: les ganglions de certains ptérygotes; deux cas différents sont figurés: à gauche, cas *a* (d'après Delphin, 1965 et Geldiay, 1959), à droite, cas *b* (d'après Johnson, 1963 et Raabe, 1965). **D**: le ganglion sous-oesophagien et les ganglions thoraciques d'un collembole entomobryomorphe (d'après Cassagnau & Juberthie, 1970). **E**: la masse nerveuse sous-oesophagienne et les ganglions de la chaîne nerveuse ventrale des diplopodes (d'après Sahli, 1966). **F**: le collier péri-oesophagien des diplopodes (seule une partie de la moitié gauche est représentée; d'après Sahli, 1966). *C C*: corps connectifs (organes neurohémaux); *ch 1*: voie de cheminement colorable à la fuchsine paraldéhyde ou à l'hématoxyline chromique; *ch 2*: voie de cheminement colorable à l'azocarmin; *cns*: cellules neurosécrétrices (fuchsine paraldéhyde ou hématoxyline chromique); *cns'*: cellule neurosécrétrice (azocarmin); *col*: collier péri-oesophagien; *conn*: connectifs; *fl 1*: flaques retenant la fuchsine paraldéhyde ou l'hématoxyline chromique; *fl 2*: flaques retenant l'azocarmin; *GG*: ganglion; *n L*: nerf de Leydic; *n lat*: nerf latéral; *n tr*: nerf transverse; *O P*: organes périsympathiques (organes neurohémaux); *tr 4*: voie cerveau-collier-masse sous-oesophagienne-chaîne nerveuse ventrale, ainsi désignée chez les diplopodes.

D'après les *affinités tinctoriales* il convient de séparer deux cas: dans le premier, le produit qui chemine retient l'azocarmin de l'azan de Heidenhain (Fig. 3*A* et *B*) et se trouve dépourvu d'affinités pour la fuchsine paraldéhyde ou l'hématoxyline chromique; dans le second cas, le produit retient énergiquement la fuchsine paraldéhyde ou l'hématoxyline chromique (Fig. 3 *C*, *D*). Dans le premier cas, le produit est stocké dans des organes neurohémaux appelés "*organes périsympathiques*". De pareils organes sont inconnus chez les diplopodes.

La *voie empruntée* est elle-même diverse. Tantôt le produit emprunte le sympathique impair: nerf de Leydic (= nerf médian; Fig. 3 *A*, *D*); tantôt, le produit emprunte les connectifs (Fig. 3 *B*, *C*). Pour les types morphologiques présentés par les organes périsympathiques, on se référera notamment aux données de Grillot, Provansal, Baudry & Raabe (1971) et de Raabe, Baudry, Grillot & Provansal (1971).

La destinée du produit qui retient la fuchsine paraldéhyde ou l'hématoxyline chromique est mal connue. Selon les cas, ce produit emprunte, soit le nerf de Leydic, soit les connectifs. Mais sur le trajet de cette voie on ne connaît pas, chez les insectes, d'organes neurohémaux individualisés stockant le produit de neurosécrétion fuchsinophile (fuchsine paraldéhyde), organes qui seraient analogues aux organes périsympathiques accumulant le produit de neurosécrétion azocarminophile. De tels organes d'accumulation d'un produit fuchsinophile (fuchsine paraldéhyde) existent par contre chez certains diplopodes: ce sont les corps connectifs, où se déverse une partie du produit de neurosécrétion cheminant le long de la voie cerveau—chaîne nerveuse ventrale (c'est-à-dire le long de la voie que nous appelons la voie *TR 4*). On peut se demander s'il ne s'agit pas là d'une voie très primitive, ayant comporté des organes neurohémaux métamériques pairs, pour la plupart disparus au cours de la phylogénèse et dont ne subsisteraient actuellement que les corps connectifs.

RESUMÉ ET CONCLUSIONS

1. Jusqu'à plus ample informé, nous considérerons les organes de Gabe ("glandes cérébrales") comme de simples organes neurohémaux, histophysiologiquement équivalents de ceux des collemboles, des campodés, de certains machilides et des symphyles.

2. Divers arguments suggèrent l'hypothèse que les corps paraoesophagiens des diplopodes constituent éventuellement l'équivalent des corpora cardiaca des insectes aptérygotes.

3. Les corps connectifs sont, comme les organes de Gabe, des organes neurohémaux céphaliques. Ils s'en distinguent notamment par

l'absence d'un produit acidophile après coloration à la fuchsine paraldéhyde-picroindigocarmin de Gabe ou l'hématoxyline chromique-phloxine de Gomori.

4. Les corps connectifs sont en rapport avec une voie de sécrétion cerveau-chaîne nerveuse ventrale, primitive, où chemine un produit colorable à la fuchsine paraldéhyde ou à l'hématoxyline chromique. Une voie comparable existe chez certaines annélides et chez certains insectes; mais, à notre connaissance, chez ceux-ci elle ne comporte pas d'organes d'accumulation individualisés histologiquement comparables aux corps connectifs des diplopodes.

5. Les corps connectifs sont peut-être des vestiges d'un dispositif primitif ayant comporté des organes neurohémaux métamériques pairs. Centres d'accumulation d'un produit de neurosécrétion fuchsinophile, cyanophile ou azocarminophile selon les méthodes utilisées, les corps connectifs sont, par leurs affinités tinctoriales, non pas homologues mais analogues aux organes périsympathiques des insectes ptérygotes, centres d'accumulation métamériques d'un produit de neurosécrétion fondamentalement azocarminophile.

Bibliographie

Cassagnau, P. & Juberthie, C. (1967a). Structures nerveuses, neurosécrétion et organes endocrines chez les Collemboles (I.). Le complexe cérébral des Poduromorphes. *Bull. Soc. Hist. nat. Toulouse* **103**: 178–222.

Cassagnau, P. & Juberthie, C. (1967b). Structures nerveuses, neurosécrétion et organes endocrines chez les Collemboles (II.). Le complexe cérébral des Entomobryomorphes. *Gen. comp. Endocr.* **8**: 489–502.

Cassagnau, P. & Juberthie, C. (1970). Structures nerveuses, neurosécrétion et organes endocrines chez les Collemboles. Neurosécrétion dans la chaîne nerveuse d'un Entomobryomorphe, *Orchesella kervillei* Denis. *C.r. hebd. Séanc. Acad. Sci., Paris* **270**: 3268–3271.

Cazal, P. (1948). Les glandes endocrines rétro-cérébrales des Insectes. *Bull. biol. Fr. Belg.* (suppl.) **32**: 1–227.

Delphin, E. (1965). The histology and possible functions of neurosecretory cells in the ventral ganglia of *Schistocerca gregaria* Forskal (Orthoptera: Acrididae). *Trans. R. ent. Soc. Lond.* **117**: 167–214.

Gabe, M. (1952). Sur l'emplacement et les connexions des cellules neurosécrétrices dans les ganglions cérébroïdes de quelques Chilopodes. *C.r. hebd. Séanc. Acad. Sci., Paris* **235**: 1430–1432.

Gabe, M. (1954). Emplacement et connexions des cellules neurosécrétrices chez quelques Diplopodes. *C.r. hebd. Séanc. Acad. Sci., Paris* **239**: 828–830.

Gabe, M. (1967). *Neurosécrétion*. Paris: Gauthier-Villars.

Geldiay, S. (1959). Neurosecretory cells in ganglia of the roach *Blaberus craniifer*. *Biol. Bull. mar. biol. Labs Woods Hole* **117**: 267–274.

Gersch, M. (1958). Neurohormone bei wirbellosen Tieren. *Verh. dt. zool. Ges.* **22**: 40–76.

Glaser, R. (1958). *Histologische Untersunchungen über das neurosekretorische System bei* ***Polydesmus testaceus*** Ltz. (Diplopoda). Diplôme Univ. Jena.

Grillot, J.-P., Provansal, A., Baudry, N. & Raabe, M. (1971). Les organes périsympathiques des Insectes Ptérygotes. Les principaux types morphologiques. *C.r. hebd. Séanc. Acad. Sci., Paris* **273**: 2126–2129.

Johnson, B. (1963). A histological study of neurosecretion in aphids. *J. Ins. Physiol.* **9**: 727–739.

Joly, R. (1966a). *Contribution à l'étude du cycle de mue et de son déterminisme chez les Myriapodes Chilopodes.* Thèse Sci. Lille No. 132: 1–110.

Joly, R. (1966b). Sur l'ultrastructure de la glande cérébrale de *Lithobius forficatus* L. (Myriapode, Chilopode). *C.r. hebd. Séanc. Acad. Sci., Paris* **263**: 374–377.

Juberthie-Jupeau, L. (1963). Recherches sur la reproduction et la mue chez les Symphyles. *Archs Zool. exp. gén.* **102**: 1–172.

Juberthie-Jupeau, L. (1967). Données sur le système endocrinien de quelques Diplopodes Oniscomorphes (Myriapodes). *C.r. hebd. Séanc. Acad. Sci., Paris* **265**: 1525–1529.

Juberthie, C. & Cassagnau, P. (1971). L'évolution du système neurosécréteur chez les Insectes; l'importance des Collemboles et des autres Aptérygotes. *Revue Ecol. Biol. Sol* **8**: 59–80.

Nguyen Duy-Jacquemin. (1971a). Etude préliminaire sur la neurosécrétion céphalique chez le Diplopode Pénicillate *Polyxenus lagurus* (Myriapodes). *C.r. hebd. Séanc. Acad. Sci., Paris* **272**: 1984–1986.

Nguyen Duy-Jacquemin. (1971b). Mise en évidence de glandes cérébrales chez le Diplopode Pénicillate *Polyxenus lagurus* (Myriapodes). *C.r. hebd. Séanc. Acad. Sci., Paris* **272**: 2195–2196.

Palm, N. B. (1955). Neurosecretory cells and associated structures in *Lithobius forficatus* L., *Ark. Zool.* (2) **9**: 115–129.

Prabhu, V. (1959). Note on the cerebral glands and a hitherto unknown connective body in *Jonespeltis splendidus* Verh. (Myriapoda, Diplopoda). *Curr. Sci.* **28**: 330–331.

Prabhu, V. (1961). The structure of the cerebral glands and connective bodies of *Jonespeltis splendidus* Verh. (Myriapoda, Diplopoda). *Z. Zellforsch. mikrosk. Anat.* **54**: 717–733.

Prabhu, V. (1962). Neurosecretory system of *Jonespeltis splendidus* Verh. (Myriapoda, Diplopoda). *Mem. Soc. Endocr.* **12**: 417–420.

Prabhu, V. (1964). Regeneration of connective bodies of the neurosecretory system of *Jonespeltis splendidus* Verh. (Myriapoda, Diplopoda). *Indian J. exp. Biol.* **2**: 5–8.

Provansal, A. (1971). Caractères particuliers des organes périsympathiques de la larve de *Diprion pini* L. (Hyménoptère, Symphyte, *Diprionidae*). *C.r. hebd. Séanc. Acad. Sci., Paris* **272**: 855–858.

Raabe, M. (1965). Etude des phénomènes de neurosécrétion au niveau de la chaîne nerveuse ventrale des Phasmides. *Bull. Soc. zool. Fr.* **90**: 631–654.

Raabe, M. Baudry, N., Grillot, J.-P. & Provansal, A. (1971). Les organes périsympathiques des Insectes Ptérygotes. Distribution. Caractères généraux. *C.r. hebd. Séanc. Acad. Sci., Paris* **273**: 2324–2327.

Sahli, F. (1958). Quelques données sur la neurosécrétion chez le Diplopode *Tachypodoiulus albipes* (C. L. Koch). *C.r. hebd. Séanc. Acad. Sci., Paris* **246**: 470–472.

Sahli, F. (1961). Sur une formation hypocérébrale chez les Diplopodes Iulides. *C.r. hebd. Séanc. Acad. Sci., Paris* **252**: 2443–2444.

Sahli, F. (1962). Sur le système neurosécréteur du Polydesmoïde *Orthomorpha gracilis* C. L. Koch (Myriapoda, Diplopoda). *C.r. hebd. Séanc. Acad. Sci. Paris* **254**: 1498–1500.

Sahli, F. (1966). *Contribution à l'étude de la périodomorphose et du système neurosécréteur des Diplopodes Iulides.* Thèse Sci. (Bernigaud & Privat), Dijon, No. 94: 1–226.

Scheffel, H. (1965). Elektronenmikroskopische Untersuchungen über den Bau der Cerebraldrüse der Chilopoden. *Zool. Jb.* (Physiol.) **71**: 624–640.

Seifert, G. (1971) Ein bisher unbekanntes Neurohämalorgan von *Craspedosoma rawlinsii* Leach (Diplopoda, Nematophora) *Z. Morph. Tiere* **70**: 128–140.

Seifert, G. & El-Hifnawi, E. (1971). Histologische und elektronenmikroskopische Untersuchungen über die Cerebraldrüse von *Polyxenus lagurus* (L.) (Diplopoda, Penicillata). *Z. Zellforsch. mikrosk. Anat.* **118**: 410–427.

Seifert, G. & El-Hifnawi, E. (1972). Die Ultrastruktur des Neurohämalorgans am Nervus protocerebralis von *Polyxenus lagurus* (L.) (Diplopoda, Penicillata). *Z. Morph. Tiere* **71**: 116–127.

Discussion

Juberthie-Jupeau: Sahli has homologized the cerebral glands with neurohaemal organs and the paraoesophageal bodies (corps paraoesophagiens, CPO) with the corpora cardiaca of insects, but this means dissociating the organs of Gabe and the CPO. The neurohaemal part of the insect corpora cardiaca is extremely well developed and is intimately associated with the glandular part. I think that *both* parts, the neurohaemal part and the CPO, are *together* homologous with the corpora cardiaca.

Sahli: I understand your point, but in this case you are not in agreement with Juberthie & Cassagnau (1971)*, who show a separate corpus cardiacum alongside a neurohaemal organ in Collembola.

Juberthie-Jupeau: I beg your pardon, I was speaking of the corpora cardiaca of the higher insects.

Sahli: The homologies I am making apply principally to the collembolans, not pterygotes. The same homologies cannot be made if pterygotes are considered, I agree.

Strasser: Are the differences you observe of use to systematics in relation to phylogeny?

Sahli: Possibly they could be useful. For example, there are no corps connectifs in *Polydesmus*; they are present only in exotic forms. Also, in relation to our previous discussion on phylogeny, consideration of the endocrine

* See list of References p. 228.

system would suggest to me that the Symphyla are more closely related to the Diplopoda than was suggested by Dr Manton's scheme.

STRASSER: Do you think that it will be possible eventually to complete your table of comparison and to establish clearer relationships?

SAHLI: Yes—but various other diplopods, especially the exotic diplopods (Spirobolidae, Spirostreptidae, etc.) will have to be studied.

Symp. zool. Soc. Lond. (1974) No. 32, 231–235.

SPERMIOGENESIS OF *LITHOBIUS FORFICATUS* (L.) AT ULTRASTRUCTURAL LEVEL

M. CAMATINI, A. SAITA and F. COTELLI

Istituto di Zoologia dell' Università di Milano, Italy

SYNOPSIS

Among a large number of invertebrate forms, the principal variations in sperm structure are seen to be adaptations to the conditions of fertilization.

In *Lithobius forficatus* (L.), the sperm is transferred to the female by means of spermatophores; accordingly, the sperm head becomes elongated and the motor apparatus modified for propulsion in a viscous medium.

There are some cytological (Descamps, 1969a, c; Joly & Descamps, 1969) and cytochemical (Descamps, 1969b) researches on spermiogenesis of *Lithobius forficatus* (L.) and a recent ultrastructural analysis of spermatogonia and spermatocytes (Descamps, 1971). The present study gives some preliminary observations on differentiation of the spermatids of this animal.

MATERIAL AND METHODS

Lithobius forficatus (L.) adults, collected in autumn and spring, during the germ cell maturation period, were fixed in 3% glutaraldehyde in cacodylate buffer, postfixed in 1% osmium tetroxide phosphate-buffered, and embedded in Epon 812-Araldite. During dehydration the specimens were prestained in uranyl acetate (1% in 90% ethanol for 1 h). Sections were obtained with an Ultrotome LKB and stained with lead citrate, according to Reynolds (1963). They were observed with an Hitachi HS-7 and Hitachi HU 11ES electron microscope.

RESULTS

It has not been possible to follow all the various structures during the differentiation process. In the early spermatid there is widespread chromatin in small granules; during nucleus elongation the chromatin appears in the form of dense filaments with a diameter of about 150–200 Å (Fig. 1), which become regularly arranged and compactly packed, in a fibrillar manner. Near the internal rim of the nuclear envelope an amorphous nuclear material is present at the apical portion and forms vesicles at the basal region of the nucleus. This material disappears during nucleus elongation and this is probably a consequence of the nuclear manchette action.

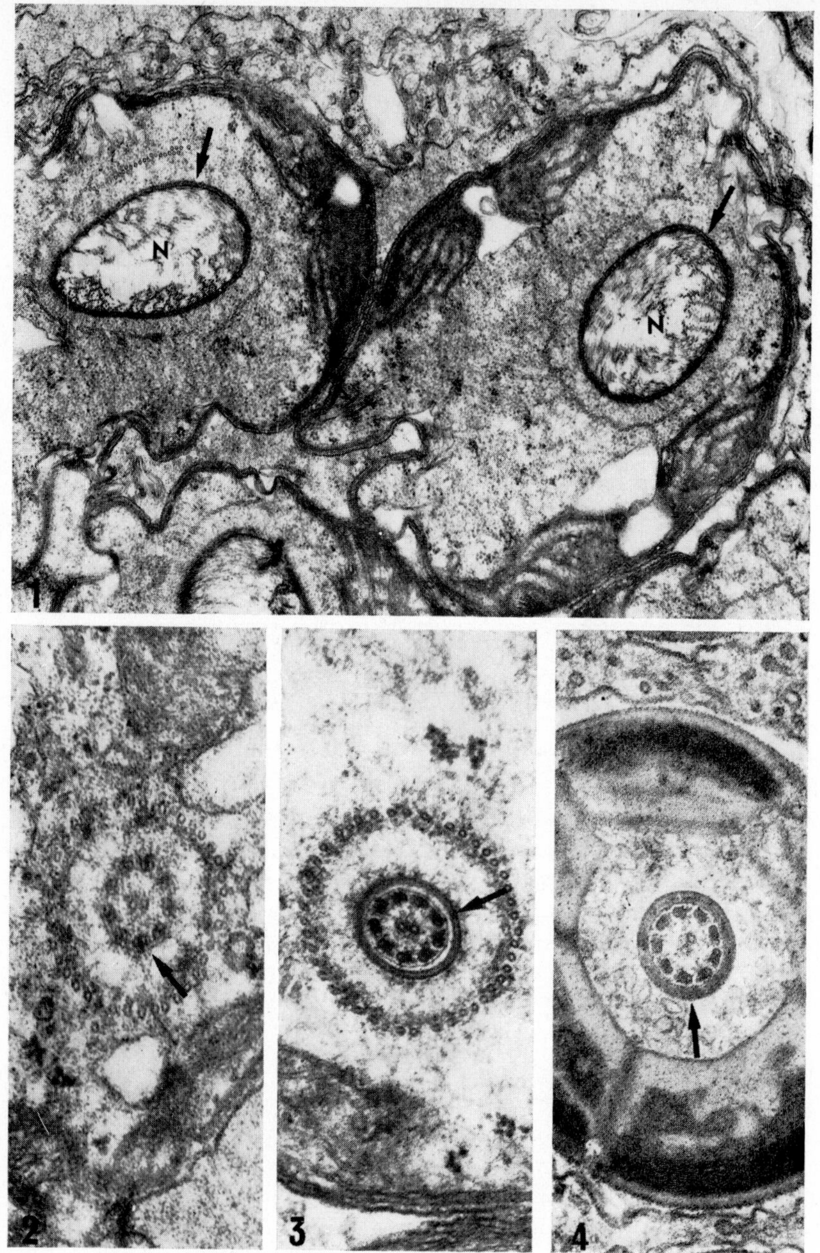

FIG. 1. Transverse section of spermatids with manchette (↗) around the nucleus (N). Cytoplasmic microtubules form a ring around the nucleus. × 32 000.

FIG. 2. Transverse section of an early spermatid flagellum. Cytoplasmic microtubules surround the central axial apparatus (↗). × 48 000.

[*Continued at foot of next page*

From the beginning to the end of spermiogenesis, microtubules follow the contour of the nucleus during changes in configuration.

The acrosome is a crescentic body at the apex of the nucleus and forms a cap-like structure covering the anterior part of the nucleus. It is composed of two coaxial caps; the internal one is more electron transparent, the external one is more opaque (Fig. 5).

In the cytoplasm of the head and connecting piece, microtubules are always present in a longitudinal array. In the early spermatid, cytoplasmic microtubules are irregularly distributed; later they constitute a ring positioned at a constant distance from the nuclear envelope (Fig. 1). In the connecting piece cytoplasmic microtubules surround the central axial apparatus of flagellar fibres (Figs 2 & 3). In this region, between the ring of cytoplasmic microtubules and flagellar fibres (Figs 2 & 3), there appears a granular material that may be composed of fibres not regularly distributed. Later a dense ring becomes evident and it forms the fibrous cylinder covering the flagellum near the outer doublets (Fig. 4). When this cylinder is completely formed, the microtubules disappear (Fig. 4).

The axial apparatus of the tail consists of a typical flagellar structure. It is positioned in the central axis of the sperm and runs from the basis of the nucleus along the connecting piece and the principal piece up to the tip of the tail (Fig. 5). The axial flagellum has the basic 9+2 arrangement of tubular fibres, and it is surrounded by a fibrous cylinder ("streitenzylinder" of Horstmann, 1968).

Between the membrane and the axial apparatus the principal piece contains a large submembranous cleft with the "mantle of the principal piece". It is characterized by four spiral septa, running in the longitudinal direction, and by numerous other septa (Fig. 6). Thus the mantle of the principal piece is subdivided into many sections containing mitochondrial derivatives (Fig. 4). The photographs of the early spermatid of *Lithobius* reveal that these structures result from a fusion of mitochondria, followed by a metamorphosis in which the typical mitochondrial morphology becomes completely lost. The matrix area of the original mitochondria contains a system of parallel membranes, spaced 45 Å centre to centre, probably of phospholipidic nature.

Continued from previous page]

FIG. 3. Transverse section of a spermatid flagellum. A dense ring (↗) constitutes the fibrous cylinder covering the flagellum. Cytoplasmic microtubules appear regularly arranged. ×48 000.

FIG. 4. Transverse section of a sperm at the connecting piece level. The fibrous cylinder around the flagellum is completely formed (↗) and cytoplasmic microtubules disappear. × 45 600.

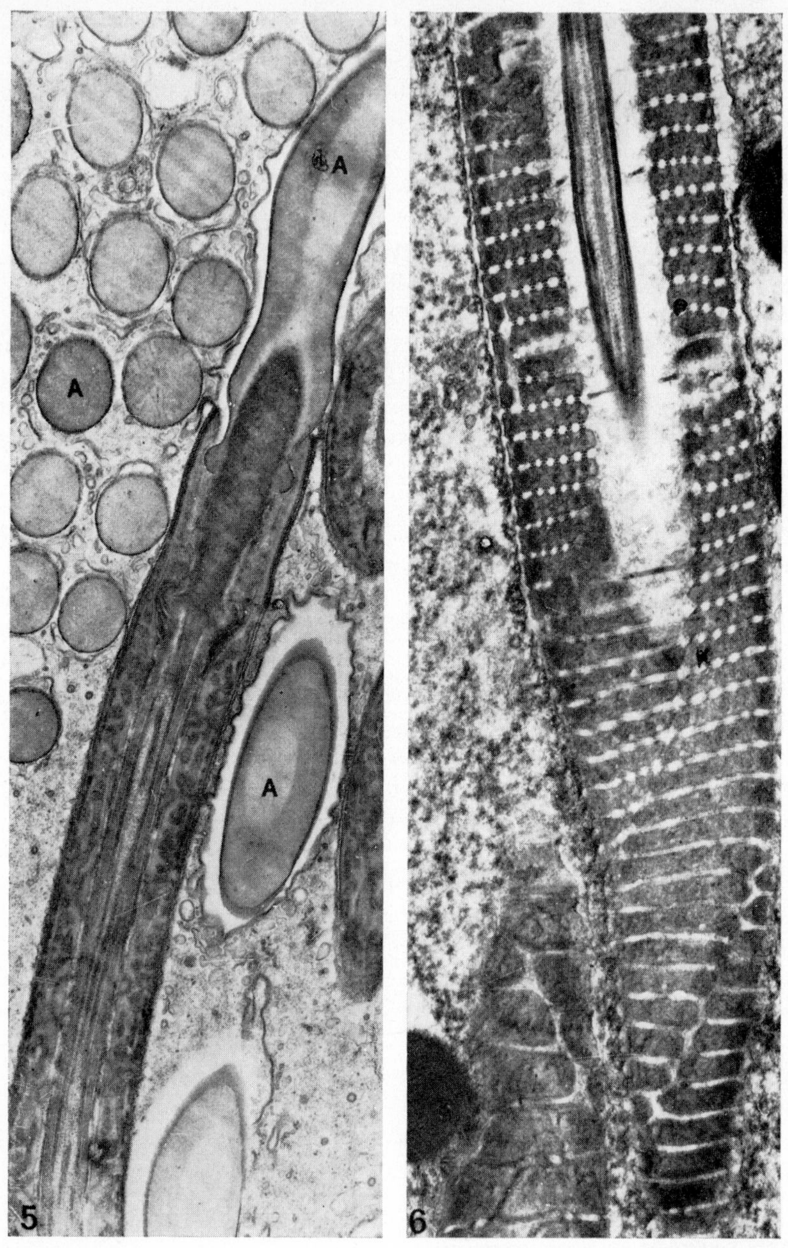

FIG. 5. Longitudinal section of the head, the connecting piece and part of the principal piece. The acrosome (A) covers the anterior part of the nucleus and presents two coaxial caps, which differ in electron transparency. × 25 600.

FIG. 6. Oblique section of the sperm principal piece. The mantle between the external membrane and the axial apparatus presents numerous septa. × 32 000.

The highly differentiated structures and their great regularity are similar to those observed in *Geophilus linearis* (Horstmann, 1968). At present more studies of spermiogenesis in Chilopoda are necessary in order to correlate ultrastructural findings with their physiological significance.

References

Descamps, M. (1969a). Étude cytologique de la spermatogénèse chez *Lithobius forficatus* L. (Myriapode Chilopode). *Archs Zool. exp. gén.* **110**: 349–361.

Descamps, M. (1969b). Étude cytochimique de la spermatogénèse chez *Lithobius forficatus* L. (Myriapode Chilopode). *Histochimie* **20**: 46–57.

Descamps, M. (1969c). Sur la présence de deux mitoses non réductionnelles après la phase de croissance spermatocytaire chez *Lithobius forficatus* L. (Myriapode Chilopode). *C. r. hebd. Séanc. Acad. Sci., Paris* **268**: 1942–1944.

Descamps, M. (1971). Étude ultrastructurale des spermatogonies et de la croissance spermatocytaire chez *Lithobius forficatus* L. (Myriapode Chilopode). *Z. Zellforsch. mikrosk. Anat.* **121**: 14–26.

Horstmann, E. (1968). Die Spermatozoen von *Geophylus linearis* Koch (Chilopoda). *Z. Zellforsch. mikrosk. Anat.* **89**: 410–429.

Joly, R. & Descamps, M. (1969). Évolution du testicule, des vésicules séminales et cycle spermatogénétique chez *Lithobius forficatus* L. (Myriapode Chilopode. *Archs Zool. exp. gén.* **110**: 341–348.

Reynolds, E. (1963). The use of lead citrate at high pH as an electron opaque stain in electron microscopy. *J. Cell. Biol.* **17**: 208–213.

Symp. zool. Soc. Lond. (1974) No. 32, 237–247.

ÉTUDE CYTOCHIMIQUE ET ORIGINE DES ENVELOPPES OVOCYTAIRES CHEZ *LITHOBIUS FORFICATUS* (L). (MYRIAPODE, CHILOPODE)

C. HERBAUT

Laboratoire de Biologie Animale, et L.A. no. 148 associé au C.N.R.S., Université de Lille, Villeneuve d'Ascq, France

SYNOPSIS

In *Lithobius forficatus* (L.) there are two types of ovocyte membranes: the primary membrane observable during the whole period of ovocyte growth and the secondary membrane or chorion which replaces the primary membrane at maturation.

The primary membrane appears at the beginning of previtellogenesis and grows to a thickness of 3–4 μm during vitellogenesis. It originates as a condensation of fibrillar material included in an amorphous substance, which is secreted by the cells of the cord joining the ovocyte to the ovarial wall. The primary membrane disappears at the end of vitellogenesis.

The secondary membrane, 10–15 μm thick, is formed from a homogeneous ground substance in which particles of glycoprotein fuse together in concentric layers. It is secreted by the epithelial cells of the ovary which is evaginated in the form of cupules. The glycoprotein of the chorion is elaborated by cells between the cupules whilst the ground substance is secreted by the cells of the cupules.

INTRODUCTION

Les enveloppes ovocytaires ont fait l'objet de nombreuses études dans diverses classes animales (Nørrevang, 1968). Le problème de leur origine a surtout été envisagé chez les insectes; selon les auteurs, elles prendraient naissance soit dans le cytoplasme ovocytaire, soit dans les cellules folliculaires (voir revue de Favard-Sereno, 1971). Nos recherches sur l'ovogénèse chez *Lithobius forficatus* (L.) nous ont permis d'apporter quelques précisions sur la mise en place des enveloppes ovocytaires chez ce chilopode.

MATÉRIEL ET TECHNIQUES

Nous avons utilisé des *Lithobius forficatus* femelles matures. Pour l'étude en microscopie photonique, les ovaires sont fixés *in situ* par le liquide de Smith. Les pièces sont incluses dans la celloïdine-paraffine et amollies au Mollifex BDH. Les coupes (7μm environ) sont colorées par l'hématoxyline de Groat et le picro-indigo-carmin. Les techniques cytochimiques utilisées ont été exposées dans une note précédente (Herbaut, 1972a).

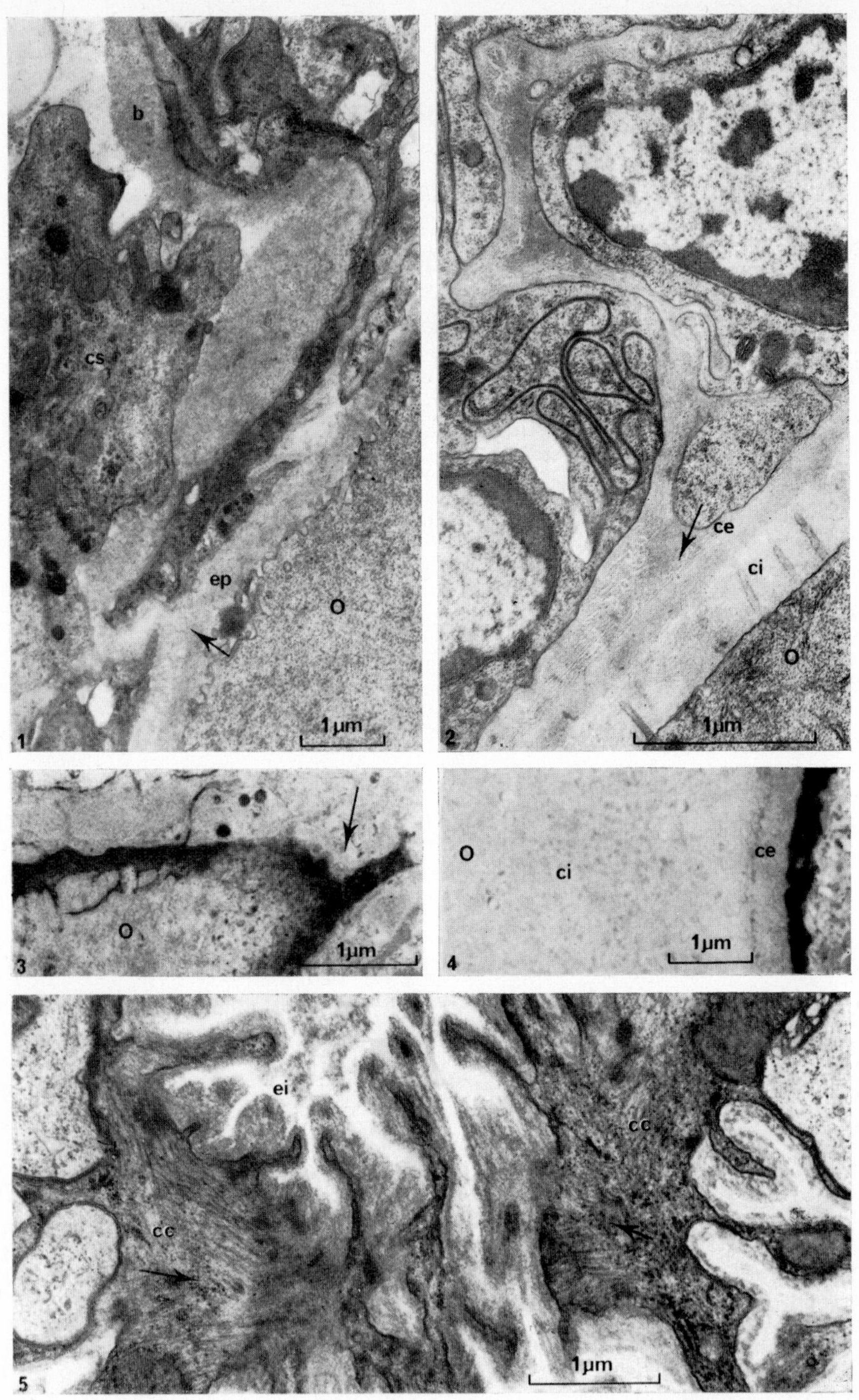

Captions on opposite page

Pour l'étude ultrastructurale, les ovaires sont disséqués et subissent une double fixation par le glutaraldéhyde à 6·25 p.100 dans le tampon phosphate à pH 7·2 et par l'acide osmique à 1 p.100. Après inclusion dans l'épon, les coupes sont contrastées par l'acétate d'uranyle et le citrate de plomb; elles sont observées au microscope électronique Hitachi H11E. La nature chimique des enveloppes a été précisée par les techniques de mise en évidence des glycoprotéines (Rambourg, 1967; Seligman modifiée par Thiéry, 1969).

RÉSULTATS

Chez *Lithobius forficatus*, il existe deux types d'enveloppes ovocytaires; l'enveloppe primitive, observable pendant la prévitellogenèse et la vitellogenèse, est remplacée, au cours de la maturation ovocytaire, par l'enveloppe définitive ou chorion.

L'enveloppe primitive

Structure

L'enveloppe primitive, absente à la périphérie des cellules en préméiose ou en début de prévitellogenèse, apparaît sous forme d'une couche de matériel homogène (Figs 1, 3), qui s'insinue entre les cellules folliculaires et l'ovocyte. Son épaisseur s'accroît progressivement et, lorsqu'elle atteint de 400 à 500 Å, il est possible de distinguer deux couches (Figs 2, 4):

1. une couche externe qui garde la structure et l'épaisseur de la couche initiale,

---

FIG. 1. Mise en place de l'enveloppe primitive (ep) chez un jeune ovocyte. Noter la continuité (flèche) avec la lame basale ovarienne (b). cs: cellule sanguine; 0: ovocyte. × 12 000.

FIG. 2. Enveloppe primitive d'un ovocyte âgé. Noter la relation de la couche externe avec le matériel extracellulaire (flèche). ce: couche externe; ci: couche interne; 0: ovocyte. × 25 000.

FIG. 3 et 4: Nature glycoprotéique de l'enveloppe primitive chez un jeune ovocyte (Fig. 3, × 16 000) et chez un ovocyte âgé (Fig. 4, × 11 000). Noter la forte colorabilité du matériel intercellulaire (flèche). ce: couche externe; ci: couche interne; 0: ovocyte.

FIG. 5: Elaboration du matériel extracellulaire qui sera à l'origine de l'enveloppe primitive. Noter les faisceaux de fibrilles convergeant vers des zones de rupture de la membrane plasmique (flèches). cc: cellules des cordons reliant les ovocytes à la paroi ovarienne; ei: espaces intercellulaires. × 18 000.

Figures 1, 2 et 5: epon, acétate d'uranyle—citrate de plomb. Figures 3 et 4: glycolméthacrylate, technique de Rambourg.

2. une couche interne, constituée d'un matériel fibrillaire comparable à celui de la couche externe et baignant dans une substance de densité électronique très faible. L'épaisseur de la couche interne augmente pendant toute la prévitellogenèse et atteint finalement 3 à 4 μm.

Durant la vitellogenèse, l'enveloppe primitive est le siège d'un transit de précurseurs vitelliniques, visibles sous forme de granules denses (Herbaut, 1972b). A la fin de cette phase, la membrane plasmique ovocytaire se rompt et le matériel de l'enveloppe primitive semble passer dans le cytoplasme de la cellule germinale.

Nature

En microscopie photonique, l'enveloppe primitive prend fortement les colorants des protéines: Millon, bleu de bromophénol mercurique. Sa réactivité au PAS, même après traitement à l'amylase salivaire,

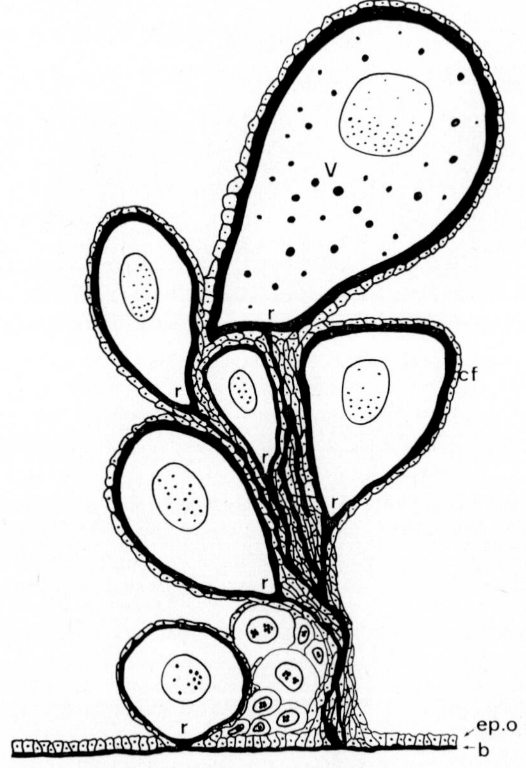

Fig. 6: Genèse de l'enveloppe primitive. r: relations avec la lame basale ovarienne (b) ou avec le matériel situé entre les cellules reliant les ovocytes à l'épithélium ovarien (ep.o). cf: cellules folliculaires; V: ovocyte en vitellogénèse.

révèle la présence de groupements vic-glycols. Pendant la vitellogenèse, les précurseurs vitelliniques lui confèrent les affinités tinctoriales du vitellus protéique.

En microscopie électronique, la couche initiale chez les jeunes ovocytes ainsi qu'une zone périphérique peu épaisse de la couche externe chez les cellules plus âgées sont fortement colorées après utilisation de la technique de Rambourg (1967) (Figs 3, 4).

L'enveloppe primitive est donc essentiellement glycoprotéique.

Origine

Les ovocytes les plus jeunes sont disposés sur la paroi ventrale de l'ovaire; les plus âgés sont reliés à cette paroi par des cordons cellulaires (Fig. 6). Dans les ovocytes jeunes, la couche initiale de l'enveloppe primitive est parfois en continuité avec la lame basale ovarienne (Figs 1, 6). Pour les ovocytes plus âgés, un matériel amorphe, comparable à la fois à la lame basale de l'ovaire et aux couches de l'enveloppe primitive, est fréquemment visible entre les cellules des cordons (Figs 2, 3, 6). Ces dernières présentent un cytoplasme riche en ribosomes libres ou associés en polysomes, avec des faisceaux de fibrilles qui semblent converger vers des zones où la membrane plasmique est interrompue (Fig. 5). Les fibrilles et les substances qu'elles pourraient acheminer paraissent constituer le matériel amorphe extracellulaire. Ce matériel s'étale à la périphérie des ovocytes, au niveau de leur zone d'attache avec les cordons cellulaires; par la suite, il pourrait être à l'origine de la couche ovocytaire interne.

L'enveloppe définitive ou chorion

Structure

Vers la fin de la phase de maturation, l'ovocyte migre vers la région postérieure de l'ovaire; son enveloppe primitive disparaît. Il est alors entouré par l'enveloppe définitive ou chorion, dont l'épaisseur atteint de 10 à 15 μm et qui est constituée de particules denses, qui peuvent fusionner en couches concentriques (Figs 7, 8).

Nature

Le chorion se délamine lors de sa confection des coupes histologiques et sa nature chimique est difficile à déterminer; il est toutefois coloré par le réactif de Millon.

La nature glycoprotéique des particules denses et des couches concentriques est mise en évidence, en microscopie électronique, par la technique de Rambourg (1967) (Fig. 8).

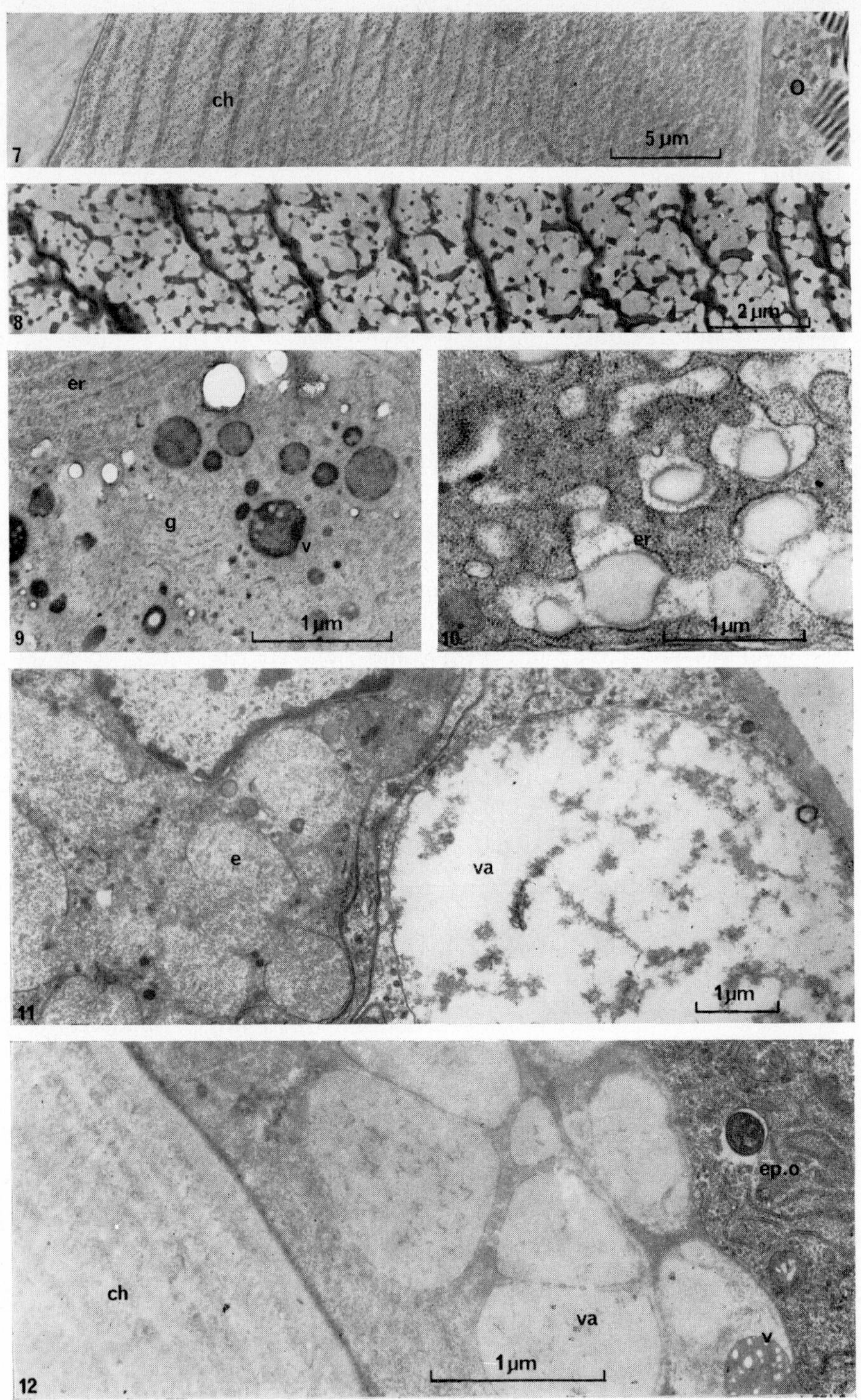

Captions on opposite page

Origine

Le chorion est sécrété par les cellules de l'épithélium ovarien, qui présente des évaginations cupuliformes (Fig. 13). En phase de repos, le cytoplasme des cellules épithéliales renferme des mitochondries, un réseau ergastoplasmique bien développé et quelques dictyosomes de faible taille. Au début de la phase sécrétoire, les cellules situées entre les cupules successives (type 1, Fig. 13) présentent des lames d'ergastoplasme épaissies, à contenu fibrillaire. Les dictyosomes sont plus nombreux et de taille plus importante. Leur face de formation est souvent en relation avec des citernes ergastoplasmiques (Fig. 13). De leur face de maturation s'échappent des vésicules qui confluent et constituent des enclaves à contenu dense et vacuolaire, dont la nature glycoprotéique est révélée par la technique de Rambourg (1967) (Fig. 9). Le matériel dense est libéré par la cellule et il est observable dans la cavité ovarienne (Fig. 12).

Les cellules épithéliales de l'ovaire présentent de profondes modifications suivant leur emplacement dans les cupules. Les cellules des parois cupulaires latérales (type 2, Fig. 13) ou apicales (types 3 et 4, Fig. 13) sont très riches en corps denses, contenant des granules ou des amas lamellaires, et de nature lysosomiale (technique de Gomori appliquée à la microscopie électronique). Ces corps denses sont des vacuoles autophagiques, selon la terminologie d'Ericsson (1969); ils semblent constitués à partir d'enclaves cytoplasmiques entourées par une lame de réticulum (Fig. 13) et donneront naissance à d'importants corps résiduels. Simultanément, l'ergastoplasme se dilate considérablement (Figs 10, 13); les citernes renferment une substance homogène, entourée d'un matériel fibrillaire.

Dans les cellules des types 3 et 4, les citernes ergastoplasmiques fusionnent pour constituer d'importantes enclaves (Figs 11, 13) qui, incomplètement entourées par une membrane unitaire, ne contiennent

---

Fig. 7. Chorion (ch) d'un ovocyte mur (o) × 3000.

Fig. 8 et 9. Mise en évidence des glycoprotéines dans le chorion d'un ovocyte mûr. ×7000.

Fig. 9: Vésicules de glycoprotéines (v) sécrétées par l'appareil de Golgi (g) dans une cellule de type 1; er: ergastoplasme. × 19 000.

Fig. 10. Dilatation de l'ergastoplasme (er) dans une cellule de type 2.

Fig. 11. Formation d'enclaves (e) (à gauche, cellule de type 3) et de vacuoles (va) (à droite, cellule de type 4). × 11 000.

Fig. 12. Fusion des vacuoles (va) dans la cavité ovarienne; ch: chorion; ep. o: épithélium ovarien; v: vésicule de glycoprotéines. × 23 500.

Figures 7, 10, 11 et 12: epon, acétate d'uranyle—citrate de plomb.

Figures 8 et 9: glycolméthacrylate, technique de Rambourg.

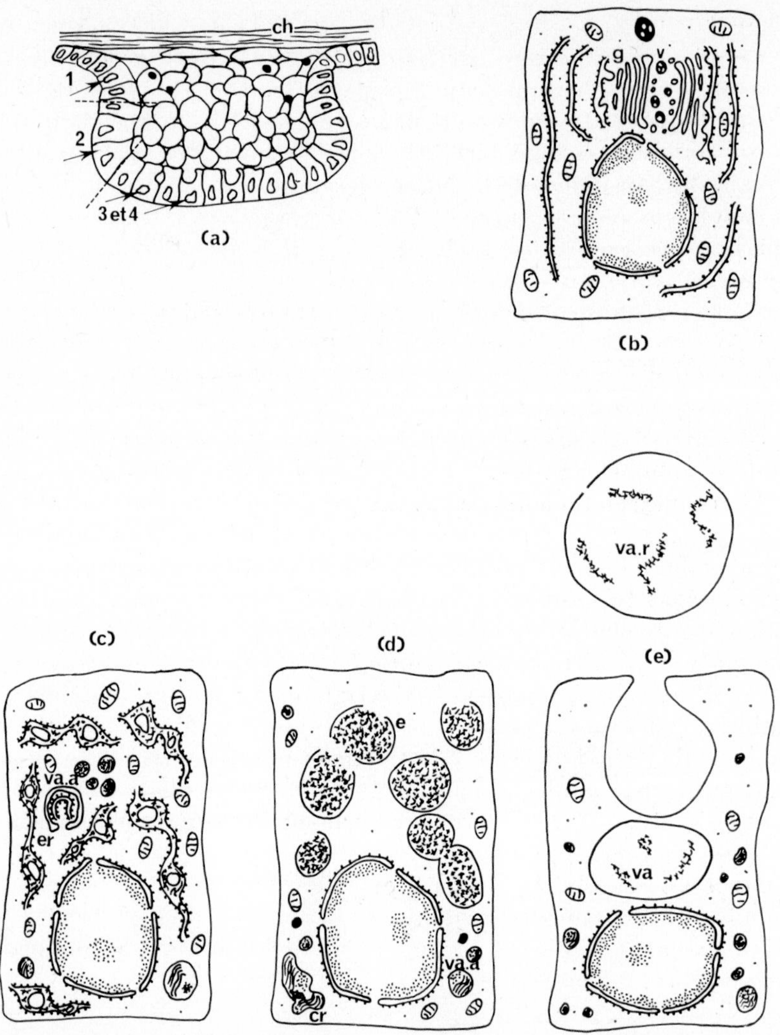

FIG. 13. Evolution cytologique des cellules cupulaires de l'épithélium ovarien. (a) cupule épithéliale avec les cellules de type 1, 2, 3 et 4; ch: chorion. (b) cellule de type 1. v: vésicule de glycoprotéines synthétisées par l'appareil de Golgi (g). (c) cellule de type 2. er: citernes ergastoplasmiques dilatées; va.a: vacuole autophagique. (d) cellule de type 3. cr: corps résiduel; e: enclave à contenu fibrillaire; va.a: vacuole autophagique. (e) cellule de type 4. va: vacuole intracellulaire; va.r: vacuole rejetée dans la lumière ovarienne.

plus que du matériel de type fibrillaire. Les zones centrales deviennent moins denses aux électrons et, après coalescence de plusieurs enclaves contiguës, il ne persiste qu'une seule grande vacuole avec une substance peu abondante et floconneuse (Fig. 11), qui est rejetée directement dans la cavité ovarienne ou dans les espaces intercellulaires (Figs 12, 13).

Le matériel glycoprotéique issu des cellules de l'épithélium ovarien en dehors des zones cupulaires (type 1) s'associe à celui des grandes vacuoles dans la cavité ovarienne (Fig. 12). Toutes ces vacuoles fusionnent et constituent le chorion des ovocytes mûrs.

DISCUSSION

Chez *Lithobius forficatus* (L.), les enveloppes ovocytaires ne sont pas élaborées par les ovocytes mais par des cellules ovariennes. Elles correspondent donc à des enveloppes secondaires, comme cela semble être le cas chez les insectes (Favard-Sereno, 1971); notons qu'elles pourraient être de type primaire et secondaire chez les poissons (Afzelius, Nicander & Sjödén, 1968).

D'autre part, chez *L. forficatus*, les cellules folliculaires n'interviennent pas dans l'élaboration des enveloppes, contrairement aux observations effectuées le plus généralement (Favard-Sereno, 1971). L'enveloppe primitive est sécrétée par des cellules situées aux pôles ovocytaires et morphologiquement différentes des cellules folliculaires, malgré leur origine probablement commune. Quant au chorion, il est sécrété par l'épithélium ovarien.

L'enveloppe primitive pourrait avoir un rôle dans le transfert des précurseurs vitelliniques (Herbaut, 1972b). Par contre, son rôle protecteur est difficile à préciser puisqu'elle disparaît en fin de vitellogenèse; notons à cet égard que les enveloppes ovocytaires successives persistent généralement et que le chorion, lorsqu'il existe, leur est superposé (Raven, 1961; Favard-Sereno, 1971).

Chez *L. forficatus*, la protection de l'oeuf est assurée par le chorion, dont la synthèse met en oeuvre deux processus simultanés. Les glycoprotéines des particules denses et des couches concentriques sont élaborées par l'appareil de Golgi des cellules situées entre les cupules de l'épithélium ovarien, à partir de précurseurs issus de vésicules ergastoplasmiques; ce processus est comparable à celui qui a été décrit dans l'épithélium intestinal du rat et de la souris (Thiéry, 1969). Quant à la substance fondamentale du chorion, elle ne semble synthétisée que par l'ergastoplasme, dont le réseau s'hypertrophie et aboutit à la formation de grosses vacuoles; une telle génèse a été décrite lors de la synthèse des lipoprotéines hépatiques (*cf.* Goldblatt, 1969). L'orientation nouvelle des synthèses cellulaires au niveau de l'épithélium cupulaire expliquerait la formation d'abondantes vacuoles autophagiques; il est en effet bien connu que les cellules qui subissent un changement de leur métabolisme sont le siège d'une autolyse importante (*cf.* Ericsson, 1969).

Bibliographie

Afzelius, B. A., Nicander, L. & Sjödén, I. (1968). Fine structure of egg envelopes and the activation changes of cortical alveoli in the river lamprey, *Lampetra fluviatilis*. *J. Embryol. exp. Morph.* **19**: 311–318.

Ericsson, J. L. E. (1969). Mechanism of cellular autophagy. In *Lysosomes in biology and pathology* **2**: 345–394. Dingle, J. T. & Fell, H. B. (eds). Amsterdam & London: North-Holland Publishing Company.

Favard-Sereno, C. (1971). Cycles sécrétoires successifs au cours de l'élaboration des enveloppes de l'ovocyte chez le Grillon (Insecte, Orthoptère). Rôle de l'appareil de Golgi. *J. Microsc.* **11**: 401–424.

Goldblatt, P. J. (1969). The endoplasmic reticulum. In *Handbook of molecular cytology*: 1101–1129, Lima de Faria A. (ed). Amsterdam & London: North-Holland Publishing Company.

Herbaut, C. (1972a). Etude cytochimique et ultrastructurale de l'ovogenèse chez *Lithobius forficatus* L. (Myriapode Chilopode). Evolution des constituants cellulaires. *Wilhelm Roux Arch. EntwMech. Org.* **170**: 115–134.

Herbaut, C. (1972b). Nature et origine des réserves vitellines dans l'ovocyte de *Lithobius forficatus* L. (Myriapode Chilopode). *Z. Zellforsch. mikrosk. Anat.* **130**: 18–27.

Nørrevang, A. (1968). Electron microscopic morphology of oogenesis. *Int. Rev. Cytol.* **23**: 114–186.

Rambourg, A. (1967). Détection des glycoprotéines en microscopie électronique; coloration de la surface cellulaire et de l'appareil de Golgi par un mélange acide chromique-phosphotungstique. *C.r. hebd. Séanc. Acad. Sci., Paris* **265**: 1426–1428.

Raven, C. P. (1961). *Oogenesis: the storage of developmental information*. London: Pergamon Press.

Thiéry, J. P. (1969). Rôle de l'appareil de Golgi dans la synthèse des mucopolysaccharides, étude cytochimique. I. Mise en évidence de mucopolysaccharides dans les vésicules de transition entre l'ergastoplasme et l'appareil de Golgi. *J. Microsc.* **8**: 689–708.

Discussion

Dohle: Did you find follicular cells?

Herbaut: Yes, around the ovocyte. The "primitive envelope" insinuates itself between the follicular cells and the ovocyte. It does not exist around the ovocyte initially but surrounds it during its growth and vitellogenesis.

Dohle: Do the follicular cells take part in vitellogenesis?

Herbaut: That is difficult to ascertain; personally I do not think they do—at least they are not directly involved in synthesizing the yolk protein which probably originates outside the ovum. There are no signs of synthesis within the follicular cells at ultrastructural level. Possibly as in *Calliphora* the follicular cells synthesise certain precursors as Anderson & Telfer (1969, 1970)* showed. Certainly an important part of the yolk protein is derived outside the ovum.

SAITA: You have shown that new material comes from the follicular cells?

HERBAUT: No. The dense particles in the primitive envelope are extra-oval in origin. I do not think they are elaborated by the follicular cells.

JUBERTHIE-JUPEAU: What happens to the follicular cells?

HERBAUT: They degenerate around the ovocyte at the end of vitellogenesis.

JUBERTHIE-JUPEAU: And nothing remains?

HERBAUT: They are occasionally observed in the ovarian lumen where they are taken up by blood cells. They begin to degenerate in contact with the ovocyte.

* Anderson, L. M. & Telfer, W. H. (1969). A follicle cell contribution to the yolk spheres of moth oocytes. *Tissue Cell* **1**: 633–644.

* Anderson, L. M. & Telfer, W. H. (1970). Trypan blue inhibition of yolk deposition. A clue to follicle cell function in the *Cecropia* moth. *J. Embryol. exp. Morph.* **23**: 35–52.

Symp. zool. Soc. Lond. (1974) No. 32, 249–259.

CONTRIBUTION À L'ÉTUDE DE L'APPAREIL GÉNITAL MÂLE ET DE LA SPERMATOGENÈSE CHEZ *POLYDESMUS ANGUSTUS* LATZEL, MYRIAPODE DIPLOPODE

JEANNINE PETIT

Laboratoire de Biologie Animale,
Faculté des Sciences, Université de Picardie, Amiens, France

SYNOPSIS

The anamorphic post-embryonic development of *Polydesmus angustus* requires eight stadia. From stadium IV (in which the secondary sexual characters appear) there is a progressive sexualization of the gonads from the posterior towards the anterior part; the testes are formed from a germinative zone situated in the posterior segments. At the same time, the deferent canals progress towards the penis.

In the adult stadium (VIII) the scalariform genital apparatus extends from the third to the penultimate segment; in the posterior two thirds there are about 20 pairs of testes of variable diameter.

Histological study reveals the existence of an antero-posterior gradient of spermatogenetic maturation: the small anterior testes (those which were formed earliest) terminated their evolution in the course of stadium IV, V and VI whilst the larger testes of the middle region only commence their spermatogenesis in stadium VII; the smaller posterior testes contain only spermatogonia until the animal has paired.

INTRODUCTION

Le développement post-embryonnaire du diplopode *Polydesmus angustus* est du type anamorphe; il comporte huit stades dont le dernier est le stade adulte.

Durant les trois premiers stades, les jeunes larves sont morphologiquement identiques. Les caractères sexuels secondaires apparaissent chez le mâle lors du stade IV; la huitième paire de pattes régresse et elle est remplacée par des bourgeons gonopodiaux.

La morphologie de l'appareil génital mâle de *P. complanatus* a fait l'objet de descriptions peu nombreuses et souvent succinctes (Fabre, 1855; Effenberger, 1909; Bessière, 1948). La spermatogenèse a été étudiée en microscopie photonique (Bessière, 1948; Tuzet, Bessière & Manier, 1957) et plus récemment la spermiogenèse d'un polydesme a été décrite en microscopie électronique (Reger & Cooper, 1968). Toutefois ces travaux ont été réalisés uniquement chez des adultes.

A notre connaissance, le développement post-embryonnaire des gonades et la nature du cycle spermatogénétique des stades immatures n'ont été que peu étudiés chez les myriapodes, citons cependant les recherches chez le chilopode *Lithobius forficatus* (Zerbib, 1966; Descamps, 1971).

Nous nous sommes intéressée à l'organogenèse de l'appareil génital mâle de *P. angustus* et parallèlement, nous avons envisagé l'étude de l'évolution spermatogénétique.

MATERIEL ET TECHNIQUES

L'examen morphologique des appareils génitaux des stades VII et VIII a été fait sous loupe binoculaire après dissection. Pour les stades larvaires plus jeunes, nous avons effectué des reconstitutions dans l'espace après observation de coupes transversales.

L'étude histologique des gonades a été réalisée soit sur des appareils génitaux isolés (stades VII et VIII), soit in situ sur des individus complets (stades IV à VI). Dans les deux cas, nous avons utilisé une fixation au glutaraldéhyde à 2 p. 100 dans le tampon cacodylate (0,1M)-sucrose (5 p. 100). Après lavage rapide dans le tampon cacodylate-sucrose, les pièces sont post-fixées au tétroxyde d'osmium (solution à 2 p. 100 dans le même tampon). L'inclusion est faite dans un mélange d'Araldite et d'Epon (Voelz & Dworkin, 1962).

Les coupes semi-fines sont colorées, soit au Bleu Azur B en milieu basique et à chaud, soit au Bleu de Toluidine à pH 4 à chaud; elles sont examinées au microscope photonique.

Les coupes ultrafines subissent une double coloration à l'acétate d'Uranyle alcoolique à 2 p. 100 et au citrate de Plomb selon Reynolds (1963); elles sont observées au microscope électronique Siemens Elmiskop IA.

RESULTATS

Nous réunirons les résultats relatifs à l'évolution de l'appareil génital au cours des différents stades post-embryonnaires et les résultats relatifs à l'évolution histologique des testicules.

Stade III

Lors de ce stade, morphologiquement non sexué, la gonade se présente sous l'aspect de deux bandes cellulaires situées dans les derniers segments larvaires.

Stade IV

Les gonades sont en position latéro-ventrale entre la chaîne nerveuse et l'intestin. Elles s'étendent du septième segment jusque dans la zone terminale de la larve. Dans la plupart des cas, on observe une sexualization progressive de la partie postérieure vers la partie antérieure.

Dans la région postérieure, les gonades ont un aspect comparable à celui qui est observé chez les larves du stade III: elles se présentent sous la forme de deux bandelettes germinatives non segmentées (Fig. 1a, b). Ces bandes se morcellent antérieurement pour donner naissance à des testicules d'un diamètre réduit (35 μm environ) (Fig. 2) et dont le nombre est peu élevé, quatre ou cinq. Les canaux déférents de faible diamètre (5 μm) sont parfois visibles.

La spermatogenèse est peu avancée; on trouve fréquemment des spermatogonies reliées entre elles par des ponts cytoplasmiques (Fig. 3). Dans certains cas, il est possible d'observer des spermatides en formation dans la partie la plus antérieure.

Stade V

De nouveaux testicules (trois ou quatre) apparaissent en arrière des testicules formés au stade IV. Dès ce stade, il existe une différence de

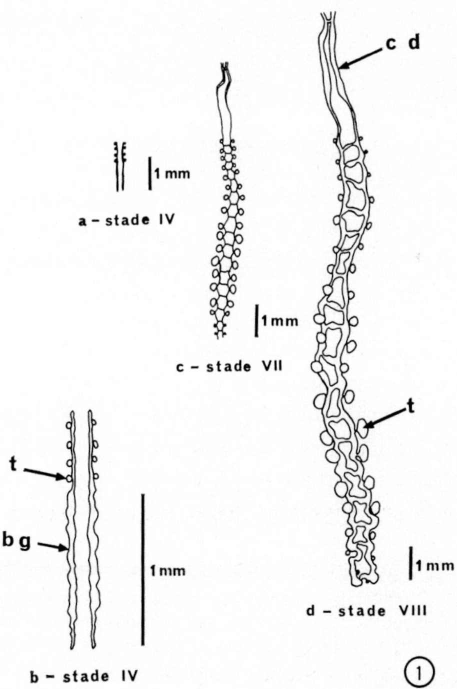

Fig. 1. Appareils génitaux mâles de *P. angustus*. a et b, stade IV; c, stade VII; d, stade VIII (adulte); b g, bandelette germinative; c d, canal déférent; t, testicule.

maturation des cellules germinales dans les deux générations testiculaires; les testicules les plus antérieurs, apparus au stade précédent, contiennent quelquefois des spermatides, voire des spermatozoïdes, alors que les testicules postérieurs ne renferment que des spermatogonies.

Stades VI et VII

L'augmentation du nombre de segments entraine la formation de testicules nouveaux dans la région postérieure.

Au stade VII, on peut dénombrer une quinzaine de testicules (Fig. 1c): les plus antérieurs et les plus postérieurs, bien qu'histologiquement différents, ont une taille comparable, toujours inférieure à celle des testicules de la région moyenne. En arrière du septième segment, des anses transversales relient les deux canaux déférents entre deux paires de testicules consécutives. Au niveau du troisième segment, une liaison antérieure unit les spermiductes avant leur débouché dans les deux pénis. Le diamètre des canaux déférents est toujours réduit et aucun spermatozoïde n'est visible dans leur lumière.

Les testicules antérieurs, les plus anciens, renferment constamment des spermatozoïdes associés par deux; de proche en proche, les testicules de la région moyenne présentent des figures de spermatogenèse.

Stade VIII = Adulte

L'appareil génital est définitivement constitué chez les jeunes adultes, avant le premier accouplement (Fig. 1d).

Le diamètre des canaux déférents a considérablement augmenté; il peut dépasser une centaine de μm. Une vingtaine de paires de testicules sont visibles dans les deux-tiers postérieurs de la gonade; leur taille et leur maturité varient selon la localisation. Nous n'avons pas mis en évidence de glandes accessoires.

Les testicules les plus antérieurs sont petits (50 μm environ de diamètre); ils sont généralement vides de leur contenu. Les cellules de leur paroi paraissent en état de dégénérescence et aucune cellule goniale n'y est visible. A leur voisinage, les canaux déférents contiennent de

---

Fig. 2. Testicule en formation. Coupe transversale d'une larve du stade IV. Coloration au bleu Azur B en milieu basique et à chaud. Ch n, chaîne nerveuse; m v, muscles ventraux; t, testicule; t a, tissu adipeux. × 550.

Fig. 3. Ponts cytoplasmiques (flèches) entre les gonies du stade IV. N, noyau. × 15 000.

Fig. 4. Spermatophores en formation (flèches) dans la "lumière" testiculaire au stade VIII. Sp, spermatozoïde. × 32 000.

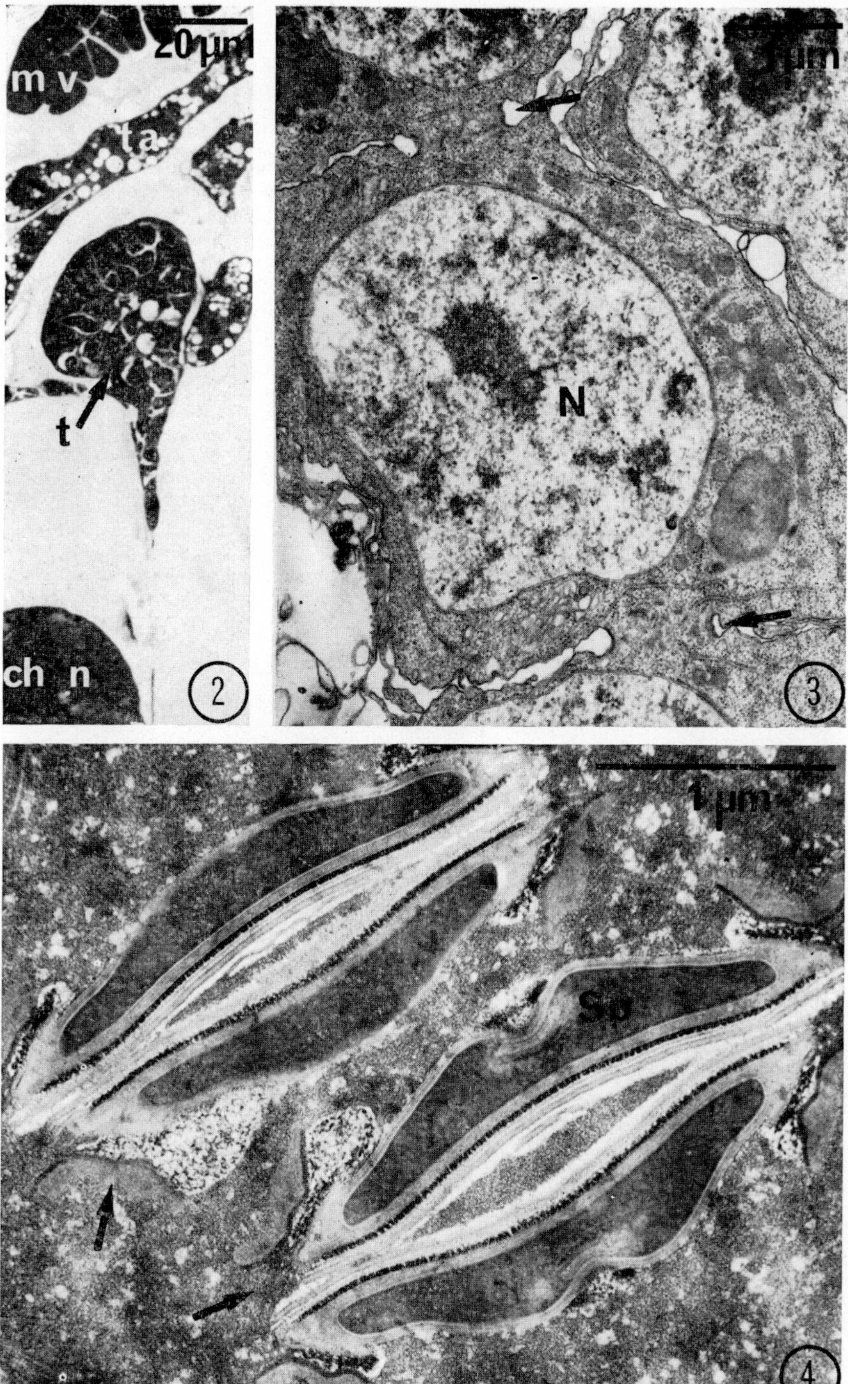

Captions on opposite page

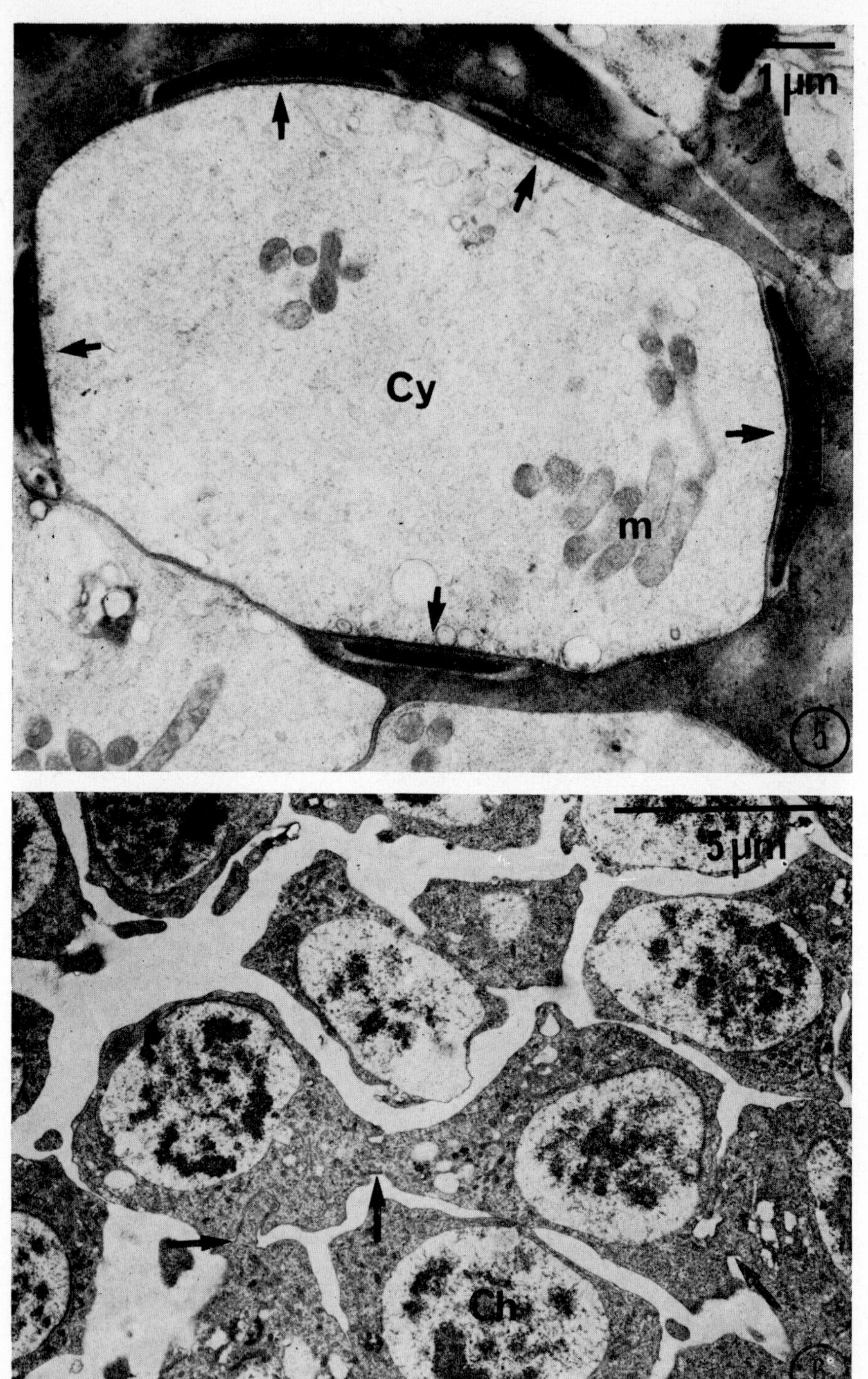

Captions on opposite page

nombreux spermatozoïdes toujours associés par deux et noyés dans une substance dense. En arrière des premières paires, la taille des testicules augmente progressivement jusqu'au niveau de la quinzième paire environ; elle atteint alors 250 μm. Les testicules de la région postérieure ont un diamètre plus faible (voisin de 100 μm).

Les cellules germinales d'un testicule donné sont toutes au même stade. De la région antérieure vers la région postérieure, les testicules renferment successivement des groupes de spermatozoïdes associés par deux (Fig. 4), puis des spermatozoïdes isolés et enfin diverses figures de spermiogenèse. Dans ce dernier cas, nous avons fréquemment remarqué autour d'une même masse cytoplasmique un nombre élevé de noyaux de spermatides (souvent supérieur à quatre) (Fig. 5). Les spermatocytes I et II sont plus rares, ce qui laisse supposer que la méïose est rapide. Les testicules formés au cours de la mue imaginale ne contiennent généralement que des spermatogonies qui sont soit au repos, soit en phase de multiplication; les mitoses sont alors synchrones.

Après le premier accouplement, de nouvelles générations spermatogénétiques sont possibles. Les testicules les plus antérieurs ne sont plus fonctionnels; les testicules postérieurs présentent des figures spermatogénétiques (Fig. 6) et libreront des spermatozoïdes dans les canaux déférents.

DISCUSSION

Morphologie de l'appareil génital mâle adulte

Les descriptions antérieures des gonades de *Polydesmus* adultes restent assez sommaires (Fabre, 1855; Effenberger, 1909; Bessière, 1948). Selon ces auteurs, l'échelle testiculaire ne compterait que treize à quatorze mailles, chacune d'elles présentant latéralement une paire de vésicules d'un diamètre voisin de 100 μm; aucun d'eux ne fait mention d'une différence de taille testiculaire.

D'après nos observations, le nombre de ponts transversaux est de 20 à 21 chez l'adulte; un pont supplémentaire est également visible juste avant le débouché des canaux déférents dans les pénis. Le diamètre testiculaire n'est pas constant et varie de 50 à 250 μm; les testicules antérieurs sont les plus petits et ceux de la région moyenne (douzième à quinzième paires) les plus gros. Les premiers testicules ne sont pas antérieurs au septième segment (segment gonopodial).

Fig. 5. Noyaux spermatiques (flèches) groupés autour d'une même masse cytoplasmique. Cy, cytoplasme; m, mitochondrie. × 15 000.

Fig. 6. Spermatocytes en prophase méïotique. Noter les ponts cytoplasmiques (flèches) et le synchronisme des divisions. Ch, chromosomes. × 6000.

L'examen de schémas d'appareils génitaux de *Strongylosoma pallipes* (Seifert, 1932) et de divers diplopodes dont un polydesmide, *Ulodesmus bispinosus* (Warren, 1934), montre que ces auteurs ont remarqué, bien qu'ils ne le mentionnent pas, une différence de diamètre entre les testicules. Chez ces diplopodes, les testicules les plus antérieurs et les plus postérieurs sont également plus petits que ceux de la région moyenne.

Spermatogenèse

Notre étude sur la spermatogenèse a surtout été faite dans le but de déterminer les degrés d'évolution des cellules germinales.

Nos observations sur les phases méïotiques confirment celles des auteurs antérieurs (Bessière, 1948; Tuzet *et al.*, 1957). Par contre, nous n'avons pas mis en évidence une complexité comparable au cours des étapes préméïotiques; pour Tuzet *et al.* (1957), en effet, l'augmentation du nombre de gonies se ferait par trois endomitoses conduisant à des gonies polyvalentes qui subiraient ensuite trois amitoses donnant des gonies de dernière génération. Chez *P. angustus*, il existe à l'intérieur des jeunes testicules une multiplication mitotique des gonies. Celles-ci ne se séparent pas complètement en fin de division; elles restent unies par des ponts cytoplasmiques ressemblant à ceux qui ont été décrits chez le gardon (Clérot, 1971). Comme le signale cet auteur, ces ponts pourraient avoir une influence sur le synchronisme des mitoses; en effet, à l'intérieur d'un même testicule, toutes les gonies sont au même stade de division.

Une étude récente de la spermiogenèse réalisée en microscopie électronique chez un polydesme (Reger & Cooper, 1968) a mis en évidence l'existence d'un nombre limité de noyaux (un ou deux) entourant une même masse cytoplasmique lors de la formation des spermatozoïdes. Chez *P. angustus*, le nombre de noyaux de spermatides est supérieur, observation qui confirme les images photoniques de Tuzet *et al.* (1957). Selon Reger & Cooper (1968), les spermatozoïdes sont directement associés par deux. En fait, il existe à l'intérieur des testicules une phase pendant laquelle les spermatozoïdes sont isolés bien que matures, et ce n'est que secondairement qu'ils s'associent deux à deux à l'intérieur des testicules. Il s'agit vraisemblablement, comme cela a été montré chez d'autres diplopodes (Horstmann & Breucker, 1969), de la constitution d'un spermatophore dans lequel les spermatozoïdes peuvent rester plusieurs mois sans subir aucune altération. Notons à cet égard la présence constante de spermatozoïdes groupés en spermatophores dès le stade VI et ne devenant fonctionnels qu'au stade VIII; de même, des spermatophores d'aspect tout à fait normal sont décelables dans les réceptacles séminaux vulvaires quinze jours après l'accouplement.

L'existence de spermatophores à l'intérieur des testicules pose le problème de leur édification. Outre une substance de fond assez opaque aux électrons, résultant vraisemblablement de la transformation des résidus cytoplasmiques après la spermiogenèse, nous avons observé des phases de sécrétion de substances très denses à l'intérieur des cellules de la paroi des testicules âgés; ces substances, qui pourraient jouer un rôle dans l'édification des spermatophores sont exudées dans la lumière testiculaire au niveau de nombreuses microvillosités.

Développement post-embryonnaire des gonades mâles

L'étude du développement post-embryonnaire des gonades mâles de *P. angustus* met en évidence l'existence d'une segmentation à la partie antérieure d'une zone germinative située dans les derniers segments larvaires. Au fur et à mesure de leur formation, les testicules présentent des figures de spermatogenèse, ce qui explique une maturation plus avancée dans les testicules les plus antérieurs. La zone germinative est moins développée chez les larves plus âgées.

L'existence d'une zone germinative a été décrite chez d'autres arthropodes tels que les crustacés amphipodes (Charniaux-Cotton, 1959). Sa présence ne semble pas générale chez tous les myriapodes; en particulier, Zerbib (1966) n'en a pas observé chez le chilopode, *Lithobius forficatus*. Chez les symphyles adultes, la formation de nouveaux testicules s'effectue à partir d'une zone germinative et les testicules apparaissent au cours des mues subies par les adultes (Juberthie-Jupeau, 1963).

Chez *P. angustus*, le nombre de testicules semble fixe chez l'adulte, et chaque phase spermatogénétique est marquée, non pas par l'apparition de nouveaux testicules, mais par l'activité de testicules jeunes restés jusqu'alors au repos. L'absence de mues chez l'adulte parait compensée par l'accumulation d'un nombre important de testicules.

Il faut également rappeler que les testicules renferment des spermatozoïdes dès le stade VI ou même dans de rares cas dès le stade V; toutefois, les spermatozoïdes ne sont visibles dans les canaux déférents que chez l'adulte. Chez *Lithobius forficatus*, la spermiogenèse n'est réalisée que chez les individus matures (Descamps, 1971).

Enfin, chez *P. angustus*, les cellules germinales sont au même stade dans chaque testicule, mais atteignent un degré de maturité de plus en plus avancé vers la région antérieure; cette observation permet de suggérer l'existence d'un gradient antéro-postérieur de maturation spermatogénétique.

Bibliographie

Bessière, Cl. (1948). La spermatogenèse de quelques Myriapodes Diplopodes. *Archs Zool. exp. gén.* **85**: 149–236.

Charniaux-Cotton, H. (1959). Etude comparée du développement post-embryonnaire de l'appareil génital et de la glande androgène chez *Orchestia gammarella* et *Orchestia mediterranea* (Crustacés Amphipodes). Autodifférenciation ovarienne. *Bull. Soc. zool. Fr.* **84**: 105–115.

Clérot, J. C. (1971). Les ponts intercellulaires du testicule du Gardon: organisation syncitiale et synchronie de la différenciation des cellules germinales. *J. Ultrastruct. Res.* **37**: 690–703.

Descamps, M. (1971). Le cycle spermatogénétique chez *Lithobius forficatus* L. (Myriapode Chilopode). I. Evolution et étude quantitative des populations cellulaires du testicule au cours du développement post-embryonnaire. *Archs Zool. exp. gén.* **112**: 199–209.

Effenberger, W. (1909). Beiträge zur Kenntnis der Gattung *Polydesmus*. *Jena. Z. Naturw.* **44**: 527–586.

Fabre, L. (1855). Recherches sur l'anatomie des organes reproducteurs et sur le développement des Myriapodes. *Annls Sci. nat.* (Zool.) (4) **3**: 257–315.

Horstmann, E. & Breucker, H. (1969). Spermatozoen und Spermiohistogenese von *Graphidostreptus* spec. (Myriapoda, Diplopoda). I. Die reifen Spermatozoen. *Z. Zellforsch. mikrosk. Anat.* **96**: 505–520.

Juberthie-Jupeau, L. (1963). Recherches sur la reproduction et la mue chez les Symphyles. *Archs Zool. exp. gén.* **102**: 1–172.

Reger, J. F. & Cooper, D. P. (1968). Studies on the fine structure of spermatids and spermatozoa from the Millipede *Polydesmus* sp. *J. Ultrastruct. Res.* **23**: 60–70.

Reynolds, E. S. (1963). The use of lead citrate at high pH as an electron opaque stain in electron microscopy. *J. Cell. Biol.* **17**: 208–212.

Seifert, B. (1932). Anatomie und Biologie des Diplopoden *Strongylosoma pallipes* Oliv. *Z. Morph. Ökol. Tiere* **25**: 362–507.

Tuzet, O., Bessière, Cl. & Manier, J. F. (1957). La spermatogenèse des Polydesmes: *Plagiodesmus oatypus* Ch. et *Polydesmus complanatus* L. *Bull. Inst. r. Sci. nat. Belg.* **33** (18): 1–11.

Voelz, H. & Dworkin, M. (1962). Fine structure of *Myxococcus xanthus* during morphogenesis. *J. Bact.* **84**: 943–952.

Warren, E. (1934). On the male genital system and spermatozoa of certain Millipedes. *Ann. Natal Mus.* **7**: 351–402.

Zerbib, Ch. (1966). Etude descriptive et expérimentale de la différenciation de l'appareil génital du Myriapode Chilopode *Lithobius forficatus* L. *Bull. Soc. zool. Fr.* **91**: 203–216.

Discussion

Demange: What happens before the appearance of the secondary sexual characters?

PETIT, J.: My husband and I have work in progress on stadium III where the gonad is not yet segmented and the form of the testes has not developed. We shall shortly report on this work.

SAHLI: What is the condition of the testes in individuals which have already paired?

PETIT, J.: What is remarkable is the fact that the testes regress immediately after forming the spermatophores and become non-functional. In an older animal which has paired several times, all of the anterior part of the testis is empty, but towards the posterior part, one finds early phases of spermatogenesis. In a younger animal which has not paired, there are no phases of spermatogenesis, merely a reserve of spermatogonia.

Symp. zool. Soc. Lond. (1974) No. 32, 261–272.

STUDIES ON THE MALE REPRODUCTIVE PATTERN IN SOME INDIAN DIPLOPODA (MYRIAPODA)

RANI KANAKA* and B. N. CHOWDAIAH

Department of Zoology, Bangalore University, Bangalore 1, India

SYNOPSIS

This paper presents a comparative account of the morphology of the male reproductive pattern in six species, belonging to three different families, of Diplopoda (Myriapoda): *Spirostreptus asthenes*, *Harpurostreptus robustior*, *Ktenostreptus costulatus*, *Aulacobolus variolosus*, *Aulacobolus levissimus* and *Arthrosphaera craspedota*.

The typical male genital system in the species investigated consists of tubes forming a simple ladder with regularly or irregularly spaced rungs and an approximately equal number of roughly paired lateral lobes. In some genera, the reticular system is replaced by a ladder system and these studies suggest a gradual evolution in the basic pattern of the male genital apparatus from a more primitive and complex type to a simpler and more advanced type.

INTRODUCTION

The contributions to the understanding of the male reproductive system in Diplopoda have been very scanty. Among the earlier workers, Newport (1841) confined his study to the family Julidae and Fabre (1855) discussed a single species from each of the families Julidae, Glomeridae, Polyxenidae and Polydesmidae. Fabre's description of the gonopods was restricted to their external anatomy. Von Rath (1890) extended the information on the male reproductive system of *Polydesmus complanatus* and other diplopods. Attems (1894) gave a more detailed description of the male gonopods of several millipedes but did not discuss the internal anatomy of these structures. The subsequent publications concerning the male reproductive system of some millipedes include mainly those of Miley (1927, 1930), Seifert (1932), Warren (1934), West (1953), Chowdaiah (1966) and Krishnan (1967).

The present paper is a contribution towards a comprehensive inquiry into the comparative morphology of the reproductive system in the male of six species of Diplopoda belonging to three families.

MATERIALS AND METHODS

The species investigated include *Spirostreptus asthenes*, *Harpurostreptus robustior*, *Ktenostreptus costulatus*, *Aulacobolus variolosus*,

* Present address: Department of Physiology, St. John's Medical College, Bangalore 560034, India.

Aulacobolus levissimus and *Arthrosphaera craspedota*, all collected from different parts of South India. They occur in abundance during the rainy season between June and September.

Several specimens of different age groups were dissected in dilute commercial formalin in order to study the variation in their reproductive pattern. For a detailed observation, the entire male reproductive system was stripped out of the dissected specimen and floated on a slide. The whole mount was made permanent by treating with tertiary-butyl alcohol and mounting in Euparol. Drawings were made from the dissections of the entire reproductive system *in situ* and the scale measurement of the same was taken in millimeters. The material for histological study was fixed in Allen's modified Bouin's fluid; serial sections were cut at thicknesses varying from 5 μm to 10 μm and stained in Heidenhain's iron haematoxylin.

OBSERVATIONS

Family: Harpagophoridae

1. *Spirostreptus asthenes* (Pocock).

The internal male genital organs which run nearly the entire length of the animal extending from the second segment almost up to the anal segment are situated ventral to the gut and immediately above the nerve-cord (Fig. 1a). The two vasa deferentia running on either side of the gonad end blindly posteriorly. These are connected all along their length by a tubular system which is variable in different regions of the gonad. The entire genital system can morphologically be differentiated into three main regions. These regions are demarcated for descriptive convenience by the letters A, B, C and D.

Anteriorly, in the region between A and B, the simple tubular system consists of a series of irregularly spaced anastomosing tubes (Fig. 1b). In the next region, between B and C, reaching almost the middle of the gonad, the tubular system forms a fairly fine mesh with an appearance of a lace ribbon (Fig. 1c). On the other hand, in the posterior half including the region between C and D, the tubular system forms a complicated reticulum by an anastomosis of a network of tubes and their branches. As a result, the identity of the longitudinal vasa deferentia is completely lost. However, a series of roughly paired stalked structures somewhat dorsally situated are found connected to the lateral edges of the reticulum (Fig. 1a). These globular lobed bodies are called lateral lobes by Warren (1934) and testicular vesicles by West (1953). These lateral lobes which are conspicuously absent in the

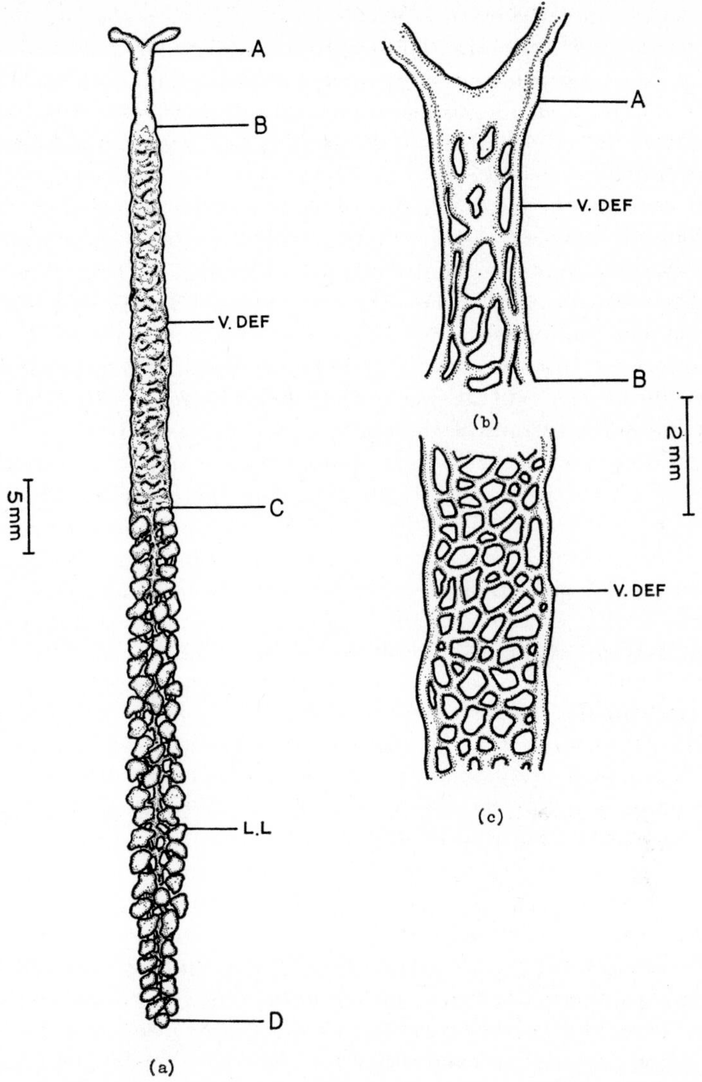

FIG. 1. *Spirostreptus asthenes*: (a) the male reproductive system of *Spirostreptus asthenes in situ*; (b) region between A and B enlarged; (c) part of the system between B and C enlarged.

anterior half are mainly spermatic lobes. Externally these lobes do not exhibit any marked difference except that they diminish in size as well as in number both at the anterior and posterior ends of the series.

2. *Harpurostreptus robustior* (Humbert).

In a general way the male genital system of *Harpurostreptus robustior* resembles that of *Spirostreptus asthenes*, but from a comparative point of view it offers several facts of interest.

The two longitudinal vasa deferentia are highly convoluted extending from the second segment almost up to the anal segment. They are not of uniform morphology throughout their length but exhibit variations at different regions of the genital system. As a result the genital system can easily be divided into four regions (Fig. 2a).

In the most anterior region, the two vasa deferentia join to form a single median common duct running between A and B which, however, forks anteriorly into two short ducts before opening independently to the exterior on the ventral side of the second segment (Fig. 2b).

The reticular system of the short region between B and C is very much reduced and the two vasa deferentia are joined together by a number of unequally spaced transverse connexions or bands forming the rungs of the typical ladder. The number of these connexions, however, is constant, being five only in the present instance (Fig. 2b).

The reticular system of the next region, between C and D, is almost the same as that of the testicular region except that the tubes of the reticular system are slightly larger and more widely spaced (Fig. 2a, c). In addition, this region and those more anterior are completely devoid of the lateral lobes.

The posterior third of the gonad includes mainly the testicular region bearing the stalked lateral lobes between D and E, commencing from the 40th segment. The right and left vasa deferentia are connected by an irregular network of fine tubules forming a closely spaced reticular system (Fig. 2a, d).

3. *Ktenostreptus costulatus* (Attems).

The male genital organs extend from the second segment almost up to the anal segment. The two vasa deferentia which end blindly at a little distance from the posterior extremity are highly convoluted and are closely approximated to each other for most of their length (Fig. 3a). However, they are connected by a series of transverse ducts throughout their length, the number of such transverse tubes and the spacing in between them being variable in different regions of the gonad.

The paired vasa deferentia in the most anterior region between A and B reveal 11 closely but unequally spaced transverse connexions (Fig. 3b). A gradual increase in the thickening of vasa deferentia is seen in the next region between B and C extending up to the 18th segment. In addition, spindle shaped swellings at different intervals

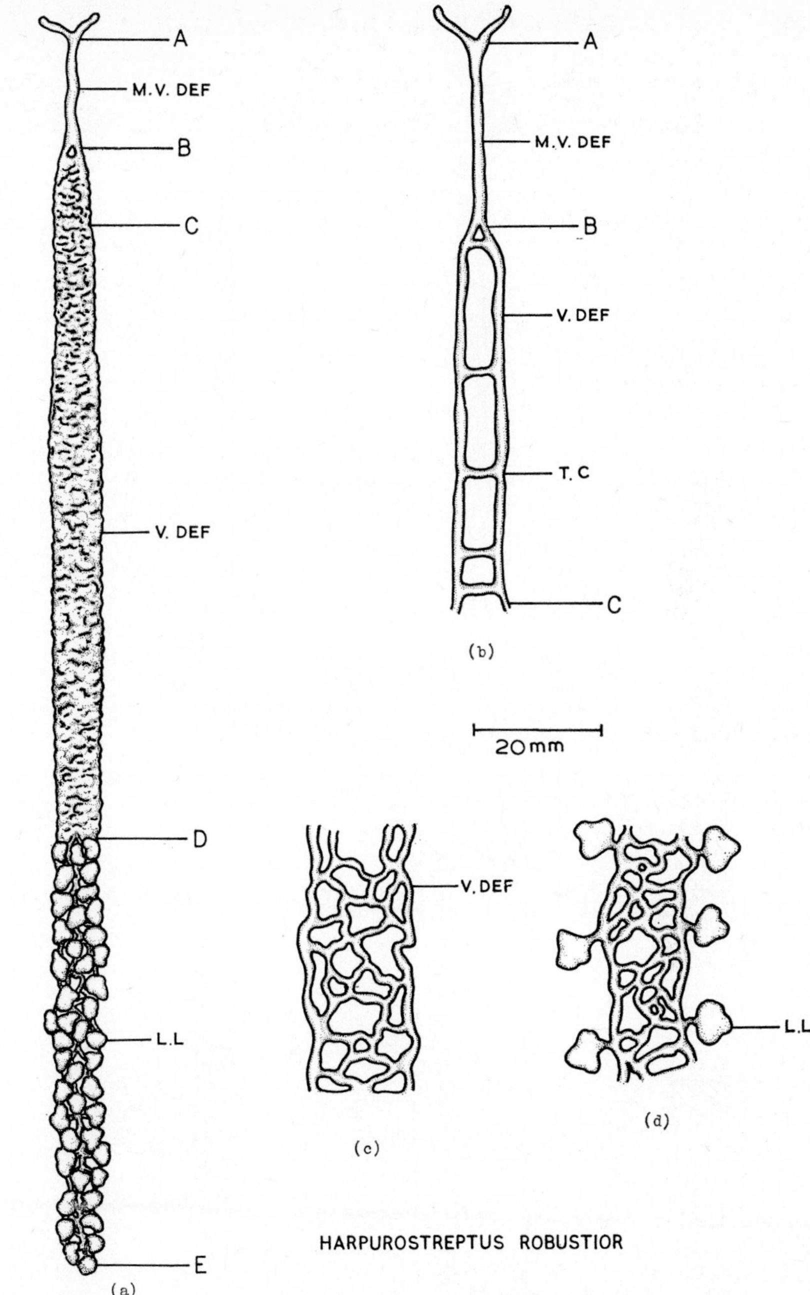

FIG. 2. *Harpurostreptus robustior*: (a) the male reproductive system of *Harpurostreptus robustior in situ*; (b) region between A and C enlarged to show the details; (c) part of the system between C and D enlarged; (d) part of the testicular region between D and E enlarged.

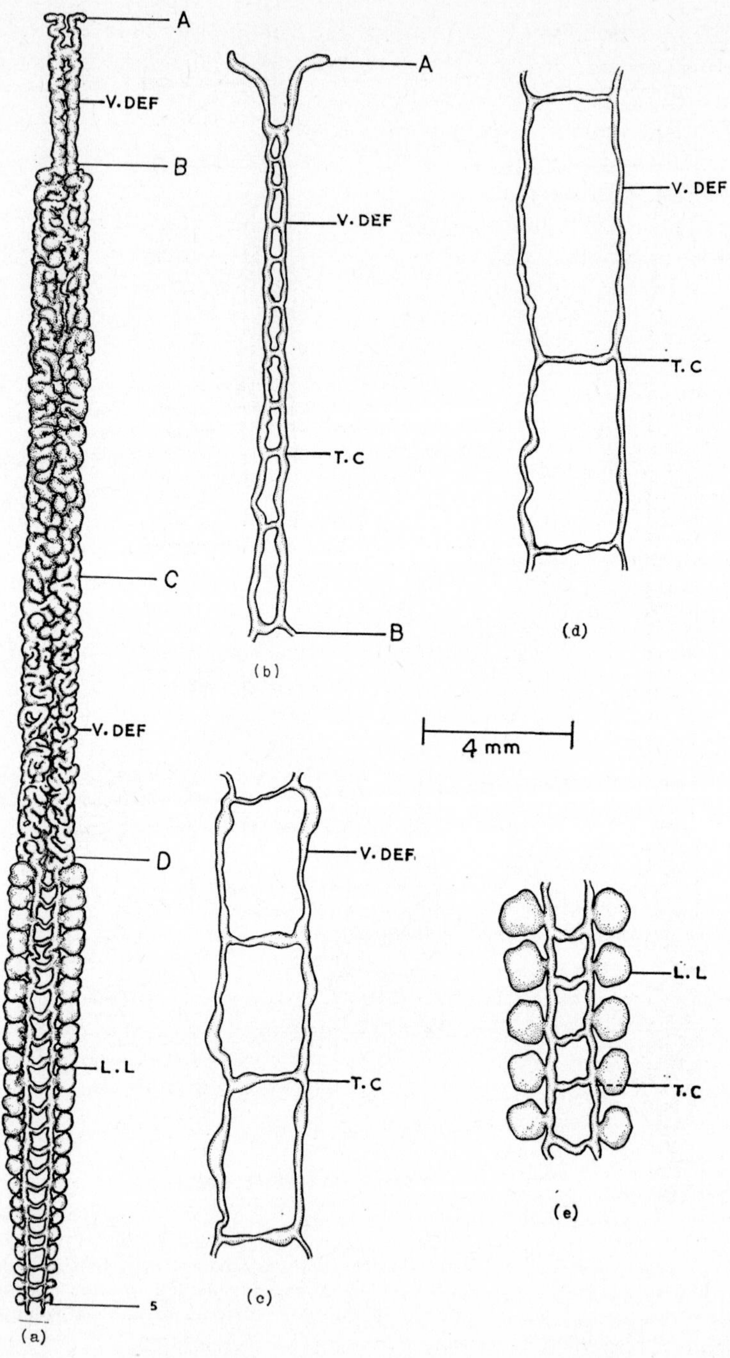

Caption on facing page

characterize both the lateral ducts and the equally spaced transverse ducts of this region (Fig. 3c). The two vasa deferentia narrow down uniformly in the region that follows, between C and D (Fig. 3d).

The last region of the gonad, between D and E, is the testicular region commencing from the 25th segment (Fig. 3e). The two vasa deferentia of this region, which run almost straight, bear the stalked lateral lobes. These lobes number approximately 22 pairs corresponding to the same number of equally and closely spaced transverse connexions which form the rungs of the typical ladder (Fig. 3a, e).

Family: Pachybolidae

1. *Aulacobolus variolosus* (Silvestri).

The basic plan of the male reproductive pattern in *Aulacobolus variolosus* (Fig. 4) is a considerable simplification of the typical ladder system of *Ktenostreptus costulatus*.

The two vasa deferentia, which retain their identity throughout, are thin and delicate in the anterior region; between the second and the seventh segments there are six closely and equally spaced transverse connexions. In the next region, between the 7th and 13th segments, the two vasa deferentia are slightly dilated and convoluted, and there are four unequally spaced transverse connexions. In the testicular region that follows, the two vasa deferentia, which are much dilated and less convoluted, are further connected by two more transverse ducts. Finally they narrow down posteriorly ending blindly to form the last region of the gonad which is devoid of the lateral lobes.

2. *Aulacobolus levissimus*, Attems.

The male reproductive pattern of *Aulacobolus levissimus* represents further simplification in the progression of evolutionary changes from the reticular-type to the ladder-type (Fig. 5).

The paired vasa deferentia run closely parallel to each other from one end to the other; they are uniformly narrow and straight except in the pre-testicular region where they are highly convoluted. There are just two transverse connexions, one in the third and another in the fourth segment, which represent the remnants of the typical ladder-system.

FIG. 3. *Ktenostreptus costulatus*: (a) the male reproductive system of *Ktenostreptus costulatus in situ*; (b) region between A and B enlarged, showing the transverse connexions; (c) part of the system between B and C enlarged, showing the vasa deferentia with irregular spindle shaped swellings and transverse connexions; (d) a part of the system between C and D enlarged; (e) part of the testicular region between D and E enlarged, showing the transverse connexions and the lateral lobes.

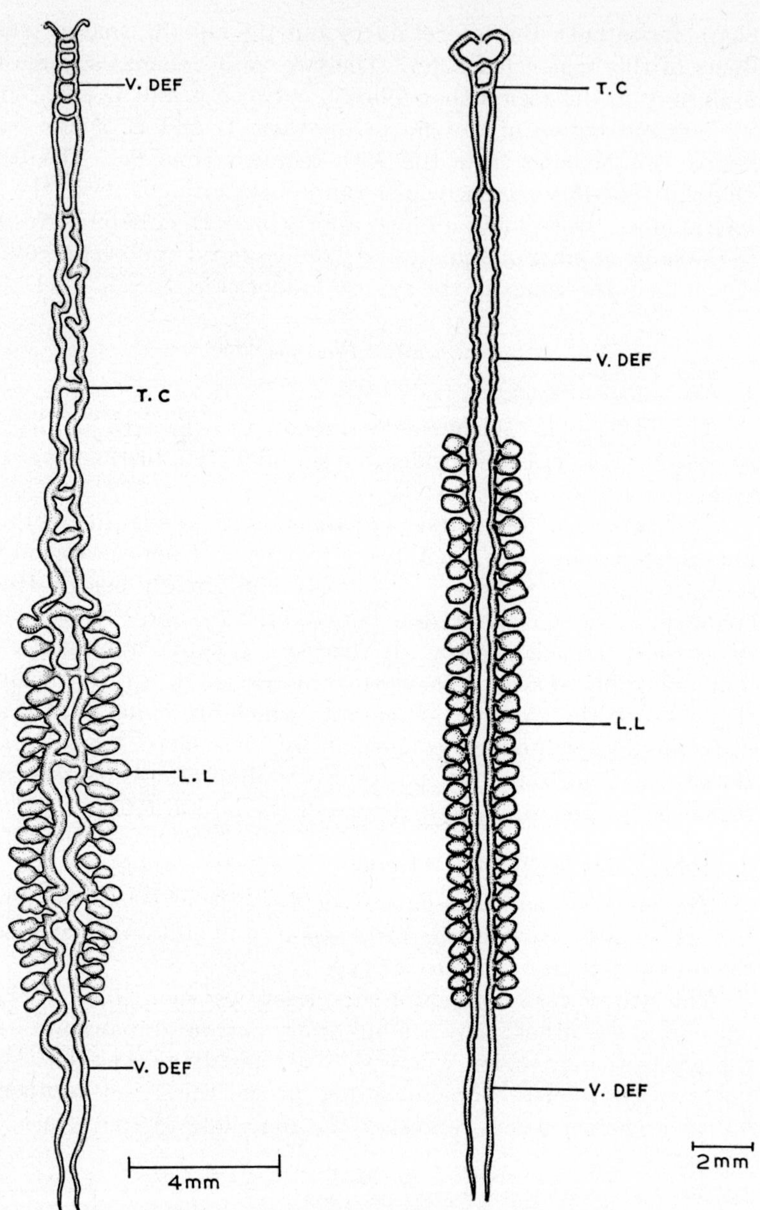

FIG. 4. *Aulacobolus variolosus*: the male reproductive system *in situ*.

FIG. 5. *Aulacobolus levissimus*: the male reproductive system *in situ*.

The testicular region commences from the 25th segment and extends up to the 40th segment. Each of the vas deferens in the testicular region bears laterally 30 delicately stalked oval lateral lobes. The post-testicular region of the genital system extending from the 40th up to the 45th segment includes the two tapering vasa deferentia devoid of the lateral lobes and ending blindly.

Family: Sphaerotheriidae

1. *Arthrosphaera craspedota*, Attems.

The internal genital system of the male reveals the usual basic plan consisting of the two vasa deferentia uniting anteriorly in a short median duct which forks into two long transverse branches running to the external apertures. These two ducts which run almost up to the last segment are widely separated and end blindly posteriorly (Fig. 6). The transverse connexions which represent the ladder-system are completely absent. The testes lobes which are fewer and usually number 16 pairs are all similar externally. They are globular and lobulated being attached to the vasa deferentia by short and thick stalks. However, they vary in size; the anterior ones are the larger and there is a gradual reduction in size towards the posterior region (Fig. 6).

DISCUSSION

The studies of Warren (1934) clearly reveal the presence of three basic types of male reproductive patterns among the six species of millipedes investigated; a reticulate-type as in the case of *Odontopyge* sp. and *Poratophilus diplodontus* where the two vasa deferentia running one on either side of the testes are connected by a series of irregularly spaced transverse tubular reticulae; a ladder-type as in the case of *Chersastus ruber* Attems and *Ulodesmus bispinosus* Attems, where the two longitudinal vasa deferentia are connected by a series of regular transverse tubes. In the case of *Sphaerotherium punctulatum* Brandt, the two vasa deferentia become exceedingly expanded in width. They are nevertheless connected posteriorly, but otherwise the rungs of the ladder are absent. This arrangement which is the characteristic of the genus *Arthrosphaera* in the family Sphaerotheriidae can be designated as the arthrosphaeroid-type.

The present studies also clearly establish the existence of these three basic types along with the intermediaries which act as the connecting links (Table I). Thus a typical reticulate-type characterizes the male reproductive system of *Spirostreptus asthenes* (Fig. 1). The

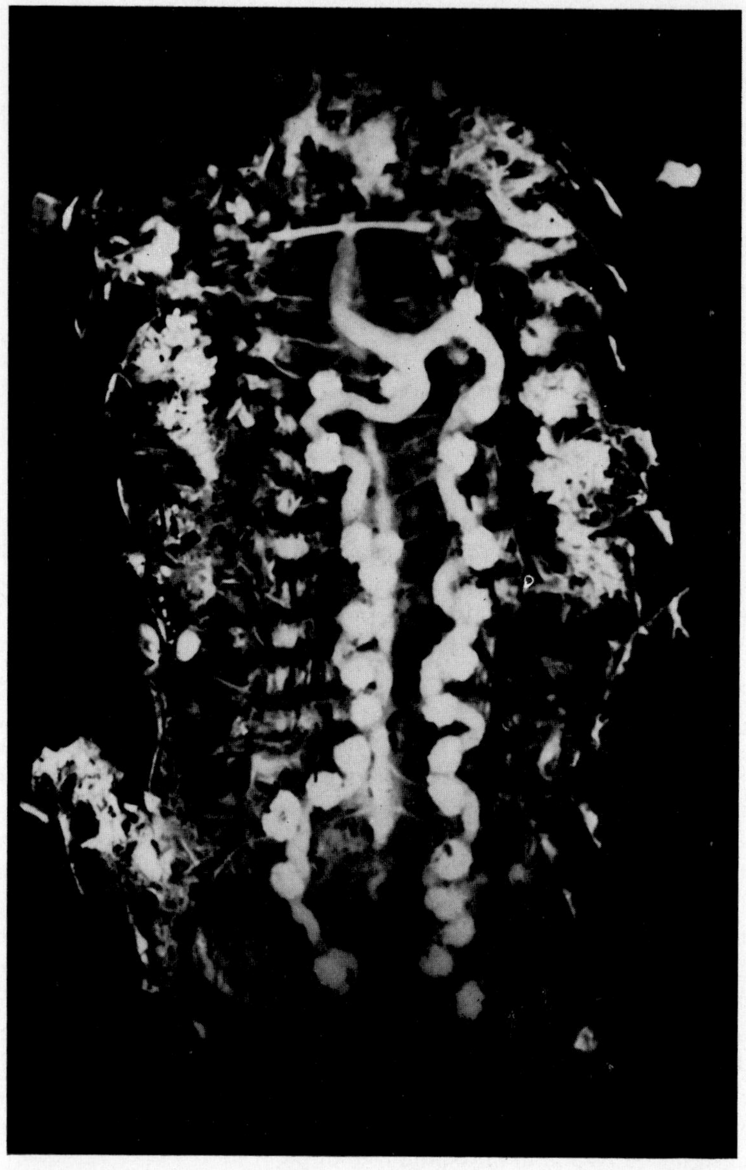

FIG. 6. *Arthrosphaera craspedota*: the male reproductive system *in situ*.

TABLE I

Comparison of the types of reproductive patterns in the species investigated

Species	Reticulate type	Ladder type	Arthro-sphaeroid type
Family: Harpagophoridae			
1. *Spirostreptus asthenes*	+		
2. *Harpurostreptus robustior*	+	+	+
3. *Ktenostreptus costulatus*		+	
Family: Pachybolidae			
4. *Aulacobolus variolosus*		+	
5. *Aulacobolus levissimus*		+	
Family: Sphaerotheriidae			
6. *Arthrosphaera craspedota*			+

ladder-type is confined to *Ktenostreptus costulatus* (Fig. 3), *Aulacobolus variolosus* (Fig. 4) and *Aulacobolus levissimus* (Fig. 5). On the other hand the arthrosphaeroid-type which is completely free from transverse connexions has been proved at least to be a generic character in the family Sphaerotheriidae. However, it is significant that the male reproductive pattern of *Harpurostreptus robustior* represents a combination of all the three basic types described (Fig. 2). Thus the family Harpagophoridae probably represents an heterogeneous group.

These studies clearly reveal that the typical male genital system in the majority of Diplopoda consists of tubes forming a simple ladder with regularly or irregularly spaced rungs and an approximately equal number of roughly paired testes lobes. There is an interesting correlation among the different types of reproductive patterns of all the species so far studied. However, it is clearly seen that the ladder-type itself has undergone considerable changes during the course of evolution from a more complicated to a simpler type. At one end of the evolutionary series as in *Ktenostreptus costulatus* the ladder-system is highly developed with a large number of transverse connexions between the two vasa deferentia all along their length of which 22 are found in the gonad-bearing region alone. In the second series which is represented by *Aulacobolus variolosus* and *Aulacobolus levissimus* a gradual reduction in the number of transverse connexions has been observed. Thus in the former there are 13 transverse connexions covering the entire genital system of which only three are confined to the gonad-bearing region. On the other hand, in the latter there are only two transverse connexions at the anterior extremity, while the remaining region of the genital system including the gonad-bearing region is completely free from any transverse connexions.

These observations and those of Warren (1934) suggest that during the course of evolution of the male reproductive system in Diplopoda, the more primitive and complex reticulate-type has been replaced by the ladder-type which in turn has given rise to the simpler and more advanced arthrosphaeroid-type. Possibly, the type of simple ladderless male genital pattern with a small number of lateral lobes as exemplified by the species of *Arthrosphaera* is not more primitive, but has evolved from the reticulate-type.

References

Attems, C. G. (1894). Die Copulationsfusse der Polydesmiden. *Sber. Akad. Wiss. Wien* No. 103: 39–54.

Chowdaiah, B. N. (1966). Male reproductive pattern in Diplopoda (Myriapoda). *J. Mysore Univ.* **20**: 41–45.

Fabre, L. (1855). Recherches sur l'anatomie des organes reproducteurs et sur le developpement des Myriapodes. *Annls Sci. Nat.* (Zool.) (4) **3**: 256–320.

Krishnan, G. (1967). The millipede *Thryopygus. CSIR Zool. Mem.* No. 1.

Miley, H. H. (1927). Development of male gonopods and life history studies of a polydesmid millipede. *Ohio J. Sci.* **27**: 25–43.

Miley, H. H. (1930). Internal anatomy of *Euryurus erythropygus* (Brandt) (Diplopoda). *Ohio J. Sci.* **30**: 229–254.

Newport, G. (1841). On the organs of reproduction and the development of the Myriapoda. *Phil. Trans. R. Soc.* **1841**: 99–130.

Rath, O. von (1890). Uber die Fortpflanzung der Diplopoden (Chilognathen). *Ber. naturf. Ges. Freiburg* **5**: 1–28.

Seifert, B. (1932). Anatomie und biologie des Diplopoden *Strongylosoma pallipes* Oliv. *Z. Morph. Okol. Tiere* **25**: 362–507.

Warren, E. (1934). On the male genital system and spermatozoa of certain millipedes. *Ann. Natal Mus.* **7**: 351–462.

West, W. R. (1953). An anatomical study of the male reproductive system of a Virginia millipede. *J. Morph.* **93**: 123–176.

ABBREVIATIONS USED IN FIGURES

V.DEF	vas deferens
M.V.DEF	median vas deferens
T.C.	transverse connexion
L.L.	lateral lobe (testis lobe)

Symp. zool. Soc. Lond. (1974) No. 32, 273–287.

RÉFLEXIONS SUR LE DÉVELOPPEMENT DE QUELQUES DIPLOPODES

J.-M. DEMANGE

Muséum national d'Histoire naturelle, Laboratoire de Zoologie-Arthropodes, Paris, France

SYNOPSIS

Some aspects of the development of diplopods are reviewed. In particular, the addition of new segments and of new ocelli and the development of the gonopods are considered. A fundamental distinction is made between the succession of moults (stages of development of varying length) and the various aspects of morphology revealed by these moults. These various aspects bear direct relation to the degree of evolution of the group considered. From this point of view, the acquisition of new segments and ocelli is particularly interesting in that one is able to attribute the absence of ocelli in the adults (a neotenic character) of very contracted groups such as the Polydesmoidea, to an accelerated development, as evidenced by the small increment of diplosegments at each moult and the reduced number of rings in the adult.

INTRODUCTION

Chez les diplopodes, le développement postembryonnaire est anamorphe, c'est à dire que les diplosegments sont acquis en partie dans l'oeuf, pendant la période embryonnaire, et en partie après l' éclosion, pendant la période postembryonnaire ou vie libre des larves. Pendant cette dernière phase, et à l'occasion de mues, les anneaux, d'abord apodes, apparaissent par petits groupes rassemblant un nombre plus ou moins grand, fixe ou variable, d'éléments métamériques.

C'est chez les iules *sensu lato* que l'on rencontre le plus grand nombre de ces apodes, de plus, ce nombre demeure élevé au cours de la croissance et remarquablement variable au sein de chaque stade*. Chez les spirobolides, les chiffres restent élevés mais les variations de chaque stade sont relativement limitées. Chez les craspédosomides, chaque stade n'apporte qu'un nombre très limité d'apodes (1, 3, 4, 3, 2, mais avec un demi-diplosegment de décalage)† et chez les polydesmides le

* Actuellement, chaque stade est limité par les mues.

† Ce demi-diplosegment n'est indiqué que par les ébauches de ses appendices; il est encore confondu, pour ses autres éléments, avec le dernier anneau du corps apparemment simple, mais dont les composants sont multiples: segmentaires et telsonien. On se souviendra également des phénomènes rencontrés à ce niveau chez les glomérides et des différences constatées entre les nombres des anneaux et des paires d'appendices (ne serait-ce que les appendices copulateurs) correspondant à un "pygidium" en bouclier. C'est dans ce "pygidium" que s'élaborent les nouveaux apodes, dans une zone de croissance ventrale.

nombre est encore plus réduit (2. 3. 2. 1). Chez les pénicillates, le nombre des diplosegments apodes acquis à chaque stade est le plus faible: 1 à chaque mue.

La première conséquence de cette variabilité dans l'acquisition des nouveaux métamères est la réduction du nombre définitif de segments chez l'adulte pour un nombre de stades postembryonnaires qui est, en réalité, sensiblement le même chez ces diplopodes : 8 (Polydesmoidea, Penicillata), 9 (Craspedosomoidea), à 8, 10 ou 12* (spirobolides, iulides). Il est évident que le nombre d'apodes présenté par la larve à la fin du stade pupoïde, nombre variable suivant les groupes, et parfois fort élevé (17–18 chez *Pachybolus*, 25 chez le Stemmiuloidea, *Diopsiulus progressus*), doit être un élément essentiel de calcul et de comparaison qui a jusqu'à présent été négligé.

Notre intention est d'étudier le développement postembryonnaire de quelques diplopodes, d'une manière générale, afin d'essayer de comprendre ou de déceler quelques uns des mécanismes conduisant à la réduction segmentaire présentée par ces myriapodes. Les mécanismes sont fort complexes; en 1968 nous avons eu l'occasion de concevoir une réduction segmentaire d'ordre évolutif et une réduction d'ordre écologique (oligomérie écologique) qui est soumise aux conditions d'existence (Demange, 1968).

Dans le second cas, la condensation segmentaire peut s'opérer en fonction de conditions écologiques bonnes ou mauvaises, réduisant plus ou moins un développement postembryonnaire déterminé; des individus à plus petit nombre de segments font donc ainsi leur apparition au cours d'une période donnée (mois, saison, année, etc. . . .). Les formules segmentaires réalisées dans ces conditions *entrent dans le cadre des variations segmentaires de l'espèce* telle qu'elle nous apparaît actuellement.

Il nous faut donc faire au moins quatre constatations fondamentales à partir de ce qui est connu.

La première constatation, importante, relève du fait que tout diplopode à nombre d'anneaux peu élevé présente, chez l'adulte, des formules segmentaires fixes (Craspedosomoidea) ou relativement fixes (par exemple Polydesmoidea).† Ce nombre est de l'ordre d'une vingtaine chez les Polydesmoidea, une trentaine chez les Craspedosomoidea, une douzaine chez les Pentazonia; par contre, tout diplopode à nombre

* Il s'agit des stades produisant des anneaux.

† Nous avons déjà eu l'occasion d'attirer l'attention, à plusieurs reprises et notamment en 1970 (Demange, 1970: 30), sur le fait que le nombre d'anneaux n'est pas fixe en réalité mais variable dans des limites extrémement étroites, de l'ordre de quelques anneaux seulement (un ou deux généralement).

élevé de diplosegments présente, chez l'adulte, une grande variabilité, par exemple chez les iulides.

La seconde constatation est que le nombre des diplosegments acquis à chaque stade est peu élevé chez les espèces à formules fixes, ou quasi fixes, alors qu'il est beaucoup plus important chez les formes à grand nombre d'anneaux.

Un troisième fait, corollaire de ce qui précède, est que, à toute acquisition segmentaire faible, correspond une fixité du nombre des apodes alors qu'un grand nombre d'apodes développé à chaque stade s'accompagne d'une variabilité du nombre de ceux-ci pour chacun de ces stades.

Enfin, à formule fixe de diplosegments chez l'adulte, se rapporte un nombre de stades moins élevé, de l'ordre de huit, généralement fixe lui aussi.

Autrement dit: tout diplopode à formule segmentaire fixe, ou quasi fixe, acquiert à chaque stade, un petit nombre d'apodes (chiffre non variable également) pour un nombre quasi fixe et peu élevé des stades postembryonnaires (phase épimorphe exclue); le résultat étant un nombre total peu élevé et non variable d'anneaux chez l'adulte.

A ces diverses possibilités d'acquisition métamériques s'oppose un réveil sexuel apparent quasi fixe pour tout le groupe des diplopodes puisqu'il semble intervenir sensiblement à un même stade III.

Le rappel de ces faits met l'accent sur certaines valeurs prises comme références, en particulier sur la limite établie entre un développement embryonnaire et un développement postembryonnaire ainsi que sur le découpage de ce dernier en un certain nombre de mues.

DÉVELOPPEMENT EMBRYONNAIRE ET DÉVELOPPEMENT POSTEMBRYONNAIRE

Considérée du point de vue strictement métamérique, la distinction établie entre développement embryonnaire et développement postembryonnaire, chez les myriapodes anamorphes, paraît superflue, car si l'on considère le développement embryonnaire des épimorphes (géophilomorphes par exemple) chez lesquels *tous* les segments sont *acquis dans l'oeuf*, on ne peut dire que le développement de *Lithobius* par acquisition segmentaire postembryonnaire est *fondamentalement différent* du développement embryonnaire du précédent quant au résultat morphologique.

Il semble que si l'on veut comprendre et surtout analyser et comparer les développements des diplopodes (il n'est pas seulement question ici de développement embryonnaire et de développement postembryonnaire) on doive faire abstraction de cadres trop rigides et en tout cas définir le sens de ceux que l'on emploie. Il semble que l'on doive considérer les *phases d'intermue*, non comme des processus immuables et fixes pour l'ensemble des diplopodes, donnant naissance à des caractères fondamentaux donnés et définitifs. Ce sont les séquences d'un développement donné et fixé, dont l'aboutissement est l'obtention d'une série de caractères précis (parmi lesquels se range le nombre total terminal des segments), fragmentant le dit développement en une série de portions, de tranches, de clichés, l'ensemble de ces portions représente le développement total final différent pour chaque groupe de diplopodes.

En d'autres termes, la série des mues que nous connaissons, phénomène actuel, se déroulant sous nos yeux, ne peut correspondre en aucun cas à la série ancestrale. Une des périodes d'intermue quelconque ne peut, encore bien moins, représenter l'une en particulier du développement ancestral. La forme actuelle que nous connaissons porte les marques d'une longue évolution, dont les empreintes sont fixées dans le patrimoine héréditaire déterminant ses caractéristiques morphologiques et biologiques; cette forme subit un certain nombre de mues pour arriver à réaliser ces caractéristiques. Ce sont ces dernières que révèlent les tranches découpées dans le développement postembryonnaire dont la longueur et le déroulement sont fixés, eux aussi, par les acquisitions successives des ancêtres phylétiques, développement plus ou moins contracté en nombre de mues et en temps.

Il y a donc lieu de ne pas confondre mues et périodes d'intermue, phénomènes actuels et révélations d'images morphologiques provoquées par ces mêmes mues car les premières sont des séquences plus ou moins labiles, sensibles à certaines conditions, les seconds quelques clichés d'une scène tout entière déjà profondément modifiée et condensée par l'évolution et enregistrée dans le patrimoine. On se trouve ainsi à des niveaux très différents subissant chacun des influences diverses et des poussées évolutives, notamment celles tendant à contracter le développement postembryonnaire.

Le nombre des mues ou des séquences de développement *est plus ou moins grand suivant les formes* pour un *aboutissement précis fixé dans le patrimoine héréditaire et les tranches découpées par ces mues sont plus ou moins épaisses*. Les répétitions de ces mues sont alors plus ou moins rapprochées, ce qui morcelle le développement (d'une façon d'ailleurs irrégulière) en des fractions plus ou moins importantes (tranches plus ou moins épaisses).

Les caractères acquis, à chacune de ces mues, sont donc très différents non seulement d'une mue à une autre en fonction des répétitions plus ou moins rapides, mais aussi d'une forme à une autre en fonction de leur degré évolutif de base.

Le nombre des mues étant sensiblement le même chez les différents diplopodes, ce qui change c'est la durée de ce développement postembryonnaire. Donc, peu importe à ce point de vue, le nombre des mues; l'important est ce qu'elles révèlent. Or, si l'on découpe le développement d'un iulide, dont la durée d'acquisition segmentaire est de plusieurs années, en dix tranches par exemple et que l'on compare les résultats en procédant à la même opération pour un polydesmide, dont les huit stades s'étalent pendant six mois ou un an, on retrouve nécessairement les différences fondamentales existant entre les deux groupes que tout diplopodologiste connaît. Il ne faut donc pas confondre les niveaux c'est à dire le niveau morphologique représenté par chaque stade et le niveau morphologique général représenté par la forme adulte; l'un est fragmentaire, l'autre un ensemble, un aboutissement; seuls deux niveaux de même nature sont comparables entre eux.

L'acquisition segmentaire devient ici un témoin très précieux de même que celle des ocelles et des caractères morphologiques des gonopodes les uns et les autres représentant des caractères éminemment variables dans le groupe des diplopodes. Il est en tout cas un fait à ne pas négliger: les myriapodes aveugles possèdent un petit nombre de diplosegments, ce nombre étant fixe si l'on néglige les variations de trés faible amplitude, alors que les oculés présentent un nombre relativement variable et élevé de ces diplosegments. Nous reviendrons ultérieurement sur le problème très particulier de ces formes aveugles.

Jusqu'à présent, c'est en se plaçant uniquement au point de vue de la morphologie des stades définis par les mues que l'on a cherché à comprendre le déterminisme des variations des formules segmentaires, la diversité des formules d'apodes présentée par les larves d'un stade donné (sanctionné par des mues), le phénomène de tassement des formules métamériques, etc.; en somme à étudier *les variations constatées en fonction des stades*, c'est à dire à faire une étude *du déroulement du développement postembryonnaire*. Mais si l'on supprime les mues par la pensée et que l'on envisage l'ensemble du développement postembryonnaire comme non fragmenté (comme le développement embryonnaire) les problèmes se présentent sous un aspect tout différent. Le développement est *continu**, marqué peut être seulement par les

* Il suffit pour s'en convaincre d'étudier la zone de croissance pour constater la continuité et la progressivité de la formation des anneaux (c.f. Krug, 1905; Pflugfelder, 1932; Demange, 1972).

variations segmentaires que l'on peut constater d'un embryon à un autre (développement plus ou moins lent)*. On n'a jamais établi, pour ces embryons de tableaux compliqués, de formules segmentaires diverses ni de calculs statistiques; le fait retenu n'était pas le déroulement du développement mais son aboutissement seul c'est à dire la réalisation d'un embryon à x segments, sa croissance terminée.

Examinons un peu plus en détail les conséquences d'intermues ainsi conçues comme fragments plus ou moins importants d'un développement donné. Au point de vue segmentaire, une longue section (tranche épaisse de développement) donnera naissance à une larve ayant acquis *un grand nombre d'anneaux* à l'inverse d'une courte section, la croissance segmentaire étant envisagée comme continue†. Une accélération du développement par diminution de la longueur de cette section produira une larve hâtive avec un plus petit nombre de segments acquis. Ce qui revient à dire que théoriquement telle larve hâtive devra muer $x+n$ fois pour atteindre le même nombre de diplosegments que telle autre larve retardée (par rapport à la première) n'ayant mué que x fois. Cela est évidemment un point de vue, un aspect de la question, valable pour des espèces comme *Pachybolus ligulatus* dont les tranches sont d'épaisseurs égales (croissance régulière suivant un rythme connu) dont le nombre des apodes acquis à chaque stade est le même pour tous les individus de ce stade; mais non pour d'autres espèces ou groupes d'espèces dont l'épaisseur des tranches, donc le nombre des anneaux acquis, peut varier dans des proportions parfois considérables.

On atteint ici le coeur du problème en énonçant que *le nombre des apodes acquis à chaque mue par un diplopode de type iule représente les caractéristiques d'une tranche de développement* et que plus le chiffre est petit, plus le développement est accéléré; d'une manière plus générale, qui dit accélération, donc raccourcissement, dit privation de plus en plus grande des possibilités de variations au niveau de chaque stade. Un temps plus long donne plus de liberté à la zone de croissance pour exprimer ses diversités segmentaires.‡

* Et dont l'aboutissement est la formation d'un individu semblable à ceux de la même population, au sein de laquelle l'on constate néanmoins des variations individuelles autour d'une formule préférentielle dont les origines sont d'ailleurs multiples; c.f. Saudray (1961); Sahli (1969); Demange (1972: lignées larvaires).

† Ce que l'on constate chez l'embryon (augmentation continue du nombre des anneaux par acquisition plus ou moins régulière des nouveaux segments).

‡ Ce qui ne veut pas dire, néammoins, que lorsqu'un groupe de larves d'une formule donnée donne naissance à plusieurs groupes de larves de formules différentes, le résultat de l'opération ne soit pas le témoin d'une image évolutive dont l'originalité essentielle est que la formule segmentaire est fondamentalement variable et le processus

[*Continued at foot of next page*

On comprend alors que pour une mue donnée, 2e, 3e, 4e, etc. . . . , *le degré de développement livré à l'observation peut n'être pas* le même pour toutes les espèces et bien plus encore pour tous les *groupes de diplopodes. Deux mues d'un même ordre ne donnent pas nécessairement des résultats identiques chez tous* au point de vue morphologique, même si le développement procède par un même nombre de mues; seul l'ordre est identique et la mue considérée peut révéler un degré morphologique de l'animal complètement différent de celui d'un autre animal; parallèlement, le caractère découvert n'est pas nécessairement et strictement à une même phase morphologique.

Le développement des gonopodes est, sous ce rapport, riche en enseignements et illustre peut-être le mieux ce qui vient d'être dit.

On sait que les gonopodes, appendices transformés, se modifient progressivement à partir des pattes ambulatoires jusqu'à l'organe le plus spécialisé. Il semble évident que le processus évoqué plus haut doive se traduire chez les mâles par des gonopodes à des dégrés de développement différents suivant les groupes. C'est ainsi que chez *Pachybolus ligulatus*, à développement très progressif, on constate des modifications morphologiques très progressives jusqu'à l'organe le plus complet tandis que chez d'autres groupes à développement accéléré, les images révélées par les mues seront très différentes les unes des autres, avec des sauts morphologiques brusques d'une mue à l'autre et plus particulièrement à la fin du développement, où l'on constate un bouleversement brusque assez spectaculaire d'un stade au suivant.

Cela illustre d'une façon éclatante la disparité existant entre les mues, phénomène actuel dont le nombre seul est plus ou moins fixé dans le patrimoine héréditaire, et les conséquences de l'évolution d'un groupe donné; mais il n'en reste pas moins que dans un cas comme dans l'autre, à développement court correspondent des images morphologiques brusques et discontinues tandis qu'à développement plus lent répondent des images plus graduelles et plus continues.

Il reste encore à évoquer et à comprendre le problème plus particulier de l'existence, chez une même espèce, de larves à morphologies segmentaire* et gonopodiale multiples.

A un même stade (mue prise comme référence) on trouve chez de très nombreux diplopodes du type iule des larves présentant plusieurs

Footnote continued from previous page]

ainsi fixé dans le génome; c'est-à-dire que la formule segmentaire est imparfaitement fixée et sujette à des variations mais ces variations ne peuvent s'exprimer que lorsque le développement est peu contracté. Elles sont dues à des avances ou à des retards de croissance segmentaire propres aux larves des lignées larvaires. Il n'en reste pas moins qu'à la fin du développement avances et retards peuvent s'annuler ou s'additionner.

* c.f. deuxième note infrapaginale, p. 273.

formules segmentaires différant par le nombre d'apodes. D'une part la possibilité offerte par chaque stade peut être restreinte (une formule ou deux) ou plus large (jusqu'à trois ou quatre possibilités). D'autre part, les organes copulateurs du mâle ne se présentent pas tous de la même manière; leur morphologies varient dans un cadre assez restreint il est vrai, mais d'une façon sensible.

Cela est le résultat du développement plus ou moins rapide des larves d'une même population; ce rythme retentit évidemment sur les images morphologiques révélées par les mues. On a conscience que dans ces conditions les morphologies diverses révélées ainsi, ou les formules découvertes, représentent *des étapes* intégrées dans un cadre beaucoup plus général du développement proprement dit de la lignée larvaire. C'est pourquoi nous avons déjà mis l'accent (Demange, 1968) sur l'existence *d'une série continue* (*ou presque*) *de nombres* (anneaux apodes) du premier stade au dernier; *ce sont des séries métamériques verticales* qui s'intègrent à une seule série segmentaire fondamentale propre à une lignée larvaire donnée. En outre, il existe également, pour chaque stade, *une série horizontale* de nombres d'apodes; elle représente les variations possibles chez les larves d'un stade donné (cf à ce propos les séries principales et secondaires de Sahli, 1969).

On constate alors pour chaque espèce ou pour chaque groupe d'espèces, l'existence d'une stabilisation plus ou moins grande du développement suivant les diversités segmentaires offertes et, par suite, des combinaisons terminales présentées par les divers adultes. Mais ce qui compte, en fait, c'est le nombre définitif des métamères réalisés au stade adulte et non les variations segmentaires au cours du développement, variations qui ne sont que des fluctuations, des *images passagères,* sans rapport avec *le résultat final à images successives intégrées.*

Ce qui précède explique les diverses combinaisons morphologiques présentées par les diplopodes et plus particulièrement les caractères morphologiques de la segmentation, des gonopodes, des ocelles, de même que la périodomorphose que nous considérons, plus que jamais, comme un phénomène biologique sans rapport direct avec celui procédant à l'acquisition des segments au cours du développement.

Il est un point important qui reste à préciser; c'est le fait que la zone de croissance exerce son activité pendant toute la croissance jusqu'à l'adulte ou tout au moins pendant la plus grande partie de la croissance;* mais à ce stade adulte, ou un peu avant comme chez *Pachybolus ligulatus*†, cette activité cesse. On constate néammoins que des segments

* Hémianamorphose des glomérides.

† Certains cas chez les iulides également.

en cours d'élaboration existent en un état de différenciation correspondant aux embryosomites et aux éosomites (parfois mésosomites). Ils sont donc en puissance mais ils ne verront jamais le jour, bloqués par l'arrêt de la croissance segmentaire qui est fixée dans le patrimoine héréditaire et sous l'influence des potentiels évolutifs du groupe ou de l'espèce.

On voit donc, par cet exemple, que la réduction segmentaire constatée chez les différents groupes de diplopodes *n'est pas le fait d'un arrêt précoce de la croissance dû à une forte accélération du développement actuel* car, dans ce cas, on devrait trouver chez l'adulte des diplopodes à nombre peu élevé de diplosegments, quelques anneaux apodes arrêtés dans leur croissance; ce qui n'est pas le cas.*

Le problème envisagé sous cet angle plus restreint ne veut pas dire que, dans le sens général, la réduction segmentaire évolutive des diplopodes n'est pas le résultat d'une condensation du développement, bien au contraire, mais la condensation constatée chez les diplopodes actuels *est déjà le résultat de l'évolution du groupe;* les différents groupes de ces diplopodes actuels nous font entrevoir, à leur tour, par les résultats qu'ils représentent, quelques clichés de la condensation segmentaire évolutive en général, des diplopodes dans leur ensemble.

C'est de ce point de vue que l'on doit étudier le problème posé par quelques diplopodes aveugles.

CONDENSATION DE LA CROISSANCE ET DÉVELOPPEMENT DES OCELLES

On ne peut éviter d'évoquer, dans le cadre de cette conception, la croissance ocellaire et le devenir des ocelles chez les formes dont le développement est plus particulièrement altéré par les phénomènes d'accélération de croissance, envisagés aussi bien du point de vue évolutif que du point de vue de la croissance proprement dite.

Tout d'abord, constatons que les diplopodes à petit nombre de segments sont généralement aveugles et que le développement d'un diplopode oculé (*Pachybolus ligulatus*, iulides, etc. . . .) comprend deux phases:

1. Une phase aveugle intrachorionale, l'embryon lui-même étant aveugle, et extrachorionale (ou larvaire) du fait que la larve 1 est souvent aveugle, ou tout au plus pourvue d'une simple tache oculaire.

* Ces faits parlent, une fois de plus, en faveur de la conception exposée plus haut, d'après laquelle mues actuelles et révélations morphologiques dues à ces mues sont indépendantes et non superposables; la seconde marquant par des clichés morphologiques le degré évolutif du groupe.

2. Une phase comportant des ocelles (un en général, 1 RO): c'est au cours de cette phase qu'intervient l'individualisation des lignées larvaires chez les diplopodes à habitus d'iule, donc à grand nombre d'anneaux. Cette particularité semble d'ailleurs être la caractéristique de cette phase.

Ce n'est qu'à partir du stade à un ocelle que les éléments visuels sont acquis rangée par rangée selon des processus parfois compliqués, s'étalant en triangle (Demange, 1972) et d'une façon plus ou moins régulière suivant les groupes.

Du point de vue pratique on peut concevoir que les polydesmides (aveugles), à petit nombre d'anneaux, ont subi une accélération de développement telle que la première phase seulement de son développement a pu voir le jour: celle correspondant à l'embryon où le premier ocelle n'est pas encore apparu. Les polydesmides seraient donc hautement néoténiques *puisque l'absence d'ocelle est un caractère embryonnaire ou de très jeune larve.* Mais, en outre, du point de vue segmentaire, l'accélération de développement, et par conséquent la contraction de ce dernier, n'apporte finalement qu'un très petit nombre d'anneaux par comparaison avec un iulide à développement remarquablement moins accéléré. On conçoit dans ce cas que la totalité des mues de ce groupe intéresse la seule première phase du développement, phase sans ocelle, et ne développe que les anneaux ou une partie seulement des anneaux de cette phase dont le nombre peut être finalement assez élevé si l'on en juge par les pupoïdes et les larves de *Pachybolus* et des Stemmiuloidea* (c'est-à-dire 17/18 et 25).

Le développement le plus progressif actuellement connu, et pour prendre un système de référence comme base d'un raisonnement, est celui de *Pachybolus ligulatus*.

Les différentes phases du développement d'un diplopode pourraient donc se rapporter à un type ***Pachybolus ligulatus*** *dont, en outre, l'acquisition ocellaire est régulière, elle aussi, rangée par rangée.*

Ce qui revient à dire que les polydesmides aveugles ne développeraient que l'équivalent maximum des 17 ou des 18 apodes du début du développement (première larve) de *P. ligulatus* pendant leur croissance segmentaire totale, On a une confirmation de ce point de vue par le nombre total de 19 à 22 anneaux rencontrés chez ce groupe (polydesmides).

L'accélération du développement conserve naturellement certains caractères larvaires devenus classiques depuis les travaux de Brölemann (1922, 1932) comme, par exemple, les soies en clavette communes aux

* Le pupoïde est aveugle chez ce groupe, au moins chez *Diopsiulus*.

larves et aux formes à plus petit nombre d'anneaux (19), les trichômes, poils transformés des pénicillates (11 à 13 anneaux) rappelant ceux des polydesmides larvaires, la volvation impossible chez les polydesmiens comme les larves des diplopodes* mais ajoutons désormais, évidemment, pour les polydesmides, le caractère néoténique capital de la non apparition des ocelles.

Ce qui vient d'être dit est confirmé chez d'autres groupes comme les Craspedosomoidea et les Stemmiuloïdea† à développement également plus ou moins accéléré, où l'on rencontre des larves à un seul ocelle pendant plusieurs stades (les trois premiers chez les craspedosomides, les deux premiers chez les Stemmiuloïdea). L'accélération, moins grande dans ce cas, tend à escamoter la première phase aveugle pour développer la tranche correspondant au début de la deuxième phase à un ocelle et naturellement révéler une quantité segmentaire en rapport avec cette accélération.

Il y a lieu, toutefois, de bien distinguer les formes ou les groupes aveugles entre eux: il ne faut pas confondre les groupes généralement aveugles à l'origine avec ceux qui ne le sont point. En d'autres termes, il faut distinguer ceux dont la disparition des ocelles est définitivement acquise, des groupes (Craspedosomoïdea, Chordeumoïdea par exemple), chez lesquels certaines espèces montrent, d'une façon très irrégulière, quelques ocelles à côté d'autres espèces complètement aveugles, le caractère paraissant incomplètement fixé.

La question délicate de l'intervention de la condensation du développement dans la disparition des ocelles n'aurait pas été abordée ici si l'éminent myriapodologiste qu'était Brölemann (1922, 1932) n'avait déjà proposé une explication. Les "êtres de plus en plus contractés n'avaient, à chaque étape, que les organes correspondant aux stades où ils se trouvaient fixés, des organes imparfaitement développés . . . , ils n'ont pu atteindre à un degré de croissance permettant un fonctionnement normal et, *comme tout organe inutilisable, ils ont été finalement éliminés*".

On voit que si nous admettons, avec Brölemann (1922, 1932), l'importance de la condensation du développement dans la disparition des ocelles, cette dernière est envisagée d'une façon totalement différente et à l'aide de documents très différents dont l'auteur ne pouvait évidemment avoir connaissance, puisque quelques uns n'ont été découverts que récemment (Demange, 1972). Il semble que les nouvelles notions aient un caractère beaucoup moins théoriques désormais.

* Rappelons que cette faculté d'enroulement est de moins en moins grande chez les larves de plus en plus jeunes.

† Première larve de *Diopsiulus progressus* avec 25 apodes.

Le déroulement de ces processus tend donc à établir des décalages dans le développement des différents groupes. C'est ce que l'on peut constater, d'une facon précise, pour les groupes les mieux connus, c'est-à-dire les pachybolides, les iulides, les polydesmides et les craspedosomides et pour les caractères rigoureusement mesurables comme le nombre des anneaux.*

Par exemple, et en comparant strictement les mues entre elles et dans leur ordre croissant, une mue x de *Pachybolus* est égale, du point de vue segmentaire, à une mue $x+n$ de polydesmide; par contre une mue $x+n$ de ces mêmes pachybolides ne représente plus rien chez les polydesmoïdes dont la croissance segmentaire en nombre est terminée depuis plus ou moins longtemps.

On constate alors que les pachybolides contiennent *la totalité des stades postembryonnaires des polydesmoïdes* (*aveugles*) dans leur toute première phase aveugle *quant à leur potentiel de développement segmentaire.*† Il en est de même en ce qui concerne les Stemmiuloïdea (25 apodes) dont la première phase segmentaire *contient l'ensemble* du développement segmentaire des craspedosomoïdes. On peut constater ici, dans les deux cas, une accélération du développement importante mais inégale chez les deux groupes; celle des Polydesmoïdea est la plus accentuée, puisqu'elle reste contenue dans la phase aveugle alors que celle des Craspedosomoïdea l'est moins puisque apparaissent des rangées d'ocelles, au moins la première; cette dernière demeure néanmoins présente pendant les trois premiers stades signe d'accélération plus intense que les stades suivants à croissance ocellaire régulière.

CONCLUSIONS

En conclusion, on doit envisager les problèmes posés par le développement segmentaire postembryonnaire des diplopodes en distinguant fondamentalement deux niveaux dans les phénomènes.

Un niveau d'ordre très général, évolutif, représentant pour chaque groupe l'aboutissement actuel d'une série de phénomènes qui ont modelé

* Les autres caractères morphologiques tirés des gonopodes, par exemple, bien qu'indicateurs précieux, sont plus subjectifs dans le sens où on ne peut les chiffrer, mais ils sont plus directement vérifiables dès l'instant que l'on démontre une relative correspondance entre les uns et les autres.

† Cette notion sera développée ultérieurement, étant bien entendu que chez *Pachybolus* la zone de croissance segmentaire *a épuisé* ses capacités avant l'avénement du stade adulte, puisque la croissance segmentaire au stade IX (Demange, 1972) c'est-à-dire deux stades avant l'adulte, alors que chez les formes plus contractées comme les craspédosomides, par exemple ou les polydesmides, l'accélération de développement *a arrêté le processus* puisque les adultes portent des anneaux apodes qui n'iront jamais plus loin la zone de croissance n'a pu épuiser complètement sa capacité de segmentation.

le groupe tel que nous le connaissons. C'est à ce niveau que se détermine le type de développement, plus ou moins contracté, tendant à une réduction fondamentale du nombre des anneaux et à la fixité plus ou moins grande des formules segmentaires.

C'est également à ce niveau, et pour ne prendre qu'un second exemple morphologique d'un autre ordre, qu'est déterminé le degré de transformation des appendices copulateurs de chaque groupe.

Le second niveau, d'un ordre plus restreint, rassemble des *phénomènes actuels* dont le plus important est la mue qui découpe le développement fixé au niveau supérieur en une série de tranches, dont le nombre, en fait, est sensiblement égal chez tous ou, en tout cas, peu différent d'un groupe à l'autre; l'épaisseur de ces tranches est variable suivant précisément la longueur de ce développement.

Les mues découvrent donc certains clichés d'un film déjà fortement bouleversé et modifié par l'évolution du groupe; il y a révélation, chaque fois, *d'un instant morphologique particulier qui est déjà un aboutissement évolutif*. Les résultats sont naturellement nombreux et de plusieurs ordres dont certains sont morphologiquement plus importants que d'autres. Du point de vue segmentaire, c'est le nombre plus ou moins grand d'apodes acquis à chaque stade: nombre élevé pour un développement plus ou moins long, nombre faible pour un développement rapide, cela en rapport avec les longueurs du développement qui sont considérablement différentes et avec les nombres de stades qui apparaissent comme sensiblement égaux.

En outre, la zone de croissance peut s'exprimer de façons diverses: soit en épuisant totalement son potentiel segmentaire chez les formes lentes, soit par un arrêt du processus dû à l'accélération du développement. En tout état de cause, une mue x de *Pachybolus* n'égale jamais en nombre de segments une mue x de polydesmoïde contracté mais une mue $x+n$; parfois même une mue $x+5$ n'a plus aucune correspondance chez les formes contractées, celle-ci ne pouvant jamais atteindre le chiffre segmentaire correspondant.

References

Brölemann, H. W. (1922). Principe de contraction contre principe d'élongation. *Bull. Soc. Hist. nat. Toulouse* **49**: 340–357.

Brölemann, H. W. (1932). La contraction tachygénétique des Polydesmiens (Myriapodes) et leurs affinités naturelles. *Bull. Soc. zool. Fr.* **47**: 387–396.

Demange, J.-M. (1968). La réduction métamérique chez les Chilopodes et les Diplopodes Chilognathes, (Myriapodes). *Bull. Mus. natn. Hist. nat. Paris* (2) **40**: 532–538.

Demange, J.-M. (1970). Myriapodes Diplopodes de Madère et des Açores. *Bolm Mus. munic. Funchal* No. 25 (107): 5–43.

Demange, J.-M. (1972). Contribution à la connaissance du développement post-embryonnaire de *Pachybolus ligulatus* (développement segmentaire, croissance ocellaire, croissance des organes copulateurs, notion de lignées larvaires, zone de croissance). *Biologia gabon.* **8**: 127–161.

Krug, H. (1905). Beiträge zur Anatomie der Gattung *Julus*. *Jena. Z. Naturw.* **42**: 485–522.

Pflugfelder, O. (1932). Über Mechanismus der Segmentbildung bei Embryonalentwicklung und Anamorphose von *Platyrrhacus amauros* Attems. *Z. wiss. Zool.* **140**: 650–723.

Sahli, F. (1969). Contribution à l'étude du développement post-embryonnaire des Diplopodes Iulides. *Annls Univ. sarav.* **7**: 1–154.

Saudray, Y. (1961). Recherches biologiques et physiologiques sur les Myriapodes Diplopodes. *Mém. Soc. linn. Normandie* (Zool.). N.S. **2**: 1–126.

Verhoeff, K. W. (1926–1932). Diplopoda. *Bronn's Kl. Ordn. Tierreichs* Band 5. abt. 2. Teil 1–2: 1–2084.

Discussion

Dohle: (Summary by Chairman), M. Demange asked us to differentiate between two aspects of development which are fundamentally different, the moults and the morphological characters that are revealed by them. The moults are initiated by some neurosecretory influences and are dependent on these but the revealed morphological characters indicate phylogenetic ancestry. The problem is to differentiate between the ontogenetic and the phylogenetic implications of the characters of each species. Diplopods which, as we say, are more contracted, as for instance the craspedosomoids or the polydesmoids, are neotenic; they become mature at a stage when others, Iuloidea for example, or Spirostreptoidea, are still juvenile. The same sort of thing is revealed by the development of the eyes. The first stage is usually without eyes; ocelli are added moult by moult. Many of those groups we regard as contracted are blind and this accords with the theory that they are neotenic.

Biernaux: In the Blaniulidae the number of new rings added at each moult is generally three or four. By contrast, in the Iulidae, the increment is usually five or six. Can we conclude that the Blaniulidae are more highly evolved than the Iulidae?

Demange: It is not only the number of segments which allows us to decide whether an animal is more or less highly evolved than another. One must consider all the characters, not only of morphology, but of physiology and general biology.

Sahli: It is perhaps not safe to make general statements about segmentation of the Iulidae or the Blaniulidae as this depends on the species under consideration. If one turns to Brölemann's work (1923)* on cavernicolous

* Brölemann, H. W. (1923). Blaniulidae (Myriapodes) (Première série) Biospéologica. *Archs Zool. exp. gén.* **61**: 99–453.

blaniulids it appears that some species add as many segments as certain iulids. There is as much variability within these families as between them.

M. Demange, you spoke of the development of ocelli in iulids from anterior to posterior.

DEMANGE: Yes—along the antero-posterior axis.

SAHLI: And you spoke of a period, more or less long, in the development of an individual. How can this relatively short period influence the appearance of new apodous segments?

DEMANGE: It is possible. Take the second example I gave, the development of gonopods. In the Pachybolidae, which I know best, there are interesting differences between individuals. In the pre-adult stadium, or the one before, there are occasionally very important differences, for example, in the seminal groove and the seminal complex at its base, seminal vesicle and ampulla. It seems possible that a period of several days can exert an influence but this has not yet been demonstrated.

SAHLI: Is there a correlation between the appearance of the number of rings and the sexual characters?

DEMANGE: One might say that the morphological differences of the gonopod rudiments between several individuals of the same stadium represent precise phases in the differentiation of gonopods. But one considers the gonopods as a whole. They will exhibit a clear character which is different to that of another larva.

SAHLI: In this fashion, if one considers the time taken for a ring to appear, in any stadium, for example the third (= 2 ocellar rows) in *Cylindroiulus silvarum*, it is evident that there are differences within a stadium but the time between moults is very short. For example, one can observe three different segmental formulae in this stadium (17/6, 18/7 and 19/8).

DEMANGE: Yes, but let us view the matter as a whole. If one considers post-embryonic development stadium by stadium, we see accelerations and retardations which more or less cancel each other out and finally produce individuals with the same segmental formula, but in their origin they are very different.

SAHLI: I agree, it is true that adults of *Cylindroiulus teutonicus* and *Tachypodoiulus niger* in our latitude have approximately the same number of segments but have achieved them in different ways.

DEMANGE: But, as you yourself have indicated, you can give results for 1972 but you cannot be certain of having the same results for 1975.

SAHLI: Yes, variation is possible, but within limits. It is not possible, for instance, to find one day a *Tachypodoiulus niger* with 80 segments, in our latitude and at 300 m high.

Symp. zool. Soc. Lond. (1974) No. 32, 289–300.

ACTION DE LA TEMPÉRATURE SUR LE DÉVELOPPEMENT EMBRYONNAIRE DE *GLOMERIS MARGINATA* (VILLERS)

L. JUBERTHIE-JUPEAU

Laboratoire Souterrain du C.N.R.S., Moulis, France

SYNOPSIS

Eggs were submitted to constant temperatures between 4 and 28°C; the duration of development at these temperatures varied from 194 to 17 days; this minimum duration was obtained at 25°C. Thermal shocks of the order of 30°C (which prove lethal if the eggs are kept at this temperature) cause malformations, the most important of which are duplication of embryos and anomalies in the segmentation of the germ band.

These anomalies in the segmentation of the germ band lead to one, two or rarely three pairs of supplementary limbs at eclosion, with or without the loss of a pair of limb-buds and the appearance of a supplementary tergite. Thus a supernumerary segment or diplo-segment has resulted either from an abnormal segmentation of the germ band or from a perturbation of the sequence of embryonic development leading to the precocious formation of appendages and segments which are normally formed, in these anamorphic animals, during post-embryonic development.

INTRODUCTION

L'étude du développement embryonnaire des diplopodes a, de longue date, retenu l'attention des auteurs; le premier travail sur ce sujet est représenté en effet par les données de Newport (1841) sur le développement de l'iule. *Glomeris marginata* a fait l'objet de quelques publications dans ce domaine: Heymons (1897) décrit la segmentation de la bandelette germinative, Hennings (1904 a, b) donne des stades du développement embryonnaire puis le développement des organes de Tömösvary; Dohle (1964), enfin, fait une description assez complète de l'embryologie de ces intéressants diplopodes. L'auteur décrit avec précision la segmentation de l'oeuf, subégale et totale, et la formation du blastoderme qui prend naissance à partir des noyaux de segmentation parvenus à la périphérie de l'oeuf et du plasma qui les entoure; cette étude se poursuit par l'étude de la formation des feuillets, des manifestations externes du développement au cours duquel l'auteur reconnaît six stades et de l'organogenèse. La larve hexapode qui sort de l'oeuf est depuis longtemps familière aux myriapodologistes.

En ce qui concerne la tolérance des adultes de *Glomeris marginata* à différentes températures constantes nous possédons les résultats de Haacker (1968). Cet auteur montre expérimentalement que les adultes peuvent supporter sans dommage n'importe quelle température comprise entre $-3°C$ et 39°C.

Le but du présent travail est une étude de l'action de la température non sur les adultes mais sur les embryons afin de déterminer:

1. l'action d'une température constante sur le développement embryonnaire,

2. l'effet de chocs thermiques sur l'embryon au cours de son développement.

Ce travail est la suite d'expériences entreprises depuis plusieurs années (Juberthie-Jupeau, 1968, 1970, 1971) et tend à regrouper un ensemble de résultats.

Au cours de ces recherches un grand nombre de jeunes larves présentant des anomalies morphologiques ont été obtenues et il apparaît actuellement que des malformations de type prévisible peuvent être obtenues en agissant de façon élective à un stade précis du développement embryonnaire. *Glomeris marginata* constitue de ce fait un matériel favorable pour l'étude de la tératologie expérimentale. Il faut noter que de façon assez paradoxale les cas signalés d'animaux tératologiques sont rares chez les diplopodes. Balazuc & Schubart (1962) qui ont effectué l'analyse des cas connus les évaluent à un peu plus d'une centaine; en ce qui touche spécialement *Glomeris marginata* qui expérimentalement peut donner un pourcentage élevé d'animaux tératologiques, il n'a été signalé comme anomalies spontanément apparues que les monstres doubles décrits par Hubert (1968). Notons que Dohle (1964) signale de nombreuses anomalies au moment de la formation de la bandelette germinative. L'auteur les rapporte au fait que la tache germinative est "très indéterminée topographiquement et physiologiquement", la bandelette germinative apparaît en avant de la tache germinative, mais elle peut aussi, dans certains cas, apparaître anormalement aussi à l'arrière, formant ainsi deux bandelettes.

ACTION DE TEMPÉRATURES CONSTANTES

Des lots d'oeufs pondus à 17°C, ont été placés moins de 24 h après la ponte aux températures suivantes: 2°, 3°, 9°, 10°, 11°, 15°, 16°, 17°, 20°, 23°, 25°, 27°, 28°, 29°, 30°. Chaque lot compte 40 oeufs sauf à 3° où il n'y a que 28 oeufs et à 9° où il n'y en a que 22.

Les oothèques ont été ouvertes après 15 jours environ de développement afin de surveiller l'éclosion de la larve hexapode.

Durée du développement (*Fig. 1*)

A 2° le développement ne semble pas se produire de même qu'à 29°.

Ces températures extrêmes constituent des limites léthales inférieure et supérieure en deça et au-delà desquelles l'oeuf ne peut se développer.

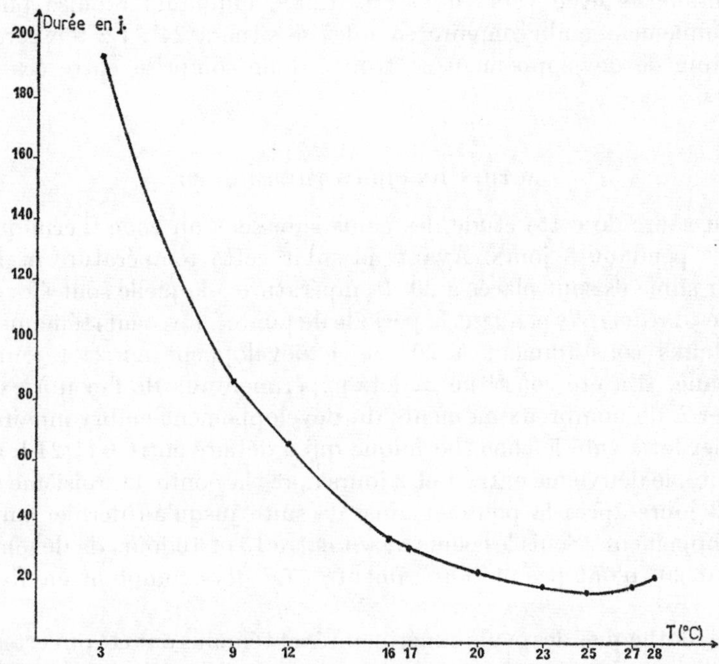

Fig. 1. Action d'une température constante sur la durée du développement embryonnaire de *Glomeris marginata*.

De 3° à 25°, on constate une accélération du développement; à 3° le développement requiert en moyenne 194 jours tandis qu'à 25° il ne dure que 17.7 jours en moyenne; c'est le temps le plus court qui ait été observé.

Au-delà de 25° il s'instaure un léger retard dans le développement, manifestation due à une température trop élevée; ainsi à 27°, le développement dure en moyenne 19 jours et à 28° il atteint 22 jours.

Pourcentage d'éclosions

Dans la gamme des températures compatibles avec le développement des oeufs, le pourcentage d'éclosion croît en même temps que la température, il atteint son maximum 100% à 16° et 17°, température au-dessus de laquelle il décroît très lentement jusqu'à 25° où il atteint 96%.

Au-dessus de cette température, il décroït rapidement pour atteindre 25% à 28°.

Ensuite de cette étude il apparaît que le pourcentage d'éclosion maximum se situe vers 16–17°; la durée minimum requise pour le développement embryonnaire complet se situe à 24°. La température optimum de développement se trouve donc comprise entre ces deux limites.

ACTION DE CHOCS THERMIQUES

Au cours de cette étude, les oeufs subissent un choc thermique de 30–31° pendant 3 jours. Avant de subir cette température et après l'avoir subie ils sont placés à 20° (température à laquelle sont élevés les couples de *Glomeris* pendant la période de ponte). Les oeufs témoins sont maintenus constamment à 20° et se développent en 24.4 jours en moyenne. Il a été constitué 22 lots expérimentaux de façon à expérimenter à de nombreux moments du développement embryonnaire. Le premier lot a subi le choc thermique qui a débuté entre 0 et 24 h après la ponte, le deuxième entre 1 et 2 jours après la ponte, le troisième entre 2 et 3 jours après la ponte et ainsi de suite jusqu'au dernier jour du développement. Seuls les oeufs ayant entre 15 et 16 jours de développement à 20° n'ont pas été expérimentés. Les lots comptent environ 40 oeufs.

Les oothèques des oeufs expérimentés et témoins ont été ouvertes une quinzaine de jours après la ponte pour repérer les éclosions.

Les résultats obtenus après les chocs thermiques sont de trois ordres:

Pourcentage d'éclosion

Le choc thermique affecte particulièrement le développement s'il est appliqué entre 0 et 24 h après la ponte; aucun oeuf ne se développe dans ces conditions. Le choc thermique a un effet de moins en moins sévère au fur et à mesure que les embryons qui y sont soumis sont plus âgés. Le pourcentage d'éclosion est en effet de 17.5% après 1–2 jours de développement à 20° et atteint un pourcentage normal quand l'action du choc thermique se situe après 6–7 jours de développement à 20°.

Durée du développement

Les oeufs soumis au choc thermique ont des durées de développement qui sont fonction du moment auquel ils ont été soumis à cette action. Le développement est ralenti lorsque le choc thermique est subi par des oeufs qui ont au maximum 10 jours de développement à 20°. Le test de Student montre un allongement de développement hautement significatif; par la suite et jusqu'à 18 jours de développement à 20°, la durée du développement est voisine de celle des animaux témoins. Lorsque les oeufs subissent le choc thermique entre 19 et 22 jours de développement à 20° leur développement est accéléré par rapport à celui des témoins, le test de Student montrant entre les deux valeurs une différence hautement significative. Au-delà de 22 jours de développement à 20°, le choc thermique n'agit plus sur la durée du développement. On constate donc un ralentissement du développement d'autant plus net que les oeufs soumis sont peu avancés dans leur développement et une accélération du développement chez les oeufs qui ont atteint environ les deux tiers de leur développement. Ce phénomène traduit une élévation de l'optimum thermique au cours du développement.

Effets morphogénétiques

L'action la plus spectaculaire des chocs thermiques est leur effet morphogénétique sur l'embryon. De nombreuses anomalies sont obtenues dont la plupart se rattachent à deux grands groupes de malformations. Ces anomalies sont prévisibles car leur apparition dépend du moment auquel l'oeuf a été soumis à l'action du choc thermique.

Lorsque les oeufs sont soumis à 30–31° après 1 à 5 jours de développement à 20° il apparaît des monstres doubles selon la fréquence suivante:

âge des embryons au début du choc	nombre d'oeufs	nombre d'éclos	nombre de monstres doubles
1–2 jours	40	7	2
2–3	41	31	3
3–4	37	27	8
4–5	40	20	1

Dans le lot soumis à 30–31° après 5–6 jours de développement à 20° aucun animal tératologique n'est apparu; 13 animaux seulement sont éclos dans ce lot.

Lorsque les oeufs sont soumis à 30–31° après 6 à 10 jours de développement à 20°, les animaux naissent avec 1, 2 ou 3 paires de pattes surnuméraires et parfois 4 paires de bourgeons de pattes au lieu de 5 paires; il apparaît parfois un tergite surnuméraire, souvent une soudure incomplète des hémi-sclérites sur la ligne médiodorsale et parfois la fusion d'hémi-sclérites décalés (hélicomérie).

La fréquence de ces malformations dans les différents lots est la suivante:

âge de l'embryon au début du choc	nombre d'oeufs	nombre d'éclos	nombre d'anormaux avec pattes surnuméraires
6–7 jours	40	35	3
7–8	40	39	25
8–9	34	24	16
9–10	42	37	19
10–11	39	37	5

Lorsque les oeufs sont soumis à 30–31° après 11 et 15 jours de développement il apparaît quelques individus anormaux en ce sens qu'ils sont très gonflés et ne présentent pas la courbure ventrale des embryons normaux; certains même ont la face ventrale qui épouse la courbure de l'oeuf. Certains présentent une évagination du tube digestif repliée sur la face dorsale ou la face ventrale. La fréquence de ces malformations est la suivante:

âge de l'embryon au début du choc	nombre d'oeufs	nombre d'éclos	nombre d'anormaux
10–11 jours	39	37	1
11–12	45	27	7
12–13	40	32	6
13–14	40	31	6
14–15	40	32	5

Notons que dans les oeufs témoins maintenus de façon constante à 20°, il n'est apparu aucun embryon ou larve tératologique.

ETUDES DES ANOMALIES DE LA SEGMENTATION

Je rapporte ici uniquement les anomalies obtenues en ce qui concerne la segmentation des larves tératologiques, l'étude morphologique des monstres doubles ayant déjà fait l'objet de publications (Juberthie-Jupeau, 1968, 1970). Certaines malformations très rares (schistomelies ou hétérosymélies) et les malformations de la tête ne sont pas rapportées ici.

Étude anatomique externe des larves normales

Avant de décrire les anomalies obtenues il est nécessaire de rappeler les caractères d'une larve normale (Fig. 2a).

Chez *Glomeris marginata* il est bien connu que la larve qui éclôt est une larve hexapode, peu active, qui normalement demeure dans son oothèque jusqu'au stade suivant.

La tête porte 2 antennes constituées de 4 articles, 3 paires d'ocelles et 1 paire d'organes de Tömösvary.

Face ventrale, le tronc présente 3 paires de pattes semblables dont le télopodite est de 5 articles, ce dernier étant pourvu d'une griffe. En arrière de ces pattes bien développées s'observent 5 bourgeons de paires de pattes; ceux-ci sont des moignons lisses, coudés et inarticulés dont l'extrémité ne présente jamais de griffe.

Dorsalement le tronc est recouvert de 8 sclérites de taille variable.

Le premier situé en arrière de la tête est court et peu large; le deuxième très long, le plus long de tous déborde largement sur les faces latérales; les 3e, 4e, 5e et 6e, sont de taille moyenne; le 7e est très étroit et de fait il est passé jusqu'à ce jour inaperçu, le 8e et dernier est long mais plus étroit que les précédents.

Cette mise au point, en ce qui concerne le nombre des tergites, est indispensable pour juger des effets morphogénétiques de la température.

Anomalies morphologiques externes de la segmentation

Les observations qui suivent font état des 67 larves obtenues dans les expériences décrites précédemment.

Les anomalies ainsi rapportées à la segmentation portent sur le nombre de paires de pattes développées au moment de l'éclosion, le nombre des bourgeons de pattes présents, et le nombre des tergites (Fig. 2b).

Différents types d'anomalies obtenues

En fait les modifications obtenues réalisent de très nombreuses combinaisons entre le nombre de pattes et des bourgeons présents. Ainsi 11 types d'anomalies ont été dénombrés, ce sont les suivants:

1. 3 paires de pattes normales, 1 paire de petites pattes puis 4 paires de bourgeons: 6 cas,
2. 3 pattes d'un côté, 4 de l'autre, 5 bourgeons de pattes d'un côté et 4 de l'autre: 1 cas,

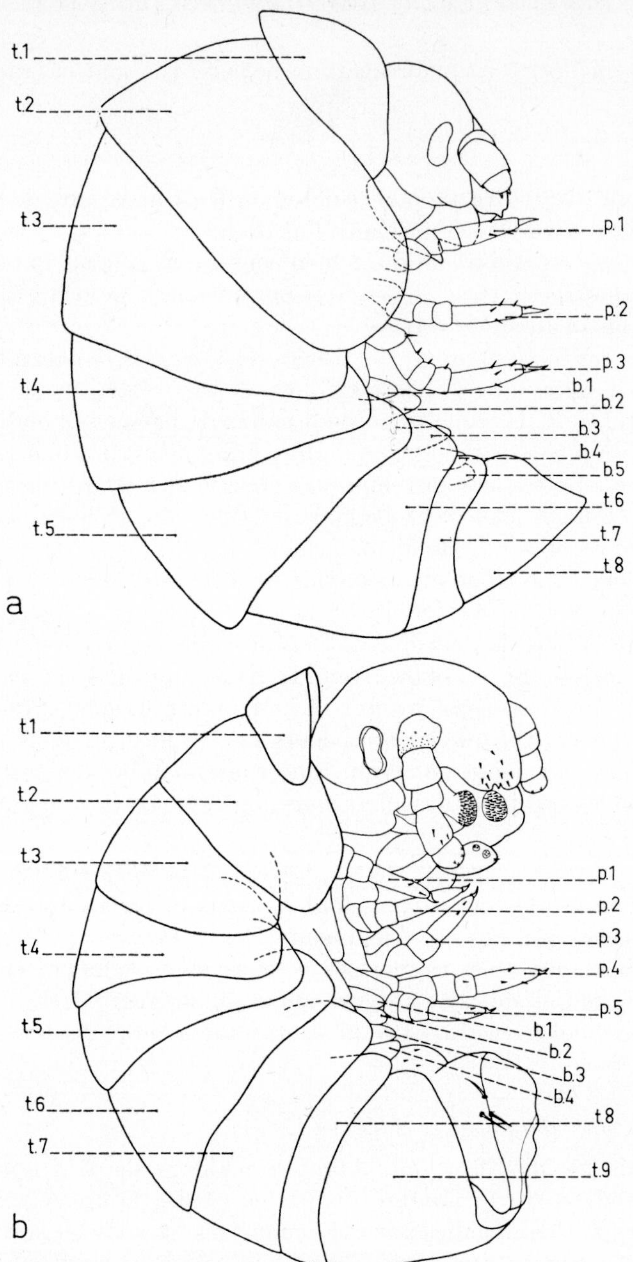

FIG. 2. *a*. Larve néonate de *G. marginata* normale (3 paires de pattes, 5 paires de bourgeons de pattes et 8 tergites); *b*. Larve néonate de *G. marginata* tératologique (5 paires de pattes, 4 paires de bourgeons de pattes et 9 tergites). *b* = bourgeon de patte, *p* = patte, *t* = tergite.

3. 4 paires de pattes, 4 paires de bourgeons de pattes et 8 tergites: 40 cas,
4. 4 paires de pattes normales, 1 paire de petites pattes, 3 paires de bourgeons, 8 tergites: 2 cas,
5. 4 paires de pattes, 4 bourgeons d'un côté, 5 de l'autre: 2 cas,
6. 4 paires de pattes, 5 paires de bourgeons, 8 tergites: 7 cas,
7. 4 paires de pattes, 5 paires de bourgeons, 9 tergites: 1 cas,
8. 5 paires de pattes, 4 paires de bourgeons, 8 tergites: 2 cas,
9. 5 paires de pattes, 4 paires de bourgeons, 9 tergites: 3 cas (Fig. 2b),
10. 5 paires de pattes, 5 paires de bourgeons, 9 tergites: 2 cas,
11. 6 paires de pattes, 4 paires de bourgeons, 9 tergites: 1 cas.

Dans cette étude l'examen de quelques larves présentant des ébauches plus ou moins apparentes entre le gnathochilarium et la première paire de pattes n'est pas envisagé.

Comparaison avec les animaux témoins

Il est à remarquer en premier lieu que les degrés d'anomalies affectant les animaux tératologiques obtenus sont plus ou moins prononcés.

En ce qui concerne les pattes surnuméraires il en apparaît soit une paire très peu développée, soit une, deux ou trois paires de taille normale et bien constituées. La symétrie entre le côté droit et le côté gauche n'est pas toujours rigoureusement respectée.

Quant au nombre de paires de bourgeons de pattes il est soit de 4 paires et dans ce cas il est réduit d'une paire, soit de 5 paires comme chez les animaux normaux. Dans les cas les plus anormaux la somme du nombre de paires de pattes et de bourgeons de paires de pattes n'est donc pas 8 comme chez les animaux normaux et tous les témoins, mais elle atteint 9 et même 10, dépassant ainsi d'une, voire de deux unités, les cas normal.

Le nombre des tergites est affecté mais dans une proportion assez faible d'animaux. Il apparaît un tergite supplémentaire chez certains animaux présentant une augmentation de la somme de leur nombre de paires de pattes et de leur nombre de paires de bourgeons. Ainsi un tergite en surnombre apparaît chez quelques animaux présentant 4 paires de pattes et 5 bourgeons de paires de pattes, 5 paires de pattes et 4 paires de bourgeons, 5 paires de pattes et 5 paires de bourgeons, 6 paires de pattes et 4 paires de bourgeons. Ainsi, un neuvième tergite s'est individualisé uniquement chez les animaux dont la somme, nombre de paires de pattes et nombre de bourgeons, dépasse le nombre normal.

Chez bon nombre d'animaux, porteurs d'anomalies de la face ventrale avec ou non un tergite supplémentaire, on observe des malformations des tergites qui se traduisent principalement sur la face médio-dorsale par des anomalies dans la soudure des hémi-tergites. Des degrés différents de soudure des hémi-tergites sont observables, qui vont de la simple encoche médiane postérieure d'un ou plusieurs tergites, à la présence d'hémi-tergites complètement séparés. Les tergites 3, 4 et 5 sont le plus fréquemment affectés. Dans quelques cas les hémi-tergites se soudent non avec leur homologue mais avec l'hémi-tergite suivant réalisant ainsi une hélicomérie dorsale.

Anomalies morphologiques internes

L'examen histologique de larves présentant à l'éclosion 4 paires de pattes, 5 paires de bourgeons de pattes et 8 tergites montre que dans ce cas il y a apparition d'un ganglion nerveux surnuméraire. En effet aux 4 paires de pattes normalement développées correspondent 4 ganglions nerveux identiques innervant chacun une paire de pattes, aux 5 paires de bourgeons correspondent 5 ganglions en formation.

CONCLUSIONS

Le développement des oeufs de *Glomeris marginata* peut s'effectuer entre 3 et 28°C, la température optimale de ce développement se situant entre 16 et 24°C.

Des chocs thermiques de l'ordre de 30–31°C appliqués pendant 3 jours à différents moments du développement peuvent engendrer des perturbations se traduisant par:

1. Une diminution du pourcentage d'éclosion chez les oeufs en début de développement.
2. Un allongement de la durée du développement pour les oeufs qui sont soumis au choc thermique avant 10 jours de développement à 20°C; un raccourcissement du développement pour les oeufs subissant l'action du choc thermique entre 19 et 22 jours.
3. Une action tératogène.

Les oeufs soumis au choc thermique au début de leur développement c'est-à-dire entre 1 et 5 jours après la ponte donnent naissance à des monstres doubles dans la proportion de 8% par rapport au nombre d'oeufs expérimentés, et 16% par rapport au nombre de larves écloses.

Il semble que dans ce cas l'action du choc thermique a lieu avant l'apparition de la tache germinative.

Des anomalies de la segmentation apparaissent lorsque les oeufs sont soumis au choc thermique entre 6 et 11 jours après la ponte. Il semble qu'à ce moment d'après les travaux de Dohle (1964) la bandelette germinative soit en train de s'édifier. Les perturbations apportées, touchent 34% des oeufs mis en expérience et 38% des larves écloses.

Des anomalies segmentaires telles que celles qui sont rapportées ici n'ont jamais été signalées. On doit remarquer d'ailleurs que tous les diplopodes naissent avec trois paires de pattes, *Polyzonium* excepté. Les résultats obtenus doivent être rapportés à une perturbation de la bandelette germinative au cours de son édification. L'action du choc thermique est parfaitement reproductible, et a été confirmée par des expériences de contrôle. On constate donc chez ces animaux dont le développement ultérieur est de type anamorphique qu'il y a une certaine possibilité d'action sur la bandelette germinative au début de sa formation.

Contrairement à l'hypothèse déjà envisagée d'une anticipation du développement ultérieur qui pourrait être due à un retard de l'éclosion, il y a lieu de remarquer que ce retard se produit également lorsqu'il apparaît une anomalie toute différente, celle qui produit des monstres doubles. Par ailleurs, on ne connaît pas ce type d'anticipation de développement dans le règne animal et certains des éléments surnuméraires qui apparaissent ne sont pas parfois exactement identiques aux normaux. En effet, toutes les larves normales ont 5 paires de bourgeons de pattes pratiquement identiques et ils donnent naissance à l'exuviation suivante à 5 paires de pattes de taille semblable et constituées comme les précédentes; dans le cas d'animaux tératologiques les pattes surnuméraires peuvent avoir une forme et une taille normale mais elles peuvent aussi être plus petites et compter moins d'articles, tous les stades intermédiaires existant entre une patte surnuméraire normale et un bourgeon de patte.

Les anomalies signalées sur la face dorsale proviennent d'anomalies de migration des plaques latérales vers la face dorsale. La présence d'hémi-tergites séparés et les hélicoméries constatées sont donc le résultat de processus embryonnaires.

En conclusion, chez *Glomeris marginata* l'action tératogène de chocs thermiques est certaine; il reste à déterminer le niveau d'action de ce facteur.

References

Balazuc, J. & Schubart, O. (1962). La tératologie des Myriapodes. *Année biol.* (4) **1**: 145–174.

Dohle, W. (1964). Die embryonal Entwicklung von *Glomeris marginata* (Villers) im Vergleich zur Entwicklung anderer Diplopoden. *Zool. Jb.* (Anat.) **81**: 241–310.

Haacker, U. (1968). Deskriptive, experimentelle und vergleichende Untersuchungen zur Autökologie rhein-mainischer Diplopoden. *Oecologia* **1**: 87–129.

Hennings, C. (1904a). Zur Biologie der Myriopoden II. *Biol. Zbl.* **24**: 252–256.

Hennings, C. (1904b). Das Tömösvarysche Organ der Myriopoden I. *Z. wiss. Zool.* **76**: 26–52.

Heymons, R. (1897). Mittheilungen über die Segmentirung und den Körperbau der Myriopoden. *Sber. preuss. Akad. Wiss.* **1897**: 915–923.

Hubert, M. (1968). Sur des cas de monstruosités doubles observés chez des larves de *Glomeris marginata* (Villers) (Myriapode, Diplopode, Oniscomorphe, Glomeridia). *Bull. Soc. zool. Fr.* **93**: 443–450.

Juberthie-Jupeau, L. (1968). Production expérimentale de monstres doubles chez *Glomeris marginata* (Villers), Myriapode, Diplopode. *C. r. hebd. Séanc. Acad. Sci., Paris* **266**: 1610–1612.

Juberthie-Jupeau, L. (1970). Action tératogène de la température sur l'embryon de *Glomeris marginata* (Villers) (Myriapode). *Bull. Mus. natn. Hist. nat. Paris* (2) **41**: 79–84.

Juberthie-Jupeau, L. (1971). Modification expérimentale du nombre des segments au cours du développement embryonnaire chez *Glomeris marginata* (Villers) (Myriapode, Diplopode). *C. r. hebd. Séanc. Acad. Sci., Paris* **273**: 1991–1994.

Newport, G. (1841). On the organs of reproduction and the development of the Myriapoda. *Phil. Trans. R. Soc.* **1841**: 99–130.

Discussion

Brade-Birks: If the optimum temperature for the development of *Glomeris marginata* is around 25°C is this reflected by what we know of its geographic distribution?

Juberthie-Jupeau: The duration of development at 25°C is minimal, but the maximum percentage hatch from the eggs is between 16° and 17°C. At 25°C fewer animals developed than at 17°C. It is difficult to say just what is the ideal temperature for the species.

Strasser: In the subsequent development of the abnormal larvae is there any compensation?

Juberthie-Jupeau: I can only give evidence up to the second larval stadium. All of the teratological animals which had more than eight limbs and rudiments, i.e. nine at least, had nine pairs of legs after the second moult; all the rudiments developed into limbs in the second stadium. There is no regulation.

Strasser: But this subsequent development was at normal temperatures without further thermal shock?

Juberthie-Jupeau: The thermal shock was applied only during embryonic development, and only for three days. If given for longer, the eggs do not develop. The eggs and young were reared at 20°C.

Symp. zool. Soc. Lond. (1974) No. 32, 301–315.

SUR LES MODALITÉS DE LA CROISSANCE ET LA RÉGÉNÉRATION DES ANTENNES DE LARVES DE *POLYDESMUS ANGUSTUS* LATZEL

GÉRARD PETIT

Laboratoire de Biologie animale, Université de Picardie, Amiens, France

SYNOPSIS

Morphological observations on the development of the antenna of *Polydesmus angustus* indicate that the change from six to seven articles results from the division of the second article at the third moult (to stadium IV). Biometric analysis of the post-embryonic growth of the antenna shows that only the second article is affected. Histological observation reveals an anatomical peculiarity at the base of the second article which could condition the segmentation.

The antenna, in course of segmentation, regenerates at all levels of amputation. The differentiation of the parts regenerated is a function of the time of amputation. There is evidence of a critical zone for regeneration. It corresponds to the articulation between the second and third articles. Amputation above and below this zone leads to regenerated antennae of the adult or juvenile condition respectively. If the regeneration proceeds from this critical zone the second article of the antenna either divides or remains undivided. The integrity of the second article is not necessary for segmentation to proceed, but sometimes disturbance to the apex of this article inhibits segmentation. The causes of the modification of the morphogenetic state of the second article are considered.

INTRODUCTION

La morphologie de l'antenne des diplopodes a été examinée au cours des différents stades du développement post-embryonnaire anamorphe (Verhoeff, 1928; Seifert, 1932; Brölemann, 1935). L'observation comparée de ces appendices a montré que l'antenne acquière de nouveaux articles au cours des premiers stades de la vie larvaire. Elle a également permis d'une part, d'établir les homologies entre les articles antennaires à divers stades larvaires et d'autre part d'envisager des segmentations d'articles dans les premières phases de la croissance.

A notre connaissance, les processus de la subdivision d'articles antennaires n'ont pas été étudiés chez les diplopodes. Nous essayerons d'en préciser les modalités dans le cas de l'accroissement de l'antenne lors du passage de la troisième à la quatrième forme larvaire chez un polydesmide: *Polydesmus angustus* Latz.

L'étude de la régénération d'appendices mettant en cause de tels processus de segmentation a été abordée chez les araignées (Vachon, 1967) et chez un diplopode (Nguyen Duy-Jacquemin, 1972). Ces auteurs ont surtout mis en évidence les relations existant entre l'aspect des régénérats et le moment d'amputation par rapport à la mue. Nous avons entrepris chez la larve du troisième stade de *P. angustus*, l'étude de la régénération de l'antenne en voie de segmentation; nous examinerons l'influence du niveau d'amputation sur la qualité des antennes régénérées, particulièrement son incidence arthrogénétique.

MATÉRIEL ET TECHNIQUES

Notre étude porte essentiellement sur des larves du stade III. Les animaux, récoltés dans la nature, sont mis en élevage séparément dans des terrariums maintenus à 10°C selon les méthodes préconisées par Saudray (1961) et Sahli (1966). La mortalité est très importante, environ 80 p. 100 des larves meurent avant d'avoir mué. La durée de l'intermue correspondant au troisième stade larvaire est très variable, elle peut dépasser quatre mois. L'exuviation est dans la plupart des cas suivie à travers la paroi des terrariums. La capture et la manipulation de larves en période préparatoire à la mue est extrêmement délicate.

Les mesures des articles antennaires sont effectuées sous loupe binoculaire à l'aide d'un micromètre. L'observation précise des antennes normales et régénérées est réalisée à partir de photographies prises après extirpation des appendices.

L'examen histologique des antennes des stades II, III et IV a été effectué à partir de coupes sériées semi-fines colorées au Bleu azur B.

Les larves du stade III sont maintenues sous loupe binoculaire et amputées d'une ou deux antennes à des niveaux variés sans anesthésie préalable. La précision du niveau d'amputation est vérifiée quelques temps avant la mue post-opératoire par appréciation de l'importance de la nécrose située sous le front de section.

RAPPEL CONCERNANT LE DÉVELOPPEMENT ET LA MORPHOLOGIE ANTENNAIRE

Le développement post-embryonnaire de *P. angustus* est du type anamorphe. Il comporte sept stades larvaires et un stade adulte. A l'approche de la mue, les animaux fabriquent une loge à l'aide de leurs déjections, puis ils s'immobilisent à l'intérieur de cette loge et effectuent leur exuviation.

La première forme larvaire possède des antennes formées de quatre articles. Au cours du développement, la segmentation de ces appendices se poursuit. Les larves des stades II et III ont des antennes de six articles. La morphologie à sept articles est acquise au stade IV et restera inchangée au cours de la croissance. Le dernier article antennaire porte distalement des quilles sensorielles insérées sur une formation cuticulaire réduite, assimilée parfois à un huitième article. Il est admis que les antennes des larves du stade III subissent au cours de la troisième mue une segmentation de leur second article.

RÉSULTATS

Modalités de la croissance de l'antenne des larves du stade III

Etude biométrique de la croissance relative des articles antennaires

A partir des valeurs moyennes obtenues et selon la méthode préconisée par Teissier (1948), nous avons établi les allométries de croissance des différents articles antennaires des animaux des stades II à VIII. Les résultats sont résumés dans le Tableau I et illustrés à la Fig. 1.

TABLEAU I

Caractéristiques biométriques de la croissance relative des articles antennaires de ***P. angustus*** *(stades II à VIII)*

Articles antennaires	Nombre de mesures n	Coefficient de corrélation r	pente p	Aspect de la croissance	
a_1	137	0·984	1·004	Isométrie	
a_4	73	0·882	1·14	Allométrie légèrement majorante	point critique à la 5e mue
	67	0·963	1·60	Allométrie fortement majorante	
a_5	72	0·931	1·27	Allométrie légèrement majorante	
	65	0·968	1·57	Allométrie fortement majorante	
a_6	135	0·983	1·07	Isométrie	
a_7	139	0·941	0·85	Allométrie minorante	
a_{23}	33	0·902	2·04	Allométrie très fortement majorante	Discontinuité
a_2	105	0·985	1·35	Allométrie majorante	
a_3	108	0·990	1·39	Allométrie majorante	

a_1, a_{23}, a_2, a_3, a_4, a_5, a_6, a_7: articles antennaires (voir Figs 2, 3).

Notre choix dans l'homologie des différents articles se trouve justifié par les valeurs très élevées des coefficients de corrélation.

Les lois de croissance sont continues pour a_1, a_6 et a_7 ainsi que pour les articles a_4 et a_5, avec cependant dans ces deux cas, un point d'inflexion à la cinquième mue. La croissance de l'article a_{23} est nettement accélérée avant la troisième mue puis présente une discontinuité. Les articles a_2 et a_3 ont des lois de croissance continues, à pentes parallèles et inférieures à celle de a_{23}.

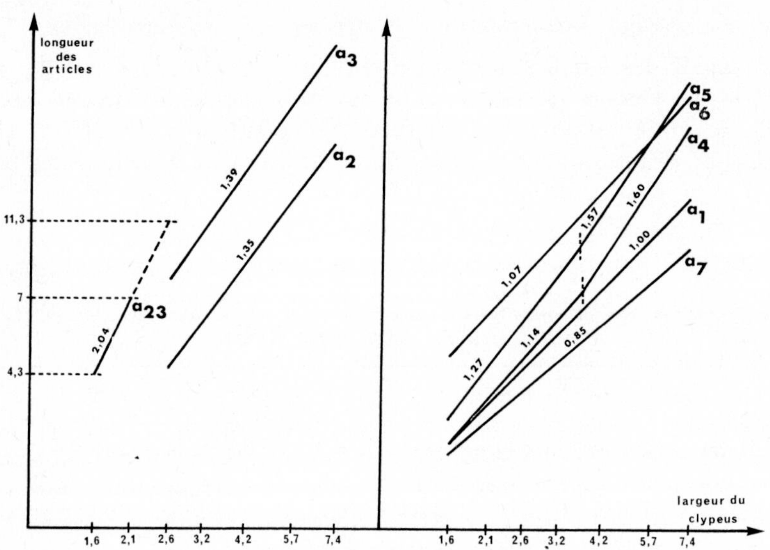

FIG. 1. Représentation graphique de la croissance relative des articles antennaires de *P. angustus* (stades II à VIII).

a_1, a_{23}, a_2, a_3, a_4, a_5, a_6, a_7 : articles antennaires (voir Figs 2, 3). La valeur de la pente (p) est indiquée sur chaque droite. Les coordonnées logarithmiques sont exprimées en unités arbitraires.

Morphologie et histologie comparées des deuxième et troisième articles antennaires des stades larvaires II, III et IV

Aucune différenciation cuticulaire ne permet d'envisager le niveau de segmentation de l'article a_{23} de l'antenne du stade III; néanmoins, la base de cet article est dépourvue de pilosité (Fig. 2). La taille du second article a_2 de l'antenne du stade IV est approximativement égale à la moitié de celle du troisième article a_3 (Fig. 3).

Au deuxième stade larvaire, les jonctions myoépidermiques des faisceaux musculaires du second article (a_{23}) se situent à la base de l'article, près de l'articulation de a_1 avec a_{23}. La longueur des faisceaux musculaires est d'environ 80 μm.

Au stade III (Fig. 4), la partie basale de ce second article est libre de toute insertion musculaire, elle n'est constituée que d'un manchon hypodermique traversé par le nerf antennaire; la longueur des faisceaux musculaires reste sensiblement égale à celle que l'on observe chez les larves du stade II, elle est approximativement de 90 μm.

Dans l'antenne de la larve du stade IV, une disposition basale des jonctions myoépidermiques s'observe à la fois dans le deuxième article (a_2) et dans le troisième (a_3).

Aspects morphologiques de la régénération de l'antenne des larves du stade III

Régénérats entièrement différenciés

Lorsque les amputations sont effectuées au moins un mois avant l'exuviation, nous obtenons la différenciation complète du blastème de régénération. Pour un niveau déterminé, des temps de régénération deux, trois ou quatre fois plus longs conduisent à un régénérat dont le degré de différenciation est identique. Les articles régénérés sont de taille réduite et ont une structure typiquement reconnaissable.

L'examen global des résultats (Tableau II) montre que de part et d'autre de l'apex du deuxième article, ou plus précisément, d'une zone correspondant à l'articulation du deuxième article avec le troisième article (Fig. 4), les amputations conduisent à des antennes régénérées qualitativement différentes. Distalement, on obtient des antennes de type adulte à sept articles (a_{23} s'est subdivisé en a_2 et a_3); proximalement, les antennes régénérées sont de type juvénile à six articles (a_{23} reste indivis) (Fig. 5). Lorsque les lésions sont faites dans la zone d'articulation, on obtient de manière égale l'élaboration de l'un ou l'autre des deux types d'antenne (Fig. 6). Nous qualifions cette partie apicale de l'article a_{23} de zone critique de régénération.

L'antenne régénérée de type juvénile possède souvent des articles globuleux. La quantité régénérée parait plus faible lorsque la régénération procède à partir des niveaux distaux. Les régénérats de type juvénile obtenus au stade IV après amputation faite dans des niveaux proximaux donnent, à la quatrième mue (18 cas) des antennes de sept articles par subdivision du deuxième article (Fig. 7). Sept de ces antennes ont été suivies jusqu'à la cinquième mue (stade VI); leurs articles tendent à reprendre un rapport de taille voisin de celui de l'antenne normale.

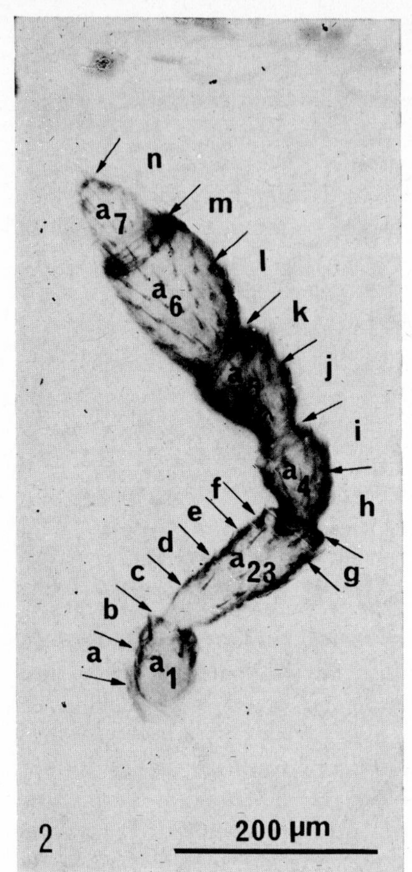

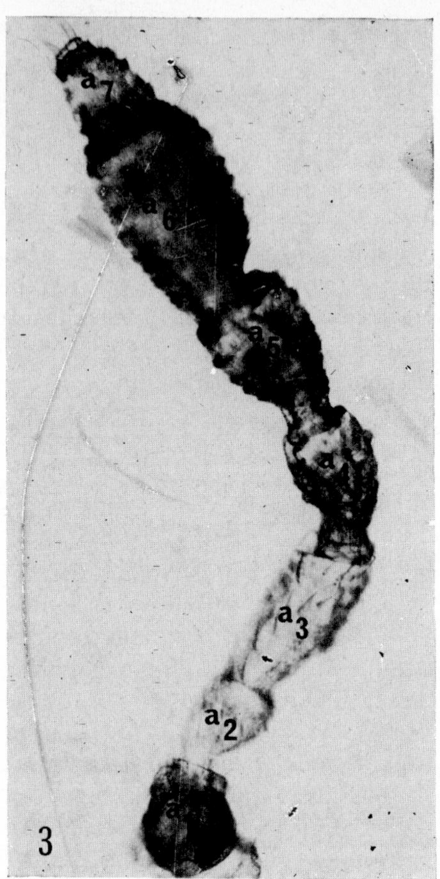

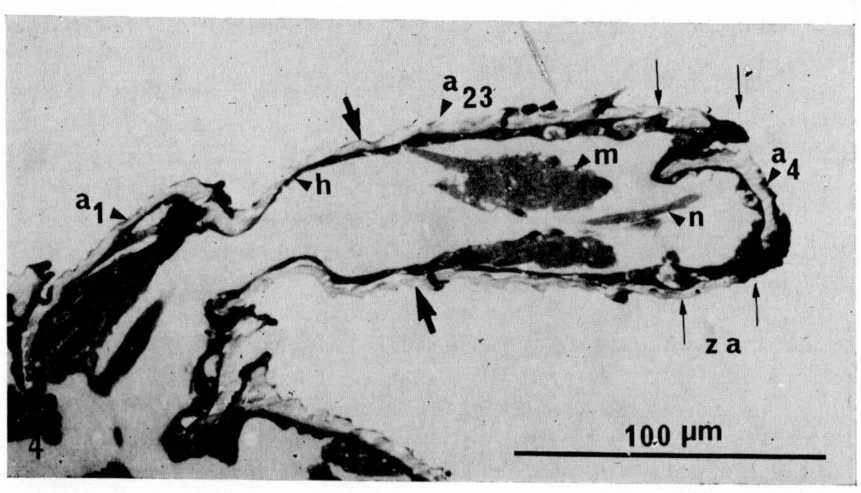

Caption on facing page

Régénérats à différenciation incomplète

Les amputations tardives effectuées 15 à 20 jours avant l'exuviation conduisent à des régénérats peu différenciés et d'autant plus réduits que l'amputation est proche de l'exuviation. On obtient une gamme de régénérats allant de la simple papille à des formations dont la segmentation est plus ou moins ébauchée. Les articles laissés en place sont bien différenciés.

Toutes les amputations distales par rapport à la zone critique permettent la segmentation du deuxième article en a_2 et a_3 (Figs 8, 9). Les sections effectuées en dessous de la zone critique conduisent à des antennes régénérées très rudimentaires dont le deuxième article, quand il est individualisé, est toujours indivis. Les amputations réalisées dans la zone apicale du second article sont suivies ou non de la segmentation de cet article.

Les régénérats incomplètement différenciés, obtenus après résection de parties distales de l'antenne montrent à la mue suivante des articles distaux parfaitement développés (six cas).

Lorsque les antennes régénérées obtenues à partir de niveaux proximaux sont limitées à une formation correspondant aux articles a_1, a_{23} et une papille distale, à la mue suivante (stade V) l'antenne est de type juvénile à six articles (cinq cas) (Fig. 10). Au stade ultérieur (stade VI) la segmentation du second article a_{23} s'effectue (deux cas).

Régénérats atypiques

Six cas d'antennes régénérées présentant des anomalies de segmentation, méritent d'être examinés. Elles résultent toutes d'amputations passant par la zone critique ou à son voisinage immédiat.

Chez trois larves du stade IV, les articles a_4 et a_5 d'une antenne régénérée ne sont représentées que par un seul article (Fig. 11). Il s'agit

---

Figures 2–4. Morphologie des antennes de *P. angustus* avant et après la segmentation du second article.

FIG. 2. Antenne de la larve du stade III. L'articulation du second article (a_{23}) avec le troisième (a_4) est bien visible. Les niveaux d'amputations inscrits au Tableau II sont représentés entre les flèches. a_1 . . . . . a_7: nomenclature des six articles antennaires.

FIG. 3. Antenne de la larve du stade IV à sept articles. Même échelle que précédemment.

FIG. 4. Coupe histologique d'une antenne de larve du stade III. a_1, a_{23}, a_4: articles antennaires; h, hypoderme; m, muscles; n, nerf; z a, zone de l'articulation de a_{23} avec a_4 (entre les petites flèches). La limite basale des insertions musculaires est indiquée par les grosses flèches.

TABLEAU II

Résultats de l'amputation des antennes chez la larve du stade III de ***P. angustus***

Antenne normale stade III	Régénération complète t = 30 à 130 jours			Régénération incomplète t = 15 à 20 jours			Régénération atypique t = 40 à 120 jours		
	Nombre de cas stade IV			Nombre de cas stade IV			Nombre de cas stade IV	Nombre de cas stade V	
Niveaux d'amputations	Total	Type juvénile 6 art.	Type adulte 7 art.	Total	a_{23} indivis	a_2, a_3 séparés	Fusion a_4–a_5	Segmentation supplémentaire dans a_4	Segmentation supplémentaire dans a_2
n	1		1						
m	2		2						
l	8		8	1		1			
k	4		4						
j	16		16	3		3			
i	23		23	1		1			
h	54		54	7		7	1		
g	32	15	17	4	1	3	2	1	2
f	18	18		1	1				
e	21	21		5	5				
d	36	36		6	6				
c	16	16		4	4				
b	4	4							
a	2	2							

Les niveaux d'amputations (a – n) sont représentés à la Fig. 2. t = temps séparant l'amputation de la mue post-opératoire; a_{23}, a_2, a_2, a_4, a_5: articles antennaires (voir Figs 2, 3).

de la fusion des articles a_4 et a_5 ou de la perte d'un de ces deux articles. Les autres articles antennaires sont présents et ont un aspect normal, en particulier a_2 et a_3.

D'autre part, trois antennes présentant des articulations surnuméraires ont également été obtenues. Elles proviennent d'un lot de six antennes régénérées après section dans la zone critique et qui ont pu être suivies jusqu'au stade V. Trois d'entre elles (deux de type juvénile et une de type adulte) ont évolué normalement en antennes de sept articles (type adulte); par contre, en ce qui concerne les trois autres, deux antennes régénérées de type adulte, ont conduit, au stade V, à des appendices présentant une segmentation nouvelle de l'article a_2 (Fig. 12) et une antenne régénérée de type juvénile a différencié en plus de l'articulation normale qui sépare a_2 et a_3 une articulation supplémentaire dans a_4 (Fig. 13).

DISCUSSION

Segmentation du deuxième article antennaire

L'aspect biométrique de la croissance, nos observations morphologiques et histologiques confirment chez *P. angustus* l'existence d'une segmentation du deuxième article antennaire lors de passage stade III–stade IV.

Ce mode d'accroissement procédant par segmentation d'articles est apparemment différent de celui qui a été étudié chez certains insectes (Bugnion, 1920; Haase, 1968); dans le cas de la blatte, deux mécanismes de segmentation assurent la croissance des antennes. Il se rapproche plutôt des processus décrits par Vachon (1945) au cours de l'arthrogenèse des pattes de limules.

Nous avons montré chez *P. angustus* que la segmentation du deuxième article antennaire au cours de la troisième mue s'effectue après une phase de croissance importante de cet article à la mue précédente. Cette segmentation n'a aucun retentissement sur la croissance des articles contigus. Il s'agit donc d'une morphogenèse qui n'intéresse que le deuxième article.

Il semble que le manchon hypodermique basal du deuxième article (a_2) de l'antenne constitue une prédisposition anatomique à la segmentation. Au cours de la troisième mue, le tiers basal de cet article se différencierait en article a_2 pendant qu'une nouvelle articulation s'établirait. Ces processus morphogénétiques permettant la segmentation de a_{23} en a_2 et a_3 seraient donc essentiellement localisés au tiers basal de l'article en voie de segmentation.

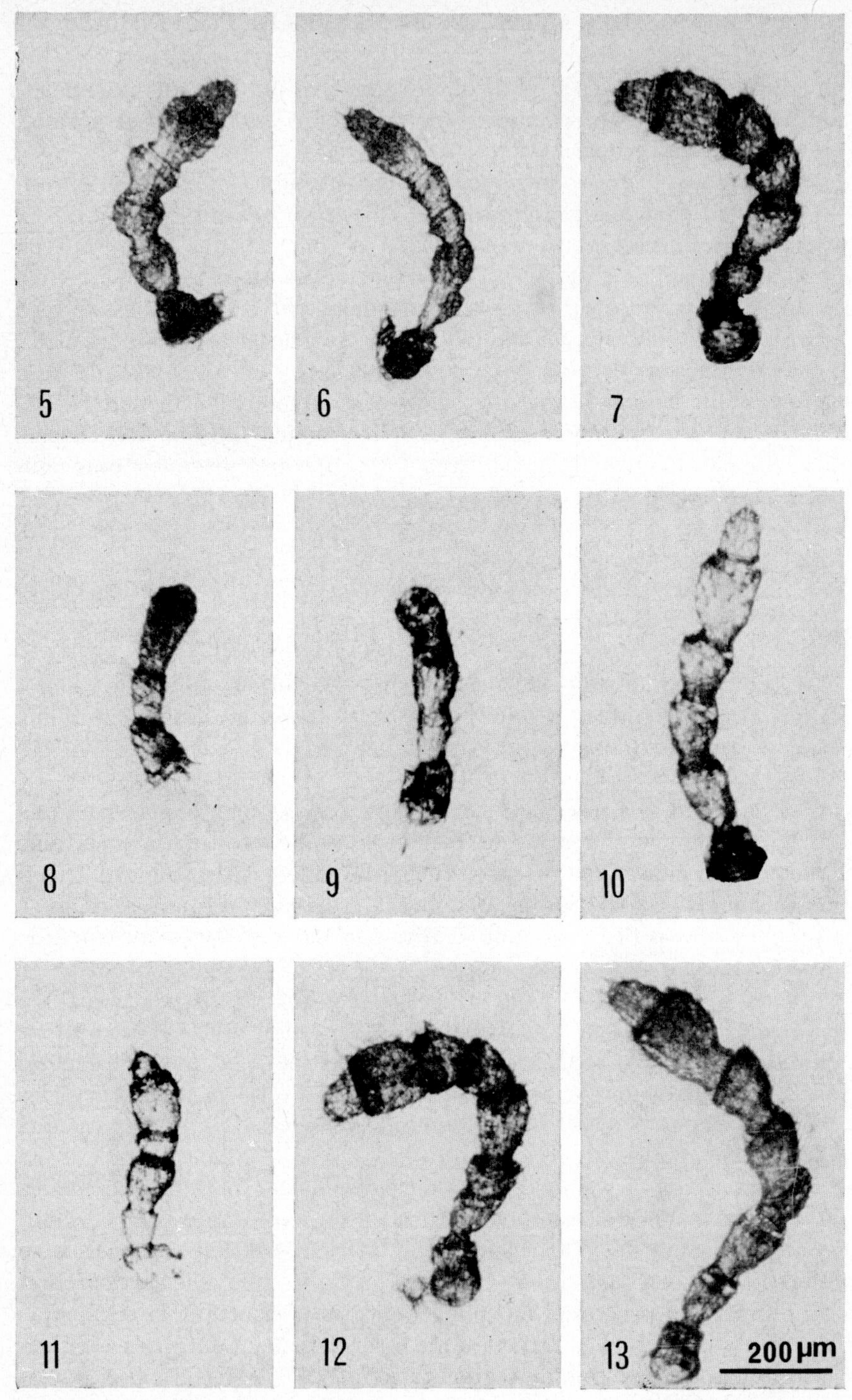

Caption on facing page

Facteurs conditionnant la régénération antennaire

En étudiant la régénération des pattes d'araignées, Vachon (1967) a bien précisé le rôle du moment d'amputation au cours du cycle d'intermue sur le polymorphisme des régénérats. Il attribue aux régénérats les qualités indifférenciés, larvaires et nymphaires. L'aspect de nos régénérats incomplets (qui correspondent aux régénérats indifférenciés) confirment l'importance du moment d'amputation sur l'aspect du régénérat. Néanmoins, nos observations ne nous ont pas permis de vérifier que les phases de différenciation du blastème étaient en conformité avec celles du développement ontogénique. Des amputations effectuées 25 à 30 jours avant l'exuviation ainsi que des expériences portant sur des larves plus jeunes (stade I, stade II) nous permettraient sans doute de retrouver les résultats de Vachon (1967) sur notre matériel, car l'observation de quelques larves du stade III portant des antennes régénérées, a montré que ces appendices se présentaient sous la forme d'antenne à quatre articles (type stade I).

Quoi qu'il en soit, le polymorphisme des appendices observés par Vachon (1967) et l'aspect de nos régénérats incomplètement différenciés correspondent au blocage de la différenciation du blastème de régénération au moment de l'exuviation.

Vachon (1967) suggère également que la qualité des régénérats peut varier en fonction du niveau d'amputation. Nguyen Duy-Jacquemin (1972) pense par contre que la régénération des antennes de *Polyxenus lagurus* ne parait pas dépendre du niveau d'amputation. Verhoeff (1928) a obtenu chez quelques larves âgées de diplopodes, des régénérats antennaires différents par le nombre des articles (six ou sept); il suppose que la présence du deuxième article est fondamentale pour

---

Figures 5–13. Qualité segmentaire d'antennes régénérées après amputations réalisées chez la troisième forme larvaire de *P. angustus*. *t*, temps séparant l'amputation de la mue post-opératoire (jours); t^1, durée de l'intermue suivant (jours); na, niveau d'amputation (voir Fig. 2).

Fig.	Stade	t	t^1	na
5	IV	40		f
6	Même animal que précédemment			g
7	V	40	110	f
8	IV	15		g
9	IV	20		h
10	V	20	90	e
11	IV	100		g
12	V	42	60	g
13	V	135	75	g

l'obtention des régénérats à sept articles. Chez les insectes, le rôle prépondérant du niveau d'amputation sur l'aspect des appendices régénérés a été bien étudié (Bart, 1971; Bullière, 1967; Urvoy, 1970).

Nos résultats montrent que la qualité juvénile (six articles) ou adulte (sept articles) des régénérats antennaires des larves du stade III de *P. angustus* dépend exclusivement du niveau d'amputation.

Le facteur temps conditionne l'état de différenciation du blastème antennaire; le niveau d'amputation impose à l'antenne régénérée une qualité segmentaire qui est toujours apparente, au moins après la première mue post-opératoire. L'influence de ces deux facteurs est simultanée dans la régénération antennaire.

Relation régénération et segmentation

La particularité de notre matériel (antenne en voie de segmentation), l'appréciation d'une zone critique de régénération réduite et bien localisée, nous permettent d'envisager les relations existant entre la morphogenèse régénératrice qui s'établit à l'apex du second article ou les processus qui lui ont donné naissance et l'inhibition de la segmentation de ce second article.

Nous avons constaté que dans tous les cas où le second article reste entier, la régénération des parties distales, même si elle s'effectue à partir d'un reste de tissus de la base du troisième article, n'affecte jamais la segmentation. Notons à cet égard, que ces résultats s'accordent à ceux de Locke (1967) et Bart (1971) qui ont montré "qu'une membrane articulaire constitue un centre relativement neutre permettant la croissance maximale des régions voisines", dans notre cas, la région sous-articulaire.

Nous avons également remarqué, qu'après obtention d'antennes de type juvénile (à différenciation complète) résultant d'amputations qui touchent profondément le second article ou qui l'éliminent, le second article régénéré se segmente toujours à la mue suivante. Par contre, la segmentation se trouve reportée à une seconde mue lorsque les articles distaux de l'antenne régénérée ne sont représentés qu'à l'état d'ébauches; ce qui laisse supposer que, dans ce cas, la segmentation du second article ne peut s'effectuer qu'après la différenciation des articles distaux.

En ce qui concerne maintenant les résultats obtenus après lésion de la zone apicale du second article, il est remarquable de constater que la morphogenèse régénératrice et l'arthrogenèse peuvent coexister dans le deuxième article, que la mise en place d'un blastème de régénération à l'apex de cet article conduit à l'élaboration de deux types d'antennes régénérées dont la qualité est liée à la réalisation ou à l'inhibition d'une formation articulaire. Ces deux modalités de régénération à arthrogenèse immédiate ou différée dépendent du comportement morphogénétique du second article, qui subit, au moins dans le deuxième cas envisagé, une modification de l'état physiologique des cellules normalement impliquées dans l'arthrogenèse; il conviendra d'examiner l'aspect cytologique de cette modification.

Dans les conditions de nos expériences, il nous est permis d'envisager deux causes possibles de l'apparition de cette modification du second article antennaire:

1. La blessure de l'apex du second article qui entraînerait une perturbation en rapport avec l'importance de la nécrose.
2. L'élimination de la zone différenciée de l'articulation du second article avec le troisième, qui conditionnerait la régénération et l'état des tissus sous-jacents.

Signalons qu'à partir d'expériences variées chez *Carausius morosus*, Bart (1970, 1971) explique le déclenchement d'une morphogenèse trochantérienne par une activation due aux lésions et par une modification de la différenciation fémorale. Le problème qui nous préoccupe (inhibition de la segmentation) est vraisemblablement très voisin.

Cependant, il ne nous est pas possible actuellement de choisir entre les deux hypothèses que nous avons proposées, pas plus que d'interpréter à partir de l'une d'elles, la genèse de niveaux anormalement différenciés qui font apparaître des segmentations anormales.

De nouvelles observations et des expériences complémentaires s'avèrent nécessaires afin de parvenir à une meilleure compréhension des mécanismes qui sont à l'origine de l'arthrogenèse antennaire.

Bibliographie

Bart, A. (1970). Blessures et transformations trochantériennes chez *Carausius morosus* (Br.). *Annls Embr. Morph.* **3**: 379–398.

Bart, A. (1971). Morphogenèse trochantérienne provoquée par une régénération fémorale chez le phasme *Carausius morosus* (Br.). *J. Embryol. exp. Morph.* **25**: 301–320.

Brölemann, H. W. (1935). Myriapodes Diplopodes. *Faune Fr.* No. 29: 1–368.

Bugnion, E. (1920). Accroissement des antennes chez *Empusa egena. Mém. Soc. zool. Fr.* **27**: 127–137.

Bullière, D. (1967). Etude de la régénération sur un insecte Blattoptéroïde, *Blabera craniifer*. Influence du niveau de la section sur la régénération d'une patte métathoracique. *Bull. Soc. zool. Fr.* **92**: 523–536.

Haase, H. J. (1968). On the epigenetic mechanisms of patterns in the Insect integument. *Int. Rev. gen. exp. Zool.* **3**: 1–51.

Locke, M. (1967). The development of patterns in the integument of Insects. *Adv. Morph.* **6**: 33–88.

Nguyen Duy-Jacquemin, N. (1972). Régénération antennaire chez les larves et les adultes de *Polyxenus lagurus*. *C.r. hebd. Séanc. Acad. Sci., Paris* **274**: 1323–1326.

Sahli, F. (1966). Contribution à l'étude de la périodomorphose et du système neurosécréteur des Diplopodes Iulides. *Thèse Sci. doct. Etat.* **94**: 1–226. Dijon: Bernigaud et Privat.

Saudray, Y. (1961). Recherches biologiques et physiologiques sur les Myriapodes Diplopodes. *Mém. Soc. linn. Normandie* **2**: 1–126.

Seifert, B. (1932). Anatomie und Biologie des Diplopoden *Strongylosoma pallipes* (Oliv.). *Z. Morph. Ökol. Tiere* **25**: 362–507.

Teissier, G. (1948). La relation d'allométrie. Sa signification statistique et biologique. *Biometrics* **4**: 14–53.

Urvoy, J. (1970). Etude des phénomènes de régénération après section d'antenne chez le phasme *Sipyloïdea sipylus* (W.). *J. Embryol. exp. Morph.* **23**: 719–728.

Vachon, M. (1945). Remarques sur les appendices du prosoma des Limules et leur arthrogenèse. *Archs Zool. exp. gén.* **84**: 271–300.

Vachon, M. (1967). Nouvelles remarques sur la régénération des pattes chez l'araignée *Coelotes terrestris* (Wid.). *Bull. Soc. zool. Fr.* **92**: 417–428.

Verhoeff, K. W. (1928). Diplopoda. *Bronn's Kl. Ordn. Tierreichs* **5**: 1–2084.

Discussion

Jubethie-Jupeau: Does amputation have any effect on the intake of food or on the duration of the inter-moult?

Petit, G.: I do not think that amputation of the antennae of the third stadium in *Polydesmus angustus* has any effect on the onset of the moult but I have no direct evidence. I have work on the amputation of antennae of older larvae in progress; here there does seem to be an effect on the onset of moulting.

Herbaut: Professor Joly (Joly & Lehouelleur, 1972)* observed a reduction in the duration of the inter-moult after amputation of antennae of *Lithobius forficatus*. The reduction was greater when a larger part of the antenna was removed.

Juberthie-Jupeau: You said that the amputation was done 15 days before the moult; you counted from the day of operation and noted the animal moulted 15 days later?

Petit, G.: Yes, of course, it does not seem possible to know in advance—I take an individual that has built or is just building its moulting chamber and I know from experience that this will moult in 15–20 days.

Juberthie-Jupeau: When you amputate earlier, you cannot determine exactly when the next moult will be?

* Joly, R. & Lehouelleur, J. (1972). Effet de la section antennaire sur le déclenchement de la mue chez *Lithobius forficatus* L. (Myriapode Chilopode). *Gen. comp. Endocr.* **19**: 320–324.

PETIT, G.: No, but this is not important; what is important, is to note the date the animal did moult, so as to know the period of regeneration; to know the period of blastema formation and of its differentiation. In *Polydesmus* all operations which were followed by a moult within a month, always produced perfectly differentiated regenerates.

DOHLE: Is there, as in the legs of *Leucophaea*, a definite area which must remain at the base, for regeneration to occur?

PETIT, G.: It is difficult to determine the precise field of regeneration, but the articulation between the head and basal article is sufficient for regeneration to occur.

SAHLI: What is true for the antennae need not be true for the limbs; I have had cases where complete extirpation of a limb was not followed by regeneration.

PETIT, G.: I have a paper in preparation in which I analyse the precise territory necessary for regeneration of the limbs of *Polydesmus*.

SAHLI: Did you obtain similar results to mine?

PETIT: Slightly different, but not absolutely contradictory.

NGUYEN DUY: You said that it is only the article A_{23} which divides in normal development but you have an abnormal case in which you said A_{45} divided during regeneration. How do you explain this?

PETIT, G.: This problem interests me particularly. I would like to have an explanation of these abnormal cases because these would no doubt serve also to explain normal segmentation. Possibly the three cases of abnormal segmentation I obtained resulted from interactions between zones which were abnormally differentiated during regeneration.

NGUYEN DUY: There was no division between the first and second stadium?

PETIT, G.: Careful measurements of the antennae of stadium I are necessary to establish what segmentation occurs and where it occurs. I have tried to do this but will have to repeat the measurements.

Symp. zool. Soc. Lond. (1974) No. 32, 317–328.

PATTERNS OF COMMUNICATION IN COURTSHIP AND MATING BEHAVIOUR OF MILLIPEDES (DIPLOPODA)

ULRICH HAACKER*

Zoologisches Staats Institut und Museum, 2 Hamburg 13, Von-Melle-Park 10, Germany

SYNOPSIS

In many millipedes (Diplopoda), mating behaviour includes communicative patterns, often associated with special structural devices, the function of which has previously been unknown or misinterpreted. Mechanical signals range from simple touching (*Graphidostreptus*, Blaniulidae) over drumming on the ground (*Chordeuma*) to stridulation (*Loboglomeris*, *Sphaerotherium*). Chemical signals are produced by coxal (Julidae, *Polyxenus*), dorsal (*Chordeuma*), or postgonopodial glands (Glomeridae, Glomeridellidae) The main functions of communicative behaviour seem to be to impart information about mating intentions, to induce adequate body postures and to identify the mate's species. Almost nothing is known about receptors or about the physical or chemical qualities of the signals. The necessity of a physiological approach to these problems is stressed.

INTRODUCTION

Recent studies in mating behaviour of Diplopoda have shown an unexpected diversity of patterns associated with courtship, mating and sperm transfer. These studies contribute to our understanding of the function of structures which have been known for a long time without any understanding of their rôle in the sexual processes. Mating behaviour in millipedes frequently involves some mechanism of communication—as in many other animals with sexual reproduction. Communication may be brought about by behavioural displays which may or may not be associated with structures that are specialized for the production of signals between mates.

A survey of all behavioural patterns so far known to be associated with communication seemed useful to me, in order to draw the attention of investigators to this special aspect. The study of millipede mating behaviour may throw new light on the morphology, physiology, ecology and evolution of this group. Systematic units may be characterized by certain behavioural properties, and species may be genetically isolated from others by specific signals used in pre-mating mechanisms of communication.

Identification of conspecific mates as well as their manoeuvring and stimulation in order to obtain an adequate position for sperm

* The death of Dr Ulrich Haacker took place on 10th September, 1972.

transfer seem to be the main functions of communication in Diplopoda. A communication system consists of a sender, who—often by structural devices and behavioural patterns—produces a signal, and a receiver who perceives this signal with the aid of adequate receptors. In most cases to be dealt with below, the male acts as sender, the female as receiver; it is probable, however, that signals are produced by the female, too; but these seem to be less apparent.

The different types of signals may serve as criteria to classify the observed patterns.

COMMUNICATION BY MECHANICAL SIGNALS

In soil-living animals which are highly sensitive to tactile stimuli it is obvious that these might easily serve as signals in intra-specific encounters. The most primitive form of communication, indeed, is touching. The following example shows that even in such a simple way definite information can be transmitted. In most species of the orders Spirobolida, Spirostreptida and Julida the males initiate mating by mounting on the female's back, clinging to her with the aid of leg-pads. The female can well distinguish this kind of contact from that of a male simply walking over her back without sexual motivation. This can clearly be seen in species displaying a special defensive behaviour. In the julid *Leptoiulus simplex* Verh. a female which is not ready to mate begins to wind her body vigorously and to secrete quinones out of her repugnatorial glands as soon as the male mounts on her back. A female of the spirostreptid *Alloporus circulus* Att. extends her head in a prognathous position which facilitates copulation, when the male has just mounted on the hindpart of her body; if, however, she is not ready to mate, she coils up her forepart thus making copulation impossible. Both examples prove that tactile stimuli connected with mounting and the action of leg-pads contain the information that there is a male which is ready to mate. This is a very primitive kind of communication in so far as transmission of information is only a subsidiary result of a behavioural pattern serving another, non-communicative, purpose. It might not have been mentioned here, if it were not probable that all higher forms of communication have evolved from such preliminary stages.

In many arthropods, tapping with body appendages is a common feature between meeting individuals. In Chilopoda, like *Scolopendra*, tapping with the antennae seems to act as a signal initiating sperm transfer (Klingel, 1957). Demange (1959) has observed similar behaviour in the spirostreptid millipede *Graphidostreptus*, Mauriès (1969) in

Typhloblaniulus and Kinkel (1955) in *Blaniulus* (both Julida); in all three cases the male taps the female with his antennae prior to copulation. A more highly developed pattern, using mechanical stimuli for communication, has evolved in the genus *Chordeuma*. Its main advantage is that it does not require body contact between the sender and the receiver of the information. Chordeumidae are fast running, omnivorous animals. When encountering a conspecific, they normally turn around about 180° and run away. Hence, a display that acts over some distance and avoids the flight reaction released by direct touching will be most favourable for communication among potential mates. Drumming, as observed in *Ch. silvestre* Koch, is a display of this type (Haacker, 1971a). When meeting a female, the male lifts the forepart of his body and drums on the ground with the frontal side of his head, with a frequency of more than five strokes per sec. The female's reaction (she actively undertakes the next step leading to copulation—see below) suggests that—over a distance up to 1 cm—she has received information about the other animal's nature and "intentions". The drumming male presumably produces vibratory stimuli; in many insects, similar signals are perceived by leg receptors. Nothing is known at present about the existence of such receptors in the legs of *Chordeuma*. However, since Rilling (1960) has described several types of sense organs from the legs of the chilopod *Lithobius*, leg receptors might account for the perception of vibratory signals in Diplopoda, too.

Communication by vibration seems to be widely distributed in terrestrial arthropods (Markl, 1969). In Diplopoda, there is at least one further example, represented by the genus *Loboglomeris*. Here, the communication system includes not only specific displays, but also structures which are specialized for communication. There is a stridulatory apparatus in the male, consisting of longitudinal ribs situated on the telopod femora and the posterior margin of the pygidium (Haacker, 1970a).

> *Note:* The femoral ribs were well-known to systematists, who used them to characterize the genus (Verhoeff, 1906; Attems, 1926); they overlooked their biological significance, because they were not trained to see and to think in terms of function.

The male of *L. pyrenaica* Latzel grasps one antenna and one vulva of the female by aid of his telopods; thus holding the female, he rubs the striated pygidial margin against his femoral ribs in a sideways movement. As the mates are in very close body contact, the female probably perceives the vibratory signals, resulting from stridulation, directly via her antenna and/or vulva. Receptors are unknown. The signals seem to contain information as to whether the male belongs to the right species or not. The areas of the two Pyrenean species *L. rugifera* Verh. and *L. pyrenaica* enter in contact at several points. I observed both species at the same place and without any ecological separation near the Col du Tourmalet (Central Pyrenees). Hence, a premating isolation mechanism of ethological nature would be very reasonable. A simple experiment gives evidence that stridulation is such a mechanism. I placed five males of *L. pyrenaica* together with five females of the same species and five further females of *L. rugifera* in a Petri dish (ϕ: 8 cm) on humid filter paper and recorded their sexual behaviour. Complete copulations only occurred between conspecifics—heterospecific copulation attempts were interrupted by the female (by coiling up her forepart) as soon as the male had stridulated (Table I).

TABLE I

Isolating effect of stridulation (36 *h-experiment*)

♂♂*L. pyrenaica*	♀♀*L. pyrenaica*	♀♀*L. rugifera*
Complete copulations	32	—
Copulation attempts interrupted after stridulation	—	7

The vibratory signals of *pyrenaica*-males obviously are the wrong information for *rugifera*-females; they distinguish them well from those produced by males of their own species.

Note: The fact that males of *L. pyrenaica* more frequently try to copulate with conspecific than with heterospecific females, may be explained by the existence of a further isolating mechanism: preliminary investigations in other species indicate that glomerid males may recognize conspecific females by chemically coded information.

In contrast to the stridulatory organ of *Loboglomeris*, those of another group of pill-millipedes, the Sphaerotheriidae, have early been recognized as sound producing organs. It even seems doubtful whether all so-called stridulatory structures, as reviewed by Dumortier (1963), are really used for stridulation. In the South African genus *Sphaerotherium*, the organ consists of a row of ribs on each of the male's telopods and two corresponding tuberculous fields on the inner side of the pygidium. In *S. dorsale* Gervais and related species stridulation forms part of the courtship behaviour (Haacker, 1969c)*.

When meeting a female (including one which has coiled up into a ball) the male lifts his pygidium and rubs his telopod ribs against the tuberculous fields, producing a faint scratching noise, audible to the human ear over a distance of more than 10 cm. Stridulation of *S. punctulatum* may even be heard over more than 5 m.

As in *Loboglomeris*, no sense organs which might account for the perception of the presumably vibratory signals are known; the temporal organs, long regarded as sound receptors (Verhoeff, 1928; Meske, 1961) seem to be chemoreceptors, judging from their fine structure (Bedini & Mirolli, 1967). One function of the male's stridulation may be to inform the female of his presence and his sexual motivation, since females have been observed to leave their coiled up position when a male stridulated nearby. Whether stridulation contains information about the species of the male, must still be checked. An analysis by oscillograms shows the signals of two morphologically nearly identical species to be very different (Table II).

TABLE II

Differences between stridulatory signals in ***Sphaerotherium*** *species*

	S. spec. 1	*S.* spec. 2	Difference
Mean distance between sequences (sec)	39 $n = 16$ $s = 8$	2·1 $n = 3$ $s = 0{\cdot}17$	Significant $p < 0{\cdot}001$
Number of series per sequence	1–4	9–72	Not tested
Mean distance between series (sec)	2·7 $n = 26$ $s = 0{\cdot}53$	0·28 $n = 22$ $s = 0{\cdot}13$	Significant $p < 0{\cdot}001$
Number of pulse groups per series	?–7–?	1–34	Not tested
Mean distance between pulse groups (msec)	39 $n = 6$ $s = 8{\cdot}5$	32 $n = 68$ $s = 13{\cdot}3$	Not significant $p > 0{\cdot}05$

* Unfortunately, the material on which these observations were based was heterogeneous, including *S. dorsale* from Natal and at least one very similar species from Tsitsikama Forest. Since nearly all characters used for separation of *Sphaerotherium* species remain very doubtful, an exact identification will only become possible after a taxonomic revision of the whole genus.

COMMUNICATION BY CHEMICAL SIGNALS

Schömann (1956), in his monograph on the biology of *Polyxenus*, was the first to discover a pheromone in Diplopoda. *Polyxenus lagurus* L. is the only millipede known to have an indirect transmission of sperm without any contact between the "mates". The *Polyxenus* male deposits his sperm on a web and, when leaving it, produces two parallel threads, about 2 cm long and regularly garnished with droplets of a secretion produced in coxal glands of the 7th segment. A female coming upon these threads obviously receives information, as she will follow them immediately and thus find the sperm. The male's secretion can be classified as a trail pheromone, comparable to those observed in ants. Unlike the rather complicated performance of *Polyxenus*, the behaviour of the sender animal is much less conspicuous in the other millipedes so far known to use chemical signals. No special behaviour at all can be found in females of the family Julidae; nevertheless, males identify conspecific females by antennal contact. Males of *Cylindroiulus punctatus* (Leach) distinguish well between their own females and those of *C. nitidus* Verh.; females of the former species are by far more frequently mounted than those of the latter (Haacker & Fuchs, 1970). The signal seems to be a chemical one, as males without antennae do not mount. Diffuse glandular cells of the hypodermis might produce the substance, since all parts of the female's body carry the information.

In all the following examples the pheromone–producing glands are localized; they either form a complex of independent bicellular units (*Chordeuma*) or they appear as a multicellular gland with one common efferent duct (*Glomeris*, see Juberthie & Tabacaru, 1968, and *Julus*).

In the male of *Chordeuma silvestre* we find single glandular cells each combined with one canal-forming cell distributed all over the dorsal hypodermis. In the (diplo-) segments 12–14 these glandular cells are aggregated to form single-layered fields; in segments 15–18 they build a multi-layered complex, but each glandular cell still has its own efferent duct, opening through the dorsal cuticle. Actually, *Chordeuma* shows how localized glands evolved from single glandular cells. Both types seem to produce the same kind of secretion. The male, after drumming, turns his back to the female, who starts licking at the forepart but soon concentrates upon the glandular region around tergite 16, which is marked by two paramedian cones. The pheromone may have a stimulative function, as described for some beetles (Matthes, 1962), but this is still to be proved; there is evidence, however, that it functions as an "arrestant" (terminology see Dethier, Barton Browne & Smith, 1960); during licking, the female takes up a position allowing the male to grasp her and to introduce his gonopods.

The same function of a chemical signal is evident in *Julus* (Haacker, 1969b). The males of *J. scandinavius* Latzel and *J. scanicus* Lohmander lift their forepart when encountering a female and turn their ventral sides towards her head, thus displaying the long coxal processes of the second pair of legs. *J. scanicus* often displays without the female's presence and may lie motionless for minutes, with his forepart upside-down. The coxal processes carry a secretion which is produced by large coxal glands; Verhoeff (1928) thought it might have a purely mechanical function helping the male to adhere to the female. In reality, the male offers this secretion to the female who takes it as a signal for copulation; only after the female has started licking the coxal processes, does the insertion of the gonopods become possible. As the areas of *J. scandinavius* and *J. scanicus* overlap, we might also consider an isolating function of the pheromone. Preliminary experiments on this question have failed because copulations occur very rarely in these two species. To emphasize the incompleteness of the actual knowledge it should be mentioned that coxal glands are known from many julid species and that in the spirobolid genus *Rhinocricus*, too, the female licks coxal processes of the male's anterior legs (Haacker, 1970b). Use of pheromone-like coxal secretions during courtship or mating may be widely spread among Diplopoda.

In 1968, Juberthie & Tabacaru described a large, postgonopodial gland in the males of the glomerid genera *Glomeris*, *Trachysphaera*, and *Glomeridella*. They assumed that it might play a part in copulation. I tried to solve the question by observing mating *Glomeris marginata* Villers. Two functions seemed possible: the secretion could help to facilitate sperm transmission by "greasing" the telopods or it could serve as a pheromone bearing some information or stimulation for the female. The latter turned out to be correct (Haacker, 1969a). As soon as a telopod displaying male has grasped one antenna and the opposite vulva of the female, he extends the skin around the opening of the gland by haemolymph pressure, and secretion pours out just beneath the female's mouthparts. While she reacts by licking, he releases her antenna and grasps the second vulva, thus establishing the definite mating position. The main function of the secretion seems to be that of an arrestant, as in *Chordeuma* and *Julus*. Whether it also may indicate the species to which the male belongs, remains doubtful. Such a function might be expected, since *G. marginata* lives syntopically with congeneric species, like *G. intermedia* Latzel, *G. conspersa* Koch, *G. pustulata* Latreille and others, over large parts of its area. However, in simultaneous choice experiments with *G. marginata* and *G. intermedia*, heterospecific copulations are frequent and often take a normal course. Interruptions by the female, if they occur at all, do not show any correlation with secretion offering and licking; they mostly occur at later stages of copulation.

Recent investigations in *Trachysphaera pyrenaica* Ribaut and *T. noduligera* Verh. indicate that in this genus the function of the postgonopodial gland is the same as in *Glomeris*; at least, the behavioural pattern of presenting the secretion and licking is entirely identical. This is not the case in *Glomeridella minima* Latzel, where the female starts licking after the male has established the mating position, and continues licking even after sperm transmission.

There is ethological evidence that a postgonopodial gland exists in still another glomerid genus: *Haploglomeris*. However, the pattern in which it takes part is again different from that of *Glomeris*. Females of *H. multistriata* Koch, when encountering telopod displaying males, start licking intensely at the grooved margin of their pygidium, where a postgonopodial skin region is extended by haemolymph pressure; eventually, the male may initiate the copulation.

Recent field observations in South Africa indicate that secretion offering by the male and licking by the female also occur in *Sphaerotherium ancillare* Att., a remarkable example of behavioural convergence. Presentation of a postgonopodial secretion by the male and licking by the female may have different functions in the cited genera. An identification of the secretion's chemical nature in related species of the same genus as well as in different genera would facilitate a better understanding of its significance.

The communicative function of a secretion and related behavioural patterns is much more apparent in the last examples to be dealt with: they concern the use of defensive secretions, which normally serve to repel predators. I observed this peculiar phenomenon in two species of the orders Julida and Spirostreptida, already referred to above. A female of *Leptoiulus* which is not ready to mate tries to prevent copulation attempts by winding her body vigorously. Nevertheless, a male will cling to the female's back trying to overcome her resistance and to fix her by his hook-shaped first legs. However, he will immediately release the female if she uses her defensive quinones as a signal of her unreadiness. Similarly, in 50% of observed copulae, females of *Alloporus* gave a signal to end the copulation by twisting their bodies round the male's forepart and pouring out defensive secretion, exactly at those diplosegments which were nearest to the male's head.

CONCLUSIONS

The aim of this contribution is to show that there is a great diversity of communicative patterns in Diplopoda, worth more detailed investigations. The examples which have been discovered so far open more new questions than they answer. Functions are often not clear; the properties of the signals and the nature of the receptors involved are in no one case exactly known. There is no doubt that still more examples of communication will be described in future, as ethological study in millipedes has just begun. It should be emphasized, therefore, that millipede species may prove to be suitable objects for the study of general problems of communication, by application of more sophisticated physiological methods.

Acknowledgements

I wish to thank, without enumerating their names, the many colleagues and friends who gave help in my own approaches to the subject, by offering working facilities, by sending or determining material, and by discussing open questions. Without their world-wide co-operation we would know still less than we actually do.

The Deutsche Forschungsgemeinschaft supported this work by the grants Ha 547/1–4.

References

Attems, C. (1926). Diplopoda. *Handb. Zool., Berl.* **4**: 29–238.

Bedini, C. & Mirolli, M. (1967). The fine structure of the temporal organ of a Pill millipede, *Glomeris romana* Verhoeff. *Monitore zool. ital.* (N.S.) **1**: 41–63.

Demange, J.-M. (1959). L'accouplement chez *Graphidostreptus tumuliporus* (Karsch) avec quelques remarques sur la morphologie des gonopodes et leur fonctionnement. *Bull. Soc. ent. Fr.* **64**: 198–207.

Dethier, V. G., Barton Browne, L. & Smith, C. N. (1960). The designation of chemicals in terms of the responses they elicit from insects. *J. econ. Ent.* **53**: 134–136.

Dumortier, B. (1963). Morphology of sound emission apparatus in Arthropoda. In *Acoustic behaviour of animals:* 277–373. Busnel, R.-G. (ed.). Amsterdam, London, New York: Elsevier Publ. Co.

Haacker, U. (1969a). Spermaübertragung von *Glomeris* (Diplopoda). *Naturwissenschaften* **56**: 467.

Haacker, U. (1969b). An attractive secretion in the mating behaviour of a millipede. *Z. Tierpsychol.* **26**: 988–990.

Haacker, U. (1969c). Das Sexualverhalten von *Sphaerotherium dorsale* (Myriapoda, Diplopoda). *Verh. dt. zool. Ges.* **32**: 454–463.

Haacker, U. (1970a). Der Stridulationsapparat von *Loboglomeris* und seine Funktion im Sexualverhalten. *Vie Milieu* (C) **20**: 57–64.

Haacker, U. (1970b). Das Paarungsverhalten von *Rhinocricus padbergi* Verh. (Diplopoda, Spirobolida). *Rev. Comport. Anim.* **4**: 35–39.

Haacker, U. (1971a). Trommelsignale bei Tausendfüsslern. *Naturwissenschaften* **58**: 59–60.

Haacker, U. (1971b). Die Funktion eines dorsalen Drüsenkomplexes im Balzverhalten von *Chordeuma* (Diplopoda). *Forma Functio* **4**: 162–170.

Haacker, U. & Fuchs, S. (1970). Das Paarungsverhalten von *Cylindroiulus punctatus* Leach. *Z. Tierpsychol.* **27**: 641–648.

Juberthie-Jupeau, L. & Tabacaru, I. (1968). Glandes postgonopodiales des Oniscomorphes (Diplopodes, Myriapodes). *Revue Ecol. Biol. Sol* **5**: 605–618.

Kinkel, H. (1955). Zur Biologie und Ökologie des getüpfelten Tausendfusses *Blaniulus guttulatus* Gerv. *Z. angew. Ent.* **37**: 401–436.

Klingel, H. (1957). Indirekte Spermatophorenübertragung beim Scolopender (*Scolopendra cingulata* Latreille; Chilopoda, Hundertfüsser). *Naturwissenschaften* **44**: 338.

Markl, H. (1969). Verständigung durch Vibrationssignale bei Arthropoden. *Naturwissenschaften* **56**: 499–505.

Matthes, D. (1962). Excitatoren und Paarungsverhalten mitteleuropäischer Malachiiden. *Z. Morph. Ökol. Tiere* **51**: 375–546.

Mauriès, J.-P. (1969). Observation sur la biologie (sexualité, périodomorphose) de *Typhloblaniulus lorifer consoranensis* Brölemann. (Diplopoda, Blaniulidae). *Annls Spéléol.* **24**: 495–504.

Meske, Ch. (1961). Untersuchungen zur Sinnesphysiologie von Diplopoden und Chilopoden. *Z. vergl. Physiol.* **45**: 61–77.

Rilling, G. (1960). Zur Anatomie des braunen Steinläufers *Lithobius forficatus* L. (Chilopoda). Skelettmuskelsystem, peripheres Nervensystem und Sinnesorgane des Rumpfes. *Zool. Jb.* (Anat.) **78**: 39–128.

Schömann, K. (1956). Zur Biologie von *Polyxenus lagurus* (L. 1758). *Zool. Jb.* (Syst.) **84**: 195–256.

Verhoeff, K. W. (1906). Zur Kenntnis der Glomeriden (zugleich Vorläufer einer *Glomeris*-Monographie). *Arch. Naturgesch.* **72**: 107–226.

Verhoeff, K. W. (1928). Diplopoden. *Bronn's Kl. Ordn. Tierreichs* Band **5**, II, **2**: 1–1072.

Discussion

Blower: Have you tried to play back the stridulation?

Haacker: No, not yet. I think it is not airborne sound, but vibrations through the substrate which form the real signal. It is difficult to imitate this, but we are trying now with material we have collected in South Africa.

Blower: How are you going to arrange for the signal to get across—how are you going to transmit?

Haacker: We shall try with plates on which the females are placed. I have the assistance of a colleague who is designing the equipment to transfer the recording to the plate.

Kraus: In your first Table you have females of two species of *Loboglomeris* and males of one of these species and I was surprised that you record so few inter-specific attempts at copulation. Perhaps there may be another mechanism which operates before partners come into contact and stridulation begins?

HAACKER: Yes; I have evidence of such a mechanism, not in *Loboglomeris* but in *Glomeris*. The reaction of the male to extend his telopods in display is released by antennal contact with the female. Females of the same species release this reaction to a significantly greater extent than females of another species.

STRASSER: Several times I have found specimens of *Glomeris* which appeared, from their morphology, to be hybrids of two species. In particular, I have found hybrids between *G. conspersa* and *G. undulata* in the Southern alps and between *G. conspersa* and *G. verhoeffi* or *G. guttata* in Calabria.

HAACKER: I have some doubts about the hybrids between *G. undulata* and *G. conspersa;* these two forms are not sympatric; possibly they are really sub-species, in which case it would not be surprising to have hybrids between them.

JEEKEL: You mentioned the scobinae in *Rhinocricus*. Did you investigate these structures; are they sound producing?

HAACKER: I have looked for some function but could find none. I found no evidence that they had any significance in sexual behaviour.

JEEKEL: I have often found males of *Cylindroiulus*, apparently rather old, which had lost a number of legs on the anterior segments. May they have lost them during copulation? Does the female bite off the legs?

HAACKER: It seems possible. In *Iulus scandinavius* the distal parts of the coxal processes of the second pair of legs are often missing. Verhoeff (1928)* suggested that the female bites through them during copulation. In *Cylindroiulus*, especially *C. nitidus*, there are very large glands on the coxae of the second pair of legs in the male. But in *C. punctatus* which I have observed in detail, I have seen nothing of the sort you mention.

ENGHOFF: Did you observe whether the inter-specific copulations between *Glomeris* species ever lead to fertile eggs or larvae?

HAACKER: I did not try because the females were always collected from the field and I did not know whether they had previously mated with males of their own species.

RAJULU: (CHAIRMAN). We have tried to locate the sense organs in the male of *Cingalobolus bugnioni* which perceive signals from the female. When the apical segment of the antenna or the distal margin of the gnathochilarium is removed, males fail to mount the female preparatory to mating. Electrophysiological recordings of an intact male show a marked response in the presence of a female. When the apex of the antenna is removed there is still a response but it is less marked. Similarly, when the gnathochilarium alone is operated on. When the apices of both antennae and the

* See list of references, p. 326.

gnathochilarium are removed there is no response whatsoever. Thus both are required for full perception of the female signal. The signal may be a sound, or a pheromone, or both.

HAACKER: Was the female in contact or just in the vicinity?

RAJULU: In the vicinity, 1–2.5 cm away, not in contact.

HAACKER: Which suggests some odour.

RAJULU: Yes. Something slightly volatile.

HAACKER: Did you try males?

RAJULU: Yes. There was no reaction. Also with a juvenile there was no response, nor with an older female—only with a female of the same size and age.

Symp. zool. Soc. Lond. (1974) No. 32, 329–346.

A CONSIDERATION OF THE CHEMICAL BASIS OF FOOD PREFERENCE IN MILLIPEDES

W. N. SAKWA

*Department of Zoology, The University, Manchester, England**

SYNOPSIS

The acceptance of decomposing materials by millipedes correlates with the nitrogen content, the carbohydrate content and the moisture content. The microbiota act to release available food substances and to decrease the polyphenol and tannin concentration. It is suggested that the release of simple sugars from complexes and polyoses serves to increase the palatability of litter leaves and also to act as an index for the nutritiveness of the food. The polyphenol and tannin content of the leaves oppose the effect of sugars upon palatability.

INTRODUCTION

Feeding preference

It has been postulated by Dethier (1937) and Thorsteinson (1953a quoted by Dadd, 1963) that the preference of insects for food plants depends upon the presence of common specific substances which provide olfactory or gustatory stimuli without which feeding fails to occur. Thorsteinson (1960 quoted by Dadd, 1963) however suggests that while feeding is primarily stimulated by universally distributed food substances, including both nutrients and water, various degrees of oligophagy result from the presence of factors which inhibit feeding. Thus food preference involves a balance between acceptance and rejection.

Many authors (Schmidt, 1952; Barlow, 1957; van der Drift, 1965) have shown that millipedes exhibit preferences in their choice of food. They have also shown that the most preferred food varies with the species of millipede under consideration although this preference may be masked by a need on the part of the millipede for a variety of foodstuffs. Schmidt (1952) notes that the inter-specific differences in food preference allow of a maximal transformation of available foods.

Physical and chemical features of preference

Although many authors have attempted to isolate the factors involved in millipede food preferences, it has only been possible to correlate chemical and physical factors of the food with its palatability, without isolating the causative factor.

* Present address: Walsall & Staffordshire Technical College, Walsall, Staffs., England.

Lyford (1943) working with *Diploiulus londinensis* found a correlation between the calcium content of litter leaves and the preference of the millipede. He concluded that leaf species was the important factor in determining the millipede's preference.

Murphy (1953), however, found that the calcium content was inversely related to the carbon : nitrogen (C : N) ratio, while Wittich (1943, 1953) found that the C : N ratio was the controlling factor in determining the rate of decomposition of leaf litter. He determined also that the nitrogen rich leaves were the most readily consumed.

Bocock (1964) found that the nitrogen content of litter increases with its age, and this can be correlated with Kheirallah's (1966) finding that the palatability of leaves increases with age. Similarly Schmidt (1952) established that the palatability of different species of dead leaves depended not only upon the species but upon the age of the leaf.

Dünger (1958) in his examination of the effects of mechanical features of leaves found that these were of little significance with regard to palatibility although Barlow (1957) points out that a certain degree of softness is required to overcome the limitations of the chewing mechanism of millipedes.

Both Dünger and Barlow feel that the moisture content is important although not the limiting factor. This is of interest since Brade-Birks (1930) observed that millipedes will eat a great variety of substances and suggested that they do this for the sake of the moisture they can obtain.

Cloudsley-Thompson (1951) postulates that millipedes attack crop plants during periods of drought for the sake of moisture. He also hypothesizes that the sugars present in the sap restrain the millipedes from returning to their normal diet thus exacerbating the severity of damage.

In his view the attractive properties of rotting substances are due only to sugars, provided that suitable contact and moisture stimuli are present. Thorsteinson (1958a quoted in Dadd, 1963) also notes that the presence of sugar commonly induces insects to feed.

We thus find that three main features of the litter leaf have been correlated with acceptance by the millipede, viz:—nitrogen content, sugars, and moisture.

The role of the microbiota

The increase in palatability of leaves with age has been explained by several authors in terms of a decrease in the concentration of a rejection factor. For example Edwards in Minderman & Daniels (1967) suggests that this is brought about by the leaching of water soluble polyphenols from the leaf.

Satchell & Lowe (1967) found a strong relationship between the polyphenol content of leaves and their palatability to *Lumbricus terrestris* but Thiele (1959) found that soaking leaves in a 2% tannin solution did not affect their palatability to millipedes. He did however find that fresh litter was rejected while litter weathered for a year was acceptable.

Other workers have attempted to explain the changes in palatability with age of litter leaves by implicating the microbiota in the breakdown of the water soluble polyphenols and of the cell contents in general. Minderman & Daniels (1967) found that newly fallen leaves had little attraction for most litter-eating animals but that leaves which had been lying on the ground for some time were readily consumed. These authors suggest that a certain concentration of bacteria and fungi has to develop in the litter before it becomes palatable. The microflora is seen as the most important source of nutriment for the litter eaters with the leaf tissue acting mainly as a sub-stratum.

Saudray (1961) also found that the nutritive value of the food eaten depends upon the extent of its alteration by fungi and bacteria both of which increase the amount of lipid, carbohydrate, and protein.

A major insight into the role of the microbiota can be derived from Handley's (1954) finding that the cell contents of litter leaves are combined into protein complexes which vary in the ease of their decomposition. The function of the microbiota is to render the cell contents accessible to the litter eaters by incorporating these complexes into their own substance. Minderman & Daniels (1967) point out that bacterial and fungal protein is more readily available to the fauna than the residual leaf protein. Similarly Striganova (1967) concludes that the decomposition of leaf litter is a microbial opening-up of the food and that the rate of this is dependent upon the structural properties of the species of leaf.

The importance of this can be gauged from the finding of Gere (1956) that the amount of litter consumed by millipedes and the amount utilized depends upon its state of decomposition when eaten. Millipedes eat more and use less of undecomposed material. Schmidt (1952) also found an intimate relationship between the extent of decomposition and the uptake of litter leaves, and concludes that the determinant of choice is unquestionably the degree of decomposition.

Further evidence for this view comes from Nielsen's (1962) finding that microbial enzyme systems are much more efficient than the enzymes of other invertebrates in the decomposition of the structural polysaccharides of leaves. Thus he envisaged the primary decomposition of litter leaves as being due to the microbiota. This was supported by his finding that there is a poor correlation between the type of diet and the carbohydrases present in a range of litter eating animals.

Although Brade-Birks (1930) suggested that millipedes do not discriminate on a chemical basis in their choice of food substances, Kheirallah (1966) was able to show that the quantity of food eaten depends upon its nutritive value. Larger amounts of the less nutritive substances were eaten if the animal was offered no choice.

Millipedes are therefore able to assess the nutritive value of their food and this may be related to the palatability. The palatability may depend on the presence of rejection and/or acceptance factors.

It is conceivable that the food of the litter feeding millipedes has little attraction when fresh owing to the presence of inhibitory factors (perhaps the polyphenols and tannins). It may become more palatable as a result of the microbial activity in reducing these factors, and also because of an increase in the concentration of attractive factors (perhaps available food substances).

Since sugars commonly induce feeding in insects (Dadd, 1963) the work in this paper has been directed at elucidating the response of millipedes to various carbohydrates. An attempt has also been made to discover the role of polyphenols and tannins in affecting palatability.

Materials and methods

The organisms studied were *Tachypodoiulus niger* (Leach 1815), *Ommatoiulus sabulosus* (Linné 1758), *Polymicrodon polydesmoides* (Leach 1815), *Polydesmus angustus* Latzel 1884, and *Glomeris marginata* (Villers 1789), collected from sycamore woods (*Acer pseudoplatanus*) in Derbyshire.

BEHAVIOURAL EXPERIMENTS

Establishment of scale of choice

The assessment of the response of millipedes to a dissolved substance is difficult because of the absence of an easily quantifiable response. It is not possible to observe the extrusion or unrolling of the mouthparts as used by Minnich (1921, 1926), Mcindoo (1934), Dethier & Wolbarsht (1956), and Hodgson (1957) for the "haustellate" insects. Nor is it possible to employ the "biting" response used by Thorpe, Crombie, Hill & Darrah (1947) for wireworms since the weak mouthparts of millipedes cannot mark filter paper (Cloudsley-Thompson, 1951).

Because of the absence of a suitable method for estimating the feeding response it is not possible to assess the feeding responses separately from the orientating responses. As a result the technique finally evolved measured a combination of the orientation responses and the feeding responses. This was a preference experiment based on the techniques used by Wigglesworth (1941), Thorpe *et al.* (1947), Cloudsley-Thompson (1951), Brookes (1963), and Fairhurst (1968).

Methods

Initially the apparatus used was that of Fairhurst after Brookes using a substrate of air dried peat after Cloudsley-Thompson. This substrate was however found to be unsuitable since the animals tended to burrow into the peat. This was perhaps a light avoidance reaction or more probably a response to the strong thigmotaxis which millipedes exhibit (Crozier & Moore, 1923). To avoid these effects, a 14 cm diam. glass Petri dish, with a low central partition of celluloid, was substituted. The substrate employed was Whatman No. 1 filter paper.

This apparatus was easy to clean between experiments and also avoided the various apparatus effects enumerated by Graham & Waterhouse (1964). Thus end and corner effects were obviated by using round dishes to prevent the collection of millipedes in corners in their efforts to maximize tactile stimulation.

Preliminary experiments were performed to detect and avoid the influence of external effects upon the apparatus and also to determine the optimal time for the experiments. Dadd (1963) suggests that insects respond to feeding stimuli only when hungry, and accordingly various periods without food were tried to obtain a consistent response. Twenty four hours' starvation on moist cotton wool prior to the experiment was found to be sufficient.

A block of 12 of these modified Petri dishes was set up, according to a random selection procedure, with de-ionized water moistened filter paper on one side and filter paper moistened with test solution on the other side. Six starved animals were introduced into each dish and their position was noted at two-minute intervals for 20 min.

A statistical analysis (Anovar) justified summing the replicates. These totals were then represented graphically and the concentration at which the behavioural response became significant at the 5% level (obtained by means of the Chi squared test) was taken to be the behavioural threshold for that substance (see Table I).

Rationale

Cloudsley-Thompson (1951) concluded from his experiments that the response of millipedes to contact chemical stimuli was purely orthokinetic. He also found no evidence for a "biting" response as distinct from an "orientating "response.

TABLE I

Threshold of behavioural preference as compared to water

Substance	*T. niger*	*O. sabulosus*	*P. polydesmoides*	*P. angustus*	*G. marginata*
D(+)Glucose	0·0970	0·0555	0·1645	0·1460	0·1645
D(−)Fructose	0·0925	0·0535	0·0843	0·0936	0·075
D(+)Galactose	Not significant	Not significant	Not significant	Not significant	Aversive
D(+)Mannose	Aversive	Not significant	Not significant	Not significant	Not significant
L(−)Sorbose	0·25	0·39	0·675	—	0·265
D(+)Xylose	0·655	0·28	Not significant	—	Not significant
D(−)Arabinose	Not significant	Aversive	Aversive	—	Aversive
Sucrose	0·0135	0·013	0·076	0·0634	0·074
Maltose	0·00995	0·00595	0·02	0·029	0·0425
Lactose	Not significant	Not significant	Not significant	Not significant	Not significant
Mannitol	Aversive	Not significant	Not significant	Aversive	Not significant
D(+)Glucosamine hydrochloride	Not significant	Aversive	Aversive	Aversive	Not significant
Sodium chloride	Aversive 0·855	Aversive 0·0825	Aversive 0·072	Aversive 0·081	Not significant
Potassium nitrate	Aversive 0·18	Aversive 0·11	Aversive 0·0755	Aversive 0·0046	Aversive 0·088

Concentration in moles per litre.

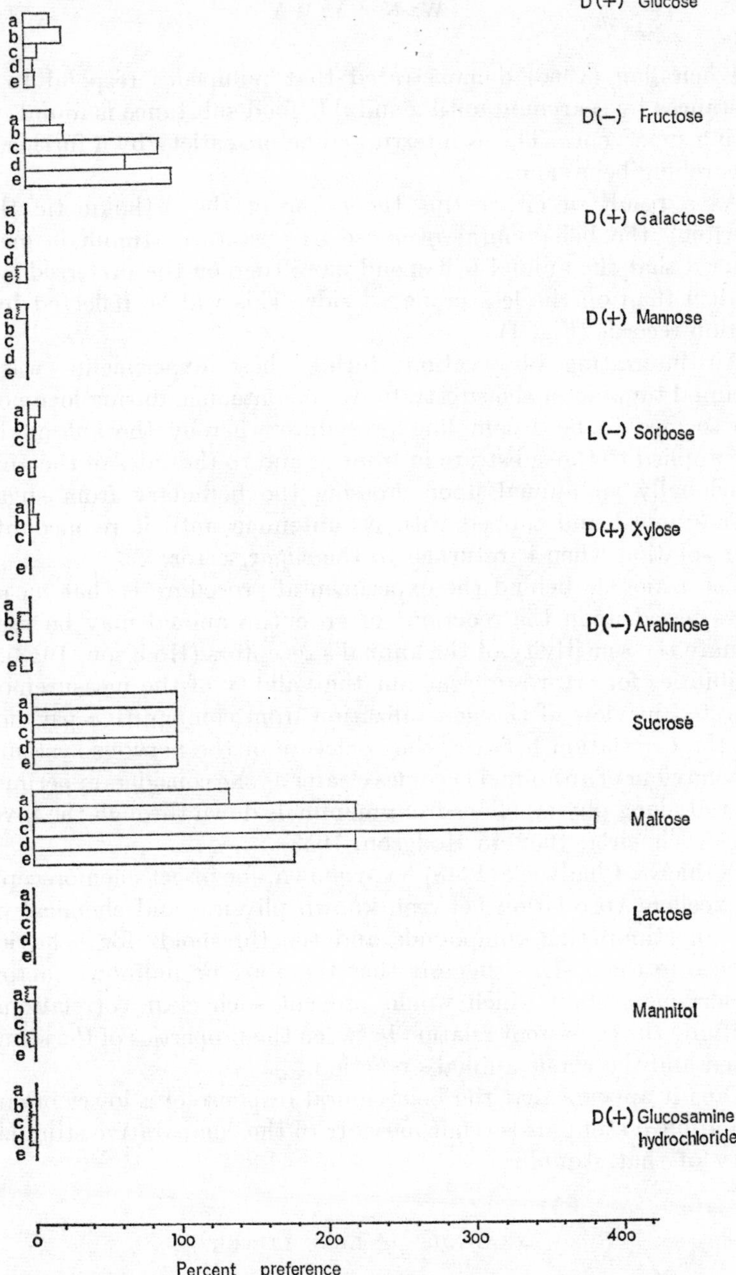

FIG. 1. Comparative behavioural preference to sugar. The histograms represent the preference of five species for a series of sugars compared with sucrose at 100%. The five species against each sugar are, from top to bottom, a–e: *Tachypodoiulus niger*, *Ommatoiulus sabulosus*, *Polymicrodon polydesmoides*, *Polydesmus angustus*, and *Glomeris marginata*.

Kheirallah (1966) demonstrated that millipedes respond to food substances by searching until a suitable food substance is found, when feeding supervenes. This is interrupted before satiety by a further bout of searching behaviour.

As a result of either this theory or of the orthokinetic theory describing the behavioural response to gustatory stimuli it may be deduced that the animal will spend more time on the preferred side of the dish than on the less preferred side. This will be reflected by the position records (Fig. 1).

An interesting observation during these experiments was the continual tapping of the substrate by the antennae during locomotion. This seemed to be a sampling procedure whereby the antennal tips were applied to the substrate in front of and to the sides of the animal. Occasionally an animal upon crossing the boundary from sugar to water stopped and tapped with its antennae until it re-encountered sugar solution when it returned to the sugar sector.

The rationale behind the experimental procedure is that measurements based upon the reactions of an entire animal may be used to estimate the sensitivity of the animal's receptors (Hodgson, 1953). The possibilities for error are clear but the validity of the measurement is suggested in view of the generalization from comparative psychology that the correlation between the structure of the nervous system and the behaviour of an animal becomes clearer as one considers experimental material along phyletic lines from mammals down through the invertebrates (Schneirla, 1952, in Hodgson, 1953).

Dethier & Chadwick (1948) have shown, for insect chemoreception, the excellent correlation between known physical and chemical properties of stimulating compounds and the thresholds for behavioural response to them. This suggests that there are no unknown factors in the nervous system which would prevent such clear correlations by modifying the consistent relation between the properties of the stimulus applied and the whole animal's reaction.

Thus it appears that the behavioural response of a lower organism to an applied stimulus is a fair measure of the comparative stimulating ability of that stimulus.

ANALYSIS OF LEAF LITTER

In an attempt to determine the relevance of these behavioural responses to the normal life of the millipedes the sycamore (*Acer pseudoplatanus*) leaf litter which they inhabited was analysed using thin layer chromatography to identify the sugars present. Colorimetric methods were used to assess quantitatively the simple carbohydrate and the glucose present. Behavioural responses to fractions of the leaf extracts were also determined.

Methods of analysis of litter leaf extract

Extraction

The laminae of freshly collected litter leaves were blended in 80% ethanol and this extract was left overnight at 60°C to ensure dissolution of the carbohydrates. The insoluble material was then removed by vacuum filtration. This solvent extracts only low molecular weight carbohydrates up to, at most, pentasaccharides (H. E. Grant pers. comm.), thus it does not extract starch or hemicellulose.

This alcoholic solution was then evaporated to dryness under reduced pressure at less than 60°C and the residue was taken up in de-ionized water. This solution was then decolorized by shaking with diethyl ether followed by ethyl acetate to remove lipids, polyphenols, and tannins.

Total carbohydrate assay

Three ml of chilled freshly prepared orcinol reagent were thoroughly mixed with 1 ml of the litter extract. The mixture was then heated on a water bath at 100°C for 10 min to develop the orange/yellow colour reaction. The solution was then cooled to 5°C and the absorption of light at 500 mμ was determined. The concentration of carbohydrate present was found from a reference graph calibrated from solutions containing 0 to 25 μg/ml of glucose. The results were expressed in terms of the quantity of carbohydrate obtained from 100 g dry weight of leaf laminae (Fig. 2).

Glucose assay

The glucose content of the litter extract was assayed using Modder's (1965) method, although the trichloro acetic acid was omitted since no protein or enzymes were present. This avoids the possibility of hydrolysis of the relatively unstable disaccharides such as sucrose or maltose (Schmidt, 1955).

The glucose concentration was obtained from a reference graph calibrated for glucose standard solutions containing 5–25 mg/1 of glucose in deionized water. The concentration of glucose was expressed in g/100 g dry weight of leaf laminae (Fig. 2).

Thin-layer chromatography

The remaining extract was de-ionized and the eluent reduced in volume to about 20 ml at reduced pressure. Two ml of *n*-propanol was added to prevent fungal and bacterial growth allowing of storage for a few days at 0°C.

The method used was derived from Stahl (1965). Inactive Kieselgel G plates 50 microns in depth were used together with methyl ethyl

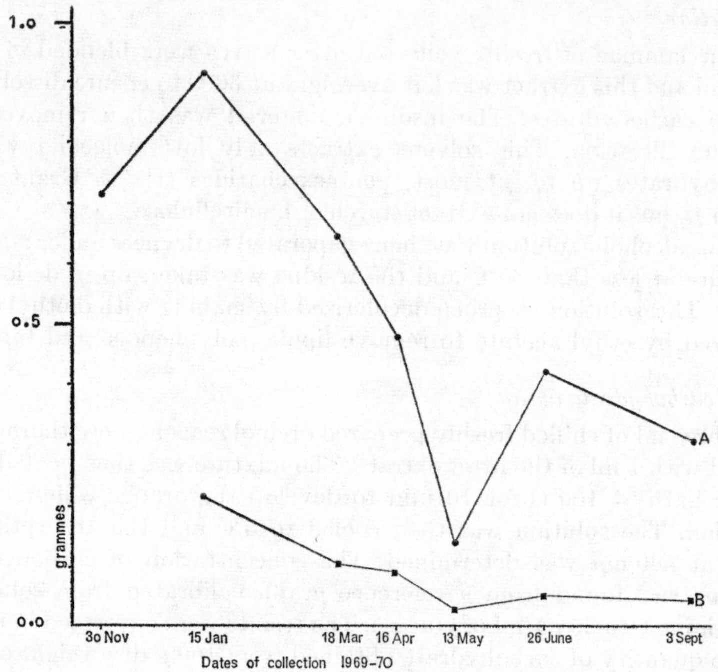

Fig. 2. Analysis of sycamore litter extract. Dry weight in g of (A) total carbohydrate and (B) glucose; both in 100 g dry weight of litter.

ketone, acetic acid (glacial) and methanol 6 : 2 : 2 as the solvent. The plates were developed with either aniline hydrogen phthallate or anisaldehyde. The Rf values were calculated for the test and reference spots and in this way the unknown sugars were identified (Table II).

Plates were also run at right angles with a second solvent, (two-dimensional chromatography), *n*-butanol, acetone, water 4 : 5 : 1 to produce a greater separation, especially of glucose and fructose, thus permitting a more accurate characterization of the sugars.

Behavioural response to fractions of extract

Since it was thought that the polyphenol and tannin content of litter leaves plays a part in determining their palatability the behavioural responses of millipedes to these substances were assessed.

TABLE II

Qualitative analysis of sycamore litter leaf extract

Date collected	Sugars present in extract					
30–11–1969	Glucose	Xylose	Rhamnose			
15–1–1970	Glucose	Xylose	Rhamnose	Raffinose		
18–3–1970	Glucose	Xylose		Raffinose	Trehalose	Arabinose
16–4–1970	Glucose	Xylose	Rhamnose	Raffinose	Trehalose	Galactose
13–5–1970	Glucose	Xylose	Rhamnose	Raffinose	Trehalose	Arabinose
26–6–1970	Glucose	Xylose	Rhamnose		Trehalose	
3–9–1970	Glucose	Xylose	Rhamnose	Raffinose	Trehalose	

Three fractions were obtained from the litter leaf extract.

1. An ethereal fraction containing lipids and certain polyphenols.
2. An ethyl acetate fraction containing other polyphenols and tannins.
3. The remaining alcoholic solution containing the carbohydrates.

These were applied to filter paper and all traces of organic solvent were removed. The papers were then moistened with de-ionized water and were tested for preference against de-ionized water as in the previous behavioural experiments. The alcoholic fraction was also compared against the ether fraction and against the ethyl acetate fraction (Table III).

DISCUSSION OF RESULTS

There is a general preference for glucose, fructose, sucrose, maltose, and sorbose compared to water while *T. niger* and *O. sabulosus* also show a preference for xylose. Galactose, mannose, D(-) arabinose, lactose, mannitol, and D(+) glucosamine hydrochloride were either avoided or gave no differential response. The results, showing avoidance of sodium chloride and potassium nitrate, indicate that the responses to sugars are not due solely to osmotic conditions (Table I).

Thus those sugars found by Hassett, Dethier & Fans (1950), Nielsen (1962) and Wyatt (1967) to be the ones most likely to be nutritive and utilizable were also the ones to which these millipedes showed the greatest preference. A correlation seems to exist between the sugars found in the leaves and those most preferred. Haslinger (1935) suggests that substances without nutritive value are tasteless while those with a nutritive value may or may not be tasted. It therefore seems likely that the palatability of sugars may be an index to their nutritive value and since sugars commonly induce feeding this may be an index to the nutritive value of the litter leaf.

There is a strong correlation between the sugars found in the leaf litter extract and those found by Roelofson (1959) to be the main components of the plant cell wall (the major part of the litter leaf). He found the following carbohydrates:

1. Cellulose.
2. Lignin.
3. Hemicelluloses containing in complexes and polyoses; *d*-mannose, *d*-glucose, *d*-galactose, *d*-xylose, *d*-arabinose, and *l*-rhamnose.

These are similar to the ones which Mehta, Duboch & Deuel (1961) found in the soil. Mannose, however, was absent in the present work

TABLE III

Behavioural response to fractions of sycamore litter leaf extract

	Control water/water	Ethereal/ water	Ethyl acetate/ water	Alcoholic/ water	Alcoholic/ ethereal	Alcoholic/ ethyl acetate
T. niger	88/92	97/83	89/91	96/84	83/97	119*/61
O. sabulosus	95/85	92/88	92/88	94/86	108*/72	87/93
P. polydesmoides	86/94	90/90	83/97	93/87	90/90	84/96
P. angustus	94/86	97/83	96/84	96/84	104*/76	120*/60
G. marginata	88/92	91/89	86/94	93/87	108*/72	88/92

Ethereal fraction contains phenols.
Ethyl acetate fraction contains phenols and tannins.
Alcoholic fraction contains sugars.

* = significant, $p < 0{\cdot}05$.
Expected ratio for chi-square = 90/90.
5 % range = 103/77.

from the extract and the trehalose found was probably derived from fungi (Nielsen, 1962).

Nielsen (1962) found that these complexes and polyoses of the plant cell wall are much more commonly hydrolysed by bacteria, fungi, and the lower herbivorous animals than are cellulose or chitin. He found that the enzymes required for the primary decomposition of litter were found mainly in the soil protozoa, snails, slugs, and in *Polydesmus*. However *Glomeris* which lacked the enzymes for primary decomposition was able to utilize the products of this decomposition, unlike *Polydesmus*.

The results from the behavioural tests with the three litter fractions are difficult to explain since none of the fractions gives a significant response when compared to de-ionized water, although there is a consistent but slight preference for the alcoholic (sugar containing) fraction (Table III). However, when the alcoholic fraction is compared to the diethyl ether fraction (containing lipids and polyphenol) *O. sabulosus*, *P. angustus* and *G. marginata* show a significant preference for the alcoholic fraction. In the case of a comparison between the alcoholic fraction and the ethyl acetate fraction (polyphenol and tannin), *T. niger* and *P. angustus* show a marked preference for the alcoholic fraction.

Thus there would seem to be some support for the view that polyphenols and tannins have an opposite effect on palatability to carbohydrates. Leaves which have a high content of polyphenols and tannin, and yet have not been subjected to the microbial action that increases the level of available carbohydrate, will be less palatable.

A preference for the alcoholic fraction over the ethereal fraction and over the ethyl-acetate fraction is not shown by *Polymicrodon polydesmoides*. However, this is a species which is found in large numbers in autumn, winter and spring when the sycamore leaves would be relatively undecomposed. It is therefore not surprising that phenol and tannin have less effect on it than on the other species.

As a result of the two factors of the primary decomposition of litter by the microflora and the need for common specific stimuli it is possible to advance an explanation for the variation of the palatability of litter leaves with age. Thus Kheirallah (1966) found that *Iulus scandinavius* rejected fresh sycamore leaves but that these became palatable after one month. The palatability then decreased over the next four months and then increased again until they were 14 months old.

This may be a result of the activity of the soil microbiota since Saudray (1961) postulated that the nutritive value of the litter depended upon the extent of its alteration by fungi and bacteria.

However a simple increase in the amount of available food substance is insufficient to explain Kheirallah's findings since they do not correlate with the carbohydrate and glucose content from my analysis. The increased carbohydrate in June could be due to the fall of green plant material and of insect frass (Bocock, 1964), especially since the sycamore trees later developed a very high aphid population, but it could also be the result of increased temperature affecting microbial activity.

Although Thiele (1959) was unable to make litter leaves less palatable to Diplopoda by soaking them in tannin, Satchell & Lowe (1967) found a strong relationship between the polyphenol content of leaves and their palatability, for *Lumbricus terrestris*. They also found a rapid loss of polyphenol in the first six weeks of weathering which they explained as a result of microbial action. This loss of polyphenol could explain the increase of palatability of fallen sycamore leaves over the first month in Kheirallah's work.

CONCLUSION

It is probable that the palatability of the sycamore leaves is dependent upon the microfloral activity since Schmidt (1952) finds that food selection is unquestionably determined by the degree of decomposition of the food material. The microflora may act either by decreasing the polyphenol content, by increasing the amounts of nutritive substances or by both of these processes. The palatability should then correlate with environmental conditions since these affect the microbial activity.

References

Barlow, C. A. (1957). A factorial analysis of distribution in three species of diplopods. *Tijdschr. Ent.* **100**: 349–426.

Bocock, K. L. (1964). Changes in the amounts of dry matter, nitrogen, carbon, and energy in the decomposing woodland leaf litter in relation to the activities of the soil fauna. *J. Ecol.* **52**: 273–284.

Brade-Birks, S. G. (1930). Notes on Myriapoda (XXXIII): The economic status of Diplopoda and Chilopoda and their allies. Part II. *Jl S.-east. agric. Coll. Wye* **27**: 103–146.

Brookes, C. (1963). *Some aspects of the life histories and ecology of* ***Proteroiulus fuscus*** *and* ***Isobates varicornis*** *with information on other blaniulid millipedes.* Thesis. University of Manchester.

Cloudsley-Thompson, J. L. (1951). On the responses to environmental stimuli and the sensory physiology of a millipede. *Proc. zool. Soc. Lond.* **121**: 253–277.

Crozier, W. J. & Moore, A. R. (1923). Homostrophic reflex and stereotropism in Diplopoda. *J. gen. physiol.* **5**: 597–607.

Dadd, R. H. (1963). Feeding behaviour and nutrition in grasshoppers and locusts. *Adv. Insect Physiol.* **1**: 47–97.

Dethier, V. G. (1937). Gustation and olfaction in lepidopterous larvae. *Biol. Bull. mar. biol. Labs Woods Hole* **72**: 7–23.

Dethier, V. G. & Chadwick, L. E. (1948). The stimulating effect of glycols and their polymers on the tarsal receptors of blowflies. *J. gen. physiol.* **32**: 139–151.

Dethier, V. G. & Wolbarsht, M. L. (1956). The electron microscopy of sensory hairs. *Experientia* **12**: 335–337.

Drift, J. van der (1965). The effects of animal activity in the litter layer. In *Experimental pedology:* 227–235. Hallsworth, E. G. & Crawford, D. V. (ed). London: Butterworths.

Dünger, W. (1958). Über die Zersetzung der Laubstreu durch die Bodenmakrofauna in Auenwald. *Zool. Jb.* (Syst.) **86**: 139–180.

Fairhurst, C. P. (1968). *Life cycles and activity patterns of schizophylline millipedes.* Thesis, University of Manchester.

Gere, G. (1956). The examination of the feeding biology and humificative function of Diplopoda and Isopoda. *Acta biol. hung.* **6**: 257–271.

Graham, W. M. & Waterhouse, F. L. (1964). The distribution and orientation of *Tribolium* on inclines and the concept of controls on gradient experiments. *Anim. Behav.* **12**: 368–373.

Handley, W. R. C. (1954). Mull and mor formation in relation to forest soils. *Bull. For. Commn, Lond.* No. 23.

Haslinger, F. (1935). Über den Geschmacksinn von *Calliphora erythrocephala* Meigen und über die verwertung von Zuchern und Zuckeralkoholen durch diese Fliege. *Z. vergl. Physiol.* **22**: 614.

Hassett, C. C., Dethier, V. G. & Fans, J. (1950). A comparison of nutritive values and taste thresholds of carbohydrates for the blowfly. *Biol. Bull. mar. biol. Labs Woods Hole* **99**: 446–453.

Hodgson, E. S. (1953). Chemoreception in aqueous and gas phases. *Biol. Bull. mar. biol. Labs Woods Hole* **105**: 115–127.

Hodgson, E. S. (1957). Electrophysiological studies of arthropod chemoreception. II. Responses of labellar chemoreceptors of the blowfly to stimulation by carbohydrates. *J. ins. Physiol.* **1**: 240–247.

Kheirallah, K. M. (1966). *Studies on the feeding behaviour of the millipede* ***Iulus scandinavius*** (*Latzel, 1884*). Thesis, University of Manchester.

Lyford, W. H. (1943). The palatability of freshly fallen forest leaves to millipedes. *Ecology* **24**: 252–284.

Mcindoo, N. E. (1934). Chemoreceptors of blowflies. *J. Morph.* **56**: 445–475.

Mehta, N. C., Duboch, P. & Deuel, H. (1961). Carbohydrates in the soil. *Adv. Carbohyd. Chem.* **16**: 335–355.

Minderman, G. & Daniels, L. (1967). Colonisation of newly fallen leaves by microorganisms. In *Progress in soil biology:* 3–9. Graff, O. and Satchell, J. E. (eds.) Amsterdam: North-Holland publishing Co.

Minnich, D. E. (1921). An experimental study of the tarsal chemoreceptors of two nymphalid butterflies. *J. exp. Zool.* **33** : 173–203.

Minnich, D. E. (1926). The chemical sensitivity of the tarsi of certain muscid flies. Biol. Bull. *mar. biol. Labs Woods Hole* **51**: 166–178.

Modder, W. W. D. (1965). *Quantitative aspects of carbohydrate digestion and utilisation by* ***Blaberus craniifer*** *Burmeister*. Thesis, University of Manchester.

Murphy, P. W. (1953). The biology of forest soils: with special reference to the mesofauna or meiofauna. *J. Soil Sci.* **4**: 155–193.

Nielsen, C. Overgaard (1962). Carbohydrases in soil and litter invertebrates. *Oikos* **13**: 200–215.

Roelofson, P. A. (1959). The plant cell wall. In *Encyclopedia of plant anatomy.* Linsbauer, K. (ed.). Berlin: Gebrüder Borntraeger.

Satchell, J. E. & Lowe, D. G. (1967). Selection of leaf litter by *Lumbricus terrestris.* In *Progress in soil biology:* 102–119. Graff, O & Satchell, J. E. (eds). Amsterdam: North Holland Publishing Co.

Saudray, Y. (1961). Recherches biologiques et physiologiques sur les myriapodes diplopodes. *Mém. soc. Linn. Normandie* (N. S. sect. Zool.) **2** (mem. no. 1): 3–126.

Schmidt, H. (1952). Nahrungswahl und Nahrungsverarbeitung bei Diplopoden (Tausendfüsslern). *Mitt. naturw. Ver. Steierm.* **81/82**: 42–66.

Schmidt, J. (1955). *A textbook of organic chemistry.* (7th Edition edited by Campbell, N.). London: Oliver and Boyd.

Stahl, E. (ed.) (1965). *Thin layer chromatography.* New York and London: Academic Press.

Striganova, B. R. (1967). Über die Zersetzung von Überwinterer Laubstreu durch Tausendfüssler und Landasseln. *Pedobiologie* **7**: 125–134.

Thiele, H.-U. (1959). Experimentelle untersuchungen über die Abhängigkeit boden-bewohnender Tierarten von Kalkgehalt des Standorts. *Z. angew. Ent.* **44**: 1–21.

Thorpe, W. H., Crombie, A. C., Hill, R. & Darrah, J. H. (1947). The behaviour of wireworms in response to chemical stimulation. *J. exp. Biol.* **23**: 234–266.

Wigglesworth, V. B. (1941). The sensory physiology of the human louse, *Pediculus corporis* de Geer. *Parasitology* **33**: 67.

Wittich, W. (1943). Untersuchungen über den verlauf der streuzersetzung auf einem Boden mit Mullzutstand. *Forstarchiv* **19**: 1–18.

Wittich, W. (1953). Der heutige stand unseres wissens von Humus und neue Wege zür Losüng des Rohhumus problems im Walde. *SchrReihe forstl. Fak. Univ. Göttingen* **9** Bd 4.

Wyatt, G. R. (1967). The biochemistry of sugars and polysaccharides in insects. *Adv. Insect physiol.* **4**: 287–360.

Discussion

Sundara Rajulu: You have suggested that the sugars present in the leaves are the phagostimulants, but Lyford (1943)* reported that calcium salts were the phagostimulants.

Sakwa: I have mentioned that the rate of decomposition of leaves depends on a low C : N ratio, which according to Murphy (1953)* is inversely related to the calcium content. It is probably more meaningful to refer the higher palatability to this low C : N ratio rather than to the calcium content.

Gabbutt: Does degradation result in an increased carbohydrate content?

Sakwa: Yes, according to Saudray, 1961*.

Gabbutt: But your graph showed a decrease later on, didn't it?

* See list of references, p. 344.

SAKWA: In Fig. 2 the first point on the graph (November) may represent the original quantity of carbohydrate present. The level drops overwinter, perhaps because of a lower microbial activity, and it then rises again during May, perhaps because of higher temperatures.

GABBUTT: The fall in the curve would then represent a loss of the originally present carbohydrate before degradation gets under way.

SAKWA: I would think so, but since I didn't analyse the leaves immediately after leaf fall—some degradation may have occurred before the first point on Fig. 2—the initial carbohydrate content may conceivably have been even lower than this.

GABBUTT: But that would suggest that the degradative processes produce only half as much carbohydrate again as was present at leaf fall.

SAKWA: Since I do not know precisely how much was originally present I cannot say.

GABBUTT: But 30th November is not long after leaf fall.

SAKWA: Yes, but another complicating factor is that the microflora present is utilizing, as well as producing, carbohydrate, and the fauna will be consuming some also. In any case my figures refer only to the simpler oligosaccharides present and not to the total carbohydrate content of the leaves; this is an important distinction.

MEIDELL: Have you analysed the faecal pellets to see which parts of the intake are utilized?

SAKWA: No—but I refer you to the paper by Bocock (1964)* which I have quoted.

* See list of references, p. 343.

Symp. zool. Soc. Lond. (1974) No. 32, 347–364.

A COMPARATIVE STUDY OF THE ORGANIC COMPONENTS OF THE HAEMOLYMPH OF A MILLIPEDE *CINGALOBOLUS BUGNIONI* AND A CENTIPEDE *SCUTIGERA LONGICORNIS* (MYRIAPODA)

G. SUNDARA RAJULU

*Department of Zoology, Madras University, Madras-5, India**

SYNOPSIS

Quantitative and qualitative analyses of the protein, carbohydrate and lipid components of the haemolymph of a millipede *Cingalobolus bugnioni* and a centipede *Scutigera longicornis* have been conducted. The average values for the haemolymph of the millipede were: protein 197·5 mg/100 ml, carbohydrate 86·4 mg/100 ml and lipid 1553 μg/ml; and for the centipede: protein 175·5 mg/100 ml, carbohydrate 82·9 mg/100 ml and lipid 1616 μg/ml.

Sexual dimorphism is seen in the number of protein components in the haemolymph of the millipede; the males have six protein components while females have seven components; the additional component resembling serum albumin. However, male and female centipedes only have six protein components. A fraction similar to fibrinogen, responsible for the coagulation of haemolymph, is seen in the millipede. Trehalose is the principal sugar and phospholipids are the main lipids in the haemolymph of both the myriapods.

The concentrations of protein, carbohydrate and lipid are directly proportional to size in both the myriapods and females always have higher values than males.

The results obtained were discussed with reference to the phylogenetic relationship of the Myriapoda with other groups of arthropods.

INTRODUCTION

Carpenter (1903), Borner (1909) and Crompton (1928) reported that Myriapoda and Insecta resemble Crustacea in the possession of mandibles, structure of the head-capsule and in the retention of segmental organs in the head. The compound eye of Crustacea is similar to those in insects and the more primitive Crustacea possess an elongated heart of the myriapod—insect type. On such evidence the Crustacea are conceived to be very close relatives to Myriapoda and Insecta. But Tiegs & Manton (1958) and Manton (1964) on the basis of an impressive array of embryological and morphological evidence have suggested that Onychophora, Myriapoda and Insecta may be phylogenetically closely related to one another and that they might have evolved independently of Crustacea and Arachnida.

* Present address: *Department of Zoology, University Extension Centre, P.S.G. College of Technology Campus, Coimbatore—641004, India.*

In recent years much emphasis has been placed on the biochemical and physiological characteristics, especially of the blood, as an index of inter-relationship among animal groups (Florkin, 1949, 1960, 1971). Therefore it may be of interest to know what light physiology and biochemistry of haemolymph can throw on the relationship of different groups of arthropods.

Sutcliffe (1963) finds that in a millipede *Julus scandinavius*, sodium accounts for only 29% and chloride 25% of the blood osmolar concentration and in this respect the blood of the myriapod is quite different from that of Crustacea and Arachnida. The total concentrations of free amino acids in the haemolymph are similar in the Onychophora, Myriapoda and Insecta but quite different from those of Crustacea and Arachnida (Sundara Rajulu, 1970; Sundara Rajulu & Ramanujam, 1972; Sundara Rajulu & Santhanakrishnan, 1972; Sundara Rajulu & Kulasekarapandian, 1972). However, it is not known how far the other organic components of the blood such as protein, carbohydrate and lipid are comparable in these groups. While information on these components in Insecta, Crustacea and Arachnida is available (Florkin, 1936; George & Nichols, 1948; Dall, 1964; Connor & Gilbert, 1968; Naidu, 1966) little is known about them in Onychophora and Myriapoda. The present investigation has been undertaken to throw some light on these components in Myriapoda.

It has been shown recently that size and sex affect the composition of the blood of crustaceans and arachnids (Gilbert, 1959; Naidu, 1966; Sundara Rajulu & Santhanakrishnan, 1972). It would be of interest to extend such a study to myriapods.

MATERIALS AND METHODS

The species investigated were the millipede *Cingalobolus bugnioni* Carl and the centipede *Scutigera longicornis*. These myriapods were collected from the Alagar Koil forest region. After collection they were brought to the laboratory and maintained as detailed elsewhere (Sundara Rajulu, 1966). Only those animals which did not show signs of moulting (Sundara Rajulu, 1969a) were used. The animals were not fed one day prior to the commencement of the experiment, to eliminate the variation in the blood constituents due to differential feeding.

Collection of haemolymph

The haemolymph was collected as described by Sundara Rajulu (1970) and pooled. The pooled haemolymph was centrifuged at 5°C for 15 min at 2000 rev/min to separate haemocytes. The supernatant was collected and used for further studies. As the haemolymph of *Cingalobolus bugnioni* was found to clot, 0·5% sodium citrate was used as anticoagulant.

Quantitative estimation of protein

The protein concentration of the haemolymph was estimated by the Biuret method of Gornall, Bardaivell & David (1949). The calibration curve obtained with bovine serum albumin was used as the standard.

Electrophoretic analyses of protein

The protein fractions of the haemolymph were analysed by paper electrophoresis using Laboi type-E electrophoretic equipment. About 15 to 20 μl of the haemolymph or 30 to 40 μl of protein extract was used for every run. The electropherograms were run on Whatman No. 1 paper strips (2 × 45 cm) under the following conditions: Veronal acetate buffer of pH 8·6 and ionic strength 0·1; constant voltage 15 V/cm; duration of run 150 min and cooling temperature for the apparatus 10–12°C. Amidoblack was used for visualizing protein bands. The curves for the pherograms were obtained with Carl-Zeiss extinction recording apparatus model ERI-10.

Quantitative estimation of amino acids of proteins

Aliquots containing 1 to 2 mg of protein were used for quantitative estimation of amino acids by means of an automatic analyser similar to one described by Spackman, Stein & Moore (1958). Quantitation was based on the ninhydrin colouring intensity of the effluent from ion-exchange columns. The absorbency of the colour developed was estimated at 570 μm and 440 μm in a spectrophotometer. The peaks on the recorded curve were integrated for loads varying from 0·1 to 3·0 μm for each amino acid.

Study of carbohydrates

To estimate free sugar value, 0·1 ml of haemolymph was taken and deproteinized by the method of Stein & Moore (1954). The sugar in the deproteinized haemolymph was estimated by the method outlined by Parvathy (1970) using an Eel photoelectric colorimeter.

Protein-bound sugar was estimated as follows: The protein was precipitated by adding 3 volumes of absolute alcohol to one volume of haemolymph. The protein precipitate was hydrolysed in N HCl for 15 h to liberate bound sugar (Dall, 1964) and the liberated bound sugar was estimated by the anthrone method described by Parvathy (1970).

For qualitative analysis of the sugars, about 10 ml of haemolymph was hydrolysed with 3 vol. of 6 N HCl for 6 h at 100°C and analysed by paper chromatography (Block, 1952) using butanol-acetic acid-water (by vol. 4 : 1 : 5) as irrigating solvent. The chromatograms were developed as given elsewhere (Sundara Rajulu & Krishnan, 1967). The sugars in the chromatograms were identified by their Rf values and by comparison with standards.

Study of lipids

After a puncture in the cuticle, the haemolymph was collected in a centrifuge tube held in ice and containing 1–2 μm reduced glutathione per ml haemolymph. The mixture was then centrifuged at 3000 rev/min for 15 min and the supernatant used for lipid analyses.

Five ml of the supernatant were extracted with 50 ml of a mixture of isopropanol-hepatane – 1 M sulphuric acid (by vol. 40 : 10 : 1) as suggested by Dole (1956). For quantitative estimation, the extract was then evaporated to dryness *in vacuo* and the residue was dissolved in 10 ml of scintillation fluid (5·5 g Packard M in 1 litre of toluene). A scintillation spectrophotometer (Model 3375) was used for quantitative estimations.

For fractionation, the extracted lipid was analysed by thin-layer chromatography, using a solvent system of hexane-diethyl ether-acetic acid (by vol. 80 : 20 : 2) (Beenakkers & Gilbert, 1968). These plates were sprayed with Rhodamine 6 G and viewed under ultra-violet light.

Appropriate regions of thin plates were removed and extracted with chloroform-methanol-water (by vol. 13 : 8 : 1) and the extract was evaporated to dryness *in vacuo*. The dry lipid fractions were dissolved in appropriate quantities of scintillation fluid and their quantities estimated by a scintillation spectrophotometer as described earlier.

RESULTS

Protein

Figure 1 presents results of quantitative estimations of protein in the haemolymph of various sizes and both sexes of *Scutigera longicornis*. Females have a higher quantity of protein than males; average values are 189 and 162 mg per 100 ml respectively. It is interesting to note that a large male weighing 3·8 g has only 168 mg of protein while a small female weighing only 1·5 g has 172 mg protein per 100 ml. Among both males and females, size seems to have a profound effect on haemolymph protein concentration; the relationship being directly proportional.

Figure 1 also shows that the concentration of protein in males of the millipede *Cingalobolus bugnioni* is remarkably similar to that in males of *S. longicornis*. But female millipedes have a higher concentration than female centipedes; the average value for the former is 233 mg/100 ml while for the later it is only 189 mg/100 ml.

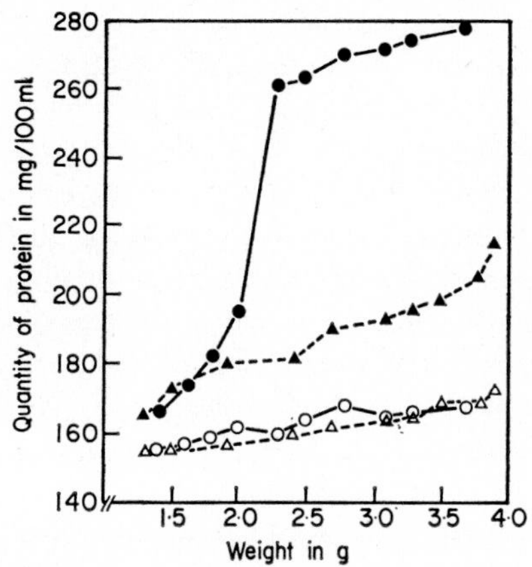

Fig. 1. Influence of size and sex on the concentration of protein in the haemolymph of *Scutigera longicornis* and *Cingalobolus bugnioni*. Closed and open circles stand for females and males, respectively, of *C. bugnioni;* closed and open triangles stand, respectively, for females and males of *S. longicornis*. Abscissa: Weight of the animals in g; Ordinate: Protein in mg per 100 ml of haemolymph.

Among millipedes size and protein concentration are directly proportional in both the sexes. In males the increase in protein concentration is gradual, but in females, there is a point at which the amount of protein rises steeply; from 195 mg/100 ml in individuals weighing 2·0 g, to 261 mg/100 ml in individuals of the next size group, 2·3 g (Fig. 1).

Electrophoretic analyses of the protein component from the haemolymph of *S. longicornis* of both the sexes and different sizes show that in all the instances it is of the same nature. Figure 2 shows a typical electrophoretic pattern of the haemolymph protein of *S. longicornis* compared to that of human plasma. The protein of the haemolymph resolves into six components. The protein bands have been numbered 1 to 6 starting from the one with the fastest anodic migration, and they will be referred to as proteins 1, 2, 3, . . . , etc.

Proteins 2, 3 and 4 show anodic mobility similar to serum albumin, a_1 and a_2 globulins, respectively. Protein 5 is cathodic and corresponds with γ-globulin. Proteins 1 and 6 do not have any counterparts in human plasma and there are no proteins in the haemolymph of the centipede corresponding to B_1 and B_2 globulins and fibrinogen.

An interesting feature of protein 1 is that it is faintly indigo-blue in colour even in unstained electropherograms, and it is decolourized when exposed to the vapours of ammonium sulphide, which suggests that it may be haemocyanin (Wilson, 1901; Sundara Rajulu, 1969b). This lends support to an earlier observation that the blood of *S. longicornis* may contain haemocyanin (Sundara Rajulu, 1969b). In order to verify this further, the protein 1 was extracted from 25 strips of electropherograms and hydrolysed, and the amino acids estimated quantitatively. The results are recorded in Table I; the data from Stewart, Dingle & Odense (1966) for the haemocyanin of a crustacean *Homarus americanus* have been included for comparison. It may be seen that the protein in question has only 17 of the amino acids found in the haemocyanin of the crustacean and the quantities of the constituent amino acids also differ.

TABLE I

Amino acid composition of haemocyanin from ***Scutigera longicornis****, compared to that of crustacean haemocyanin, in mole %*

No.	Amino acids	Haemocyanin from *S. longicornis*	Haemocyanin from Crustacea*
1	Alanine	7·12	6·22
2	Arginine	3·63	4·95
3	Aspartic acid	10·48	13·64
4	Cystine	2·46	0·48
5	Glutamic acid	12·32	10·75
6	Glycine	7·12	6·03
7	Histidine	5·38	7·69
8	Isoleucine	3·61	4·82
9	Leucine	5·05	7·48
10	Lysine	3·11	5·09
11	Methionine	4·23	2·23
12	Phenyl alanine	5·01	5·75
13	Proline	5·76	4·61
14	Serine	—	4·17
15	Threonine	7·02	5·35
16	Tryptophane	1·12	0·06
17	Tyrosine	5·12	3·86
18	Valine	5·44	6·78

* From Stewart *et al.* (1966).

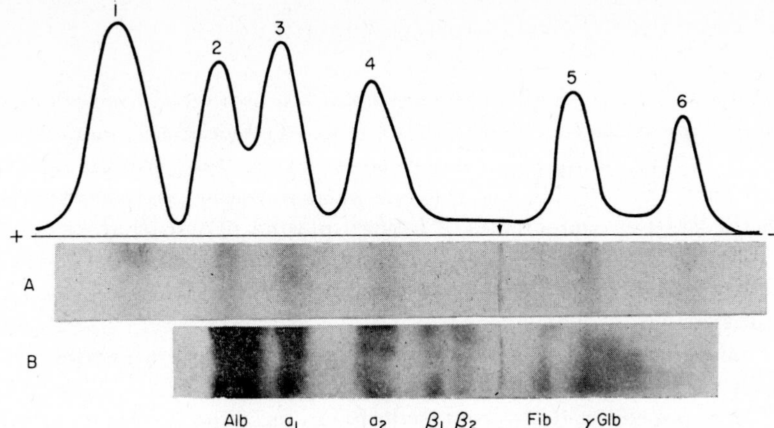

Fig. 2

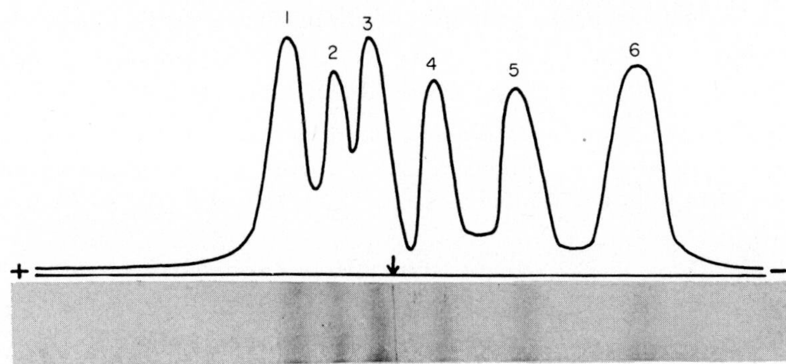

Fig 3

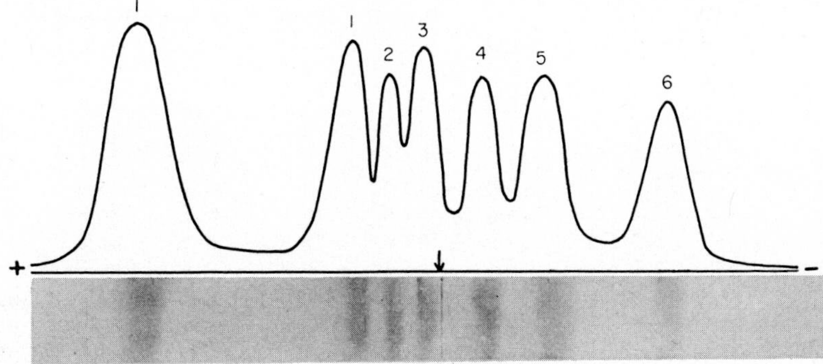

Fig. 4

FIG. 2. Electrophoretic pattern of the haemolymph proteins of *Scutigera longicornis* (A), compared with that of human plasma (B).

FIG. 3. Electrophoretic pattern of the haemolymph proteins from males of *Cingalobolus bugnioni*.

FIG. 4. Electrophoretic pattern of the haemolymph proteins from the females of *Cingalobolus bugnioni*.

While sex does not affect the nature of the protein component of the haemolymph in the centipede, it has a marked effect on the haemolymph protein in the millipede. Males have six protein components, of which three are anodal and the rest cathodal (Fig. 3). A comparison with the protein components of human plasma reveals that proteins 1 and 2 correspond to B_1 and B_2 globulins and proteins 4 and 5 to fibrinogen and γ-globulin, respectively. Proteins 3 and 6 have no counterparts in human plasma.

Females have seven protein components. The extra component is anodal and it corresponds to serum albumin (Fig. 4).

The millipede and the centipede have dissimilar protein patterns; the most important differences being the presence of a protein corresponding to fibrinogen in the millipede, but not in the centipede, and a haemocyanin-like protein fraction in the centipede but not in the millipede.

Carbohydrate

The total amounts of free and bound sugar in the haemolymph of *Scutigera longicornis* is given in Table II and Fig. 5. Females have a higher titre of sugar than males, and in both sexes, the concentration

TABLE II

Results of quantitative estimations of the carbohydrate content of the haemolymph of ***Scutigera longicornis****, in mg glucose/100 ml*

		Male			Female		
No.	Weight of the animal in g	Free sugar	Bound sugar	Total sugar	Free sugar	Bound sugar	Total sugar
1	1·2	44·3	33·9	78·2	45·1	34·1	79·2
2	1·4	44·9	33·4	78·3	46·0	35·4	81·4
3	1·7	44·7	33·8	78·5	46·4	36·1	82·5
4	1·9	45·2	33·3	78·5	47·6	36·3	83·9
5	2·4	45·5	33·6	79·1	48·8	36·6	85·4
6	2·7	45·9	33·4	79·3	50·2	36·9	87·1
7	3·1	46·4	33·4	79·8	51·6	38·0	89·6
8	3·4	47·1	33·1	80·2	52·1	37·5	89·6
9	3·6	48·6	32·1	80·7	54·7	38·0	92·7
10	3·9	47·8	33·3	81·1	55·5	37·8	93·3
	Average	46·0	33·3	79·3	49·8	36·7	86·5
	Average for males and females—82·9						

TABLE III

Results of quantitative estimations of the carbohydrate content of the haemolymph of ***Cingalobolus bugnioni****, in mg glucose/100 ml*

No.	Weight of the animal in g	Male			Female		
		Free sugar	Bound sugar	Total sugar	Free sugar	Bound sugar	Total sugar
1	1·6	46·2	35·1	81·3	49·3	33·8	83·1
2	1·7	48·4	33·2	81·6	51·5	34·0	85·5
3	1·9	47·5	34·5	82·0	52·9	34·8	87·7
4	2·2	46·3	33·9	80·2	55·1	33·5	88·6
5	2·5	47·6	33·5	81·1	57·2	33·2	90·4
6	2·6	47·1	33·5	80·6	58·8	34·1	92·9
7	2·8	49·2	32·5	81·7	60·3	33·0	93·3
8	3·0	48·8	32·5	81·7	63·4	31·8	95·2
9	3·3	49·7	32·6	82·3	64·0	32·0	96·8
10	3·5	48·9	33·7	82·6	66·5	32·9	99·4
	Average	48·0	33·5	81·5	57·9	33·3	91·3

Average for males and females—86·4

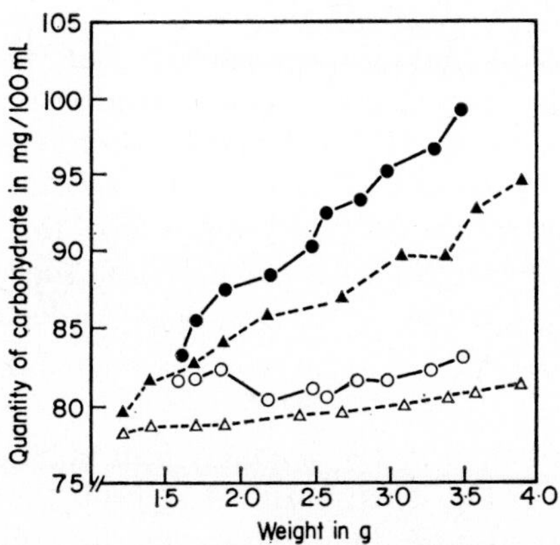

FIG. 5. Influence of size and sex on the concentration of carbohydrate in the haemolymph of *Scutigera longicornis* and *Cingalobolus bugnioni*. Closed and open circles stand for females and males, respectively, of *C. bugnioni;* closed and open triangles stand, respectively, for females and males of *S. longicornis*. Abscissa: Weight of the animals in g; Ordinate: Carbohydrate in mg per 100 ml of haemolymph.

of free sugar is higher than that of bound sugar. A similar relation is seen in the sugar values of the haemolymph of *Cingalobolus bugnioni* (Table III and Fig. 5).

TABLE IV

Sugars in the haemolymph of ***Cingalobolus bugnioni*** *and* ***Scutigera longicornis***

Sugars	*C. bugnioni*	*S. longicornis*
A. Free sugars		
Mannose	+	+
Glucose	+	+
Galactose	+	+
Fucose	+	—
Trehalose	+ +	+ +
B. Bound sugars		
Mannose	+ +	+ +
Glucose	+	+
Fucose	+	—
Trehalose	+ +	+ +

Key: – negative reaction, + positive reaction, + + intensely positive reaction.

Chromatographic analysis of the haemolymph sugars of different sizes and both sexes of the millipede reveals that the relation between the free and bound sugars remains similar, and the results recorded in Table IV are applicable to all the individuals of the species. Mannose, glucose, galactose, fucose and trehalose are present as free sugars, and, except galactose, are found as bound sugars also. But for the absence of fucose, the centipede has the same haemolymph sugars as the millipede (Table IV). Trehalose is the commonest sugar in both (Table IV).

Lipid

The quantities of lipid in the haemolymph of *S. longicornis* and *C. bugnioni* are shown in Fig. 6. In both the species, the size and the lipid content are directly proportional. However, in both the myriapods, sex has a profound effect on the haemolymph lipid content. The largest female contains almost double the amount found in a male of the same size. In both sexes, there is a slightly higher concentration of lipid in the centipede than in the millipede (Fig. 6).

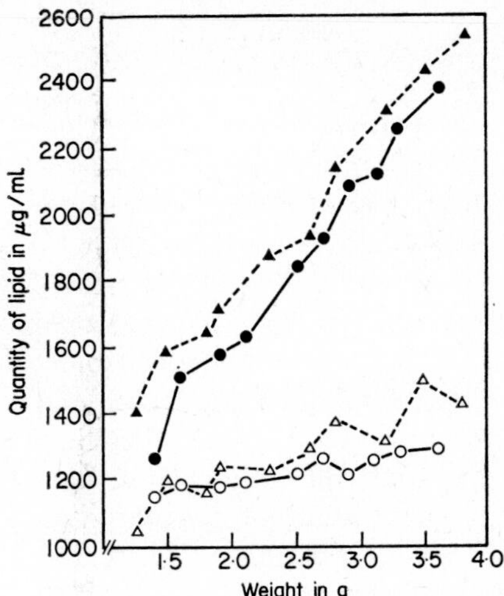

Fig. 6. Influence of size and sex on the concentration of lipid in the haemolymph of *Scutigera longicornis* and *Cingalobolus bugnioni*. Closed and open circles stand for females and males, respectively, of *C. bugnioni;* closed and open triangles stand, respectively, for females and males of *S. longicornis*. Abscissa: Weight of the animals in g; Ordinate: Lipid in µg per ml of haemolymph.

Thin-layer chromatographic analyses of the haemolymph samples from individuals of various sizes of both sexes and both species reveal that the lipid fractions are free fatty acids, mono-, di- and triglycerides and phospholipids, and that they are found in similar proportions in both the myriapods (Table V). Phospholipids and diglycerides account for 74% and 80% of the total lipid of the millipede and centipede respectively.

TABLE V

Results of quantitative estimations of lipid fractions from the haemolymphs of ***Cingalobolus bugnioni*** *and* ***Scutigera longicornis*** *as percentage of total lipid content*

Lipid fraction	*C. bugnioni*	*S. longicornis*
Free fatty acids	1·7	1·9
Monoglycerides	18·2	12·9
Diglycerides	32·2	35·3
Triglycerides	6·3	5·2
Phospholipids	41·6	44·7

DISCUSSION

It is known that crustaceans and arachnids have a much higher concentration of protein in the haemolymph than insects; the crustaceans like *Palinurus elephas* and *Carcinus maenas* have 1427·6 mg and 1622·5 mg of protein per 100 ml of haemolymph, respectively (Damboviceanu, 1932; Drilhon-Courtis, 1934a,b), and the arachnids like *Limulus polyphemus* and *Heterometrus fulvipes* have 2500 mg and 5119 mg/100ml, respectively (Dailey, Fremont-Smith & Carrol, 1931; Naidu, 1966). But the highest value of haemolymph protein from insects is only 832 mg/100 ml for honey bees and the lowest value is 166 mg/100 ml for primitive Orthoptera (Florkin, 1936). It is observed in the present study that the millipede *Cingalobolus bugnioni* has 197·5 mg of protein per 100 ml of haemolymph and the centipede *Scutigera longicornis* has 175·5 mg/100 ml. These values recall those recorded for the orthopterous insects.

This similarity of the haemolymph protein concentration of myriapods to that of insects is reflected also in the number of protein components. Crustaceans and arachnids have usually only three and rarely four components of protein in the haemolymph (Clark & Brunet, 1942; Kurup, 1965). But in insects the number of protein fractions is usually five and the maximum number is 11 (Wyatt, 1961). Therefore the presence of six fractions in the haemolymph of both male and female centipedes and also in male millipedes, and seven fractions in the female millipedes, is comparable to the condition present in insects. An interesting feature of the protein pattern of the haemolymph of millipedes is its remarkable similarity to that reported for the larvae of insects such as those of the silk worm *Bombyx mori* (Sridhara & Ananthasamy, 1963).

The presence of an extra protein fraction in the haemolymph of the female millipede recalls a similar feature known from several insects. Telfer (1954) noted an extra protein fraction, which he calls "female protein", in the haemolymph of females of the silk worm, *Cecropia*. The investigations of Telfer (1960) and Coles (1965) on *Hyalophora cecropia* and *Rhodnius prolixus* show that the extra protein fraction found in the haemolymph of females may be of the nature of albumin. The additional protein component from the females of the millipede *Cingalobolus bugnioni* also corresponds to serum albumin.

A feature of interest of the carbohydrate component of the haemolymph of *C. bugnioni* and *S. longicornis* is the presence of trehalose as the principal sugar. This again recalls a similar feature in insects where this non-reducing disaccharide is the main sugar (Evans &

Dethier, 1957; Wyatt & Kalf, 1956). In addition the sugar levels of the haemolymph from both the myriapods are also markedly similar to those in insects.

May (1935) observes that the orthopteroid insects have 74 to 108 mg sugar per 100 ml, while in the larvae of the solitary bee *Arthopera* sp. the value of sugar reaches the phenomenal concentration of 6554 mg/100 ml (Duchateau & Florkin, 1959). But crustaceans and arachnids have only 20–50 mg sugar per 100 ml (Damboviceanu, 1929; Kisch, 1929; Naidu, 1966). Therefore the concentration of sugar in the haemolymph of both the species of myriapods is comparable to that in orthopteroid insects; the centipede has 82·9 mg/100 ml and the millipede has 86·4 mg/100ml.

While the concentration of protein and carbohydrate components of the haemolymph of the myriapods studied are similar to those in insects, the lipid content is lower than that in insects. Sridhara & Bhat (1965) found 3 mg lipid/ml haemolymph in the larvae of *Bombyx mori* and in the larvae of the wax moth the value is 16 mg/ml (Wlodawer & Wisniewska, 1965). But the millipede *C. bugnioni* has only 1·55 mg/ml and the centipede *S. longicornis* has 1·62 mg/ml haemolymph. Despite this difference in the quantity of total lipid concentration a remarkable feature is the abundance of diglycerides in the haemolymph of the myriapods as in insects (Chino & Gilbert, 1965). In contrast to this, crustaceans virtually lack mono- and di-glycerides in the haemolymph (Bligh & Scott, 1966; Connor & Gilbert, 1968). Since little is known of the lipid concentration of the haemolymph of Arachnida, no comparison is possible.

Despite the differences in the lipid concentration, the organic constituents of the haemolymph of the two myriapods studied are generally similar to those of primitive orthopteroid insects and are markedly different from those of Crustacea and Arachnida. It is already known that the free amino acid concentration of the haemolymph of a millipede *Spirostreptus asthenes* and a centipede *Ethmostigmus spinosus* is similar to that in primitive hemimetabolous insects (Sundara Rajulu, 1970). Hence, these results would substantiate the observations of Tiegs & Manton (1958) and Manton (1964) that myriapods may not be related to the crustacean stock.

A comparison of the constituents of the haemolymph of the millipede with those of the centipede, shows that the concentrations of protein and carbohydrate are slightly higher in the former, but that the latter has more lipid. However, the differences may not be significant.

It is also noted that size and sex have a profound effect on the concentrations of protein, carbohydrate and lipid of the haemolymph in both the myriapods; the size and the concentration of these components are directly proportional in both males and females, and females always have a higher titre of these components than males. This again is in contrast to the condition reported for Crustacea and Arachnida where the males are said to have a higher concentration of organic substances in the haemolymph than the females (Naidu, 1966; Sundara Rajulu & Kulasekarapandian, 1972; Sundara Rajulu & Santhanakrishnan, 1972). The significance of this feature is not clear and this merits further study.

References

Beenakkers, A. M. T. & Gilbert, L. I. (1968). The fatty acid composition of fat body and haemolymph lipids in *Hylophora cecropia* and its relation to lipid releases. *J. Insect Physiol.* **14**: 418–494.

Bligh, E. G. & Scott, M. A. (1966). Blood lipids of the lobster *Homarus americanus*. *J. Fish. Res. Bd Can.* **23**:1629–1631.

Block, R. J. (1952). *Paper chromatography, a laboratory manual.* New York and London: Academic Press.

Borner, C. (1909). Neue Homologien Zwischen Crustaceen und Hexpoden. *Zool. Anz.* **34**:100–112.

Carpenter, G. H. (1903). On the phylogeny of mandibulate arthropods. *Proc. R. Ir. Acad.* **24**: 36–44.

Chino, H. & Gilbert, L. I. (1965). Lipid release and transport in insects. *Biochim. Biophys. Acta* **98**: 94–110.

Clark, E. & Brunet, F. M. (1942). The application of the serological method to the study of Crustacea. *Aust. J. exp. Biol. med. Sci.* **20**: 89–95.

Coles, G. C. (1965). Haemolymph proteins and yolk formation in *Rhodnius prolixus* Stal. *J. exp. Biol.* **43**: 425–431.

Connor, J. D. & Gilbert, L. I. (1968). Aspects of lipid metabolism in crustaceans. *Am. Zool.* **8**: 529–539.

Crompton, G. C. (1928). Homology of compound eyes of Crustacea and Hexopoda. *Can. Ent.* **60**: 14–22.

Dailey, M. E., Fremont-Smith, F. & Carrol, M. P. (1931). The relative composition of sea water and of the blood of *Limulus polyphemus*. *J. biol. Chem.* **93**: 17–28.

Dall, W. (1964). Studies on the physiology of a shrimp *Metapenaeus mastersii* (Haswell). I. Blood constituents. *Austr. J. mar. Freshwat. Res.* **15**: 146–161.

Damboviceanu, A. (1929). Recherches sur les constantes physico-chimiques du plasma des Invertebrés à l'état normal et en course d'immunisation. 1. Constantes du plasma de quelques crustaces Decapodes à l'état normal. *Archs roum. Path. exp. Microbiol.* **2**: 5–38.

Damboviceanu, A. (1932). Composition chimique et physicochimique du liquide cavitaire chez les crustaces Decapodes. *Archs roum. Path. exp. Microbiol.* **5**: 239–309.

Dole, V. P. (1956). Relation between non-esterified fatty acids in plasma and the metabolism of glucose. *J. clin. Invest.* **35**: 150–154.

Drilhon-Courtis, A. (1934a). De la regulation de la composition minerale de l'haemolymphe des Crustaces. *Annls Physiol. Physicochim. biol.* **10**: 377–414.

Drilhon-Courtis, A. (1934b). Étude des éléments mineraux du milieu intérieur de *Telphusa fluviatilis* (Latr.) et son adaptation aux changements de salinité. *Bull. Inst. océanogr. Monaco* No. 644: 1–12.

Duchateau, G. & Florkin, M. (1959). Sur la trehalosemine des insectes et sa signification. *Archs int. Physiol.* **67**: 306–314.

Evans, D. R. & Dethier, V. G. (1957). The regulation of taste thresholds for sugar in the blowfly. *J. Insect Physiol.* **1**: 3–17.

Florkin, M. (1936). Nouvelles données sur la teneur en protéines du plasma sanguin des insectes. *C.r. Séanc. Soc. Biol.* **123**: 1024–1026.

Florkin, M. (1949). *Biochemical evolution.* New York and London: Academic Press.

Florkin, M. (1960). Blood chemistry. In *The Physiology of Crustacea* **1**: 141–159. Waterman, T. H. (Ed.). New York and London: Academic Press.

Florkin, M. (1971). Evolution as a biochemical process. In *Biochemical evolution and the origin of life* **1**: 366–380. Schoffeniels, E. (Ed.): Amsterdam: North Holland Publishing Company.

George, W. C. & Nichols, J. (1948). A study of the blood of some Crustacea. *J. Morph.* **83**: 425–440.

Gilbert, A. B. (1959). The composition of the blood of the shore crab, *Carcinus maenas* (Pennant), in relation to sex and size. *J. exp. Biol.* **36**: 356–362.

Gornall, A. G., Bardaivell, C. J. & David, M. (1949). Determination of serum proteins by means of the biuret reaction. *J. biol. Chem.* **177**: 751–766.

Kisch, B. (1929). Der Gehalt des Blutes einiger Wirbelloser an reduzierenden substanzen. *Biochem. Z.* **211**: 292–294.

Kurup, P. A. (1965). Investigation on haemolymph of the South Indian scorpion *Heterometrus scaber*, especially the nature of carbohydrates, proteins and amino acid compounds. *Curr. Sci.* **34**: 663.

Manton, S. M. (1964). Mandibular mechanisms and the evolution of arthropods. *Phil. Trans. R. Soc.* (B.) **247**: 1–183.

May, R. M. (1935). Reducing substances and chloride in the blood of Orthoptera. *Bull. Soc. chim. biol.* **17**: 1045–1053.

Naidu, B. P. (1966). Ionic composition of the blood and the blood volume of the scorpion *Heterometrus fulvipes*. *Comp. Biochem. Physiol.* **17**: 157–166.

Parvathy, K. (1970). Blood sugars in relation to chitin synthesis during cuticle formation in *Emerita asiatica*. *Mar. Biol., Berl.* **5**: 108–112.

Spackman, D. H., Stein, W. H. & Moore, S. (1958). Automatic recording apparatus for use in chromatography of amino acids. *Analyt. Chem.* **30**: 1190–1220.

Sridhara, S. & Ananthasamy, T. S. (1963). Blood proteins of the silk worm *Bombyx mori*. *Proc. Indian Acad. Sci.* **48**: 19–27.

Sridhara, S. & Bhat, J. V. (1965). Lipid composition of the silk worm *Bombyx mori* L. *J. Insect Physiol.* **11**: 449–462.

Stein, W. H. & Moore, S. (1954). The free amino acids of human blood plasma. *J. biol. Chem.* **211**: 915–926.

Stewart, J. E., Dingle, J. R. & Odense, P. H. (1966). Constituents of the haemolymph of the lobster *Homarus americanus* Milne Edwards. *Can. J. Biochem.* **44**: 1447–1459.

Sundara Rajulu, G. (1966). Cardiac physiology of *Scolopendra morsitans*, a chilopod. *J. Anim. Morph. Physiol.* **13**: 114–120.

Sundara Rajulu, G. (1969a). Moult cycle of a millipede *Spirostreptus asthenes* (Diplopoda: Myriapoda). *Sci. Cult.* **35**: 483–485.

Sundara Rajulu, G. (1969b). Presence of haemocyanin in the blood of a centipede *Scutigera longicornis* (Chilopoda: Myriapoda). *Curr. Sci.* **38**: 168–169.

Sundara Rajulu, G. (1970). A comparative study of the free amino acids in the haemolymph of a millipede *Spirostreptus asthenes* and a centipede *Ethmostigmus spinosus* (Myriapoda). *Comp. Biochem. Physiol.* **37**: 339–344.

Sundara Rajulu, G. & Krishnan, K. R. (1967). The chemical composition of the cuticle of the earthworm *Megascolex mauritii*. *Indian J. exp. Biol.* **5**: 55–56.

Sundara Rajulu, G. & Kulasekarapandian, S. (1972). A study of the free amino acids from the haemolymph of Crustacea. *Z. Zool. Syst. EvolForsch.* **10**: 137–145.

Sundara Rajulu, G. & Ramanujam, R. M. (1972). Free amino acids in the haemolymph of *Eoperipatus weldoni* (Onychophora). *Experientia* **28**: 87.

Sundara Rajulu, G. & Santhanakrishnan, G. (1972). An investigation of the free amino acids in the haemolymph of Arachnida. *Zool. Anz.* **189**: 222–232.

Sutcliffe, D. W. (1963). The chemical composition of haemolymph in insects and some other arthropods, in relation to their phylogeny. *Comp. Biochem. Physiol.* **9**: 121–135.

Telfer, W. H. (1954). Immunological studies of insect metamorphosis. II. The role of a sex limited blood protein in egg formation by the cecropia silk worm. *J. gen. Physiol.* **37**: 539–558.

Telfer, W. H. (1960). The selective accumulation of blood proteins by the oocytes of saturniid moths. *Biol. Bull. mar. biol. Lab., Woods Hole.* **118**: 338–351.

Tiegs, O. W. & Manton, S. M. (1958). The evolution of the Arthropoda. *Biol. Rev.* **33**: 255–337.

Wilson, W. H. (1901). Reactivity of haemocyanin to chemical reagents. *Rec. Egypt. Govt Sch. Med.* **1**: 151–162.

Wlodawer, P. & Wisniewska, A. (1965). Lipids in the haemolymph of waxmoth larvae during starvation. *J. Insect Physiol.* **11**: 11–20.

Wyatt, G. R. (1961). The biochemistry of insect haemolymph. *A. Rev. Ent.* **6**: 75–102.

Wyatt, G. R. & Kalf, G. F. (1956). The chemistry of insect haemolymph. II. Trehalose and other carbohydrates. *J. gen. Physiol.* **40**: 833–847.

Discussion

KRAUS: You have studied one centipede which is highly evolved (*Scutigera*), and one millipede. Do you think that these are sufficiently representative of both groups? In several cases you had quite large differences even between the sexes of the one species.

SUNDARA RAJULU: Certainly we cannot come to a general conclusion from just one representative of each, but this is only a preliminary study. We would welcome any other laboratories entering this field and giving more information.

STRASSER: Can you amplify your remarks on the phylogenetic relationships?

SUNDARA RAJULU: It is known from embryology, comparative and functional morphology that Crustacea and Arachnida resemble each other and are quite distinct from the Onychophora, Myriapoda and Insecta. Now we find that there are differences in the contents, for example, of free amino acids and lipids, both qualitative and quantitative, between the Crustacea and Arachnida on the one hand and the Myriapoda and Insecta on the other.

SAKWA: How did you obtain the haemolymph samples?

SUNDARA RAJULU: The blood was drawn out from the mid-dorsal line by a hypodermic needle which had a large bore so as not to crush the haemocytes and contaminate the haemolymph by their organic components.

MANTON: I would like to say that never at any time have I or Professor Tiegs maintained that there was a close relationship between Crustacea and Arachnida. In fact, the modern view is that they represent two entirely separate phyla. I do not think one should divide the Arthropoda into two groups and link the Crustacea with the Chelicerata which differ so fundamentally.

KRAUS: Do you think it practical to use biochemical characters as measures of phylogenetic similarity? May they not merely reflect adaptations to different ways of life?

SUNDARA RAJULU: I agree that morphological and embryological differences are more valid data on which to base our phylogeny. However, although the information from biochemistry is meagre, it supports the conclusions of the morphologists and embryologists and never contradicts. Since biochemistry gives us supporting evidence, we should accept it.

DOHLE: I think that what you have said does support the close relationship of Myriapoda and Insecta but I think it is a theoretical mistake to believe that dissimilarities would indicate non-relationship. If this were so, the Rhizocephala, for example, would not be considered as Cirripedia.

SUNDARA RAJULU: When the dissimilarities are of a smaller magnitude, they may not indicate non-relationship, but when they are conspicuous they might indicate non-relationship.

MANTON: I recall the impact of the discovery of differences in the biochemistry of the nitrogen cycle which seemed, at the time, to fit in quite well with the classification of animals based on their morphology, and the ensuing realization that this biochemical evidence was not as simple as it first appeared. One has to treat this type of evidence with great care.

KRAUS: In comparison with our great experience of morphological criteria, our knowledge of how to interpret biochemical characters is still very rudimentary.

SUNDARA RAJULU: I agree. Whilst I think it really must be accepted that the relationships between different groups of vertebrates can best be established by using biochemical and physiological criteria, we are only just beginning to understand these things in the Arthropoda. However, such data as we possess seem to support the conclusions from morphology.

EASON: Have you investigated the differences between, say, marine, freshwater and terrestrial crustaceans?

SUNDARA RAJULU: Yes. I have a paper published on the subject (Sundara Rajulu & Kulasekarapandian, 1972)*. I have studied some crabs from all three habitats and some prawns which live in brackish water during the early part of their life cycle and later migrate to the sea. There are certain differences but fundamentally they are all similar. In some molluscs the amino acid taurine is present in marine representatives, decreases in amount in estuarine and freshwater forms, and is totally absent in terrestrial forms (Simpson, Allen & Awapara, 1959).† I have found a similar relationship in crabs from these various habitats but all of them are alike in having about 50 mg of free amino acids per 100 ml of haemolymph, irrespective of their habitat conditions. By contrast, the concentration of free amino acids in the blood of myriapods and insects ranges from 245–2500 mg per 100 ml.

* See list of references.

† Simpson, J. W., Allen, K. & Awapara, J. (1959). Free amino-acids in some aquatic invertebrates. *Biol. Bull. mar. biol. Lab. Woods Hole* **117**: 371–381.

Symp. zool. Soc. Lond. (1974) No. 32, 365–382.

THE SPIRACLE STRUCTURE AND RESISTANCE TO DESICCATION OF CENTIPEDES

ALAN CURRY

*Zoology Department, University of Manchester, Manchester, England**

SYNOPSIS

The structure of the spiracles in *Lithobius forficatus*, *Lithobius variegatus*, *Haplophilus subterraneus*, *Geophilus insculptus*, and *Cryptops hortensis* is described. Basically, the spiracles consist of an opening, which leads into an atrium. The walls of the atrium are invariably covered in cuticular outgrowths (trichomes). Tracheae arise from the base of the atrium. No spiracular closing mechanisms have been found in any of the centipedes examined, although muscles, which may facilitate breathing movements, have been found adjacent to the spiracular atrium in *C. hortensis*.

Desiccation experiments have been carried out at 0% relative humidity at 19 ± 2°C on *L. forficatus*, *L. variegatus*, and *G. insculptus*. There is a simple linear relationship between the body weight lost as water, and time, in both species of *Lithobius*. It is suggested that the cuticle is as important as the spiracles as a site of water loss. Evaporation of water from *G. insculptus* has been found to be more complex. The rate of water loss falls initially, then increases, and falls again. A tentative explanation for the shape of this curve is put forward.

INTRODUCTION

Previous workers have seldom attempted to relate the resistance to desiccation shown by centipedes to either the structure of the cuticle, or the character of the spiracular openings.

It has been suggested by Lewis (1963), that the size of the spiracular opening and the degree of development of trichomes in the spiracle atrium may be important in controlling the rate of water loss in geophilomorph species. Verhoeff (1941) states that chilopods possess a well developed tracheal system with spiracular closing devices. Lewis (1963) also states that lithobiomorph spiracles possess a closing device, but that the spiracles of geophilomorphs and *Cryptops*, (though not all scolopendromorphs), have no closing device. Blower (1955) suggests that despite their hydrofuge cuticle, most myriapods lose water rapidly through their spiracles; most rapidly in those centipedes which have imperfect spiracular closing devices. Kaufman (1962) has stated that *Lithobius forficatus* has imperfect spiracular closing devices, and that *Cryptops sp.* (Kaufman, 1964) has a mechanism for

* Present address: *Public Health Laboratory, Withington Hospital, Manchester, England.*

opening its spiracles. However, Kaestner (1968)* states that there are no mechanisms or muscles for closing the spiracles in centipedes.

The aim of this study has been to clarify the spiracle structure of centipedes, and to investigate their response to desiccation.

MATERIALS AND METHODS

Collection of animals

Lithobius forficatus, *Geophilus insculptus* and *Haplophilus subterraneus* were collected from beneath stones on wasteland and gardens in Manchester. *Lithobius variegatus* was collected from Great Shacklow Wood, near Ashford, Derbyshire (G.R. SK696179). *Cryptops hortensis* was collected from leaf litter in mixed deciduous woodland at Esher, Surrey (G.R. TQ126632).

Desiccation experiments

It is extremely difficult to determine the precise humidity conditions to which an animal is exposed, when kept in a closed vessel in "still" air over a humidity controlling substance (Ramsay, 1935). Therefore, it was decided to use a completely dry atmosphere (anhydrous calcium chloride in a desiccator) at 19 ± 2°C. Adequate control of the conditions had to be weighed against the fact that centipedes are not likely to be exposed to such extreme conditions in nature.

Animals were placed into specially made desiccation tubes, each consisting of a 3in. by 1in. glass tube, with the bottom removed, and a piece of fine meshed gauze held across one end by means of an elastic band. The mesh was used not only to restrain the animal, but also to reduce the possibility of a localised atmosphere forming in the bottom of the tube. To remove moisture from the gauze, these tubes were left in the desiccator for at least 24 h before an experiment was due to begin. The tubes were then weighed. A single animal, unfed for at least 24h, was placed in each tube, and the tubes reweighed to determine the initial weight of the animal. The tubes were then placed in the desiccator for 30 min and weighed again. This process was repeated at 30 min intervals for the duration of the experiment. Damaged animals were included in the results to determine whether there were any significant differences between the desiccation rates of damaged and undamaged animals.

* Though Kaestner gives a bibliography at the end of each chapter, he does not indicate which author is responsible for the information given. However, the above information probably came from Füller (1960).

Histological technique

L. forficatus, *L. variegatus*, *G. insculptus*, and *H. subterraneus* were fixed in Bouin's fluid. *C. hortensis* was fixed in neutral formalin solution. Fixed animals were cut into small pieces and placed into Diaphanol for a week, to soften the cuticle (for details of action, see Kennaugh, 1957). After dehydration the pieces of animal were embedded in wax, and 10 μm serial sections cut, using a Cambridge Rocking Microtome. Sections were cut in transverse, frontal and sagittal planes of most of the centipedes. Sections were stained in Haematoxylin and alcoholic Eosin.

Some animals were cleared in benzyl alcohol for a few days, in order to trace the paths of the tracheae.

RESULTS

Description of spiracles

Lithobius

The spiracular openings are borne on the elliptical stigmatopleurites in the pleural region of pedigerous segments 3, 5, 8, 10, 12 and 14. A pair of spiracles occurs on each of these segments. The openings are slit-shaped and run antero-ventral to postero-dorsal. The spiracular openings are directed slightly backwards.

In profile, the apex of a spiracle of *L. forficatus* has a triangular appearance (Fig. 10), whereas that of *L. variegatus* is curved (Fig. 14). The size of the anterior spiracular openings in adult *L. forficatus* is about 300 μm. According to Füller (1960), the size of these openings varies evenly down the body; the first stigma being twice as long as the last. The size of the anterior spiracular openings of adult *L. variegatus* is about 200 μm. It is not known whether the size of these varies down the body. The stigmatopleurites have sense hairs on them. These are usually three or four in *L. variegatus* and, according to Füller (1960), four to six in *L. forficatus* (although a diagram of Rilling's (1960), shows nine sensory hairs on the stigmatopleurite of *L. forficatus*).

Each spiracle has an atrium (Figs 1, 2, 8, 9, 10, 14, 15, 17). The atrium has the appearance of a flattened funnel, with the tapering end running inwards. The mouth of the atrium has two lips (Figs 10, 11), which are bordered by large cuticular outgrowths (Figs 10, 14, 15). The inner surface of the atrium is covered with small cuticular outgrowths (trichomes). In both *L. forficatus* and *L. variegatus*, these trichomes diminish in length from the mouth to the base of the atrium (Figs 1, 2, 15).

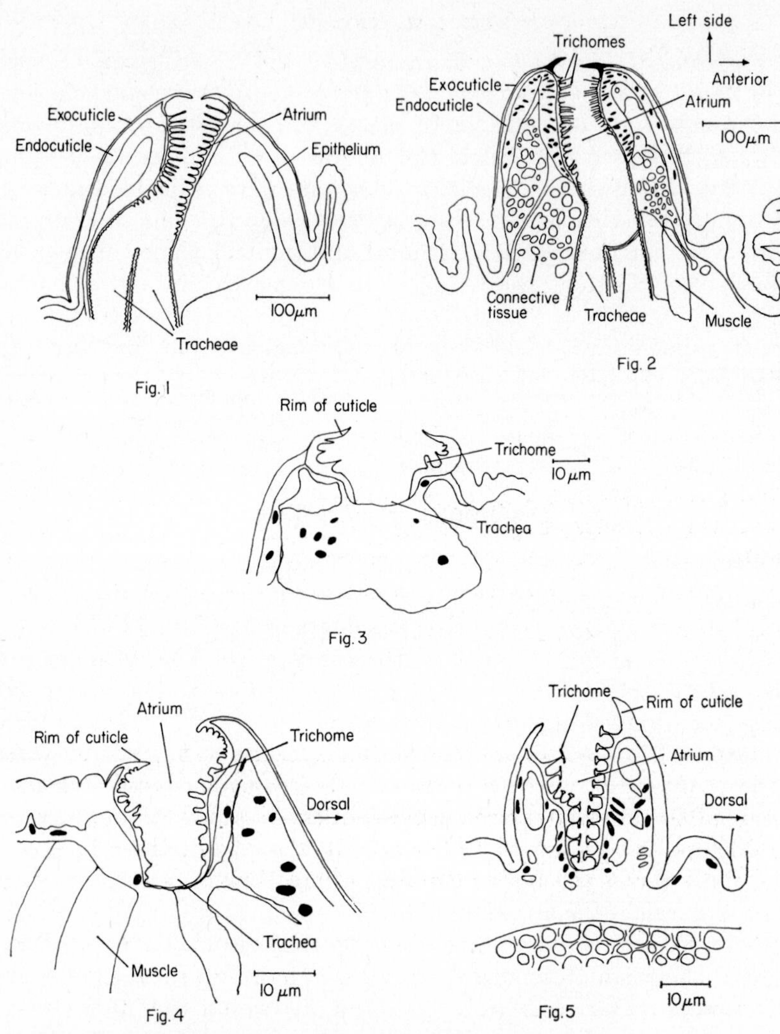

FIG. 1. Longitudinal section through a spiracle of *L. forficatus*.

FIG. 2. Longitudinal section through a spiracle of *L. variegatus*.

FIG. 3. Median vertical section through a spiracle of *G. insculptus*.

FIG. 4. Median vertical section through a spiracle of *H. subterraneus* (anterior region).

FIG. 5. Median vertical section through a spiracle of *C. hortensis*.

The trichomes of *L. forficatus* are covered in exocuticle with a small cone of endocuticle extending up the centre (Fig. 6). Some of the trichomes have a spatulate appearance. In *L. variegatus* there appear to be two sizes of trichomes lining the atrium; large ones interspersed with smaller ones. This is probably an optical effect, and the structure of the trichomes is interpreted to have the form shown in Fig. 7, where each trichome is surrounded by a short tube of exocuticle, and the endocuticle extends through and beyond this tube.

Near the base of the atrium of *L. forficatus* some trichomes fuse to form a mesh-like network (Fig. 8).

The cuticle of the stigmatopleurite has a thick exocuticle. The pleural membrane, which surrounds the stigmatopleurite, has a thin colourless exocuticle.

The epidermis of the stigma consists of columnar cells, with elongate nuclei (Figs 6, 7). The cavity between the epithelium of the stigmatopleurite and the epithelium of the atrium is filled with connective tissue, which contains spherical vesicles (Figs 2, 9, 15, 17).

In some sections, muscles have been seen near to the base of the stigmatopleurite (Figs 2, 15). It seems probable that these muscles are part of the body musculature, and have no function in closing the spiracle, but may facilitate ventilation and/or restrict the air passage.

Haplophilus and *Geophilus*

The spiracles of geophilids are borne on the stigmatopleurites in the pleural region of all pedigerous segments, except the first and the last. The diameter of the anterior spiracular openings of *H. subterraneus* is about 30 to 50 μm, whereas that of the posterior spiracles is approximately 20 μm. The spiracular openings along the length of *G. insculptus* have not been measured, but those of the anterior end have a diameter of about 20 to 30 μm.

In both species the spiracular openings are circular and are surrounded by a rim of sclerotised cuticle (Figs 4, 18). The atrium of *H. subterraneus* is funnel-shaped, tapering inwards (Figs 4, 18), whereas that of *G. insculptus* is disc-shaped (Fig. 3). In both species the inner surface of the atrium is covered with trichomes, which are mainly cone-shaped in *G. insculptus* (Fig. 3), but are elongate plates in *H. subterraneus* (Fig. 4).

The cuticle of the stigmatopleurite of both species is sclerotized, particularly in *G. insculptus*, where the endocuticle is very thin.

In geophilids, the tracheae from the spiracles are interconnected into a system, which runs down the entire length of the body (Fig. 19). The atria and tracheae of both *H. subterraneus* and *G. insculptus* are surrounded by body musculature.

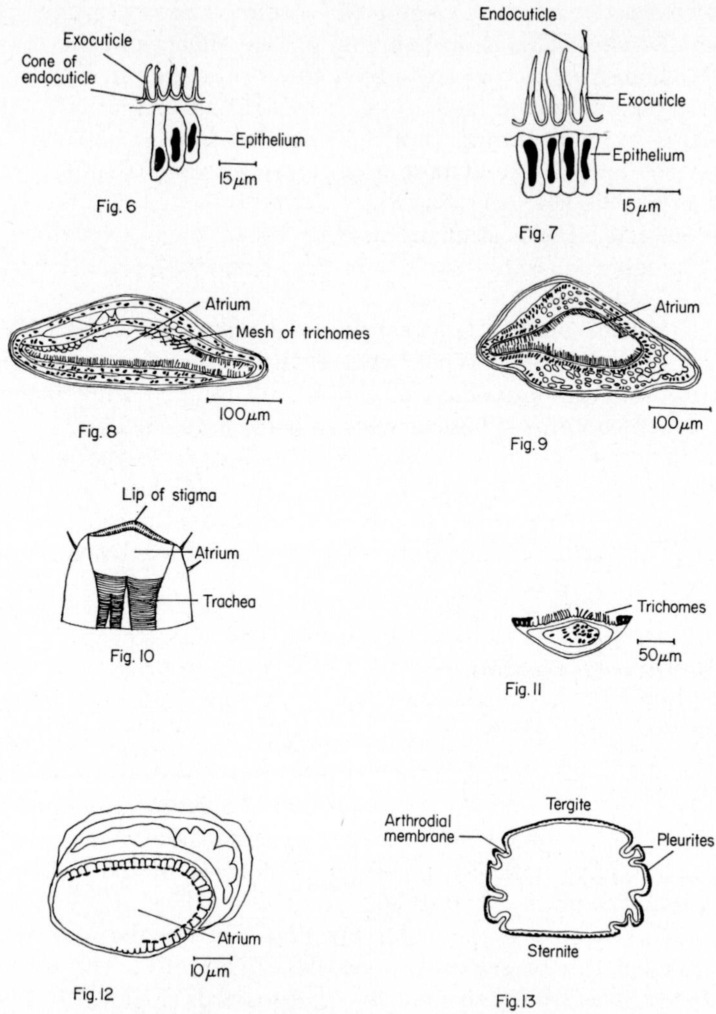

FIG. 6. Atrial trichomes of *L. forficatus*.

FIG. 7. Atrial trichomes of *L. variegatus*.

FIG. 8. Cross section through the atrium of a spiracle of *L. forficatus*.

FIG. 9. Cross section through the atrium of a spiracle of *L. variegatus*.

FIG. 10. Diagrammatic sketch of a spiracle of *L. forficatus*.

FIG. 11. Cross section through a lip of a spiracle of *L. forficatus*.

FIG. 12. Cross section through the atrium of a spiracle of *C. hortensis*.

FIG. 13. Transverse section of *H. subterraneus* (from Blower, 1951).

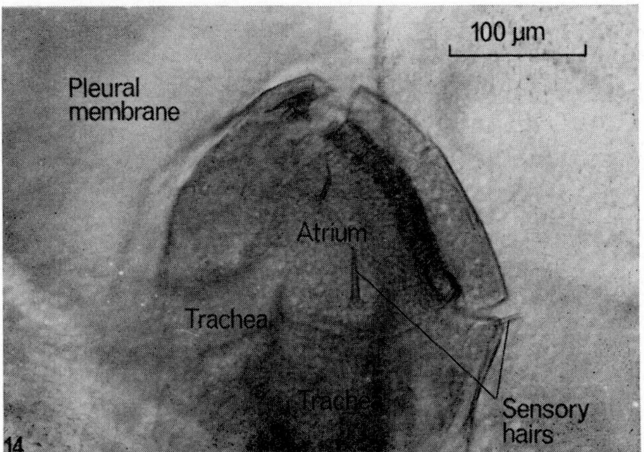

FIG. 14. A cleared spiracle of *L. variegatus*.

Cryptops

The body of *C. hortensis* consists of 21 leg bearing segments. The spiracles are situated in the pleural region of segments 3, 5, 8, 10, 12, 14, 16, 18, and 20. The external opening of the spiracle is elliptical, with its major axis running parallel to the lateral margin of the tergite. The atrium is funnel-shaped, tapering inwards. Sections show the entire inner surface of the atrium to be covered with trichomes, which have the appearance of columns, with broad flattened ends (Figs 5, 12).

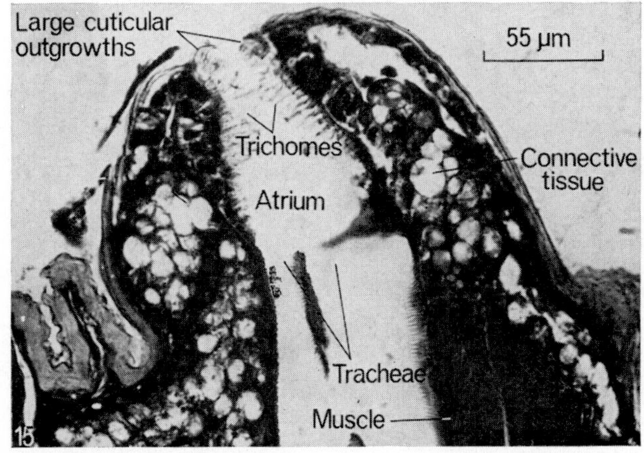

FIG. 15. A longitudinal section through a spiracle of *L. variegatus*.

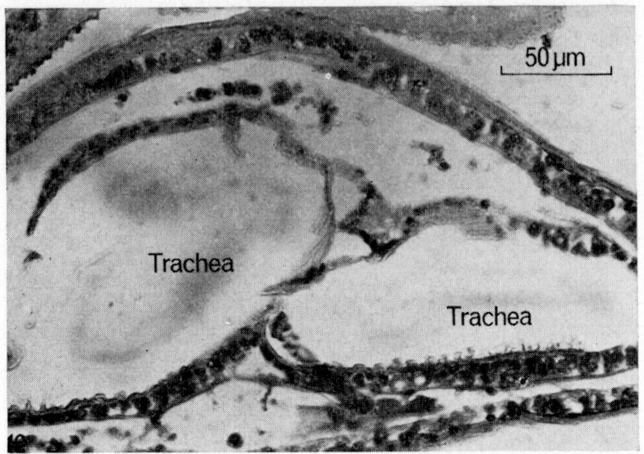

FIG. 16. A transverse section through the atrial base of a spiracle of *L. forficatus* showing tracheae.

The cuticle of the stigmatopleurites and the atrium is heavily sclerotized in *Cryptops*. This is shown by the extreme thickness of the amber coloured exocuticle, and the apparent absence of an endocuticular layer in this area.

The tracheae of *Cryptops*, in contrast to *Geophilus* and *Lithobius*, have a tendency to form vesicular swellings, especially at branching points (Kaufman, 1964). In *Cryptops* there is a transverse connection

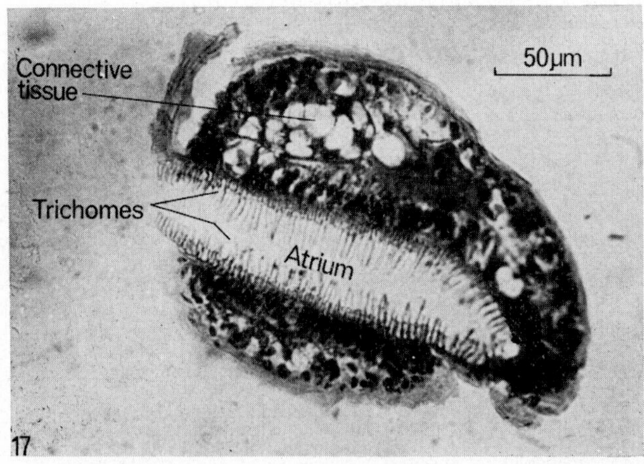

FIG. 17. A transverse section through the top of the atrium of a spiracle of *L. variegatus*.

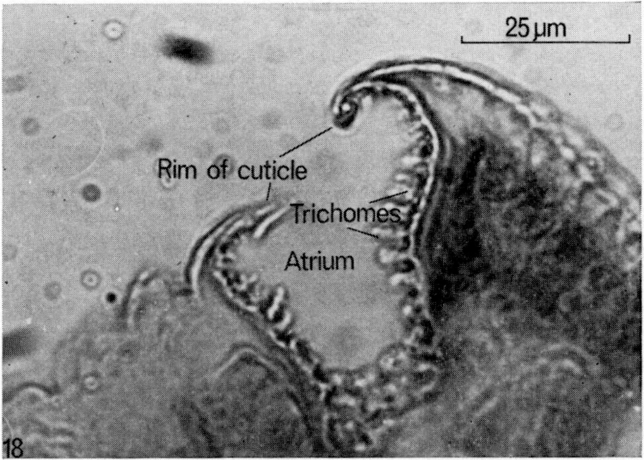

FIG. 18. A median vertical section through a spiracle of *H. subterraneus*.

between spiracles of the same segment. In larger scolopendromorphs, for example *Scolopendra*, longitudinal links are also established.

The spiracles of *Cryptops* do have muscular attachments, but these are not inserted in the atria, as one would expect if there was any closing device, but in the walls of the tracheae (see also Kaufman, 1964 and Füller, 1960).

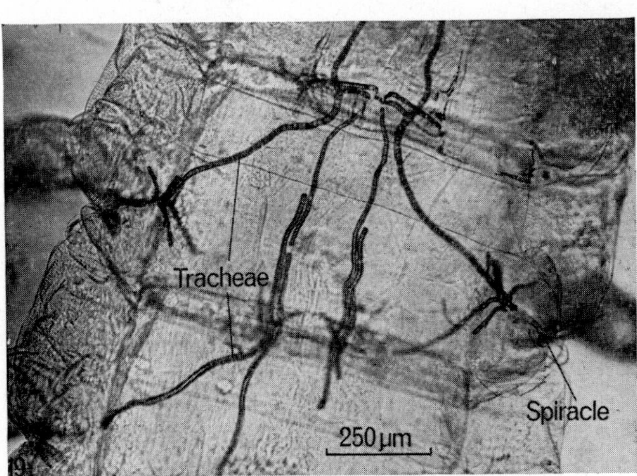

FIG. 19. The main trunks of the tracheal system in the mid-trunk region of *H. subterraneus*.

Desiccation experiments

Desiccation experiments have been performed on *L. forficatus*, *L. variegatus*, and *G. insculptus*. The results of the experiments on both *L. forficatus* and *L. variegatus* are shown in Fig. 20, and the results of *G. insculptus* in Fig. 21.

Both species of *Lithobius* attempt to run up the sides of the glass desiccation tubes. This behaviour, which is probably a response to low humidity, ceases, and the centipedes go into a period of quiescence, when 20–30% of their body weight has been lost. The lithobiids no longer move on being prodded with a blunt seeker when about 40% of their body weight has been lost. Lewis (1963) regarded the desiccated centipedes in this condition as dead. This, however, is not the point of death, as at least 50% recover if placed, dorsal surface down, on damp filter paper.

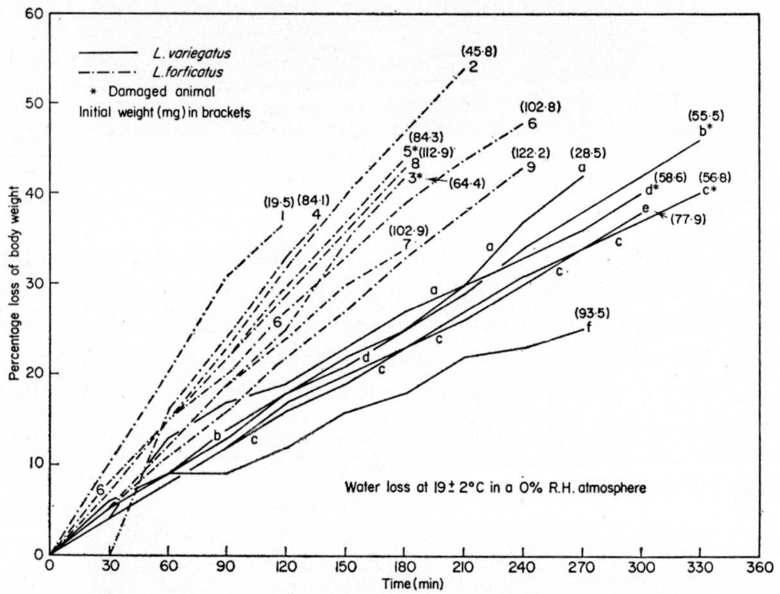

FIG. 20. Water loss of *Lithobius variegatus* and *L. forficatus*.

From Fig. 20, it appears that there is a simple linear relationship between time and percentage loss of body weight. There is little or no overlap between the rates of evaporation of water when comparing both species; the rate for *L. forficatus* being greater than for *L. variegatus*. Within each species, it will be seen that, in general, the smaller the initial weight, the greater the rate of water loss, and the greater the initial weight the less the rate of water loss, the variation being related to surface area.

There seems to be little or no difference between the desiccation rates of intact animals and those with some legs missing.

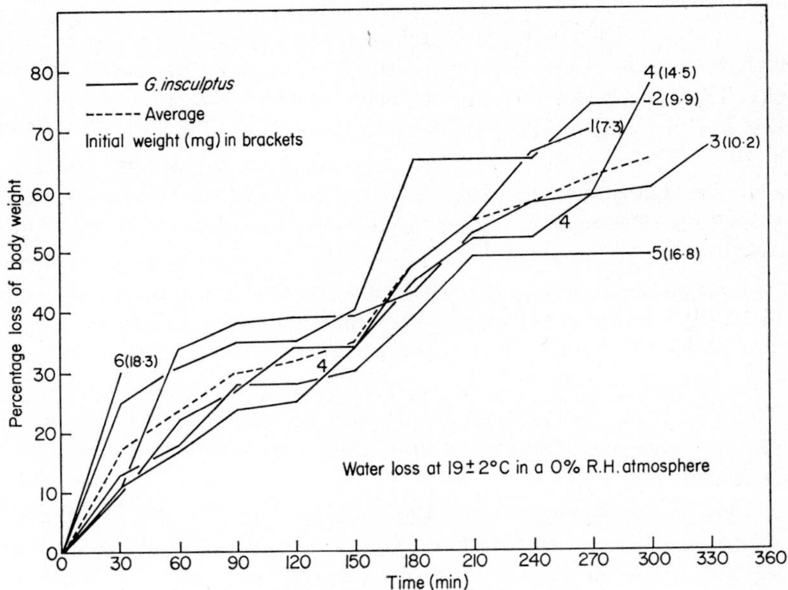

FIG. 21. Water loss of *Geophilus insculptus*.

The process of desiccation in *Geophilus* appears to be more complex than that in *Lithobius*. From Fig. 21 it can be seen that the rate of water loss is divisible into two phases, each initially steep, then flattening. On being placed in the desiccator *Geophilus* walks around the bottom of the tube. When about 50–60% of the body weight of *Geophilus* has been lost, and the animals are stimulated, they are observed to walk backwards dragging the anterior end.

DISCUSSION

The five centipede species examined have a similar spiracle structure. Each spiracle consists of an opening which leads into an atrium. The walls of the atrium are invariably covered with trichomes. Kaufman (1962) regards the trichomes as protective devices, which stop the respiratory passages from becoming clogged. It seems more likely, however, that they prevent the spiracle from closing during the activity of the centipedes, thus preventing the cutting off of the local air supply. Manton (1965) is also of this opinion. This is important as, in lithobiomorphs, the tracheae coming from the spiracles have many branches, but do not connect with tracheae from other spiracles (Haase, 1884). Thus if a spiracle is blocked, air from other spiracles cannot easily diffuse to the area cut off.

The anterior third of a geophilid does more muscular work in forming a burrow than do most of the segments further along the body (Manton, 1965). The musculature in the anterior of the body is therefore more bulky, and respiratory exchanges must go on here at a higher level than elsewhere. These respiratory demands are met in part by the larger size of the spiracular openings in anterior segments (Manton, 1965). Constriction of a spiracle when a segment becomes short and thick, (as in burrowing), does not prevent ventilation. Exchange will continue via the air spaces between the trichomes. Even if the spiracular opening is blocked by the substratum, respiration can continue, as air can diffuse along the length of the animal through the longitudinal trunks of the tracheal system.

Geophilids penetrate more deeply into the soil than lithobiomorphs, where they are more likely to encounter the problem of dilution of body fluids. The cuticle of geophilids is more heavily water-proofed than that of *Lithobius* (Blower, 1951, 1955). The size and frequency of trichomes is much the same in geophilids of different body sizes, suggesting that the dimensions of the trichomes may be related to physical properties, such as surface tension of water and wetting of cuticular surfaces (Manton, 1965). As Manton has pointed out it is possible that the trichomes could act as a physical gill under the wet conditions which are likely to occur in the deeper layers of the soil. Evidence of spiracles being used as physical gills may be found in the littoral centipede *Hydroschendyla submarina*, which is found at or below high-tidemark (Eason, 1964).

Füller (1960) failed to find muscles attached to the spiracles in *G. longicornis* or *L. forficatus*, but did find them, as in this study, inserted in the walls of the tracheae of *C. hortensis*. These may facilitate breathing movements independent of body movements. However, Dubuisson (1928) could not distinguish any ventilation movements in *L. pyrenaicus*, *G. carpophagus*, or *C. hortensis*; but a ventilation due to the pulsations of the heart was noted in *G. carpophagus*. In lithobiids and geophilids, body movements may be important in ventilation, whereas in times of quiescence gaseous exchange must occur by diffusion, or by the activity of the heart and other organs.

The desiccation experiments have yielded some interesting results. The rate of water loss in individual lithobiids appears to be constant, for the duration of the experiments. However, the smaller the animal, the greater the rate of evaporation. The importance of size was recognised by Kennedy (1927), who first pointed out that all other parameters remaining the same, the water content is a function of the volume of the animal, whereas evaporation is a function of surface area. Thus the smaller the organism, the more will its water content be depleted through a particular duration of exposure to a given humidity.

One would have expected damaged animals to lose water more rapidly than undamaged animals, but since the results show that there is little or no difference, the indication is that the cuticle may be as important as the spiracles as a site of water loss. Immobilized specimens of *L. forficatus* and *L. variegatus* were placed between slides, covered with anhydrous cobalt chloride paper, to attempt to determine the sites of water loss. If the spiracles constituted the main site of water loss, the area of cobalt chloride paper overlying the edge of the tergites and sternites would surely hydrate first. However, the entire area of the paper covering the sternites and tergites hydrated simultaneously, indicating that the cuticle was losing water at least as rapidly as the spiracles.

Compared with the cuticle of *Lithobius*, the cuticle of geophilids is heavily impregnated with lipids (Blower, 1951). This evidence suggests that the cuticle of *Lithobius* is less waterproof than that of geophilids. The present study supports this theory.

As the results of the desiccation experiments have shown, the rates of water loss from *L. forficatus* and *L. variegatus* do not overlap. This discrepancy may be related to differences in the areas of these two animals, (that is, *L. forficatus* may have a greater surface area per volume than *L. variegatus*, and hence lose water more quickly under identical conditions); alternatively or in addition, it may be related to the fact that the permeability of the cuticle of *L. forficatus* is greater than that of *L. variegatus*. These two parameters, surface area and permeability, must be reflected by the slope of the graph, for a given size of animal.

Thus:—

$$A.x \text{ is proportional to } \tan\theta$$

where A = the surface area of the animal; x = the average permeability of the cuticle (rate of evaporation per unit area) of that animal (this average including water lost from the mouth, spiracles and anus); $\tan\theta$ = the slope of the desiccation curve of that animal.

Unfortunately it is extremely difficult to determine the surface area of *Lithobius* (or any chilopod), accurately. Attempts have been made to determine the surface area of weighed, preserved specimens. The circumference of the body was estimated and multiplied by the length. To this sum the area of the antennae and the legs was added. Calculations were based on the assumption that these structures were tubular. By these means an approximate figure for the area was reached. The slope of the graph, corresponding to the weight of the animal whose area had been estimated, was then noted, and the values substituted into the above equation. From these values, it was found that the average permeability of the cuticle of *L. forficatus* was greater than that of *L. variegatus*. If the values for the permeability had been similar, then the differences in the desiccation rates could have been attributed to surface area alone. However, until a more accurate method of measuring the surface area is found, and the weights of live animals are used, the above result must remain speculative.

Auerbach (1951) suggested that moisture plays a role in the differential distribution of centipedes, and that resistance to desiccation may be important in this respect. *L. variegatus* and *L. forficatus* are surface dwelling animals, occupying similar ecological niches, and frequently occur together (Lewis, 1965). Roberts (1956), working in woodland near Southampton, found that *L. variegatus* was uncommon in a dry sweet-chestnut wood, where *L. forficatus* was common; the opposite being the case in the damper Burley wood. The distribution of animals is a very complex subject, but the above observation indicates that *L. forficatus* is more resistant to desiccation than *L. variegatus*, which has not been shown in this study.

The rate of water loss from *G. insculptus* is more complex than that found in *Lithobius*, and a tentative explanation is put forward to explain the complex form of Fig. 21. A transverse section of a fully hydrated geophilid (Fig. 13) shows that the arthrodial membrane is exposed between the sclerotized pleurites. Water probably evaporates faster through the arthrodial membrane than through the sclerotized pleurites (J. G. Blower, pers. comm.). As the body loses water, it will surely tend to shrink, thus pulling the pleurites together, and so eliminating exposure of the arthrodial membrane. Hence the initial drop in the rate of water loss. Evidence for this has so far been inconclusive.

The accelerated rate of water loss, after the initial fall, is intriguing. It is as if there is a sudden breakdown in some internal water retaining mechanism, of which the epidermis may be a part, allowing more water to be lost rapidly. This is, however, purely speculative.

The second fall in the rate of water loss is also difficult to explain; but, as has been noted above, the posterior end of the animal drags the anterior around when 50–60% of the body weight has been lost. (Lewis, 1963, also noticed this.) This may indicate that at this stage the major sites of water loss are the mouth and spiracles in the anterior region, or that the anterior has a greater surface/volume ratio. As the head and anterior of the animal dry out, the rate of water loss again decreases. Thereafter, the main sites of water loss are the smaller spiracles of the rear two-thirds of the animal, and the cuticle. An alternative but less likely explanation is that there is a change in the property of the cuticle about the "transition point". Characteristic of the Geophilomorpha is a change in the structure of the trunk, which occurs before the mid-point of the body, at about the junction of the anterior two-fifths with the posterior three-fifths. This change is spread over several segments in *Geophilus* (Eason, 1964).

Little is known of the ecology of *G. insculptus*, but it is primarily a subterranean animal.

As Beament (1961) has pointed out, land animals, which survive in humidities substantially below saturation, must conserve water. Four possible sites of water loss can be distinguished in chilopods:

1. defecation,
2. the spiracles,
3. the mouth, and
4. the integument.

It is possible to eliminate water loss by excretion almost entirely, by the use of insoluble, low-hydrogen compounds of nitrogen, which can be excreted dry (Edney, 1957). Centipedes excrete uric acid and therefore defecation is probably not important in water loss. This error was eliminated from the experiments by starving the animals for 24 h before the beginning of an experiment.

This study has indicated that the cuticle of centipedes is not resistant to desiccation, and as Edney (1957) points out, there is little point in greatly reducing water loss from the spiracles in the absence of an impermeable cuticle. The present study has yielded no evidence of the closing mechanisms that have been suggested by some previous authors (Verhoeff, 1941; Blower, 1955; Kaufman, 1959, 1962, 1964; Lewis, 1963).

The importance of the mouth as a site of water loss in geophilids was recognized by Lewis (1963), and the findings of this study tend to support this theory.

The hypothesis that the atrial trichomes may be important in controlling the rate of water loss in Geophilomorpha (Lewis, 1963) is still unproved. But, as lithobiomorphs and *Cryptops* have a similar spiracle structure, this hypothesis may justifiably be extended to these groups. The trichomes may reduce water loss through the spiracles, but more probable functions of these structures are:

1. to act as a filter mechanism, as suggested by Kaufman (1962),
2. to prevent the spiracle closing during muscular exertion, and
3. possibly to act as a physical gill in wet situations, especially in geophilids.

Much more work needs to be done therefore on the humidity aspects of the ecology of these species, and indeed on centipedes in general.

SUMMARY

No spiracular closing mechanisms have been found and indeed none are necessary as the cuticle seems to be permeable.

Acknowledgements

I am grateful to Dr D. W. Yalden, Mr N. Sakwa and Mr G. R. Hosey, for their help in the collection of animals. My thanks also to Dr J. Kennaugh for his advice on histology, Dr J. G. E. Lewis for his helpful critism, Mr R. F. Grayson for his helpful comments and suggestions, and to Miss J. A. Albinson for her help in preparation of this manuscript. My most grateful thanks go to Mr J. G. Blower for his extremely helpful advice and encouragement, and for making his own slide preparations available to me.

References

Auerbach, S. I. (1951). The centipedes of the Chicago area with special reference to their ecology. *Ecol. Monogr.* **21**: 97–124.

Beament, J. W. L. (1961). The water relations of insect cuticle. *Biol. Rev.* **36**: 281–320.

Blower, J. G. (1951). A comparative study of the Chilopod and Diplopod cuticle. *Q. Jl. microsc. Sci.* **92**: 141–161.

Blower, J. G. (1955). Millipedes and Centipedes as soil animals. In *Soil zoology*: 138–151. Kevan, D. K. M. (ed.) London: Butterworth.

Dubuisson, M. (1928). Recherches sur la ventilation trachéenne chez les Chilopodes et sur la circulation sanguine chez les Scutigères. *Archs Zool. exp. gén.* **67** (Notes et Rev.): 49–63.

Eason, E. H. (1964). *Centipedes of the British Isles.* London: Warne.

Edney, E. B. (1957). *The water relations of terrestrial arthropods.* Cambridge: University Press.

Füller, H. (1960). Untersuchungen über dem Bau des Stigmen bei Chilopoda. *Zool. Jb.* (*Anat.*) **78**: 129–144.

Haase, E. (1884). Das Respirationssystem der Symphylen und Chilopoden. *Zool. Beitr.* **1**: 65–96.

Kaestner, A. (1968). *Invertebrate zoology* **2**: 1–472 Arthropod relatives, Chelicerata, Myriapoda. Translated and adapted from the German by H. W. Levi & L. R. Levi. New York: Interscience Publishers (a division of John Wiley and Sons).

Kaufman, Z. S. (1959). Spiracular structure in *Geophilus proximus*, C. L. Koch. *Dokl. Akad. Nauk SSSR* **129**: 698–701.

Kaufman, Z. S. (1962). Structure and development of stigmata in *Lithobius forficatus* L. (Chilopoda, Lithobiidae). *Ent. Obozr.* **41**: 223–225.

Kaufman, Z. S. (1964). Structure of the tracheal system in *Cryptops* sp. (Chilopoda, Scolopendromorpha, Cryptopidae). *Ent. Obozr.* **43**: 167–169.

Kennaugh, J. (1957). Action of diaphanol on arthropod cuticles. *Nature, Lond.* **180**: 238.

Kennedy, C. H. (1927). Some non-nervous factors that condition the sensitivity of insects to moisture, temperature, light and odors. *Ann. ent. Soc. Am.* **20**: 87–106.

Lewis, J. G. E. (1963). On the spiracle structure and resistance to desiccation of four species of geophilomorph centipede. *Entomologia exp. appl.* **6**: 89–94.

Lewis, J. G. E. (1965). The food and reproductive cycles of the centipedes *Lithobius variegatus* and *Lithobius forficatus* in a Yorkshire woodland. *Proc. zool. Soc. Lond.* **144**: 269–283.

Manton, S. M. (1965). Functional requirements and body design in Chilopoda *J. Linn. Soc.* (Zool.) **45**: 251–501.

Ramsay, J. A. (1935). Methods of measuring the evaporation of water from animals. *J. exp. Biol.* **12**: 355–372.

Rilling, G. (1960). Anatomie des braunen Steinlaufers *Lithobius forficatus* L. (Chilopoda). Skelettmuskelsystem, peripheres Nervensystem und Sinnesorgane des Rumpfes. *Zool. Jb.* (*Anat.*) **78**: 39–128.

Roberts, H. (1956). *An ecological study of the arthropods of a mixed beech-oak woodland, with particular reference to Lithobiidae.* Ph.D. thesis, University of Southampton.

Verhoeff, K. W. (1941). Zur Kenntnis der Chilopodenstigmen. *Z. Morph. Ökol. Tiere* **38**: 96–117.

Discussion

Wallwork: Referring to your graph of water loss of *Lithobius forficatus* and *L. variegatus*. If you bear in mind the surface/volume ratio surely you would expect a higher rate of water loss (per unit body weight) in the smaller species?

Curry: The water loss is indeed greater, the smaller the stadium (of one species). I haven't tried any of the smaller species of *Lithobius*.

WALLWORK: I wonder whether there may be some form of waterproofing notwithstanding the absence of an obvious lipid layer on the cuticle. In terrestrial isopods there's some suggestion now that there may be a water barrier of some kind, perhaps even in the endocuticle. It is perhaps worth looking for?

CURRY: Blower (1951)* records that lipids are poured onto the surface of the chilopod cuticle; this layer is more obvious in geophilomorphs than lithobiomorphs. Noting the inability of these animals to retain water, he suggested that the function of the surface lipid was to prevent the ingress of water.

WALLWORK: I wonder whether there is a difference between the cuticles of *Lithobius forficatus* and *L. variegatus* which might explain your graph?

CURRY: I would like to examine these under the EM but I have not got around to this yet.

BLOWER: The difficulty in comparing these two lithobiids stems from their different shapes which cannot yet be taken into account when estimating surface and volume.

CURRY: The equation which I have used will have to be improved eventually.

* See list of references p. 380.

Symp. zool. Soc. Lond. (1974) No. 32, 383–404.

CONTRIBUTION TO THE MORPHOLOGY OF CHILOPOD EYES

R. R. BÄHR

Zoologisches Institut der Universität Münster/Westf., Germany

SYNOPSIS

The ultrastructural organization of the ocelli of *Lithobius forficatus* (L.) in general and under conditions of light- and dark-adaptation is presented. As accepted by Hesse (1901), the ocellus contains pigmented visual cells of two different morphological types with characteristically distributed cell-elements. In normal and dark-adapted distal receptors forming the closed part of the rhabdom, an extended perirhabdomal Schaltzone (intercalary zone) (Hesse, 1901) can be observed, consisting of an accumulation of endoplasmic cisternae, palisades and vacuoles of variable size and diameter.

Illumination of low intensity changes expansion as well as structure of the Schaltzone and is accompanied by migration of pigment and mitochondria. It is suggested that in the ocellus of *Lithobius*, which corresponds to the apposition type, reversible variations of Schaltzonen and pigment migration represent a pupil-like mechanism, influencing the rate of absorption within the rhabdom. The morphological properties described are discussed in relation to the unknown system responsible for the adaptive structural movements.

INTRODUCTION

Anatomical and histological investigations on the eyes of a few selected species of chilopods using light microscopy have been made by Graber (1880), Grenacher (1880), Carrière (1885), Willem (1892), Hesse (1901) and Hanström (1934). With respect to the supposed primitive condition of the chilopod eye, these authors found remarkable differences between species as well as between genera in the number of ocelli, the level of aggregation and the histological organization. For example, the scolopendrids have two ocelli on each hemisphere and their eye-cup has a crystalline cone. *Lithobius forficatus*, one of the best-known species, shows an accumulation of ocelli ("gehäufte Punktaugen") which appear to have a two-layered receptor epithelium (Grenacher, 1880; Willem, 1892; Hesse, 1901). Only the Scutigeromorpha have a well-developed faceted eye which is homologous to the compound eyes of other arthropod groups. *Scutigera coleoptrata* has up to 250 ommatidium-like ocelli (Bähr, in prep.), containing crystalline cones, a smaller number of distal and proximal cells and two kinds of pigment cells (see Hanström, 1934).

In addition to these investigations on the structure of the ocellus and its constituents, the function of these visual sense organs was studied in behavioural experiments carried out by Plateau (1886), Verhoeff (1925), Klein (1934), Scharmer (1935), Görner (1955) and Meske (1960). From these results I was led to examine the ocelli of chilopods, especially those of *Lithobius forficatus*, with regard to their functional significance, using electrophysiological methods. The electroretinograms I obtained indicated that illumination causes depolarization of the sensory cells dependent on the intensity, duration and frequency of the stimulus. I also found that light sensitivity increases during dark-adaptation (Bähr, 1965, 1967).

In contrast to our growing knowledge of the ultrastructure of insect and crustacean compound eyes, little has been published on the fine structure of chilopod ocelli. In his work on *Limulus* eyes Miller (1957) mentioned the rhabdom configuration of *Scutigera* which he found similar to that of the bee. The first ultramicroscopical investigation was made by Bedini (1968) on the lithobiid *Polybothrus fasciatus*. Bedini was unable to reconcile her findings on this species with existing histological descriptions of the ocelli of *Lithobius forficatus*. The differences are possibly due to the different methods used, different planes of sections and also the variation of homologous structures within species. In 1969 Joly published electron-microscopical data on *Lithobius forficatus*. He was unable to confirm the differences within the receptor layer described by Grenacher (1880), Willem (1892) and Hesse (1901).

In this paper I give the results of my examination of *Lithobius* ocelli by electron microscopy, attempt to correlate these details of fine structure with the histology as revealed by the light microscope and describe the cytological changes occurring during light- and dark-adaptation.

MATERIAL AND METHODS

Adult specimens of *Lithobius forficatus* (L). of 2·5 cm body length, freshly collected and dark-adapted over more than 15 h, were used. Under dim red light the heads were cut off and fixed in the mixture of Fahrenbach (1969), or in glutaraldehyde 2–3%, buffered with sodium cacodylate or phosphate buffer (0·1 M) after Sörensen at pH 7·2 for 1 h. Other fixatives have also been used (see Bähr, 1971). After postfixation in buffered osmium tetroxide (1%) objects were dried over alcohol and acetone and embedded in Araldite. Thin sections were made by an LKB ultratome, double stained with uranyl acetate and lead citrate and observed under a Zeiss EM 9A electron microscope. Ultrastructural changes were studied in specimens illuminated for various periods ranging up to 3 h. Illumination intensity ranged from 200 Lux to sunlight

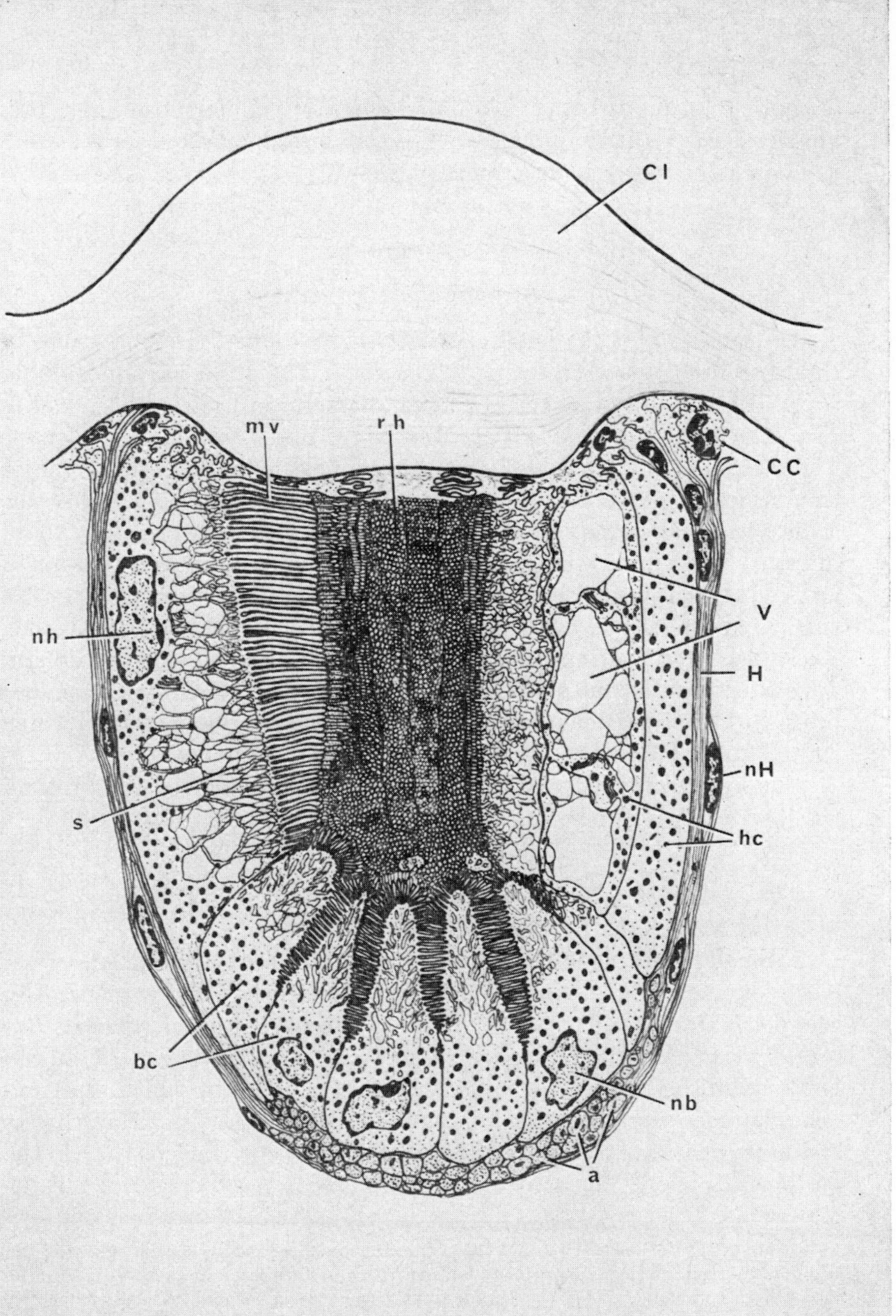

Fig. 1. An ocellus of *Lithobius forficatus* reconstructed from longitudinal sections at different levels. a: axons; bc: basal cell; cc: corneagenous cell; cl: corneal lens; H: covering cell (Hüllzelle); hc: hair cell (distal receptor); mv: microvilli; nb: nucleus of basal cell; nH: nucleus of covering cell; nh: nucleus of hair cell; rh: rhabdom; s: intercalary zone (Schaltzone); v: vacuole of intercalary zone.

of more than 6×10^4 Lux. To find reversible cellular structures, previously dark-adapted animals were exposed, dark-adapted for a certain time and then fixed as described above.

RESULTS

General organization

On each side of the head of *Lithobius forficatus* there is a group of single ocelli arranged in five to seven rows. The ocelli vary in number (up to 40 in stadium maturus senior), diameter of their corneae and in size. The most dorsal ocelli are the larger and have elliptical lenses. The ocellus at the lateral edge of the eye field is the largest; its major axis is nearly perpendicular to those of its neighbours but it has the same histological organization as they do; its rhabdom measures 75μm in transverse sections (major axis). A general view of the constitution of an ocellus, as revealed by electron microscopy, is shown in Fig. 1. The single eye consists of (1) a corneal lens (cl), (2) corneagenous cells (cc), (3) Hüllzellen (covering cells) (H), which cover the whole eye-cup and appear as modified and sunken epithelial cells, and (4) a variable number of up to 110 visual cells, discernible by their sensory structures, which contribute to either the distal or the proximal rhabdom.

Small pigment cells, which Grenacher (1880) believed to surround the internal cuticle between the lenses, have not been discovered.

The fine structure of the components of the ocellus

Corneal lens (cl)

From different longitudinal sections it is clear that all lenses are biconvex (normally 50 μm in diam. and 35 μm thick) with symmetrically rounded external and in most cases a tub-like internal surface. Externally a predominantly homogenous epicuticular layer of 2–3 μm can be distinguished from the underlying endocuticle, in which electron-dense laminae alternate with less dense interlaminar layers. These layers are more separated from each other in the middle of the lens than in the peripheral region. The lamellated structure of the dioptric apparatus

---

FIG. 2. (a) Endocuticular part of the corneal lens at high magnification, showing the alignment of microfibres (arrows) in laminae (l) and interlaminar layer (il). (b) the structure of corneagenous cell membrane dilatation between corneal lens and the central rhabdom. cc: corneagenous cell; en: endocuticular part of the lens; md: interdigitating membranes of the corneagenous cells; rh: rhabdom. (c) Slightly transverse section between the interlens space, showing the distal part of hair cells including nuclei and corneagenous cell bodies. hc: hair cell; mH: membranes of covering cells (Hüllzellen); nc: nucleus of corneagenous cell; nh: nucleus of hair cell; pi: pigment grains.

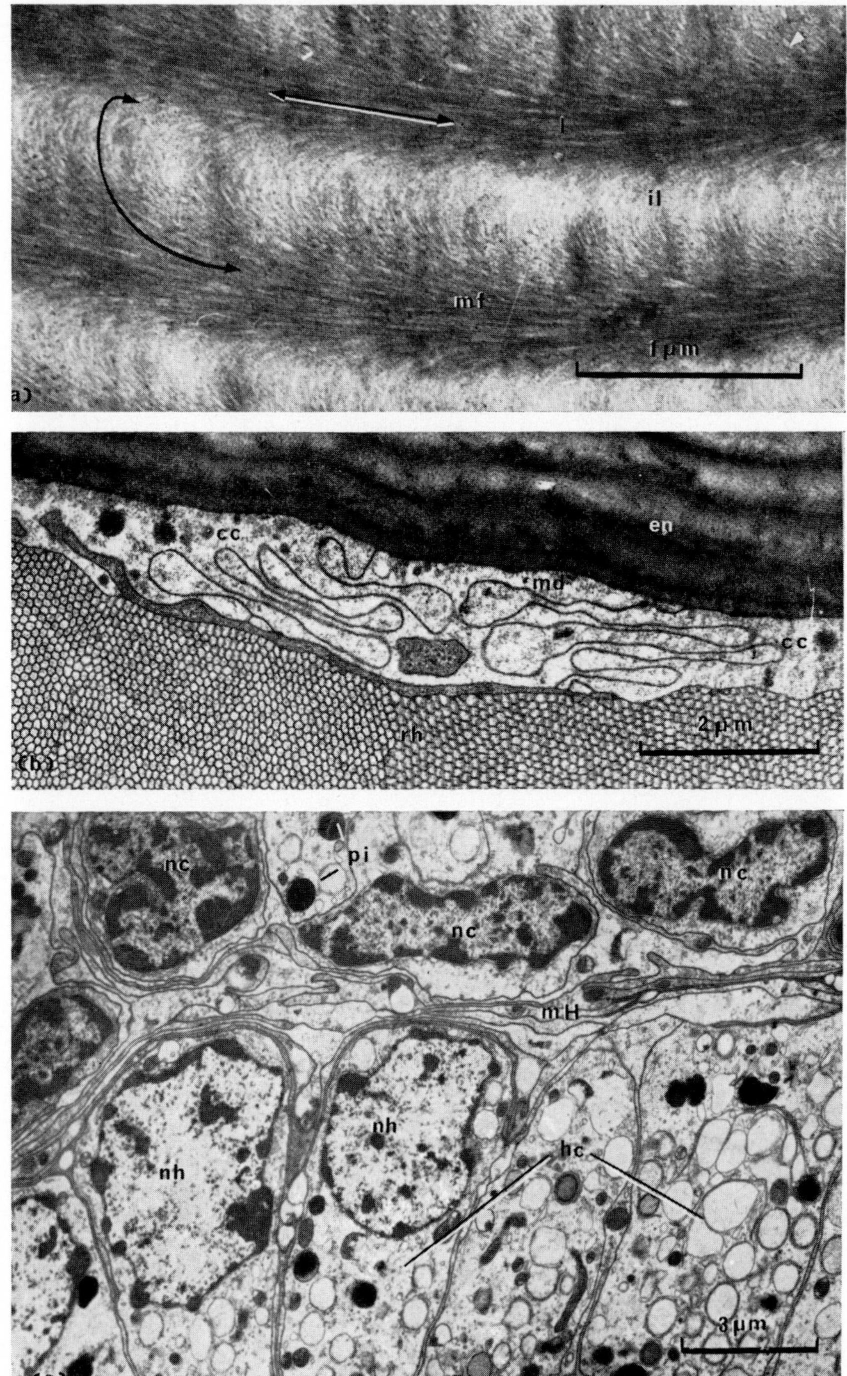

Caption on opposite page

can be explained by the different alignment of microfibres running parallel in the laminae and changing their direction by approximately 180° within the interlaminar layer (Fig. 2a).

Corneagenous cells (*cc*)

In contrast to the observations of Bedini (1968) in *Polybothrus* the corneagenous layer in *Lithobius* does not have a regular appearance. Usually the cell bodies including the flattened nuclei (5–11 μm diam.) are located lateral to the internal vault of the lens (Fig. 2c). Within the central part, i.e. the optical axis, one can find, with few exceptions, only three to nine membranous dilatations of the corneagenous cells which interdigitate (Fig. 2b). Nuclei apparently situated in the midline as drawn by Grenacher (1880) result from tangential sectioning, or possibly from displacement of the optical axes of lens and eye-cup in relation to each other. The cytoplasm of the corneagenous cells appears very osmiophilic in different ocelli, containing a rough endoplasmic reticulum and numerous free ribosomes. This is presumed to be a sign of the metabolic activity of these cells. The high content of cytoplasmic microtubules is noteworthy. Other organelles are occasionally observed, mitochondria, a very few small pigment grains, cisternae of ribosome-free ER and also Golgi structures. Intercellular connections occur at the membranes close to the internal surface of the lens as well as between the corneagenous and receptor cells. My investigations (Bähr, 1971) do not support Grenacher's (1880) observation of peripherally located pigment cells. The corneagenous cells are too heterogenous and have insufficient pigment grains to serve as screening cells.

Hüllzellen (*covering cells*) (*H*)

The Hüllzellen (cellules bordantes, Joly & Herbaut, 1968; satellite cells, Bedini, 1968) near the corneagenous cells exhibit morphological properties transitional between the corneagenous cells and the cells detached from the peripheral layer which cover the eye-cup. Their structural aspects and distribution suggest three possible functions:

1. secretion of the endocuticular part of the lenses as shown in Fig. 3;
2. the morphological and mechanical separation of the ocelli from each other; and
3. the isolation of one or more axons.

Hüllzellen covering the sensory epithelium form a flat epithelium with prolonged cell bodies and spindle-shaped nuclei (2×9 μm). In longitudinal sections their cell membranes are attached, and intercellular spaces are not always recognizable. From this evidence one might expect that the Hüllzellen also serve as a physiological barrier between the haemolymph space and the receptor layer; this is confirmed by ferritin incubation experiments, which were found negative (Bähr, 1971). Cytoplasmic inclusions consist of mitochondria, lipoid droplets, small vesicles, glycogen particles and microtubules in a proximal-distal direction parallel to the cell membranes. The ER is poorly developed. Dense bodies and multivesicular bodies of the fenestrated type are present.

The photoreceptor cell layer

Grenacher (1880) was the first to notice the difference between the composition of the eye-cup proximally and distally. He discriminated:

> "Ein Kranz grosser prismatischer Zellen, die, keilförmig gestaltet, sich zu einem dickwandigen, durch und durch pigmentierten Hohlcylinder zusammenfügen",

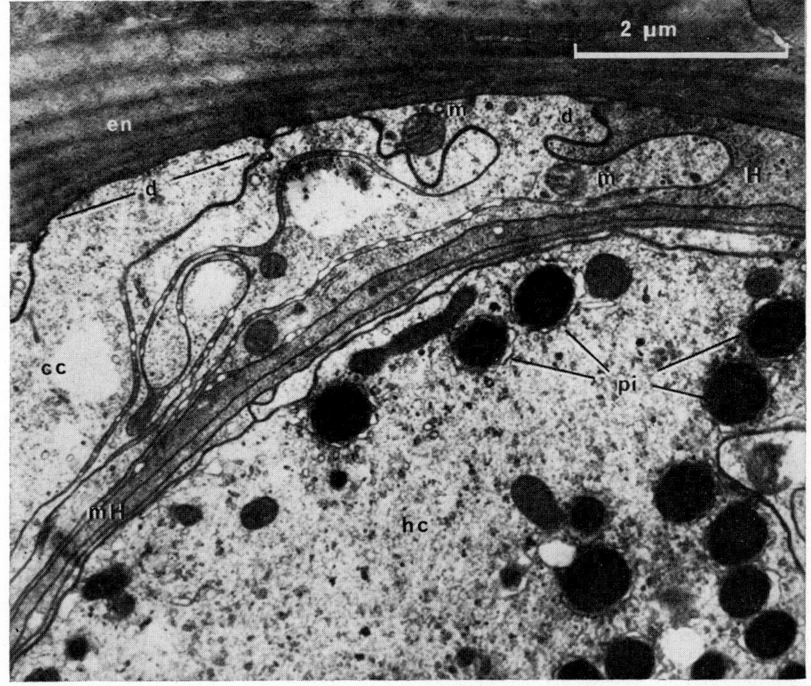

FIG. 3. Longitudinal section through the distal part of the eye-cup covering. cc: corneagenous cell; d: desmosome; en: endocuticle; hc: hair cell; m: mitochondrium; mH: distal processes of the covering cells (Hüllzellen) (H), connected with the lens internal surface; pi: pigment grains of hair cell.

and (p. 443):

> "Den hinteren Theil des Auges bildet wieder die halbkugelige, hier nur von einer geringen Zahl von Zellen (einigen zwanzig nach meiner Schätzung) gebildeten Retina . . .".

This spatial separation and distribution of the receptor cells led Hesse (1901) to the term "beginnende Zweischichtigkeit" and in agreement with Willem (1892) he identified Grenacher's "Haarzellen" as sensory

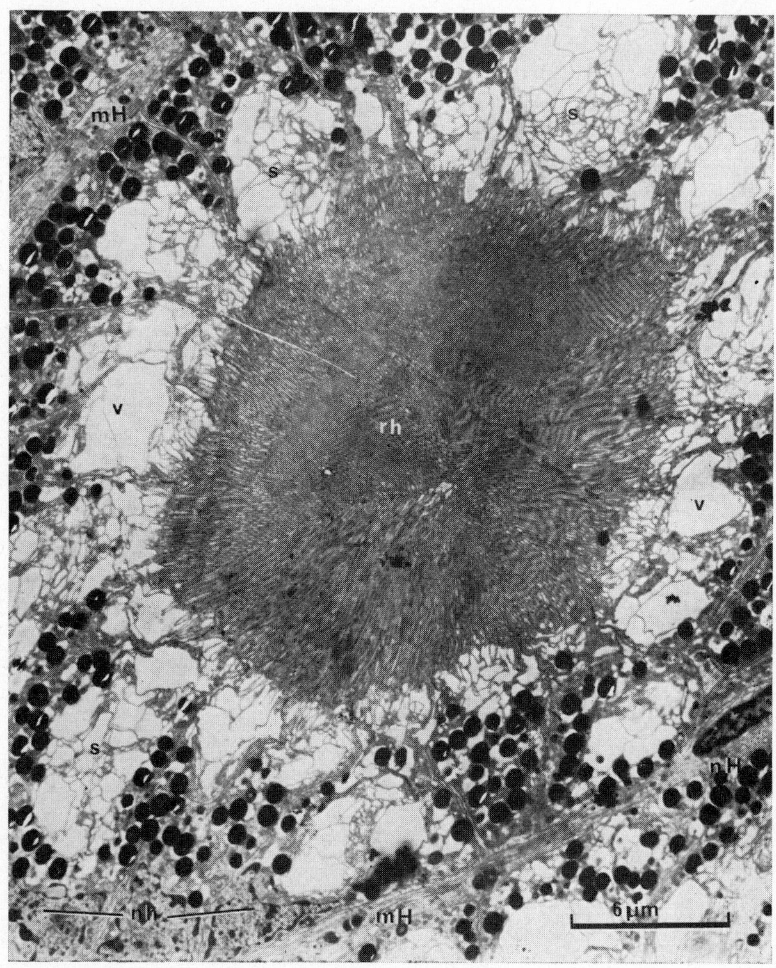

FIG. 4. Transverse section of a dark-adapted ocellus. mH: membranes of covering cells; nH: nucleus of covering cell; nh: nucleus of hair cells; rh: fused rhabdom; s: intercalary zone (Schaltzone); v: large vacuole space.

and, moreover, visual cells. Recent investigations confirm the existence of two kinds of receptor cells, distinguished by their ultrastructural properties. Bedini (1968), however, has clearly demonstrated that *Polybothrus fasciatus* possesses only one type of sensory cell which, so far as I can determine, corresponds to the basal cells of *Lithobius*.

The distal receptors or hair cells (Grenacher, 1880) (*hc*). The distal and greater part of the eye-cup is occupied by about 25 to 70 large, somewhat prismatic sensory cells (Fig. 5.) Their symmetrical arrangement is shown in Fig. 4 where they are dark-adapted. The hair cells of the dark-adapted state show a succession of sectors because of the characteristic distribution of their cell elements from the axis outwards: (1)

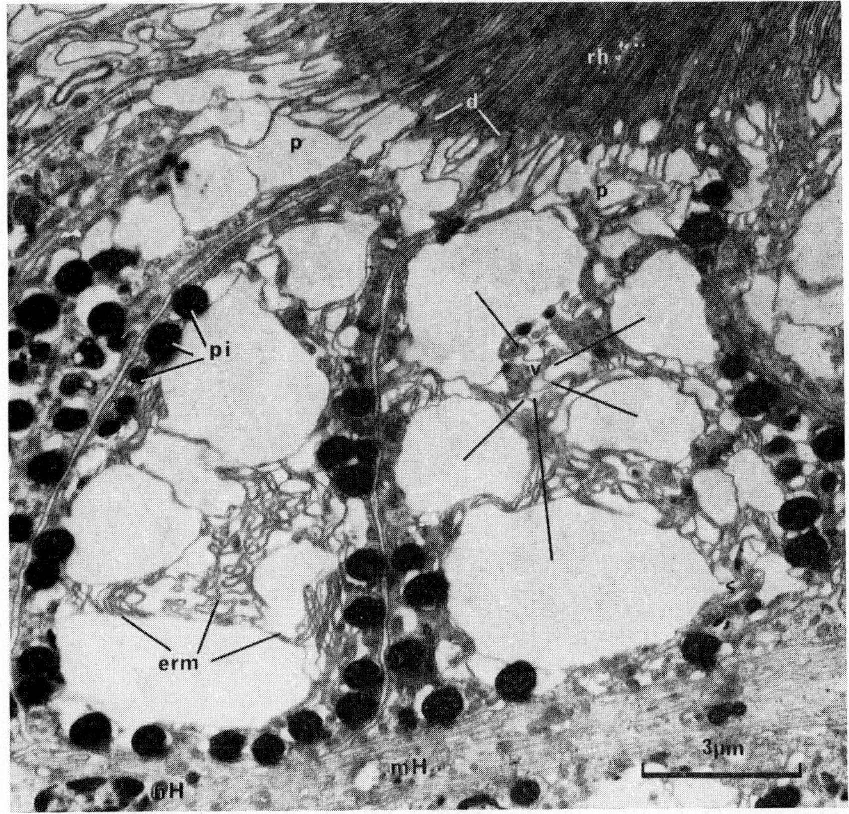

FIG. 5. Hair cells (dark-adapted) in transverse section at high magnification. d: desmosome; erm: endoplasmic membrane system; mH: membranes of covering cells; nH: nucleus of covering cell; p: palisade structure; pi: pigment; rh: rhabdomeres; v: large vacuoles of the intercalary zone (Schaltzone).

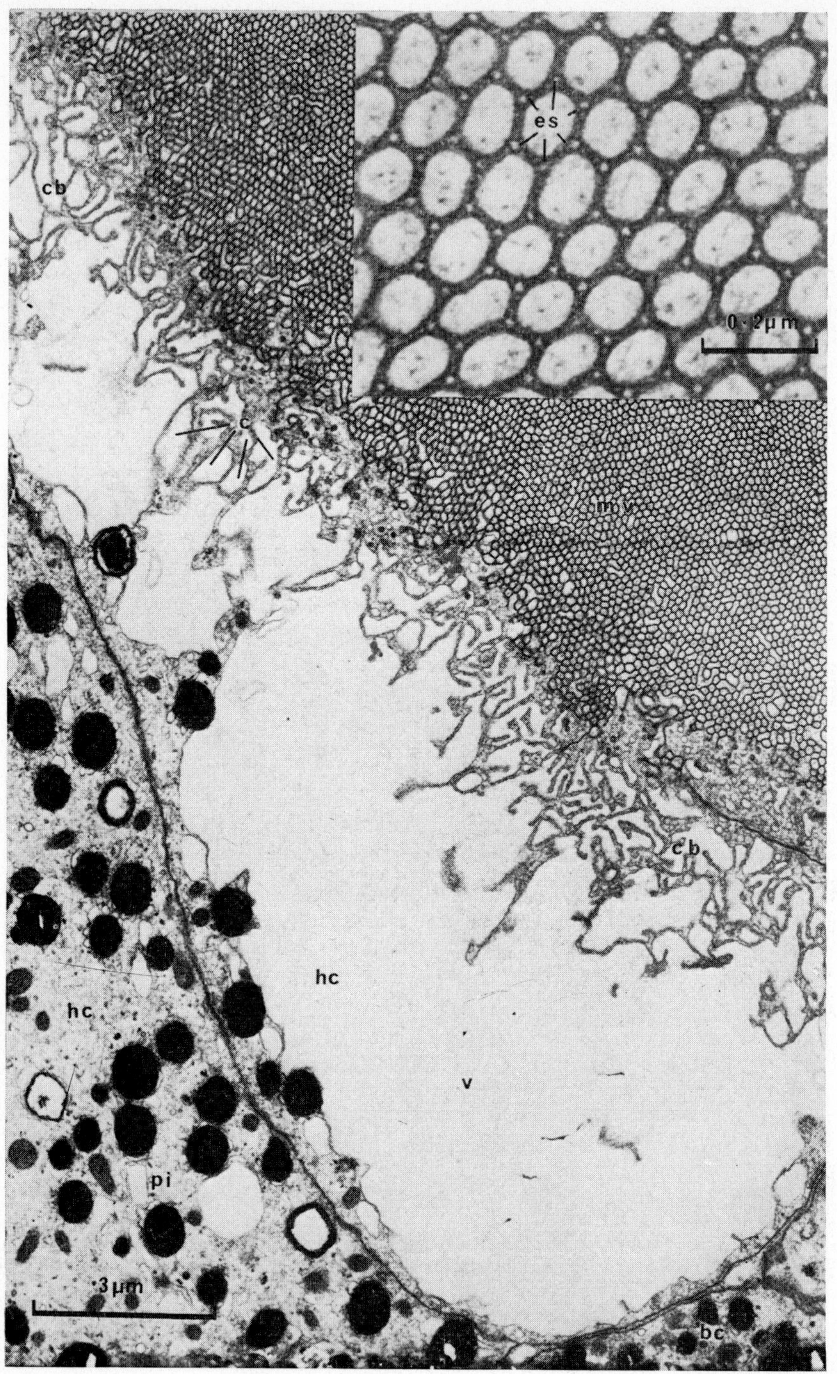

Caption on opposite page

the rhabdomere (rh); (2) the so-called Schaltzone (intercalary zone) (s); (3) the main cytoplasmic part with cell organelles; and (4) the nerve fibre process. As known from other photoreceptors, the rhabdomeres are composed of fine tube-like membrane formations, the microvilli (see Fig. 6 inset). Their insertion, perpendicular to the optical axis, occurs only at the centrally oriented membrane for a distance of 25–30 μm, thus forming the fused part of the rhabdom (Fig. 1). Compared with other arthropod visual cells, the microvilli of *Lithobius* are of large size (1–17 μm in length and 750–2500 Å in diam.). Because of their dimensions they were already known from light microscope studies and therefore called "hairs" or "hair-border" (Grenacher, 1880).

Beyond the microvilli extending towards the eye-cup periphery is a clear, bright region, the Schaltzone. The term was first used by Hesse (1901), to describe a zone between the cell body and the microvilli, and did not then include any physiological significance, as shown below. Furthermore, Hesse mentioned that small fibres which he called neurofibrils run from the dilated base of the microvilli through the Schaltzone and the cell body to the axon. As shown in Fig. 4, the Schaltzonen of the distal receptors may completely surround the fused rhabdom, depending on the state of dark-adaptation. Electron microscopic findings reveal that the Schaltzone consists of cytoplasmic bridges which surround small cisternae of the smooth endoplasmic reticulum, and, further from the centre, of vacuoles with low electron density (Figs 4, 5, 6). The alignment of the plasmic bridges and lengthened cisternae (also called "palisades" by Horridge & Barnard, 1965), led Hesse (1901) to accept the presence of fine microvilli processes. The less developed Schaltzonen in Hesse's illustration and those shown electron-microscopically by Joly (1969) seem to be at an intermediate stage in the process of dark-adaptation. Under these conditions the Schaltzone is extended around the rhabdom over the whole vertical cell axis and thus determines the distribution of the cytoplasmic cell organelles (Fig. 6). I observed a large number of several kinds of multivesicular bodies, large multilamellated bodies (MLB or onion body) as a part of the specialized smooth endoplasmic reticulum, a high content of pigment grains, Golgi structures (mostly located in the cytoplasm neighbouring the Schaltzone), mitochondria, ribosomes, lysosomes and microtubules. In animals dark-adapted over more than 12 h the nuclei

---

FIG. 6. Hair cells (dark-adapted) in oblique longitudinal section at different cell levels. bc: basal cell; c: cisternae of the endoplasmic reticulum; cb: cytoplasmic bridges; es: extracellular space between the microvilli; hc: hair cell; mv: microvilli (at high magnification see inset); pi: pigment grains; v: large vacuole space within the intercalary zone (Schaltzone).

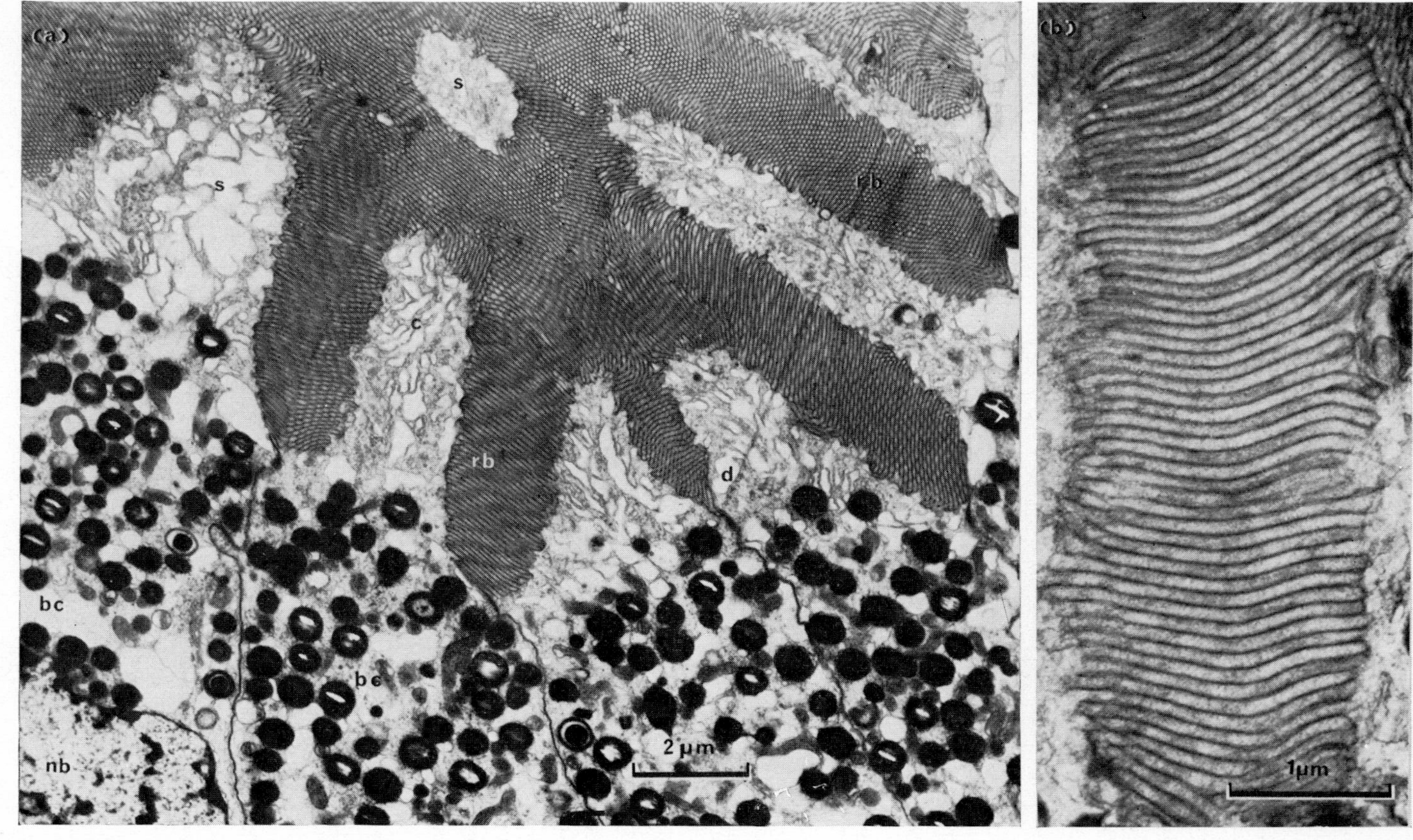

FIG. 7. (a) Proximal part of the ocellus in slightly oblique longitudinal section. bc: basal cell; c: cisternae of the intercalary zone; d: desmosome; nb: nucleus of basal cell; rb: rhabdomeric structure of the basal cells; s: intercalary zone (Schaltzone). (b) Single rod-like structure of the interdigitating rhabdomeric microvilli.

of the hair cells are found to be situated predominantly in the distal region of the cell (compare Fig. 2c). From the level of the lowest microvilli the oblique inner membranes converge at the cell base which becomes small and forms the nerve fibre. The axons (0·4–0·5 μm in diam.) occur at the periphery of the eye-cup and are single or grouped, surrounded by Hüllzellen. From the ocellus nerve appearing between the

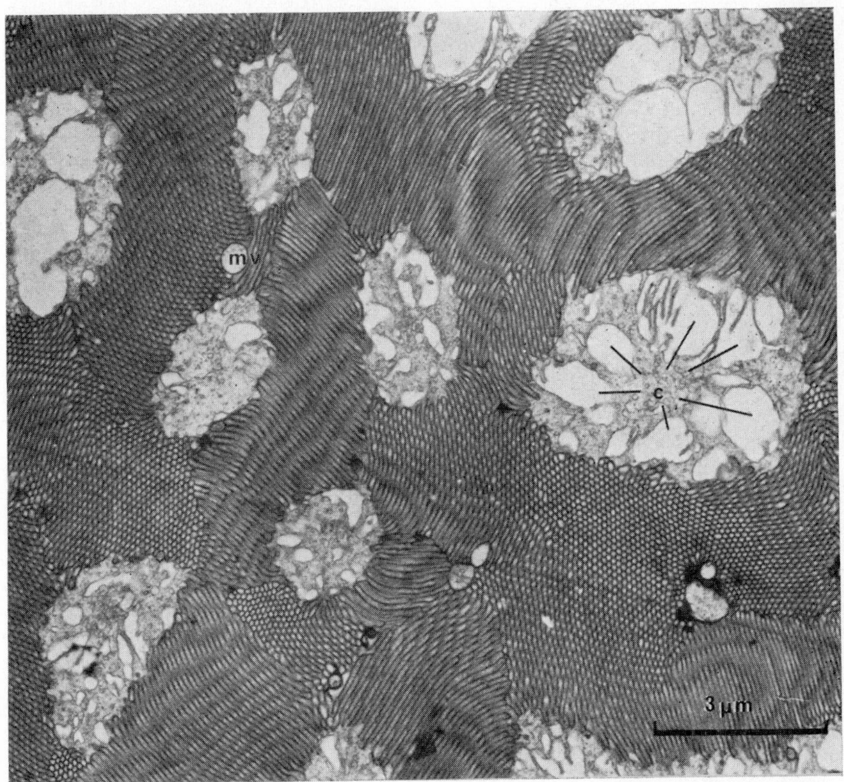

Fig. 8. Transverse section through a dark-adapted ocellus below the fused rhabdom at the level of the intercalary zones of basal cells. c: enlarged cisternae of the endoplasmic reticulum; mv: microvilli of basal cells at several planes.

ocelli one can estimate by counting the axons the number of sensory cells which form the eye-cup. Components of the axoplasm are mitochondria, neurotubules, vesicles of 800–1000 Å and some of 4000 Å diam., and rare neuronal multivesicular bodies.

The basal cells (*retina*) (*bc*). Longitudinal and transverse sections through the proximal part of the ocellus show quite a different rhabdomeric organization to that of the distal part. In this context, one should refer to the findings of Grenacher (1880), who surveyed the so-called "Stäbchen" within the basal eye-cup and noticed a fine platelet-like construction. Like Willem (1892), Hesse (1901) was able to show the arrangement of the microvilli of the basal cells. The relation between

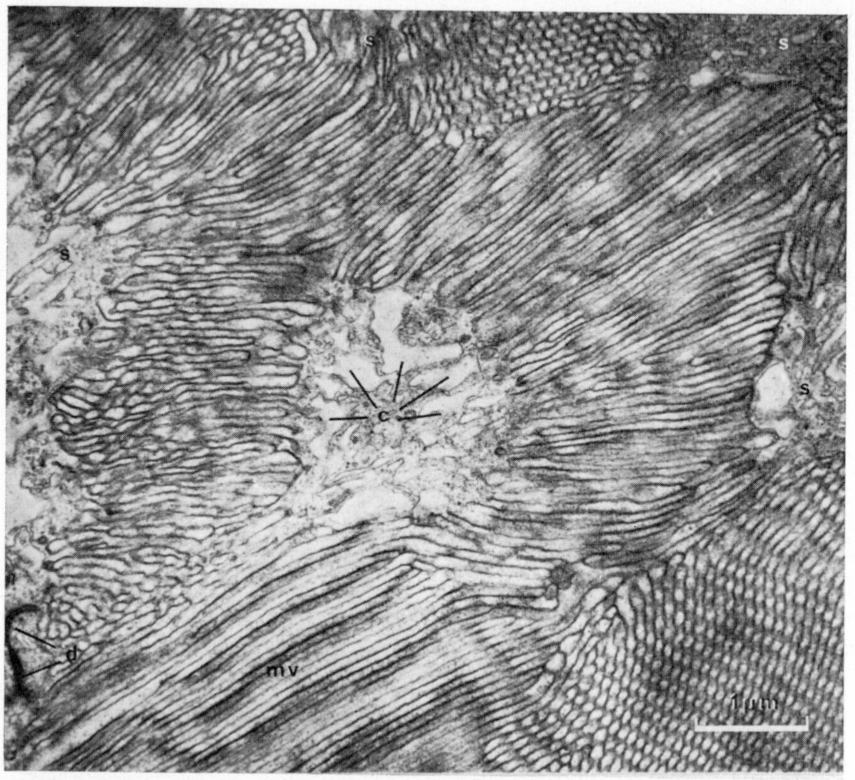

FIG. 9. Single basal cell with microvilli arising mainly from diametrically opposite poles. c: cisternae of endoplasmic reticulum; d: desmosome; mv: microvilli; s: intercalary zones of neighbouring basal cells.

the proximal sensory cells and their contribution to the formation of the "Sehstäbchen" is shown in Figs 7 and 8. While the microvilli of the hair cells are perpendicular to the axis of the ocellus, those of the basal cells are aligned approximately perpendicular to the long diameter of the cell body. The microvilli are inserted on all sides of the distal and apical cell membranes, but most of them lie on opposite sides. Microvilli of neighbouring basal cells are not separated by rhabdomeric radii but they interdigitate. From this spatial arrangement it is easy to see how the stilt- and platelet-like formation of the rhabdomeric structure in the retinal part of the eye-cup is derived. Another feature of the basal cells is their Schaltzone, quite unlike the distal Schaltzonen which completely surround the fused rhabdom. Here the prominent structure is localized between the membranes edged with microvilli (Figs 7a, 8, 9). The ultrastructural aspect of the Schaltzone components is quite similar to that described for the distal receptor type. Sometimes an axial cytoplasmic rope from which derive small plasmic bridges, curving to the base of the microvilli, can be found. Large vacuolar spaces rarely appear except in the basal cells which are directly connected to the hair cells. The content of cytoplasmic cell organelles resembles that of the distal receptors. Pigment grains are distributed over the cytoplasm up to the level of the desmosomes (Fig. 7a). The description just given refers to the dark-adapted state of the ocellus. In the next paragraph I discuss the evidence for structural changes (a) during illumination and (b) during darkness preceded by illumination.

Light-induced changes of the visual cells

In earlier experiments it was stated that the duration of darkness influences the rate of the Schaltzone extension. Changes should also occur if ocelli which have been dark-adapted for a long period are illuminated. Some results of these studies, using a constant illumination intensity of 250 Lux for serial tests, are now reported.

After exposure for 5 min (250 Lux) the distribution of Schaltzone elements is already reduced to less than half their original extent. The cytoplasmic sector has enlarged and the pigment grains and mitochondria have moved against the rhabdomere (Fig. 10). This effect is increased by prolonged illumination and depends also on light intensity. After 25 min illumination the peripheral part of the Schaltzone is occupied by pigment grains forming a screening ring around the fused rhabdom, which seems to be complete after 60 min exposure (Fig. 11). During movements of pigment and mitochondria, the fine structure of the Schaltzone also changes. Under 250 Lux the Schaltzone cross-sectional area is reduced from 35–42% to 6–7% (ocellus area = 100%). In most cases the perirhabdomal layers of the palisade are diminished or disappear, while the connexions of the larger vacuoles to the rhabdomeric evaginations become more numerous. As shown by Horridge & Barnard (1965) and Menzel & Lange (1971) the endoplasmic cisternae could be dispersed within the cytoplasm, but there is no evidence in *Lithobius* of the manner in which the cisternae and vacuoles, prominent in the dark-adapted state, are dispersed into the cytoplasm by illumination.

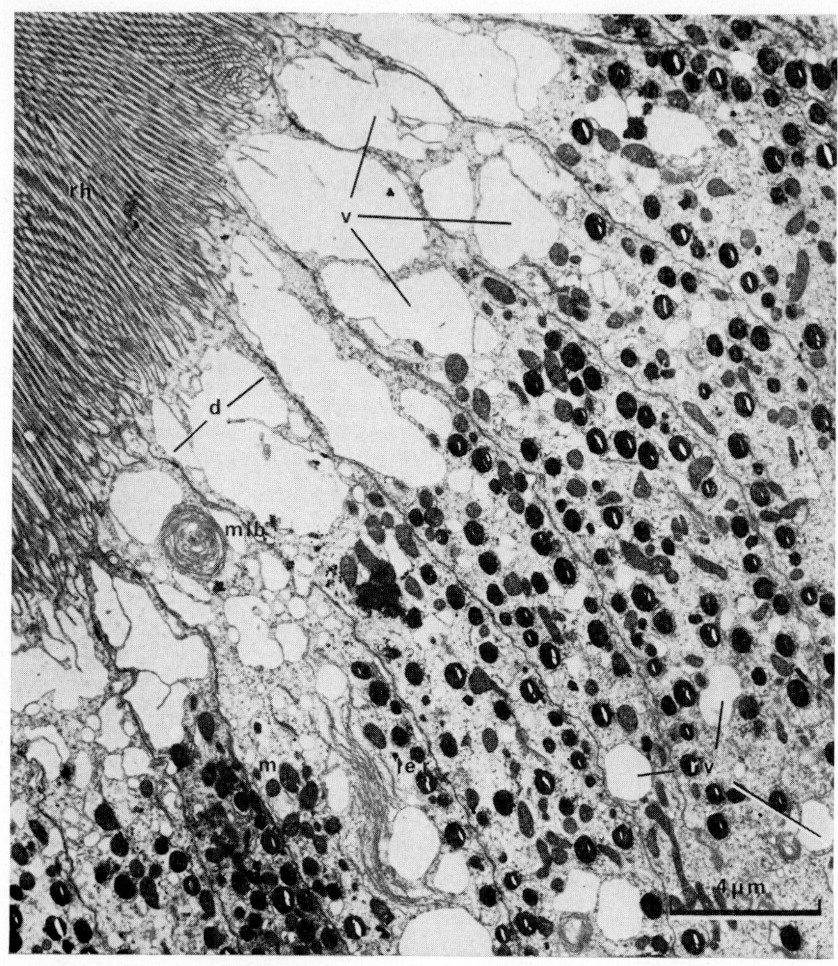

Fig. 10. Hair cells of ocellus illuminated for 5 min with 250 Lux in transverse section. d: desmosome; ler: lamellar endoplasmic reticulum; m: mitochondria; mlb: multilamellar body; rh: rhabdom; rv: resting vacuoles of the original intercalary zone; v: vacuoles.

Possibly some of the membranes surrounding the endoplasmic cisternae and vacuoles could be transformed by illumination to form the lamellated ER-membranes which may be the precursors of the multilamellar bodies. However, at this time, it is impossible to explain the change in distribution of endoplasmic constituents induced by illumination (Figs 10, 13). Intense or prolonged illumination has another effect, concerning the arrangement and organization of the microvilli. The parallel alignment characteristically found in dark-adapted animals is affected by light. From the base of the microvilli as well as in the central area of the rhabdom the microvilli membranes undergo structural

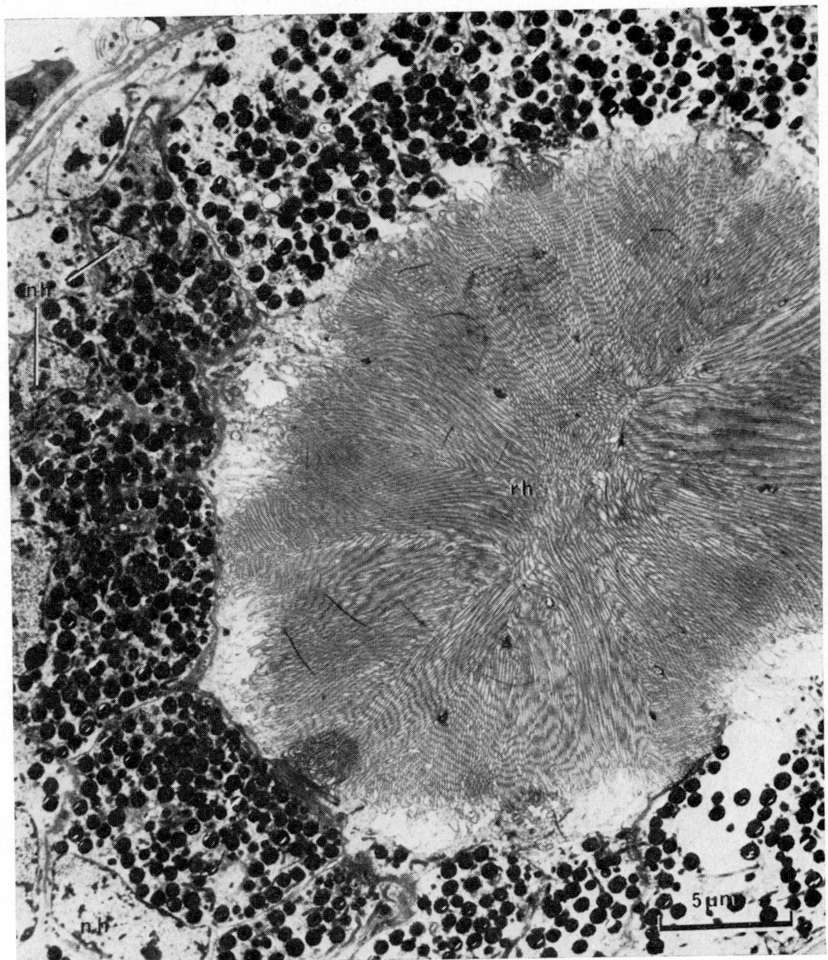

FIG. 11. Transverse section of a *Lithobius* ocellus illuminated for 1 h at 250 Lux, showing the translocation of pigment around the fused rhabdom.

changes. There are indications of a possibility that in slightly illuminated states, elements of the palisade endoplasmic reticulum can reconstitute parts of the microvilli membranes. Furthermore, I have found that multilamellar bodies seem to be transferred to the rhabdomere and are used as material for the microvilli membrane system.

There is also nothing known about the mechanism of pigment migration, which could be mediated by a passive transport within the cytoplasm or by a structure-bound system involving endoplasmic membranes, microtubules or the microfilamentous organelles.

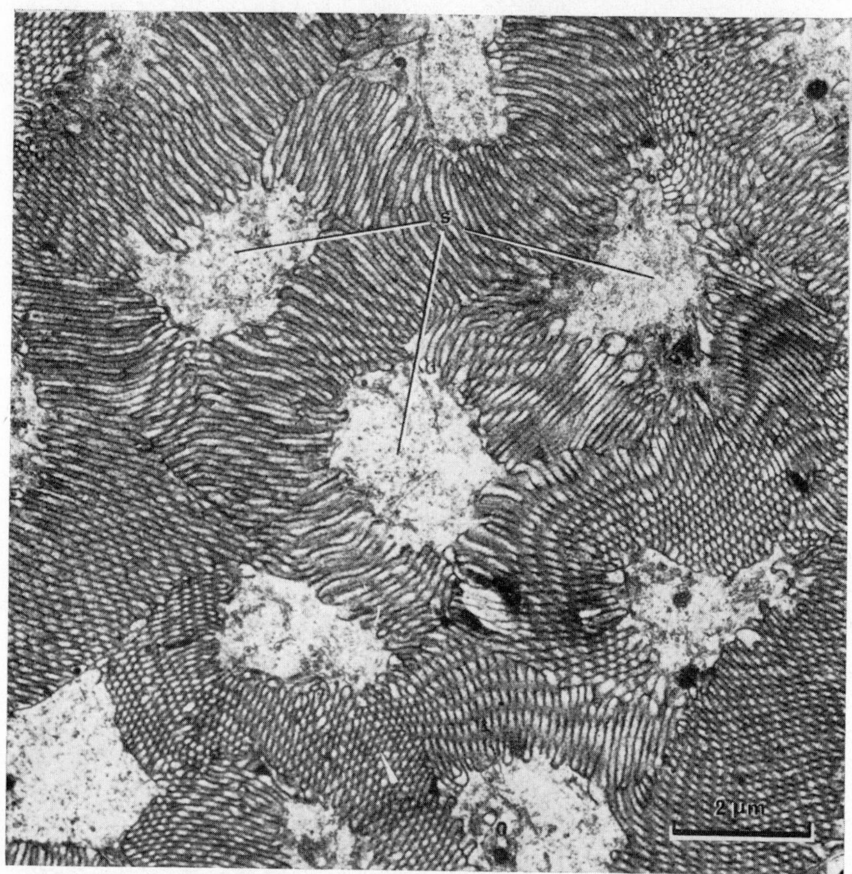

FIG. 12. Transverse section through the ocellus below the fused rhabdom as in Fig. 8 after 60 min illumination (250 Lux). Note the cytoplasmic structure in which the cisternae have disappeared (compare Fig. 8). s: intercalary zones (Schaltzonen) of neighbouring basal cells.

FIG. 13. Appearance of complex multilamellar bodies after 90 min illumination (250 Lux). (a) Transverse section through hair cells at proximal level. (b) Multiple formation of multilamellar bodies in basal cells. c: cisternae of endoplasmic reticulum; cy: central cytoplasmic portion within the mlb; mH: membranes of covering cells; mlb: multilamellar bodies; mv: microvilli undergoing light-induced structural changes; nh: sunken nucleus of hair cell.

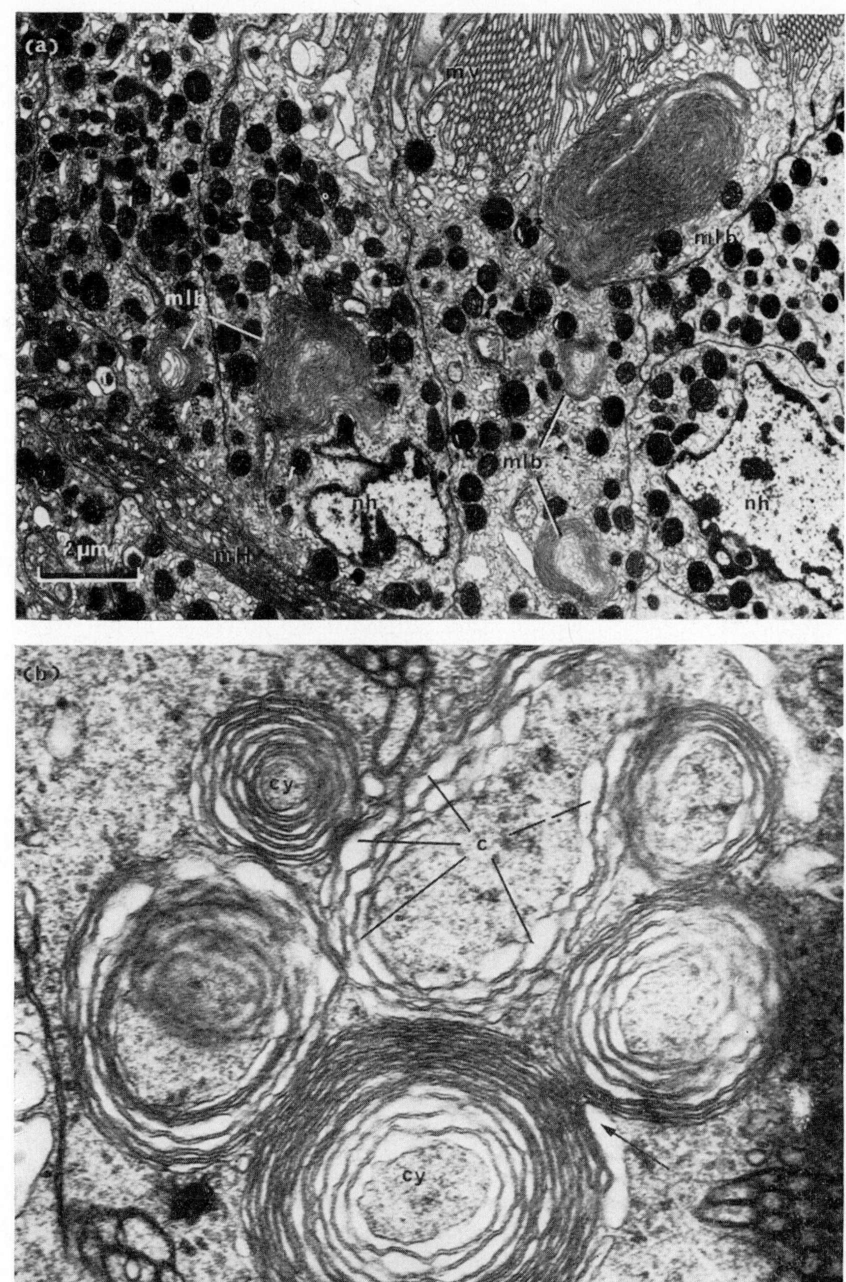

Caption on opposite page

The migration of pigment grains within the basal cells is not influenced in the same manner as is the pigment distribution in the hair cells. The position of pigment in the dark-adapted cells and the slight movement of grains between the "Sehstäbchen" during illumination suggest that pigment translocation had already occurred while the animal was handled under dim red light. The enlarged endoplasmic cisternae of the basal receptors as shown in Figs 7a and 8 diminish or disappear when illuminated (Fig. 12).

Although there exist no exact correlations between duration of illumination and adaptational changes in all specimens, nevertheless it is possible to show whether ocelli were excited by light or not.

Hence, the question arises, are Schaltzone changes and pigment migration reversible on return to darkness after illumination? I have tested some animals and found that the large vacuoles extend, and are surrounded by vesicles of 0·5–0·9 μm in the cytoplasm bordering the vacuoles. After 60 min exposure mitochondria and pigment have also moved towards the ocellus periphery. The reconstitution of the fine cytoplasmic bridges within the palisades is however a long-term process, more pronounced in the distal than in the basal cells.

From the results reported here we might assume that in *Lithobius forficatus* the reversible changes of Schaltzonen and pigment distribution resemble an adaptive mechanism regulating the rate of incoming light, functionally similar to those described for *Tachycines* (Tuurala & Lehtinen, 1967a), *Oniscus* (Tuurala & Lehtinen, 1967b), *Musca* (Kirschfeld & Franceschini, 1969), *Calliphora* (Seitz, 1970), and *Formica* (Menzel & Lange, 1971).

Acknowledgements

I wish to acknowledge the valuable technical assistance given to me by Miss Barbara Blum.

References

Bähr, R. (1965). Ableitung lichtinduzierter Potentiale von den Augen von *Lithobius forficatus* L. *Naturwissenschaften* **52**: 459.

Bähr, R. (1967). Elektrophysiologische Untersuchungen an den Ocellen von *Lithobius forficatus* L. *Z. vergl. Physiol.* **55**: 70–102.

Bähr, R. (1971). Die Ultrastruktur der Photorezeptoren von *Lithobius forficatus* L. (Chilopoda: Lithobiidae). *Z. Zellforsch. mikrosk. Anat.* **116**: 70–93.

Bähr, R. (1972). Licht- und dunkeladaptive Änderungen der Sehzellen von *Lithobius forficatus* L. (Chilopoda: Lithobiidae). *Cytobiologie* **6**: 214–233.

Bedini, C. (1968). The ultrastructure of the eye of a centipede *Polybothrus fasciatus* (Newport). *Monitore zool. ital.* (N.S.) **2**: 31–47.

Carrière, J. (1885). *Die Sehorgane der Thiere*. München und Leipzig: R. Oldenbourg.

Fahrenbach, W. H. (1969). The morphology of the eyes of *Limulus*. II. Ommatidia of the compound eye. *Z. Zellforsch. mikrosk. Anat.* **93**: 451–483.

Görner, P. (1955). Optische Orientierungsreaktionen bei Chilopoden. *Z. vergl. Physiol.* **42**: 1–5.

Graber, V. (1880). Über das unicorneale Tracheaten-Auge. *Arch. mikrosk. Anat.* **17**: 58–93.

Grenacher, H. (1880). Über die Augen einiger Myriapoden. *Arch. mikrosk. Anat.* **18**: 415–467.

Hanström, B. (1934). Bemerkungen über das Komplexauge der Scutigeriden. *Acta Univ. lund.* (N.F.) (Avd. 2) **30**: 1–14.

Hesse, R. (1901). Untersuchungen über die Organe der Lichtempfindung bei niederen Thieren VII. Von den Arthropoden-Augen, 2. Die Augen der Myriapoden. *Z. wiss. Zool.* **70**: 347–473.

Horridge, G. A. & Barnard, P. B. T. (1965). Movement of palisade in locust retinula cells when illuminated. *Q. Jl microsc. Sci.* **106**: 131–135.

Joly, R. (1969). Sur l'ultrastructure de l'oeil de *Lithobius forficatus* L. (Myriapode Chilopode). *C.r. hebd. Séanc. Acad. Sci., Paris* **268**: 3180–3182.

Joly, R. & Herbaut, C. (1968). Sur la régéneration oculaire chez *Lithobius forficatus* L. (Myriapode Chilopode). *Archs zool. exp. Gén.* **109**: 591–612.

Kirschfeld, K. & Franceschini, N. (1969). Ein Mechanismus zur Steuerung des Lichtflusses in den Rhabdomeren des Komplexauges von Musca. *Kybernetik* **6**: 13–22.

Klein, K. (1934). Über die Helligkeitsreaktionen einiger Arthropoden. *Z. wiss. Zool.* **145**: 1–38.

Menzel, R. & Lange, G. (1971). Änderung der Feinstruktur im Komplexauge von *Formica polyctena*. *Z. Naturforsch.* **26b**: 357–359.

Meske, C. (1960). *Untersuchungen zur Sinnesphysiologie und zum Verhalten von Chilopoden und Diplopoden*. Dissertation Zool. Inst. Universität Münster.

Miller, W. H. (1957). Morphology of the ommatidia of the compound eye of *Limulus*. *J. biophys. biochem. Cytol.* **3**: 421–428.

Plateau, F. (1886). Recherches sur la perception de la lumière par les Myriopodes aveugles. *J. Anat. Physiol.* Paris **22**: 431–457.

Scharmer, J. (1935). Die Bedeutung der Rechts-Links-Struktur und die Orientierung bei *Lithobius forficatus*. *Zool. Jb.* (Zool.) **54**: 459–506.

Seitz, G. (1970). Eine Pupillenreaktion im Auge der Schmeißfliege. *Zool. Anz.* Suppl. **33**: 169–174.

Tuurala, O. & Lehtinen, A. (1967a). Zu den photomechanischen Erscheinungen im Auge der Gewächshausheuschrecke *Tachycines asynamorus* Adel. *Commentat. biol.* **30**: 1–4.

Tuurala, O. & Lehtinen, A. (1967b). Über die Wandlungen in der Feinstruktur der Lichtsinneszellen bei der Hell- und Dunkeladaptation im Auge einer Asselart *Oniscus asellus* L. *Suomal. Tiedeakat. Toim.* (A.IV) **123**: 3–7.

Verhoeff, K. (1902–1925). *Chilopoda. Bronn's Kl. Ordn. Tierreichs* Band 5 Abt. 2: 1–725.

Willem, V. (1892). Les ocelles de *Lithobius* et de *Polyxenus*. *Bull. Séanc. Soc. malac. Belg.* **27**: 69–71.

Discussion

DOHLE: Can you be more precise as to why you describe the eyes of *Scutigera* as pseudo-faceted eyes and homologous with those of Insecta and Crustacea?

BÄHR: Structures are homologous when they are derived phylogenetically and ontogenetically from the same rudiments. The term "pseudo-faceted eye" comes from the literature (Adensamer, 1893*). The Scutigeridae are the one recent group of chilopods which possess compound eyes formed from a group of conical ocelli. It is known that each pseudommatidium consists of a crystalline cone (also found in *Scolopendra*), distal and proximal pigment cells and a two-layered visual epithelium. These elements of the ocellus of *Scutigera* are certainly homologous with those of the ommatidium of the compound eyes of Insecta and Crustacea. Whether the eye of *Scutigera*, in regard to the known primitive characters of this group, is to be regarded as highly evolved (by convergence), I cannot decide without knowledge of its ontogeny and fine-structure. I would be interested to hear from the taxonomists whether fossil lithobiids possess compound eyes and whether such eyes have been investigated.

* Adensamer, T. (1893). Zur Kenntnis der Anatomie und Histologie von *Scutigera coleoptrata*. *Verh. zool.-bot. Ges. Wien* **43**: 573–578.

Symp. zool. Soc. Lond. (1974) No. 32, 405–410.

PAUROPODA FROM ARABLE SOIL IN GREAT BRITAIN

ULF SCHELLER

Lundsberg, S-68080 Storfors, Sweden

SYNOPSIS

Seven species of Pauropoda (Myriapoda) are listed from arable soil in Great Britain. Three of them, *Allopauropus milloti, Pauropus lanceolatus* and *Polypauropus duboscqi* are new to Britain and one more, *Allopauropus multiplex*, is new to arable soil there.

INTRODUCTION

The Pauropoda are surely richly represented in Great Britain but have only sporadically been reported from there. The number of known species is low and their distribution is unknown. Most reports are from secondary forests, parks, gardens and other localities modified by man, only very few are from arable soil.

Bagnall was the first to report them from fields: *Allopauropus vulgaris* (Hansen) in 1909, *Pauropus huxleyi* Lubbock in 1911 and *Allopauropus cuenoti* (Remy) and *A. gracilis* (Hansen) in 1935. Remy (1961) has also mentioned *A. broelemanni* Remy and *A. millotianus* Leclerc from fallow fields at Bridgwater in Somerset, together with *A. cuenoti*, *A. gracilis* and *A. vulgaris*. The occurrence of Pauropoda in arable soil has also been shown by Edwards, Thompson & Lofty (1967) in a paper on the effect on soil invertebrates of organophosphorus insecticides and chlorinated hydrocarbons.

This study lists partly the species which were in a collection sent to me by Mr H. J. Gough of Jealott's Hill Research Station, Berkshire, and collected by him and the staff of the Ecology section there, and a collection made by Dr C. A. Edwards of Rothamsted Experimental Station, Harpenden. The list includes three of the species mentioned above and four more (*Allopauropus milloti, A. multiplex, Pauropus lanceolatus, Polypauropus duboscqi*). The latter are all new to arable soil in Britain and three of them (*A. milloti, P. lanceolatus, P. duboscqi*) are here reported for the first time from the British Isles.

SYNOPSIS OF THE SPECIES

Family Pauropodidae

Subfamily Pauropodinae

Genus ***Allopauropus*** *Silvestri*, 1902.

1. *Allopauropus* (*Decapauropus*) *cuenoti* (Remy, 1931) (Fig. 1b). *Archs Zool. exp. gén.* **78**: 67–83: Figs 1–12.

Material examined. 3 specimens.

Distribution. **Somerset,** Cannington, 18.II.1958, 1 subad.* 8 (♀), (Coll. Edwards). **Gloucestershire,** Tarlton, limestone soil, depth 0–15 cm, 25.II.1970, 1 ad. 9 (♀), and depth 2–4 cm, 16.X.1970, 1 ad. 9 (♀), (Coll. Gough).

2. *Allopauropus* (*D.*) *gracilis* (Hansen, 1902) (Fig. 1c). *Vidensk. Meddr dansk naturh. Foren.* **1901**: 395–397: pl. V, Fig. 3a–3f.

Material examined. 26 specimens.

Distribution. **Somerset,** Bridgwater, 18.II.1958, 1 ad. 9 (♂), (Coll. Edwards); Cannington, 18.II.1958, 1 ad. 9 (♀), (Coll. Edwards); Cole, Draycott, 13.VII.1954, 1 ad. 9 (♀), (Coll. Edwards); Bath, 11.XI.1954, 1 ad. 9 (♀), (Coll. Edwards). **Hampshire,** Frensham, sandy soil, depth 0–15 cm, 6. IX.1971, 2 ad. 9 (♂,♀), 2 subad. 8 (♂♂), (Coll. Gough). **Berkshire,** Jealott's Hill, Broadricks, arable clay loam, 2.VI.1969 depth 20–22 cm, 1 ad. 9 (♀), depth 26–28 cm, 1 ad. 9 (♀), and 9.III.1970, depth 20–22 cm, 1 ad. 9 (♀), depth 22–24 cm, 2 ad. 9 (♀♀), and 9.XII. 1970, depth 10–12 cm, 1 ad. 9 (♀), depth 14–16 cm, 1 ad. 9 (♀), depth 18–20 cm, 1 ad. 9 (♀), depth 26–28 cm, 1 ad. 9 (♀), 1 juv. 5, depth 28–30 cm, 2 ad. 9 (♀♀), 1 juv. 6; same place, CP1, clay loam, depth 0–15 cm, 8.X.1971, 1 ad, 10 (♀), (Coll, Gough). **Gloucestershire,** Tarlton, limestone soil, depth 0–2 cm, 12.XII.1970, 1 ad. 9 (♀), (Coll. Gough). **Lincolnshire,** Thurlby Fen, peat soil, depth 0–15 cm, 25.II. 1966, 2 ad. 9 (♂,♀), (Coll. Gough).

3. *Allopauropus* (*D.*) *milloti* Remy, 1945 (Fig. 1d). *Mém. Mus. natn. Hist. nat., Paris* **21**: 136–137: Fig. 4.

Material examined. 2 specimens.

Distribution. **Gloucestershire,** Tarlton, limestone soil, depth 20–22 cm, 17.XII.1970, 2 ad. 9 (♀♀), (Coll. Gough).

The species has not previously been reported from Great Britain, neither from primary habitats nor from arable land.

4. *Allopauropus* (*D.*) *multiplex* Remy, 1936 (Fig. 1f). *Zool. Anz.* **116**: 315–316: Fig. 3.

Material examined. 30 specimens.

Distribution. **Berkshire,** Jealott's Hill, Broadricks, clay loam, 9.III.1970, depth 12–16 cm, 1 ad. 10 (♀), depth 22–24 cm, 2 ad. 10 (♀♀),

* Abbreviations: ad. . . . and subad. . . ., an adult or a sub-adult specimen with the number of pairs of legs indicated; juv. . . ., a juvenile specimen with the number of pairs of legs indicated.

1 ad. 9 (♀), and 9.X.1970, depth 2–6 cm, 1 ad. 9 (♀), and 9.XII.1970, depth 10–12 cm, 1 ad. 10 (♀), depth 22–24 cm, 1 ad. 9 (♀); same place, CP1, depth 0–15 cm, 8.X.1971, 5 ad. 9 (♀♀), (Coll. Gough).

There is only one previous record of this species in Britain, in Wales (Remy, 1956).

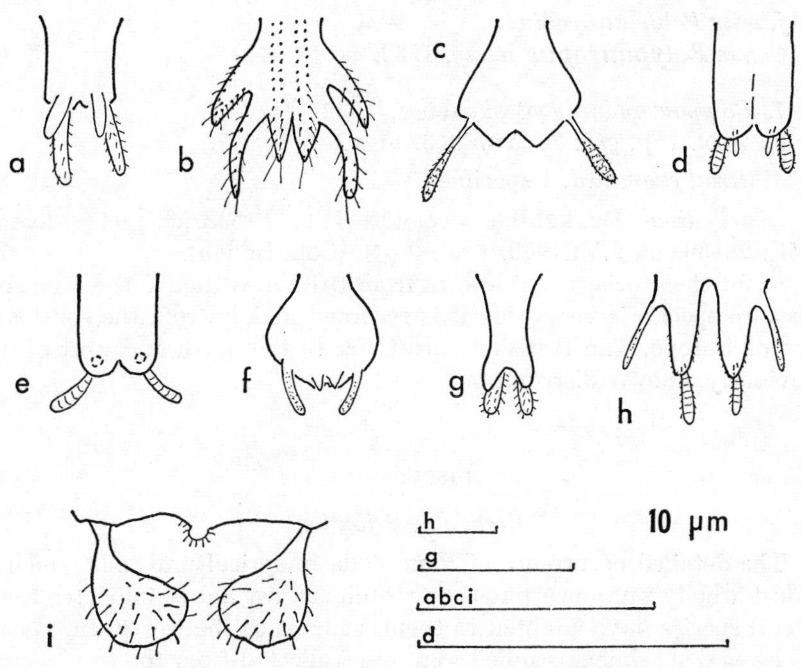

Fig. 1. Anal plates of British Pauropoda from arable soil (a,d,e,h,i, tergal view; b,c,f,g, sternal view). a, *Allopauropus broelemanni*. b, *A. cuenoti*. c, *A. gracilis* (here var. *sequanus*). d, *A. milloti*. e, *A. millotianus*. f, *A. multiplex*. g, *A. vulgaris*. h, *Pauropus lanceolatus*. i, *Polypauropus duboscqi*.

5. *Allopauropus* (*D.*) *vulgaris* (Hansen, 1902) (Fig. 1g).
Vidensk. Meddr dansk naturh. Foren. **1901**: 392–395: pl. V, Fig. 2a–2g.

Material examined. 3 specimens.

Distribution. **Somerset,** Cannington, 18.II.1958, 1 juv. (stad.?), (Coll. Edwards); Cole, Draycott, 13.VII.1954, 1 ad. 9 (♀), (Coll. Edwards); **Wiltshire,** Berwick St. James, chalk soil, depth 0–2 cm, 6.IX.1971, 1 ad. 9 (♀), (Coll. Gough).

Genus **Pauropus** *Lubbock*, 1867.

6. *Pauropus lanceolatus* Remy, 1956 (Fig. 1h).
Suomen hyönt. Aikak. 1937, **3**: 141–144 and *Mém. Inst. scient. Madagascar*, 1956, Sér. A, **10**: 109.

Material examined. 1 specimen.

Distribution. **Somerset,** Winsley, 10.XI.1957, 1 stad.?

The species has not previously been reported from Britain.

Subfamily Polypauropodinae

Genus **Polypauropus** *Remy*, 1932.

7. *Polypauropus duboscqi* Remy, 1932 (Fig. 1i).
Archs Zool. exp. gén. **74**: 287–303, Figs 1–8.

Material examined. 1 specimen.

Distribution. **Berkshire,** Jealott's Hill, Broadricks, clay loam, depth 28–30 cm, 2.VI.1969, 1 ad. 9 (♂), (Coll. Gough).

So far the species is not known from Britain. Although the range of this rare species is very wide it is recorded mainly from the southern part of Europe. The British record is far to the north and west of its previously known distribution.

DISCUSSION

Vertical distribution

The detailed occurrence of Pauropoda in agricultural soils remains almost wholly unknown but the preliminary list above indicates that several species have adapted to them, at least to the top 30 cm. Most species and specimens studied were also collected from the top 15 cm, but *Allopauropus milloti* and *Polypauropus duboscqi* were below 20 cm. These two species may belong to a group of Pauropoda penetrating deep into the soil. Schäfer (1951) is of the opinion that *P. duboscqi* is most often near the ground-water table and the present author has collected it from a depth of 50–55 cm; *A. milloti* occurs at least at 40–45 cm deep (Scheller, 1973). However, pauropods occur in even deeper layers; the author has collected them from a depth of 70–75 cm.

General distribution

Altogether 10 species of Pauropoda have been collected from arable soil in Britain. Most of them are very widespread or even subcosmopolites or cosmopolites. So far only *Allopauropus broelemanni*, *A. multiplex* and *Pauropus huxleyi* seem to have more restricted ranges, but they are all found in several parts of the western Palaearctic.

References

Bagnall, R. S. (1909). Notes on some Pauropoda from the counties of Northumberland and Durham. *Trans. nat. Hist. Soc. Northumb.* (N.S.) **3**: 462–465.

Bagnall, R. S. (1911). A synopsis of the British Pauropoda. *Trans. nat. Hist. Soc. Northumb.* (N.S.) **3**: 654–660.

Bagnall, R. S. (1935). On *Thalassopauropus remyi* gen. et sp. n., a halophilus pauropod, and on the genus *Decapauropus* Remy. *Scott. Nat.* **1935**: 79–82.

Edwards, C. A., Thompson, A. R. & Lofty, J. R. (1967). Changes in soil invertebrate populations caused by some organophosphorus insecticides. *Proc. 4th British insect and fungic. Conf.* **1967**: 48–55.

Remy, P. A. (1956). Quelques stations de Symphyles et de Pauropodes dans les Iles Britanniques. *Ann. Mag. nat. Hist.* (12) **9**: 287–288.

Remy, P. A. (1961). Sur la microfaune du sol de Grande-Bretagne. I. Pauropodes. *Ann. Mag. nat. Hist.* (13) **4**: 149–153.

Schäfer, H.-W. (1951). Über die Besiedlung des Grundwassers. *Verh. int. Verein. theor. angew. Limnol.* **11**: 324–330.

Scheller, U. (1973). Pauropoda and Symphyla from the Pyrenees. *Revue Écol. Biol. Sol.* **10**: 131–149.

Discussion

Hüther: I have studied pauropods from arable land in Germany; it would be interesting to know which species are common to Britain and Germany. I never found *Polypauropus duboscqi* in arable land in Germany; other species are only found in arable land and never in forests. As regards vertical distribution, *Allopauropus vulgaris* lives in the upper layers of most soils but in arable soils it is only found deeper than 10 cm.

Scheller: I cannot comment precisely since the collections I have studied were purely for taxonomic purposes. I think species probably occur at varying depths. The larger species of the genera *Pauropus* and *Stylopauropus* probably do not penetrate deeper than 10–20 cm, or so.

Hüther: I agree, but there are differences in the depths of penetration according to whether the soil is arable or not, and I think this has been established for the other smaller arthropods such as Collembola.

Pierrard: Are pauropods sensitive or resistant to insecticide residues?

Scheller: Edwards, Thompson and Lofty* have studied this problem and stated that pauropods are very susceptible to chlorinated hydrocarbons and organophosphorus insecticides. More so than mites and collembolans. (See also Edwards, this Symposium, pp. 645–655.)

Hüther: I have an example of the use of carbon disulphide; after a year no pauropods survived, so they must be very sensitive to this.

Scheller: Edwards is of the same opinion.

* See list of references above.

MANTON: Of all the myriapods whose locomotory mechanisms I have investigated, pauropods are most susceptible to everything; they die at the slightest provocation, too much light, too much moisture, too little moisture; they have to be treated with the greatest care. I am not surprised that they are susceptible to insecticides.

JEEKEL: Dr Latzel* remarks in his book that pauropods could be found beneath stones but I have never found them. What is your experience?

SCHELLER: When I began my studies I looked for pauropods beneath stones for nearly six months without success. Knowing that they ought to be there I eventually looked in Dr Hansen's localities on Moen Island in Denmark. I found them immediately. Most specimens I have collected in the course of the years are from the undersides of stones. However, I now use a water flotation method.

GOUGH: I have obtained a few specimens from a simple Tullgren funnel but I also use the flotation method of Salt and Hollick. The pauropods in my samples from arable soil form only 1–2% of the total number of animals extracted.

SCHELLER: It is worth adding that these minute animals, very thin and 0·5–1 mm long, can never be easy to find. The best method to spot them on the underside of a stone is to blow gently over the surface. They can be recognized immediately by the manner in which they run. Although superficially like the slow moving collembolans, pauropods (the common Pauropodidae) run rapidly forwards, they stop, they run backwards or twist their bodies in many directions.

MANTON: I have found *Pauropus* in considerable number in decaying logs. Tiegs found them in decaying *Dixonia* (Australian Tree Ferns) but they do occur in astonishing places. Professor Hinton produced *Pauropus* for me in his cellar; they were walking up the walls. The relative humidity was just right for them.

SCHELLER: It is easy to get a large number of pauropods by flotation. I once had more than 200 specimens from a 192 cm^3 soil sample.

JUBERTHIE-JUPEAU: Do you find the same species and in similar proportions, under stones as you extract from soil?

SCHELLER: In southern Sweden and Norway I have had species by flotation extraction which I have never found in any other way, e.g. *Allopauropus tenellus* Scheller†. The proportions of the various species are different and also the age structures may be very different.

* Latzel, R. (1884). *Die Myriopoden der Österreich-Ungarischen Monarchie. 2. Die Symphylen, Pauropoden und Diplopoden.* Wien.

† Scheller, U. (1971). Two new Pauropoda species from northern Europe. *Ent. scand.* **2**: 304–308.

Symp. zool. Soc. Lond. (1974) No. 32, 411–421.

ZUR BIONOMIE MITTELEUROPÄISCHER PAUROPODEN

W. HÜTHER

Sammlungen der Abteilung für Biologie,
Ruhr-Universität, D-463 Bochum, Germany

SYNOPSIS

The occurrence of pauropods in six different biotopes in the district between the Rhine and the Saar, and in the neighbourhood of Bochum and Bräunschweig, is described. Over 900 sample units of 500 ml have been taken and the animals extracted by Tullgren funnels. Constancy of pauropods is lowest in coniferous forest (17·6 %) and highest in arable land and vineyards (39 % and 41 %). More than half of the sample units were taken in vineyards and a separate analysis of the frequencies of pauropods in seven different soil types is given.

The lowest density of pauropods was in arable land and vineyards (2·6 and 2·1 per unit) and the highest was in deciduous forest (4·6 per unit). There were usually less than three or four species per unit. Differences in ecological preferences were observed; both stenotopic and eurytopic species were found. Two forms of *Decapauropus gracilis* var. *amaudruti* are probably ecological vicariants. Five species were found to prefer a particular depth of soil. In nearly all species both sexes were captured but females were generally more numerous than males.

EINLEITUNG

Die Ökologie der Pauropoden ist noch äußerst wenig bekannt. Der einzige, der speziell dieses Problem bearbeitete, ist Starling (1944). Als weitere Arbeiten, die die Pauropoden nicht nur als Gruppe zitieren, sondern ökologische Angaben zu einzelnen Arten bringen, sind noch zu erwähnen: Dunger (1968), Leruth (1938), Loksa (1966), Remy & Husson (1938) und Remy & Condé (1961).

In der folgenden Darstellung wird der Versuch gemacht, elf von etwa 40 in Mitteleuropa gefundenen Arten ökologisch zu charakterisieren (siehe Tabelle V). Trotz einer verhältnismäßig großen Zahl von Proben (über 800 Bodenproben und zahlreiche direkte Aufsammlungen) aus verschiedenen Biotopen können diese Ausführungen jedoch nur als vorläufige Ergebnisse angesehen werden. In Anbetracht der meist sehr geringen Abundanz und oft auch Frequenz der Pauropoden wäre ein Vielfaches der Zahl an Proben nötig, um gesicherte Aussagen machen zu können.

METHODE

Das Material wurde größtenteils aus Bodenproben von 500 ml mit Berlese-Tullgren-Trichtern gewonnen (seit 1955). In den Weinbergen handelt es sich um Mischproben bis zu einer Bodentiefe von 50 cm (genaue Angaben über diesen Biotop siehe Hüther, 1961; die Proben aus der total begifteten Parzelle sind im folgenden nicht mit berücksichtigt). In den übrigen Biotopen wurden größtenteils einheitliche Proben bis zu einer Tiefe von 7–10 cm genommen (das Volumen dieser Proben ist nicht immer exakt 500 ml). Die einzelnen Standorte liegen hauptsächlich im Gebiet zwischen Rhein und Saar, sowie in der weiteren Umgebung von Bochum und Braunschweig.

ERGEBNISSE

Effektivität der Berlese-Tullgren-Methode

Die Berlese-Tullgren-Methode ist für die Pauropoden nicht sehr geeignet. Besonders die großen *Pauropus*- und *Stylopauropus*-Arten werden nur zum Teil erfaßt, wie ich durch direktes Aussuchen von Vergleichsproben feststellte. Das gleiche gilt für *Decapauropus distinctus,* den ich nie mit dieser Methode fing, sondern nur unter Steinen und Holz fand. Bei anderen Arten scheinen die Larven wesentlich empfindlicher zu sein als die Adulten, da ihre Zahl im Verhältnis zu den Erwachsenen ziemlich gering ist (besonders bei *Decapauropus cuenoti* und *D. vulgaris*). Von *Decapauropus viticolus* und *Cauvetauropus rhenanus* wurden dagegen ausschließlich Larven gefunden. Es lassen sich daher vorläufig keine Vergleiche über die Besiedlungsdichten verschiedener Arten durchführen, weshalb ich auf Dominanzberechnungen verzichtet habe. Vergleichbar sind nur die Populationen einer Art von verschiedenen Standorten.

Konstanz und Abundanz

Die Besiedlung der einzelnen Biotope ist recht verschieden (Tabelle I; offenes Gelände = Wiesen, Steppenheiden, Parks, kleine Feldgebüsche und ähnliches). Die Konstanz ist am geringsten im Nadelwald, am größten merkwürdigerweise in den Feldern und Weinbergen. Dagegen ist in diesen die Abundanz am niedrigsten, während sie im Laubwald am höchsten ist.

Eine genauere Analyse ergibt, daß zwischen den einzelnen Beständen eines Biotops, in diesem Fall der Weinberge, erhebliche Unterschiede bestehen können (Tabelle II). Der Sandboden ist sehr arm an Pauropoden, dagegen erreicht die Frequenz vor allem in den etwas feuchteren, durch Staunässe in der Tiefe gekennzeichneten Lehmböden (Parabraunerde und Pseudovergleyte Parabraunerde) ziemlich hohe Werte, jedoch bleibt die mittlere Abundanz sehr niedrig. Die höheren Werte

TABELLE I

Konstanz und Abundanz (bezogen auf 500 ml Substrat) der Pauropoden in den einzelnen Biotopen

Biotop	Nadelwald	Mischwald	Laubwald	Offenes Gelände	Felder	Weinberge
Probenzahl	51	16	162	95	23	566
Konstanz (%)	17·6	25·0	32·1	32·5	39·2	40·8
Mittl. Abundanz	3·4	3·3	4·6	3·0	2·6	2·1
Maxim. Abundanz	5	17*	170*	59*	9	15

* Bei der mittleren Abundanz nicht berücksichtigt

TABELLE II

Frequenz und mittlere Abundanz der Pauropoden in den Weinbergen

Bodentyp*	Sbr.	Pbr.	PsPbr.	Ku.	Mr.	VPr.	Pr.
Probenzahl	64	130	64	121	64	64	59
Mittl. H_2O-Geh.†	10·4	22·2	21·7	15·4	17·0	20·1	21·7
Mittl. Temp. °C	9·9	8·0	9·6	10·4	9·4	10·1	8·1
Frequenz	7·8	48·5	54·8	50·3	29·7	37·5	37·4
Mittl. Abundanz	1·0	1·7	1·8	3·3	2·2	1·4	1·5

* Sbr. = Sandbraunerde, Pbr. = Parabraunerde, PsPbr. = Pseudovergleyte Parabraunerde, Ku. = Kultosol, Mr. = Mullrendsina, VPr. = Verbraunte Pararendsina, Pr. = Pararendsina.

† In % des Trockengewichts.

TABELLE III

Abundanz einzelner Pauropoden-Arten

	Felder + Weinberge		Übrige Biotope	
	Mittl. Abund.	Maxim. Abund.	Mittl. Abund.	Maxim. Abund.
D. gracilis	2·0	13	3·0	16 (95)*
D. cuenoti	1·4	5	1·4	4
D. vulgaris	1·0	1	3·0	5 (20)*
A. danicus	1·0	1	3·0	8
D. viticolus	1·6	5	—	—
C. rhenanus	1·4	4	—	—
P. huxleyi	—	—	1·4	5
P. lanceolatus	—	—	c.3·0	c.40 (170)*

* Einmalige Höchstwerte.

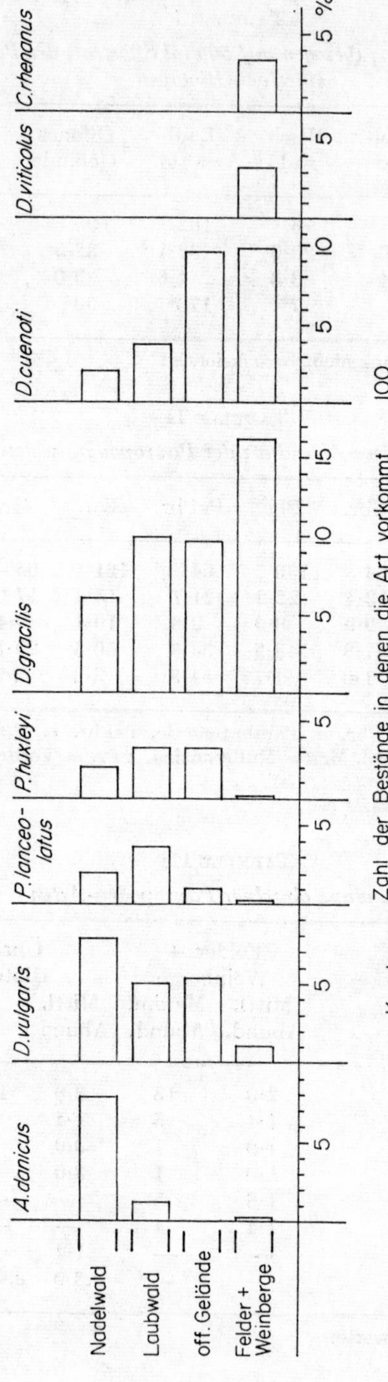

ABB. 1. Konstanz einiger Pauropoden-Arten in verschiedenen Biotopen.

des trockeneren Kultosol wurden durch verschiedene Bodenbedeckungen bedingt (Hüther, 1961). In einem anderen, ziemlich trockenen Weinberg betrug die Frequenz sogar 61% bei 36 Proben, die Abundanz jedoch auch nur 1.4 (da diese Proben nicht einzeln qualitativ untersucht wurden, werden sie im folgenden nicht berücksichtigt).

Die einzelnen Arten verhalten sich dabei unterschiedlich (Tabelle III). Auffallend ist die gleichmäßig geringe Abundanz in allen Biotopen von *Decapauropus cuenoti*. In sehr hoher Populationsdichte wurden nur einmal *Pauropus lanceolatus* und *Decapauropus gracilis* gefunden.

Erwähnt sei noch, daß die Art-Abundanz meist unter drei bis vier liegt. Das Maximum beträgt sieben Arten bei 14 Individuen, jedoch ist das Volumen dieser Probe nicht genau bekannt.

Biotopbindung

Verteilung der Pauropoden in verschiedenen Biotopen

Interessant ist die Verteilung der Pauropoden in den einzelnen Biotopen (Abb. 1). Die Angaben für die beiden *Pauropus*-Arten sind allerdings unsicher, da bei ihnen, wie eingangs erwähnt, der methodische Fehler besonders groß ist und außerdem zu wenige Proben aus verschiedenen Gegenden vorliegen. Es scheint jedoch so zu sein, daß *P. lanceolatus* gegenüber *P. huxleyi* etwas feuchtere Standorte bevorzugt. Ich fand ihn zum Beispiel regelmäßig und oft zahlreich (vergleiche Tabelle III) in einem Blattkomposthaufen in Laubwald, wo letzterer nie gefunden wurde. *Allopauropus danicus* besiedelt möglicherweise bevorzugt zerfallendes Holz. Da ich derartige Proben nur in geringer Zahl untersuchte, kann sich auch bei ihm das ökologische Bild ändern; eine Bevorzugung von Nadelwald halte ich jedoch für wahrscheinlich.

Für die übrigen Arten können die Angaben als gesicherter angesehen werden. Demnach bewohnt *Decapauropus vulgaris* vornehmlich Biotope mit geschlossener Pflanzendecke bei einer Bevorzugung von Nadelwald, während *D. cuenoti* vorwiegend außerhalb des Waldes vorkommt. Die beiden Arten *D. viticolus* und *Cauvetauropus rhenanus* sind ausgesprochen stenotop. Nur *D. gracilis* ist in allen Biotopen relativ häufig. Doch handelt es sich bei ihm sehr wahrscheinlich um eine Sammelart, wofür einige Beobachtungen sprechen.

Decapauropus gracilis gilt als sehr variabel, es sind auch einige Varietäten von ihm beschrieben worden. Von der var. *amaudruti* fand ich nun zwei morphologisch verschiedene Formen, die sich ökologisch auszuschließen scheinen (Tabelle IV). Die form A wurde (mit einer Ausnahme) nur im Laubwald gefunden, die Form B dagegen nur im offenen Gelände und einmal unter Steinen in Laubwald. In den

TABELLE IV

Vorkommen zweier Formen von ***Decapauropus gracilis v. amaudruti***

	Laubwald			Offenes Gelände		
	Proben -Zahl	Konstanz	Indiv. -Zahl	Proben -Zahl	Konstanz	Indiv. -Zahl
Form A	7	4·3	23	1	1·1	1
Form B	1	0·6	2	4	4·2	66

Weinbergen kommt ebenfalls die Form B vor, sowie wahrscheinlich noch eine dritte Form.—Das Material ist noch zu gering, um sichere Aussagen machen zu können. Es gibt jedoch einen Hinweis, daß in *D. gracilis* s.l. zumindest verschiedene Ökotypen vereinigt sind, die meiner Meinung nach gute Arten darstellen.

Verteilung der Pauropoden innerhalb eines Biotops

Ähnlich wie bei der Abundanz bestehen auch hier zwischen den Beständen eines Biotops beträchtliche Unterschiede, wie am Beispiel der Weinberge gezeigt werden kann (Abb. 2). *Decapauropus vulgaris* kann seiner geringen Frequenz und Abundanz nach wohl kaum als biotopeigene Art angesehen werden. *Decapauropus cuenoti* ist zwar in fünf der sieben Bodentypen vertreten, jedoch nur in dreien regelmäßig. Die beiden schon erwähnten stenotopen Arten *D. viticolus* und *Cauvetauropus rhenanus* sind auch innerhalb der Weinberge wiederum stenotop, kommen an ihrem typischen Standort jedoch das ganze Jahr über regelmäßig vor. Für *D. gracilis* gilt das bereits oben Gesagte.

Korrelationen zwischen Bodeneigenschaften und dem Vorkommen der Pauropoden konnten nicht festgestellt werden. Abgesehen von den Weinbergsböden wurden allerdings auch keine genauen Bodenanalysen durchgeführt.

Vertikalverteilung

Die Tiefenverteilung wurde nur in den Weinbergen untersucht (Abb. 3). Eine bestimmte Vorzugstiefe der einzelnen Arten ist deutlich zu erkennen. Am geringsten äusgeprägt ist eine solche Bevorzugung bei *Cauvetauropus rhenanus*, am stärksten bei *Decapauropus viticolus*, der bei 20–30 cm seine größte Frequenz und Abundanz hat, und oberhalb von 10 cm überhaupt nicht gefunden wurde. Interessant ist das Verhalten von *D. vulgaris*, der in diesem für ihn ungünstigen Biotop

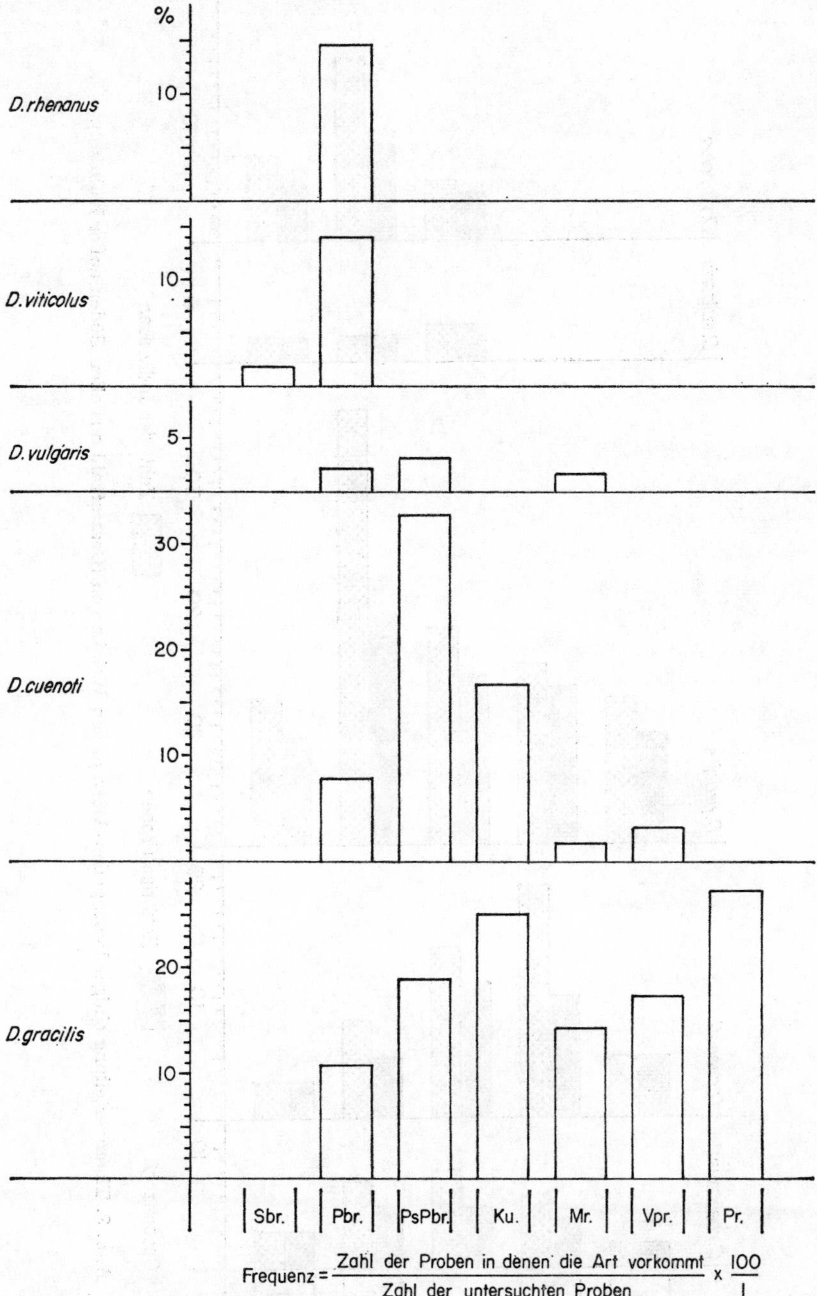

ABB. 2. Frequenz einiger Pauropoden-Arten in den Weinbergsböden (Sbr. = Sandbraunerde, Pbr. = Parabraunerde, PsPbr. = Pseudovergleyte Parabraunerde, Ku. = Kultosol, Mr. = Mullrendsina, VPr. = Verbraunte Pararendsina, Pr. = Pararendsina).

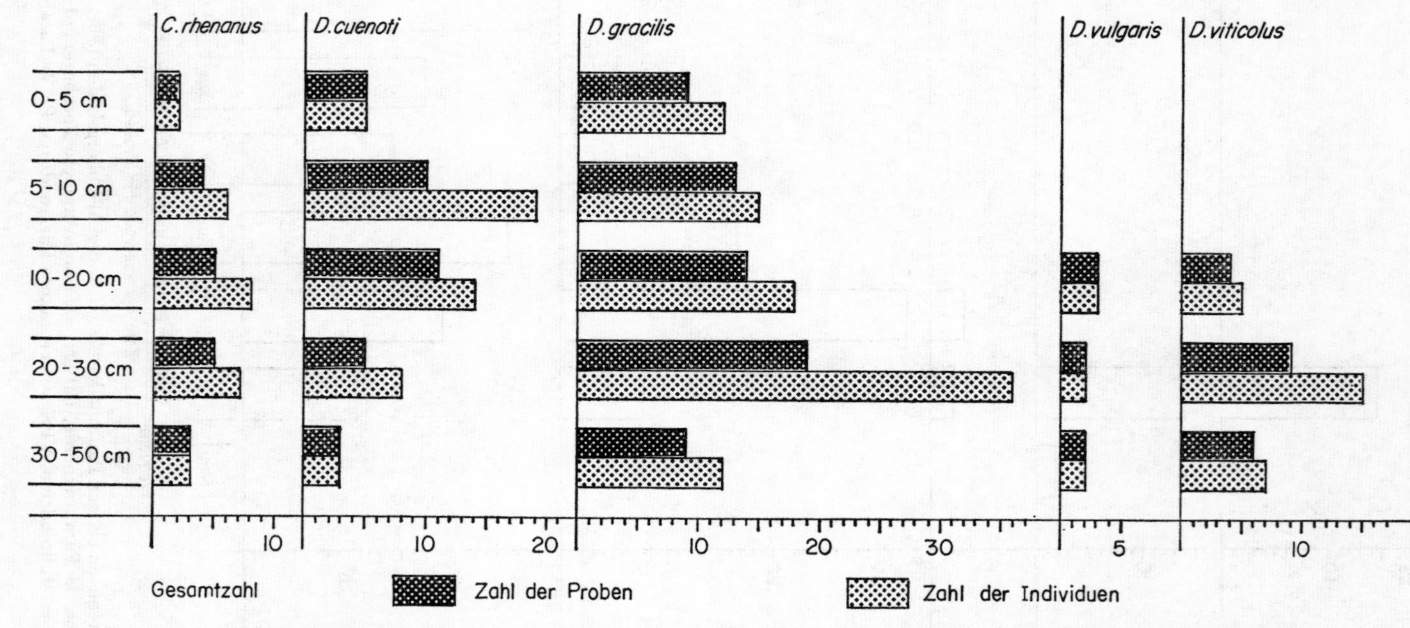

ABB. 3. Tiefenverteilung einiger Pauropoden-Arten in den Weinbergen (Gesamtzahl aus allen Böden außer Kultosol).

ebenfalls nur im Tiefenbereich vorkommt, während er in anderen Biotopen auch die obersten Schichten des Bodens besiedelt. Im einzelnen bestehen noch gewisse Unterschiede zwischen den verschiedenen Böden, die jedoch hier nicht weiter erörtert werden sollen.

Unterschiede in der Tiefenverteilung der einzelnen Entwicklungsstadien oder zu verschiedenen Jahreszeiten wurden nicht beobachtet.

Phänologie

Jahreszeitliche Unterschiede im Auftreten der Pauropoden sind nicht sicher zu ermitteln. Dies gilt sowohl für die Gesamtzahl als auch für die einzelnen Arten und Entwicklungsstadien. Inwieweit dies auf die methodischen Mängel zurückzuführen ist, läßt sich vorläufig nicht entscheiden.

Zahlenverhältnis der Geschlechter

Eine Zusammenstellung der gefundenen Männchen und Weibchen zeigt, daß von allen Arten, von denen adulte Individuen bekannt sind, beide Geschlechter vorkommen (Tabelle V). Von *Decapauropus viticolus* und *Cauvetauropus rhenanus* wurden als älteste Stadien nur Individuen mit acht Beinpaaren gefunden. Da bei diesen keine männlichen Geschlechtsorgane zu erkennen sind handelt es sich mit ziemlicher

TABELLE V

Zahlenverhältnis der Geschlechter einiger Pauropoden-Arten

	♂	♀
Stylopauropus pedunculatus (Lubbock)	1	7
St. pubescens Hansen	5	1
Pauropus huxleyi Lubbock	5	6
P. lanceolatus Remy	8	13
Allopauropus danicus (Hansen)	7	7
Decapauropus cuenoti Remy	6	56
D. gracilis (Hansen)	85	160
*D. g.*v. *amaudruti* (Remy) Form A	8	8
*D. g.*v. *amaudruti* (Remy) Form B	2	44
D. distinctus (Remy)	2	5
D. vulgaris (Hansen)	14	28
D. viticolus Hüther*	—	7
Cauvetauropus rhenanus Hüther	—	3

* Hüther, W. (in press). Ein neuer *Decapauropus* aus der Pfalz. *Revue Ecol. Biol. Sol.*

Sicherheit um Weibchen (bei allen anderen Arten sind die äußeren männlichen Geschlechtsorgane in diesem Larvenstadium bereits ausgebildet).

Auffallend ist das Überwiegen der Weibchen bei den meisten Arten, besonders aber bei *D. cuenoti* und der Form B von *D. gracilis* v. *amaudruti*. Diese Tatsache ist im Vergleich zu *D. gracilis* von Interesse, da sie ebenfalls auf eine Selbständigkeit dieser Form hinweist.

DISKUSSION

Vergleichbare Angaben lassen sich nur sehr wenige aus der Literatur entnehmen, da in den meisten faunistischen und systematischen Arbeiten die Fundortbeschreibungen sehr ungenau sind. Mit den Ergebnissen von Starling (1944) stimmt nur die mittlere Abundanz überein, auch er fand (unterhalb der Streu, die sehr schwach besiedelt war) zwei bis drei Individuen/500 ml Boden. Dagegen stellte er sowohl Korrelationen zwischen Bodenfeuchtigkeit und Individuenzahl fest, als auch jahreszeitliche Populationsschwankungen. Allerdings waren an den von ihm untersuchten Standorten die klimatischen Schwankungen erheblich größer (Proben von 1–50% Wassergehalt) als in den Weinbergen. In diesen lag der Wassergehalt fast immer in dem von Starling als optimal festgestellten Bereich (Tabelle II). Man muß dabei jedoch auch berücksichtigen, daß sich seine Untersuchungen auf völlig andere Arten beziehen.

Vergleichsuntersuchungen zwischen sehr verschiedenartigen Biotopen und über die Tiefenverteilung liegen noch nicht vor. Einige Angaben scheinen jedoch meine Befunde zu bestätigen, daß *Decapauropus vulgaris* vorwiegend in offenem Gelände und Wald, *D. cuenoti* vorwiegend in offenem Gelände und in Ackerland und *D. gracilis* in allen Biotopen vorkommt (Chalupský, 1967; Dunger, 1968; Loksa, 1966; Remy & Condé, 1961). *Decapauropus gracilis* v. *amaudruti* fand Dunger (1968) nur an zwei Standorten in einzelnen Exemplaren. Es wäre interessant festzustellen, zu welcher der beiden hier unterschiedenen Formen (A oder B) diese Tiere gehören.—Für die unterschiedliche Verteilung läßt sich hier wie bei anderen Bodentieren keine Erklärung finden. Diese Frage kann nur dürch experimentelle Untersuchungen gelöst werden.

Über die Verteilung der Geschlechter machen Remy & Husson (1938) nähere Angaben. Sie fanden von *Pauropus lanceolatus* in allen Höhlen wesentlich mehr Männchen als Weibchen. Aus zahlreichen faunistischen und systematischen Arbeiten geht hervor, daß die Geschlechterverteilung bei den Pauropoden sehr verschieden sein kann.

Ob dies mit geographischen, ökologischen oder jahreszeitlichen Gegegebenheiten in Zusammenhang steht, bleibt zu untersuchen.

Literatur

Chalupský, J. (1967). Bohemian Pauropoda III. *Věst. csl. Spol. zool.* **31**: 121–132.

Dunger, W. (1968). Die Entwicklung der Bodenfauna auf rekultivierten Kippen und Halden des Braunkohletagebaues. *Abh. Ber. naturk. Mus. Görlitz* **43**: (2): 1–256.

Hüther, W. (1961). Ökologische Untersuchungen über die Fauna pfälzischer Weinbergsböden mit besonderer Berücksichtigung der Collembolen und Milben. *Zool. Jb.* (Syst.) **89**: 243–368.

Hüther, W. (1971). Zwei interessante Pauropoden aus dem Oberrheingebiet. *Mitt. Pollichia Pfälz. Ver. Naturk. Nat Schutz* (III. Reihe), **18**: 170–177.

Leruth, R. (1938). Contribution à l'étude de la faune endogée et saproxylophile. I. Les Pauropodes en Belgique. *Bull. Soc. r. Sci. Liège* **1938**: 381–387.

Loksa, I. (1966). *Die bodenzoozönologischen Verhältnisse der Flaumeichen-Buschwälder Südostmitteleuropas.* Budapest: Verlag der Ungarischen Akademie der Wissenschaften.

Remy, P. & Condé, B. (1961) Sur la microfaune du sol de Grande-Bretagne. *Ann. Mag. nat. Hist.* (13) **4**: 149–154.

Remy, P. & Husson, R. (1938). Les Pauropodes des galeries de mines et des cavernes naturelles. *C.r. prem. Congr. Lorrain Soc. Sav. Est France* **1938**: 1–19.

Starling, J. H. (1944). Ecological studies of the Pauropoda of the Duke Forest. *Ecol. Monogr.* **14**: 291–310.

Discussion

Haacker: You have shown that males are less common than females. Do males occur seasonally?

Hüther: No. Males are less numerous than females throughout the year.

Enghoff: You think that both forms of *Decapauropus gracilis* var. *amaudruti* are good species, but *amaudruti* itself may be only a sub-species!

Hüther: The form *amaudruti* was first described as a good species by Remy. Later he found that the typical *gracilis* was very variable and occurred together with *amaudruti* and so he then considered it as a variant of *gracilis*. In fact, *amaudruti* is morphologically distinct. I first separated the two forms of *amaudruti* by morphological characters and was surprised to find there were also distinct ecological differences. Clearly the group needs revision!

Enghoff: I suppose breeding experiments are not easy to carry out?

Hüther: Up to now it is impossible to do this under controlled conditions.

Symp. zool. Soc. Lond. (1974) No. 32, 423–431.

THE ECOLOGY OF CENTIPEDES AND MILLIPEDES IN NORTHERN NIGERIA

J. G. E. LEWIS

Dover College, Dover, Kent, England

SYNOPSIS

The life histories of eight polydesmoid millipedes and four scolopendromorph centipedes in northern Nigeria are discussed with particular reference to survival during the dry season.

The adults of most millipede species are short-lived and the species survive the dry season as diapausing larvae in moulting chambers in the soil. Diapausing stadium VII larvae give rise to adults which appear at the beginning of the rains. Earlier stadia which have diapaused appear later, possibly because they spend the dry season deeper in the soil below the depth reached by the early rains.

Of the centipedes studied, *Scolopendra amazonica* is atypical in that it is surface active throughout the year. The other three species remain active after the rains have finished but eventually disappear from surface habitats presumably entering crevices in the soil. The presence of a substantial soil fauna enables them to continue to grow through the dry season.

INTRODUCTION

Most of the work on the ecology of myriapods has been carried out in temperate regions of the world where the major factor restricting or inhibiting activity is probably the low temperature experienced in the winter months. In the dry tropics the major factor restricting the activity of these animals is the occurrence of one or more dry seasons when the rainfall is negligible or absent.

The work to be reviewed in this paper was carried out on eight species of polydesmoid millipedes and four species of scolopendromorph centipedes in the northern guinea savanna vegetation belt at Zaria, northern Nigeria. Zaria has a mean annual rainfall of 112 cm. The rainfall is seasonal, there being five months (November to March) which have less than 25 mm of rain. The rains in April and May tend to be spasmodic so that the soil may dry out several times during this period (Fig. 1) but after the end of May the rains are normally continuous and remain so until the end of September (Lewis, 1971b).

MILLIPEDES

The millipedes studied belong to four families which exhibit marked differences in structure and habit.

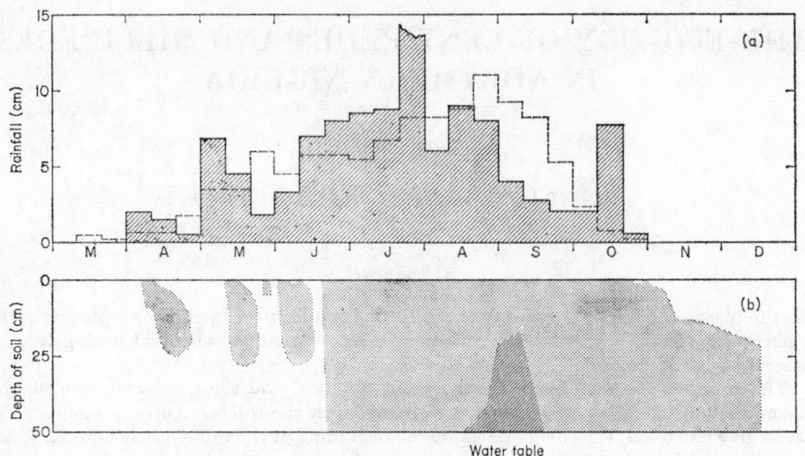

Fig. 1. Soil moisture and rainfall in Zaria from March to December 1969. (From Lewis, 1971b.) (a) Rainfall for Zaria in 10-day periods. Hatched columns: 1969 rainfall. Pecked line: mean rainfall for 1928 to 1965. (b) Soil moisture conditions. Damp soil stippled; waterlogged soil hatched.

Gomphodesmidae

Two gomphodesmids (*Tymbodesmus falcatus* Karsch and *Sphenodesmus sheribongensis* Schiøtz) have been investigated (Lewis, 1971a). Both are relatively large millipedes, measuring 22 to 48 mm in length,

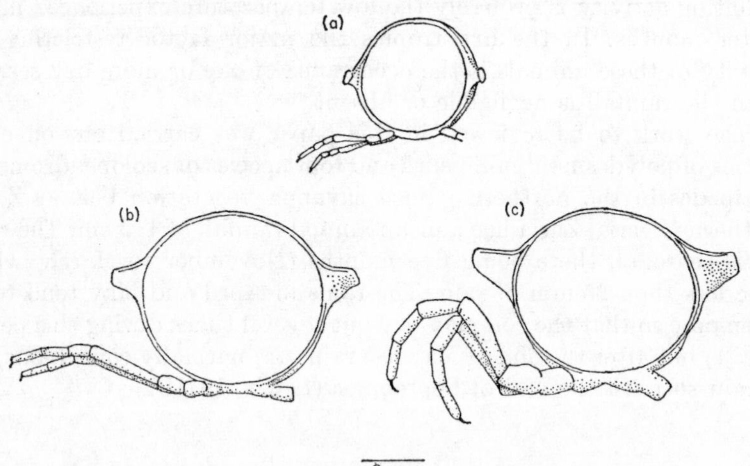

Fig. 2(a)–(c). The tenth body ring of *Tymbodesmus falcatus*, posterior aspect. (From Lewis, 1971a.) (a) Tenth body ring of stadium VI male. (b) Tenth body ring of mature female. (c) Tenth body ring of mature male.

with moderately developed lateral keels (Fig. 2b,c). The adults of both species are surface active and feed on soil and litter. The seven larval stadia of *S. sheribongensis* and the first six larval stadia of *T. falcatus* are soil dwelling but stadium VII *T. falcatus* are frequently found on the soil surface. The soil dwelling stadia have poorly developed lateral keels (Fig. 2a).

T. falcatus has a two-year life cycle. Newly moulted adults appear on the soil surface soon after the rains have begun in April or May. Egg-laying takes place in June and the adults have virtually disappeared by the end of July. One-year-old stadium VII larvae appear on the soil surface several weeks after the adults.

During the rainy season the larvae moult without producing moulting chambers but at the end of the rains, thick walled moulting chambers are produced and the larvae diapause in these through the dry season. The moulting chambers are probably important in water conservation at this time.

Paradoxosomatidae

The three paradoxosomatids studied (*Habrodesmus duboscqui* Brölemann, *Xanthodesmus* sp. (near *penicularius* (Attems)) and *Xanthodesmus physkon* (Attems)) are soil surface living forms feeding on cryptogams. They are rounded and have poorly developed lateral keels (Fig. 3a,b). The larvae form large swarms comprising a thousand or more specimens.

H. duboscqui and *Xanthodesmus* sp. are characteristic of open habitats and, like gomphodesmids, first appear as adults in April or May. One-year-old stadium VII larvae appear several weeks after the adults. Egg-laying takes place in June and adults are rare by the end of the month; the resulting larvae become rare in August or September, some time before the rains have ended. *Xanthodesmus* sp. larvae diapause in thick-walled moulting chambers and it is probable that *H. duboscqui* does so as well.

X. physkon is found in wooded areas and first appears as larvae at the beginning of June. These larvae reach adulthood and lay eggs in July and August. The larvae resulting from these eggs can still be found on the soil surface in October, presumably because favourable conditions prevail for a longer period in shaded habitats.

Pterodesmidae and Oxydesmidae

The remaining two families investigated contain litter-dwelling species, the pterodesmids *Aporodesmus zaria* Hoffman and *A. aestivus* Hoffmann and the oxydesmid *Coromus* sp. Both genera have very well

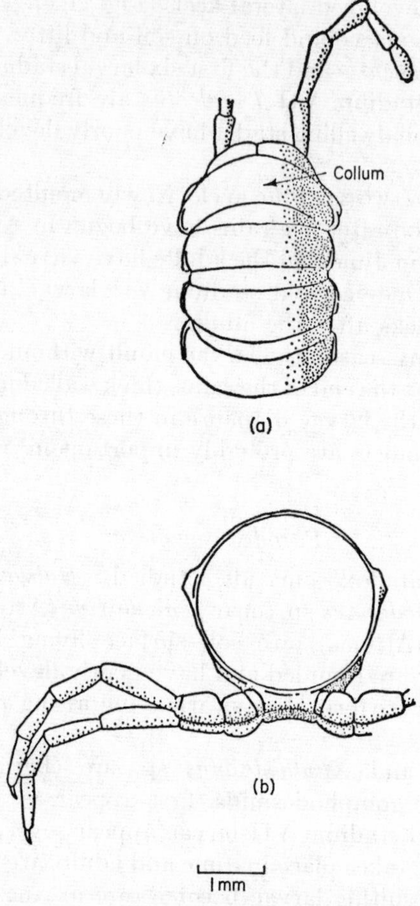

FIG. 3(a), (b). A male *Xanthodesmus physkon* from Zaria, Nigeria. (From Lewis, 1971b.) (a) Dorsal view of the head and first four segments (the legs are not shown); (b) Posterior view of the tenth body ring.

developed lateral keels characteristic of litter dwelling polydesmoids (Fig. 4).

Apart from a brief appearance of adult *A. zaria* on the soil surface prior to egg-laying in May, *A. zaria* and *A. aestivus* do not appear in the litter until the middle of the rainy season. The two *Aporodesmus* species have differently phased life histories. *A. zaria* lays eggs in the soil at the beginning of the rains: the resulting stadium VI larvae migrate into the litter in July and by late August have produced adults which will spend the dry season in the soil. *A. aestivus*, however, passes

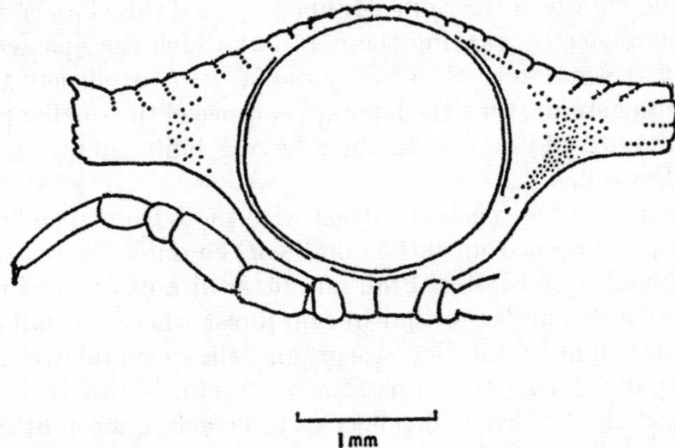

FIG. 4. Posterior view of the eleventh body ring of a male *Aporodesmus zaria* from Zaria, Nigeria.

the dry season as larvae in the soil and these larvae give rise to short-lived adults which appear in the litter at the end of June. The eggs laid by these adults give rise to larvae which remain in the litter until the end of the rainy season (Lewis, in prep.).

Adult *Coromus* sp. appear in the middle of the rainy season and probably pass the dry season as larvae in the soil (J. G. E. Lewis, unpublished data).

Conclusions regarding savanna millipedes

Seven of the eight species studied in Zaria have short-lived adults and survive the dry season as larvae in the soil. Lewis (1971b) has pointed out that this is unusual in that most temperate zone millipedes have long-lived adults which overwinter. He has suggested that the reason why adults do not live through the dry season in savanna is that they cannot produce moulting chambers and are thus denied the necessary protection against desiccation that these afford. Adult millipedes, because of their large size, are the most desiccation resistant active stage. This may be the reason why they are the first stadium to appear at the beginning of the rainy season.

It is probable that rain is the factor causing the termination of larval diapause. It has been shown that early in the rainy season water penetrates the soil relatively slowly. For example, in 1969 the first rains fell in mid-April but by mid-June they had only penetrated to a depth of 30 cm (Lewis, 1971b). In that year, therefore, animals passing the dry season deep in the soil were cut off from the surface by a layer of dry soil until the middle of the rains. Millipedes could therefore control their time of emergence by varying the depth at which they passed the dry season. If, as seems likely, the smaller stadia are more efficient burrowers than the larger stadia, then the later appearance of the smaller paradoxosomatid stadia could be due to their having built moulting chambers deeper in the soil.

The litter dwelling species contrast with those from open habitats in that they do not appear until the middle of the rains. There are no data on the seasonal pattern of leaf fall and litter disappearance in savanna but it is probably similar to that in rain forest where leaf fall reaches a peak at the height of the dry season and the accumulated litter disappears rapidly during the rains (Madge, 1965). If this is so then one would expect the litter dwelling species to be active when litter is most abundant, that is at the beginning of the rainy season. Lewis (in prep.) has suggested that the reason why litter dwelling species are not active at this time is that food shortage could often be caused by the complete removal of litter by the fires which are characteristic of the latter part of the dry season. In such a situation the litter feeding species would be dependent for food upon trees which shed their leaves during the rainy season.

The most noticeable characteristic of many of these tropical polydesmoids is their high degree of surface activity. Lewis (1971a) has suggested that surface activity is beneficial in that it enables the species to search for a mate in a two dimensional rather than a three dimensional habitat. The surface also offers a richer source of food than the soil. The attendant risk is that of higher predation and it may be significant that there is a correlation between the degree of surface activity and fecundity.

CENTIPEDES

Four scolopendromorph centipedes have been studied in Zaria, namely, *Scolopendra amazonica* (Bücherl), *Rhysida nuda togoensis* Kraepelin, *Ethmostigmus trigonopodus* (Leach) and *Asanada sokotrana* Pocock.

S. amazonica is unusual in that it is active throughout the dry season finding high humidities and abundant food in cow dung. The life cycle is completed in a year and young appear in March and again in October (Lewis, 1970). The other three species appear as soon as the first rains have fallen.

R. nuda togoensis occurs in litter and under stones. The young appear in May and the life cycle is completed in one year. Specimens can be found in damp habitats some time after the rains have ended but become rare in December.

E. trigonopodus is common in deserted *Trinervetermes* mounds. The young appear in April and the life cycle is probably completed in one year, although some specimens may live for two years. Specimens are rare after the end of October (Lewis, 1972a).

A. sokotrana is invariably found in deserted *Trinervetermes* mounds (Lewis, 1973) where the young appear in March and, as is the case with *R. nuda togoensis* and *E. trigonopodus*, the life history is completed in one year.

The three species continue to grow through the dry season although they are absent from surface habitats. Growth rates are high during the rainy season so that the young have reached a large size by September or October.

DISCUSSION

It has been shown that with one exception all paradoxosomatids and gomphodesmids have entered diapause in moulting chambers in the soil by the end of September. The exception is *X. physkon*, larvae of which are still active in October. These, however, occur only in shaded habitats where the ground remains damp longer than elsewhere and hence the rains are effectively prolonged. Most, if not all species are inactive in the dry season, many of them diapausing in moulting chambers.

The centipedes studied contrast markedly with the millipedes in a number of respects. They appear earlier in the rains than millipedes and remain active after the rains have ended. The young appear at the beginning of the rains and have reached a large size by September or October and thus their resistance to desiccation is increased. This resistance and their mobility enable them to search for favourable habitats in which they can continue to feed but they become less common on the surface as the dry season progresses. As scolopendromorphs do not burrow actively or deeply (Manton, 1952) they must enter the soil through crevices which appear as it dries out. There is evidence that the soil contains a substantial fauna through the dry season (Lewis, 1972b) and this enables carnivores such as centipedes, though not herbivores such as millipedes, to continue their activity throughout the year. The crevices which form during the dry season will fill up with water at the beginning of the rains due to run-off. This brings about the rapid appearance of the adult centipedes on the surface the following year.

References

Lewis, J. G. E. (1970). The biology of *Scolopendra amazonica* in Nigerian Guinea savannah. *Bull. Mus. natr. Hist. nat., Paris* **41**: Suppl. no. 2: 85–90.

Lewis, J. G. E. (1971a). The life history and ecology of the millipede *Tymbodesmus falcatus* (Polydesmida: Gomphodesmidae) in northern Nigeria with notes on *Sphenodesmus sheribongensis*. *J. Zool., Lond.* **164**: 551–563.

Lewis, J. G. E. (1971b). The life history and ecology of three paradoxosomatid millipedes (Diplopoda: Polydesmida) in northern Nigeria. *J. Zool., Lond.* **165**: 431–452.

Lewis, J. G. E. (1972a). The life history and distribution of the centipedes *Rhysida nuda togoensis* and *Ethmostigmus trigonopodus* (Scolopendromorpha: Scolopendridae) in Nigeria. *J. Zool., Lond.* **167**: 399–414.

Lewis, J. G. E. (1972b). The population density and biomass of the centipede *Scolopendra amazonica* (Bücherl) (Scolopendromorpha: Scolopendridae) in sahel savanna in Nigeria. *Entomologists' mon. Mag.* **108**: 16–18.

Lewis, J. G. E. (1973). The taxonomy, distribution and ecology of centipedes of the genus *Asanada* (Scolopendromorpha: Scolopendridae) in Nigeria. *Zool. J. Linn. Soc.* **52**: 97–112.

Lewis, J. G. E. (in preparation). *The biology of* ***Aporodesmus zaria*** *and* ***A. aestivus*** *(Polydesmida: Pterodesmidae) in Nigeria, with a discussion of the adaptations of millipedes to the savanna habitat.*

Madge, D. S. (1965). Leaf fall and litter disappearance in a tropical forest. *Pedobiologia* **5**: 273–288.

Manton, S. M. (1952). The evolution of arthropodan locomotory mechanisms. Part 3. The locomotion of Chilopoda and Pauropoda. *J. Linn. Soc.* (Zool.) **42**: 118–166.

Discussion

WALLWORK: What is the duration of the life cycle in the centipedes you studied?

LEWIS: Certainly a year in three species, one or two years in the fourth species.

WALLWORK: Do the life histories of the species of millipedes in savanna differ from those living in woodland?

LEWIS: All the habitats are savanna, some of them more wooded than others. There is a species, *Xanthodesmus physkon*, which is only found in the more wooded areas. This has a life history which differs from that of its close relatives in more open localities.

WALLWORK: Does it have a longer adult stage?

LEWIS: No. The species in the open habitats appear as adults as the rains begin; the species in the more wooded habitat appear as larvae shortly after the rains begin and their activity is prolonged into the beginning of the dry season—but this is because rainy season conditions continue longer in shaded habitats.

EASON: You mentioned a species of millipede which survives the dry season in crevices. It cannot build a moulting chamber, does it just lie completely torpid, not feeding, not growing?

LEWIS: I have not found it during the dry season. The adults disappear at the end of the rainy season and reappear six months later. I suppose they pass the time in deep crevices.

JEEKEL: In South America, Guiana, in the wet season, large areas of the tree roots are flooded and millipedes move up the trees and stay there a long time. Did you observe anything like this?

LEWIS: I have seen *Xanthodesmus physkon* a few feet up a tree over a period of one or two days when the soil was very wet indeed.

HAACKER: The construction of a moulting chamber in which they pass the dry season is a very striking feature of the ecology of these millipedes. Did you say that only one stadium behaved in this way?

LEWIS: No. Several stadia do this.

HAACKER: And in the rainy season?

LEWIS: No. Not in the rainy season. With a two year life-history where there is an overlap of instars, those same instars which build moulting chambers in the dry season do not build them in the rainy season.

HAACKER: If you could transfer instars from the rainy season into the dry season conditions, would they construct a moulting chamber?

LEWIS: This would be my guess. Yes!

SAKWA: Do you know if these millipedes take refuge in termitaria in the dry season?

LEWIS: I have looked in inactive mounds and only found an occasional specimen.

Symp. zool. Soc. Lond. (1974) No. 32, 433–462.

THE LIFE HISTORY OF THE MILLIPEDE *GLOMERIS MARGINATA* (VILLERS) IN NORTH-WEST ENGLAND

J. HEATH

Biological Records Centre, Institute of Terrestrial Ecology, Monks Wood Experimental Station, Abbots Ripton, Huntingdon, PE17 2LS England

K. L. BOCOCK

Soil Ecology Section, Institute of Terrestrial Ecology, Merlewood Research Station, Grange-over-Sands, Lancashire LA11 6JU, England

and

M. D. MOUNTFORD

Biometrics Section, Institute of Terrestrial Ecology, 19 Belgrave Square, London, SW1X 8PY, England

SYNOPSIS

The life-history of the millipede *Glomeris marginata* in north-west England is described on the basis of the following data collected in the field and laboratory:

1. Observations on the timing and pattern of mating and egg-laying, the number of eggs laid and the location of egg-laying sites.
2. Seasonal changes in the condition of the ovary.
3. Observations on the occurrence and rate of development of the eggs, the five anamorphic stadia and up to ten epimorphic stadia. The latter were defined by analysing the population size-class structure and ageing the size-classes using growth data, live weight being used as the main criterion of size.

Glomeris lays up to 86 eggs per year mainly during April, May and June. Embryonic development takes about 1·5 months under field conditions and the second anamorphic stadium emerges from the egg-capsule 2–2·5 months after egg-laying. The second to fourth anamorphic stadia are reached by the first winter and the last anamorphic to second epimorphic stadia by the second winter. The males become mature in the second epimorphic stadium when they are two to three years old. Females mature after three to four years when they are in the third epimorphic stadium. Both sexes can live for up to 10 or 11 years.

INTRODUCTION

Information on the life-histories of *Glomeris* species is scattered throughout the literature published during the past 136 years. Some aspects of the life-history have been dealt with fully, for example, the more obvious morphological differences between the sexes of various species including *G. marginata* were described by Chalande (1905), Haacker (1964), Schubart (1934), Verhoeff (1906, 1916, 1928) and vom Rath

(1890) and an account of mating behaviour was given by Haacker (1964) following the earlier observations of Humbert (1872), Verhoeff (1906, 1916, 1928) and vom Rath (1890, 1891). Chalande (1905), Evans (1910, 1911), Fabre (1855), Gervais (1837), Haacker (1964), Hennings (1904), Juberthie-Jupeau (1967a), Verhoeff (1910, 1928) and vom Rath (1890, 1891) gave details of egg-laying and construction of the egg-capsule by *Glomeris* whilst Dohle (1964), following Heymons (1897), described the embryology of *G. marginata* very fully. Descriptions of part or the whole of the development or of some or all of the anamorphic stadia were given by Bocock, Mountford & Heath (1967), Chalande (1905) and Hennings (1904) for *G. marginata*, Latzel (1884) for *G. hexasticha* and *G. multistriata*, vom Rath (1890, 1891) for *G. conspersa*, *hexasticha* and *pustulata* and Verhoeff (1910) for *G. pustulata* and *G. marginata*. Verhoeff (1928) included details for *G. hexasticha* and summarized previous work. Attems (1926), Blower (1958) and Schubart (1934) also provided general summaries of the life-history of *Glomeris*.

Information on many other aspects of the life-history can be found in papers by Blower & Gabbutt (1964), Haacker (1968), Juberthie-Jupeau (1967b, 1967c, 1968, 1970, 1971) and Verhoeff (1932) as well as in the papers already mentioned. However, coverage of the life-history of *G. marginata* is incomplete partly because many of the published data were obtained in the laboratory or under unspecified conditions or for species other than *G. marginata*. It is the purpose of this paper to give a description of the life-history based on our studies of *G. marginata* in woodland (Bocock, 1963; Bocock & Heath, 1967; Bocock *et al.*, 1967) emphasizing in detail those aspects which have previously been covered inadequately or on which conflicting opinions exist.

FIELD SITES

Two main field sites were used in these studies, Heaning Wood (National Grid Reference SD(34)398805) and Eggerslack Wood (National Grid Reference SD(34)408796) both near Grange-over-Sands, north Lancashire. Twenty-nine of the presence records for the anamorphic stadia (Fig. 5) were for a third site, Roudsea Wood National Nature Reserve (National Grid Reference SD(34)333823, 8 km WNW of Grange-over-Sands.

All the sites were in mixed deciduous woodland on shallow stony base-rich soil overlying Carboniferous limestone. This soil is glacial drift of Silurian shaly flags and grits with admixture of variable quantities of limestone rubble and residue from limestone weathering. Bocock

& Gilbert (1957) and Bocock, Gilbert *et al.* (1960) described briefly the soil and vegetation of the Roudsea site. Nicholson, Bocock & Heal (1966) described part of the Eggerslack site. The remaining part is similar to Nicholson's site but coppiced trees of hazel (*Corylus avellana* L.) and ash (*Fraxinus excelsior* L.) dominate the vegetation. The soil and vegetation in Heaning Wood are very similar to those in Eggerslack Wood. Both sites are on steep hillsides but Heaning has a westerly aspect whereas Eggerslack faces east. The Roudsea site is on the top and gently sloping sides of a limestone ridge running north to south.

Data collected outside the tree canopy indicate that the climate on all three sites is similar, with a mean annual rainfall of approximately 125 cm and a mean annual air temperature of about 9·5°C. Under the tree canopy about 85% of the rain reaches the soil and the annual mean soil temperature is about 1°C lower than the value for the air.

SEX RATIO

Little information is available on the sex ratio. When data for seven species of *Glomeris* (Verhoeff, 1928) are pooled, a ratio of 0·617 (477 males : 773 females) is obtained, but the efficiency of the sampling method, apparently hand-collection, is unknown. During our studies, 15 samples of epimorphic stadia of *Glomeris*, ranging in size from 88–1197 animals, were collected by hand-sorting the top 10–20 cm of soil in the field from April to October inclusive (Fig. 1). From our knowledge of the behaviour of *Glomeris marginata*, for example its vertical movements in the soil (Bocock & Heath, 1967), we suspected that the sampling method would give representative samples of the *Glomeris* population when the millipedes were active near the soil surface. However, joint studies with D. C. Pickering (Liverpool Polytechnic), to be described in detail elsewhere, indicated that the method recovers only 71·4% of males and 86·7% of females present. Accordingly, the sex ratio obtained for our samples, 0·8194 (1960 males : 2392 females), was corrected to a mean value of 0·995 with a standard error of 0·145 which includes a component associated with the variance of the recovery data.

In an experiment in which 134 *Glomeris* were grown from egg to the first epimorphic stadium (refer to pp. 437 & 446), the resulting sex ratio was 1·197. In joint work with O. J. W. Gilbert, one of us (Bocock) obtained a sex ratio of 1·249 (231 males : 185 females) for *Glomeris marginata* collected in pitfall traps (Fig. 1). These two values are not significantly different ($P \geqslant 0{\cdot}05$) from the value of 0·995 obtained by hand-sampling. A chi-square test indicated that two of the three values, 0·995 and 1·197, were not significantly different from unity ($P \geqslant 0{\cdot}05$). The sex-ratio for pitfall-trapped animals changed significantly with season, unlike the ratio obtained for the hand-sampled animals (Fig. 1), because it depended on the change in relative activity of the two sexes during the year.

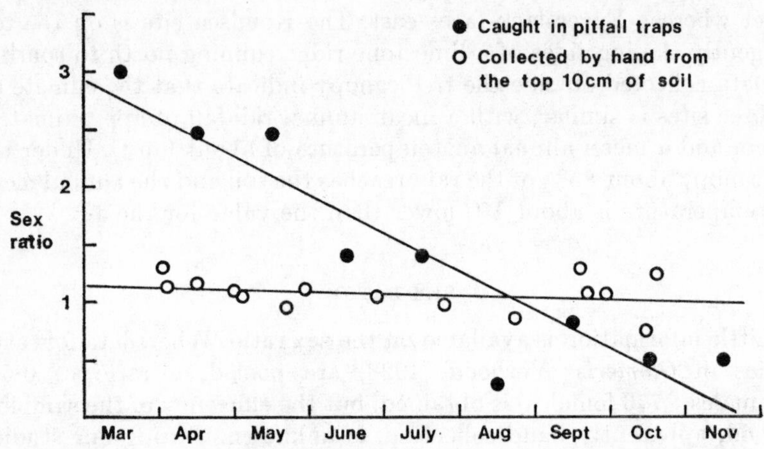

FIG. 1. Seasonal variation in the sex ratio of *Glomeris marginata*. Total numbers of animals per sample ranged from 6–145 in the pitfall samples and 88–1197 in the hand-collected samples. Linear regressions: pitfalls $y = 2{\cdot}938 - 0{\cdot}0110x$, slope significantly different from zero ($P < 0{\cdot}001$); hand collections, $y = 1{\cdot}143 - 0{\cdot}000563x$, slope not significantly different from zero ($P > 0{\cdot}25$).

Published sex ratios for other millipedes vary greatly within a range of about 0·2 to 2·0 and appear to depend in part on the sampling method used. For example, Blower (1970) found a ratio of approximately unity for *Iulus scandinavius* Latzel when he extracted mature animals from soil, but obtained values of 1·404 and 1·735 in successive years when he pitfall trapped the same species.

The sex ratio of millipedes also varies with the difference in mortality patterns of the sexes. The sex ratio of *Brachydesmus superus*, for example, is about unity at the time of hatching and declines to about 0·33 during subsequent stadia because males have a higher mortality than females (Stephenson, 1960). In contrast, the sex-ratio of epimorphic *Glomeris* is about unity throughout the epimorphic phase and the sexes have a similar maximum life-span (Table III) and mortality pattern (K. L. Bocock, M. D. Mountford & J. Heath, in prep.).

MATING AND EGG-LAYING

Collection of data

Data were gathered in three main ways:

Observation of the activities of **Glomeris** *in the field*

During all the field-work involved in our *Glomeris* studies, careful notes of the activities of the various stages of the millipede were kept to discover the timing of reproduction.

Laboratory and field breeding experiments

These studies, which have been outlined previously (Bocock *et al.*, 1967), were begun primarily to produce information on the number of eggs laid and the pattern of egg-laying.

In the laboratory study, 15 pairs of mature millipedes, with live weight ranges of 24–73 mg (males) and 64–295 mg (females), were collected from the field from 2nd April–8th May 1963. When collected, the animals in each pair were mating or were in very close proximity. They were placed in covered glass jars (5 cm high, 10 cm in diam.) with moist soil and decaying tree leaves and kept under a naturally fluctuating light and temperature regime (temperature range during the April–October period 3·5°–20·5°C). They were examined every few days, egg capsules produced since the previous examination were counted and transferred to other similar jars, and mating, egg-laying and moulting were noted.

In the field experiment, 24 mature *Glomeris* of each sex, with live weight ranges 25–104 mg (males) and 66–245 mg (females), were kept from spring until autumn 1964 in eight open-topped field enclosures as described previously (Bocock *et al.*, 1967). Three males and three females were placed in each enclosure. The contents were sorted by hand in early June and late August, any mating, egg-laying or similar events were noted and egg-capsules counted, removed and their contents examined by dissection under a low-powered microscope. In the laboratory and field experiments, checks for egg-capsules were discontinued when all the mature females had moulted.

Examination of the ovaries

In most months from April 1963 to May 1964 (Fig. 3), approximately 100 female *Glomeris* in the epimorphic stages were collected from the field. After two to three days starvation under moist conditions, they were weighed and preserved individually in 70% alcohol.

Each animal was dissected within a few weeks of fixation, the ovary was described briefly, and all the "mature" and "developing" eggs were counted. In this study, the term "mature" is used to describe the eggs which were lying freely in the sac-like oviduct (Fig. 2). When fresh, these eggs were almost spherical with minimum and maximum diameters of 0·9 mm and 1·0 mm respectively. The term "developing" is used to describe the eggs which were enlarging and accumulating yolk but which were still inside ovarioles. These eggs were ovoid in shape with a length of 0·5–1 mm (Fig. 2). In ovaries which had been fixed in alcohol, the large eggs (= "mature" and "developing" eggs) assumed a pale yellowish colour and were therefore clearly distinguishable from the white non-yolky oocytes.

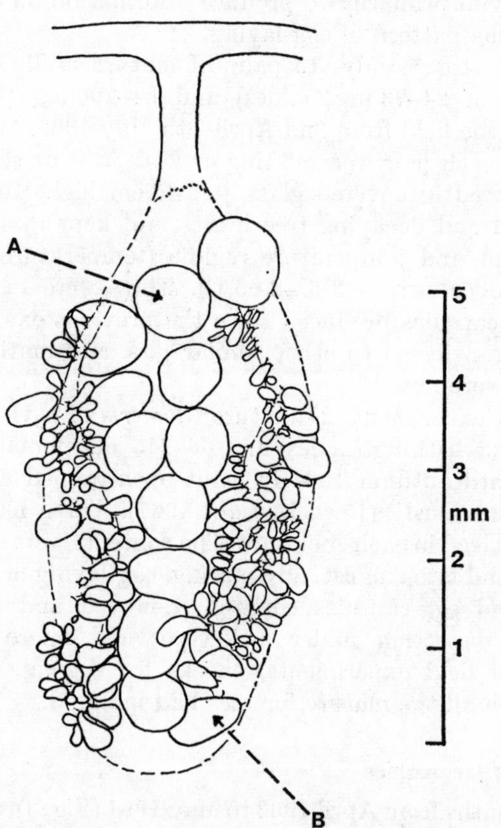

FIG. 2. Appearance of the ovary of epimorphic stadia of *Glomeris marginata* as seen typically in mature stadia, that is VIII and beyond. A = a "mature" egg, B = a "developing" egg.

Results and discussion

Changes in the ovary

Three main types of ovary condition were recognized in *Glomeris:*

1. Ovary so little developed that it was undetectable under a low-power microscope.

2. Ovary about 3–5 mm in length and with two series of similarly-sized follicles.

3. Ovary longer than about 5 mm and usually containing all sizes of follicles up to, but not necessarily including some of "mature" egg size; very rarely, with only large follicles present (Fig. 2).

Condition (1) was common among the definitely immature females of less than 50 mg live weight. Occasionally, particularly in the middle of the moulting season in August, ovaries were not detected in obviously mature animals. Condition (2) was characteristic of immature animals at all times of the year but was found frequently, particularly in spring and autumn, in animals which had only just become mature (54–90 mg live weight). It was also common in mature animals during the moulting period. Condition (3) was characteristic of the ovaries of mature females. Five animals in the moult recovery stage (Verhoeff, 1937) examined in August had ovaries in condition (3); one of these animals contained four "developing" eggs.

Females whose body cavities were heavily infected with gregarines or which contained a *Hexamermis* sp., respectively 1·40% and 1·22% of all females collected, usually contained atypical ovaries. For example, the ovaries of parasitized mature animals in spring either contained less than the average number of large eggs or were in condition (2).

The general patterns of change in the mean numbers of "mature", "developing" and large eggs in the ovary is clear from Fig. 3. When the common logarithms of the sample variances were plotted against the common logarithms of the sample means a line with a slope of about 1·26 was produced so we concluded that it was necessary to add 0·375 to each of the data and to apply a square root transformation before analysing the data statistically (Jeffers, 1959). The general pattern of changes in numbers in Figs 3 and 4 is similar. Several significant differences ($P < 0{\cdot}05$) in large eggs (Fig. 4) are worth noting: the mid-August value is lower than any of the other values; the significant rise in January–Febuary 1964 after three months of no significant change; the fall in March–April 1964 after a month of no significant change; the significant rise in numbers from April to May 1963. The latter was due in part to the sample being unrepresentative of the

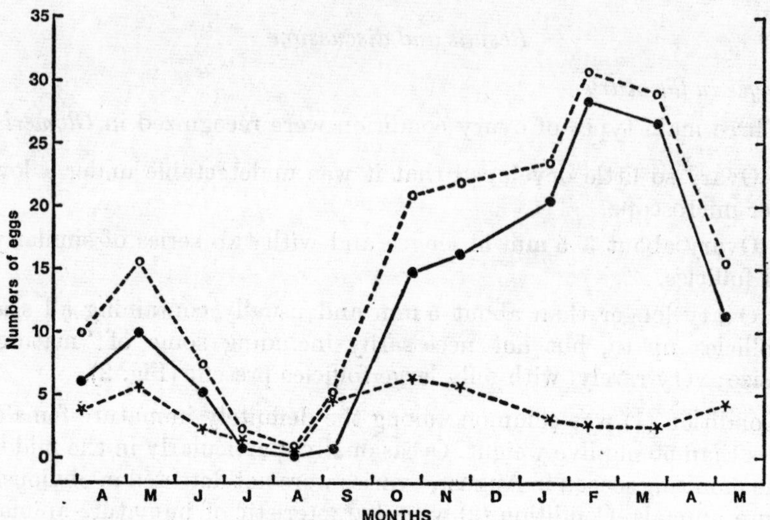

FIG. 3. Changes in the mean numbers of eggs in the ovary of mature *Glomeris marginata* from April 1963–May 1964. Solid circles, stars and open circles indicate respectively "mature" eggs, "developing" eggs and total large eggs. For convenience 54 mg, which is approximately the lower 95% confidence limit of the live weight distribution for stadium VIII and also the upper 99·8% limit for a similar distribution for stadium VII, was taken to be the minimum live weight for mature *Glomeris*. The means are based on all data for animals above this weight.

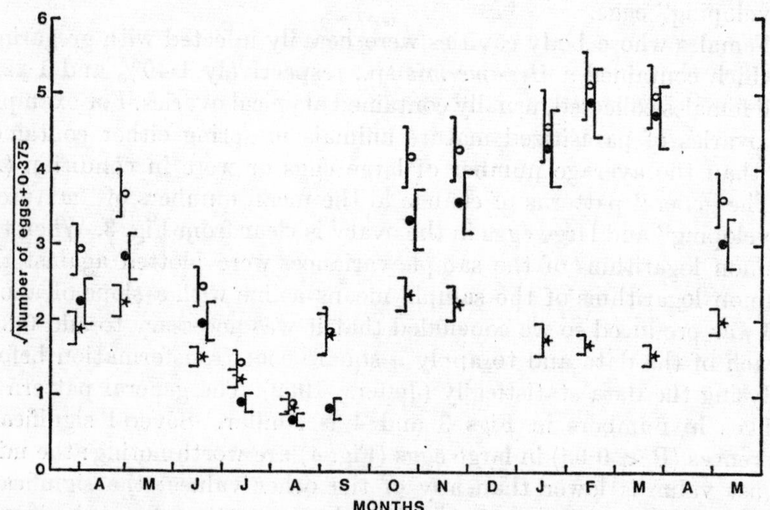

FIG. 4. Seasonal changes in the number of eggs in the ovary of *Glomeris marginata*. The means and 95% confidence limits are based on transformed data. Solid circles, stars and open circles indicate respectively "mature" eggs, "developing" eggs and total large eggs (see pp. 437–438 in text).

whole population. Females which produce the largest numbers of eggs, that is animals over about 155 mg (Fig. 7), were not found in April 1963. The long and exceptionally severe winter of 1962–1963 may also have influenced the number of large eggs in the ovaries in spring 1963.

The patterns of change in numbers of "mature" or "mature" plus "developing" eggs parallel each other as the numbers of "developing" eggs are, in general, low. However, a significant rise in numbers of the latter occurred in August–September in newly moulted females.

Timing of breeding

From the change in the number of large eggs in the ovary it appears that egg-laying could occur between the beginning of March and the middle of July (Figs 3 and 4). However, it is unusual for *Glomeris* to become active near the soil surface before early March (Bocock & Heath, 1967) and from 3–30 days elapse between copulation and egg-laying (Attems, 1926; Haacker, 1964; vom Rath, 1890). Our data for the first occurrences of mating and egg-laying indicate that this interval may be about a week but because of doubts about the exact value an interval has not been indicated in Fig. 5.

Our first and last records of egg-laying in the field are for 23rd March and 8th July but mating was recorded on 17th March and 16th July. In a laboratory breeding experiment (see p. 437), the last egg-laying occurred in the period 9th–31st July. We conclude that the breeding season on our sites extends from mid-March to mid-July with a peak in mid-May. This conclusion is in broad agreement with the comments made by other authors (Gervais, 1837; vom Rath, 1890, 1891; Hennings, 1904; Chalande, 1905; Evans, 1910; Attems, 1926; Verhoeff, 1928; Schubart, 1934; Dohle, 1964; Haacker, 1964, 1968; Hubert, 1968). However, Chalande (1905) states that egg-laying extends to the beginning of September.

Mating of *Glomeris* in autumn does not appear to have been recorded previously although Verhoeff (1928) implies that glomerids as well as iulids and polydesmids pair in the autumn and produce young but only in exceptional circumstances. Vom Rath (1891) states that he did not observe mating and egg-laying in *Glomeris* in September, October and November in spite of much field-work and Haacker (1968) comments similarly. We observed mating only twice in autumn although we spent several hours in the field in most weeks under various conditions and in several years. Neither the population-size structure (Bocock *et al.*, 1967) nor the ovary condition (Fig. 4) provide evidence of breeding in the autumn.

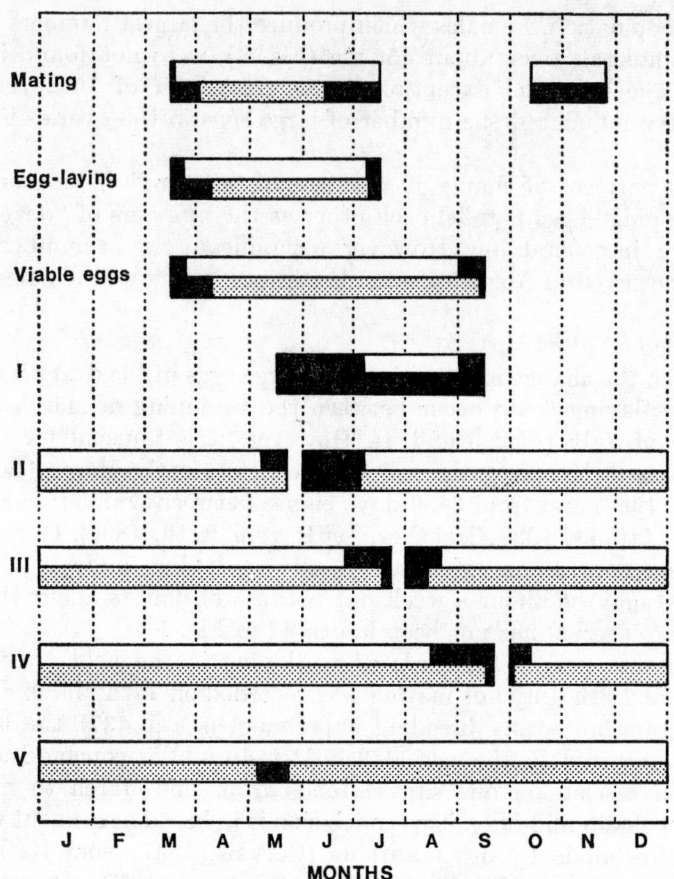

FIG. 5. The early life-history of *Glomeris marginata*. The open areas indicate presence of a particular activity or a stadium in Roudsea Wood (May 1956–September 1957 and in December 1964) or in Eggerslack and Heaning Woods (April 1963–April 1966). Similar data from growth studies (see pp. 437 or 446 in text) are summarized by the stippled areas. The black areas indicate periods for which the presence of an activity or stadium is strongly suspected (see text for supporting arguments). Stadium numbers are given at the left-hand side of the diagram. Viable eggs = egg capsules containing live eggs or stadia I and II.

Pattern of breeding

Data from individual animals kept in the laboratory (Fig. 6) or from animals kept in threes in the field clearly indicate that egg-laying is an intermittent process extending for individuals over a period of from 4–14·5 weeks. Matings occur repeatedly during this period. This confirms Evans's (1910) and Haacker's (1968) observations and opposes vom Rath's (1890) views on the pattern of egg-laying.

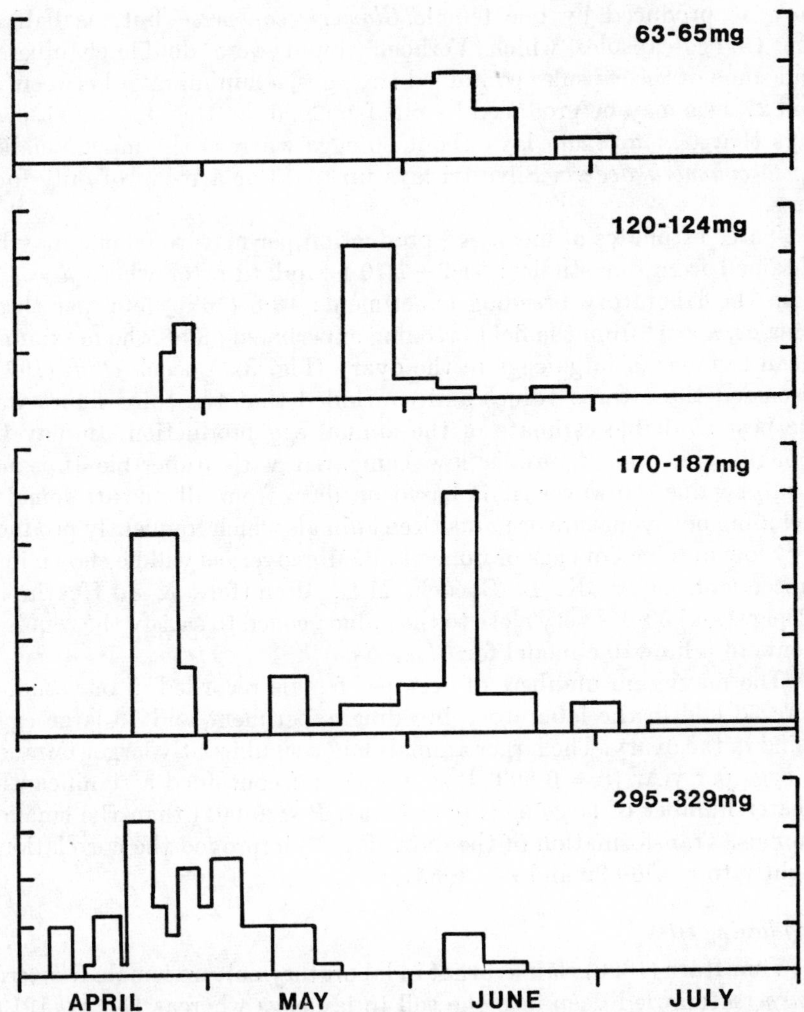

FIG. 6. Some examples of the pattern of egg-laying by *Glomeris marginata*. The live weight ranges of the females during egg-laying are given at the right-hand side of each diagram. The vertical scales indicate number of eggs laid, each sub-division representing five eggs.

Numbers of eggs laid

Several authors, for example Verhoeff, 1928; Halkka, 1958; Blower, 1969, 1970, have provided data on the number of eggs laid by iulid and polydesmid millipedes but data for glomerids are scanty. Schubart (1934), probably quoting Verhoeff (1910), states that up to 18 eggs may be produced by one female *Glomeris conspersa* but, as half of the 18 egg-capsules which Verhoeff found were double-chambered and some of the capsules contained two eggs, a minimum of between 18 and 27 eggs may be produced by one female. Juberthie-Jupeau (1967c) says that *G. marginata* lays about 50 eggs whereas the much smaller *Spelaeoglomeris doderoi* Silvestri lays up to 13 or a mean of only four eggs.

Three estimates of mean egg production per mature female may be obtained from our studies: $41{\cdot}2 \pm 7{\cdot}70$ (standard error where $n = 15$) from the laboratory breeding experiment; 18·8 ($56{\cdot}5 \pm 5{\cdot}92$ per three females, $n = 8$) from the field breeding experiment; 30·8, the maximum mean number of large eggs in the ovary (Fig. 3). Bocock *et al.* (1967) discussed these three values and concluded that the third figure was the best available estimate of the annual egg production. It may be thought that this figure is low compared with Juberthie-Jupeau's (1967c) value. However, it is based on data from all mature females including newly mature or parasitized animals which frequently produce very low numbers of eggs or none at all. Moreover, as will be shown in a subsequent paper (K. L. Bocock, M.D. Mountford & J. Heath, in preparation) 30·8 is very close to the value needed to satisfy the requirements of a life-table model for *Glomeris*.

The maximum numbers of eggs per female recorded in our studies were 86 laid in the laboratory breeding experiment and 76 large eggs found in the ovary. The larger animals laid a significantly larger number of eggs per year ($r = 0{\cdot}803$, $P < 0{\cdot}001$) and contained a significantly greater number of large eggs ($r = 0{\cdot}531$, $P < 0{\cdot}001$) than the smaller animals. Transformation of the data (Fig. 7) improved the correlations slightly to $r = 0{\cdot}832$ and $r = 0{\cdot}551$.

Egg-laying sites

Vom Rath (1890) claimed that in laboratory culture female *Glomeris conspersa* crawled deep into the soil to lay eggs whereas Evans (1910) and Hennings (1904) maintained that in the field *G. marginata* laid eggs at or very near to the soil surface under moss or dead leaves. Our field and laboratory observations plus the occurrence of most of the adult and near-adult *Glomeris* in the top 5 cm of the mineral soil during the breeding season (Bocock & Heath, 1967) support the previous views on *marginata*. However, occasionally, we have found groups of egg capsules at a depth of 5–10 cm in cavities between stones in soils formed on scree slopes.

The location of the main egg-laying areas within our sites appeared to be related to slope, type of soil, vegetation cover and the amount of

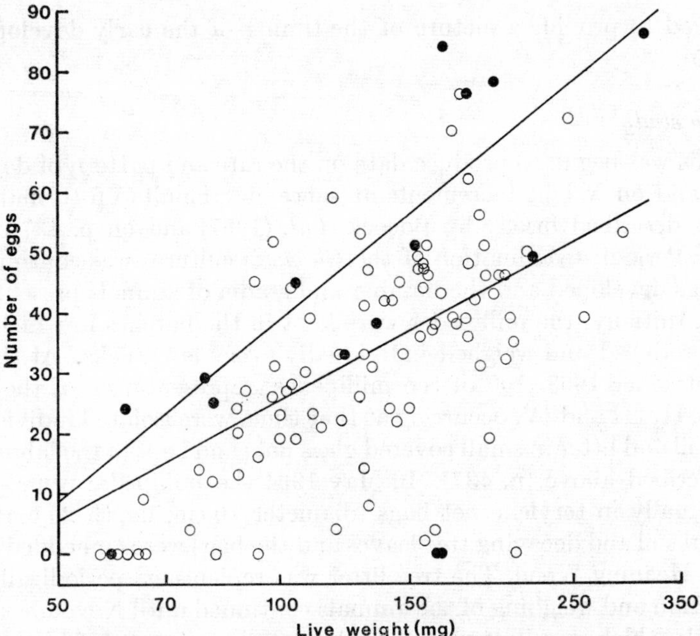

FIG. 7. Number of eggs laid by *Glomeris marginata* per year in relation to live body weight. The solid circles and upper regression line, $y = 96{\cdot}071 \log x - 153{\cdot}609$, refer to egg production by animals in the laboratory (see text p. 437). The open circles and lower line, $y = 65{\cdot}687 \log x - 105{\cdot}798$, refer to the numbers of large eggs in the ovaries of animals collected from the field in February 1964 (text pp. 437, 438). The analysis did not include any $y = 0$ data because the latter were associated with immature animals or with animals which were heavily parasitized.

plant remains on the soil surface. For example, in Heaning Wood, the favoured positions were the small level areas immediately on the up-slope side of coppiced hazel bushes. These areas were frequently well-shaded by overhanging hazel branches and the herb and ground layers were sparse. The soil had a litter layer extending into the coppice stool and beneath this a stony well-drained mineral horizon rich in organic matter which seemed always to be moist but not wet. In such areas the highest densities of all *G. marginata*, including anamorphic stadia, were found. The aggregation of *Glomeris* will be examined in another paper.

DEVELOPMENT OF THE EGG AND ANAMORPHIC STADIA

Sources of information and methods

Direct observation of **Glomeris** *in the field*

Records were kept of all *Glomeris* eggs or anamorphic stadia found in the field and these data, several hundred in all, were then arranged to provide a picture of the timing of the early development (Fig. 5).

Growth study

This was begun to produce data on the rate and pattern of development and on weight increments at successive moults up to maturity. It was described briefly by Bocock *et al.* (1967) and on p. 437 in this paper. Periodic examination of the *Glomeris* cultures was continued as the eggs developed and the number and stadia of animals present were noted. Initially, the millipedes were kept in the batches in which they were produced and weighed individually every few weeks. At the end of September 1963, 100 of the millipedes, representative of the three stadia, II, III and IV, occurring at that time, were isolated individually with soil and litter in small covered glass pots and kept in the laboratory as described above (p. 437). In May 1964 the millipedes were placed individually in terylene net bags (diameter 10 cm, depth 20 cm) with mineral soil and decaying tree leaves and the bags were embedded in the soil in Heaning Wood. The tree litter was replenished periodically and inspection and weighing of the animals continued until November 1965. Animals which were not placed in the pots in autumn 1963 were used to replace any of the 100 animals which died. They were kept in batches in the laboratory until they were transferred to large terylene net bags (diam. 15 cm, depth 30 cm) embedded in the field soil in either December 1963 or May 1964.

Examination of the contents of egg capsules

This was undertaken to facilitate description of the development from egg to second larval stage. In all, 741 capsules, collected from the field from July to September and 114 egg capsules produced in laboratory cultures of *Glomeris* kept as described above (p. 437) were examined under a low-powered binocular microscope. Only a selection of the results is given in Table I.

Results and discussion

Development and characteristics of the anamorphic stadia

In a previous paper (Bocock *et al.*, 1967), we gave a table summarizing the main morphological features which can be used to distinguish the five anamorphic stadia. On one point only, the number of apparent tergites in stadium I, we differed from all other authors except Dohle (1964) and more recently Juberthie-Jupeau (1971). Dohle indicated that the tergites of the seventh and eighth

TABLE I

Contents of egg capsules of ***Glomeris marginata***

Number of capsules examined	Embryos with eight tergites	Stadium I with eight tergites	Stadium II with eight tergites + exuvium from I	Capsule with exit hole and exuvium from I	Stadium I with brownish gut contents	Stadium II with brownish gut contents
75*	0	36	14	25	1	8
39†	8	28	3	0	0	0
35‡	0	33	2§	0	3	0

* Laboratory cultures, 29–30th July 1965.
† Laboratory cultures, 10th August 1964.
‡ The field, 11–13th August 1964.
§ Number of exuvia not checked.

body segments become separated from the proliferation zone in the sixth embryonic stadium but that these segments have a common lateral plate.

We confirmed that the late embryo had eight tergites and found that both stadium I and stadium II had eight tergites after examination of the contents of egg capsules (Table I). This examination also revealed other new information. First, after stadium I hatches from the egg it contains much yolk food reserve as indicated by other authors but nevertheless it sometimes feeds on the embryonal membranes or on the wall of the egg-capsule (Table I). Occasionally we found the remains of the tough brownish chorion. If stadium II was present or if it had emerged from the egg-capsule the remains of an exuvium, mainly the cast skin of the head and mouth parts, were always present (Table I). The exuvium and embryonal membranes therefore do not always form the first food of stadium II as suggested by Hubert (1968) but this stadium does eat the wall of the egg-capsule as it emerges (Table I) as observed by other authors, for example Fabre (1855), Hennings (1904) and Hubert (1968). Unlike Dohle (1964) we found that yolky material from the egg was frequently present in the body of the young stadium II but that it disappeared rapidly about the time of emergence of the animal from the capsule.

As both stadium I and stadium II have eight tergites, eight pairs of jointed legs, five antennal articles in each antenna and three ocelli on each side of the head other distinguishing characteristics must be sought. Hubert (1968) has stressed the marked difference in size by quoting head widths (0·4–0·5 mm and 0·82 mm) and body length measurements (about 1·5 and 2·18–2·20 mm). To this we can add live weight: I—0·4038 mg ($n = 1120$), II—1·3638 (range 1·1–1·8 mg, $n = 138$) and the lengths of components of the fourth to eighth pairs

TABLE II

Lengths (μm) of the segments of legs four to eight in stadia I and II of ***Glomeris marginata*** *dissected from egg-capsules*

Stadium	n	Pre-femur	Femur	Post-femur	Tibia	Tarsus
I	3	60	100	20	20	140
II	28	80	120	50	50	200

n = the number of animals examined.

of walking legs, particularly the tibia and tarsus which are considerably larger in II than in I (Table II). The fourth to eighth pairs of legs appear to be fully segmented and are clearly not the unsegmented parapodia described by vom Rath (1890) or Verhoeff (1928) for other species of *Glomeris*.

Timing of development

In a previous paper (Bocock *et al.*, 1967), we outlined the approximate timing and main pathways of development through the anamorphic and early epimorphic stadia (Fig. 8). Here we have attempted to elucidate the timing of development in greater detail to provide a sound basis for estimation of production of *Glomeris* (K. L. Bocock, M. D. Mountford & J. Heath, in prep.).

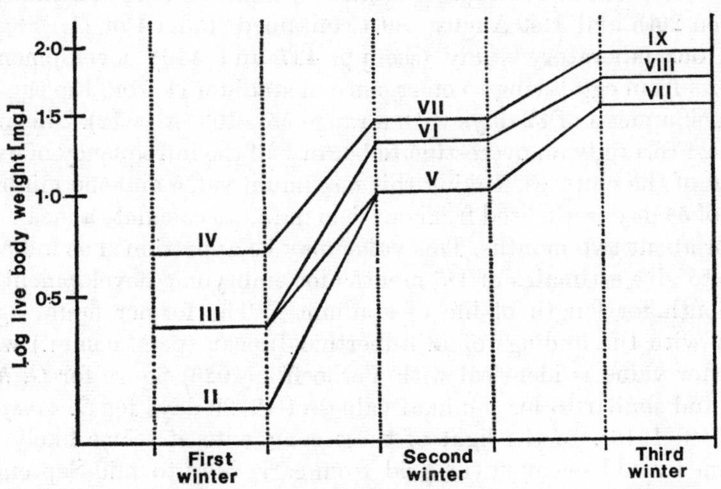

FIG. 8. The main pathways of development of female *Glomeris marginata* from egg to maturity (re-drawn from Bocock, Mountford & Heath, 1967). Stadium numbers are given above the paths.

Freshly laid eggs occurred in the field from mid-March until mid-July, but egg-capsules containing viable eggs or stadia I or II were found until 31st August (Fig. 5). Some capsules in a similar condition were found in laboratory cultures on 12th September (Fig. 5).

The embryonic development of *Glomeris* is said to take 30 days (Attems, 1926), three to four weeks (Schubart, 1934) or one month (Hennings, 1904). These estimates are almost certainly derived in part from vom Rath (1890) who quotes four weeks and 30 days for *G.*

conspersa or Chalande (1905) who, apparently independently of vom Rath and Hennings, obtained a figure of 30 days for *G. marginata* and *G. pyrenaica*. More recently, for *G. marginata* Dohle (1964) and Hubert (1968) quoted respectively 21–22 days for animals kept under unspecified conditions, probably in the laboratory, and three to eight weeks (minimum and maximum calculated from the dates given) for animals at 15°C in the laboratory. Chalande (1905), Dohle (1964) and Hubert (1968) all indicate that stadium I lives for about ten days. This is equivalent to about one quarter of the total time which the egg plus stadium I is said to spend inside the capsule.

From the difference between the recorded end of the egg-laying period and the last occurrence of egg capsules containing viable larvae or eggs in the field (Fig. 5), we estimate that development in the egg-capsule takes 54 days. This is probably an underestimate because 30 of this last batch of 173 capsules collected from the field and examined between 20th and 31st August 1964 contained stadia I or II.

In our laboratory study (see pp. 437 and 446) development of *Glomeris* from egg-laying to emergence of stadium II from the egg-capsule took a mean of 82 days with a range 55—103 ($n = 12$). This mean is almost certainly an overestimate because of the infrequency of examination of the cultures. Taking this maximum value and the minimum mean of 54 days estimated from our field data, we calculate a mean of 68 days or about two months. This value may be apportioned as indicated above to give estimates of 1·5 months for embryonic development and 0·5 month for length of life of stadium I. The former figure agrees closely with the findings of L. Juberthie-Jupeau (pers. comm.) whilst the latter value is identical with Verhoeff's (1928) figure for *G. hexasticha* and similar to his minimal values of 12–20 days for *G. conspersa* (Verhoeff, 1910). In the light of these comments, it seems likely that stadium I could occur in the field from early May to mid-September. However, development of eggs laid in March will be considerably slower than development of those laid in the warmer months so mid-May is a more acceptable first date (Fig. 5). If this is correct and stadium I lives about 0·5 month before moulting, II may be present from about early June, five weeks before our first field and laboratory records (Fig. 5). This is not unreasonable as stadium II had emerged from 116 (41·6%) of 279 egg capsules which we collected in the field on 6th–13th July 1964. However, Hubert (1968) first found II in laboratory cultures at 15°C on 24th June following egg-laying at the end of April onwards but on our sites egg-laying began one month earlier. Blower & Gabbutt (1964), working on a slightly warmer site than ours, found "the first stadia", presumably I, from May to August.

The lengths of life of stadia II, III, IV and V are very variable because animals can overwinter in any of these stadia (Fig. 8), no moults occurring from about early November until at least early April. Verhoeff's (1928) data for stadium II for *G. hexasticha* indicate a duration of about 80 days ending on 21st September. Our presence data for II and III indicate durations for II of 45 days based on first records (6th July–20th August) and 51 days (6th May–26th June) based on last records (Fig. 5). Comparable but entirely independent data from the growth study are 39 days (4th July–12th August) and 56 days (22nd May–14th July) (Fig. 5). However, if our estimated starting date for II is used in these calculations, the above estimates based on first records become 80 and 72 days for the presence and growth data respectively. A mean value of 62 days, or approximately two months duration, seems therefore to be a reasonable figure to use for non-overwintering stadia II. Overwintering II in the growth study took about nine months to develop to stadium III (Fig. 5). Verhoeff (1928) found stadium III of *G. hexasticha* first on 21st September. Vom Rath (1890) states that stadium III of *G. conspersa* and other unstated species of *Glomeris* occur in November, whereas Verhoeff (1910) found this stadium of *G. pustulata* from the beginning of August onwards.

Our presence and growth data indicate that III occurs from the middle of August and during the following complete year (Fig. 5). If the duration of non-overwintering stadia II is about two months III can be expected to occur from the beginning of August onwards. Our estimates of the duration of non-overwintering stadium III are as follows: 73 days (1st August–12th October) based on presence records for larva IV and suspected first presence for III; 53 days (20th August–12th October) and 62 days (17th July–17th September) based on the growth study. A mean value of 60 days or about two months can therefore be taken. Overwintering III and IV may live for up to eight or nine months before moulting.

Stadium IV can be first expected at the beginning of October and this timing is achieved in the growth studies (Fig. 5). None of these animals moult to give V until the following year (Figs 5 and 8). Our estimates of the duration of non-overwintering stadium IV are 51 and 62 days based on respectively the first and last occurrences of III and IV in the growth study, and 53 and 50 days from respectively the first and last presences of III and IV in the field. A mean of 54 days, very approximately two months, can therefore be used.

Stadia V were found in all months of the year (Fig. 5) so calculation of durations of this stage must be based solely on observation of the growth of individual animals. Such data indicate that V lives for two to three months if it does not overwinter or about nine months including overwintering.

THE EPIMORPHIC STADIA

Outline of development

The following is a summary of our current information on the epimorphic development. This topic will be dealt with more fully in another paper now in preparation.

Verhoeff (1928) and other authors were aware that *Glomeris* had one or more immature epimorphic stadia and that the adult moulted several times after the onset of maturity. However, except for Verhoeff's attempts to separate the early epimorphic stadia (refer to p. 456), no progress was made in analysing the size- or age-class structure until the work reported in our earlier paper (Bocock *et al.*, 1967). In Fig. 3 of that paper, a frequency histogram of live-weights of female *G. marginata*, we indicated that four fairly distinct size-classes were followed by up to 16 overlapping indistinct classes. A similar picture was found in small population samples and during analysis of growth data, so we concluded that the classes were real size-classes and not artefacts and we attached the numbers 12(1)–12(20) to them. The mean weights and ages of classes beyond 12(4) were determined from the growth data (Bocock *et al.*, 1967). The size-class structure of males was similar to that for the females, except that the weight range extended only up to about 130 mg. Only the first three epimorphic stadia were recognizable and the remaining size-class distributions overlapped even more than in the females. Our current interpretation of the relation of these classes to developmental stadia is given in Table III.

Moulting

Seasonal occurrence

Hennings (1904) said that *Glomeris* moults during the summer months in the weeks subsequent to mating and egg-laying. Vom Rath (1890), who commented similarly, found fully grown animals moulting especially in July and August, whereas Verhoeff (1928) stated that moulting of diplopods in general occurs from May–October but particularly in September. Our observations (Figs 5 and 9) support and extend the above comments. Moulting animals were not seen outside the months of May-October inclusive but it is possible that some moulting occurred in late April or early November (Fig. 9).

TABLE III

Live weights ± standard errors of the early epimorphic stadia of **Glomeris marginata**

	MALES			FEMALES		
Stadium*	Stadium†	Live weight (mg)	Age in years	Stadium†	Live weight (mg)	Age in years
12 (1)	VI	18·2 ± 3·9	2	VI	18·2 ± 3·9	2
12 (2)	VII	33·1 ± 6·8	2 or 3	VII	33·1 ± 6·8	2 or 3
12 (3)	VIII	38·7 ± 2·6	3 or 4	VIII	72·0 ± 8·9	3
12 (4)				IXA	95·8 ± 4·9	3 or 4
12 (5)	IX	47·0 ± 4·5	4 or 5	IXB	111·4 ± 10·3	4
12 (6)				XA	135·2 ± 7·1	4 or 5
12 (7)	X	55·3 ± 5·8	5 or 6	XB	145·9 ± 13·1	5
12 (8)				XIA	169·7 ± 10·8	5 or 6
12 (9)	XI	63·6 ± 6·9	6 or 7	XIB	170·6 ± 16·6	6
12 (10)				XIIA	194·4 ± 14·8	6 or 7
12 (11)	XII	68·7 ± 7·8	7 or 8	XIIB	195·4 ± 19·5	7
12 (12)				XIIIA	219·2 ± 18·0	7 or 8
12 (13)	XIII	73·8 ± 8·7	8 or 9	XIIIB	220·1 ± 22·0	8
12 (14)				XIVA	243·9 ± 20·6	8 or 9
12 (15)	XIV	78·9 ± 9·4	9 or 10	XIVB	244·9 ± 24·2	9
12 (16)				XVA	268·7 ± 22·1	9 or 10
12 (17)	XV	84·0 ± 10·2	10 or 11	XVB	269·6 ± 26·2	10
12 (18)				XVIA	293·4 ± 25·1	10 or 11

The mean weights and ages of each stadium were determined by analysis of growth data and population size–class structure (Bocock, Mountford & Heath, 1967).

* Stadium numbering of Bocock, Mountford & Heath (1967).

† Conventional stadium numbering.

The first four anamorphic stadia moult during the periods indicated in Figs 5 and 8. Stadium V can be found moulting at any time during May–October inclusive but the main moulting activity occurs in spring and autumn. Figure 8 indicates that female stadium VI and VII moult

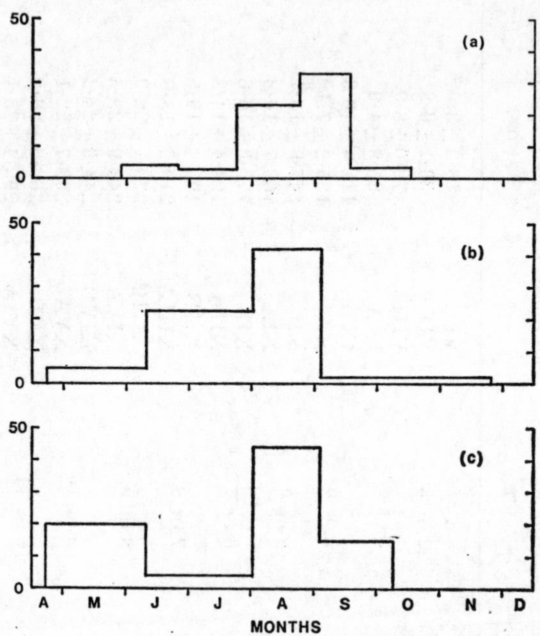

FIG. 9. Seasonal occurrence of moulting in epimorphic *Glomeris marginata*. (a) unenclosed animals of all sizes collected during population sampling, (b) males of all sizes, (c) females less than 100 mg live weight kept in large terylene bags embedded in the soil. The vertical scales indicate the number of moulting animals or exuviae per 100 animals collected.

during the May–October period and immature females in general pass through two or three moults per year. The pattern of moulting is similar for the males. Mature animals, that is, stadia VII (males) or VIII (females) and beyond, moult only once per year during the July–October period. This has been determined by frequent observation of several hundred *Glomeris* kept in the field in enclosures (Bocock *et al.*, 1967) or in terylene net bags as described above (p. 446).

Length of the moulting period for individual animals

Halkka (1958), Vannier (1966) and Verhoeff (1937) indicated that the moulting period of various individual millipedes lasted for about

2 to 3 weeks in the laboratory. In one of our field studies, in which pairs of epimorphic *Glomeris* were enclosed in terylene net bags in the soil and checked every 2 to 4 weeks for feeding activity and moulting, the minimum number of weeks from feeding to feeding with an intervening moult ranged from 2 to 6. These values must be maximum estimates because of the infrequency of examination of the animals. Re-examination of the gut content data given in Fig. 1 of Bocock & Heath (1967) suggested that the frequency of moulting, as indicated by the frequency of non-feeding animals, was normally distributed with a mean at 17th–18th August and a standard deviation of 16·7 days. The mean length of the moulting period for epimorphic *Glomeris* calculated from the area under the normal curve was about 22 days. Accordingly, from our two estimates, we conclude that the moulting period for epimorphic *Glomeris* in the field lasts about 2 to 4 weeks. The duration of the period for smaller *Glomeris* was not examined but we would expect it to be considerably less than 2 to 4 weeks because of the relatively higher metabolic activity of smaller millipedes.

The moulting process

The four stages of moulting described by Halkka (1958) and Verhoeff (1928) for certain millipedes are also found in *Glomeris*. Evans (1910), Hennings (1904) and Vom Rath (1890) concluded that *Glomeris* does not seek any special protection during moulting but that it lies freely in loose soil or under moss or dead leaves. Verhoeff (1928) states that many *Glomeris* moult in primitive moulting chambers—hollows in the soil smoothed internally by rocking movements performed by the animal.

We frequently found *G. marginata* moulting whilst lying on the mineral soil surface under moss, dead leaves or stones, in loose soil and in simple chambers in the soil which we confirm are smoothed internally by the animal rocking. We have also seen moults occurring rarely inside a thin-walled tent-like chamber similar to that described by Vannier (1966) for an African millipede. These chambers were irregular in shape, about 2 cm in diam. and constructed of blobs of mineral-soil-rich material stuck together to form continuous sheets about 1–2 mm in thickness. The sheets stretched between any convenient objects, roots and stones in the field, stones and the wall of the culture jar in the laboratory. The reason for construction of these chambers may be related to the moisture status of the soil. Before the laboratory observations were made, the soil in the culture jars had been allowed to become rather dry and the tent was constructed at the bottom of the jar 2–3 cm below the soil surface where the dry soil was probably most moist.

Achievement of maturity

The point in development at which *Glomeris* becomes mature was determined by gathering together evidence of reproductive activity for animals either of a known live weight or of a known stadium.

Millipedes of known live weight

This approach involved using the live weights of the early epimorphic stadia (Table III) as an aid in allocating an animal to a particular stadium after it has shown evidence of maturity. The approach could only be a preliminary one because of the large variation in live weight of animals belonging to an identifiable size- and age-class (Table III).

In the ovary examination (see pp. 437, 438), the live weights of the smallest females which contained large eggs ranged from 58·2–77·4 mg. The smallest animals which contained some "mature" eggs weighed 59·6–85·9 mg.

In the breeding experiment in the laboratory (see p. 437), the live weights of the three smallest females which produced eggs varied as follows during the breeding period: 75·9–82·9 mg, 63·5–65·0 mg and 89·7–92·0 mg. A female of 59·4–64·0 mg did not produce eggs but was observed mating.

The live weights of the six smallest males, which were kept with females which produced eggs, varied as follows during the breeding period: 27·6–30·0 mg, 30·3–33·2 mg, 24·1–25·5 mg, 28·4–30·0 mg, 35·2–36·5 mg and 35·0–36·6 mg. Egg production by a female collected in the spring and kept with a male is not conclusive evidence of maturity of the male because the female may have mated before collection. However, mating followed by egg-laying is more conclusive and the first two males in the above list were seen to mate.

Millipedes of known stadium

The second and more convincing method of checking for onset of maturity involved initially collecting stadia V and VI from the field in August 1964 and placing them singly with garden soil and decaying plant remains in small covered glass jars. Stadium V was identified from the number of tergites and VI by its live weight (< 20 mg) and the light pigmentation and narrowness of the penultimate tergite. The culture jars were kept in a cabinet in which the temperature fluctuated daily between about 13° and 20°C (extremes 12·5° and 22·0°C). The millipedes were checked for moulting once or twice weekly and weighed fortnightly as an additional check on growth.

The maturity of an animal was checked by placing it in a culture jar with a *Glomeris* of the opposite sex for periods of a few days to a few weeks. In all, 41 males and 19 females were tested for maturity, some individuals being checked in each of two successive stadia (Table IV). Great significance cannot be attached to the live weights of the stadia after V and VI in this study because the weight increments at successive moults may have been influenced by the laboratory conditions and frequent handling of the animals.

TABLE IV

Summary of maturity tests on ***Glomeris marginata***

Stadium during reproductive activity	Live weight (mg) during reproductive activity	Reproductive activity T*	M†	E‡	Number of animals
Males					
VII	23·1–33·8	+	+	+	4
VII	24·6–36·1	−	+	+	5§‖
VII	27·1–37·6	+	−	+	3
VII	34·4–36·0	—	+	—	1
VIII	29·6–37·6	+	+	+	4§¶
VIII	29·9 only	+	−	+	1
Females					
VIII	50·0–52·1		+	−	1§‖
VIII	58·6–64·7		+	−	1
VIII	64·1–69·1		−	+	1
VIII	68·4–81·2		−	+	1
IX	69·4–77·6		+	−	1§‖
VIII	84·8–96·8		+	−	1

* Telopods protruding in preparation for mating.
† Mating.
‡ Egg-laying by female of pair.
§ Includes one animal which showed reproductive activity in two successive stadia.
All the animals tested were in stadium V when collected except for:
‖ one animal in stadium VI when collected,
¶ two animals in stadium VI when collected.

Verhoeff (1928) claimed that in oniscomorph millipedes, there are three immature epimorphic stadia between stadium V and the fully mature animal. He based his claim apparently entirely on examination of the characteristics, particularly the pigmentation, chitinization and morphology, of the telopods of male *Gervaisia*, *Typhloglomeris*, *Onychoglomeris* and various *Glomeris* species. For females, he assumed that the same three stadia existed and suggested that characteristics of the vulvae might be useful in distinguishing the stadia, but he apparently did not examine these in detail nor did he check these stadia critically for maturity.

Data which we have collected indicate that Verhoeff's comments about onset of maturity do not apply to *Glomeris marginata*. From a combination of the data given in Table III and on p. 456 above, we conclude tentatively that males are mature in stadia VII when they are two or three years old whilst females are mature in stadia VIII and IX at three or four years old.

For females, negative results in maturity tests on all stadium VII animals tested and egg-laying by two indisputably stadium VIII animals (Table IV) confirm our tentative conclusions convincingly. However, *Glomeris* weighing 54–72 mg frequently contain no large eggs in autumn or late winter (Fig. 7) so it is probable that only part of the stadium VIII animals lay eggs in their third year. Virtually all stadium IX females (live weight approximately 86–106 mg) and all the larger VIII females (live weight approximately 72–90 mg) contained large eggs (Fig. 7) and were therefore mature.

The tests on males (Table IV) confirm our tentative conclusions concerning onset of maturity. None of the stadium VI animals tested produced evidence of maturity. It is possible that some of the males do not mate until stadium VIII but we have no convincing evidence of this. The negative results in the maturity tests were evenly spread over the whole range of live weights of males tested. A record of mating by the same male in stadia VII and VIII (Table IV) supports the generally accepted view that periodomorphosis does not occur in *Glomeris* (Halkka, 1958).

Longevity

Verhoeff (1928) studied the development of *Glomeris hexasticha* and argued that because anamorphic development takes 20·5 months and the last anamorphic stadium is about half the size of what he regarded as the mature animal the latter must be about 3·5 years old. As Verhoeff's mature stadium was stadium IX his ageing of this stage was accurate (Table III) if *G. hexasticha* and *G. marginata* develop at the same rate.

Verhoeff (1928) also argued that because some species of *Glomeris* moult several times after reaching maturity a life-span of 6·5–7·0 years is possible. Table III indicates that Verhoeff greatly underestimated the possible life-span of *Glomeris* and also that this life-span greatly exceeds the durations of about 2 to 6 years quoted for some other millipedes whose life-history has been studied (see for example Blower, 1958, 1969, 1970; Blower & Gabbutt, 1964; Halkka, 1958; Stephenson, 1960; Verhoeff, 1928).

Acknowledgements

We are very grateful to our colleagues Mrs D. M. Howard and Mrs J. Parrington whose painstaking work made our *Glomeris* studies possible. We also wish to thank Mr D. F. Spalding of the Biometrics Section of the Nature Conservancy for his expert assistance with the statistical analysis of our data and Dr. H. E. Welch of the Research Branch of the Canada Department of Agriculture for identifying *Hexamermis*.

References

Attems, C. (1926). Diplopoda. *Handb. Zool.* **4**: 239–402.

Blower, J. G. (1958). British millipedes. *Synopses Br. Fauna* **11**: 1–74.

Blower, J. G. (1969). Age structures of millipede populations in relation to activity and dispersion. *Publs Syst. Ass.* **8**: 209–216.

Blower, J. G. (1970). The millipedes of a Cheshire Wood. *J. Zool., Lond.* **160**: 455–496.

Blower, J. G. & Gabbutt, P. D. (1964). Studies on the millipedes of a Devon oak wood. *Proc. zool. Soc. Lond.* **143**: 143–176.

Bocock, K. L. (1963). The digestion and assimilation of food by *Glomeris*. In *Soil organisms*: 85–91. Doeksen, J. & Van der Drift, J. (eds). Amsterdam: North Holland Publishing Company.

Bocock, K. L. & Gilbert, O. J. W. (1957). The disappearance of leaf litter under different woodland conditions. *Pl. Soil* **9**: 179–185.

Bocock, K. L., Gilbert, O., Capstick, C. K., Twinn, D. C., Waid, J. S. & Woodman, M. J. (1960). Changes in leaf litter when placed on the surface of soils with contrasting humus types. I. losses in dry weight of oak and ash leaf litter. *J. Soil Sci.* **11**: 1–9.

Bocock, K. L. & Heath, J. (1967). Feeding activity of the millipede *Glomeris marginata* (Villers) in relation to its vertical distribution in the soil. In *Progress in soil biology*: 233–240. Graff, O. & Satchell, J. E. (eds). Braunschweig: Friedr. Vieweg. Amsterdam: North Holland Publishing Company.

Bocock, K. L., Mountford, M. D. & Heath, J. (1967). Estimation of annual production of a millipede population. In *Secondary productivity of terrestrial ecosystems* **2**: 727–739. Petrusewicz, K. (ed). Warszawa–Krakow: Państwowe Wydawnictwo Naukowe.

Chalande, J. (1905). Recherches sur les Myriapodes du Sud-Ouest de la France. *Bull. Soc. Hist. nat. Toulouse* **38**: 46–154.

Dohle, W. (1964). Die Embryonalentwicklung von *Glomeris marginata* (Villers) im Vergleich zur Entwicklung anderer Diplopoden. *Zool. Jb.* (Anat.) **81**: 241–310.

Evans, T. J. (1910). Bionomical observations on some British millipedes. *Ann. Mag. nat. Hist.* (8) **6**: 284–291.

Evans, T. J. (1911). The egg-capsule of *Glomeris*. *Zool. Anz.* **37**: 208–211.

Fabre, L. (1855). Recherches sur l'anatomie des organes reproducteurs et sur le développement des Myriapodes. *Annls Sci. nat.* (4) **3**: 257–316.

Gervais, P. (1837). Etude pour servir à l'histoire naturelle des Myriapodes *Annls Sci. nat.* (2) **7**: 35–59.

Haacker, U. (1964). Das Paarungsverhalten des Saftkuglers *Glomeris marginata*. *Natur Mus., Frankf.* **94**: 265–272.

Haacker, U. (1968). Deskriptive, experimentelle und vergleichende untersuchungen zur Autökologie rhein-mainischer Diplopoden. *Oecologia* **1**: 87–129.

Halkka, R. (1958). Life history of *Schizophyllum sabulosum* (L) (Diplopoda, Iulidae). *Suomal. eläin-ja kasvit. Seur. Van. eläin. Julk.* **19** (4): 1–72.

Hennings, C. (1904). Zur Biologie der Myriapoden. *Biol. Zbl.* **24**: 251–283.

Heymons, R. (1897). Mitteilungen über die Segmentierung und den Korperbau der Myriopoden. *Sber. preuss. Akad. Wiss.* **40**: 915–928.

Humbert, A. (1872). Etudes sur les Myriapodes. *Mitt. schweiz. ent. Ges.* **3**: 230–544.

Hubert, M. (1968). Sur des cas de monstruosités doubles observés chez des larves de *Glomeris marginata* (Villers) (Myriapode, Diplopode, Oniscomorpha, Glomeridia). *Bull. Soc. zool. Fr.* **93**: 443–450.

Jeffers, J. N. R. (1959). *Experimental design and analysis in forest research.* Stockholm: Almqvist & Wiksell.

Juberthie-Jupeau, L. (1967a). Les oothèques de quelques Diplopodes Glomerida *Révue Ecol. Biol. Sol* **4**: 131–142.

Juberthie-Jupeau, L. (1967b). Ponte et développement larvaire de *Spelaeoglomeris doderoi* Silvestri (Myriapode, Diplopode). *Annls Spéléol.* **22**: 147–166.

Juberthie-Jupeau, L. (1967c). Etude du biotope et du développement d'un Diplopode cavernicole, *Spelaeoglomeris doderoi* Silvestri. *Spelunca, Paris* **5**: 273–276.

Juberthie-Jupeau, L. (1968). Production expérimentale de monstres doubles chez *Glomeris marginata* (Villers), Myriapode, Diplopode. *C.r. hebd. Séanc. Acad. Sci., Paris* (D) **266**: 1610–1612.

Juberthie-Jupeau, L. (1970). Action tératogène de la température sur l'embryon de *Glomeris marginata* (Villers). *Bull. Mus. natn. Hist. nat., Paris* (2) **41**; Supp. No. 2: 79–84.

Juberthie-Jupeau, L. (1971). Modification expérimentale du nombre des segments au cours du développement embryonnaire chez *Glomeris marginata* (Villers), Myriapode, Diplopode. *C.r. hebd. Séanc. Acad. Sci., Paris* (D) **273**: 1991–1994.

Latzel, R. (1884). *Die Myriapoden der österreichisch-ungarischen Monarchie.* **2**: 1–414. Wien: Hölder.

Nicholson, P. B., Bocock, K. L. & Heal, O. W. (1966). Studies on the decomposition of the faecal pellets of a millipede (*Glomeris marginata* (Villers)). *J. Ecol.* **54**: 755–766.

Rath, O. vom (1890). Uber die Fortpflanzung der Diplopoden (Chilognathen). *Ber. naturf. Ges. Freiburg* **5**: 1–28.

Rath, O. vom (1891). Zur Biologie der Diplopoden. *Ber. naturf. Ges. Freiburg* **5**: 161–199.

Schubart, O. (1934). Diplopoda. *Tierwelt Dtl.* **28**: 1–318.

Stephenson, J. W. (1960). The biology of *Brachydesmus superus* (Latzel) Diplopoda. *Ann. Mag. nat. Hist.* (13) **3**: 311–319.

Vannier, G. (1966). Loge de mue d'un nouveau type construite par un diplopode africain. *Révue Ecol. Biol. Sol* **3**: 241–258.

Verhoeff, K. W. (1906). Zur Kenntnis der Glomeriden. *Arch. Naturgesch.* **72**: 107–226.

Verhoeff, K. W. (1910). Mitteilung betreffend Ökologie, Einrollungsarten und Metamorphosecharakter bei *Glomeris. Zool. Anz.* **36**: 298–315.

Verhoeff, K. W. (1916). Ist die physiologische Bedeutung der Glomeridentelopoden geklart? *Biol. Zbl.* **36**: 167–174.

Verhoeff, K. W. (1928). Diplopoda 1. *Bronn's Kl. Ordn. Tierreichs* **5** (2): 1–1072.

Verhoeff, K. W. (1932). Diplopoda 2. *Bronn's Kl. Ordn. Tierreichs* **5** (2): 1073–2084.

Verhoeff, K. W. (1937). Die Perioden der Hautungszeit bei Chilognathen. *Z. Morph. Ökol. Tiere* **33**: 290–296.

Discussion

Blower: Do you know why the distribution of the weights of the epimorphic stadia is bimodal?

Bocock: Even if animals are taken from the field and grown under constant conditions in the laboratory the distribution of their weights is bimodal. Bimodality is characteristic of the species and seems to be related to the timing of development of individual animals and the overlap of generations illustrated in Fig. 8. We are not entirely satisfied with this explanation and hope to comment further on this point in a future paper. Incidentally, bimodality is only obvious in the females where the range of weight distribution is wider than in the males.

Strasser: Verhoeff distinguished four epimorphic stages, a prematurus, pseudomaturus, maturus junior and maturus senior. Can you tell me to which of your stages these four of Verhoeff correspond?

Bocock: We think the mature stage (our stage 12(3)) is the maturus junior.

Haacker: In the female.

Bocock: Yes.

Haacker: In the male your 12(3) would be the pseudomaturus?

Bocock: Yes. the first of Verhoeff's stages was the "vorstadium" or stadium antecedens, I think. The second was the pseudomaturus. So, in the male, it is the pseudomaturus which is the first mature stage.

Dohle: You include the number of apparent body segments in your Table; I should prefer the use of the term "tergite".

Bocock: Yes. I shall use "tergite" or "apparent tergite" in the paper.

Standen: Is the figure of 40–86 eggs laid, the number laid once per season?

BOCOCK: The figure I gave was "up to 86"; this is the number laid per female per year.

CAUSEY: How many eggs are there in one clutch?

BOCOCK: This varies considerably; it may be one or two laid in one particular place, it may be 30 or 40. If the female chooses a highly favourable site initially, it may lay all the eggs in this place at one time. A smaller number will be laid if the animal is forced away from the site, by adverse weather conditions, for example.

MEIDELL: Do you have any thoughts about what factors limit the distribution of *G. marginata*?

BOCOCK: We think moisture and drainage important. The species is not common on very moist sites. Even on our sites, which are very well drained during prolonged wet weather, we have found large numbers of epimorphic stadia in the litter and amongst moss, in a very waterlogged condition—the arthrodial membranes between the tergites were exposed.

MEIDELL: This idea fits very well with the distribution in Norway; it occurs half way up our eastern coast but has not been found in the western part where the humidity is much higher.

HEATH: This waterlogging is noticeable in laboratory cultures. This is the only thing that *can* go wrong with cultures.

BOCOCK: The anamorphic stadia are very susceptible to waterlogging. There is a large mortality in the egg and first two anamorphic stadia which is probably due to waterlogging.

WALLWORK: If this is the case presumably the water is taken across the general body surface?

BOCOCK: Presumably between the tergites.

WALLWORK: Yet the species is common on dry sites in the Chilterns and this implies a waterproofing device in the cuticle.

BOCOCK: Yes, we find it also in very dry sites. Where the soil dries completely in the summer the animals respond by moving into deeper regions as they do in the winter in response to adverse temperatures. We mention that we have sometimes found them in moulting chambers both in the field and in the laboratory. I think this is the first observation of this habit in this species.

HAACKER: Is it a real moulting chamber or just a hole in the soil?

BOCOCK: Frequently they are found in a smoothed hollow. The animal performs rocking movements which smooth the depression. But I was referring to definite moulting chambers formed from material which has passed through the gut.

Symp. zool. Soc. Lond. (1974) No. 32, 463–469.

SEX RATIO AND PERIODOMORPHOSIS OF *PROTEROIULUS FUSCUS* (AM STEIN) (DIPLOPODA, BLANIULIDAE)

MAIJA RANTALA

Järvensivuntie 93, Tampere, Finland

SYNOPSIS

Proteroiulus fuscus reared from eggs in laboratory cultures began sexual differentiation usually at stadium VI. Of 906 individuals, 187 (20·6%) were males. In the field, in Finland, the proportion of males is much smaller (0·64–0·71%). Progeny from individual females included up to 100% males.

Males first attain maturity at stadium VIII, IX or even X, moulting next into the intercalary ("Schalt") condition and usually, after a second moult, attain the copulatory phase again. Only in two individuals were successive intercalary phases recorded. One of these, taken from the field as a stadium IX copulatory in 1966, passed through nine further stadia, only one of which was a copulatory phase, eventually dying in 1972 as a stadium XVIII intercalary.

INTRODUCTION

From 1966 to 1968 I studied anamorphosis and periodomorphosis in *Proteroiulus fuscus* (Rantala, 1970). From 1968 to 1972 I have continued the investigation of periodomorphosis and the sex ratio. The cultures consisted of the same individuals as in these previous studies, and their progeny.

SEX RATIO

It was shown in my earlier investigation that especially in laboratory cultures, but also in nature, there are conditions in which the proportion of males is much higher than usual. The individuals hatched in the cultures (at c. 15–18°C.) of the years of 1968–1971 were reared until their sex could be determined. Sometimes this was possible at the 5th instar (8–10 pairs of defence glands), but as a rule, as in my earlier investigations, not until the 6th instar (11–15 pairs of defence glands).

Altogether, sex was determined in 906 larvae. Of these, 187 (20·6%) were males. According to the earlier investigations (Rantala, 1970:124), of over 6600 individuals of *Proteroiulus* found in nature in Finland, 0·64–0·71% were males.

In the cultures, different *Proteroiulus* individuals have been kept isolated from each other, as far as possible. In some cases, the progeny of one female of the P generation has been followed; from this, 1–4 females have been transferred to different culture jars. In this way it has been possible to follow the progeny of the F_1 and F_2 generations separately.

The progeny in all of 45 culture jars were investigated. There were male larvae in 24 of these.

TABLE I

Sex ratios of different progenies

Size of progeny			Number of defence glands of the female	Time (years) spent by the female in culture before reproduction	Male present (+) not present (−) in the culture
♀ larvae	♂ larvae	? larvae			
7	1	—	33	1	+
14	1	7	21	1	+
11	1	—	22	1	—
12	5	—	30	1	—
13	2	—	23	1	—
23	7	—	26	1	—
12	5	—	30	1	—
7	3	—	?	1	—
11	3	—	27	2	—
23	7	—	26	1	—
15	13	—	23	0	—
5	10	10	24	1	—
5	10	10	24	1	—
5	9	4	23	1	—
5	5	4	?	1	—
0	17	2	28	2	+

In Table I those cases are presented which are known with certainty to be the progeny of one female. According to the Table, the percentage of males in the progeny may even be 100%. More observations are needed before it can be said whether correlations exist between the number of males and the other variables presented in the Table or variables not presented in the Table (such as humidity, food, etc). In three cases the sex ratios (Table II) in two successive progenies of the same female have been followed. Table III shows the sex ratios of the progenies in two successive generations.

TABLE II

Sex ratios of successive progenies of three different females

Female no.	Number of defence glands	Year of reproduction	Progeny no.	Sex ratio ♂ : ♀
1	22	1969	1	1 : 11
	26	1970	2	0 : 10
2	25	1969	1	0 : 9
	29	1970	2	0 : 13
3	38	1967	1	0 : 10
		1968	2	0 : 10

In the cases investigated, no difference worth mentioning exists between the sex ratios of the first and second progenies of the same female, whereas between the progenies of individuals of the *P* and *F* generations the percentage of males varies considerably.

In addition to the rearing experiments it will be necessary to investigate spermatogenesis before anything certain can be concluded about the causes of the different sex ratios.

TABLE III

Sex ratio in successive generations

Culture no.	Generation	Year of reproduction	Sex ratio (♂ : ♀) of the progeny and number of "mothers"
1	P	1967	0 : 10/1 ♀
	F_1	1969	0 : 24/2 ♀
2	P	1967	10 : 14/3 ♀
	F_1	1969	0 : 22/1 ♀
3	P	1968	0 : 10/1 ♀
	F_1	1970	3 : 11/1 ♀

TABLE IV

Moults of the males in rearing experiments

V	VI	VII	VIII	IX	X	XI	XII	XIII	XIV	XV	XVI	XVII	XVIII	Instar
		L	L	L	C									
		L	L	C	S	C								
	L	L	C											
	L	L	C											
	L	L	L	C	S	C								
	L	L	L	C	S	C								
	L	L	L	L	C									
	L	L	L	C										
L	L	L	L	C										
		L	L	C	S									
			L	C	?	C								
		L	L	C	S	C								
		L	L	C	S	C								
	L	L	C	?	C									
			L	C										
	L	L	C	S										
	L	L	C	S										
		L	L	C	S									
		L	L	C	S									
			L	C	S									
		L	C	?	C									
			L	C	?	C								
			L	C	?	C								
		L	L	C	S	C								
			L	C										
	L	L	L	L	C	S	C							
	L	L	L	L	C	S								

8–10	11–14	15–19	20–23	24–26	27–29	30–31	32–33	34–35	36–37	38–39	40–41	42–43	44–45	Pairs of defence glands
	L	L	L	C										
	L	L	L	C										
	L	L	L	C										
	L	L	L	C										
	L	L	L	C										
			L	C	S									
	L	L	L	C	S	?	C							
		L	L	C	S									
		L	L	C	S									
	L	L	C	S	C									
		L	C	?	C									
	L	L	L	C	S									
		L	C											
		L	C	?	C									
				C	S	S	S	S	S	S	C	S	S	
				C	S									
						C	S	C						
			C	S										
					S	C								
					C	S	C							
				C	S	C								
					C	S	C							
				S	C	S	S	C						
			C	S										
			C	S	C									
						C	?	S						
				C	S									
				C	S									
		L	L	L	C	?	C							

L = larval male C = copulatory male S = "schalt" male

PERIODOMORPHOSIS

By following the moults of over 30 males, the following results were obtained by the year 1968 (Rantala, 1970):

1. S instars were found evenly at all seasons of the year.
2. Periodomorphosis was observed in all males which moulted.
3. Males were observed to moult at most twice during one summer.
4. Moulting was more frequent in spring and autumn.
5. In many cases, S–C moults, as well as L–C moults, occurred in autumn.
6. C–C moults were never observed.
7. One S–S–S moult was observed.
8. The C or S phase was sometimes considerably prolonged.
9. The S phase sometimes lasted only one month.
10. The first C phase occurred when the number of defence glands was 22–26.

(L = larval (immature), C = copulatory (mature), S = "Schalt" (intercalary).)

In the investigations after 1968 considerably more males were available than before. The results obtained by the year 1971 are presented in Table IV.

These investigations confirm the results presented above and add some new observations:

1. In four cases the first C phase did not begin until the number of defence glands was over 26.
2. S males had been observed by 1971 at all seasons of the year, as before, but in winter 1972 they did not occur at all, whereas the number of C males was about 60.
3. Very few successive S moults were observed. Up to now they have been observed only in two specimens.
4. Not more than two C phases have so far been noted in any male.

This also applies to the specimen whose moults I have followed for almost six years. When it was taken from the wild in 1966 it was evidently in the first C phase and it had 25 pairs of defence glands. By the year 1971 it had gone through the following moults: C–S–S–S–S–S–S–C–S–S. It was at this phase when it died in 1972. It is possible that it was the oldest specimen in my material of over 10 000 *Proteroiulus*. It had 44 pairs of defence glands, which is more than in any other individual studied so far.

REFERENCE

Rantala, M. (1970). Anamorphosis and periodomorphosis of *Proteroiulus fuscus* (Am Stein) (Diplopoda, Blaniulidae). *Bull. Mus. natn. Hist. nat. Paris* (2) **41**: Suppl. no. 2: 122–128.

Discussion

Sahli: Please allow me to congratulate Mme Rantala on these excellent observations on periodomorphosis and in particular the evidence for a remarkable succession from copulatory, through six intercalary, a second copulatory and two further intercalary stadia. Has this succession been observed in one and the same male?

Rantala: Yes.

Sahli: A truly remarkable accomplishment (*applause*). Also, your observations on the appearance of adults in three stadia accord with my own in iulids.

Causey: You didn't mention parthenogenesis?

Rantala: I am more than sure that parthenogenesis does occur in this species since I have reared young from many isolated females, themselves reared in culture.

Hüther: Is it possible that in *Proteroiulus fuscus* there are females which produce only males or only females (monogenous) and others which produce both sexes (amphigenous) as in isopods?

Rantala: I cannot say, my investigations are not complete.

Symp. zool. Soc. Lond. (1974) No. 32, 471–483.

VERTICAL ORIENTATION AND AGGREGATIONS OF *PROTEROIULUS FUSCUS* (AM STEIN) (DIPLOPODA, BLANIULIDAE)

MAIJA PEITSALMI

Kotirannankatu 7, 13200 Hämeenlinna 20, Finland

SYNOPSIS

Old stumps of trees (chiefly pine and spruce) form the most important sites of *Proteroiulus fuscus* in a number of localities in Finland.

The effect of humidity upon the distribution of *P. fuscus* has been studied with the aid of two kinds of experimental arrangements simulating the various gradients occurring in natural stumps.

In the vertical gradient animals choose the saturated zone (RH 100%) irrespective of whether it is in the upper or lower part of the apparatus. If the RH is uniformly 100%, the animals all tend to go to the upper part of the apparatus, whilst in dry air they tend to collect in the lower part.

Orientation to the "preferred" humidity results in the formation of one or two clumps composed of densely grouped animals. The duration of such a clump is humidity-dependent, being clearly shorter in low humidities than if the air is saturated. If the humidity is very low such aggregations are not formed.

The responses observed have been used to interpret field observations on the seasonal distribution of *P. fuscus* in the stumps.

INTRODUCTION

Several studies have been published about the factors affecting the movements of diplopods, especially the temperature and relative humidity of the air (*cf.* Perttunen, 1953, 1955; Haacker, 1968). Observations have also been made on their habitat selection in relation to environmental factors (Blower, 1955; Kinkel, 1955; Schömann, 1956; Haacker, 1970; Fairhurst, 1970). Little is known about the vertical movements of diplopods (Blower, 1956), but recently Haacker (1967) reported observations on the diel vertical movements of some species in experimental cages filled with a suitable substratum.

Rantala (1970) described some features of the local distribution of *Proteroiulus fuscus*, a diplopod which is common in the forests of southern and central Finland (Palmén, 1949). This species is often found in large numbers in stumps of pines and other trees, in which it shows a distinct pattern of vertical distribution, thus providing an opportunity to correlate the changes in this distribution with environmental factors. I shall describe the pattern of this vertical distribution of *P. fuscus* in the field, and the results of experiments designed to show its causes.

FIELD OBSERVATIONS

Quantitative samples were taken to show the vertical distribution of *Proteroiulus fuscus* in tree stumps, at different seasons and under different weather conditions. For this purpose separate samples were taken from three levels of each stump. During the active period of the species (April–September) the specimens were more numerous in the uppermost level of the stump, which includes both the bark at this level and the moss, lichen and detritus covering the cut surface of the stump. During the summer months (June–August) most of the animals found at this level were actually in the layer covering the tops of the stumps. It is to be noted, however, that changes in climatic conditions seem to affect the microdistribution of the species. Table I shows that during dry spells the numbers of specimens found in the layer covering the stumps were much smaller (253) than during rainy periods (1123). The percentage of first-instar juveniles was only about 22 in dry spells, but nearly 50 in rainy weather. These data indicate that the animals tend to collect in the uppermost level of the stumps in damp weather and to move downwards during drought. These observations prompted me to study the behaviour of the species in a vertical gradient of humidity.

TABLE I

Numbers of specimens (in parentheses: percentages) of the different instars of ***P. fuscus*** *in the layer covering the tops of 32 similar stumps during rainy and dry periods of the summer, according to samples collected in 1966–1970*

	1st instar	2nd instar	3rd instar	4th instar	Totals
Dry periods	56	54	99	44	253
	(22·1)	(21·3)	(39·1)	(17·5)	(100)
Rainy periods	559	334	159	71	1123
	(49·8)	(29·7)	(14·2)	(6·3)	(100)

EXPERIMENTS

In order to study the relation between air humidity and the vertical movements of *P. fuscus*, two different types of apparatus were constructed. An important point was to avoid a non-transparent substratum that would interfere with observations on the behaviour of the test animals, i.e. to devise conditions where all the animals were always visible. The two arrangements are described below.

Cylindrical gradient apparatus

A cylinder of filter paper is placed inside a slightly taller and wider simple glass cylinder 0·5 m high and 3 cm in diameter. The animals are unable to climb on the glass, and so the two sides of the filter paper cylinder provide their only walking surface. Any part of the filter paper can be moistened, and if the top of the cylinder is open, the lowest air humidity, even if the whole filter paper is moist, is in the uppermost part, which communicates with the air of the laboratory. In experiments to demonstrate the movements of the animals in a cylinder where the whole filter paper is moist, the cylinder can be closed with a plug. Before such experiments the cylinder was kept closed for five hours to produce an even humidity (100% RH) within the whole cylinder.

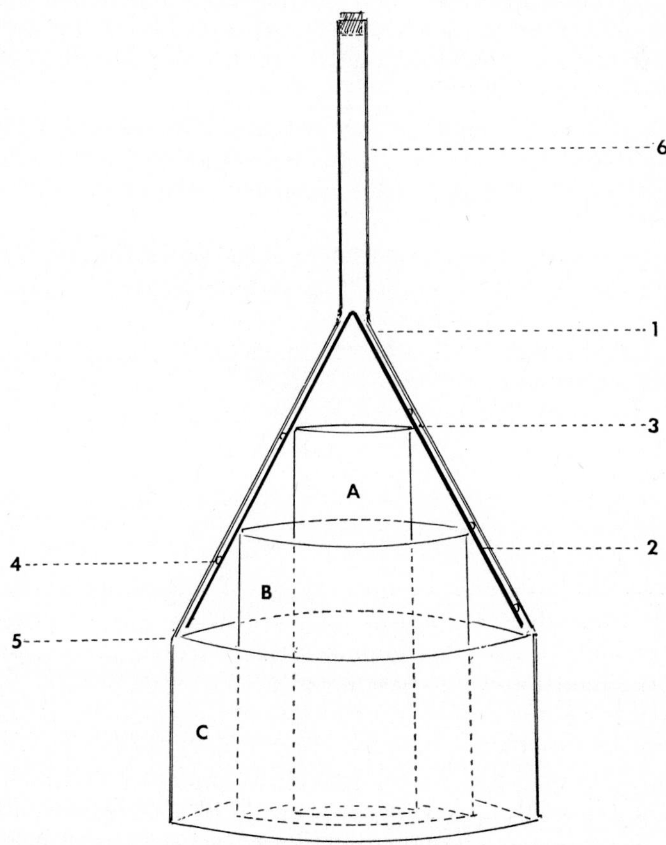

Fig. 1. Funnel gradient apparatus. For explanation, see text.

Funnel gradient apparatus

The design of this apparatus can be seen in Fig. 1. It consists of three glass jars (A, B, C) of different heights, each containing a supersaturated salt solution or water (*cf.* below) to produce the required humidity. The jars are covered with a cone made of copper wire gauze (2) exactly fitting the rim of each jar, and this cone is covered with an ordinary glass funnel (1). The mouth of the wire gauze cone is 6·6 cm in diameter and that of the glass funnel 6·8 cm. The outer surface of the cone is about 68·4 cm^2, and it provides a walking surface for the test animals, which can be dropped into the funnel through the inlet (6) of the glass funnel which is easy to plug and make air-tight. To produce an even space (3) between the funnel and the cone (about 1 mm for the adults and slightly less for the juveniles), small drops of tin (4) were soldered to the copper wire cone. During each experiment the glass funnel is tightly fixed to the glass jar C with a plastic ring (5) which also supports the copper wire cone and is taped to the jar in order to make the apparatus air-tight.

The different air humidities were obtained with the aid of the usual supersaturated salt solutions (*cf.* Gunn & Kennedy, 1936; Wigglesworth, 1941; Perttunen, 1953). In each experiment, two of the jars usually contained the same salt solution and the third contained water. By changing the order of the salt solutions and the water, the position of the dry section could be changed, i.e. it could be in either the lower or the upper part of the apparatus.

In order to avoid the disturbing effect of light, the experiments were made in darkness, i.e. the apparatus was covered with a cone made of black cardboard. The experiments were made at room temperature (about 20°C).

Sources of error

An inherent weakness of both types of apparatus is that it is impossible to produce a gradient with a distinct border between the alternative humidities. To minimize this disadvantage, the space between the funnel and cone was made as narrow as possible, i.e. just wide enough to permit free movement of the animals. It was not possible to measure the actual humidities in the different parts of either apparatus. However, because the humidities produced by the different super-saturated solutions are known and the air space involved is very small, it has been assumed that the gradient has been established within the three-hour period preceding each experiment.

RESULTS

Responses in a vertical gradient of humidity

A series of experiments was made with the aid of the funnel gradient apparatus to study the sensitivity of *P. fuscus* to air humidity and to check whether the position of the "dry" zone has any effect upon the response. In each experiment, saturated air (RH 100%) was employed as one alternative, the other alternative being one of the following humidities (theoretical values obtained with the aid of different supersaturated salt solutions, see above): 99·5, 97, 93, 87, 77, 67, 56, 34 and 0%. Several experiments showed that the positions of the humid and dry zones had no effect upon the responses, i.e. when given enough time the animals always aggregated in the more humid part of the gradient. Tables II and III show some typical examples of these experiments. It can be seen that in all cases, even if the theoretical difference in humidities between the two parts was only 0·5%, the young specimens showed a positive response towards humidity after as short a time as 15 min, and that the response was practically the same irrespective of the alternative humidities employed (Table II). In the full-grown specimens the response was less distinct (Table III): if the humidity difference was small (about 30% or less), there was no clear response during the three-hour experiments, but if the difference was greater, the response was clearly hygropositive, often even within less than one hour. If experiments with small differences in humidity were prolonged, a positive humidity response occurred within the next few hours. If full-grown animals are desiccated at room temperature for a few hours, their responses are practically the same as those of the young individuals.

The results of these experiments show that *P. fuscus* is strongly hygropositive, and that this response is independent of the position of the humid zone in a vertical gradient.

Vertical orientation in the cylindrical gradient apparatus

Since it was observed that *P. fuscus* tends to aggregate in the upper parts of the stumps during rainy weather, a series of experiments was made with the cylindrical apparatus under conditions where the filter paper within the cylinder was thoroughly moist and the top of the cylinder was covered with a lid, i.e. there was no humidity gradient. Under these conditions the specimens aggregated in the upper part of the cylinder and maintained this position for at least six days (Fig. 2d).

In other experiments with the same apparatus four different alternatives were employed, viz.:

TABLE II

Responses of young individuals of ***P. fuscus*** *in the funnel gradient apparatus to different alternative humidities*

Alternative humidities	Time in minutes											
	15	30	45	60	75	90	105	120	135	150	165	180
100: 99·5	14	*16*	16	17	18	18	19	18	19	19	19	20
100: 97	15	13	15	14	*12*	12	15	15	15	16	17	17
100: 93	18	*16*	18	18	19	19	19	20	20	20	20	20
100: 87	13	14	*16*	18	17	18	18	18	19	19	19	19
100: 77	12	*13*	17	18	18	17	18	18	19	19	19	19
100: 67	13	11	*15*	16	16	16	17	18	18	19	20	20
100: 56	*18*	17	18	19	18	19	19	20	20	20	20	20
100: 34	13	12	*16*	16	17	18	19	19	19	19	19	19
100: 0	19	*17*	16	17	17	18	18	17	18	19	19	19

The figures indicate the number of specimens (out of 20) in the humid zone of the apparatus. The young specimens showed a hygro-positive response after 15 min, irrespective of the alternative humidities employed. Figures in italics indicate the first observation of an aggregation in each experiment.

TABLE III

Responses of full-grown individuals of ***P. fuscus*** *in the funnel gradient apparatus with different alternative humidities*

Alternative humidities	Time in minutes											
	15	30	45	60	75	90	105	120	135	150	165	180
100: 99·5	9	9	9	8	10	7	8	7	8	8	9	10
100: 97	7	6	7	7	5	5	4	7	6	6	6	7
100: 93	8	6	9	6	8	7	6	5	8	8	9	9
100: 87	7	7	6	7	7	7	7	8	9	8	8	9
100: 77	7	8	9	8	7	6	8	9	10	9	10	10
100: 67	6	7	9	*10*	11	12	13	13	13	13	14	14
100: 56	12	14	9	12	*12*	13	15	16	18	18	19	19
100: 34	11	*12*	14	17	17	16	18	19	19	18	19	19
100: 0	18	*19*	18	19	19	19	19	19	19	19	19	19

The figures indicate the number of specimens (out of 20) in the humid zone of the apparatus. It can be seen that if the difference between the humid and dry zones is about 30% or more, there is a definite hygropositive response within 30–60 min. Figures in italics indicate the first observation of an aggregation in each experiment.

1. An open cylinder with the filter paper moistened at the top.
2. An open cylinder with moist strips of filter paper at the upper end of the cylinder.
3. An open cylinder with the filter paper moistened throughout.
4. An open cylinder with dry filter paper.

Figure 2 shows typical responses of *P. fuscus* under these experimental conditions. If only the upper part of the filter paper was moist the animals at first aggregated in this part, but as soon as the paper began

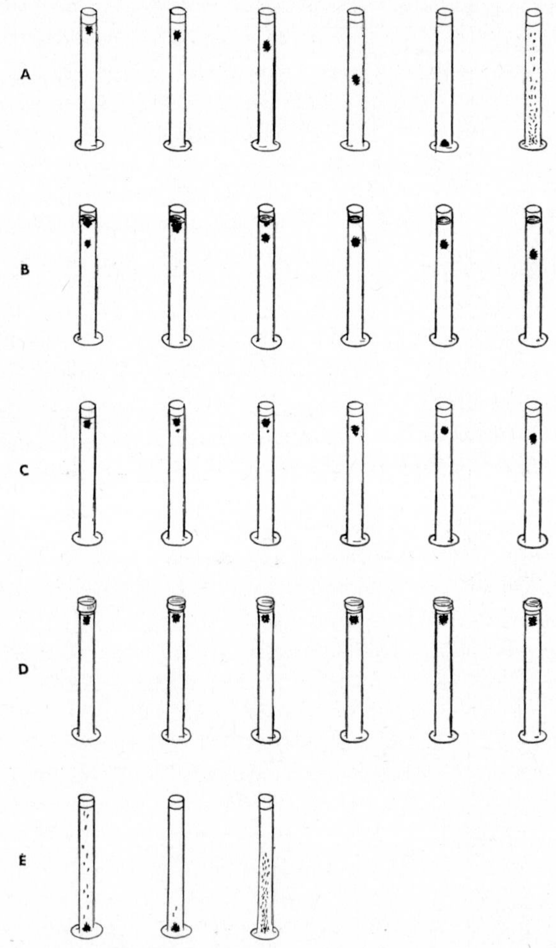

FIG. 2. Changes in the position of the aggregations under different experimental conditions. For explanation, see text.

to dry, the aggregations shifted successively downwards, and by the fifth day the animals were on the bottom of the cylinder. During the next day, they moved restlessly to and fro (Fig. 2a). If the cylinder had moist filter paper inside the upper edge, the animals aggregated at this lining and near it, where a relatively high moisture was maintained inside the cylinder, i.e. the conditions resembled those in a moist and closed cylinder (Fig. 2b). If the filter paper in the open cylinder was thoroughly moist, they aggregated in the upper part of the cylinder throughout the six days, but tended to shift downwards a little, probably owing to the drying of the upper part of the filter paper. Because the bottom of the cylinder was still moist, a high humidity was maintained in the greater part of the cylinder (Fig. 2c). If the cylinder was dry throughout, the animals usually moved restlessly around, but occasionally aggregated on the bottom for a short time, then scattering again (Fig. 2e). If left in these conditions without water, they soon perished.

These experiments show that the vertical position of *P. fuscus* is strongly affected by humidity. In high air humidity there is an upward movement, which is most clearly demonstrated if the air is saturated and the cylinder closed.

Aggregations

In the experiments described above, the specimens of *P. fuscus* showed definite thigmokinetic behaviour which was reflected in the formation of aggregations in the "preferred" humidity. An aggregation usually formed when two or more animals moving in the gradient apparatus (Fig. 3) or in the cylinder (Figs 4, 5) chanced to meet in the "preferred" zone of humidity. They then stopped and soon other specimens joined the group, and gradually a "clump" was formed which consisted of all or nearly all the specimens present in the apparatus at the same time. If young individuals and adults are kept in the apparatus simultaneously the first nucleus of the aggregation is usually formed by the young, and the adults join the clump later. Such aggregations are a regular phenomenon if the animals settle in the zone where the air is saturated, but for a short time aggregations may even form on the drier side of the gradient, particularly if the saturation deficit is not very great. The compactness of the aggregation seems to depend on the air humidity: the higher the humidity, the looser the aggregation. In saturated air such aggregations may persist for as long as a week at exactly the same place. Occasional individuals may leave the clump for a short time, but they soon turn back and join the aggregation again. It is typical of the aggregations that they scatter at a certain time of

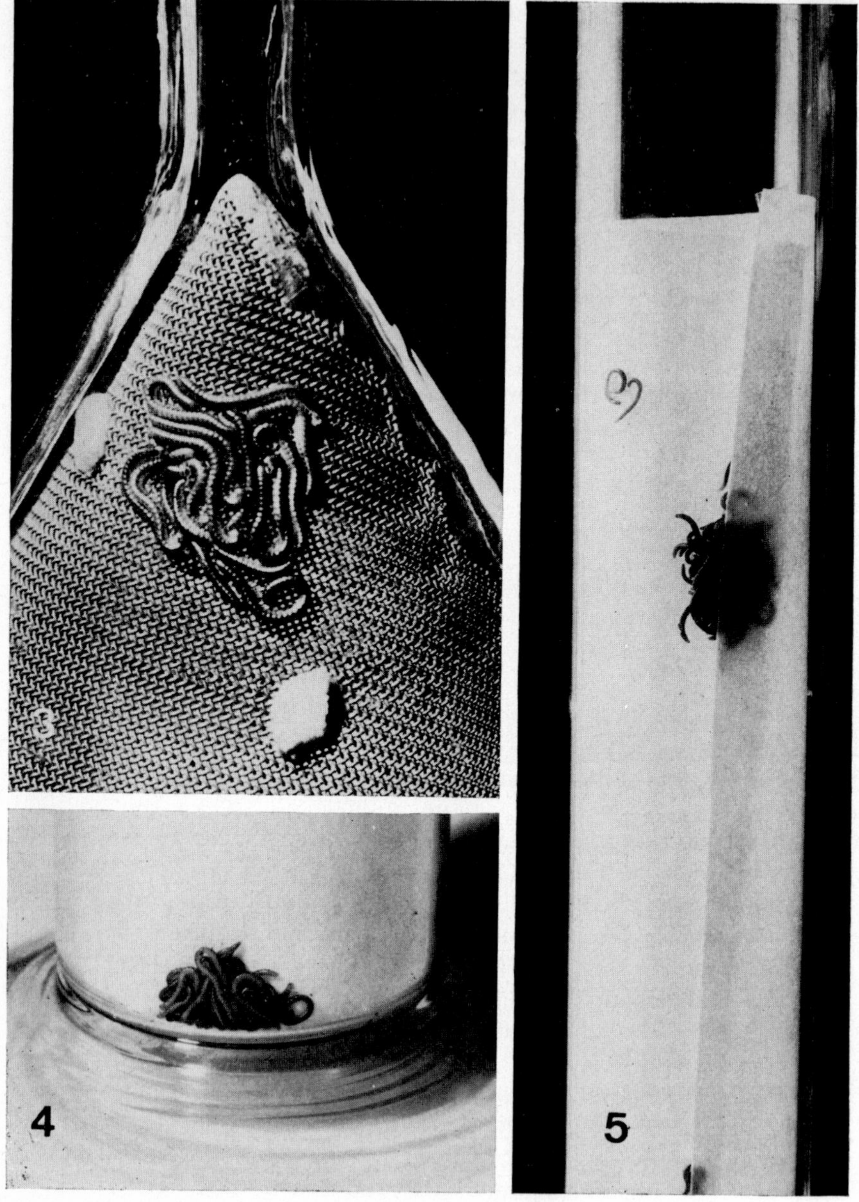

FIG. 3. A clump in the upper part of the funnel gradient apparatus.
FIG. 4. A clump on the bottom of the dry cylinder. For explanation see p. 479.
FIG. 5. A clump in the upper part of the cylinder. For explanation, see p. 479.
(Photographs by Mr Tuomo Niemelä).

day, usually around midnight. Figure 6 shows a typical case where the number of specimens present in the aggregation was always lowest after midnight, whilst the aggregation formed again during the morning. If the air is thoroughly saturated within the whole cylinder, this rhythm is practically absent, and the aggregation persists for several days.

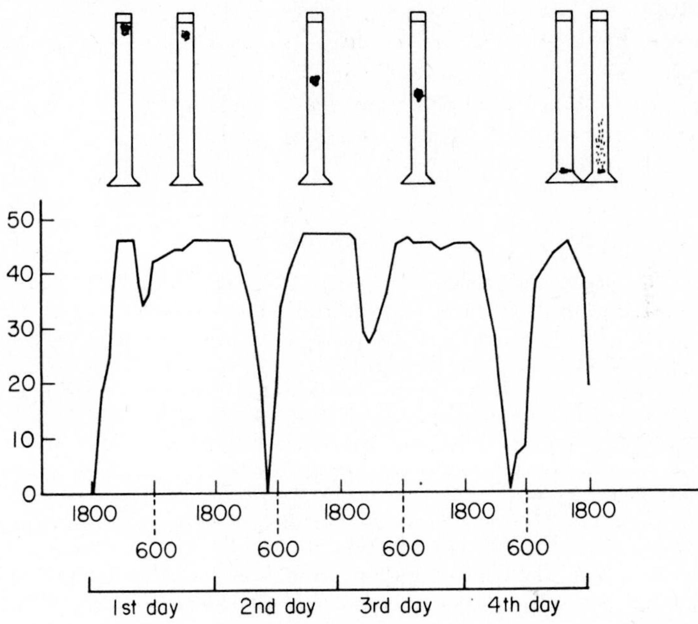

FIG. 6. Changes in the numbers of specimens in an aggregation of *Proteroiulus fuscus* during four days in an open cylinder kept moist at the top. Ordinate: number of specimens in the aggregation, abscissa: time. The diagrams show the position of the aggregation on successive days. Note the downward shift of the aggregation.

From field observations and the experiments described above it can be concluded that the actual sites of the aggregation of *P. fuscus* are largely determined by the humidity conditions in combination with possibilities for true thigmokinetic responses. If the relative humidity is maintained at the level of saturation, and suitable contact surfaces are available, the site of the aggregation in a moist stump probably does not change for a long time, because enough food is available. Drops in relative humidity, on the other hand, will induce movements that result in aggregation in more suitable places; the animals in a dry stump, for instance, will shift to the lower parts near the soil.

Acknowledgements

The present study was begun at the suggestion of Prof. Ernst Palmén, an expert on Finnish millipedes. My sincere thanks are due to him for his encouragement and for valuable help at different phases of my work.

I have pleasure in thanking Ph. D. Vilho Perttunen for his various help during my work.

Sincere thanks are due to Mrs Maija Rantala for her aid in collecting material.

Mrs Jean Margaret Perttunen has checked the English of my manuscript, for which I express my deep gratitude.

References

Blower, J. G. (1955). Millipedes and centipedes as soil animals. In *Soil zoology:* 138–151. Kevan, D. K. McE. (Ed.) London: Butterworth.

Blower, J. G. (1956). Some relations between millipedes and the soil. *Sixième Congr. Int. Sci. Sol., Paris* **3**: 169–176.

Gunn, D. L. & Kennedy, J. S. (1936). Apparatus for investigating the reactions of land arthropods to humidity. *J. exp. Biol.* **13**: 450–459.

Fairhurst, C. P. (1970). Activity and wandering in *Tachypodoiulus niger* (Leach) and *Schizophyllum sabulosum* (L.). *Bull. Mus. natn. Hist. nat., Paris* (2) **41** Suppl. No. 2: 61–66.

Haacker, U. (1967). Tagesrhythmische Vertikalbewegung bei Tausendfüsslern. *Naturwissenschaften* **54**: 346–347.

Haacker, U. (1968). Deskriptive, experimentelle und vergleichende Untersuchungen zur Autökologie rheinmainischer Diplopoden. *Oecologia* **1**: 87–129.

Haacker, U. (1970). Experimentelle Untersuchungen zur Ökologie von *Unciger foetidus* (C. L. Koch). *Bull. Mus. natn. Hist. nat., Paris* (2) **41** Suppl. No. 2: 67–71.

Kinkel, H. (1955). Zur Biologie und Ökologie des getüpfelten Tausendfüsses, *Blaniulus guttulatus* Gerv. *Z. angew. Ent.* **37**: 401–436.

Palmén, E. (1949). The Diplopoda of Eastern Fennoscandia. *Suomal. elaïn-ja kasvit. Seur. van. Julk* **13** (6): 1–54.

Perttunen, V. (1953). Reactions of diplopods to the relative humidity of the air. *Suomal. elaïn-ja kasvit. Seur. van. Julk* **16** (1): 1–64.

Perttunen, V. (1955). The effect of antennectomy on the humidity of normal and desiccated specimens of *Schizophyllum sabulosum* L. (Diplopoda, Iulidae). *Suomen hyönt. Aikak.* **21** (3): 157–162.

Rantala, Maija (1970). Anamorphosis and periodomorphosis of *Proteroiulus fuscus* (am Stein). *Bull. Mus. natn. Hist. nat., Paris* (2) **41**, Suppl. No. 2: 122–128.

Schömann, K. (1956). Zur Biologie von *Polyxenus lagurus* (L.). *Zool. Jb.* (Syst.). **84**: 195–256.

Wigglesworth, V. B. (1941). The sensory physiology of the human louse *Pediculus humanus corporis* De Geer (Anoplura). *Parasitology* **33**: 67–109.

Discussion

WALLWORK: Has Mlle Peitsalmi found aggregations in the field similar to those she has demonstrated in the laboratory?

PEITSALMI: Yes, during rainy weather, under leaves on the cut surface of stumps. I have seen 52 individuals under one little leaf (about 2 cm long). There are many more places in nature where the animals' thigmokinesis can be manifest than in my cylinder in the laboratory.

BROOKES: What do you think happens in winter when (as Rantala (1970)* reports) they go deep in the soil? Is the response of the animals to come up in the stumps negated by the low temperature?

PEITSALMI: Yes, perhaps. My observations refer to the warmer conditions of summer time.

JEEKEL: Is there any indication of how the animals of the species find each other after they have dispersed? How do they reform an aggregate—just by walking around and meeting each other by chance?

PEITSALMI: I think the olfactory sense may be involved.

* See list of references p. 482.

Symp. zool. Soc. Lond. (1974) No. 32, 485–501.

THE LIFE CYCLE OF *PROTEROIULUS FUSCUS* (AM STEIN) AND *ISOBATES VARICORNIS* (KOCH) WITH NOTES ON THE ANAMORPHOSIS OF BLANIULIDAE

CHARLES H. BROOKES

School of Biology
Department of Chemistry and Biology
Manchester Polytechnic, Manchester, England

SYNOPSIS

Anamorphosis and the life histories in Britain of two blaniulids, *Proteroiulus fuscus* and *Isobates varicornis*, are described. *P. fuscus* hatches from the egg as a white immobile pupa which is followed by six larval stadia, adulthood being achieved for the first time at stadium VII. Further moulting of adult stadia occurs so that additional adult females, albeit in declining numbers, are found from stadia VIII to XIII. *I. varicornis* hatches from the egg into a pupoid from which arises a stadium I unique among Diplopoda in that it moves, feeds and on occasion overwinters. Unlike *P. fuscus*, *I. varicornis* has only five larval stadia, maturity being achieved at stadium VI. Again further adults in progressively declining numbers are found from stadium VII to stadium XII. Reproduction in both species is parthenogenetic. Typically, *P. fuscus* breeds for the first time at the age of three years and *I. varicornis* at the age of two years. The egg laying period in *P. fuscus* occurs between May and July whereas *I. varicornis* lays eggs singly and continuously over a number of summer months. *I. varicornis*, an obligate bark inhabitant, has several features of its life cycle which are seen as adaptions to the transitory nature of its restricted habitat and to problems of dispersal. By contrast, being more generally distributed, *P. fuscus* is able to oscillate between bark and litter habitats as climatic conditions dictate. Warmer temperatures appear to advance the rate of development of *P. fuscus*. It is suggested that the duration of the life cycle may be shorter in southern latitudes but extended at the northern limits of the geographical range of the millipede.

INTRODUCTION

Most of the published literature dealing with iuliform millipedes of the family Blaniulidae has been concerned with taxonomy and morphology, relatively little being known of life cycles and ecology. Published records on the life cycle of *P. fuscus* and in particular *I. varicornis* are scarce. A notable exception is an account of anamorphosis and periodomorphosis of *P. fuscus* by Rantala (1970). Life history and ecological data even for the economically important blaniulid *Blaniulus guttulatus* are fragmentary and to some extent contradictory. However, in recent years the position for *Blaniulus guttulatus* has been improved by the work of Biernaux (1972).

SAMPLING AND METHOD

Most of the investigations described in the present paper were carried out during the years 1959–62 in Ernocroft Woodland (Grid ref. SJ/975 910), situated in Cheshire some 12 miles south-east of Manchester. Two main sites were selected; one a stand of larch and spruce, *Larix decidua* (Mill.) and *Picea sitchensis* (Bong.) which supported a population of *P. fuscus*; the other an open site of recently felled elm trees (*Ulmus* spp.) inhabited by high densities of both *P. fuscus* and *I. varicornis*.

Each month, over a period of three years, 24 litter samples of equal volume (2270 cm^3) and known but variable areas were removed from the Larch/Spruce site. For an initial period of one year samples of the humus layer were also taken but later discontinued since only very small numbers of *P. fuscus* frequented this layer. Moreover, *P. fuscus* was notably absent from soil samples which were also discontinued after one year for the same reason.

Animals were extracted from samples using a modified Tullgren apparatus (Brookes, 1963). Larch and spruce stumps and logs on the same site were also sampled, approximately equal areas of bark being removed at monthly intervals over the same three-year period.

Stumps on the Elm site were sampled similarly. Millipedes were collected from bark by hand. In addition large collections of *P. fuscus* and *I. varicornis* were obtained from other parts of the woodland for use in separation of stadia and setting of laboratory cultures.

ANAMORPHOSIS

Post embryonic development of diplopods is exclusively anamorphic in all orders except the Pentazonia. There are several larval stadia and at each moult the number of body rings and pairs of legs increases. In the Iuliformia, the number of segments added at each stadium, except for the very early stadia, is variable. Consequently in older stadia there is an overlap of the number of segments between one stadium and the next.

The immediate problem at the outset of my investigation was to delimit the post embryonic stadia of *P. fuscus* and *I. varicornis*. In all, some 13 000 specimens of *P. fuscus* and over 3000 individuals of *I. varicornis* were examined in order to establish their modes of anamorphosis. In order to categorize each stadium, a combination of characteristics was used, including the number, size and serial groupings of defence glands (after Halkka, 1958), number of leg pairs, number and arrangement of ocelli and length measurements.

Post embryonic development of **P. fuscus**

Like all diplopods, *P. fuscus* hatches from the egg as a white immobile pupa enclosed within a membrane. In common with most authors but unlike Rantala (1970), I have applied the term larval stadium I to the unenclosed hexapod stage. On this basis, anamorphosis in the female comprises a succession of six larval stadia; adulthood being achieved for the first time at stadium VII. Further moulting of adult stadia occurs so that additional adults, albeit in declining numbers, are found from stadia VIII–XIII.

Typically, more podous segments are added at moults producing instars V and VI than at any other moult. Production of segments from the proliferation zone then falls off gradually towards the end of anamorphosis until moults producing stadia X–XIII add constantly only two segments.

The main course of anamorphosis is similar for populations of *P. fuscus* on the Larch/Spruce site and the Elm site. However, it is interesting to note that in general in those stadia where a range of variations can exist, the number of segments added by individuals on the Elm site is at the higher end of the incremental range. The larger increment of segments seen in individuals begins in stadium III and is consolidated especially between stadia IV, V and VI. The larger

Table I

The commonest course of anamorphosis in **P. fuscus**

Stadium	Podous segments	Pairs of legs	Pairs of Defence Glands	Apodous segments	Ocelli
I	4	3	—	2	—
II	6	7	1	4	1
III	10	15	5	4	2
IV	14	23	9	4–5	3
V	18–19	31–33	13–14	4–5	4
VI	22–24	39–43	17–19	4	5
VII	26–28	47–51	21–23	3–4	6 +
VIII	29–31	53–57	24–26	2–3	7 +
IX	32–33	59–61	27–28	2	8 + variable
X	34–36	63–67	29–31	2	,,
XI	36–38	67–71	31–33	2	,,
XII	38–39	71–73	33–34	2	,,
XIII	40–41	75–77	35–36	2	,,

segment numbers are therefore maintained in some degree throughout the later stadia. The general course of anamorphosis and the characteristics of each stadium are presented in Table I.

Post embryonic development in ***I. varicornis***

Examination of samples of *I. varicornis* taken from under bark on the Elm site has established the course of anamorphosis. Fortunately, as in *P. fuscus*, successive increments of defence glands were separately recognizable so that the "defence gland method" of Halkka (1958) could be applied also to *I. varicornis*. In females of *I. varicornis* there are only five larval stadia, adults appearing for the first time at stadium VI. Further moults of adult stadia occur so that adults although in declining numbers are found from stadia VI–XII.

As in *P. fuscus*, the egg of *I. varicornis* hatches into a pupoid enclosed with its pupal skin. Stadium I emerges by breaking through the pupal skin, which is left behind. Only stadium I has constant characteristics. It is unique in several respects, being the only recorded case of a stadium I millipede capable of feeding independently of its stored yolk. When the yolk granules disappear, stadium I moves actively and begins feeding on soft tissues under bark. The commonest course of anamorphosis and details of individual stadia are detailed in Table II.

TABLE II

The commonest course of anamorphosis in ***I. varicornis***

Stadium	Podous segments	Pairs of legs	Pairs of Defence Glands	Apodous segments	Ocelli
I	4	3	—	2	—
II	6	7	1	4	1
III	10	15	5	4, 5	3
IV	14, 15	23–25	9, 10	5	6
V	18, 19, 20	31–35	13, 14, 15	5	10
VI	23–25	41–45	18, 19, 20	4–5	10 + variable
VII	28, 29, 30	51–55	23, 24, 25	3	10 + variable
VIII	31, 32, 33	57–61	26–28	2–3	,,
IX	34, 35	63–65	29–30	2	,,
X	36, 37	67–69	31–32	2	,,
XI	38, 39	71–73	33–34	2	,,
XII	40	75	35	2	,,

Anamorphosis in other Blaniulidae

There are considerable similarities in at least the early part of anamorphosis in seven of the blaniulids occurring in Britain. The early stages of anamorphosis of seven species of British blaniulids, based on several criteria, are recorded in Table III.

It is evident that there is particularly close resemblance between the anamorphic development of *P. fuscus*, *Nopoiulus minutus* and *Choneiulus palmatus* on the one hand and between the post embryonic development of *B. guttulatus*, *Archeboreoiulus pallidus* and *Boreoiulus tenuis* on the other.

Perhaps worthy of special attention are my observations on stadium II of *B. tenuis*. The latter is certainly peculiar among Iuliformia in possessing a pair of defence glands on *each* of segments six and seven. Consequently, the stadium is also peculiar in having nine pairs of legs. Most other iuliforms at stadium II have seven pairs of legs and one pair of prominent defence glands.

In general, Table III illustrates that in spite of variation within stadia, no overlap occurs between any of the larval stadia of any one species of blaniulid, that is, as far as stadium VI. Furthermore, it is suggested that "the defence gland method" used in the present work should prove equally effective in delimiting both larval and adult stages of all species of Blaniulidae.

Life cycle of ***P. fuscus***

Samples containing large numbers of *P. fuscus* were obtained from conifer litter on the Larch/Spruce site at approximately monthly intervals. For each sample, individual millipedes were assigned to the appropriate stadium thereby providing an estimate of the successive abundance of stadia with respect to time (Fig. 1). For each year the maximum growth period extends from the end of March to early October. Growth is then at a standstill from late October to the following spring.

Year I

Eggs are laid in litter during late May/June and July mainly by stadia VII and VIII but millipedes from stadia IX–XII contribute a small number of eggs.

The eggs hatch to give peak numbers of stadium I in midsummer. During the remaining summer and early autumn, the majority of individuals reach stadium II, while a small proportion develop as far as stadium III. No animals remain at stadium I. Growth then ceases and larvae overwinter at stadia II and III.

TABLE III

Species and sources of information

— = Undescribed	*Proteroiulus fuscus* (Brookes, 1963)	*Isobates varicornis* (Brookes, 1963)	*Boreoiulus tenuis* (Brookes, 1963)	*Archeboreoiulus pallidus* (Brookes, 1963)	*Blaniulus guttulatus* (Brookes, 1963)	*Choneiulus palmatus* (Verhoeff, 1928)	*Nopoiulus minutus* (Humbert, 1893) (Latzel, 1884) (Verhoeff, 1928, 1933, 1939)
Stadium I							
No. of podous segments	4	4	—	—	—	4	4
No. of apodous segments	2	2	—	—	—	2	2
Pairs of legs	3	3	—	—	—	3	3
No. of gland pairs	0	0	—	—	—	0	0
Ocelli	0	0	—	—	—	0	0
Stadium II							
No. of podous segments	6	6	7	6	6	6	6
No. of apodous segments	4	3–4	4	5	5	4	3–4
Pairs of legs	7	7	9	7	7	7	7
No. of gland pairs	1	1	2!	1	1	1	1
Ocelli	1	1	0	0	0	1	1
Stadium III							
No. of podous segments	10	9–10	11	11	11	10	9–10
No. of apodous segments	3–5	3–5	4–5	5–6	4–5	4	4–5
Pairs of legs	15	13–15	17	17	17	15	13–15
No. of gland pairs	5	4–5	6	6	6	5	4–5
Ocelli	2	3	0	0	0	2	2

Stadium IV							
No. of podous segments	13–15	13–15	15–16	16–17	15–16	14	13–15
No. of apodous segments	3– 5	4– 6	4– 6	5– 6	4– 5	4	4– 5
Pairs of legs	21–25	21–25	25–27	27–29	25–27	23	21–25
No. of gland pairs	8–10	8–10	10–11	11–12	10–11	9	8–10
Ocelli	3	6	0	0	0	3	3
Stadium V							
No. of podous segments	17–19	17–21	20–21	21–23	20–21	18	17–19
No. of apodous segments	3– 5	4– 6	4 –5	5– 6	4– 6	4	4– 5
Pairs of legs	29–33	29–37	35–37	37–41	35–37	31	29–33
No. of gland pairs	12–14	12–16	15–16	16–18	15–16	13	12–14
Ocelli	4	10	0	0	0	3?	4
Stadium VI							
No. of podous segments	21–24	22–26	25–26	26–29	24–27	22	21–23
No. of apodous segments	3– 5	3– 5	4– 5	4– 5	5– 6	4	3– 5
Pairs of legs	37–43	39–47	43–47	47–53	43–49	39	37–41
No. of gland pairs	16–19	17–21	19–21	21–24	19–22	17	16–18
Ocelli	5	10 +	0	0	0	4	3– 5
Stadium VII							
No. of podous segments	25–28	26–30	29–30	31–35	29–32	—	25–26
No. of apodous segments	3– 4	2– 4	3– 4	3– 4	4– 5	—	3– 4
Pairs of legs	45–51	47–55	53–55	57–65	53–59	—	45–47
No. of gland pairs	20–23	21–25	24–25	26–30	24–27	—	20–21
Ocelli	6 +	12 +	0	0	0	—	—

Year II

Growth recommences in spring and by autumn of the second year most larvae reach stadium V. A small proportion, however, only attain stadium IV but a few reach as far as stadium VI. Growth ceases so that the larvae spend their second winter as stadia IV, V and VI.

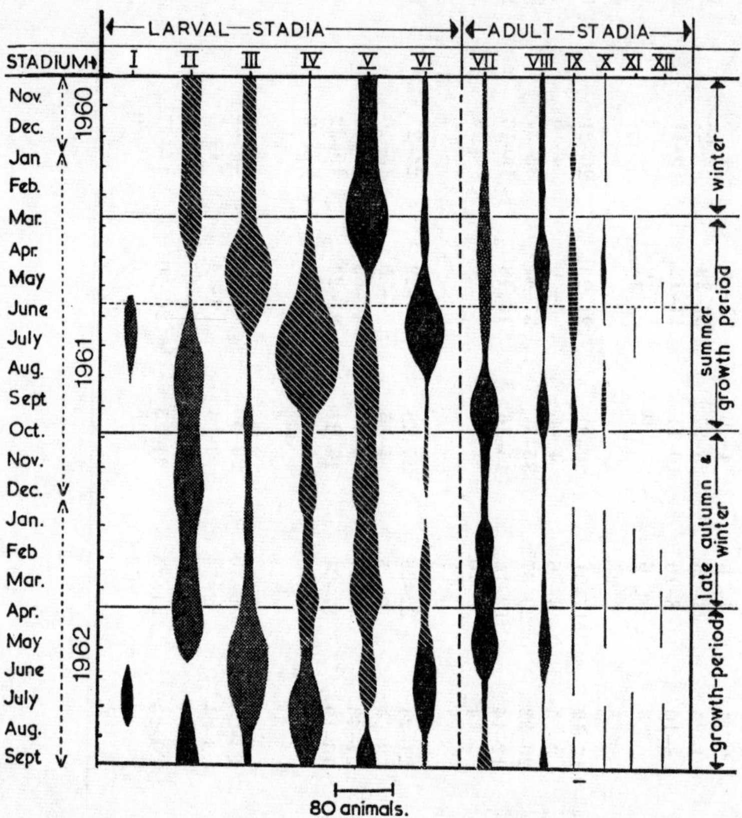

FIG. 1. The seasonal occurrence of the stadia of *Proteroiulus fuscus* in conifer litter on the Larch/Spruce site, 1960-1962. Widths of columns represent the densities of animals of each stadium in each monthly sample of 2·65m^2.

Year III

After the recommencement of growth in spring, most larvae achieve stadium VI during June and July of the third year. By October all the stadium VI have passed into stadium VII which is the first adult stadium. Growth is then arrested, when the adult stadia overwinter, but begin the production of eggs within the ovitube.

Year IV

By May in the spring of the fourth year, the adult female contains peak numbers of mature eggs. During late May, June and July, the stadium VII females deposit their eggs.

The duration of the life cycle is therefore three years from egg to egg.

After breeding, many of the adult stadium VII die. A small proportion survive and moult in late summer or autumn to overwinter at stadium VIII, prior to breeding again in the fifth year.

It is evident from the above data that, in any year, two distinct larval and one adult generation overwinter with peak numbers at stadia II, V and VII respectively.

The life cycle of *P. fuscus* on conifer bark shows essentially the same pattern as in the conifer litter, but the samples from under bark on the Elm site reveal that the developmental stages reached by both larval generations were approximately one stadium in advance of the corresponding larval generation found on the Larch/Spruce site (Fig. 2). In this figure the relative numbers of each stadium are expressed as a percentage of the total number in the sample collected each month. Results are expressed in this way to overcome at least in part the intrinsic difficulty of obtaining exactly replicate quantities of bark of similar age from such inherently heterogeneous material, which makes for less comparability between samples. Figures expressed as percentages are more easily compared and any trends resulting from them are instantly more readable.

Figure 2 shows that the generation born in 1959 under the elm bark overwintered mainly at stadium III whereas the generation hatched the previous year spent the winter at stadium VI. In addition, adults overwintered from 1959–60 with peak numbers at both stadia VII and VIII.

Eggs, pupoids and stadium I made their appearance under elm bark in June and July of the following year. Also in June of the same year, peak numbers of stadium VII adults appeared, having first moulted from the generation which had overwintered at stadium VI. It must be noted that these adult stadia which arrived early did not develop mature eggs until the following spring. The advanced state of development therefore conferred little advantage on the millipede, since individuals arrived too late to breed in the same year. Subsequent

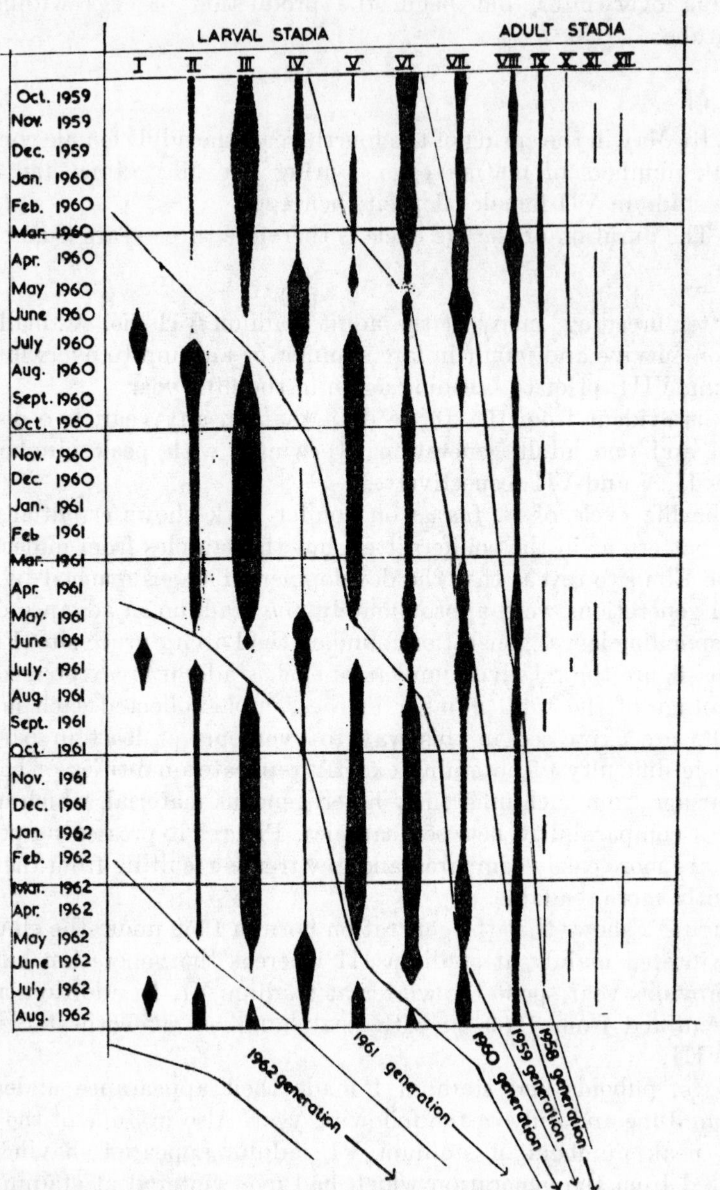

FIG. 2. The seasonal occurrence of the stadia of *Proteroiulus fuscus* under elm bark on the elm site, 1959-1962. Widths of columns represent the percentages of each stadium in each monthly sample.

development of these stadium VII adults took one of two courses. Some survived without moulting again and reproduced the following year, whereas others moulted later the same year and overwintered at stadium VIII prior to breeding in the following summer. Since mature eggs appear gradually in the ovitube during winter and early spring months, it is clear that only individuals which overwinter as adults breed in the subsequent summer.

The overall pattern of development on elm was much the same in each year from 1959–1962. Exceptionally, however, a peak of pre-adult animals overwintered mainly at stadium VI in 1959, not at stadium V as in subsequent years. The advanced state of development on elm compared with that on the conifer site can probably be explained by a consideration of the nature of the two habitats, particularly in 1959 when the stumps on the Elm site were afforded virtually no shade. Indeed daytime bark temperatures recorded on the sunlit elm stumps were consistently higher than those for stumps on the densely shaded Larch/Spruce site. These facts taken together with the exceptionally hot summer of 1959 probably account for the advanced stages of development found on elm.

Life cycle of ***I. varicornis***

I was particularly fortunate in finding a large population of *I. varicornis* upon which to base a long-term sampling programme. Normally, numbers of this millipede in any one habitat are too low to provide useful information on the life cycle of this species. Consequently the life cycle presented in the present paper is based solely upon data obtained by sampling regularly the bark of stumps and logs of elm and sycamore found on the Elm site.

The data are presented in Fig. 3 as percentages, for the reasons given above. As in *P. fuscus* growth extends from late March until October. Growth is at a standstill from November to early March. Eggs are laid singly particularly by adults of stadia VI and VII, from mid June to early August. Intensive egg laying occurs in early July. Eggs, pupoids and a few stadium I appear early in July. Most hatching of pupoids and the consequent appearance of peak numbers of stadium I takes place during mid and late July. Not until early August is egg laying completed so that hatching of eggs and pupoids continues into late August. Growth continues until late October by which time most individuals hatched the same year reach stadia II and III. A small proportion of the early hatched individuals, however, reach as far as stadium IV, whereas late hatching individuals remain at stadium I. Stadium I, being able to feed on an external food supply, is the only stadium I diplopod known to survive the winter. After cessation of growth in November of the first year larval stadia I, II, III and IV all overwinter, but mainly at stadia II and III.

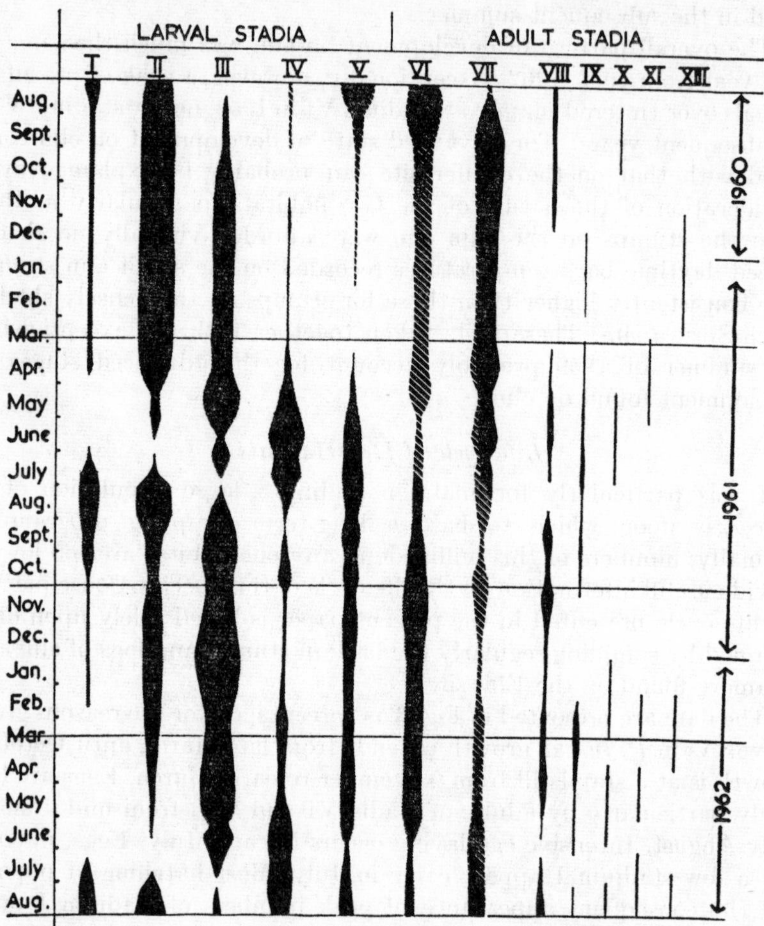

FIG. 3. The seasonal occurrence of stadia of *Isobates varicornis* under bark on the Elm site 1960-1962. Widths of columns represent the percentages of each stadium in each monthly sample.

During the second period of growth from March to October most larvae reach stadium V and mainly stadium VI. Growth ceases in autumn when stadia V and VI and occasionally small numbers of stadium VII overwinter. During the early months of the third year adult stadia begin producing mature eggs in the ovitube. Reproduction is parthenogenetic, eggs being deposited later in the same summer in soft tissues under bark.

The life cycle is therefore typically of two years' duration from egg to egg. After breeding, many of the adult stadium VI moult, particularly in August and September, into stadium VII. Newly moulted adult stadium VII overwinter to reproduce the following summer. Heavy mortality then results in only a small fraction of individuals surviving to moult into stadium VIII. Expectation of life beyond stadium VIII is very low and decreases further with each successive stadium.

DISCUSSION

It may be useful to compare the life cycles of *P. fuscus* and *I. varicornis* in relation to their respective habitats.

Individuals of both species may attain stadium VI by the autumn of the second year, but since this stadium is mature in *I. varicornis*, the generation begins breeding the following summer. Stadium VI of *P. fuscus* is not mature and it is only very rarely (e.g. in the exceptionally hot summer of 1959) that a small number reach maturity at stadium VII after only two years of growth. Typically, *P. fuscus* which attain maturity at stadium VII must wait until the next year before breeding. Thus, by advancing the attainment of maturity by one stadium, *I. varicornis*, an obligate bark inhabitant, eliminates one year from its life cycle. Clearly, the shorter life cycle of *I. varicornis* is suited to the transitory nature of the bark habitat. *I. varicornis* by maturing earlier is more likely to achieve adulthood before the bark habitat disintegrates. *P. fuscus*, although able to breed under bark, normally requires three years to complete development. Completion of a three-year life cycle under bark would be less certain. *P. fuscus*, however, whenever possible retains a population which breeds in litter. From litter, which after all is the more permanent habitat, individuals may emigrate under suitable climatic conditions in order to colonize new habitats. Chances of colonizing new habitats in this way are greatly enhanced by the fact that reproduction in *P. fuscus* is parthenogenetic, only one individual female being required to begin a new colony. In *I. varicornis* also, parthenogenesis aids colonization of new logs and stumps. Of further advantage to a sub-cortical species like *I. varicornis* is its ability to lay eggs singly and continuously over an extended period of time. The method of egg laying employed by *I. varicornis* may preclude desiccation of many eggs at one time, for example during short periods of drought. Even so, the fact that *I. varicornis* is a relatively less common species throughout its geographical range perhaps suggests that dispersal is for some reason inefficient, leading to difficulties in finding new logs or stumps. By contrast *P. fuscus*, while lacking some of the attributes of a thoroughgoing bark inhabitant, appears in large numbers in both bark and litter and is able to oscillate between the two habitats as climatic conditions change throughout the year. During the summer months, larger numbers of *P. fuscus* were found in conifer litter on the Larch/Spruce site, many of them having emigrated from conifer bark which at this time becomes dry and brittle. Clearly at these times conifer bark is less hospitable to a species like *P. fuscus*, which requires and seeks out places of high humidity. In autumn, when bark becomes moist and generally humid conditions return, conifer logs and stumps are repopulated by *P. fuscus*, in particular by newly moulted adults.

An appearance in October of each year of increased numbers of adults under bark, together with a corresponding drop in numbers in litter, indicates movement onto the inhabitable bark from litter. Also at this time more general dispersive movements of adults are indicated by the presence of large numbers of adult *P. fuscus* on the ephemeral fructifications of certain fungi, which are found sprouting from bark and litter during September, October and November and finishing with the late November frosts. It is suggested that in autumn a general dispersion of *P. fuscus* affecting mainly adult stadia leads to a marked decrease in numbers aggregated under bark and in litter during the winter months. Although the possibility suggested by Rantala (1970) that in winter numbers of adults descend deep into the soil cannot entirely be ruled out, I have found little evidence of downward movements into the humus and soil from my samples taken during 1959–1960 and from samples taken by a colleague in 1960–1961.

It is interesting to consider also the effects of temperature on the life cycle of *P. fuscus*, particularly during the period of growth between March and October. On the open Elm site with its warmer summer temperatures, development proceeds at a faster rate with bigger increments of segments between certain stadia than on the shaded and therefore cooler Larch/Spruce site. These observations raise the question of whether or not the duration of the life cycle varies in different geographical regions. Since clearly the rate of development varies between sites in Ernocroft Woodland, it is not unreasonable to suppose that the time taken to reach adulthood may be extended at the northern limits of its range (66° 45″N) but shortened in the southernmost latitudes (40° 10′ N).

The only additional observations on the life cycle of *P. fuscus* are those of Rantala (1970) made in southern and central Finland. Now, if the numbering of stadia used by Rantala (1970) is converted to the system used by the present author, it is clear that in Finland three separate co-existing generations overwinter with peak numbers at stadia II, IV and VI respectively. It will be recalled that in the present study, three distinct generations also hibernate but with peak numbers at stadia II and III; at stadium V; and at stadium VII. If we interpret the above data cited by Rantala (1970) in terms of the present account of the life cycle of *P. fuscus*, then it would appear that development in Britain is faster, being in fact approximately one stadium ahead of the population of this millipede in Finland. In general then, it would seem that warmer temperatures whether on a microclimatic or on a geographical scale serve to advance the rate of development of this species.

References

Biernaux, J. (1972). *Chorologie et étude biologique comparée de deux familles de Myriapodes-Diplopodes belges: les Blaniulidae et les Iulidae.* Dissertation originale présentée à la Faculté des Sciences Agronomiques de l'Etat, Gembloux.

Brookes, C. H. (1963). *Some aspects of the life histories and ecology of* ***Proteroiulus fuscus*** *(Am Stein) and* ***Isobates varicornis*** *(Koch) (Diplopoda) with information on other blaniulid millipedes.* Ph.D. Thesis, University of Manchester.

Halkka, R. (1958). Life history of *Schizophyllum sabulosum* (L.) (Diplopoda, Iulidae). *Suomal. eläin-ja kasvit. seur. van. eläin. Julk.* **19** (4): 1–72.

Humbert, A. (1893). *Myriapodes des environs de Genève.* Geneva.

Latzel, R. (1884). *Die Myriopoden der österreichisch-Ungarischen Monarchie.* Zweite Halfte. Die Symphylen, Pauropoden und Diplopoden. Vienna.

Rantala, M. (1970). Anamorphosis and periodomorphosis of *Proteroiulus fuscus* (Am Stein). *Bull. Mus. natn. Hist. nat., Paris* (2) **41** (Suppl. no. 2): 122–128.

Verhoeff, K. W. (1928). Diplopoden. *Bronn's Kl. Ordn. Tierreichs* **5** II Abt. 1 Teil: 1–1072.

Verhoeff, K. W. (1933). Wachstum und Lebensverlängerung bei Blaniuliden und über die Periodomorphose. *Z. Morph. Ökol. Tiere* **27**: 732–749.

Verhoeff, K. W. (1939). Wachstum und Lebensverlängerung bei Blaniuliden und über die Periodomorphose. II. *Z. Morph. Ökol. Tiere* **36**: 21–40.

Discussion

SAHLI: It is perhaps unfortunate that you do not give the limits for the numbers of pedigerous segments for each stadium. An individual of stadium X with 36 pedigerous segments (your Table I) may have been derived from an individual in stadium IX with 34 (since there are two apodous in IX); such an individual does not appear in your Table I where the highest number given for stadium IX is 33.

BROOKES: But Table I gives only the *commonest* course of anamorphosis and I must make it clear that these Tables are compiled from the actual individuals in the sample—but the history of each individual, its previous increments, have been traced back by the serial defence gland method of Halkka (1958)*.

SAHLI: (trans. Haacker). Without following individuals through, it is difficult to reconstruct the development of individuals from a sample?

BROOKES: Agreed. An individual with, say, 32 segments could belong to any of three stadia, but when traced back by the defence gland method it can be placed exactly in one stadium; in fact, its entire past history, from the egg, can be recalled.

GABBUTT: Different generations might be genetically isolated—what is the evidence for annual cyclic isolation in *Proteroiulus* and *Isobates*?

BROOKES: I have none.

GABBUTT: A proportion of adults of stadia VIII, IX and X might contribute to overcoming the inherent genetic isolation?

BROOKES: It is not possible to discover which animals belonging to which particular generations have bred together.

BLOWER: But both species are parthenogenetic?

BROOKES: The possibility of some bisexual reproduction cannot be ruled out. However, unless the unusual numbers of males in Rantala's culture (Rantala, 1970)* sometimes occur in the field (and this seems not to be so) then the 0·1% ♂♂ in Britain would presumably make an unimportant contribution.

BIERNAUX: The details of anamorphosis in the several blaniulids you find in England are very similar to those I have established in Belgium. It is certainly difficult to establish the life histories of these animals and I congratulate you on this attempt.

MEIDELL: In Norway I collected about 770 individuals of *P. fuscus* in 1967–1968 and not one male was included. In 1969–1970 I collected 86 individuals and four of these were males. 1967 had the highest rainfall for 30 years (more than 200% of normal); 1968 was one of the driest years (less than 1 mm fell in August). Possibly this alternation of extremes of climate could lead to more males occurring?

BROOKES: Perhaps. Mme Rantala reported (Rantala, 1970)* that when conditions in her cultures are not good, males seem to turn up. Perhaps males survive in culture better than the field.

* See list of References, p. 499.

FAIRHURST: In some iulids there are indications that stadium II may not be able to survive a winter. This is obviously not the case with female blaniulids, but I wonder whether males might be eliminated at this stage?

BROOKES: Unfortunately, as you know, males are not recognizable until stadium V, but it is fascinating to speculate that the males in Rantala's cultures (Rantala, 1970)* were due to the permanently *good* conditions.

Symp. zool. Soc. Lond. (1974) No. 32, 503–525.

THE LIFE-CYCLE AND ECOLOGY OF *OPHYIULUS PILOSUS* (NEWPORT) IN BRITAIN

J. GORDON BLOWER and PETER F. MILLER*

Department of Zoology, The University, Manchester, England

SYNOPSIS

Post-embryonic development of *Ophyiulus pilosus* is very similar to that of *Iulus scandinavius*. Early stadia follow precisely the same succession and not that given by Verhoeff (1928). Maturity is achieved in stadia IX, X or XI, most commonly in stadium X, but usually takes two years (compared with three in *I. scandinavius*) and this possibly accounts for its success as a colonist of countries as far apart as North America and New Zealand.

The size, age structure and fecundity of a Welsh population are given, a life-table is constructed and the standing crop and production are calculated. An average over-wintering standing crop of 260 individuals per square metre, with a fresh weight of 2·1 g produces between 2·5 and 5·0 g (fresh weight) annually. The details of the anamorphosis, life-cycle, survivorship, standing crop and production of *O. pilosus* are compared with the available data for *I. scandinavius*.

INTRODUCTION

A brief account of the anamorphosis and life-cycles of the two common British iulines, *Iulus scandinavius* Latzel and *Ophyiulus pilosus* (Newport) (=*O. fallax* Meinert) was given by Blower & Gabbutt (1964). Further details for *I. scandinavius* have been given by Blower (1970). Recently we have gathered quantitative data which enable us to give a detailed description of the post-embryonic development and life cycle of *O. pilosus*.

Interest in these two iulines is focussed on four features of their biology. Firstly, they are the commonest iulids associated with leaf litter. It has long been realized that they must contribute to litter degradation and it is important to know the scale and nature of their effect. Secondly, these two iulines are peculiar in being semelparous—in reaching maturity and then dying, unlike most of our iulids which survive for several years after first becoming mature and breed in each of these (Blower, 1969). Thirdly, *O. pilosus* appears to mature a year earlier than *I. scandinavius* and it has been suggested that this type of neotenic contraction of the life-history has conferred on the species the ability to succeed as a colonist (Blower, 1969).

* Present address: *Biological and Chemical Research Institute, Rydalmere 2116, New South Wales, Australia.*

Fourthly, *O. pilosus* has held an enigmatic position with regard to its post-embryonic development. Verhoeff (1928) has reported that its first stadium hatched from the egg with six instead of the usual three pairs of legs. In noting the great similarity of the post-embryonic development of *O. pilosus* to that of *I. scandinavius*, which hatches with the usual three pairs, Blower & Gabbutt (1964) suggested that Verhoeff had probably made an error of observation and that the early anamorphosis of *O. pilosus* would probably turn out to be typical. In addition to confirming this suspicion, the present paper pursues the matter of ecological similarity in relation to the taxonomic distinctness of these two species. Notwithstanding the fact that *O. pilosus* and *I. scandinavius* belong to different genera, it is very difficult to assign any individual, except a mature male, to one species or the other. In view of the similarities between many details of their life-cycles and ecology and their frequent co-existence, even in the same patch of leaf litter, it is essential to list those morphological differences which will facilitate practical separation of the two, and to prepare the ground for a further enquiry into the subtle ecological differences which must theoretically exist between such closely similar sympatric species.

FIELD SITES AND METHODS

Two mixed deciduous woods on carboniferous limestone have been sampled quantitatively over a period. The first, at Llethrid, Gower, South Wales, has a shallow soil and mull-like moder humus. The second, at Milldale, Derbyshire, has a deeper soil and the humus form is a mull. Both sites have rich diplopod faunas of 11 and 9 species respectively. These sites and their faunas are fully described elsewhere (Blower & Miller, in prep.).

The site at Llethrid consisted of a rectangle 10×20 m with its long side parallel to the slope. It was divided into a lower and upper square of 10 m side. For the first 4 samples, 10 units of soil and litter were taken from each of the upper and lower squares and distributed between 2 sets of funnels (type A and type B) for extraction. The numbers of *O. pilosus* did not differ significantly in the upper and lower parts of the rectangle but type A funnels yielded consistently and significantly higher numbers. Accordingly, only the figures for the lower square extracted by type A funnels were used. A further 5 samples consisted of 10 units of soil and litter from the lower site only and were extracted by type A funnels. The sample units consisted of one tenth of a square metre of litter and one twentieth of a square metre of soil beneath. The units were taken at the centre of square metres located by means of random co-ordinates within the 10 × 10 site.

At the Derbyshire site there was very little litter at the time of sampling and so the surface litter and soil were taken together as units of one twentieth of a square metre, about 10 cm deep. The 20 units were randomized within a rectangle 10 × 20 m. The long axis of the rectangle again followed the slope of the hillside but units from its upper and lower parts were not kept separate. At Llethrid, no single individual of *I. scandinavius* occurred on the site but at Milldale it was present in small numbers. At Llethrid, *I. scandinavius* did occur in the vicinity of the site; several individuals were included in hand samples from further up the hillside.

Hand samples were taken on each sampling occasion to provide material for assessing breeding condition and weight. Many other sites have been sampled by hand on one or two occasions each, usually within the period from autumn to spring when development is at a standstill and an estimate of the overwintering stadial spectrum is obtained. Each individual from the hand and funnel samples was assigned to a stadium by counting the rows of ocelli. In addition, the numbers of podous and apodous segments were counted and these figures often provided confirmatory evidence of stadium and species. Hand collected animals were weighed alive and the females dissected to determine the number and size of the eggs in the ovitube. Details of segment number have been gathered from all the material, whether collected by funnel or by hand.

POST-EMBRYONIC DEVELOPMENT

On two occasions at Llethrid, aggregates of stadia II and III, probably the contents of single nests, emerged from the funnels although the animals do not normally leave the nest until stadium III. Further information on the earlier stadia has been obtained from eggs reared in vivaria (Blower, 1974). As in *I. scandinavius* (Blower & Gabbutt, 1964) the first stadium does not leave the adherent pupoid skin, nor does it leave the egg case; thus stadium II appears to walk out of the egg capsule. This second stadium, like the first, is relatively inactive, still fully charged with yolk, and the number of pairs of legs is difficult to establish. Stadium I and stadium II are, however, like those of most iulids in having three pairs and seven pairs respectively on the first four and six rings. Thus, whilst the details given by Verhoeff (1928) are wrong, the reason for his observational errors is easily understood. In any case, Verhoeff correctly established the most important feature which distinguishes the early anamorphosis of *O. pilosus* from that of most other iulids, that is, the complete dependence of the second stadium on the yolk of the egg and the postponement of independent feeding until stadium III. This peculiarity *O. pilosus* shares with the other common British iuline, *I. scandinavius*.

The course of anamorphosis is summarized in Table I and details of the variation in increments of segments are given. As in *I. scandinavius* the secondary sexual modifications first appear in stadium VII and maturity is first achieved, in males, in stadia IX, X or XI and in females, in stadia X or XI. The general course of anamorphosis is thus almost identical in *O. pilosus* and *I. scandinavius* but the actual increments added are larger in *O. pilosus* (see Fig. 1).

TABLE I

Anamorphosis of ***Ophyiulus pilosus****; podous and apodous segments, from Llethrid, Gower*

Stadium	Podous segment*	Apodous segments*	N.
I	4	2	
II	6	5	
III	11	4	
IV	15	(5) 6, 7 (8)	51
V	(21) 22, 23	(6) 7, 8	64
VI	(28) 29, 30 (31)	(5) 6, 7 (8)	336
VII ♂	(34) 35–37 (38)	(4) 5, 6	276
VII ♀	(35) 36–38	(4) 5, 6 (7)	206
VIII ♂	(40) 41–43	(3) 4, 5	60
VIII ♀	(40) 41–44	(3) 4, 5	58
IX ♂ imm.	*(43)* 44–47 *(48)*	2, 3 (4)	19
IX ♂ mature	*(42)* *43* 45–48	(2) 3 (4)	26
IX ♀	44–47 (48)	3, 4	34
X ♂	(47) 48–51 (52)	(1) 2 (3)	113
X ♀	(47) 48–51 (52)	1, 2 (3)	118
XI ♂	51–54	2	3
XI ♀	51–54	2	5

* The segment numbers in brackets are possessed by 5% or less of the sample. Italicized bracketed numbers occur, but not yet in Gower.

THE LIFE HISTORY OF *O. PILOSUS*

Figure 2 shows the stadial composition of *O. pilosus* in the funnel samples from Llethrid. The 71 individuals of stadia II and III in the sample of 1st May all emerged from one of the 10 units and were clearly not ready to leave the nest. The presence of stadia II in the sample taken on 22nd June also indicates that these animals were probably still in the nest and that oviposition therefore extends over at least three weeks.

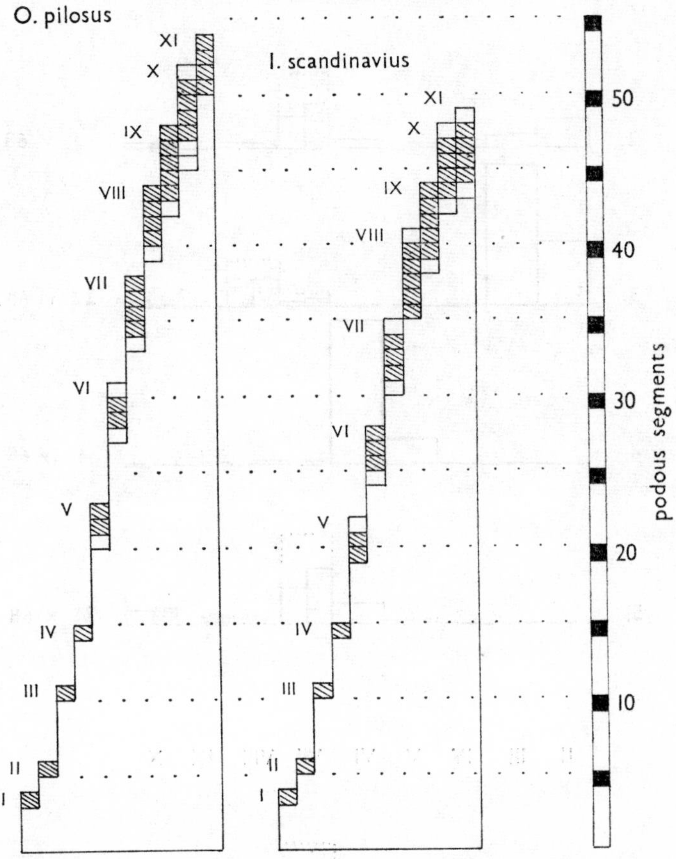

FIG. 1. Range of podous segment numbers in each stadium of *O. pilosus* compared with that of *I. scandinavius*. Open rectangles represent less than 5% of the sample.

Eggs reared in the laboratory took three weeks to reach stadium III at a temperature somewhat higher than in the field. Probably at least a month will be necessary for the first individuals of stadium III to appear in the field. Thus the eggs represented in samples 1 and 2 were probably laid from early April to late May. The generation represented by stadia II and III in May and June appears to reach stadia V and

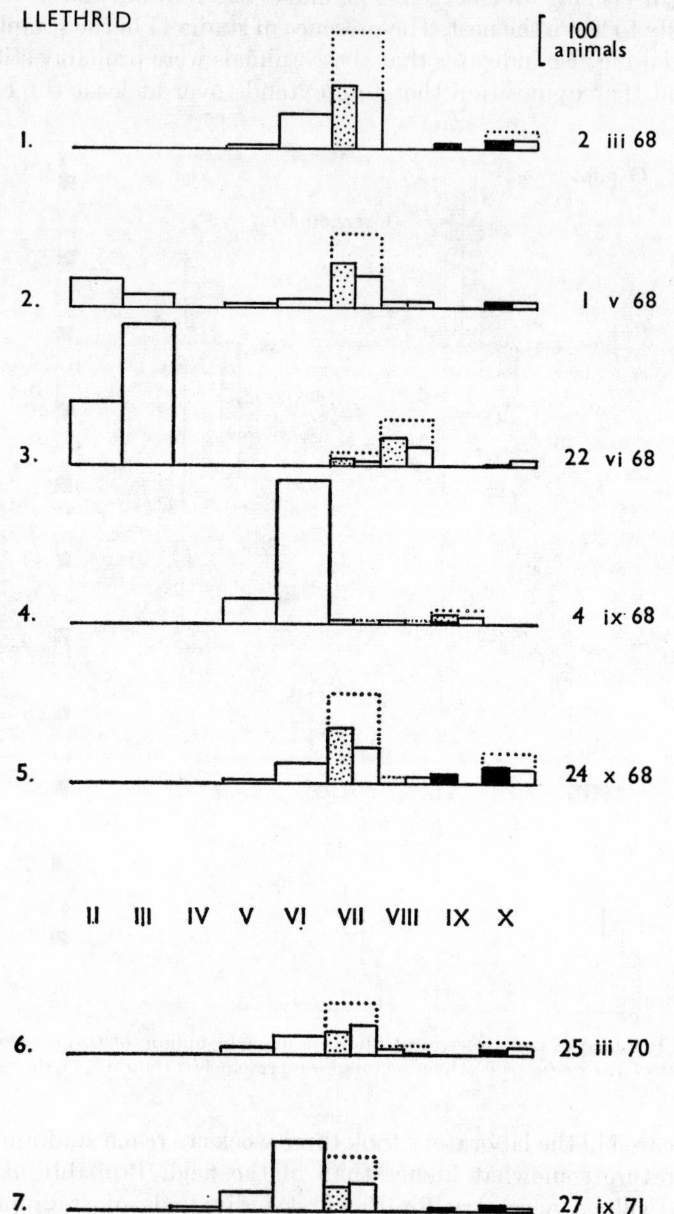

FIG. 2. Stadial composition of *O. pilosus* in the samples from Llethrid, Gower. The number and date of each sample is given. Divided columns show males on the left and females on the right. Mature males black, immature males stippled.

VI by early September and stadia VI, VII and VIII (and possibly males of stadium IX) by the end of October—the bulk of the generation achieves stadium VII. A similar stadial spectrum can be seen in samples 1 and 6 from the spring of 1968 and 1970. It is possible that the peak of stadium VI in sample 6 late in September 1970 may still have reached the usual condition of a peak at stadium VII by late October. There seems little doubt that a few advanced animals may reach stadium VIII (see samples 5 and 6) in the autumn of the year of their birth. From the laboratory cultures it seems just possible that males might reach maturity at stadium IX by the autumn following their birth. Indeed, it is noteworthy that any individuals overwintering in stadium IX are usually mature males (see samples 1, 5 and 7).

In the second growing season, the majority of stadia VI and VII which have overwintered appear to proceed to stadium X. It is noted in laboratory reared animals that both males and females of stadium VII

TABLE II

Fresh weights in mg of **Ophyiulus pilosus** *from Llethrid, Gower*

Stadium and sex	Mean	Actual Range	N	Mean wt of *I. scandinavius* (from Blower, 1970)
IV	0·76	0·6–1·0	5	
V	1·20	0·8–1·5	11	2·6
VI	3·23	2·3–5·0	27	5·2
VII ♂	4·85	2·4–6·7	31	
VII ♀	5·94	4·1–7·7	39	
VII ♂ ♀	5·40			9·9
VIII ♂	9·48	6·9–12·7	5	18·5
VIII ♀	9·80	7·7–12·5	9	26·5
VIII ♂ ♀	9·64			
IX ♂ imm.	15·05	14·6, 15·5	2	} 32·8
IX ♂ mature	17·59	12·6–23·5	16	
IX ♀	19·40	15·2–25·9	7	36·0
IX ♂ ♀	17·89			
X ♂	23·19	17·0–28·9	49	39·8
X ♀	51·18	43·2–70·0	88	} 90·9
X ♀*	53·83	43·2–70·0	34	
X ♂ ♀	38·51			
XI ♀	87·0	86·7, 87·3	2	136·1

* Mean weight of fully gravid females. This figure is used for the average of both sexes.

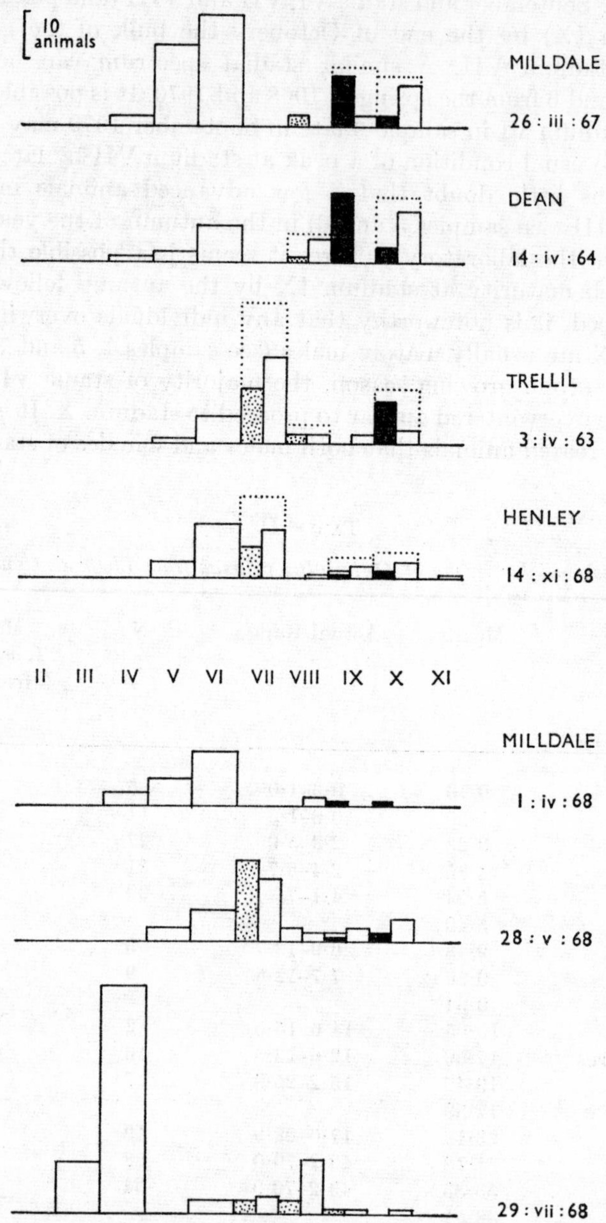

FIG. 3. Stadial composition of *O. pilosus* in the Milldale samples and from four hand samples from various localities. Divided columns show males on the left and females on the right. Immature males stippled.

take a similar time to develop from the egg. However, since mature males are much smaller than females of the same stadium (less than half the weight in stadium X, see Table II), it is possible that they will spend less time developing from stadium VII. In fact it takes only two months, in the laboratory, for a male to reach maturity at stadium IX from stadium VII whilst a female takes from 10–13 months to reach stadium X—which is the first stadium in which females can achieve maturity. Curiously, the males in the laboratory which matured in stadium X or stadium XI took a similar time to the females. Since these late maturing males spent a large part of the time inactive in moulting chambers, it appears that the retardation of development is specially designed to ensure synchronous appearance with the females. Thus, although males have the potential for much faster development that the females, this is rarely realized. As the data in Fig. 2 show, similar numbers of both sexes reach stadium X together.

Figure 3 gives the stadial composition of three funnel samples from Milldale together with the overwintering stadial spectrum of hand samples from Milldale and three other places. The hand sample from Milldale taken in the spring of 1967 agrees with the funnel sample from spring 1968 in showing that the species here only proceeds as far as stadium VI in the first growing season and a few animals get no further than stadium IV. The hand sample from the Forest of Dean also shows this pattern. The hand samples from Milldale and Forest of Dean also show a preponderance of mature males in stadium IX and a correspondingly lower number in stadium X. In view of the absence of overwintering animals in stadium VII, the small numbers of stadium VIII must be construed as late animals of the second growing season in contrast to the condition at Llethrid where they appear to be early arrivals of the first autumn. The populations at Milldale and Dean appear to develop more slowly than those at Llethrid and there is a distinct possibility that a few females may take three years to reach maturity. The hand samples from Trellil in Cornwall and Henley on Thames appear to have the same overwintering spectrum as at Llethrid.

VERTICAL DISTRIBUTION

Figure 4 shows the distribution of *O. pilosus* between the litter and the soil at Llethrid. It will be seen that never more than 35–40% of the sub-adult individuals inhabit the litter in the daytime. However, occasionally more than half of the population of adults are found in the litter.

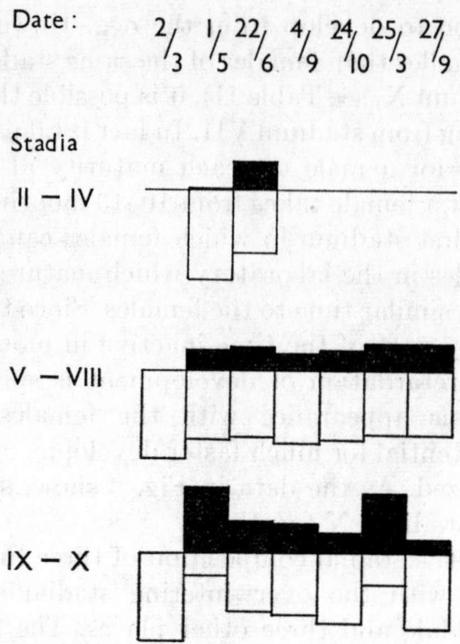

FIG. 4. Vertical distribution of *O. pilosus* in the samples from Llethrid, Gower Each column represents 100%, the black part gives the proportion in the litter.

FECUNDITY AND SURVIVAL

From Table III it can be seen that less than half of the stadium X females in autumn have full-sized eggs. By spring the proportion has risen to about 80%. A small number of those in the samples of 16th April and 15th May had already laid their eggs. Females carrying

TABLE III

Number of eggs and weights in stadium X females from Gower

	Number dissected	Proportion with eggs	Mean wt with eggs mg	Mean wt without eggs mg	Mean no. of eggs
24.10.68	41	0·40	54·18	41·96	72·10
16.4.69	14	0·77	51·10	50·31	78·25
15.5.69	18	0·83	55·10	53·20	73·40

full-sized eggs in autumn are significantly heavier than those without full-sized eggs ($p < 0{\cdot}001$) but the difference in the spring samples is less

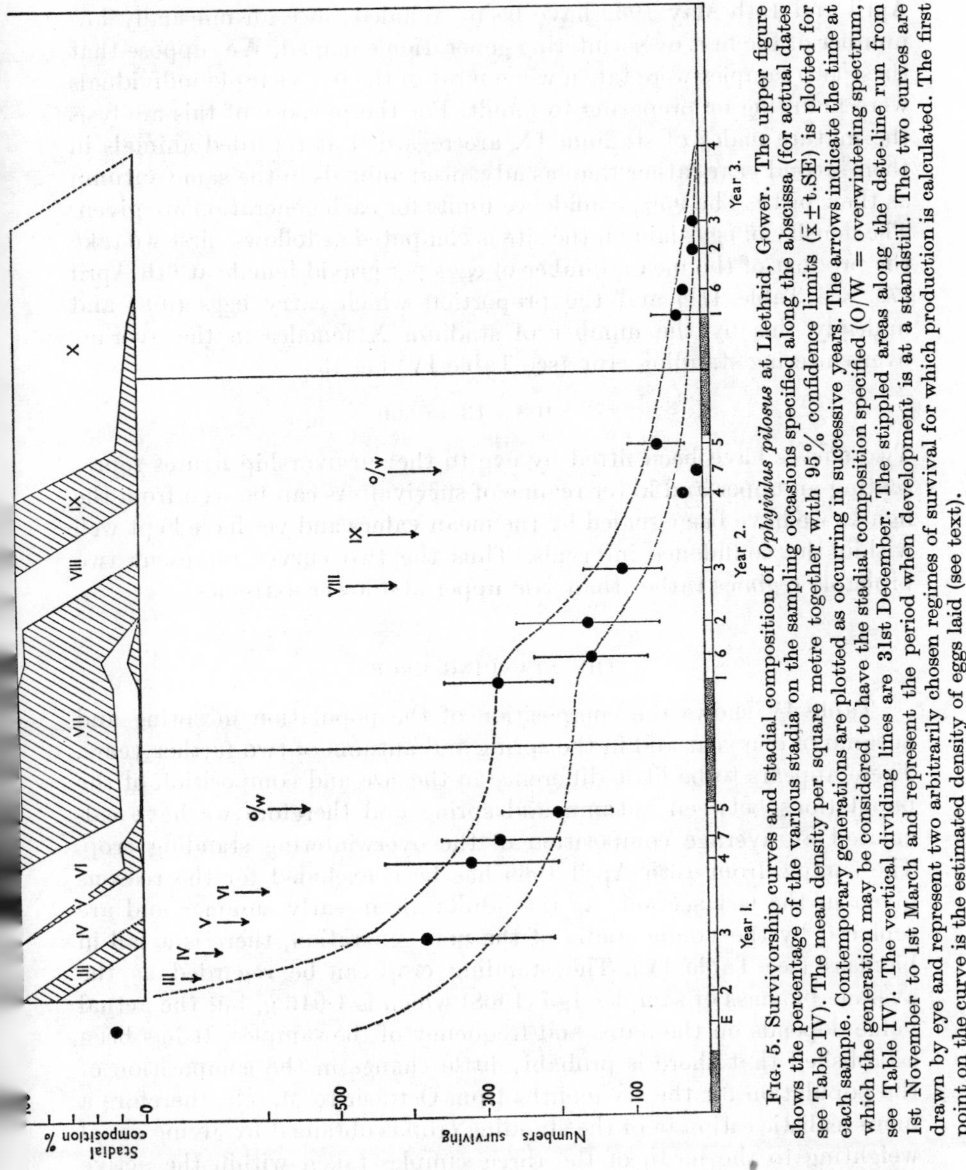

FIG. 5. Survivorship curves and stadial composition of *Ophyulus pilosus* at Llethrid, Gower. The upper figure shows the percentage of the various stadia on the sampling occasions specified along the abscissa (for actual dates see Table IV). The mean density per square metre together with 95% confidence limits ($\bar{x} \pm t.SE$) is plotted for each sample. Contemporary generations are plotted as occurring in successive years. The arrows indicate the time at which the generation may be considered to have the stadial composition specified (O/W = overwintering spectrum, see Table IV). The vertical dividing lines are 31st December. The stippled areas along the date-line run from 1st November to 1st March and represent the period when development is at a standstill. The two curves are drawn by eye and represent two arbitrarily chosen regimes of survival for which production is calculated. The first point on the curve is the estimated density of eggs laid (see text).

obvious and not significant. Of the wet weight of the females in autumn 19·15% is apparently due to eggs; this is equivalent to a fresh weight per egg of 0·131 mg.

In Fig. 5 the numbers of the separate generations of each sample are treated serially as a survivorship curve. Two samples taken on 16th

April and 15th May 1969 have been excluded since an unusually low number of the first overwintering generation emerged. We suppose that these two samples were taken when most of the 0–1 year old individuals were moulting or preparing to moult. For the purpose of this analysis the mature males of stadium IX are regarded as retarded animals in their second year rather than as advanced animals in the same autumn as their birth. The 95% confidence limits for each generation are given. The density of eggs laid on the site is computed as follows: first we take the product of the mean number of eggs per gravid female at 6th April (78, see Table III) and the proportion which carry eggs (0·8) and multiply this by the number of stadium X females in the average overwintering standing crop (see Table IV) i.e. 13

$$78 \times 0{\cdot}8 \times 13 = 806$$

Two curves have been fitted by eye to the survivorship figures representing an upper and lower regime of survival. As can be seen from the figures we have been guided by the mean values and yet have kept well within the confidence intervals. Thus the two curves represent two probable regimes rather than safe upper and lower extremes.

THE STANDING CROP

Table IV shows the composition of the population in spring and autumn of one year and in the spring and autumn of two further years. There appears to be little difference in the size and composition of the populations between autumn and spring and therefore we have calculated an average composition of the overwintering standing crop. The sample from 16th April 1969 has been excluded for the reasons given in the last section. As the adults die in early summer and are replaced by the young stadia of the new generation, there is a fall in biomass (see Table IV). The standing crop can be regarded as the average biomass of samples 1–5 (1968) which is 1·916 g, but the actual value depends on the dates and frequency of the samples. It has been established that there is probably little change in the composition of the population for the six months from October to March; therefore a more realistic estimate of the standing crop is obtained by giving equal weighting to the mean of the three samples taken within the active period (1·342 g) and of the mean overwintering population (2·778 g), which gives an overall mean of

$$(1{\cdot}342 + 2{\cdot}778)/2 = 2{\cdot}060 \text{ g.}$$

But the overall average for the overwintering standing crop is 2·108g.

Combining this with the average for the 1968 active season we have an annual mean of

$$(1{\cdot}342+2{\cdot}108)/2 = 1{\cdot}725 \text{ g.}$$

However, we feel that the standing crop in relation to production can best be defined as "that population which survives an inactive period, to produce again in the next season" i.e. the overwintering standing crop. This value is the most clearly defined; furthermore, it was the value used by Blower (1970) for *I. scandinavius* with which we are concerned to make effective comparisons. Accordingly we shall take the average value of 2·108 g.

PRODUCTION

The stadial composition of each sample is given in Fig. 5 along with the survivorship curves. From these data the times at which maximum numbers of each stadium are present are judged and marked on the graph. At no one time in the first year of growth is the entire generation represented by one stadium, however. The average stadial composition of the overwintering population (see Table IV) is marked on the curves instead of stadia VII or X. For the rest, the oversimplification is not thought to involve a large error; for although the population is never represented or is hardly ever represented by one stadium, a declining set of production values can be theoretically conceived for the separate stadia and points have been chosen to represent these.

The maximum values on the ordinate for each of the above points are read off from each curve. These are the values used in the calculation of the net annual production set out in Table V. The eggs develop to stadium III using the yolk. The production of eggs, pupoids, stadia I, II and III is therefore included in that of the adults. The fresh weights (w) for the overwintering stadia indicate the average weight per individual of this overwintering *group* of stadia. The figures in brackets are the fresh weights of stadia VII and X respectively and do not differ much from the compound values we have used in their stead.

FOOD CONSUMPTION

In Table V the age–specific food consumptions of a laboratory population of *O. pilosus* (Blower, 1974) have been added and used to compute the population food consumption—assuming that sycamore litter is the sole food. A discussion of these figures is included in the paper just quoted.

TABLE IV

Densities and biomass

A. Density per square metre of overwintering populations

Sample. no.	Date	Stadia and sex IV	V	VI	VII ♂	VII ♀	VIII ♂	VIII ♀	IX ♂ imm.	IX ♂ mature	X ♂	X ♀	Generation 0–1 yr	Generation 1–2 yr	Total
1	2.3.68		2	61	119	100				2	14	14	282	30	312
5	24.10.68		1	32	101	65		4		11	27	24	203	62	265
6	25.3.70		15	31	43	55	8	2			10	12	154	22	176
7	27.9.71	7	37	135	48	55			1		5	1	282	7	289
Means		1·75	13·75	64·75	146·5		3·5		0·25	3·25	26·75		230·25	30·25	260·5

B. Biomass of overwinter populations in mg

		0–1 yr	1–2 yrs	Total	
(i) Calculated from the mean overwintering spectrum above		1051·8	1055·8	2107·6	(i)
(ii) From individual samples	1	1382·0	1071·4	2453·4	
	5	1039·5	2062·2	3101·7	
	6	743·7	818·2	1561·9	
	7	1042·0	238·2	1280·2	
		1051·8	1047·5	2099·3	(ii)

C. Biomass of summer populations in mg

				2 + yrs	
2	1.5.68	17·3	843·2	274·3	1134·8
3	22.6.68	245·3	962·9	279·1	1487·8
4	4.9.68	939·3	436·6	—	1375·9
Mean					1332·8

The discrepancy between the total overwintering biomass at (i) and (ii) is entirely due to the different combination of the sexes of stadia IX and X.

TABLE V

Calculation of the net annual production and food consumption of a population of ***Ophyiulus pilosus*** *in South Wales*

Stadium or group of	Survival and production				Food consumption				
	Surviving	Dying	Fresh weight mg	Production mg	Eaten by stadium x mg dry	l_{x+1}	$d_x/2$		Population consumption mg
x	l_x	d_x	w	$w.d_x$	f	a	b	$a+b$	$f(a+b)$
Upper regime									
III	635	128	—	—	1·1	507	64	571	628·1
IV	507	75	0·76	57·0	3·3	432	37·5	469·5	1594·4
V	432	45	1·20	54·0	10·2	387	22·5	409·5	4176·9
VI	387	83	3·23	268·1	15·0	304	41·5	345·5	5182·5
O/W(1)*	304	144	4·57(5·4)	658·1	41·8(55)	160	72	232	9697·6
VIII	160	44	9·64	424·2	129·5	116	22	138	17871·0
IX	116	36	17·89	644·0	179·5	80	18	98	17591·0
O/W(2)*	80	80	34·90(37·2)	2792·0	247·0(262)	—	40	40	9880·0
				4897·4					66621·5

Lower regime									
III	342	38	—		1·1	304	19	323	355·3
IV	304	36	0·76	27·4	3·3	268	18	286	943·8
V	268	34	1·20	40·8	10·2	234	17	251	2560·2
VI	234	43	3·23	138·9	15·0	191	21·5	212·5	3187·5
O/W(1)*	191	107	4·57(5·4)	489·0	41·8	84	53·5	137·5	5747·5
VIII	84	28	9·64	269·9	129·5	56	14	70	9065·0
IX	56	22	17·89	393·6	179·5	34	11	45	8077·5
O/W(2)*	34	34	34·90(37·2)	1186·6	247·0	—	17	17	4199·0
				2546·2					34135·8

Data for age specific food consumption from a laboratory population (Blower, 1974).
Details of survival (l_x) are read from the two curves in Fig. 5, representing an upper and lower regime.

* O/W(1) and O/W(2) refer to the average overwintering stadial spectra from Table IV. The fresh weight and food consumption per individual within these spectra are given. The actual fresh weights and food consumption of stadia VII and X are given in brackets for comparison.

TABLE VI

Some comparative data for ***O. pilosus*** *and* ***I. scandinavius***

A. Dimension

		O. pilosus					*I. scandinavius*				
		Length mm	Diam. mm	Volume ($\pi r^2 l$) mm³	Ratio length /diam.	N	Length mm	Diam. mm	Volume ($\pi r^2 l$) mm³	Ratio length /diam.	N
IX	♂♂	15·88	1·07	14·29	14·84	18	16·2	1·50	28·63	10·80	15
	♀♀	16·31	1·24	19·67	13·15	6	18·1	1·63	37·77	11·10	17
X	♂♂	17·82	1·10	16·95	16·20	10	20·8	1·68	46·11	12·38	16
	♀♀	23·67	1·69	53·17	14·00	20	23·8	2·14	85·60	11·12	18
XI	♀♀	27·91	2·01	88·54	13·88	2	26·5	2·47	96·00	10·73	16

B. Density, biomass and production

	Overwintering standing crop density per sq. metre 0–1 yr	1–2 yr	2–3 yr	Total	Biomass (g)	Net annual production (g)
O. pilosus (South Wales)	230	30	—	260	2·1	2·5–4·9
I. scandinavius (Cheshire)	35	14	9	58	1·25	1·5–2·5

A COMPARISON OF *O. PILOSUS* AND *I. SCANDINAVIUS*

What follows is mainly based on a population of *O. pilosus* from South Wales and a population of *I. scandinavius* from Cheshire. However, we have recorded similar life-histories from Cornwall and Oxfordshire and contrasted life-histories from Cheshire and Gloucestershire. We know of no obvious correlation between life-cycle and geographical position.

The increments of new segments are usually larger in *O. pilosus*. Compared with *I. scandinavius*, an extra segment is added at the moults IV/V, V/VI and VII/VIII. At the moult from stadium VI to VII either one or two extra segments are added. Increments from stadium VIII are the same as in *I. scandinavius*. Thus stadia VIII, IX, X and XI of *O. pilosus* have, on average, four more podous segments in the male and five more in the female. Figure 1 summarises the data for the two species. It will be seen that the 90% ranges of podous segments for a given stadium do not overlap except in stadium IX at 44 segments. Even here, the sex of the individual removes the overlap; a female stadium IX with 44 podous segments will belong to *Iulus scandinavius*, a male will belong to *Ophyiulus pilosus*—in 90% of cases.

O. pilosus is proportionally longer and thinner than *I. scandinavius* (Table VI). A given stadium of *I. scandinavius* is about twice the volume and twice the weight of *O. pilosus*, males slightly more than twice, females slightly less. Compared with *O. pilosus*, a stadium X female of *I. scandinavius* is 1·75 times the volume, 1·75 times the weight and lays 1·75 times the number of eggs. Stadium X males of *I. scandinavius* are more than twice the weight of stadium X males of *O. pilosus*. The fact that *O. pilosus* is longer and thinner than *I. scandinavius* and has more segments might facilitate more efficient burrowing. *O. pilosus* is found in the soil much more frequently than *I. scandinavius* (compare Fig. 4 with Fig. 7 in Blower, 1970). These dimensional data are summarized in Table VI.

The survivorship curves of both species are compared in Fig. 6. The mortality of the early stadia of *I. scandinavius* in their first year is much greater than that of the same stadia of *O. pilosus*, but the latter suffers relatively greater mortality of its older stadia in their second year. Although far fewer adults of *I. scandinavius* survive to maturity, their larger size in part compensates and the density of eggs laid by this species is not very much smaller than that laid by *O. pilosus*.

We might hazard the suggestion that physical factors may largely account for the mortality of the delicate younger stadia of *I. scandinavius* whilst the relatively greater mortality of the older stadia of *O.*

pilosus in the second year is perhaps more likely to be due to biotic factors. From Table V it will be seen that the older stadia consume much more food and in mid-summer this is likely to be very limiting (See Miller, 1974).

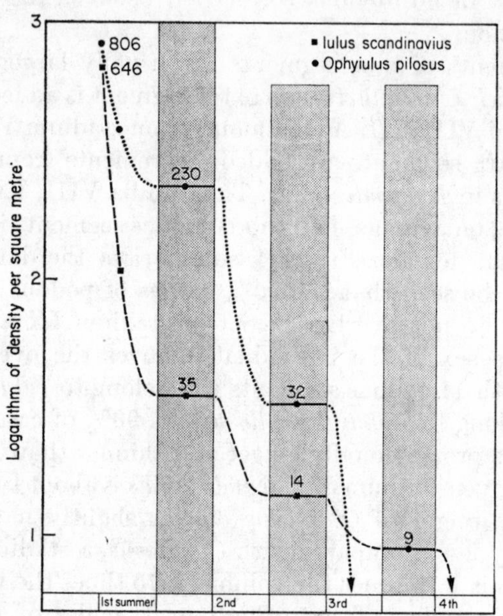

FIG. 6. Survivorship curves of *O. pilosus* from Llethrid, Gower and *I. scandinavius* from Cheshire. The horizontal parts of the curve are drawn at the average overwintering densities. The first points are the estimated egg-densities.

Population density of *O. pilosus* at the Welsh site is much higher than that of *I. scandinavius* at the Cheshire site. Our subjective impression at many other sites is that *O. pilosus* forms denser populations than *I. scandinavius*. But although the density of *O. pilosus* at the Welsh site is over four times greater than that of *I. scandinavius* in Cheshire, since a large part of the overwintering standing crop consists of animals 0–1 year old and since *O. pilosus*, stadium for stadium, is lighter than *I. scandinavius*, the biomasses of the overwintering standing populations are not so discrepant. The overwintering standing crop of *I. scandinavius* is more than half that of *O. pilosus*. The production of the two species is related in similar manner (see Table VI).

Acknowledgements

Both authors gratefully acknowledge a grant from the Natural Environment Research Council.

References

Blower, J. G. (1969). Age structures of millipede populations in relation to activity and dispersion. *Syst. Ass. Publs* No. 8: 209–216.

Blower, J. G. (1970). The millipedes of a Cheshire Wood. *J. Zool., Lond.* **160**: 455–496.

Blower, J. G. (1974). Food consumption and growth in a laboratory population of *Ophyiulus pilosus* (Newport) *Symp. zool. Soc. Lond.* No. 32: 527–561.

Blower, J. G. & Gabbutt, P. D. (1964). Studies on the millipedes of a Devon oak wood. *Proc. zool. Soc. Lond.* **143**: 143–176.

Blower, J. G. & Miller, P. F. (in prep.). *The millipedes of two limestone woodlands.*

Miller, P. F. (1974). Competition between *Ophyiulus pilosus* (Newport) and *Iulus scandinavius* Latzel. *Symp. zool. Soc. Lond.* No. 32: 563–584.

Verhoeff, K. W. (1928), Diplopoda, I. *Bronn's Kl. Ordn. Tierreichs* **5**(2): 1–1072.

Discussion

Meidell: You mentioned the number of animals on the lower and upper parts of the slope. I only have one locality for *Iulus scandinavius* in western Norway where they seem to concentrate in the lower parts where litter accumulates and probably provides a higher humidity. Did you notice anything like this?

Blower: This aspect will be discussed by Mr Miller in his paper. We do think that *Ophyiulus pilosus* favours the flushed regions at the bottoms of slopes. Whether this is due to flushing or to the depth of litter which accumulates there I cannot say, but Mr Miller will discuss this. On this Welsh site *Iulus* was rarely present in either upper or lower parts. The numbers of *Ophyiulus* were not significantly different in the two parts. There is, in fact, little distance between the upper and lower sub-divisions of this site; though we recognize such effects as you were speaking of, they were probably not apparent here.

Deshmukh: Have you carried your confidence limits through to your production figures and is there a significant difference in the production of these two species?

Blower: No, we have not, but production has been calculated by three separate methods which gave similar results; although rough and ready, we think they are meaningful.

ANDERSSON: From your survivorship curve it seems that every animal survives the winter, that there is no mortality.

BLOWER: Certainly this is what the curve suggests. There seems to be very little mortality; there are no significant differences between the population sizes before and after winter.

ENGHOFF: Is the fact that no mortality occurs during the winter due to reduced predation?

BLOWER: We do not think that predation is important but our evidence is very flimsy. We have, essentially, two concave survivorship curves corresponding with the two periods of activity. We think much mortality occurs at the moults; we observe this in the laboratory and infer this in the field. Physical factors seem to us to be more important.

JEEKEL: Could emigration influence numbers?

BLOWER: Yes, but we looked at the number of aggregates of stadia III in our 20 units; this was roughly the same as the number of ovipositing females we supposed were present. If there had been a significant gain or loss of adults, this would only affect the slope of the early part of the survivorship curve. However, the production up to stadium III is included in that of the adults. It is certainly our estimates of the numbers of ovipositing adults which are critical.

MEIDELL: I mentioned to you the schaltstadium, or what I considered to be a schaltstadium, of *Iulus scandinavius* which I found in Norway. Do you think, in view of differences between Britain and Norway, that schaltstadia of *Iulus* could occur in Norway?

BLOWER: I would think the ecological circumstances in Norway would have to be very different to account for such a fundamental difference in the organization of the life cycle of this species.

MEIDELL: But there are big differences as one goes further north. *Polyxenus lagurus* has two breeding periods in Germany (Schömann, 1956)* but only one in Norway (Meidell, 1970)†.

BLOWER: Yes, I would expect differences—but not such a profound difference.

CAUSEY: What is the nest like; is there much preparation for egg laying?

* Schömann, K. (1956). Zur Biologie von *Polyxenus lagurus* (L. 1758). *Zool. Jb.* (Syst.) **84**: 195–256.

† Meidell, B. A. (1970). On the distribution, sex ratio, and development of *Polyxenus lagurus* (L.) (Diplopoda) in Norway. *Norsk ent. Tidsskr.* **17**: 147–152.

BLOWER: We have seen nests which have been built against the transparent floor of plastic boxes. It is hemispherical and smooth on the inside. We haven't seen them being built but those we have seen are like those described by Evans (1910)‡ and could have been constructed in the manner he describes.

‡ Evans, T. J. (1910). Bionomical observations on some British Millipedes. *Ann. Mag. nat. Hist.* (8) **6**: 284–291.

Symp. zool. Soc. Lond. (1974) No. 32, 527–551.

FOOD CONSUMPTION AND GROWTH IN A LABORATORY POPULATION OF *OPHYIULUS PILOSUS* (NEWPORT)

J. GORDON BLOWER

Department of Zoology, The University, Manchester, England

SYNOPSIS

Over a period of two years several individuals of *Ophyiulus pilosus* have been reared to maturity from broods laid in vivaria. In one brood of 83 eggs, 78 hatched and 18 reached maturity; 63% of the mortality occurred in stadium III. The third stadia of two other broods were separated into groups earlier in the stadium, but although mortality in stadium III was halved (32%) only six matured (4%).

Animals were kept on non-nutrient agar plates and fed on 0–1 year old sycamore litter. Squares of leaf (2 × 2 cm) were pressed on to the agar surface and the amounts eaten by each individual of each stadium estimated by area and by residue. Area estimates were found to be adequate for later stadia. Applying the figures obtained to a field population described elsewhere (Blower & Miller, 1974, this Symposium pp. 503–525) and supposing only sycamore is eaten, the population of a square metre would consume from 34 to 67 g of leaf (dry weight) per annum. Since veins are not eaten, nearly twice this weight of intact sycamore leaves would be skeletonized.

The pattern of feeding and growth in each stadium is described, together with details of moulting. More than a tenth of the time from egg to adult is spent in moulting and most of the food taken to reach maturity is eaten in a third of the remaining time.

INTRODUCTION

Most work on millipede feeding has sought to establish the existence of and reasons for particular preferences but there have been few attempts to determine the total amount of food consumed during the animal's lifetime. Van der Drift (1951) reared the iulid *Cylindroiulus punctatus* for part of its life on leaf litter of varying ages. In recording rather a low retention percentage (assimilation efficiency) in *Glomeris marginata* of about 5%, he concluded that the total quantity of leaf litter consumed by millipedes must be considerable. Kheirallah (1966) reared *Iulus scandinavius* on various species of deciduous leaf litter in the laboratory, having previously determined by analysing gut contents that deciduous litter was the major component of the diet of this species. This last author noted that oven-dried and remoistened litter was not readily acceptable to the animals; accordingly he subjected the litter to air drying alone, to determine quantity. It is possible that millipedes may depend on the microflora and fauna of dead leaves for their food, either directly or via the effects of the micro-organisms on the availability of leaf substance. In the course of rearing *Ophyiulus pilosus* in the laboratory to clarify details of its life-history in the field (Blower & Miller, 1973) an attempt was made to rear the species on litter collected directly from the field, without any intervening drying which might interfere with the microflora and fauna. The species was reared from eggs to adults on sycamore litter alone and the weight consumed was estimated. Although Verhoeff (1928) had reared *Tachypodoiulus niger* from egg to adult and Kheirallah (1966) had reared *Iulus scandinavius* to adulthood, a fully documented rearing has not previously been achieved. Accordingly full details are given here along with the estimates of food consumed.

METHODS

Adult *Ophyiulus pilosus* of both sexes were collected from a sycamore wood at Kerridge in Cheshire on 23rd February, 1970. They were placed in a plastic box on a substrate of sifted soil covered with moist sycamore litter. Eggs were laid in nests on the bottom of the box where they were visible through the transparent floor. One nest, Nest 1, appeared on 1st March and another three, close together, on the 15th March. The contents of two of these three were grouped and are referred to hereafter as "Nest 2". When the eggs had hatched they were transferred to Petri dishes, $3\frac{1}{2}$ in. i.d., containing tap water (non-nutrient) agar. The agar maintains a constant relative humidity at or close to 100%. The agar also serves as an excellent method of supporting an area of leaf and keeping it moist. The method was developed by Brookes (1963) and Healey (1963).

The numbers of individuals introduced to each plate and details of their subsequent history from oviposition in March 1970 to the death of the last three adult females in April 1972, are given in Fig. 1. The plates were kept in the dark except when examined, or when new food was added. The temperatures at which they were kept are recorded in Fig. 3. Plates were examined and new food added where necessary, every two to three days up to stadium VII and thereafter, once or twice a week.

Selection, preparation and presentation of food

Sycamore leaves up to one year old were collected from the wood at Kerridge (and on one occasion from Ernocroft Wood, see Table I). They were used immediately or kept moist in a plastic bag until required. A leaf with little or no fenestration was chosen, rinsed in tap water to remove adherent soil and other debris and cut into squares 2×2 cm. After the first two months the five main veins were removed

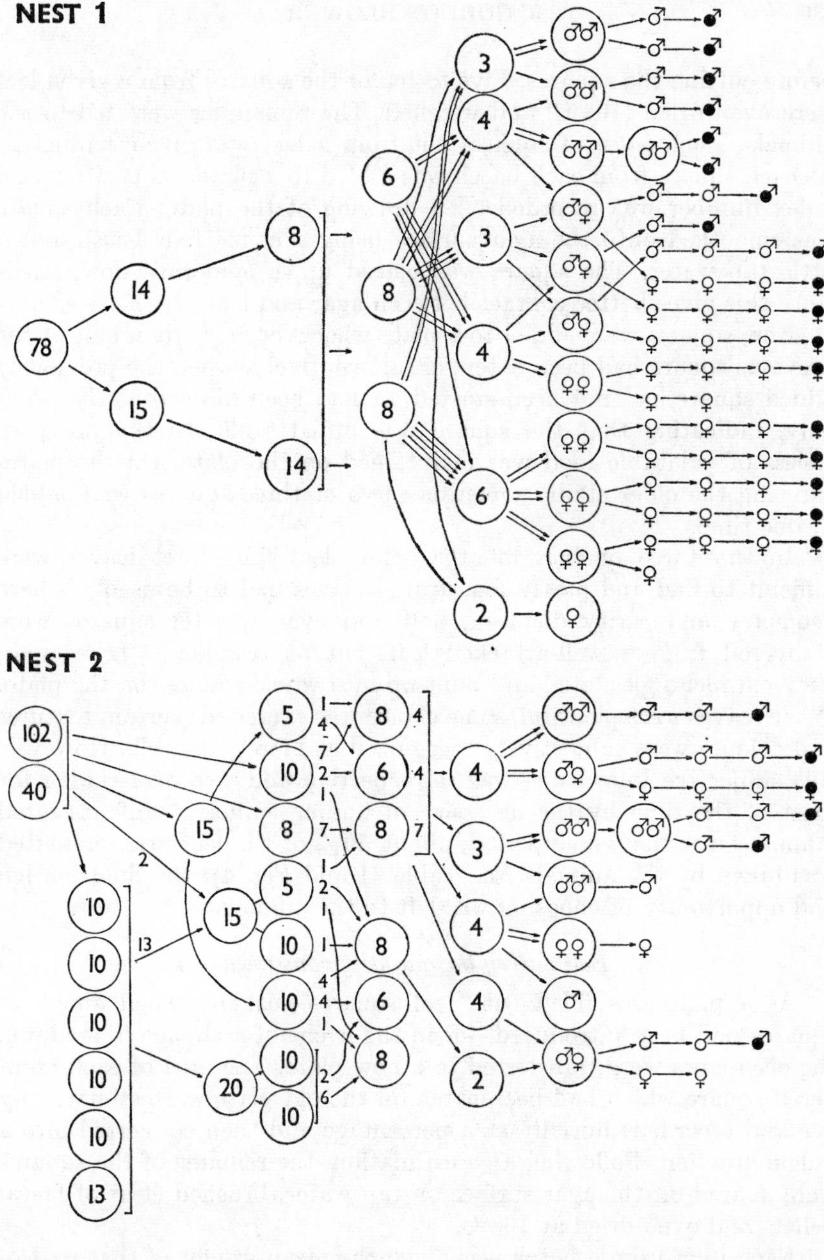

Fig. 1. The rearing scheme. The number of individuals reared on a plate is indicated within the circles representing the plates. The individuals reared singly from stadium VIII, are indicated by the appropriate symbol designating sex; the black symbols indicate maturity. The individuals of Nest 1 in the later stadia are figured in the same order in which they appear in Fig. 3.

before cutting the squares. Five to ten of the squares from a given leaf were oven dried (108°C) and weighed. The remainder were fed to the animals. Each *batch* of squares cut from a leaf was given a number. As each square from each batch was added to a plate, its position and index number was recorded on a drawing of the plate. Each square was smoothed onto the agar surface using a camel hair brush and a little tap water. The square was placed upper-epidermis downwards since this gives better contact between agar and leaf.

New squares were added to a plate whenever more than half of the previous square had been eaten, or, alternatively, when the previously added square had not been started, or had been nibbled at the edges only, indicating that the square was unpalatable. In this way, an excess of palatable food was maintained on the plate. On the plates carrying the older stadia sometimes two or three squares were added at one time.

In the three or four months before leaf fall intact leaves were difficult to find and partly fenestrated leaves had to be used. Where geometry and rarity dictated, half and even quarter squares were recovered from a well-attacked leaf, cut as triangles, "L" shapes, 1 × 2 cm rectangles, etc., and built-up into whole squares on the plate. When leaves were plentiful, some choice was exercised; certain textures and colours were subjectively recognized as more palatable. However, this subjective knowledge was retrospective and was over-ridden for most of the time by the necessity of finding sufficient unfenestrated lamina. Over the whole period, about 50% of the leaf area presented was taken by the animals (see Table II and Fig. 4); the final burden and opportunity of choosing was left to the animals.

Estimating the quantity consumed

As a plate was filled with leaf squares—usually when about 10 squares had been introduced—or in the event of a change in stadium, the occupants were transferred to a new plate. The area of each numbered square which had been eaten on the old plate was estimated by eye and recorded; initially as a percentage and then converted into a vulgar fraction. Following area estimation, the remains of the squares were floated off the agar surface on tap water, brushed clear of faecal pellets and oven dried at 108°C.

Each item from a batch was given the mean weight of that part of the batch which had been dried. The *area estimate* of food eaten for each plate was then obtained as the sum of the products of each batch mean (mg/sq.) and the area of each square eaten, expressed as a vulgar fraction. The *residue estimate* of food eaten for each plate was computed

TABLE I

Weights of leaf squares (mg oven-dry)

Period presented	Individual squares N	Individual squares x	2SD	$2\cdot23SE_{10}$	Batches of squares N	Batches of squares $\bar{x}$	Range
1970							
19/6–31/7	187	15·41	6·68	2·37	86 (19/6–23/10)	15·10	9–22
5/8–18/8	106	15·60	6·78	2·42			
24/8–23/10	143	14·79	6·50	2·33			
*6/11–8/1	125	14·90	6·88	2·46	25	14·22	9–19
1971							
21/1–31/3	185	14·52	7·48	2·71	31	14·40	10–22
1/4–11/6	150	16·29	7·06	2·51	30	16·23	10–26
16/6–2/8	124	14·69	5·18	1·85	26	14·96	9–19
5/8–30/11	119	15·37	5·80	2·07	26	15·11	11–20

* From Ernocroft Wood, Cheshire, England; the rest all from Kerridge Wood, Cheshire, England.

as the difference between the totalled batch means for each square and the dry weight of the residue.

It will be appreciated that neither *area* nor *residue* estimate is an absolute measure of consumption since the squares presented are merely replicates of those which were weighed. If the replication is good, then the *area estimates* would be more accurate than the *residue estimates* since there will be an "invisible" loss in weight due to the activity of micro-organisms which does not result in fenestration and this would give an artificially high residue estimate. However, the accuracy of the area estimate depends on a visual determination of what should ideally be a homogeneous plane surface. It is therefore necessary to establish the limits of accuracy:

1. with respect to replication, and
2. with respect to the relation between area and residue estimates.

Replication

Table I gives the means and 95% confidence limits for the weights of *all* the individual squares together with the means and actual range of the batches. The actual range of the batch means exceeds the 95% range for the individual squares which shows that most of the variance is between leaves and not within a leaf; i.e. the batches are fairly homogeneous. However, the replication is closer than these figures themselves would suggest since it depends on the random accumulation of eight to 12 squares on a plate, each square coming from a different batch. The average plate eventually contained 10 squares randomly chosen from the population and therefore the mean weight of these 10 squares is predicted by 2·23 (t) times the standard error ($SD/\sqrt{10}$). Thus, although 95% of the weights of individual squares lay in the range of the mean ± 5·0–7·5 mg, 95% of random selections of 10 squares will have a mean value within the range of the true mean ± 2·0–2·7 mg, i.e. from 11·81–18·80 mg. In fact, 70 plate means from 23.III.71–29.XI.71 were within the range 13·5–17·5 mg. The mean weights which form the base of both area and residue estimates can, therefore, be taken as having an accuracy of ± 2–2·5 mg.

The weights of the individual squares are slightly asymmetrically distributed, the mode being 1 or 2 mg below the mean. This is probably due to the distribution of veins.

Comparison of area and residue estimates

The area and residue estimates for each plate are compared in Fig. 2 and Table II. It will be seen that the area estimates for stadia III–VII are the higher but for stadia VIII–XI it is the residue estimates which are higher. This seems to be partly due to the fact that the proportion of veins taken by the millipede increases in the later stadia. A stadium III or IV eats none of the veins or veinlets, leaving a finely cleared entire skeleton. Stadia IX, X and XI leave only the main veins (actually the veins of the second order since the main veins are excluded at the outset).

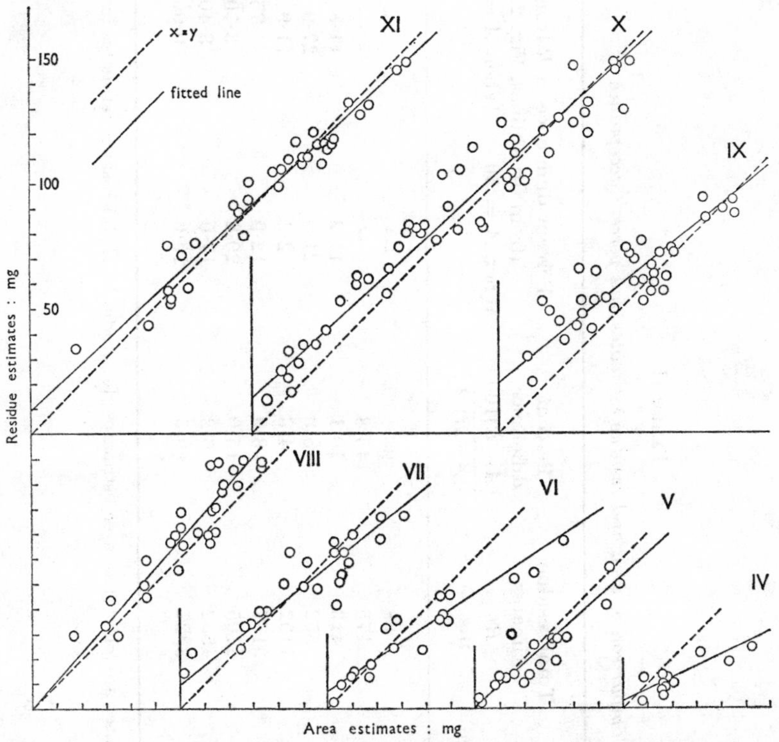

Fig. 2. Area and residue estimates compared. The area and residue estimates of food eaten in mg dry weight of sycamore leaf on each completed plate are plotted for each stadium. The fitted line is the arithmetic mean of the two regression lines, y on x and x on y.

It must be stressed that the tests of closeness of the estimates are all retrospective. Nevertheless, there is an opportunity to allow for an overestimate in one stadium in the next, albeit subconsciously; if this has happened, there has clearly been slight over-compensation from the early to the later stadia. The errors incurred are given in Table II. It will be noted in Fig. 2 that several points for stadium IX

TABLE II

Comparison of area and residue estimates and percent acceptance

Stadium	Total area estimates A mg	Total residue estimates R mg	Percent difference $\frac{(A-R)100}{(A+R)/2}$	Percent difference of R from A (from fitted lines, Fig. 2.)		Acceptance: percent eaten of leaf-squares offered
				When $A = 50$	When $A = 100$	
III	255	88	98	—	—	—
IV	279	175	45·8	—	—	—
V	457	413	10·1	11·2	11·8	62·4
VI	667	552	18·9	21·0	27·9	67·1
VII	1248	1192	4·6	2·2	11·9	55·5
VIII	2924	3197	8·9	12·9	4·75	54·4
IX	4051	4705	15·0	20·2	3·70	37·0
X	4278	4552	6·2	17·0	3·40	57·5
XI	3311	3327	0·5	10·0	0·00	72·4

For stadia III–VII residue estimates are *lower* than area estimates; for stadia VIII–XI residue estimates are *higher*.

show that the area estimate was unusually low. In this stadium (see Fig. 4) there was an unusually long period of inactivity before the moult. All the aberrant points referred to apply to the last plates in the series where little food was taken thus allowing maximum opportunity for reduction in weight by micro-organisms.

An attempt was made to measure the extent of this "invisible" loss by keeping control squares on plates without millipedes. Variable, though small, losses were recorded, but the final test involved accumulating squares on control plates in the same manner and at the same rate as in the experimental plates. Five such plates gathered from 16th April–28th May 1971, weighed on the 6th June, showed a mean loss of 6·6% (4·6–9·6%). However, it may be supposed that the organisms responsible for the "invisible" loss will be those rendering the leaf acceptable as food for the millipedes. We can then expect that millipedes will rapidly consume those squares with a potentially high rate of loss due to the activity of micro-organisms; only those squares *not* likely to lose weight by microbial attack will remain. Thus a falsely high residue estimate is unlikely when the plates carry actively feeding animals. Supporting this suggestion is the reasonable correspondence of area and residue estimates, especially for later stadia where consumption is much more rapid.

Close correspondence of area and residue estimates means that further quantitative determination might be made with area estimates alone; since much of the time and labour of servicing plates is associated with the determination of the residue estimates, their elimination would allow three times the number of plates to be serviced.

RESULTS

The achievement of maturity

The first result of note is the demonstration that *Ophyiulus pilosus* can be reared from egg to maturity on dead sycamore leaves alone. Unfortunately the adult females did not proceed to lay eggs although they were successfully mated with both reared males and field males. The whole process of mating was followed, the partners approached head on, each lifting the first part of the ventral surface towards the other as Haacker (1969) describes for *Iulus scandinavius*. Insertion of the gonopods was observed. On dissection of each female on death, only half-sized eggs were found.

Another indication that rearing conditions were not quite adequate was the number of gynandromorph "males" produced (see Bigler,

1920). Two of the nine stadium IX males from Nest 1 and the two stadium XI males from Nest 2 had vulvae in addition to gonopods.

The weights of the stadia reared accord well with those determined from the field. Having established that morphological maturation is possible on dead leaf material alone, the next point of interest concerns the time and/or the stadium at which maturity is reached. Blower & Miller (1974) have shown that males can mature in either stadium IX, X or XI and females in stadium X or XI. Seven out of nine males from Nest 1 matured in stadium IX, one in stadium X and one in stadium XI, whereas all the females matured in stadium XI. From Nest 2, three males matured in stadium X and two in stadium XI. The only female to mature was in stadium XI.

Moulting

Immediately before the moult the earlier stadia, from IV to VI, spent an average of five days (2–8) inactive, either in a specially prepared cell hollowed out in the agar, or on the surface without any apparent preparation. Stadium VII was inactive for an average of nine or ten days (2–17) before the moult. About half constructed cells, the rest moulted on the surface; a shorter period was spent inactive on the surface than when preparing and occupying a cell. Stadia VIII and IX invariably made an agar cell in which to moult and occupied this for an average of 19–20 days (11–28). After the seventh or eighth day the developing vulvae or gonopods were extruded in preparation for the moult which ensued after another seven days. A third period of about five days succeeded the moult before the cell was vacated. During this period the cast was eaten. The females moulting from stadium X to XI all remained on the agar surface and remained inactive for an average of 10 days (4–16).

In these later stadia feeding obviously ceased some seven to 30 days before excavating a cell, or resting motionless on the surface. However, the rate of food uptake falls much earlier than this—90% of the food being taken in the first third of the instar (see Fig. 4).

Of the two males from Nest 1 which did not mature in stadium IX one did not mature until stadium XI. This animal stopped feeding as a stadium IX immature male 21 days before entering its cell. It occupied this for 63 days from 7th January to 5th March 1971—57 days before moulting and a further six before finally relinquishing it. Thus the animal was inactive for 12 of the 19 weeks spent in stadium IX.

As a stadium X immature male it again entered a cell on 24th June and was forcibly extracted from it on 30th July when concern for its

TABLE III

Durations of stadia (in days)

		Nest 1 Mean days and range	*	*	N	Accumulated days	*	*	Nest 2 Mean days and range	N	Accumulated days
Egg		13				13			13		13
Pupoid		4				17			} 6		
I		2				19					19
II		5				24			3		22
III		19				43			18– 20·5– 24	22	43
IV		22				65			21– 24·5– 28	21	67
V		20– 28·5– 35			21	94			24– 25 – 25	15	82
VI	♂	14– 22 – 28			8	116			21– 22 – 24	7	104
VII	♂	21– 25 – 29	*66*	*82*	7	141	*160*	*176*	44– 69 – 87	5	173
VIII	♂	35– 45 – 53	*45*	*231*	7	186	*205*	*407*	123–157 –200	5	330
IX	♂	126–198 –241	*134*	*118*	6		*339*	*525*	98–160 –220	5	490
X	♂		*195*	*139*			*534*		83–105 –143	5	(595)
XI	♂		*36*						133	1	
VI	♀	15– 24 – 29			12	118			25– 25 – 26	3	107
VII	♀	40– 62 – 76			11	180			35– 42 – 54	3	149
VIII	♀	104–123 –138			10	303			102–170 –232	3	319
IX	♀	174–227 –264			9	530			71	1	390
X	♀	50– 79 –174			8	609			80	1	470
XI	♀	132–164 –235			9				176	1	

* Each column of italicized figures for individual males maturing at XI and X.

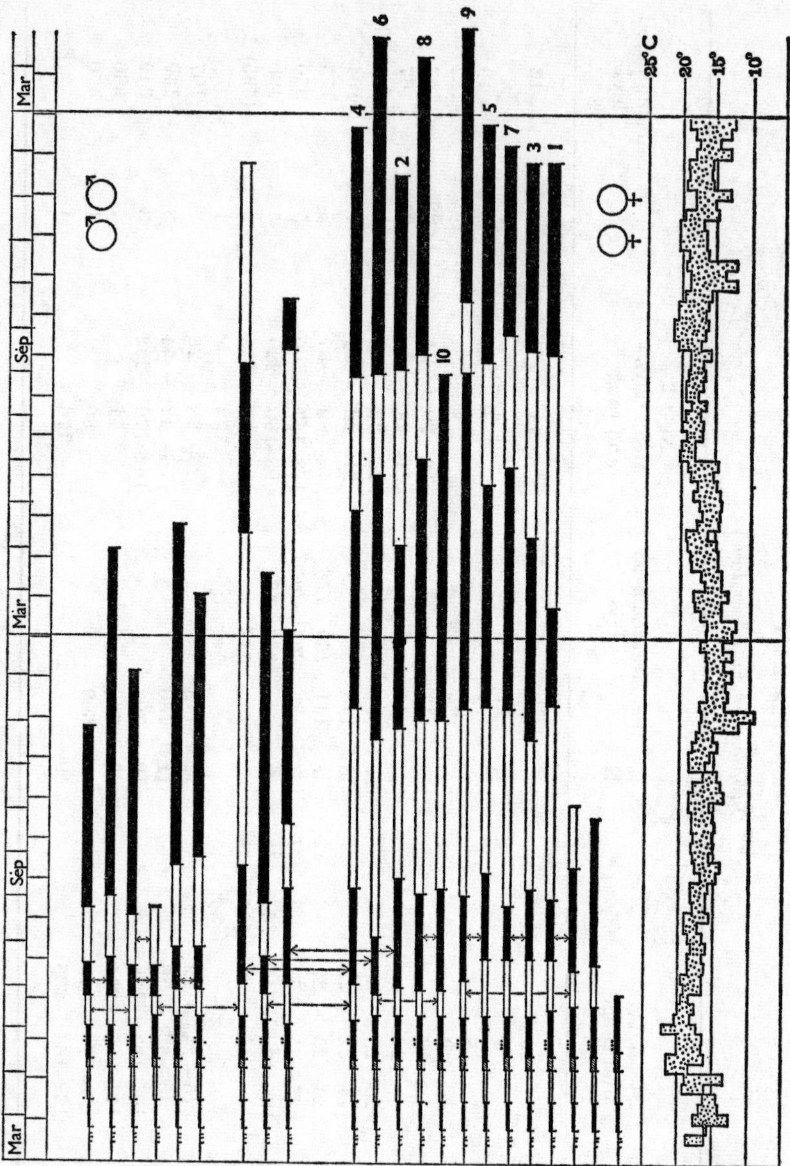

Fig. 3. The history of the individuals of Nest 1 and the temperature at which animals were reared. The nine surviving males occupy the upper part of the figure. Below are the 10 females numbered as in Fig. 4. The upper time scale, along the top, is in calendar months from March, 1970 to February, 1972 inclusive. The lower scale is marked in lunar months. Each line shows the duration of egg, pupoid, first, second and subsequent

[*Continued on opposite page*

well-being was growing—but it immediately resumed normal feeding. One of the stadium IX immature males from Nest 2, which ultimately matured at stadium X, stopped feeding after the first 52 days in the stadium and entered a cell three days later. It occupied this cell for 85 days until extracted from it. It then behaved normally and moulted into a stadium X mature male after a further 48 days; it had been inactive for 88 days in the middle of the stadium which occupied 188 days in all. Each of the two males just referred to became perfectly normal mature males. The occupation of cells for such long periods is considered in relation to the different development times of males and females in the next section.

Duration of stadia

Table III lists the average and range of the times spent in each stadium. Figure 3 shows the times for Nest 1 and also gives the prevailing temperatures. Individuals from both tests reached stadium VII in three and a half to four months (116 and 118 days for males and females of Nest 1; 104 and 107 for Nest 2). There appears to be no significant difference between the times taken by the two sexes. In the field, Blower & Miller (1974) find that a rapidly developing population in South Wales reaches stadium VII in six months and winter is spent in this stadium. Thus the rearing temperatures have accelerated development about one and half times. In the laboratory populations "winter" temperatures came later, whilst most animals were in stadium VIII or IX (Fig. 3), and it is noticeable that a greater time was spent in these stadia than in all the other sub-adult stadia.

Maturity was achieved by most of the males of Nest 1 in stadium IX in just over six months (186 days). Thus it is just conceivable that males in the field could achieve maturity in the same year as their birth. Two of the nine males of Nest 1 and all those of Nest 2 deferred maturity until stadium X or XI, taking 16–18 months to reach it. All the females from Nest 1 matured at stadium XI, in 20 months. The single surviving female from Nest 2 matured at stadium XI in 16 months. Periods of 18 and 13 months were taken by females of Nest 1 and Nest 2 respectively, to reach stadium X, which is the stadium in which maturity

---

stadia from oviposition to death. Odd-numbered stadia from III are black. The cross-hatched portions of stadium V indicate the period they spent before being separated into groups. The three groups of stadium V are indicated by one, two or three dots; the groups of 3, 3, 4, 4, and 6 stadium VI, the 10 pairs of stadium VII, and the single pair of stadium VIII are indicated by connecting dotted lines (compare with Fig. 1). The weekly maximum and minimum temperatures (summarized from daily readings) are plotted along the bottom of the figure.

is more usually achieved. In the Welsh population previously referred to, maturity of most individuals of both sexes is achieved in stadium X, in the autumn of the year following their birth (i.e. in 18 months). Thus the laboratory populations have taken as long as the field population notwithstanding a much more rapid development up to stadium VII. There appears to have been some obligatory lowering of activity in the laboratory, possibly caused by the drop in temperature which occurred when the animals were in stadia VIII and IX (see Fig. 3) This certainly appeared to be the case with some of the males which spent long periods in moulting chambers.

It is remarkable that some males matured at IX in just over six months and yet a further period of 13–14 months was necessary for maturing at X or XI. It is also remarkable that these males, in stadium X and XI, weighing 21 and 27 mg respectively (Table V) should take about the same time to develop as the females in these same stadia which are twice to three times the weight of the males (45 and 77 mg respectively). It seems possible that the long periods spent by some of these males in moulting chambers might be adaptive inactivity to synchronize the maturation of the males with that of the females.

The pattern of food consumption and growth

Consumption of food at the beginning of a stadium is greater than in the middle and final part. Where the stadium is sufficiently advanced for the individual to need several plates of food we can obtain a quantitative estimate of this trend. Figure 4 illustrates the course of consumption and growth of each female in the last three stadia.

Most of the food is taken in the first third of the instar in stadia IX and X, but there is a more even uptake in stadium XI. At the time of the initial rapid rate of feeding, a large percentage of the food offered is taken (upper curves in IX and X). As the rate of feeding declines, the animal becomes more discerning and the percentage of the food offered which is taken declines. The behaviour of individual No. 1 in stadium X emphasizes this point. There is a higher rate of feeding later in the stadium than usual and this is accompanied by a sharp rise in the percentage taken (broken line of the upper set of curves).

However, the animals become increasingly discerning *before* the decline in the rate of feeding. This is evident in stadium IX. If the first two sets of readings are compared, it will be seen that there is very little fall-off in feeding rate between the first and second plates but the percentage taken drops from 50–85% to 20–40%. This behaviour will tend to upset any experimental determination of preference based

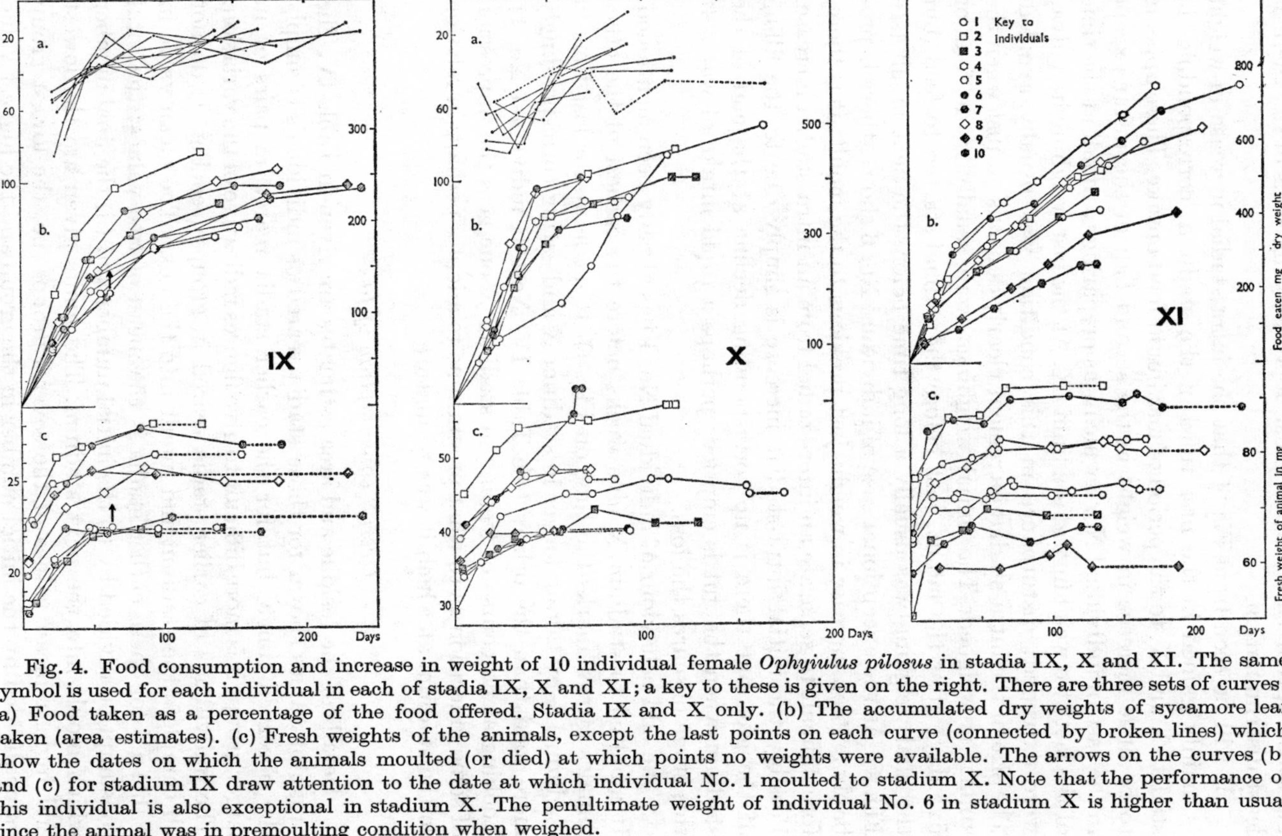

Fig. 4. Food consumption and increase in weight of 10 individual female *Ophyiulus pilosus* in stadia IX, X and XI. The same symbol is used for each individual in each of stadia IX, X and XI; a key to these is given on the right. There are three sets of curves: (a) Food taken as a percentage of the food offered. Stadia IX and X only. (b) The accumulated dry weights of sycamore leaf taken (area estimates). (c) Fresh weights of the animals, except the last points on each curve (connected by broken lines) which show the dates on which the animals moulted (or died) at which points no weights were available. The arrows on the curves (b) and (c) for stadium IX draw attention to the date at which individual No. 1 moulted to stadium X. Note that the performance of this individual is also exceptional in stadium X. The penultimate weight of individual No. 6 in stadium X is higher than usual since the animal was in premoulting condition when weighed.

on consumption since, when most food is being consumed, there is least discrimination.

It will be seen from Fig. 4 that the intra-stadial increase in weight is mainly gained in the first third of the stadium, corresponding, in stadia IX and X, to the period of maximum food uptake. There appears to be little increase in weight in the second half of the instar except in the case of individual No. 6 which was preparing to moult at the time of the last two weighings at stadium IX. At the last weighing its vulvae were extruded; disturbance in this condition was usually avoided. The size of the intra-stadial increment clearly depends on just when the animal is weighed. The initial weighings were usually made four or five days after the moult, just before the animal started to feed, but the final weighing was usually a long time before the next moult (see Fig. 4). As the exceptional case of individual No. 6 shows, there is probably a steep increase in weight just previous to the moult; it is therefore difficult to separate an inter-stadial from an intra-stadial increase. Possibly the initial intra-stadial increase is simply due to the filling of the gut, but there is no corresponding decline at the end of the stadium when the gut is emptied; perhaps a rapid intake of water at this stage obscures the loss.

The odd behaviour of individual No. 1 has already been mentioned. It moulted to stadium X nine weeks before the earliest of the other females (arrow marks the time on Fig. 4). It made up for lost time in stadium IX by taking longer in stadium X and eating correspondingly more food than the others (see Table IV, Nest 1, individual No. 1). Looking at its course of feeding in stadium X one gets the impression of some indecision as to what programme to adopt—perhaps in relation to the stadium at which it was to mature.

Overall consumption of food

Details of the residue and area estimates are given in Table IV. The range of values shown for those stadia reared as individuals is simply the observed range, but for the earlier stadia reared in pairs or in larger groups it is more difficult to give figures indicative of the variation. The individuals of earlier stadia reared in groups (see Fig. 1) do not all complete the stadium and moult into the next. The mean value in Table IV for these earlier stadia is the mean of two values: the total consumption divided by (i) the initial number and (ii) the final number which moult into the next stadium. The limits given are the lowest and highest mean values of the several groups, i.e. the lowest group value divided by the initial number in the group and the highest value divided by the final number in the group. Table II gives the percentage difference between total area and residue estimates for each stadium and the differences indicated by the fitted lines in Fig. 2. These graphs in Fig. 2 suggest that the difference between the two estimates is least at the higher values. In fact, the majority of the plates had values

TABLE IV

Food consumed in mg dry weight

Nest			
III	(2)	0·95– 1·13– 1·30	(83/68)
IV	(2)	2·7 – 3·3 – 3·6 *3·2 – 4·7 – 6·1*	(68/44)
V	(2)	7·9 –10·2 –12·5 *8·6 –11·1 –13·6*	(30/19)
VI	(1)	13·0 –15·9 –12·5 *17·0 –19·6 –22·5*	(11/10)
	(2)	13·7 –14·2 –18·7 *16·7 –19·4 –26·5*	(20)

♂♂	Nest 1 Maturing at IX	Nest 2 Maturing at X or XI	Nest 1 Maturing at X	 XI
VII	24– 25– 26(6) *25– 29– 33*	45 – 63(4) *45– 56– 68*	27 *38*	50 *57*
VIII	38– 59– 77(7) *45– 59– 82*	109–132–146(4) *106–112–114*	165 *125*	59 *64*
IX	117–220–316(6) *123–190–250*	84– 203(5) *70–132–168*	150 *137*	106 *101*
X		— — — *51–147–231*(5)	— *155*	214 *177*
XI		— *238* (1)		— *25*

TABLE IV (*continued*)

♀♀	Nest 1	Nest 1 Individual 1	Nest 2
VII	39– 53– 63(8)	53	
	41– 54– 60	*54*	*36*
VIII	120–157–200(10)	151	207
	125–147–178	*139*	*177*
IX	206–264–308(9)	128	150
	183–227–274	*122*	*119*
X	358–410–500(8)	537	286
	289–377–455	*498*	*273*
XI	*261–528–749*(8)	*409*	*238*

Lower limit, mean and upper limit are given for each stadium for each nest.
The numbers cultured follow each entry in brackets; only plates with single sex are used.
The upper means and limits are the residue estimates.
The lower figures, italicized, are the area estimates.
Missing residue means due to one or two plates for which the estimate was not available.

around 100 mg. The percentage difference between estimates is clearly less in the later stadia where more food is consumed.

The difference between the two estimates for stadium IX has been ascribed to the fact that the last plate of each series within this stadium carried individuals which had virtually stopped feeding. Higher residue estimates were therefore the result of the low quantities of food eaten and the "invisible" loss due to microbial activity. Thus the figures for stadium IX are exceptional and the reason for this is known. For the others, the differences between area and residue estimates for stadia VIII, X and XI are 9, 6 and 0·5% respectively. Taking all these considerations into account, the safest (and most conservative) set of estimates for the entire period consists of the residue estimates for stadia III–VII and the area estimates for stadia VIII to XI. These are summarized in Table V.

It has been mentioned that the pattern of consumption in the later stadia of males differs, for each stadium from VIII onwards, according to the stadium in which maturity is achieved. The food necessary for maturation at stadium IX appears to be minimal; there is good agreement between the accumulated food consumed by the male from Nest 1 maturing at stadium X and those from Nest 2 maturing in either X or XI—both taking 300 mg for stadia VII, VIII and IX (accumulated figures from Table IV using residue estimates for stadium VII). Since Blower & Miller (1974) assert that it is more usual for a population to have males in stadium X the figures for males in Nest 2 maturing at X or XI are used in the set of best estimates given in column 2 of Table V.

In Table IV one of the females from Nest 1 has been separated from the others. This is the individual No. 1 to which reference has already been made. It will be seen that its consumption in stadium IX was well below average but it compensated by taking more food in stadium X. Its overall consumption in stadia VII–XI totals 1222 mg as compared with the mean figures for the other eight females of 1333 mg. It is therefore possible that other individuals compensate between stadia in this manner. If the area estimates for each stadium are accumulated for each individual female from Nest 1 there is less variation between these values than between the sum of the extremes in Table IV (24% as against 31%).

The remaining columns of Table V relate the weight of food eaten to the weight of each stadium and to the increments in weight. In general, the stadia consume a dry weight of food equal to six to ten times their fresh weight (roughly equivalent to 18–30 times their dry weight). The exceptions are stadium III (for which the estimate is perhaps the most subject to error), stadia VII and VIII males and stadium VIII females; the higher consumption in these stadia recalls the observation of Kheirallah (1966) that the consumption of stadium VI of *Iulus scandinavius* was higher than in all the others. Perhaps the higher consumption of stadia VII and VIII is related to an ensuing period of temperature-imposed quiescence as Kheirallah suggested for *Iulus*. However, there is clearly no trend consistent with the idea that smaller individuals consume relatively more food. The relation between the weight of the consumer and the amount of food consumed appears to be simple, and not that suggested by van der Drift (1951) based on a surface derivative of weight.

A more interesting relation is that between the increment of growth and the food consumed. The cost of producing a given increment appears in Table V column 7 and the reciprocal, the gross growth efficiency, in columns 8 and 9. It will be appreciated that the dry weights of the individuals reared through are not available but from field animals we can take the dry weight to be roughly a third of the fresh

TABLE V

Food consumption in relation to animal weight

1 Stadium	2 Food eaten mg	3 Accum. food eaten mg	4 Fresh weight of animal mg	5 Food per fresh weight 2/4	6 Gain in fresh weight mg	7 Food per gain 2/6	8 Gross growth efficiency animal fresh weight per food dry 100/7	9 animal dry weight per food dry 8/3	
III	1·1	1·1	0·27	4·0	0·27	4·1	24	8	
IV	3·3	4·4	0·54	6·3	0·50	6·6	15	5	
V	10·2	14·6	1·04	9·8	1·96	5·2	19	6·4	
VI	15·0	29·6	3·00	5·0	1·90	7·9	13	4·3	♂ ♂
					2·90	5·2	19	6·3	♀ ♀
Males									
VII	56	86	4·9	11·4	2·1	26·7	3·7	1·2	
VIII	112	198	7·0	16·0	7·9	14·2	7·0	2·3	
IX	132	330	14·9	8·9	6·1	21·6	4·6	1·5	
X	147	477*	21·0	7·0	5·7	25·8	3·9	1·3	
XI	238	715	26·7*	8·9					
						17·87	5·64	1·88†	

				Females				
VII	54	84	5·9	9·2	5·4	10·0	10·0	3·3
VIII	147	231	11·3	13·0	13·1	11·2	8·9	3·0
IX	227	458	24·4	9·3	20·8	10·9	9·2	3·1
X	377	**835***	45·2	8·3	32·0	11·8	8·5	2·8
XI	528	1363	**77·2***	6·8				
						10·82	**9·28**	**3·09†**
				Males and females				
XI		1312 (eaten up to XI)	103·9			*12·6*	*7·94*	*2·65*

† The bold figures are derived from the accumulated food marked* and the final weights similarly marked. The last line shows the means of these two sets of bold figures.

weight. In this way the values in column 8 have been converted to those of column 9. The cost of producing stadia VII to XI is higher than that of the early stadia, but most noteworthy is the fact that the cost of growth in the males is nearly twice that in the females.

Blower & Miller (1974) use the stadium specific estimates of consumption of Table V together with their estimates of the stadium specific survival rates of a field population in Wales, to compute the total amount of litter consumed per year. Assuming the species takes sycamore alone, the population of one square metre is estimated to consume 34–67 g/year. Another method of estimating the population consumption is to use the estimates for production given by Blower & Miller (1974) in conjunction with the figures for the gross cost of growth from Table V here. From this Table we see that the accumulated food consumed to produce stadium XI males and females with fresh weights of 26·7 and 77·2 mg is 477 and 835 mg respectively. Dividing the sum of food consumed by the sum of the weights of the animals, we obtain a factor of 12·6. This is the cost, in units of dry weight of food, per unit of growth in animal fresh weight. (It is not possible to use the consumption figures for stadium XI since we have no measure of what is produced by this consumption.) Blower & Miller (1974) estimate a net annual production by the Welsh population, of 2·5–4·9 g fresh weight/sq. m. This should represent a food consumption of 12·6 times this figure of oven-dry litter, that is 32 to 62 g. This accords well with the estimate obtained directly from the life-table. The agreement depends on there being not too great a difference in the cost of producing the heavier later stadia.

These estimates of population food consumption are for leaf lamina alone. The larger veins and petioles which are excluded, account for nearly half the weight of a sycamore leaf. Thus it appears that some 60 –120 g of sycamore leaves could be reduced to woody-vein fragments, per square metre per year. Miller (1974, this Symposium, pp. 553–574) has estimated the leaf fall on this Welsh site to be 413 g/sq. m, of which 245 g are sycamore leaves.

DISCUSSION

All the earlier stadia were cultured in groups. It is possible that intraspecific competition for food might have reduced the consumption per individual. This possibility has been examined by Miller (1973). However, the suggested estimates are based on the consumption of successful individuals. The scale of consumption is, at all events, high, as van der Drift (1951) anticipated. At the site in Wales referred to, it seems possible that a quarter of the leaf fall is consumed by this one species. There were several other species common on the same site (Blower & Miller, in prep.). Still more striking is the realization that this large quantity of leaf litter is consumed in a very short part of the animal's life. In the first year, stadium III appears towards the end of May and has just 26 weeks before the temperature falls below its apparent threshold, at the end of October (Blower & Miller, 1974). Of these 26 weeks, two and a half are spent in moulting chambers, and of the remaining $23\frac{1}{2}$, only about a third are spent actively feeding. In the second year there is a longer period for development, 35 weeks, but more of these are spent in moulting chambers (about seven weeks). In all, eight to nine weeks of each year will be spent by the immature animals feeding. The adults, however, will be feeding for a longer period.

It is noteworthy that the reared females matured in the latest possible stadium. The observation of Kheirallah (1966) that reared females of *Iulus scandinavius* reached stadium XII (one further than ever observed in the field), is recalled. However, late maturation cannot be entirely a consequence of rearing conditions since most of the males matured in the earliest possible stadium. It is indeed intriguing to speculate about the possible environmental cues which might initiate moulting or stimulate the achievement of maturity by a given stadium. A glance at Fig. 3 reveals very few relations between temperature and culture conditions and the programme of development. There is a possible connection between the drop in temperature in December/January and the moults of the females from stadia VIII to IX which occurred within a very short time, but the same does not apply to the males or to the individuals from the other nest. The moults of females from stadia VIII to IX are the most suggestive of synchrony. It is just possible that pheromonal continuity between separately plated individuals might have been established each time new leaf squares were added, since the same brush was used to flatten the squares. However, there is very little evidence for this. The second feature illustrated by Fig. 3 is that both the males which elected to defer their maturity beyond stadium IX had been reared with females in stadium VIII. All but one of the males maturing at the earliest opportunity (at stadium IX) had been reared with the other males. However, the actual grouping depended on which animals moulted first and needed transferring. The two males proceeding beyond stadium IX evidently moulted into stadium VII later than the others and therefore were esconced with females. This might suggest that the stadium in which maturity will be achieved is already determined by stadium VII. It would be of great interest to investigate the mechanisms which lead to some of the males apparently "marking time" to ensure synchrony with the more slowly developing females.

Worthy of a final word are the implications of the rearing results for field sampling. Of the 35 weeks available for development in the field, up to three are spent in moulting chambers by stadia III–VII and seven by stadia VIII–X in the second period of growth. There is thus a probability of 0·2 of missing an animal in the extraction in the second year. The probability of securing males in the sample is perhaps less, since they either fit more moults into the year (to mature at IX) or spend more time in moulting chambers if they are to mature in a later stadium.

ACKNOWLEDGEMENTS

Dr P. D. Gabbutt and Mr P. F. Miller read through my typescript and made many suggestions which contributed greatly to the clarity of the final copy. Mr Miller tended the animals during my absence with the same care I gave to them myself and helped me in numerous other ways. To both these gentlemen I offer my most sincere thanks.

References

Bigler, W. (1920). Über einige Diplopoden aus Holstein und über einen Fall von Gynandromorphismus bei *Ophiiulus fallax*. *Festschr. Zschokke*. Nr 7: 1–14.

Blower, J. G. & Miller, P. F. (1974). The life-cycle and ecology of *Ophyiulus pilosus* (Newport) in Britain. *Symp. zool. Soc. Lond.* No. 32: 503–525.

Blower, J. G. & Miller, P. F. (in preparation). *The millipedes of two limestone woodlands.*

Brookes, C. H. (1963). *Some aspects of the life-histories and ecology of* ***Proteroiulus fuscus*** *(Am Stein) and* ***Isobates varicornis*** *(C. L. Koch) with information on other blaniulid millipedes.* Thesis: University of Manchester.

Drift, J. van der (1951). Analysis of the animal community in a beech forest floor. *Meded. Inst. toegep. biol. Onderz. Nat.* **9**: 1–168.

Haacker, U. (1969). An attractive secretion in the mating behaviour of a millipede. *Z. Tierpsychol.* **26**: 988–990.

Healey, V. (1963). *Studies on the ecology of the woodlouse* ***Trichoniscus pusillus pusillus*** *Brandt* 1833. Thesis: University of Manchester.

Kheirallah, A. M. (1966). *Studies on the feeding behaviour of the millipede* ***Iulus scandinavius*** *Latzel,* 1884. Thesis: University of Manchester.

Miller, P. F. (1973). *The consumption and utilisation of food by millipede populations.* Thesis: University of Manchester.

Miller, P. F. (1974) Competition between *Ophyiulus pilosus* (Newport) and *Iulus scandinavius* Latzel. *Symp. zool. Soc. Lond.* No. 32: 553–574.

Verhoeff, K. W. (1928). Diplopoda 1. *Bronn's Kl. Ordn. Tierreichs* **5** (2): 1–1072.

Discussion

Enghoff: When millipedes are reared on agar plates are fungi and bacteria not a problem?

Blower: No, they are no problem in practice. If a leaf-square is attacked by micro-organisms it will perhaps be especially palatable and will, therefore, be rapidly eaten by animals before the micro-organisms can develop an extensive growth.

Eason: Do the millipedes tend to eat the complete leaf-square or do they leave some behind?

Blower: I think they would eat the complete square, all except the larger veins, but before they get to this stage they have been provided with another; it has been policy to keep an excess of food on the plate.

Malcolm: You have talked about food preferences and rates of development on various types of leaf litter. Is there any other survival value for food preferences other than increased rate of development?

Blower: I did not want to imply that the reason for food preference was an increased rate of development, although that is perhaps one observable result. I do not know the selective value of food preferences; Sakwa has worked in this field; one supposes the proximal cause of food preference is some chemical or other.

Malcolm: I was wondering about chemical end-products, do the millipedes have defensive secretions?

Blower: They do, and they are often phenolic, but whether there is any connexion between these and the phenols in the dead leaves I do not know.

Symp. zool. Soc. Lond. (1974) No. 32, 553–574.

COMPETITION BETWEEN *OPHYIULUS PILOSUS* (NEWPORT) AND *IULUS SCANDINAVIUS* LATZEL

PETER F. MILLER

*Department of Zoology, University of Manchester, Manchester, England**

SYNOPSIS

The millipedes *O. pilosus* and *I. scandinavius* are both widely distributed throughout Britain and Europe. Both species inhabit the woodland floor and are litter feeders. Studies of a number of woods have revealed a degree of ecological isolation which is related to topography. *O. pilosus* is found most often in the deeper litter accumulations at the foot of slopes. *I. scandinavius* is found higher up in sparser litter where conditions are thought to be drier. Several sites have mixed populations and here competition seems probable.

INTRODUCTION

The two British iulines, *Ophyiulus pilosus* (Newport) (= *O. fallax* (Meinert) and *Iulus scandinavius* Latzel, are common animals of the woodland floor. Their life histories are described by Blower (1970) and Blower & Miller (1974) on two sites having pure populations of each species. These studies have confirmed Blower's view (1956) that they are true soil species, though often they will be found in the litter. Kheirallah (1966) and Barlow (1957) conclude that the main food of *I. scandinavius* is leaf material and this is almost certainly the case for *O. pilosus*. Recently Blower (1974) has successfully reared the latter species from stadium III to adulthood using only sycamore leaves.

Sometimes *O. pilosus* and *I. scandinavius* occur at the same place and here it is quite normal to find them feeding on the same leaf, thus they are potential competitors. The study of these mixed populations has been hindered in the past because only mature males could be distinguished with certainty. However a reliable method of stadial determination (Vachon, 1947) has enabled this problem to be resolved;

(a) because the segment number of a particular stadium is largely non-overlapping between the two species and,

(b) because *I. scandinavius* is approximately twice the size of *O. pilosus* of the equivalent stadium (see Blower & Miller, 1974).

First, I will summarise what is known of the geographical ranges of the two species and then examine their habitat preferences with

* Present address: *Biological and Chemical Research Institute, Rydalmere, New South Wales, Australia.*

particular reference to topography and litter cover. Finally, I will consider the detailed distribution of a mixed population within a small, fairly homogeneous site.

GEOGRAPHICAL DISTRIBUTION

Details of the European distribution of *O. pilosus* and *I. scandinavius* are presented in the monographs of Schubart (1934), Lang (1954) and Stojalowska (1961), and some general conclusions can be drawn from these. Their ranges overlap in central Europe between latitudes 44°N and 59°N, but *O. pilosus* extends slightly further south than *I. scandinavius* into Northern Italy and Northern Yugoslavia. *O. pilosus* is apparently absent from France, Belgium and the Netherlands though there seems to be no obvious explanation for this. Outside Europe *O. pilosus* is present in the Eastern United States (Chamberlin & Hoffman, 1958) and New Zealand (Dawson, 1958; Johns, 1962, 1967). Blower (1969) has suggested that this success of *O. pilosus* as a colonist is due to its shortened life history. It thus seems that *I. scandinavius* and *O. pilosus* are both found in regions characterized by temperate deciduous forest.

Blower (1972) lists the British species of millipedes according to the number of vice-counties from which they are recorded; *O. pilosus* and *I. scandinavius* are ranked the ninth and eleventh most common species of the 48 found.

At present their known distribution in the British Isles is merely a reflection of the areas which have been collected extensively (Miller, 1973); however, it seems certain that when coverage is more complete both species will be found to be widespread.

PREVIOUS HABITAT RECORDS

Although both species are recorded from open land and coastal areas their most common habitat is woodland, in the soil or litter. Rotting logs and the underside of stones provide occasional refuges—at least during the day, when these collections were made (Table I). Both inhabit a range of litter types (Table II). *O. pilosus* appears to occur less frequently, alone, in oak litter than the two species together or *I. scandinavius* alone.

The available myriapod records of sample size > 5 were scored for the presence or absence of *O. pilosus* and *I. scandinavius* and the results analysed to discover if there was any inter-specific association. Figures for the Lake District are given separately since these are the

TABLE I

Percentage occurrence of **O. pilosus** *and* **I. scandinavius** *in three woodland habitats*

	Litter and surface soil	Beneath stones	Rotting logs
I. scandinavius	23	14	8
O. pilosus	19	3	16
Both	25	17	3
Neither	33	66	73
Number of sites sampled	48	35	37

TABLE II

Occurrence of **O. pilosus** *and* **I. scandinavius** *in different litter types*

	Litter type			
	Beech	Oak	Sycamore	Mixed
O. pilosus	3	2	1	5
I. scandinavius	1	9	1	6
Both	3	4	4	4

result of a more extensive survey in which the sample size was larger (J. G. Blower, unpubl. data). The test of association used was that devised by Cole (1949). The results shown in Table III indicate that there is a slight positive association which probably reflects their similar habitat requirements. The limitations of this sort of test, where samples are

TABLE III

Occurrence of **O. pilosus** *and* **I. scandinavius** *in myriapod samples* ($n \geqq 5$) *from the British Isles and Lake District*

	British Isles	Lake District
Only *I. scandinavius* present	36	7
Only *O. pilosus* present	43	4
Both present	27	5
Neither present	172	20
Coefficient of Interspecific Association (Cole, 1949)	$+0{\cdot}236 \pm 0{\cdot}064$	$+0{\cdot}222 \pm 0{\cdot}136$

scored simply on a presence or absence basis, is that no allowance is made for the proportion of the two species when they occur together, nor for their age distribution. Both are important considerations when determining whether a particular habitat is the preferred one or merely an area into which animals have wandered during dispersal. It is also very important to know the size of the area sampled since patchy, mutually exclusive distributions can be masked if this is large.

To account for the absence of a species from a particular place we can invoke dispersal barriers or physical or biotic factors. Certain aspects of the British distribution may be explicable in terms of dispersal barriers (e.g. the absence of *I. scandinavius* from the Isle of Man) but these cannot adequately account for marked changes in the proportion of the two species occurring over small areas of woodland. This is the situation in the majority of the sites to be discussed and here physical or biotic factors, which may include competition, are most likely.

THE STUDY SITES AND SAMPLING METHODS

Much of the woodland remaining in Britain is situated on hillsides as these are often unsuitable for agricultural purposes. Where this was the case with the study areas two or more contrasting sites were usually chosen, a lower one and an upper one. Information on the position and topography of the study areas is provided in Fig. 1. A map of Kerridge A, which has been the subject of a more extensive survey, is provided in Fig. 2.

Three methods were used to sample the millipede fauna:—

1. Individuals were collected by hand from the litter and the interface between this and the organo-mineral layer.
2. Litter and soil, to a depth of approximately 6 cm, were collected and mechanically sieved to separate off the gross, organic debris and stones. The material passing through was then searched by hand.
3. 0·05 m^2 samples of soil, to a depth of approximately 6 cm, together with the overlying litter, were removed and the animals extracted using Tullgren funnels.

Each method has certain advantages. Hand sampling rarely yields animals of the early stadia since these are not easily seen. However, a greater area can be covered than is feasible if using Tullgren funnels, thus the adults, which are at a low density, can be obtained in larger numbers. The advantage of Tullgren funnel extraction is that, so far as we know for millipedes, it is not biased towards certain stadia. Thus a

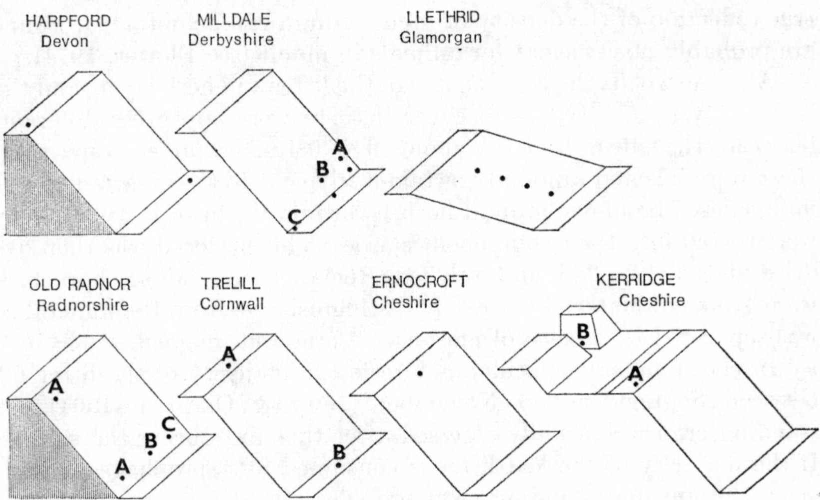

FIG. 1. Diagrammatic representation of the topography of the study areas together with the relative positions of the sites sampled.

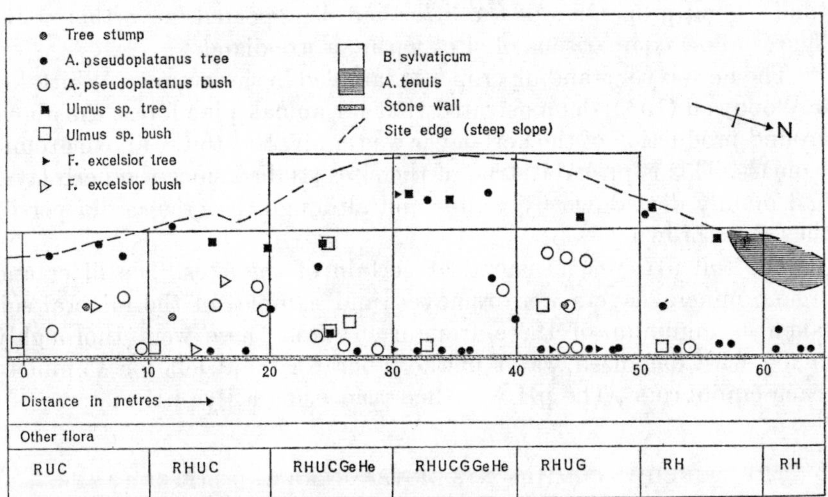

FIG. 2. Map of site at Kerridge A. Abbreviations: R, *Rubus fruticosus* L.; U, *Urtica dioica* L.; C, *Campanula latifolia* L.; H, *Heraclium sphondylium* L.; Ge, *Geranium robertianum* L.; G, *Geum urbanum* L.; and He, *Hedera helix* L. The background vegetation over most of the site is *Brachypodium sylvaticum* (Huds) Beauv., it is particularly common between 10 and 20 m. There is a small patch of *Agrostis tenuis* (Sibth) at the southern end.

true reflection of the density of each stadium can be obtained. Funnels are probably not efficient for animals in moult (see Blower, 1974).

At some woods the components of the litter and herb layers and their biomass were assessed to discover if there were any gross differences between the sites. Usually about five 0·1 m^2 samples were taken, these were selected randomly or taken adjacent to the area sampled for millipedes. The above-ground herb layer and the litter in each quadrat were sorted into their component species. This material was then oven dried at 105°C for 24 h and weighed. Remains too small to characterize were grouped. Material not easily distinguishable from the mineral soil was separated by a sieve of mesh size 4 mm and weighed.

In cool temperate deciduous forests the majority of the litter falls between September and November (Bray & Gorham, 1964). The standing crop was usually assessed after this and during the summer. If the majority of the leaf litter is consumed by saprophages within a year, the autumn standing crop provides an estimate of production. Where the litter is more persistent, litter traps, left out during leaf fall, will be superior. This latter method was used at Llethrid. The standing crop is not a good estimate of woody litter production because it persists for more than a year and the production of different years is difficult to separate. Some sites were only sampled in summer; by this time a good proportion of the litter had disappeared, nevertheless the figures allow comparisons of sites within a woodland.

The herb layer standing crop was sampled in the summer. Whittaker & Woodwell (1969) demonstrated that for an oak/pine forest the above ground production of the herb layer was equivalent to the above ground biomass. This is probably true of the sites studied since the herb layer had mainly died down by winter and although the grasses did persist they were brown.

The soil pH was measured at certain of the sites. The litter and organo-mineral layers were removed and samples of the mineral soil taken, a minimum of three from each site. These were thoroughly mixed with deionized water in the laboratory and left for 15 min to reach equilibrium. The pH was then read using a Pye pH meter.

THE PRINCIPAL COMPONENTS OF THE HERB AND LITTER LAYERS

Litter from trees and herbs is an important source of energy for saprophages inhabiting the forest floor; it can also affect the micro-climate since deep litter usually provides a damp microhabitat and also ensures that moist food will be available—millipedes will not usually eat dry litter. The nature of the tree canopy (open or closed) will modify the microclimate directly and also by influencing the herb layer and the thickness of the litter. The figures in this section refer to the autumn mean standing crop of leaf litter (except at Llethrid and Ernocroft) and the summer mean standing crop of the herb layer (Tables IV and V).

Kerridge

At Kerridge in October 1971 the mean weight of litter per m^2 was 328 g; it was a little deeper alongside the wall. Because the centre of the site is wider there was a larger area here with sparse litter than at either end. The main component was sycamore (*Acer pseudoplatanus* L.) with a little elm (*Ulmus* sp.) and ash (*Fraxinus excelsior* L.). The proportions of these were fairly constant along the length of the site (Fig. 4). They are all fairly palatable and by late August 1972 only sycamore petioles remained, apart from a few prematurely fallen 1972 leaves. In summer the herb layer (Fig. 2) was well developed (mean standing crop = 122 g/m^2) but this began to die down in September leaving the site largely bare in winter except for the leaf litter and some areas of grass. Although the floral diversity increased in the centre of the site the standing crop was similar.

Kerridge B is characterized by a very deep sycamore litter accumulation which precludes a herb layer.

Llethrid

The lowest part of this site (Llethrid lower) has been the subject of an extensive study (Blower & Miller, 1974). The litter input at Llethrid in autumn 1971 was 416 g/m^2, slightly higher than at Kerridge though there was a less well developed herb layer (72 g/m^2). No assessment was made of the litter input at the upper site in autumn but the June figures suggest that it is lower and that sycamore is absent (Table IV). Litter traps proved to be ideal for determining the autumn litter production at Llethrid because its main component, oak, tends to persist for more than one year. The June quadrat samples do include this persistent litter (Remainder column Table IV) and this explains the rather small reduction in total litter between October and June. Sycamore per m^2 had declined markedly by June but the oak litter remained similar. It may be that oak litter, since it takes longer than one year to break down, reaches a steady state. The herb layer at Llethrid upper and lower was similar, the main components being bramble and fern.

TABLE

The standing crop

Date	Location	No. of samples	Sycamore	Oak
Oct. 1971	Kerridge A			
	0·3 m from wall	7	32·00 (8·56–43·45)	0·11 (0–0·53)
	2·5 m from wall	7	26·67 (15·30–35·50)	0·09 (0–0·40)
June 1972	0·3 m from wall	7	8·01 (1·53–14·42)	0·07 (0–0·24)
	2·5 m from wall	7	1·63 (0·55–2·42)	0·08 (0–0·30)
Sept. 1972	0·3 m from wall	7	2·48 (0–3·54)	0
	2·5 m from wall	7	0·58 (0–0·81)	0
Oct. 1971	Llethrid lower	5	24·48 (13·62–46·42)	14·00 (6·08–27·52)
June 1972	Llethrid lower	5	4·93 (1·45–13·73)	14·74 (5·90–21·69)
June 1972	Llethrid upper	6	0	4·05 (2·93–5·19)
June 1972	Old Radnor lower A	5	0	15·85 (12·49–25·29)
June 1972	Old Radnor lower B	2	0	24·54
June 1972	Old Radnor lower C	5	0	9·41 (4·51–16·21)
	Ernocroft			
Autumn 1960	(Healey, 1963)		12·2	3·5
August 1972		6	3·98 (0–9·23)	0·19 (0–0·82)
Nov. 1972	Milldale A lower	3	12·30 (9·75–15·79)	0·44 (0·31–0·52)
Nov. 1972	Milldale A upper	3	16·00 (10·49–22·50)	0·10 (0–0·31)
Nov. 1972	Milldale B lower	3	49·09 (40·24–62·29)	0·03 (0–0·10)
Aug. 1972	Milldale B lower	3	4·76 (3·42–5·93)	0
Nov. 1972	Milldale B upper	3	19·54 (11·57–30·39)	1·02 (0·95–1·10)
Aug. 1972	Milldale B upper	3	0	0

The remainder column refers to the leaf litter which could not be assigned

* Component not assessed.

IV

of the litter

Mean weight of litter in g/0·1 m² (actual range is given in brackets)					
Elm	Ash	Woody tissue	Remainder	Herbs	Total leaf litter
3·43 (1·69–7·77)	1·14 (0·38–1·62)	*	0	0·01	36·69
2·95 (1·35–4·39)	0·79 (0·49–1·19)	*	0	0·04 (0–0·17)	28·84
0·05 (0–0·28)	0·01 (0–0·06)	*	0	0	8·14
0	0	*	0	0	1·71
0·14 (0–0·53)	0	10·00 (0·64–26·39)	0	0·08 (0–0·25)	2·70
0·04 (0–0·17)	0	6·27 (1·86–13·37)	0	0·43 (0·02–1·53)	1·05
2·81 (0–8·34)	0	*	*	0·32 (0–1·82)	41·61
0	0	11·12 (5·18–15·32)	18·17 (11·42–25·60)	1·03 (0·84–1·50)	38·87
0	0	17·36 (8·69–25·42)	7·73 (5·11–10·40)	1·33 (0·60–1·98)	13·11
0	0	15·98 (2·86–37·94)	7·31 (2·63–14·08)	0	23·16
17·69	0	9·52	15·44	0	57·67
0	0	8·79 (4·84–14·70)	1·01 (0–3·73)	0	10·42
0	2·5	*	*	*	18·2
0	0	5·37 (0·53–11·20)	0	0	4·17
21·29 (19·18–24·80)	0·83 (0·73–0·99)	18·31 (9·75–27·73)	0	0·47 (0–1·38)	34·86
7·96 (6·00–11·72)	0	13·32 (2·64–19·42)	0	2·04 (0–5·67)	24·06
0·51 (0·34–0·67)	0·57 (0·09–1·22)	29·40 (20·00–38·79)	0	2·58 (2·31–2·90)	50·20
0	0	16·23 (6·21–28·46)	5·00 (3·49–10·87)	0	9·76
1·11 (0·82–1·62)	2·40 (1·73–3·57)	15·92 (1·83–26·49)	0	0·36 (0–0·36)	24·05
0	0	15·39 (13·27–16·93)	0	0	0

to a species. If this was small it was added to the largest litter component at that site.

TABLE V

The midsummer standing crop of the herb layer

Location	No. of samples	Mean weight of above ground herb layer in g/0·1 m² (actual range in given in brackets)						
		Grasses	Nettles	Ferns	Bramble	Other dicotyledons	Mosses	Total
Kerridge A	14	7·69 (0–30·13)	0·48 (0–5·01)	0	1·44 (0–8·23)	2·56 (1·17–17·26)	—	12·17
Llethrid Lower	5	0	0	0·48 (0–1·87)	6·19 (1·86–8·63)	0·5	0	7·17
Llethrid Upper	5	0	0	0	5·60 (0–11·50)	1·28 (0–3·84)	1·44 (0–3·76)	8·32
Old Radnor Lower								
A	5	0	0	0	0	0·07 (0–0·33)	0	0·07
B	2	0	0	0	0	0	0	0
C	5	20·84 (8·10–36·14)	0	0	0	0·75 (0–2·39)	0	21·59
Ernocroft	5	34·11 (8·99–50·20)	0	13·68 (0–37·39)	1·50 (0–9·02)	2·40 (0–8·39)	0	51·69
Milldale B Lower	3	0	10·26 (8·58–11·20)	0	0	0	0	10·26
Upper	3	0	0	0	0	1·09 (0–3·27)	0·34 (0–1·02)	1·43

Ernocroft

The autumn litter production, mainly of sycamore, is low (182 g/m^2), however the standing crop of the herb layer, principally *Holcus lanatus* L., is the highest of all the sites studied (517 g/m^2). A detailed description of the site is given by Blower (1970).

Milldale

Milldale A and B have a thick autumn litter accumulation at the lower sites (349 and 502 g/m^2 respectively) but this is less pronounced at the upper sites (241 g/m^2 at both). The predominant species are sycamore and elm. The herb layer is not well developed at any of the sites but at Milldale B upper it is virtually absent leaving the ground bare in summer except for twigs. Milldale C has a deep layer of mixed deciduous litter.

Old Radnor

The lower sites, A and B, have a well developed litter layer of oak and mixed deciduous litter respectively. Less litter accumulates at site C and this allows an extensive herb layer to develop (cf. Ernocroft). At Radnor A upper the litter is thinner than at the equivalent lower site.

Harpford and Trelill

These places were not visited personally. At Harpford the predominant litter is oak and this is thicker at Harpford lower. The contrast between Trelill A and B is in the deeper litter accumulation at the latter site; at both it is predominantly oak.

Thus the autumn standing crop of the litter was low at Ernocroft but similar at the other sites. In all of the sites sampled in summer it was greatly reduced. At this time woody tissue was an important component. Though usually unpalatable it could provide a "food refuge" during late summer when leaf litter is sparse. The herb layer is probably not an important source of energy at the majority of the sites. However it is significant at Ernocroft and Old Radnor C owing mainly to a large standing crop of grasses.

TOPOGRAPHICAL DISTRIBUTION

The distribution of the two species is often related to the topography and its associated litter cover. Four sites had pure *I. scandinavius* populations; Harpford upper, Ernocroft, Radnor upper and Milldale B upper. These are all on, or at the top of, slopes and have a sparse litter cover, some of which is no doubt removed by wind (Fig. 3, Table VI). A few *O. pilosus* were found at Ernocroft but there were only four of these discovered in nine years of sampling (Blower, 1970). A recent sample from Ernocroft (October, 1970) has revealed that

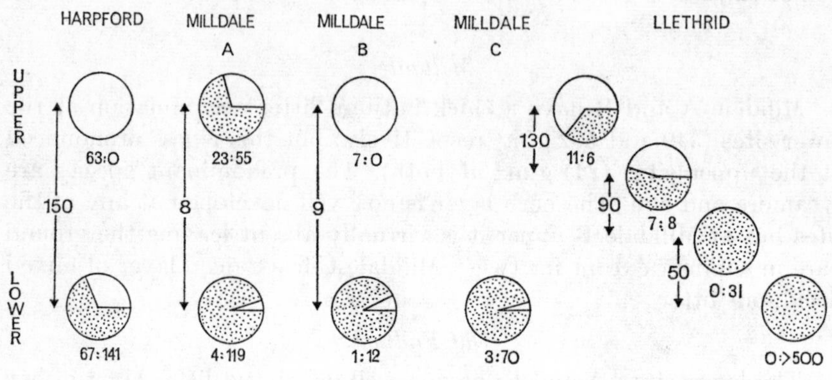

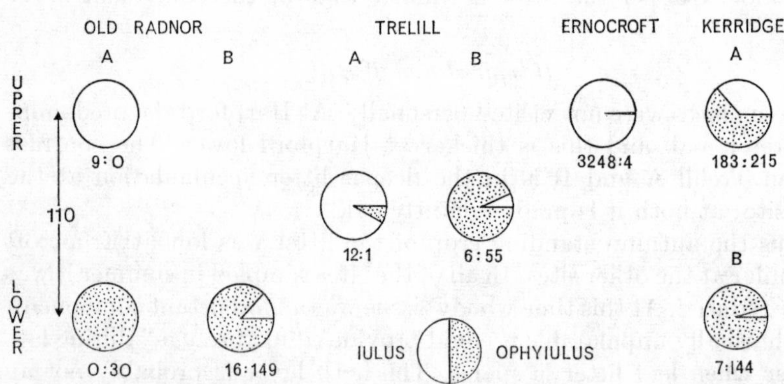

FIG. 3. Ratio of *O. pilosus* to *I. scandinavius* at the sites studied. The distance between upper and lower sites in the same wood is given in metres.

I. scandinavius is still the dominant millipede. The *I. scandinavius* density at Milldale B upper and Radnor upper is low. The former site is by no means inimical to all millipedes as *Cylindroiulus nitidus* (Verhoeff) was common, however all millipedes were scarce at the latter. The two sites with pure *O. pilosus* populations were Llethrid lower and Radnor A lower; both were at the base of slopes and had a thick litter cover.

TABLE VI

The age structure and breeding status of **O. pilosus** *and* **I. scandinavius**. *Details of the pH and canopy at the sites are also provided.*

Place	Date	Sampling method	*O. pilosus* Age 0–1 yr	*O. pilosus* Age 1+yr	*I. scandinavius* Age 0–1 yr	*I. scandinavius* Age 1+yr	Species breeding	Canopy	Ph	Number and size† of samples extracted by funnel
Harpford upper	June 1954 –May 1955	funnel	0	0	all stadia		*I*	medium	6·5	Blower & Gabbutt (1964)
Harpford lower	Feb. 1953 –May 1954	funnel	all stadia		all stadia		*O/I*	open	6·3	
Milldale A upper	April–July 1968	funnel	all stadia		15	8	*O/I*	—	6·9	J. G. Blower & P. F. Miller (unpub. data).
lower		funnel	all stadia		1	3	*O*		7·6	
Milldale B upper	Sept. 1971	funnel	0	0	7	0	*I**	—	5·6	10 × 0·05 m²
lower		funnel	10	2	0	1	*O*		6·3	10 × 0·05 m²
Milldale C lower	23.3.1964	hand	44	26		3	*O*	—	—	
Lethrid upper	26.6.72	funnel	143	12	10	7	*O/I*	medium	6·4	10 × 0·05 m²
	Sept. 1971	sieve	5	1	11	0				
middle	Sept. 1971	sieve	7	1	6	1	*O/I*	—	—	
	Sept. 1971	sieve	23	8	0	0	*O*			
lower	Mar. 1968 –Sept. 1971	funnel	all stadia		0	0	*O*	close	5·3	Blower & Miller (1974)
Radnor A upper	9.11.71	funnel/hand	0	0	3	6	*I**	medium	—	5 × 0·05 m²
lower	1971–1972	funnel	all stadia		0	0	*O*	medium		Brookes (unpubl. data)
B lower	9·11·71	hand	21	55	1	11	*O*	medium	—	2 × 0·05 m²
lower	11.5.72	funnel/sieve	34	39	1	3	*O*			
C lower		funnel	0	0	1	2	?	open	—	5 × 0·05 m²
Trelill A	Aug. 1961	hand	0	1	1	11	*I**	—	—	
B	April 1963	hand/funnel	39	16	4	2	*O*	—	—	
Ernocroft	1959–1965			4	all stadia		*I*	open	4·0–4·5	Blower (1970)
Kerridge A	1971	hand	39	194	14	124	*O/I*	medium	7·2	21 × 0·05 m²
A		funnel	23	18	41	4				
B	May 1971	hand	14	130	1	6	*O*	medium	5·2	

0–1 yr includes stadia up to VII.
1+yr includes stadia from VIII–XI
*The samples from these sites are small, nevertheless a tentative conclusion as to which species breeds there is put forward.
†The number of soil samples taken for extraction by Tullgren funnel is only given for those sites which have not been the subject of extensive surveys by other authors.

All the other sites studied in detail have mixed populations. However the presence of one or both species could be maintained by immigration from a population centre elsewhere. The absence of young stadia in an area, sampled at the appropriate time of year using a non-discriminating method, would probably indicate this. It is thought to be the situation at the Milldale lower sites and Radnor B, and perhaps at Trelill A, although the sample size is small here. Thus the millipede populations occupying these sites probably conform to expectations based on the topography, i.e. they are monospecific. However all the anomalous distributions on the sites studied cannot be resolved in this way. A wide stadial spectrum of each species suggests a true equilibrium association and this is apparently the situation at the remaining sites. At the one mixed site sampled for a number of years (Kerridge A) both species have persisted, suggesting that the association is stable.

THE SOIL pH

The soil pH could be important in determining the distribution of either species and it too could be related to the topography. Since the woods were often on slopes one might expect flushing to occur rendering the top acidic and the bottom basic. The survey of J. G. Blower (unpubl. data) gave some indication that *O. pilosus* was associated with basic areas and *I. scandinavius* with acidic ones, the former species occurring more often on limestone, the latter species occurring less commonly on limestone but sometimes in pine forests or on peat. When the pH of the sites studied in this paper are considered, however, the situation is by no means clear cut (Table VI). Whilst there is a tendency for *I. scandinavius* to occupy the more acidic sites its range (pH 4·0–6·8) overlaps markedly with that of *O. pilosus* (pH 5·2–7·6).

MIXED POPULATIONS AND SPATIAL SEPARATION WITHIN THEM

Coexistence has been noted at Kerridge A and Harpford lower, and probably Llethrid upper, Trelill B and Milldale B upper (Fig. 3, Table VI). Two of these sites (Llethrid and Kerridge) have been studied further and at the latter the distribution within the site has been investigated.

At Llethrid hand samples, taken in March 1970, at 20 m intervals up the slope from the lower site, showed that *I. scandinavius* was present at the top. The numbers of *O. pilosus* and *I. scandinavius* in these samples were 7 and 0, 4 and 0, 12 and 0, 4 and 0, 0 and 2 and 3 and 2. A later hand sample revealed a similar trend (Fig. 3). The presence of young stadia of the two species at the upper site suggested oviposition by both and a further sample was taken in June 1972 to confirm this (Table VI). A similar sample was taken from the lower site to enable a comparison to be made. Early stadia of *I. scandinavius* and *O. pilosus* were found at the upper site (Table VI) but *I. scandinavius* constituted a much lower proportion than would have been predicted from the hand sampling data. The hand samples were not taken from precisely the same place as the funnel samples and this may account for the discrepancy; or possibly *I. scandinavius* was in moult from stadium III to IV. Nevertheless oviposition does seem to occur. The other species of millipede found at the two sites were similar (Table VII) but the overall millipede density was lower at the upper site; this is perhaps attributable to the reduced litter layer (Table IV).

TABLE VII

The density of the millipedes at Llethrid upper and lower in June 1972

	Lower		Upper	
	Number per 0·05 m^2	$\frac{s^2}{m}$	Number per 0·05 m^2	$\frac{s^2}{m}$
Iulus scandinavius Latzel	0		1·70	1·18
Ophyiulus pilosus (Newport)	35·40	23·55	15·50	13·18
Microchordeuma scutellare (Ribaut)	13·20	3·09	6·50	5·03
Chordeuma proximum Ribaut	6·40	6·29	4·10	3·66
Brachydesmus superus Latzel	4·60	2·57	1·10	
Glomeris marginata (Villers)	1·40		2·30	5·03
Tachypodoiulus niger (Leach)	0·30		1·00	
Polydesmus spp.			0·20	
Polymicrodon polydesmoides (Leach)	0·10		0	
Cylindroiulus punctatus (Leach)	0		0·10	
TOTAL	61·40		32·50	

s/m^2 was not calculated if the density per 0·05 m^2 was less than 2·00.
s^2/m was high for *O. pilosus* because of the aggregated pattern of the newly hatched millipedes of stadium III.

Both *I. scandinavius* and *O. pilosus* have been collected from Kerridge A for a number of years. They are the two commonest species, though the total millipede density is small (Table VI) and

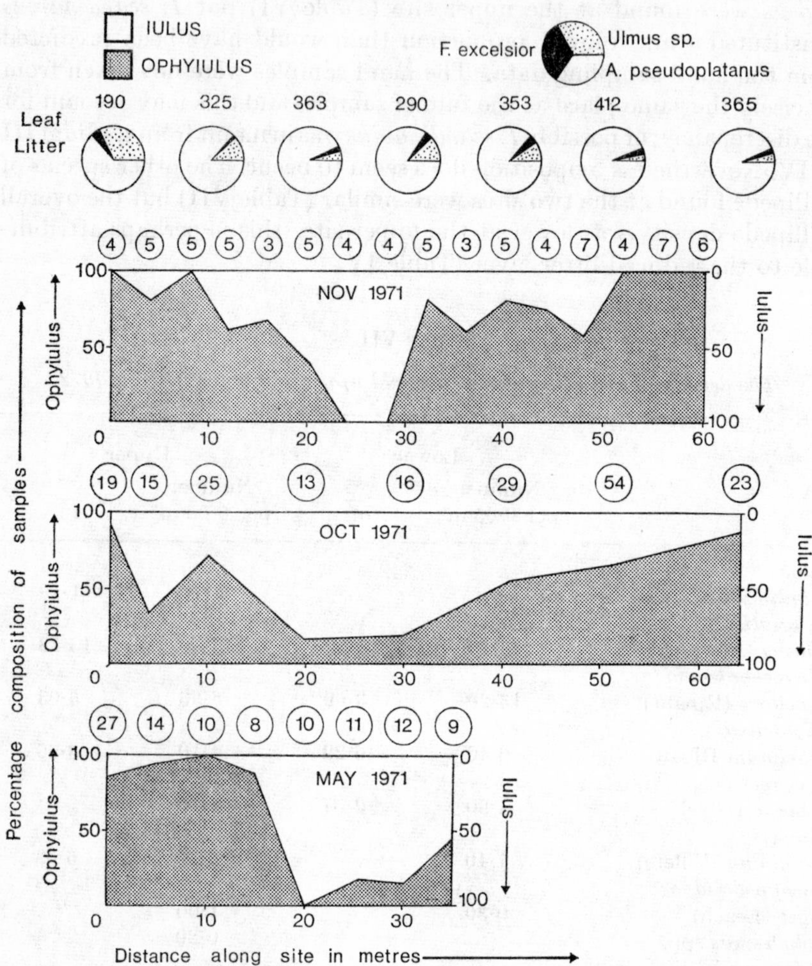

Fig. 4. The leaf litter and distribution of *O. pilosus* and *I. scandinavius* at Kerridge site A. The numbers in circles are the sample sizes. The weight of the litter in g/m² is given above each pie diagram.

contrasts markedly with that at Llethrid lower at the same time of year. The litter fall is not so different (Table IV) and it may be that some litter is removed from the exposed Kerridge A site by wind.

A transect made in May 1971 gave the first indication that *O. pilosus* and *I. scandinavius* were not distributed evenly over the site. *O. pilosus* was more common in the samples from the southern end of the site and *I. scandinavius* predominated in the middle (Fig. 4).

Further hand samples were collected in October and November 1971, and at the same time as the October samples two quadrats of litter were collected from each sampling position. These later transects, which include samples from between 40 and 60 m, indicate that *I. scandinavius* is common at the centre and *O. pilosus* at either end, though they do overlap. The herb layer is more diverse at the centre (but the standing crop is similar) the pH is lower and the deeper litter accumulation alongside the wall slightly less prominent because of the increased width of the site. These differences do not seem marked enough to account for the unusual distribution of animals.

Ground litter is very sparse in late summer (Table IV) and palatable litter may be scarce at other times of the year. It is possible that the distribution of *O. pilosus* and *I. scandinavius* is a response to this food shortage, and that competition emphasizes the microclimatic differences along the site, i.e. the niche is restricted by competition.

DISCUSSION

The main constituent of the food found on the forest floor is litter from trees; the herb layer contributes on average only 9% of the total biomass (Bray & Gorham, 1964). This seems to hold true for the majority of the sites studied although at Radnor C and Ernocroft the herb layer is more important. Numerous saprophages, among them millipedes, utilize portions of this energy and although some have become specialized (e.g. *Cylindroiulus punctatus* (Leach) feeds mainly on woody tissue) the majority probably consume leaf litter. Milne (1961) defines competition as "the endeavour of two animals to gain the same particular thing or to gain the measure each wants from the supply of a thing when that supply is not sufficient for both". The non-accumulation of leaf litter with time in deciduous woodland is circumstantial evidence that food may be limiting. This is very obviously the case at Kerridge A where the ground is virtually bare in midsummer despite an extensive autumn litter fall. Moreover the presence of leaf litter on the ground does not necessarily mean that it is available for consumption. Usually a period of "weathering" is required to render it palatable. Also neither *I. scandinavius* nor *O. pilosus* will eat dry litter.

Thus saprophages may experience an early check on their numbers due to the unpalatability of the apparently abundant leaf litter and a later check during August and September when the majority of the litter has been consumed.

It is, of course, possible that saprophages are limited by other factors, e.g. predators or climate, and that litter is destroyed by the microbiota, though Edwards & Heath (1963) suggest otherwise. They found that litter broke down extremely slowly if soil invertebrates were excluded. Indeed the implication of these experiments is that if the nutrients "locked up" in the litter are to become available rapidly the leaf-eating saprophages must be food limited!

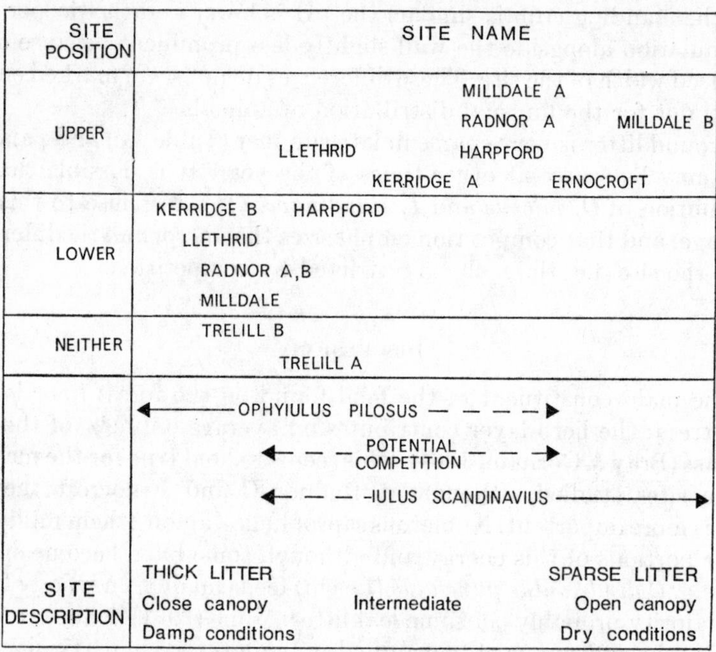

SITE POSITION
SITE NAME
UPPER
MILLDALE A
RADNOR A
MILLDALE B
LLETHRID
HARPFORD
KERRIDGE A
ERNOCROFT
LOWER
KERRIDGE B
HARPFORD
LLETHRID
RADNOR A,B
MILLDALE
NEITHER
TRELILL B
TRELILL A
OPHYIULUS PILOSUS
POTENTIAL COMPETITION
-IULUS SCANDINAVIUS
SITE DESCRIPTION
THICK LITTER
Close canopy
Damp conditions
Intermediate
SPARSE LITTER
Open canopy
Dry conditions

FIG. 5. Summary of the range of *O. pilosus* and *I. scandinavius* in woodland.

At some of the sites studied *I. scandinavius* and *O. pilosus* often occupy different areas of the same woodland and thus there is no need to invoke competition. *I. scandinavius* is found more commonly at the top of slopes where the litter cover is usually thinner and conditions probably drier. *O. pilosus* is often lower down in the deeper litter accumulations (Fig. 5). Laboratory culture of the animals suggests that *I. scandinavius* may be more susceptible to drowning in damper conditions (Miller, 1973). The vulnerable stage appears to be during moult when the intersegmental membrane is exposed. *I. scandinavius* may be excluded from deep, wet litter accumulations for this reason. O'Neill (1969) found that larger millipedes were more resistant to desiccation because of their smaller surface area in relation to volume and this may exclude the small *O. pilosus* from drier sites. A differing response to damp conditions could explain the distribution at Kerridge A, though here the microclimatic differences are probably more subtle. Ecological isolation, however, is not perfect for at some sites mixed populations flourish and frequently individuals of both species are found feeding on the same leaf. The ability of these species to coexist has been further investigated in the laboratory (Miller, 1973).

References

Barlow, C. A. (1957). A factorial analysis of distribution in three species of diplopods. *Tijdschr. Ent.* **100**: 349–426.

Blower, J. G. (1956). Some relations between millipedes and the soil. *Congrès Int. Sci. Sol* **6**: 169–176.

Blower, J. G. (1969). Age structures of millipede populations in relation to activity and dispersion. *Publs Syst. Ass.* No. **8**: 209–216.

Blower, J. G. (1970). The millipedes of a Cheshire wood. *J. Zool., Lond.* **160**: 455–496.

Blower, J. G. (1972). The distribution of British millipedes as known at the end of 1969. *Bull. Br. Myriapod Gr.* **1**: 9–38.

Blower, J. G. (1974). Food consumption and growth in a laboratory population of *Ophyiulus pilosus* (Newport). *Symp. zool. Soc. Lond.* No. 32: 527–551.

Blower, J. G. & Gabbutt, P. D. (1964). Studies on the millipedes of a Devon oak wood. *Proc. zool. Soc. Lond.* **143**: 143–176.

Blower, J. G. & Miller, P. F. (1974). The life cycle and ecology of *Ophyiulus pilosus* (Newport) in Britain. *Symp. zool. Soc. Lond.* No. 32: 503–525.

Bray, J. R. & Gorham, E. (1964). Litter production in the forests of the world. *Adv. ecol. Res.* **2**: 101–157.

Chamberlin, R. V. & Hoffman, R. L. (1958). Checklist of the millipedes of North America. *Bull. U.S. natn. Mus.* No. 212.

Cole, L. C. (1949). The measurement of interspecific association. *Ecology* **30**: 411–424.

Dawson, E. W. (1958). Exotic millipedes (Diplopoda) in New Zealand. *N.Z. Ent.* **2**(3): 1–5.

Edwards, C. A. & Heath, G. W. (1963). The role of soil animals in breakdown of leaf material. In *Soil organisms:* 76–84, Doeksen, J. and Drift, J. van der (eds.) Amsterdam: North Holland Publ. Co.

Healey, V. (1963). *Studies on the ecology of the woodlouse* ***Trichoniscus pusillus pusillus*** *Brandt* 1833. Ph.D. Thesis, Manchester University.

Johns, P. M. (1962). Introduction to the endemic and introduced millipedes of New Zealand. *N. Z. Ent.* **3**: 38–46.

Johns, P. M. (1967). A note on the introduced millipedes of New Zealand. *N. Z. Ent.* **3**: 60–62.

Kheirallah, A. M. (1966). *Studies on the feeding behaviour of the millipede* ***Iulus scandinavius*** *Latzel 1884*. Ph.D. Thesis. Manchester University.

Lang, J. (1954). Mnohožky-Diplopoda. *Fauna C S R* **2**: 1–183.

Miller, P. F. (1973). *The consumption and utilization of food by millipede populations.* Ph.D. thesis, Manchester University.

Milne, A. (1961). Definition of competition among animals. *Symp. exp. Biol.* No. 15: 40–61.

O'Neill, R. V. (1969). Adaptive responses to desiccation in the millipede *Narceus americanus* (Beauvois). *Am. Midl. Nat.* **81**: 578–583.

Schubart, O. (1934). Tausendfüssler oder Myriapoda I. Diplopoda. *Tierwelt Dtl.* **28**: 1–318.

Stojalowska, W. (1961). *Krocionogi (Diplopoda) Polski.* Warsaw: Polska Akad. Nauk. Inst. Zool.

Vachon, M. (1947). Contribution à l'étude du développement post-embryonnaire de *Pachybolus ligulatus* Voges. Les étapes de la croissance. *Annls Sci. nat.* (Zool.) (11) **9**: 109–121.

Whittaker, R. H. & Woodwell, G. M. (1969). Structure, production and diversity of oak-pine forest at Brookhaven, New York. *J. Ecol.* **57**: 155–174.

Discussion

Sakwa: Might the differences in food preferences between the two species depend not on the substances most preferred but on the range of preference? Perhaps *Iulus scandinavius* prefers a wider range of foods or will accept a wider range because it is less sensitive to polyphenols?

Miller: Yes, I have had the idea that *Iulus* will take food which *Ophyiulus* would find unpalatable; this was in my mind when I set up both species with a range of foods in a competitive situation. I thought that more unpalatable food might have been taken rather earlier than when the two species were reared apart. I have the results of this experiment but they need interpreting; they are to be published elsewhere.

Dohle: You suggested that available food was limiting but I am not quite convinced. If competition occurs in an area with similar food, A is present at one end, B at the other and both together in between, the density will be the same over the whole area. But the numbers you gave from your samples are so different, from nine to 1000.

Miller: But the sample sizes were different, the figures I gave were not densities, some are from hand samples, some from funnels; the object was merely to determine the ratio between the two species, to show that in some areas just one or the other occurs and in other areas, both are associated.

Dohle: Therefore, from the numbers you give, the idea of competition is not supported?

Miller: Agreed, but where the two species occur together on the same site, in good numbers, each with a full stadial spectrum, I think there is the possibility of competition. Where they occur separately there is evidently no possibility of competition.

DOHLE: The fact of co-existence does not necessarily indicate that there may be competition.

MILLER: I think that the observed distribution at Kerridge is most probably explained in terms of competition; I think that either species would occupy the whole site in the absence of the other.

LEWIS: I am not sure why you confine your attention to food, there may be so many other factors involved.

MILLER: I think that food is an important factor, it is the factor I am interested in. I agree there are other things such as availability of sites for oviposition but I have not been looking at these, my work is on food consumption.

BOCOCK: Have you looked at the temperature gradients over some of these sites?

MILLER: No—but I should point out that I am aware that there are many ecological reasons for the differential distributions of the two species, several of which I discuss at greater length in my paper. I was not suggesting all the details of distribution are explainable in terms of competition.

WALLWORK: Evidence is accumulating that in the case of many detritus feeding soil animals, the primary regulator of population size is the effect of climatic conditions and breeding performance, and that food is rarely a limiting factor. The fact that *I. scandinavius* and *O. pilosus* occur together would indicate that food is not limiting, i.e. they may both be selecting the same food, but may not be competing for it, because it is present in virtually unlimited quantities.

MILLER: But we would then expect food to accumulate, for example, on the mull sites. At Kerridge in August we find it difficult to get food for our animals. There is no leaf lamina on the ground, just petioles.

WALLWORK: They could possibly change their diet, say to fungi?

BLOWER: Kheirallah (1966)* established that algae and fungi contributed a very small part of the diet of *Iulus scandinavius*.

WALLWORK: At that time of the year?

BLOWER: At most times. I think Kheirallah recorded a very small percentage of algae and fungal spores in November or December, or perhaps a little earlier.

HAACKER: You did not say much about other diplopods in the areas, you mentioned *Cylindroiulus nitidus* but were there other litter feeding species?

* See list of references, page 571.

MILLER: Yes, there were. *Blaniulus guttulatus* occurred quite commonly at one site and there were other litter feeding species. I suppose they parcel out the food supply somehow.

FAIRHURST: One would expect different species to feed on different aspects of the environment; are the blaniulids, for instance, restricted to the mould rather than the litter?

BROOKES: *Blaniulus* is more often in the soil than the litter, but surely we are attempting the impossible—why do these animals live where they live. We have heard one aspect of the question, the effect food may have; I don't think we are justified in considering all other possibilities.

STANDEN: In my experience these deep accumulations of litter are unsuitable habitats for many animals, particularly the smaller stadia. I found that the small woodlouse *Trichoniscus pusillus* drowned readily in these deep accumulations. Have you any evidence that young stadia of *Iulus scandinavius* are more susceptible to drowning? Yesterday Mr Bocock mentioned that eggs and young stadia of *Glomeris* were easily waterlogged in culture.

MILLER: Yes, this is a possibility. Also I think that *Iulus* can survive better than *Ophyiulus* in drier habitats with sparse litter.

CAUSEY: Have you tried any kind of "seeding" experiments?

MILLER: No. We did consider this but were put off by the numbers which would be required to provide a reasonable field transfer—something like 10,000 for quite a small area. Possibly we could try with just adults.

Symp. zool. Soc. Lond. (1974) No. 32, 575–587.

THE ADAPTIVE SIGNIFICANCE OF VARIATIONS IN THE LIFE CYCLES OF SCHIZOPHYLLINE MILLIPEDES

COLIN FAIRHURST

*Department of Science, Stockport College of Technology, Stockport, England**

SYNOPSIS

Published work has accumulated on variations in size and number of stadia with geographical implications. Variation in the duration of the life cycle is now considered in *Tachypodoiulus niger*, with locality and year, within Britain. Similar differences are found in *Ommatoiulus sabulosus* over the range of the species. An explanation is advanced for the limits of the species in the light of a restricted season for the first season's development. Another point to emerge is that the onset of sexual dimorphism may occur at a certain time rather than at a certain stadium.

Within an area on Anglesey, the life cycles of *Ommatoiulus* have been followed in several contrasting habitats. The differences may amount to a delay of one year in the onset of maturity and an apparent reduction in the number of breeding males in the dense forest region.

The differences in schizophylline life-cycles are at present attributed to climatic influences.

INTRODUCTION

The Schizophyllinae is a sub-family of the Iulidae, the snake millipedes. There are only two British species, *Tachypodoiulus niger* (Leach) and *Ommatoiulus sabulosus* (L.). The latter has previously been known as *Schizophyllum sabulosum* but see Jeekel (1968). These species are two of the largest millipedes found in Britain and may reach a length of two inches.

Both species have a distinctive appearance. *Ommatoiulus* is brown with two yellow dorso-lateral stripes and *Tachypodoiulus* has a black body with white legs. The species have wide European and wide habitat distributions and have a remarkably vagile way of life, which sometimes leads to mass wanderings (Fairhurst, 1968).

Probably linked with these features is the phenomenom of periodomorphosis which makes possible a succession of male adult stadia (Blower & Fairhurst, 1968). The behavioural adaptations which are responsible for the wide ranging activity are now established (Fairhurst, 1970) and all the evidence points to the species being highly adapted for a dispersive role (Blower, 1969).

* Present address: *Department of Biology, The University, Salford, England.*

All these facets of the ecology of the Schizophyllinae have given rise to a great deal of published data throughout Europe so that it is now possible to consider the variations in the life cycle within differing geographical regions and in contrasting habitats.

THE LIFE CYCLE OF *TACHYPODOIULUS NIGER*

Nests are found in the field mainly in April, while animals kept in an unheated room lay eggs up to a month earlier. Pupoids and Ist stadia each take about a week to develop. Therefore, the active IInd stadium would be expected to appear in the field in early May. In fact, this stadium may be found until August. The pitfall captures at two sites in Cheshire and at sand-dunes in Lincolnshire indicate that the first winter is spent as the Vth stadium with occasional representatives of the IVth. The plot sampling carried out during the period 1959–1963 in a Cheshire wood (Blower, 1970) showed that in the 1959/1960 winter, IVth and Vth stadia were present. The following winter was passed in the Vth stadium. However, plot sampling in a Lancashire wood in this winter showed IIIrd and IVth stadia. In the cold winter of 1962/1963, only IVth stadium individuals were found at either site.

In the second season the IVth and Vth stadial larvae emerging from overwintering cavities soon moult to the VIth stadium which is sexually dimorphic. The IVth stadium is rarely found after April and the Vth is not common after May. The VIth stadium may last through to the autumn before moulting, but it can also moult into the VIIth in May. These early VIIth stadial animals moult in July/August to produce the first mature individuals. The VIIIth stadium males may copulate in the autumn activity peak but ripe eggs have not been found in the females. The function of copulation at this time is still in doubt.

In the third spring, the VIIth and VIIIth stadia moult to the VIIIth and IXth for the breeding season. As far as the males are concerned, only 2·5% of the VIIIth stadium have been found to be immature, and so the majority of these two-year-old males will be able to breed. As Sahli (1966) indicates, males may mature as late as the Xth stadium, so that some may mate for the first time three years from the egg stage.

With the females, the IXth stadium may contain eggs developed in the previous moult, but the VIIIth females will not hold eggs and therefore these animals may not breed for a further year. The ratio of VIIIth to IXth females appears to change from year to year, with either predominating.

In conclusion, we see an example of variation in life cycle from year to year within Britain and even within the same area. The likeliest causative agent would seem to be some climatic influence. The difference may lead to either a two or three year life cycle for the females although the males may not be so much affected.

THE LIFE CYCLE OF *OMMATOIULUS SABULOSUS*: LARVAE

In Anglesey, the author has been working on life cycles and other aspects of sand dune millipedes since 1964. The area under study, Newborough Forest and Nature Reserve, comprises a 3500 acre sandy region. Approximately half is a Forestry Commission plantation started in 1947.

In 1966, egg-laying appeared to commence at the end of May and continued until mid July. In the laboratory, adults were maintained in culture jars from 29th June, egg-laying carrying on until the 16th July. The rearings were carried out at 18°C in a dark room. Halkka (1958) also carried out clutch rearings and commented on the variation in rate of development with temperature. At 18–22°C, the eggs hatched in 7–18 days, while at 15°C only two clutches hatched after 17–19 days. At 27–29°C, the eggs hatch after eight days and at this temperature, the pupoid stage is passed through within the egg, so that the larvae were at the first stadium on hatching.

Table I shows the duration and occurrence of larvae raised in the laboratory and also those found in the field.

Halkka (1958) found that the pupoid stage lasted from five to seven days and the Ist stadium for three to six days. From her figures

TABLE I

Life history of larval ***Ommatoiulus sabulosus***

Stage	Duration	Laboratory occurrence	Field occurrence
Egg	17–20 days	1st July–16th July	
Pupoid	8–12 days	10th July–2nd Aug.	
I	12–16 days	23rd July–14th Aug.	
II	16–26 days	1st Aug.–20th Sept.	
III		27th Aug.–	10th Aug.–28th Sept.
IV			12th Sept.–14th Dec.
V			23rd Nov.–21st April
VI			7th April–July
VII			2nd June–

it would seem that the season during which the young can moult is comparatively short in Finland. This is shown in Fig. 1. In Finland, the larvae pass the first winter in the IIIrd stadium, while in this country and in Belgium (Breny & Biernaux, 1966), they first overwinter in the IVth and Vth stadia. Similarly, in Finland the second overwintering stage is the VIIth (with a few at VI and VIII), whilst in England and Belgium, it is the VIIIth with a few VIIth and IX stadial animals.

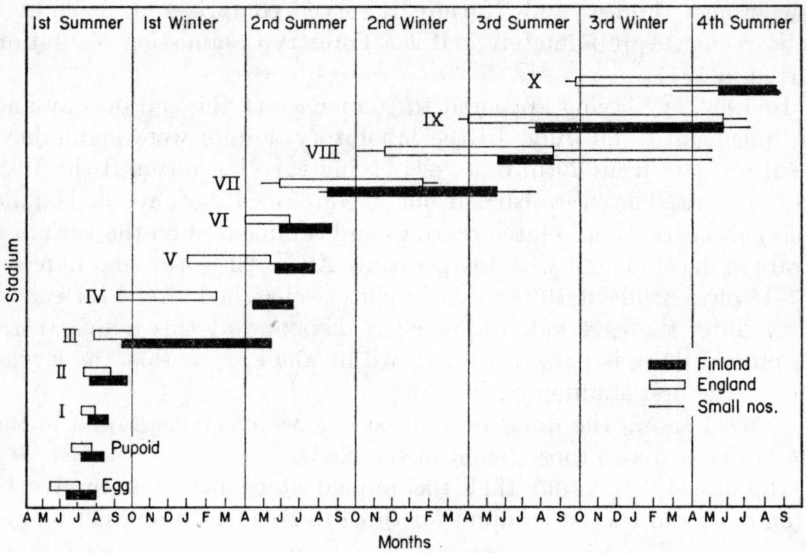

FIG. 1. Seasonal occurrence of stadia of a single generation of *Ommatoiulus sabulosus*.

Another interesting point emerges from the comparison. In Finland sexual dimorphism occurs at the IVth stadium whilst in this country it occurs at the VIth. In both cases, this represents the first moult of the spring, so that sexual dimorphism may occur at a set time rather than at a certain stadium.

When one looks at the climatic differences between the various sites from which stadial data is available, it appears that a mean temperature of 9°C may be necessary before full activity starts. As egg-laying usually occurs six weeks after this date, the time between egg-laying and overwintering will vary greatly with climate. Halkka mentions that it is highly probable that the IInd stadium is not capable of overwintering and the present author's results show that the

IIIrd stadium takes an average of nine weeks to develop from the new laid egg. Hence, from the "development time" calculated in Table II (and derived from Fig. 2), it seems likely that the northern limit of the species is governed by there being too short a period between egg-laying and the end of the active period for the IIIrd stadium to develop.

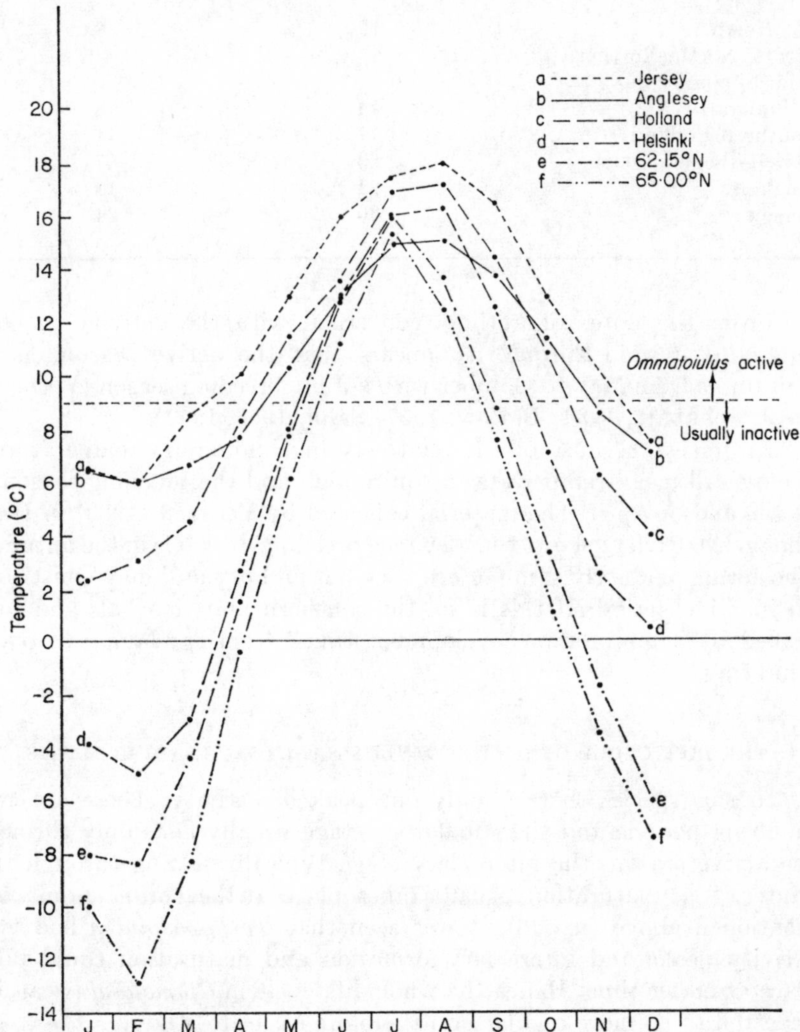

Fig. 2. Mean monthly temperatures in the different parts of Europe in which *Ommatoiulus* stadia have been characterized.

TABLE II

Activity times for the sites mentioned in Fig. 2

Location	Approximate number of weeks of activity	Number of weeks from egg laying to end of activity
65° North	12	6
62° 15′ North (Northern limit of species in Finland)	14	8
Southern Finland	17	11
Dutch–Belgian Coast	23	17
Anglesey	24	18
Jersey	30	24

From the notes of authors concerned with the activity of *Ommatoiulus* in mid-Europe, it appears that the active season, as in Britain, is from May to October with a June breeding season (Verhoeff, 1913; Schubart, 1934; Barlow, 1957; Sahli, 1955, 1957).

In Jersey, because of the relatively high minimum temperatures, the overall mean temperature is quite high and the breeding season is at the end of April. The material collected by Verhoeff (1933) in Italy and on the Riviera shows that the true breeding season is in the autumn. The spring and early summer are very hot and dry and may constitute an inactive period. If this is so, the southern limit may also be controlled by a minimum "development time" from egg-laying to overwintering.

THE LIFE CYCLE OF *OMMATOIULUS SABULOSUS*: ADULT MALES

In this species there is only one peak of activity. There are two moulting periods and the copulatory stage usually lasts only through the active season, the intercalary stage typically lasting through the winter. Also maturation usually takes place at the spring moult. As mentioned above (p. 576), it was seen that *Tachypodoiulus* had two activity peaks and intercalary formation and maturation could take place at either time. Hence the whole life cycle in *Ommatoiulus* can be expected to be more regular and therefore easier to study.

The previous work on the life cycles of *Ommatoiulus* has now been extended by sampling populations in different habitats. At the Anglesey area mentioned previously, "home sites" have been studied by various methods. A "home site" is defined as an area in which the immature stadia are fully represented (Fairhurst, 1970).

(a) Sand dune transect with pitfall jars.
(b) Litter at the edge of forest, sampled by hand and pitfall jars.
(c) Dune transect with young trees eight feet high sampled by pitfall jars.
(d) Pitfall transect in old forest.
(e) Plot sampling in old forest.

There is evidence to suggest that there are distinct behavioural differences between immatures and adults and between copulatory and intercalary males and between the sexes. There is not yet any reason to assume behavioural differences between copulatory male stadia and so the stadial numbers can be compared (see Table III).

TABLE III

Stadial percentages of copulatory males from June to August in the sites mentioned above

Site	VIII	IX	X	XI	XII	XIII	XIV	Total number
(a)	5·4	*76·3*	4·3	*10·8*	2·1	1·1	0	93
(b)	2·3	*75·0*	*11·4*	4·6	2·3	2·3	2·3	44
(c)	0	*30·4*	*30·4*	*26·1*	13·0	0	0	23
(d)	0	*20·0*	*24·0*	*44·0*	8·0	4·0	0	25

Italicized figures in each row comprise over 85% of the population and indicate the stadia in which the copulatory phase is most usually achieved.

In the transect area (a), home sites are very localized, and there is a marked alternation of peaks which signifies a strict periodomorphosis. Ninety-six percent of these males had moulted on emergence in late spring and we can expect the overwintering stadia to be mainly VIIIth immatures and Xth intercalaries. The stadial spectrum indicates that the males mature as early as possible, predominantly at the IXth stadium.

In the second area (b), the males are maturing for the first time at the IXth and Xth stadium and one can expect similar proportions of immature VIIIth and IXth stadia in the preceding winter. This is verified in material collected from this site in winter litter collections.

At site (c), there is a sheltered dune hollow with young trees and a ground cover of creeping willow. Here the stadial spectrum indicates that the animals are maturing equally at the IXth and Xth stadia and also apparently at the XIth. This trend is carried a step further in site (d) and it now becomes apparent that the majority of males are becoming mature in their third spring instead of in their second as in the warren area.

Two more examples can be cited to illustrate further the trends outlined above. Both the ratio of mature to immature IXth stadial males (Table IV), and the ratio of intercalary to copulatory males (Table V) indicate the reduction of breeding males in the afforested regions as compared to the dunes. The values for the forest plot sampling will be biased because this material represents the total population instead of the active population. As the copulatory males will be more active than immatures or intercalaries these latter animals will be better represented in the plot samples.

TABLE IV

Percentages of mature and immature IXth stadial males

Home site	June to August samples % immature	% mature	Total number
(a)	2·7	97·3	73
(b)	15·4	84·6	39
(c)	50·0	50·0	14
(d)	44·5	55·5	9
(e)	93·9	6·1	114

TABLE V

Proportions of intercalary and copulatory males

Home site	June to August samples % intercalary	% copulatory	Total number
(a)	0	100	94
(b)	0	100	43
(c)	8	92	25
(d)	16·7	82·3	30
(e)	67·3	32·7	25

The high percentage of intercalaries at home site (e) (Table V) must involve the moult sequence of intercalary to intercalary. This is the moult pattern which previous workers have found common when following the life history of the species in the laboratory (Halkka, 1958; Sahli, 1966), but which was not expected to be common in the field.

THE LIFE CYCLE OF *OMMATOIULUS SABULOSUS*: ADULT FEMALES

In the section on males, reasons were given for taking the adult stadium best represented as the one in which most animals mature first. The pitfall captures of females show similar trends to those of males from like habitat. In the dune area 38% of the females are in the Xth stadium which is probably formed in the third spring. In the forest transect, however, the stadial spectrum indicates that, like the males, the females mature mainly at the XIth stadium (39%). It is thus possible that in the forest, the life cycle is four years, whereas in the dune area it is three.

DISCUSSION

In *Tachypodoiulus*, the first winter may be spent as the IIIrd, IVth or Vth stadium. The difference is found in different years and in different habitats. This variation may lead to either the VIIth or VIIIth stadium being reached by the second winter. In both cases maturity can be reached in the following spring by the males but the females may have their maturity delayed a year in the case of the slower

TABLE VI

Stadia at which male maturity is first achieved

Species	Stadia: VII	VIII	IX	X	XI
Tachypodoiulus	rare	+ + + + +	+		
Ommatoiulus at:					
Sand dune		+	+ + + + +	rare	
Forest edge		rare	+ + + + +	+	
Young forest			+ + +	+ + +	rare
Older forest			+ +	+ +	+ +

moult sequence. The stadial onset of maturity does not appear to vary. In *Ommatoiulus*, by contrast, males take either two or three years to reach maturity, mainly as stadium IX or X, depending on the habitat (see Table VI); females mature in three or possibly four years.

Thus, we can trace the effects on the life-cycle of the different environments that the species inhabits as home sites. The warren may represent a relatively "hostile" environment in that it greatly restricts the home sites and provides little protection against desiccation. The rules governing seasonal occurrence of copulatory males and the onset of maturation may be more strictly enforced here. The forest with moist litter and fungus layers on loose sand may provide a more congenial habitat, with plentiful food and protection from climatic extremes. Early maturation here may not be so vital.

An alternative hypothesis can be erected along similar lines to the work on *Proteroiulus fuscus* (Brookes, 1963). The dune area, without the insulation layer of a tree canopy, may achieve a higher average summer temperature than the sheltered forest floor. The difference may lead to a more rapid larval development in the dunes.

A third hypothesis is that these variations are a response to population pressure. The species becomes very numerous in certain years and exhibits mass movements, especially in this area. These fluctuations in density are apparently generated within the forest although the movements are more apparent in the more open areas.

In conclusion, it appears that variations in climate may account for the variations in the life cycle of the schizophylline millipedes that occur from year to year, from place to place, or from one country to another. For *Ommatoiulus*, the possibility of additional connections between habitat variations in life cycle and population pressure is the subject of present work on the species.

Acknowledgements

It is a pleasure to acknowledge the invaluable advice and encouragement of Mr J. G. Blower throughout the course of this work. I am also indebted to Professor R. Dennell and Professor A. R. Gemmell for the provision of laboratory facilities at the Universities of Manchester and Keele, to Mr K. Bocock and Mr R. Arthur of the Nature Conservancy and Mr B. Griffiths and the Forestry Commission for access to property and material. Thanks are also due to my wife Dr J. M. Fairhurst for her help whilst carrying out a parallel study on the carabid beetles of the sites mentioned in the text.

References

Barlow, C. A. (1957). A factorial analysis of distribution in three species of diplopods. *Tijdschr. Ent.*, **100**: 349–426.

Blower, J. G. (1969). Age structures of millipede populations in relation to activity and dispersion. *Syst. Ass. Publs* No. 8: 209–216.

Blower, J. G. (1970). The millipedes of a Cheshire wood. *J. Zool., Lond.* **160**: 455–496.

Blower, J. G. & Fairhurst, C. P. (1968). Notes on the life-history and ecology of *Tachypodoiulus niger* (Diplopoda, Iulidae) in Britain. *J. Zool., Lond.* **156**: 257–271.

Breny, R. & Biernaux, J. (1966). Diplopodes Belges: position systématique et biotopes. *Bull. Annls Soc. r. ent. Belg.* **102**: 269–326.

Brookes, C. H. (1963). *Some aspects of the life histories and ecology of* ***Proteroiulus fuscus*** *(Am Stein) and* ***Isobates varicornis*** *(Koch) with information on other blaniulid millipedes.* Ph.D. thesis. University of Manchester.

Fairhurst, C. P. (1968). *Life cycles and activity patterns in schizophylline millipedes.* Ph.D. thesis. University of Manchester.

Fairhurst, C. P. (1970). Activity and wandering in *Tachypodoiulus niger* (Leach) and *Schizophyllum sabulosum* (L.). *Bull. Mus. natn. Hist. nat. Paris* (2) **41**: Suppl. No. 2: 61–66.

Halkka, R. (1958). Life history of *Schizophyllum sabulosum* (L.) (Diplopoda, Iulidae). *Annls zool. Soc. zool.-bot. fenn. Vanamo* **19**: 1–72.

Jeekel, C. A. W. (1968). The generic and sub-generic names of the European Iulid generally referred to as *Schizophyllum* (Verh. 1895) (Diplopoda, Iulidae). *Ent. Ber., Amst.* **28**: 49–51.

Sahli, F. (1955). Diplopodes de Sarre. *Annls Univ. sarav.* (Sci.) **4**: 357–366.

Sahli, F. (1957). Diplopodes de Sarre (deuxième contribution). *Annls Univ. sarav.* (Sci.) **6**: 280–283.

Sahli, F. (1966). *Contribution à l'étude de la périodomorphose et du système neuro-sécréteur des diplopodes Iulides.* Thèse Sc. Dijon.

Schubart, O. (1934). Tausendfüssler oder Myriapoda, 1. Diplopoda. *Tierwelt Dtl.* **28**: 1–318.

Verhoeff, K. W. (1913). Erscheinungszeiten und Erscheinungsweissen der reifen Tausendfüssler Mitteleuropas und zur Kenntnis der Gattungen *Orobainosoma* und *Oxydactylon. Verh. zool.-bot. Ges. Wien* **63**: 334–381.

Verhoeff, K. W. (1933). Wachstum und Lebensverlängerung bei blaniuliden und über die Periodomorphose. *Z. Morph. Ökol. Tiere* **27**: 732–748.

Discussion

MEIDELL: What do you think is the function of schalt males?

FAIRHURST: M. Mauriès has some unpublished work on spermatogenesis; apparently sperm formation proceeds in the intercalary stages *and* in the stadium before maturity. In the copulatory stages, if I understand M. Mauriès correctly, there is no spermatogenesis. I have shown that intercalary males are much less active than copulatory males; thus it seems that the intercalary males are resting stages in which sperm is built-up for the active copulatory stage.

MEIDELL: But that is what they *do* as schalt males: why do these animals, in contrast to many others, need schalt males? What is their survival value?

FAIRHURST: In a paper on *Tachypodoiulus niger*, by Blower and myself (1968)* we argued that intercalary males were probably of importance in dispersal. With such vagile animals it is essential that males are represented in every group of dispersing animals (without periodomorphosis the sex-ratio becomes biased to females since these survive and the males cannot).

MEIDELL: Yes, I see, compared with *Proteroiulus fuscus* where a dispersing female, being parthenogenetic, can build up a population herself, dispersing schizophyllines must have accompanying males.

DOHLE: Is the correlation with meteorological data sufficiently well-established to say that a longer life cycle is due to a lower temperature?

FAIRHURST: That temperature affects the length of the life-cycle seems highly probable but accurate temperature records in each of the four places I mentioned would be necessary to ascertain whether temperature was directly and solely responsible for the observed variations in duration of life cycle.

SAHLI: In *Tachypodoiulus* you have copulatory males in stadium VII?

FAIRHURST: Yes, Mr. Blower records just one.

SAHLI: Verhoeff mentioned that he had found a single individual.

MEIDELL: Do you know any other animals, except diplopods, that show periodomorphosis? I think there are some spiders?

FAIRHURST: There is a crayfish, *Cambarus*, which alternates between copulatory and non-copulatory phases. Some species of *Carabus* have a physiological rather than a morphological intercalary stage.

KRAUS: Most, or even all "bird-spiders" (Orthognatha) to the dismay of the taxonomists, have females which continue to moult after first reaching maturity. The late Dr Büchli had the idea that males might moult as well, but there is no verification of this.

DESHMUKH: Were the other workers who determined details of life cycle in relation to latitude working with similar habitats to yourself?

FAIRHURST: Halkka* gave very clear descriptions of her sites but more accurate data on the rest would be useful.

JEEKEL: Are the population figures for each site taken over a year or a shorter period?

FAIRHURST: Over a period of one year, and these correlate closely with at least two subsequent years' sampling.

* See list of References, p. 585.

JEEKEL: But I wonder whether you have taken into account the fact that the population in the dune area may be just wandering animals? In Holland during May, you see them emerging from the vegetation and moving into the bare sandy places where they perish in large numbers.

FAIRHURST: Yes. The areas I have considered are home sites, that is, sites where all young stadia are fully represented. I have several other pitfall trap transects designed to sample *just* the migrations. It is these mass movements, of which you speak, which primarily interest me.

HAACKER: How can we exclude the possibility of immigration affecting the stadial spectrum of adult males in the forest site?

FAIRHURST: I am certain that the males from the forest site have not come from other habitats. Site (d) is well within 1000 acres of even-age conifers. The site on the edge of the forest could be affected in the way you suggest.

HAACKER: The migrating animals are not just old males or young males, but consist of all stadia?

FAIRHURST: This depends how far away from the home site we sample. In the fore-dune pitfall trap transect, some distance from the nearest home site, the stadial spectrum has a mode at stadium XI or XII; the youngest males here are stadium X.

Symp. zool. Soc. Lond. (1974) No. 32, 589–602.

A POPULATION OF *CYLINDROIULUS LATESTRIATUS* (CURTIS) ON SAND DUNES

M. J. COTTON

Department of Biology, The Polytechnic, Sunderland, England

and

P. F. MILLER

*Department of Zoology, The University, Manchester, England**

SYNOPSIS

The distribution and activity of *Cylindroiulus latestriatus* in marram dunes on the east coast of Scotland (Tentsmuir) is described. The numbers falling into pitfall traps from 1966–1969 and the sex and stadial composition of the catch from October 1967 to November 1968 are given. Stadia were determined by the ocular field method which appears to be reliable for this species. Apart from a few animals of stadium VI, most were mature males of stadium VII and VIII and occasionally IX, and mature females of stadia VII to X (mainly VIII and IX). More males were trapped than females. The numbers of males gradually increased from March to October. Females began falling into traps in March but high numbers did not appear until September and October. The females are presumed to be less active in the spring because they are preoccupied with ovipositing.

Most animals were trapped in areas of stable microclimate, the semi-fixed and fixed dunes. Very few occurred on embryo dunes or in the slacks and then only later female stadia. On the fore-dunes there are higher numbers in the marram tussocks but on the more stable dunes they also occur in open sandy places. The life-cycle of the population at Tentsmuir appears to be different to that of a population previously described in Devon.

INTRODUCTION

Few studies have been made on the ecology of arthropods on dune systems although such systems represent one of the best understood communities with respect to formation and vegetational succession. During embryo dune formation there are only three species of grasses which constitute the major part of the flora. A brief examination of the area gives one the impression of little animal life, yet sampling shows an active and diverse arthropod fauna throughout the entire year. The two British millipedes associated with sand dune systems are *Cylindroiulus latestriatus* (Curtis) (= *C. frisius* Verhoeff) and *Ommatoiulus* (= *Schizophyllum*) *sabulosus* (L.). The latter is also a fairly common woodland species; the former is sometimes found in coastal woodland (Blower & Gabbutt, 1964; Barlow, 1958; Blower & Fairhurst, 1968).

* Present address: *Biological and Chemical Research Institute, Rydalmere 2116, New South Wales, Australia.*

Salisbury (1952) describes sand dunes as "temperate deserts the arid condition of which is conditioned not by climatic circumstances but by lack of capacity of the dune soil to retain water. The dune is therefore technically an edaphic desert". However, even on a hot dry day the sand immediately beneath the surface may be moist and cool, especially in the tussocks. This arises from "internal dew formation", warm moisture-laden air being drawn into pore spaces as a result of the upward convection currents maintained on the dune crest after dusk. The moist air impinges on the cold sand grains and dew is deposited. Relative humidity near sand level during the summer night approaches saturation both within the tussock and in the bare sand areas, although it is always higher in the latter as a result of the sand having become much hotter during the day and cooling more during the night. The high relative humidity will however be maintained longer during the following day in the tussock.

The present study, commenced in September 1966 by one of the authors (M.J.C.), followed the temporal and spatial distribution of arthropods through stages of dune succession in relation to micro-climatic conditions and vegetation colonization. In particular the marram dune system and the habit of tussock formation by marram grass with respect to arthropod distribution was investigated.

AREA AND METHODS

The study area

The dune system chosen formed part of the Tentsmuir National Nature Reserve to the south of the estuary of the River Tay in Scotland. The present sand dunes are of fairly recent origin (about 30 years) having accumulated in the first instance against a line of concrete blocks which formed part of the east coast defences during the 1939–1945 war. For this reason, then, they do not form a very extensive system and stretch little more than 80 m in an east–west direction, between the beach and the dune slack. The vegetation is typical of an acidic dune system. It can be divided, somewhat arbitrarily in this case, into mobile, semi-fixed and fixed dunes, to the east of which lies an extensive slack, each zone having its characteristic pattern of vegetation. Typically the dominant species on the dunes is *Ammophila arenaria* which is found in early stages of colonization on the mobile dunes together with scattered patches of *Agropyron junceiforme* and some *Elymus arenarius*. Somewhat atypically the tussocks of *Ammophila arenaria* reach a stage of maximum development some distance to landward of the dune ridge. It is this region, in this case, which shall be referred to as "semi-fixed" dune for it is here that one first sees the establishment of a more varied flora with bryophytes, lichens and algae stabilizing the sand surface against wind erosion. From this point onward as far as the dune slack, the dense tussocks of *Ammophila* become dissociated and their dominance is gradually eroded as *Festuca rubra* var. *maritima* becomes established. Thus in the semi-fixed dunes *Festuca* is co-dominant with *Ammophila*, and in the fixed dune *Ammophila*, while still abundant, has become of secondary importance to *Festuca*.

The zonation so far has been fairly indistinct since the transition between one region and the next is gradual. Between the fixed dune and the dune slack, however, the plant community changes within the space of a few yards from one dominated by two species of grass to one in which the dominant species are *Carex arenaria*, *Salix repens*, *Lotus corniculatus* and *Juncus* spp. A more detailed account of the dune slack flora of Tentsmuir is given by Crawford & Wishart (1966).

Behind the slack to the west is a line of dense alder carr in association with some birch and buckthorn and indeed *Hippophae rhamnoides* is found in clumps at the slack–dune boundary. Beyond this again is a region of dune heath and then marsh. Finally, the whole reserve is fringed by a Forestry Commission plantation consisting mostly of pine. A feature of interest is the colonization of the fixed dunes and slacks by pine seedlings as a result of the proximity of the plantation.

The water table is maintained by a small burn which runs into the dunes after passing through the plantation.

Methods of study

Pitfall traps have been used by a number of workers in the past, e.g. Gilbert (1956), Greenslade (1964), Murdoch (1966), and their advantages and disadvantages have been well enumerated. However, provided their limitations are fully recognized, they remain an extremely convenient technique for sampling the ground-living arthropod fauna. A grid of pitfall traps was set between the strand line and the centre of the dune slack; a distance of 100 m. In the first year the grid comprised only 18 traps set in six lines parallel to the shore. Traps in each line were 10 m apart and the line corresponded to a distinct zone, characterized by topography, vegetation cover and plant diversity. In subsequent years the grid was increased to 24 traps, each of the six zones having four traps. In those zones in which marram grass was the dominant plant species two of the traps sampled were from within marram tussocks and two from the exposed open sand. The small number of traps used throughout was necessitated by the fact that all arthropod groups were collected, sorted and identified, and the volume of material collected was at times so great that more traps could not be properly managed. In addition there was the problem that over-sampling might seriously damage the populations present in the Nature Reserve. In order to avoid the effects of trampling it was necessary to move the grid slightly along the dunes each year, although the entire sampling area over the full study period was only 50 m wide and the nature of the zones remained as far as possible the same from year to year. The traps contained phenyl mercury acetate as a preservative; they remained open all the time and were emptied each week.

Microclimate data was collected from various stations across the dunes. Temperature was recorded by means of a "Grant" Multipoint Recorder and relative humidity by means of cobalt thiocyanate paper (Solomon, 1957). The microclimate studies will be published in full elsewhere.

Determination of stadia

Determinations of sex and stadia were made by one of us (P.F.M.). Stadia were determined by counting the rows of ocelli, the stadium being equal to the number of rows plus one (Vachon, 1947). Blower & Gabbutt (1964) determined stadia of *C. latestriatus* by probability analyses of lengths; Blower & Fairhurst (1968) and Blower (1970) maintained that the ocular field of this species was too irregular for the use of Vachon's method. However, the individuals of *C. latestriatus* from the Tentsmuir traps were extended by the preservative and measurement of lengths was not possible, therefore the possibility of using Vachon's method was re-examined. Although the eye-rows are more difficult to discern in *C. latestriatus* than in most other iulids, the attempt was made and Table I indicates the measure of success achieved. In this Table the number of podous and apodous segments of a sub-sample of the characterized animals is compared with the figures given by Blower & Gabbutt (1964). Some of the females with 40 or more podous segments which were placed in stadia IX and X may belong to stadia XI or XII since it is difficult to determine the later additions of ocelli to the ocular field. However, such misplaced animals would be very few in number.

DISTRIBUTION AND ACTIVITY OF *CYLINDROIULUS LATESTRIATUS*

Seasonal activity

The only millipede to occur in the study area at Tentsmuir is *C. latestriatus* although *O. sabulosus* occurs in the marram dunes at

St. Cyrus Nature Reserve only 20 miles to the north. There is an overall increase in the seasonal activity of *C. latestriatus* from March until September and October when there is a distinct autumn peak (Fig. 1).

TABLE I

Segment numbers of stadia from Tentsmuir determined by the ocular field method compared with those recorded from Harpford Wood by Blower & Gabbutt (1964) for stadia determined by probability analysis

		Harpford		Tentsmuir		
		Podous	Apodous	Podous	Apodous	No.
♂♂	VII	30–35	2–4	30–36	2–4(5)	33
	VIII	32–36	2–4	31–36	2–4	23
	IX	34–37	2–3	39	2	1
♀♀	VI	24–29	2–5	26 and 28	4 and 5	2
	VII	30–34	2–4	30–32	3–4	4
	VIII	32–38	1–3	33–39	1–3	28
	IX	35–38	1–2	35–42	1–2(3)	15
	X	36–40	1–2	37–42	1–2	6
	XI	38–41	1–2			
	XII					
	XIII	(43)	1			

Numbers in brackets refer to one specimen.

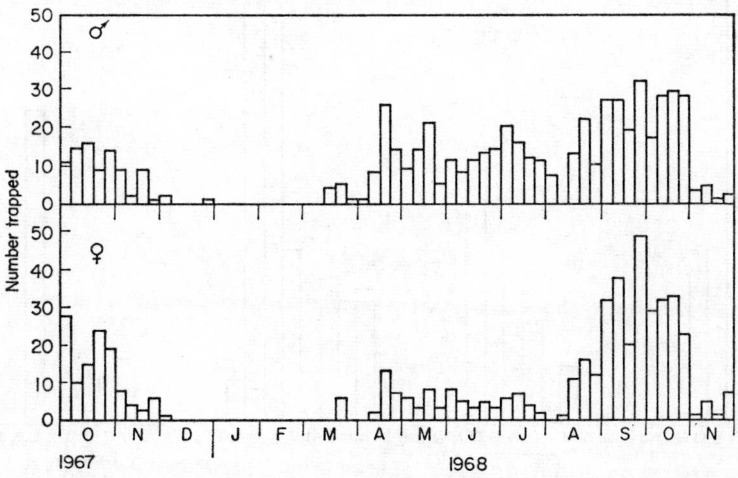

FIG. 1. Seasonal activity of male and female *C. latestriatus* in the marram dunes.

Activity falls rapidly in November and very few individuals are caught in winter months. This pattern is shown particularly by males but females are less active in spring and produce a sudden activity peak in September (Fig. 1). Although both sexes are active in autumn females outnumber males and the sex ratio (Fig. 2) does indeed change significantly over the year ($P < 0{\cdot}001$).

During three complete years of the sampling programme (1966–1969) four periods of peak activity were studied (Fig. 3). Since the number of traps used each year varied slightly the results are expressed

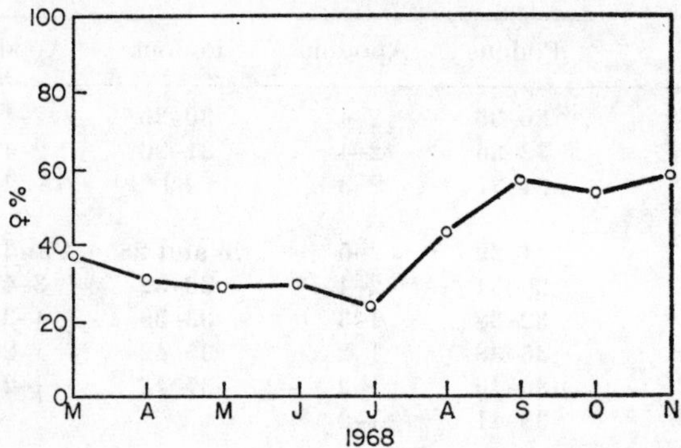

Fig. 2. The monthly sex ratio of *C. latestriatus* expressed as percentage change in number of females.

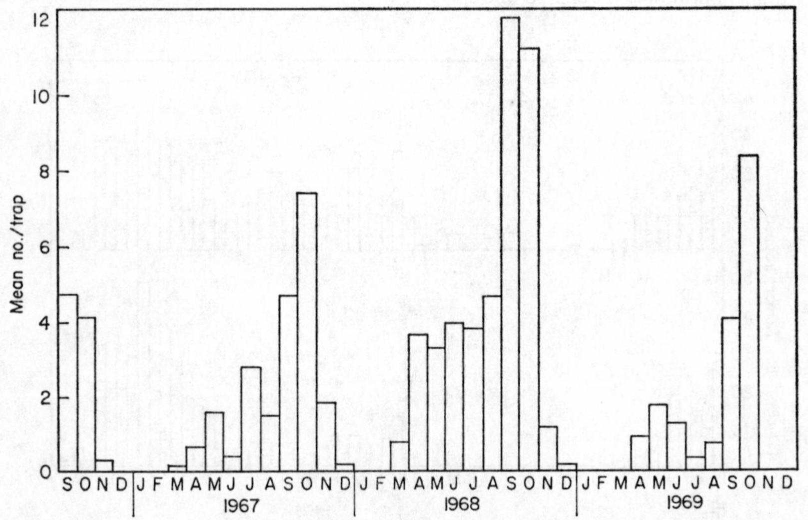

Fig. 3. Monthly activity of *C. latestriatus* from 1966–1969.

as a mean value for each trap. In the first year no traps were set in the dune slack but since in later years only 3% of the total catch was from the slack these few individuals have been ignored and the results refer only to those millipedes found in the marram dunes. There is good agreement in the seasonal activity pattern between years especially at the time of the autumn peak, when it is thought a true reflection is given of the population density. Differences appear in the degree of activity shown in spring and summer months, especially during 1969. This may result from the prolonged hot and dry weather in that year which limited activity in the exposed sand areas between marram tussocks. However, with a decline in temperature in September and higher relative humidities, especially at night, activity increased and the autumn peak was similar to previous years.

Life history and sexual activity

In the Devon wood population of *C. latestriatus* described by Blower & Gabbutt (1964) it was suggested that both sexes develop as far as stadia VII and VIII by the autumn of the year following their birth. Many of the animals in these stadia were mature but some males remained immature and moulted to stadia VIII and IX by the next autumn. Thus adult males were present as stadia VII, VIII and IX but represented two separate generations. Mature females, unlike the males, can moult repeatedly. Females belonged to stadia VII to XIII and represented three or more generations, first VII and VIII, then VIII and IX and so on. Most belonged to stadium VIII since this contained representatives of animals in their first and second years of maturity.

At Tentsmuir, over 95% of the animals of stadium VII and above were mature. Males were present as stadia VII and VIII but only a small number as stadium IX. Females were present mainly as stadia VIII and IX, there being very few in stadium VII (Fig. 4). Although most animals were mature, we know there must have been many immature males in stadium VII since the mature males of stadium VIII can only have been derived from immature animals. The fact that immature animals of these stadia did not fall into the traps suggests that the surface activity has sexual significance. However, we have no reason to believe that mature animals of different stadia are differentially active and thus the numbers trapped must bear a simple relation to their density. The seasonal activity is shown in Fig. 4. Since there are more males active in stadium VII than in stadium VIII, we presume that most matured in stadium VII, fewer in stadium VIII and hardly any in stadium IX, which contrasts markedly with the population in Devon where equal numbers of males were present in these three stadia.

By the same argument, the low number of females active in stadium VII probably reflects their low density. Thus the majority of the females at Tentsmuir do not appear to reach maturity until stadium VIII, a stadium later than the Devon population. Again, since most of the females trapped were in stadia VIII and IX, we presume that only one moult follows maturity compared with the three or more suggested in Devon. Thus the Tentsmuir population appears to mature later and to have fewer broods in successive years (less marked iteroparity).

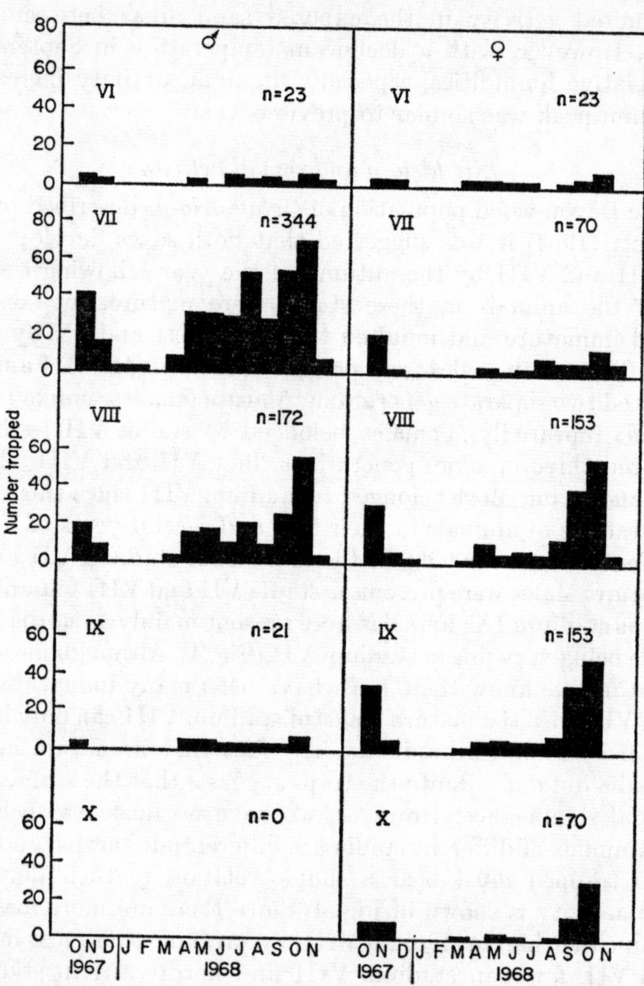

FIG. 4. Monthly activity of each stadium of the millipede population.

The peak activity in autumn might indicate mating and/or dispersal and selection of suitable overwintering sites. Iulids do pair in autumn (Blower & Fairhurst, 1968) and *C. latestriatus* caught alive in the traps at Tentsmuir were often found *in copula*. The activity of males in spring and summer may indicate pairing in this season also. It is possible that some females overwinter as virgins and pair in spring whilst others have already been inseminated in autumn. The selection of subterranean ovipositing sites by the females in spring and summer may account for their low surface activity relative to the males. Oviposition probably occurs in the marram tussocks since we have obtained very young stadia by sieving sand and litter from the base of these.

Spatial distribution in relation to dune succession

The species shows relatively even distribution from the marram ridge (zone 1a) to the more fixed dunes bordering the slack (zone 2b) (Fig. 5). In the first year of the study peak numbers were found in the semi-fixed dunes (zone 2a) but in 1968–1969 the peak was observed in the more fixed dunes (zone 2b) characterized by a dense scrub vegetation of sea buckthorn (*H. rhamnoides*) and good ground cover provided by *Festuca rubra*. It is of significance that this vegetation provided more equable conditions of microclimate and a higher relative humidity during a summer when the general climate was hot and dry. Millipede activity was therefore somewhat confined to this habitat. *C. latestriatus* seldom enters the slack, except for older females, nor does it commonly occur in the early embryo dunes (zone 0b). There was little overall difference in numbers collected in those traps placed within marram tussocks and those placed outside in bare sand. In 1967–1968 the total from sand traps was 510 and from tussock traps 441.

TABLE II

Number of millipedes collected in sand and tussock traps in relation to vegetational succession

Zone	Number trapped 1967–1968 Sand	Tussock	% in sand
1a	95	127	43
1b	162	112	60
2a	109	91	55
2b	144	111	58

$\chi^2 = 14{\cdot}7$
$P < 0{\cdot}01$

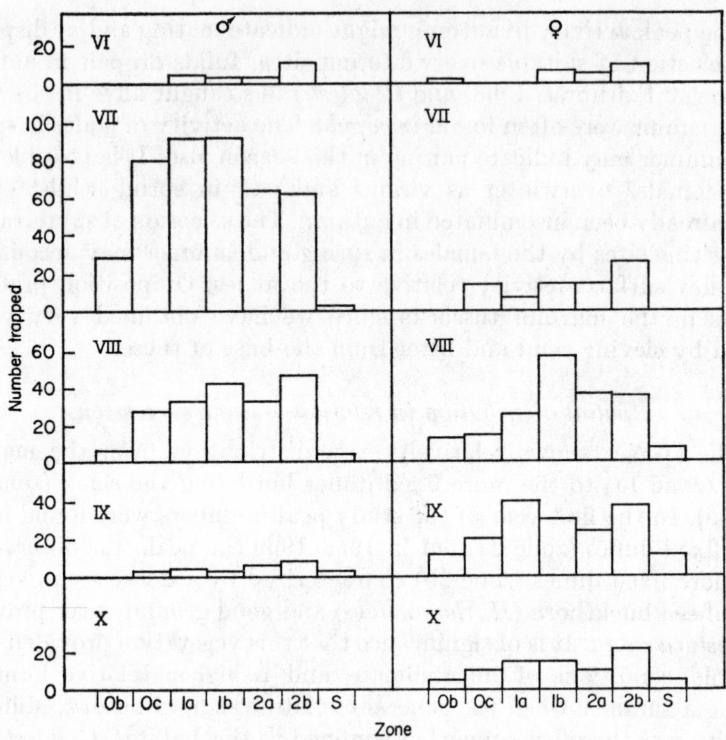

FIG. 5. Spatial distribution of each stadium in relation to zones of vegetational succession. 0b—*Agropyron* embryo dunes; 0c—*Elymus* embryo dunes; 1a—*Ammophila* dune ridge; 1b—*Ammophila* mobile dunes; 2a—*Ammophila* semi-fixed dunes; 2b—*Ammophila* fixed dunes; S—Dune slack.

However, the ratios in sand and tussock traps for the different vegetation zones were significantly different (Table II). Millipedes occurred more in the tussocks on the dune ridge (zone 1a) and in the open sand traps in the more landward stable areas (zones 1b, 2a and 2b). This presumably reflects activity since in the semi-fixed dunes individuals can wander freely at night and tend to be caught in sand traps. On the more exposed foredune ridge, at the limit of their ecological range, millipedes tend to aggregate in the dense tussocks.

The majority of the animals were trapped in the fixed and semi-fixed dunes which seem to represent the population centre. Animals, particularly older females, do disperse into the dune slack and embryo dunes though there is no evidence of population establishment here.

There is great similarity in the distribution of each stadium between tussock and sand traps (Fig. 6), and it appears that within the confines of the marram dune system *C. latestriatus* has adapted to the ecological problems associated with this habitat and that only under certain climatic conditions (as in the summer of 1969) or at the extreme limit of its range is its distribution pattern seriously affected.

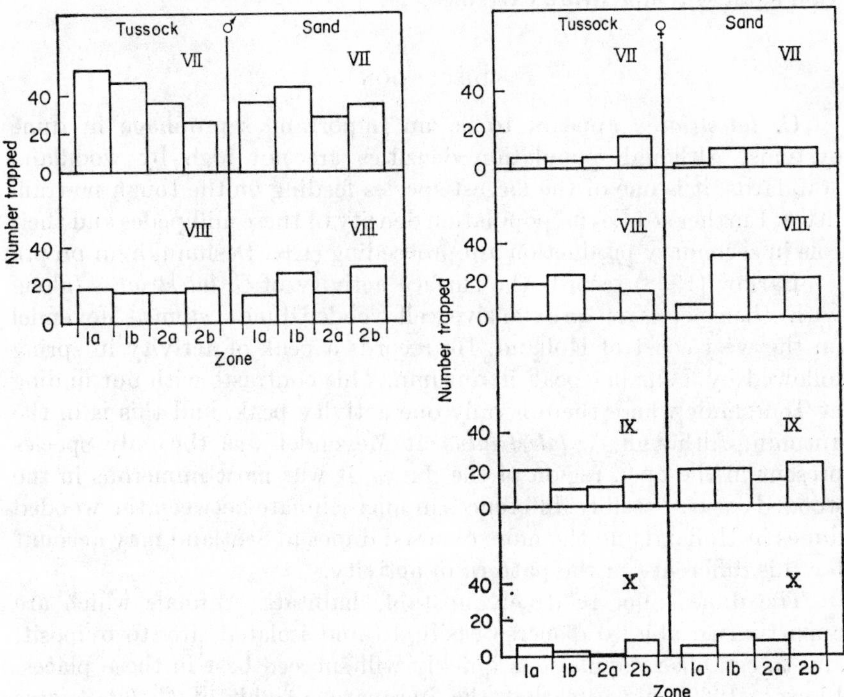

FIG. 6. Comparison of the distribution of each stadium between tussock and sand traps in relation to zones of vegetational succession. Symbols as in Fig. 5.

Blower & Gabbutt (1964) consider that the thermal properties of a sandy soil are involved in determining the rather narrow European range of this species and that its coastal distribution may reflect the more equable climate in these regions. This is substantiated by the absence of millipedes in the dune slack, the lowest part of the dune close to the water table, where temperature conditions are more extreme and water-logged conditions or occasional flooding occur. Sub-zero temperatures above and below soil level are often experienced in winter and the slack forms a "frost hollow" with freezing and compacting of the sand to a depth of 20–30 cm. During this period other arthropods often disperse from the slack into the warmer marram tussocks in the seaward parts of the dunes and microclimatic conditions within the slack appear to determine the spatial distribution of other dune species (Cotton, 1971). In summer the diurnal temperature range is most extreme in the slack with its prostrate vegetation cover and absence of a litter layer. Even in the embryo dunes a small marram tussock affords good protection against temperature extremes.

DISCUSSION

C. latestriatus appears to be an important saprophage in dune systems. Although population densities are not high by woodland standards, it is one of the largest species feeding on the tough marram litter. Further studies on population density of these millipedes and their role in secondary production are proceeding (I.K. Deshmukh, in prep.).

Barlow (1958) records the surface activity of *C. latestriatus* (along with other species) on an extensive well-wooded dune system at Meyendel on the west coast of Holland. He records a peak of activity in spring followed by a smaller peak in autumn. This contrasts with our finding at Tentsmuir where there is only one activity peak, and this is in the autumn. Although *C. latestriatus* at Meyendel was the only species present in the open region of the dunes, it was most numerous in the wooded areas. Possibly differences in microclimate between the wooded dunes in Holland and the more exposed dunes in Scotland may account for this difference in the pattern of activity.

The dunes offer relatively unstable habitats. Animals which are opportunists, able to disperse easily, to find isolated sites to oviposit, and to establish populations quickly will succeed best in these places. Blower (1969) suggests that the iteroparous habit of *C. latestriatus* adapted it for dispersal whilst its ability to mature a stadium earlier than most iulids enables it to establish quickly. However, he was referring to the population in Devon, in comparison with which the Tentsmuir population appears to be retarded and less markedly iteroparous.

Acknowledgements

The study was financially supported by a grant from the Natural Environment Research Council during the period 1968–1971 which enabled detailed recording of microclimate and employment of a research assistant, Miss M. McCormick, who provided much of the labour in both field and laboratory. The study area formed part of the Tentsmuir Nature Reserve and thanks are due to the Nature Conservancy (Scotland) for permission to work there and erect a field laboratory.

The assistance of the Reserve Warden, Mr M. Smith, was appreciated throughout. The authors are indebted to Mr J. G. Blower for many hours of discussion and help with regard to the study.

References

Barlow, C. A. (1958). Distribution and seasonal activity in three species of Diplopods. *Archs néerl. Zool.* **13**: 108–133.

Blower, J. G. (1969). Age-structures of millipede populations in relation to activity and dispersion. *Publs Syst. Ass.* No. 8: 209–216.

Blower, J. G. (1970). The millipedes of a Cheshire wood. *J. Zool., Lond.* **160**: 455–496.

Blower, J. G. & Fairhurst, C. P. (1968). Notes on the life-history and ecology of *Tachypodoiulus niger* (Diplopoda, Iulidae) in Britain. *J. Zool., Lond.* **156**: 257–271.

Blower, J. G. & Gabbutt, P. D. (1964). Studies on the millipedes of a Devon oak wood. *Proc. zool. Soc. Lond.* **143**: 143–176.

Cotton, M. J. (1971). The distribution of *Boreus hyemalis* (L.) (Mecoptera) on a sand dune system. *Entomologists' mon. Mag.* **106**: 174–176.

Crawford, R. M. M. & Wishart, D. (1966). A multivariate analysis of the development of dune slack vegetation in relation to coastal accretion at Tentsmuir, Fife. *J. Ecol.* **54**: 729–743.

Gilbert, O. (1956). The natural histories of four species of *Calathus* (Col., Carabidae) living on sand dunes in Anglesey, North Wales. *Oikos* **7**: 22–47.

Greenslade, P. J. M. (1964). Pit-fall trapping as a method for studying populations of Carabidae (Coleoptera). *J. anim. Ecol.* **33**: 301–310.

Murdoch, W. W. (1966). Aspects of the population dynamics of some marsh Carabidae. *J. anim. Ecol.* **35**: 127–156.

Salisbury, E. J. (1952). *Downs and dunes.* London: Bell.

Solomon, M. E. (1957). Estimation of humidity with cobalt thiocyanate papers and permanent colour standards. *Bull. ent. Res.* **48**: 489–506.

Vachon, M. (1947). Contribution à l'étude du développement post-embryonnaire de *Pachybolus ligulatus* Voges. Les étapes de la croissance. *Annls Sci. nat.* (*Zool.*) (11) **9**: 109–121.

Discussion

Deshmukh: I have recently obtained details of the population of *Cylindroiulus latestriatus* at Tentsmuir from direct sampling. A generation spends the first winter as stadia III and IV and the second as V and VI; maturation occurs in the third autumn after birth and is thus a year later than the Devon population studied by Blower & Gabbutt (1964)*. Is this delay due to the difference in latitude between Devon and N.E. Scotland or to the extreme nature of the sand-dune environment? We have just heard from Dr Fairhurst (this Symposium: pp. 575–587) that *Ommatoiulus sabulosus* develops more rapidly in the dunes than in the forest!

* See list of references above.

COTTON: I think the situations in the Devon oak wood and our dunes in Scotland, some 500 miles further north, will be very different and could account for the longer life-cycle at Tentsmuir. By the same token, I would not be surprised to find that if a population of *C. punctatus* is studied at the same latitude as Tentsmuir, it would have a longer life-cycle than the Devon population (four years?). Fairhurst's paper reminded us too, that factors other than latitude may alter the length of the life-cycle.

FAIRHURST: I think one of the most important principles to emerge from the papers in this section, concerns the great variation in life-cycles; we should be very cautious in generalizing life-cycle details from one study area. It must also be quite clear that any description of life-cycle should be accompanied by very detailed descriptions of the habitats involved.

Symp. zool. Soc. Lond. (1974), No. 32, 603–609.

THE DISTRIBUTION OF BRITISH MILLIPEDES AS KNOWN UP TO 1970

J. GORDON BLOWER

Department of Zoology, The University, Manchester, England

SYNOPSIS

Reference is made to a recent analysis of millipede records in Great Britain and Ireland. Up to 1970, three-quarters of the counties had fewer than half of the possible species recorded in them and 18 counties were without a single record. The 48 species occurring here are listed according to the number of vice-counties in which they have been recorded and a measure of "commonness" is thereby accorded to them. Differences in the ranking of a few of the commoner species in the different principalities suggest a geographic bias. The precise location of the 16 least common species is given.

INTRODUCTION

I have recently analysed several thousand millipede records from Great Britain and Ireland which I have gathered over the years from over 250 published papers and from collections made by colleagues and myself (Blower, 1972). It has been the habit of the older naturalists to pay more attention to the geographic location of the record than to the precise details of its habitat. We already know, in broad terms, the various habitat requirements of the commoner species—which live in dead wood, under bark, in fungal fructifications, which prefer base-rich soil, etc. Further analysis of the records merely confirms these habitat features which I have given elsewhere (Blower, 1958). It is easy to see the geographic distribution of the rare species since there are so few records but not until all the records had been marshalled were geographic patterns discernible for a few of our commoner species. The exercise has revealed the incompleteness of our present data on distribution but has also provided some quantitative measure of the usually nebulous concepts of commonness and rarity.

HISTORICAL

Named species of millipedes were first recorded from specific localities in the first half of the 19th century. Leach (1815) recorded 11 species in Britain, seven of which he himself had described as new to science. Johnston (1835) added a 12th species; he and Templeton (1836) were the first to give precise localities for their records. In the second half of the 19th century Newport, Curtis, Pocock and Evans (references in Blower, 1972) added 11 species thus bringing the total recorded up to the year 1907, to 23 species.

Between 1912 and 1922, Brade-Birks and Bagnall between them added a further 18 species and many times this number of records. These 41 species are the valid entries in the check-list drawn up by Brade-Birks (1939) and remained as the British list until 10 years after the Second World War. Four more species appeared in the list of Blower (1958) and, more recently, a further three species have been added; these are *Chordeuma proximum* (Nelson, 1964), *Chordeuma silvestre* (J. G. Blower, unpubl.) and *Leptophyllum armatum* (E. H. Eason, unpubl.).

THE COMMON AND THE RARE SPECIES

In Table I the 48 species are listed in the order of the numbers of vice-counties in which they have been recorded. The species most frequently recorded are at the head of the list and can be regarded as the commonest; those further down the list are what we may term the rarities. When naturalists refer to a given species as "common" or "rare", their judgement is usually subjective. If the measure of "wide-spreadness" provided in Table I is really the same as commonness, then we have here an objective and quantitative measure.

Within a vice-county some species are naturally more frequently recorded than others but such variations in frequency do not appear in Table I. However, over the entire country, the frequency of records as measured by the number of vice-counties in which they occur is closely similar to the frequency *within* a given vice-county. The probability of a given species being found depends, to some extent, on its size, the proportion of adults in the population and on its habits. Our smaller species, those less than 5 mm marked on Table I, will be more difficult to find, or if found, may be discounted by the uninitiated as young stadia for which a name cannot be secured. Then again, most of these smaller species are poorly pigmented, subterranean in habit and rarely surface active. However, most of the available records have been made by the initiated few and it will be seen from Table I that such considerations do not appear to have affected the position of the smaller species in the Table.

The species added in each of the four half-centuries mentioned in the last section are marked in Table I and it will be seen that there is a correlation between the time of their discovery and their position in the list. All of the 20 commonest species had been found by the turn of the century.

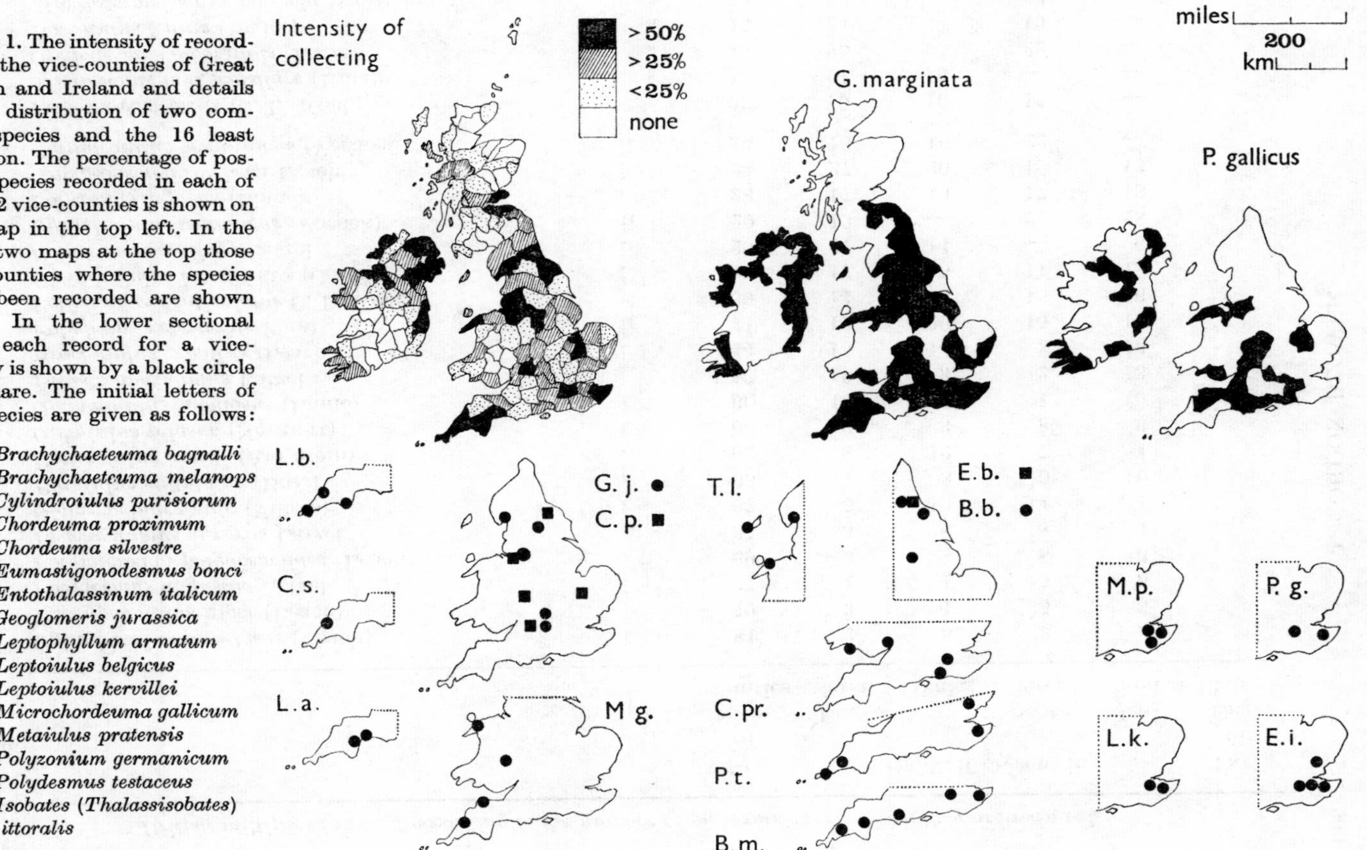

FIG. 1. The intensity of recording in the vice-counties of Great Britain and Ireland and details of the distribution of two common species and the 16 least common. The percentage of possible species recorded in each of the 152 vice-counties is shown on the map in the top left. In the other two maps at the top those vice-counties where the species have been recorded are shown black. In the lower sectional maps each record for a vice-county is shown by a black circle or square. The initial letters of the species are given as follows:

B.b. *Brachychaeteuma bagnalli*
B.m. *Brachychaeteuma melanops*
C.p. *Cylindroiulus parisiorum*
C.pr. *Chordeuma proximum*
C.s. *Chordeuma silvestre*
E.b. *Eumastigonodesmus bonci*
E.i. *Entothalassinum italicum*
G.j. *Geoglomeris jurassica*
L.a. *Leptophyllum armatum*
L.b. *Leptoiulus belgicus*
L.k. *Leptoiulus kervillei*
M.g. *Microchordeuma gallicum*
M.p. *Metaiulus pratensis*
P.g. *Polyzonium germanicum*
P.t. *Polydesmus testaceus*
T.l. *Isobates (Thalassisobates) ittoralis*

TABLE I

British millipedes ranked according to the number of vice-counties in which they are recorded

Rank	Species			No. of vice-counties	Rank, if present in: England	Wales	Scotland	Ireland	Part of England
1.	*Cylindroiulus punctatus* (Leach)	i		81	2	5	2	2	
2.	*Tachypodoiulus niger* (Leach)	i		79	3	1	5	3	
3.	*Polydesmus angustus* Latzel	i		77	1	1	5	6	
4.	*Polymicrodon polydesmoides* (Leach)	i		69	3	8	8	6	
5.	*Brachydesmus superus* Latzel	ii	s	67	5	8	8	4	
6.	*Glomeris marginata* (Villers)	i		67	5	12	14	1	
7.	*Blaniulus guttulatus* (Bosc)	i		65	7	8	3	10	
8.	*Proteroiulus fuscus* (Am Stein)	ii		62	8	12	7	4	
9.	*Ophyiulus pilosus* (Newport)	i		60	9	3	8	6	
10.	*Ommatoiulus sabulosus* (Linné)	i		60	10	8	1	10	
11.	*Iulus scandinavius* Latzel	i		52	10	3	12	13	
12.	*Brachyiulus pusillus* (Leach)	i		44	15	14	3	16	
13.	*Polydesmus coriaceus* Porat	ii		41	12	20	15	10	
14.	*Polydesmus denticulatus* C. L. Koch	ii		39	14	5	15	16	
15.	*Cylindroiulus latestriatus* (Curtis)	i		37	17	5	11	18	
16.	*Polydesmus gallicus* Latzel	ii		29	21	14	—	6	
17.	*Cylindroiulus teutonicus* (Pocock)	ii		29	13	—	22	18	
18.	*Polyxenus lagurus* (Linné)	i	ss	28	17	16	17	18	
19.	*Craspedosoma rawlinsi* Leach	i		24	27	20	12	13	
20.	*Cylindroiulus britannicus* (Verhoeff)	ii		24	16	16	22	—	
21.	*Isobates varicornis* (C. L. Koch)			23	19	16	17	—	
22.	*Microchordeuma scutellare* (Ribaut)		s	17	21	16	—	—	
23.	*Ophiodesmus albonanus* (Latzel)		ss	17	20	—	22	—	
24.	*Boreoiulus tenuis* (Bigler)		s	15	24	—	19	—	
25.	*Macrosternodesmus palicola* Brölemann		ss	14	24	—	22	—	

No.	Species	Period	Size						
26.	*Nopoiulus minutus* Brandt			14	32	—	22	15	
27.	*Archeboreoiulus pallidus* (Brade-Birks)			14	23	—	—	—	
28.	*Oxidus gracilis* (C. L. Koch)	ii		12	26	—	19	—	
29.	*Cylindroiulus nitidus* (Verhoeff)			12	27	—	19	—	
30.	*Cylindroiulus londinensis* (Leach)	i		7	29	—	—	—	
31.	*Brachychaeteuma bradeae* (Brölemann & Brade-Birks)		s	7	29	—	—	—	
32.	*Choneiulus palmatus* (Nemec)			7	29	—	—	—	
33.	*Cylindroiulus parisiorum* (Brölemann & Verhoeff)			5	33	—	—	—	
34.	*Polydesmus testaceus* C. L. Koch	ii		4	34	—	—	—	S
35.	*Chordeuma proximum* Ribaut	iv		4	36	20	—	—	S
36.	*Brachychaeteuma melanops* Brade-Birks		s	4	34	—	—	—	S
37.	*Brachychaeteuma bagnalli* Verhoeff		s	3	36	—	—	—	NE
38.	*Isobates littoralis* Silvestri			3	42	20	—	—	NW
39.	*Leptoiulus belgicus* (Latzel)			3	36	—	—	—	SW
40.	*Metaiulus pratensis* Blower & Rolfe	iv		3	36	—	—	—	SE
41.	*Entothalassinum italicum* (Latzel)			3	36	—	—	—	SE
42.	*Geoglomeris jurassica* Verhoeff	iv	ss	3	36	—	—	—	
43.	*Microchordeuma gallicum* (Latzel)	iv		2	45	20	—	—	SW
44.	*Polyzonium germanicum* Brandt			2	42	—	—	—	SE
45.	*Leptoiulus kervillei* (Brölemann)	iv		2	42	—	—	—	SE
46.	*Chordeuma silvestre* Latzel	iv		1	45	—	—	—	SW
47.	*Eumastigonodesmus bonci* (Brölemann)		ss	1	45	—	—	—	NE
48.	*Leptophyllum armatum* Ribaut	iv		1	45	—	—	—	SW

After the name and author of the species a small roman numeral indicates the period during which the species was first recorded in this country:

i—by Leach (1815) or, for *O. sabulosus*, Johnston (1835),
ii—between the years 1844 and 1907 by Newport, Curtis, Pocock and Evans,
iv—since 1956, and
the rest were recorded by Brade-Birks and Bagnall between 1912 and 1922.

s after the name means the adults are less than 10 mm long.
ss indicates less than 5 mm long.
Details for the entries from 33 to 48 are shown in Fig. 1.

GEOGRAPHIC DISTRIBUTION

It will be seen from Table I that 16 of the 19 least common species are known only from England and the other three only from England and Wales. Furthermore, 14 of the 19 are known only from a small part of England, or England and Wales. Their limited distribution is indicated in the Table and also in Fig. 1. Six of the 20 most common species appear to have geographically biased distributions. There is no evidence to suggest that the remainder will not eventually be found all over the country. Of the six species with biased distributions two of them are absent from parts of the country which have been well-worked and the bias appears clearly from a simple plot of the records (compare the plotted distribution of *Glomeris marginata* and *Polydesmus gallicus* with the plot of intensity of recording in Fig. 1). *G. marginata* is not recorded in Scotland north of the firths of Forth and Clyde. In Ireland it is the most frequently recorded species. *P. gallicus* is absent from Britain north of the county of Cheshire but is widespread in Ireland.

The biased distribution of the four other species only appears when their frequency within one of the principalities is compared with their overall frequency. *Ommatoiulus sabulosus* (= *Schizophyllum sabulosum*) is the most frequently recorded species in Scotland but is 10th overall. Similarly, *Brachyiulus pusillus* is the third most frequent in Scotland but 12th overall. *Craspedosoma rawlinsi* is ranked 19th overall but 27th in England and is clearly more frequent in the north and west. *Cylindroiulus teutonicus* is less frequent in Scotland and Ireland than in England where it is more frequent in the south and east of the country.

Of the 152 vice-counties of Great Britain and Ireland, there are records for 133 but many of these have very few species recorded in them (see Fig. 1, map showing "Intensity of collecting"). It will be seen from Table I that the highest number of vice-county records for a species (*C. punctatus*) is only 81. In terms of counties, up to 1970, three-quarters of the counties had fewer than half of the possible species recorded in them and 18 were without a single record.

CONCLUSIONS

In view of the uneven coverage of the country so far, it would be premature to discuss reasons for the few apparent patterns of distribution. We can now concentrate our efforts, for example, to check that *P. gallicus* does not occur north of Cheshire or that *Glomeris marginata* is not found north of the Forth and Clyde and then begin to seek explanations. For species such as *Ommatoiulus sabulosus* which can be expected to occur in all our counties and yet appears to be more frequent in the north, further details will be difficult to gather in a meaningful manner. Possibly the collection of records from 10 Km grid squares will reveal more useful information. Certainly the systematized collection of habitat data along with 10 Km locality records which Barber & Fairhurst (1974, this Symposium, pp. 611–619) are beginning should greatly improve our knowledge of habitat requirements and may possibly reveal some more subtle geographic patterns of distribution.

References

Barber, A. D. & Fairhurst, C. P. (1974). A habitat and distribution recording scheme for Myriapoda and other invertebrates. *Symp. zool. Soc. Lond.* No. 32: 611–619.

Blower, J. G. (1958). British Millipedes (Diplopoda). *Synopses Br. Fauna* No. 11: 1–74.

Blower, J. G. (1972). The distribution of British millipedes as known at the end of 1969. *Bull. Br. Myriapod Gr.* **1**: 9–38.

Brade-Birks, S. G. (1939). Notes on Myriapoda XXXVI. Sources for description and illustration of the British fauna. *Jl S.-east agric. Coll. Wye* **44**: 156–179.

Johnston, G. (1835). Insecta Myriapoda found in Berwickshire. *Loudon's Mag. nat. Hist.* **8**: 486–494.

Leach, W. E. (1815). A tabular view of the external characters of four classes of animals, which Linné arranged under Insecta. *Trans. Linn. Soc. Lond.* **11**: 306–400.

Nelson, J. M. (1964). *Chordeuma proximum* Ribaut, a millipede new to Britain. *Ann. Mag. nat. Hist.* (13) **7**: 527–528.

Templeton, R. (1836). Catalogue of Irish Crustacea, Myriapoda and Arachnida, selected from the papers of the late John Templeton, Esq. *Loudon's Mag. nat. Hist.* **9**: 9–14.

Symp. zool. Soc. Lond. (1974) No. 32, 611–619.

A HABITAT AND DISTRIBUTION RECORDING SCHEME FOR MYRIAPODA AND OTHER INVERTEBRATES

A. D. BARBER

Plymouth College of Further Education, Plymouth, England

and

C. P. FAIRHURST

*Department of Science, Stockport College of Technology, Stockport, England**

SYNOPSIS

Distribution studies based on the 10 km grid or other systems yield information of strictly limited value. To gain further insight into the factors influencing occurrence, some form of standardized recording of habitats is required to cover both major habitat and microsite together with other information of ecological interest. The present scheme uses a record card based on an extension of previous habitat classification systems to include those habitats most relevent to chilopods, diplopods, and freshwater and terrestrial isopods. It is designed to be used for computer analysis and to allow a bank of accessible, useful data to be built up, and at the same time to be readily usable in the field. It appears possible to extend its use, both to other countries and to other groups of invertebrates.

INTRODUCTION

By far the most conspicuous feature of the majority of the older published records of plants and animals was the lack of systematic data on both habitat and geographical distribution. The habitats of rare species were often better known than those of common ones and comparative studies were difficult to make, owing to the vagueness of the descriptions. Since then, a number of schemes have been devised to put both the distribution and habitat information on a more systematic basis.

DISTRIBUTION RECORDING

Early accounts of distribution, when they gave details, tended simply to state the area e.g. "Kew" or "Sherwood Forest" but during the nineteenth century the Watson–Praeger system of vice-counties was introduced for vascular plants in Britain and Ireland. This was subsequently extended to animals (see Balfour-Browne, 1931) and a

*Present address: *Department of Biology, The University, Salford, Lancs., England.*

number of more recent works show distribution maps on this basis, e.g. Hynes (1958) (Plecoptera), Ragge (1965) (Orthoptera s.l.). The units are based on the old administrative counties with appropriate sub-division and do not necessarily bear any relation to geological or topographical features or to urbanization. However, with a large area of country to cover with relatively few records, this method is obviously useful for gaining an initial insight into the main distribution parameters of the species concerned.

A more precise mapping technique is one based on the 10 km National Grid square standard on British Ordnance Survey maps. As well as greatly increasing precision, such a technique allows mechanical plotting from map co-ordinates of the distribution maps. The first maps based on this system were those for vascular plants (Perring & Walters, 1962). Under the auspices of the Biological Records Centre, Monks Wood, this mapping has been extended to non-vascular plants and animals (Perring, 1971). The European Invertebrate Survey (Heath, 1971a) and other similar schemes are a logical development using the U.T.M. (Universal Transverse Mercator) grid and 10/50 km squares. Provisional maps have been produced for certain British, Belgian and European invertebrate groups (Heath, 1970, 1971 a & b; Biernaux, 1971).

With most groups of invertebrates including myriapods the data available at present are so incomplete that 10 km maps show more a pattern of collectors' activities than a genuine geographical distribution. Only by making use of information built up over the course of a number of years will a more complete picture emerge.

HABITAT RECORDING

A system of habitat recording must be usable both in the sense of reducing subjective judgements to a minimum and in being sufficiently straightforward to be used by relatively inexperienced workers. It must also be so designed as to be of value in extracting the maximum useful ecological information. Any such system must therefore necessarily be a compromise between these various demands. Elton (1966) designed a habitat classification covering all possible habitat systems but with particular reference to terrestrial aspects. The latter he divided laterally into four formation types ranging from open ground to woodland and vertically into seven layers, two of which could be sub-divided as necessary (Fig. 1). To these were added various qualifiers and provision was made for edges and mixtures. A similar breakdown was used for the aquatic system and the aquatic/terrestrial transition.

Other aspects of the environment were covered by the domestic, subterranean, high air and general systems. He used the general systems for a number of specific habitats not otherwise included as such. These were "small in compass and scattered throughout the major habitats" and the main resource was "other than a green plant in healthy

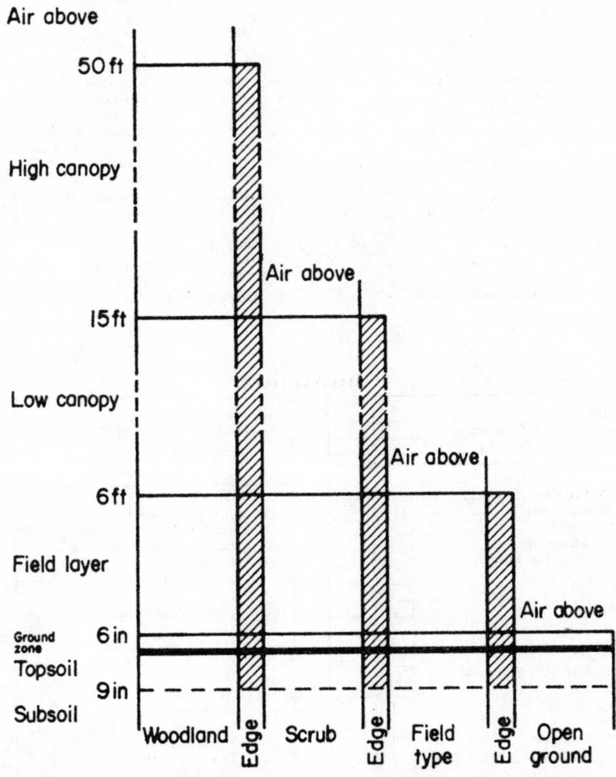

FIG. 1. Elton's Terrestrial Formation Types (from "The Pattern of Animal Communities", London, 1964).

condition". Here were included a number of habitats of importance to myriapods such as dead wood and macrofungi together with human and animal artefacts, carrion and dung. Hand-sorted punched cards for each species record with habitat information based on this system were used for the Wytham Ecological Survey.

The Society for the Promotion of Nature Reserves (S.P.N.R., 1969) designed a habitat classification for the recording of sites of biological interest (nature reserves, etc.). The importance here is in

(22–24) LOCALITY		CENTIPEDES 6200–6211 (1–4
		Date (60–64) | V.C. No. (33–3
RECORDER	DET.	(77–79) V.C.
		Alt. (73–76) | Code No. (65–6

Grid Ref. (25–32)

(5–10)
101 Brachygeophilus truncorum
201 Brachyschendyla dentata
202 monoeci
301 Chaetechelyne montana
302 vesuviana
401 Chalandea pinguis
501 Clinopodes linearis
601 Cryptops anomalans
602 hortensis
603 parisi
701 Dicellophilus carniolensis
801 Geophilus carpophagus
802 electricus
803 fucorum seurati
804 insculptus
805 osquidatum
806 pusillifrater
901 Haplophilus subterraneus
1001 Hydroschendyla submarina
1101 Lamyctes fulvicornis
1201 Lithobius agilis
1202 aulacopus
1203 borealis
1204 calcaratus

(5–10)
1205 Lithobius crassipes
1206 curtipes
1207 duboscqui
1208 erythrocephalus
1209 forficatus
1210 lapidicola
1211 melanops
1212 muticus
1213 nigrifrons
1214 piceus
1215 pilicornis
1216 tricuspis
1217 variegatus
1301 Necrophloeophagus longicornis
1401 Nesoporogaster souletina brevior
1501 Pachymerium ferrugineum
1601 Schendyla nemorensis
1602 peyerimhoffi
1603 zonalis
1701 Scutigera coleoptrata
1801 Strigamia acuminata
1802 crassipes
1803 maritima

Other species:

HABITAT DATA

A 1 tick (obligatory):

	11
Coastal <15km from sea	1
Inland >15km from sea	2

B 1 tick (obligatory):

	12
Urban	1
Suburban/village	2
Rural	3

C 1st order habitats;
1 tick (obligatory):

	13–15
Aquatic: Canal	001
River >5m wide	002
Lake >1 acre (0.4 hectare)	003
Estuary	004
Sea	005
Marsh: Fen	011
Carr	012
Bog	013
Salt marsh	014
Cave/Well/Tunnel: Threshold	021
Dark zone	022
Building: Inside	031
Outside	032
Garden: Domestic	041
Waste ground: <25% veg. cover	051
>25% veg. cover	052
Arable: Cereal crops	061
Root crops	062
Fodder crops	063
Grass ley	064
Market garden/allotment	065
Grassland: Ungrazed	071
Lightly grazed	072
Heavily grazed	073
Mown	074
Scrubland: Dense	081
Open with herbs/grass	082
Woodland: Dense	091
Open with scrub	092
Open with herbs/grass	093
Acid heath/moor: Moss/lichen	101
Grass/sedge/rush	102
Heather	103
Vaccinium (bilberry)	104
Mixed	105
Sand dune: Bare sand	201
Tussocky	202
Dense sward	203
Dune slack	204
Dune heath	205
Other: If none of above	301

D 2nd order habitats;
1 tick (where applicable):

	16, 17
Cold frame	01
Rockery	02
Flower bed	03
Lawn	04
Compost/refuse heap	05
Dung heap	11
Hay (or other) stack	12
Potato (or other) clamp	13
Hedge	21
Roadside verge	22
Embankment/cutting	23
Woodland ride/firebreak	24
Wood fence	25
Dry stone wall	31
Wall with mortar	32
Quarry face	33
Quarry floor	34
Natural cliff face	35
Rock pavement	36
Stablilised scree	37
Unstabilised scree	38
Grike	41
Road/path	51
Dry water course bed	61
'Dry' ditch	71
Wet ditch	72
Shore/water edge/strandline	81
Vegetated stream	91
Unvegetated stream	92
Puddle	93
Pond <1 acre (0.4 hectare)	94
Flood patch	95

Biological Records Centre June 1970 RA 14

the general nature of the site rather than in specific groups of animals and it was designed with British habitats in mind. For terrestrial sites eight major habitat types are defined by vegetation, soil reaction, and water regime with appropriate modification for various types.

HABITAT DATA

E MICROSITE (animal actually found under, on or in); 1 tick (obligatory): 36, 37

Microsite	Code
Stones	01
Shingle	02
Soil/sand	03
Litter	11
Tussocks	21
Bark (living trees or shrubs)	31
Dead wood	32
Dung	33
Carrion	34
Bracket fungi	35
Ant colony (specify if possible)	41
Bird/mammal nest (specify below)	51
Rock	61
Stone or brick work	62
Shore line jetsam	71
Human rubbish/ garbage	81
Other (specify below if possible)	91

F HABITAT QUALIFIERS; 1 tick in each section, where applicable:

(a) Building: Cellar	38: 1
Inhabited/public	2
Uninhabited/outbuilding	3
Ruin	4
Greenhouse (heated)	5
Greenhouse (unheated)	6
(b) Shore: Intertidal	39: 1
Splash zone	2
Between splash zone and 100m	3
100–1000m above H.W.M.	4
(c) Encrustations: Moss	40: 1
Lichen	2
Pleurococcoids	3
(d) Waterspeed: Fast	41: 1
Slow	2
Standing	3
(e) Watercourse bed: Rocks	42: 1
Pebbles	2
Sand	3
Silt	4
Peat	5

G LIGHT LEVEL; 1 tick (obligatory):

Full daylight	43: 1
Half-light/dusk/dawn	2
Dark	3

Other information e.g.
Abundance,..................(number) collected per..................(please state unit of time/ space/volume) (53) Aspect/degree of slope (54) Behaviour (55) Food (56) Predators and Parasites (57) Age structure and Sex ratio (58) etc. (59)

H SOIL/LITTER DETAILS, terrestrial habitats only; 1 tick in each section where applicable: 44, 45

(a) Litter mainly: Oak	01
Beech	02
Birch	03
Sycamore	04
Mixed deciduous	05
Coniferous	11
Mixed decid./conif.	21
Gorse	31
Hawthorn	32
Heathers	33
Sea Buckthorn	34
Litter/veg. mainly: *Carex*	41
Molinia	42
Dactylis	43
Festuca	44
Bromus	45
Brachypodium	46
Grass—species unknown	47
Mixed grass/herbs	51
Nettles	61
Reeds (*Phragmites*)	62
Juncus	63
Bracken	64
Other (specify below)	71
(b) Litter age: Fresh	46: 1
Old	2
Both	3
(c) Litter cover: Exposed	47: 1
Protected by thin veg.	2
Protected by thick veg.	3
(d) Soil/exposed rock: Calcareous	48: 1
Non-calcareous	2
(e) Soil: Heavy clay	49: 1
Clayey	2
Peat	3
Loam	4
Sandy	5
Pure sand	6
(f) Humus type: *Mull*	50: 1
Mor	2

I LOCATION OF ANIMAL; 1 tick in each section where applicable:

(a) Horizon: >3m above ground	51: 1
<3m above ground	2
On ground surface	3
In litter	4
<10cm in soil	5
>10cm in soil	6
(b) Position: In open	52: 1
In crevice	2

FIG. 2. Habitat record card for centipedes, reverse side; opposite page: obverse side.

Only in the cases of woodland and agricultural sites are provisions made for indicating dominant plant species. There is a similar system for aquatic habitats and "Habitat Modifiers" are used as required. Habitat information is then recorded with other details on a habitat record card. Such cards are designed to be used in conjunction with separate species cards of a standard type. Although such a system of recording is useful for the analysis of particular habitats, the use of separate habitat and species cards is in many ways a cumbersome procedure and habitat cards designed for this purpose may lack the precision required for ecological studies of specific groups of animals.

Species cards are now in use for a variety of invertebrate groups. The basis of each card is a space for locality, date and species with, in some cases, space for sex and stage. By themselves these cards can be used for distribution but not habitat studies.

THE ISOPOD/MYRIAPOD SURVEY SCHEMES

The majority of terrestrial invertebrates with the exception of pterygote insects tend to occur in a restricted range of habitats within any formation type; that is soil, surface layers, and other systems. Hence either of the recording schemes mentioned previously is of limited value for either an isopod or myriapod study. Much more specific microhabitat information is required and much greater coverage of domestic sites.

The record card (Fig. 2) originated from the demands for a habitat card for freshwater and terrestrial isopods on which both distributional and ecological information could be recorded. It was modified in the light of experience and to incorporate Chilopoda and Diplopoda; three versions are in use, identical apart from the different species lists for the three groups. The cards are designed for computed usage of the data.

As well as having the standard locality/date/species list format of other schemes, the card incorporates an easily used and comprehensive habitat recording system as an integral part. This system incorporates aspects of the preceding schemes with selection and extension of relevant sections. Most conspicuously the general and domestic systems of Elton are incorporated into the rest of the terrestrial system and expanded to cover a variety of specialized microhabitats. It also incorporates certain sections designed to deal with specific features of the ecology of these animals such as (A) Coastal/Inland, (B) Urban/Suburban or village/Rural, (G) Light level, (I) Location i.e. Horizon and Position.

First Order Habitats (C) represent 12 major habitat types with allowance for other zones and incorporate the terrestrial, aquatic and domestic systems with their qualifiers/modifiers making a total of 41 categories, one of which must be indicated. Second Order Habitats (D) are utilized where relevant and these represent a variety of specialized habitats found in gardens (01–05), farmland (11–13), various sites of particular interest, and aquatic habitats (91–95). The aquatic habitats are for freshwater isopods. Microsites (E) are obligatory and indicate the likely habitat of the species concerned. They include the majority of cryptozoic habitats. Soil/Litter Details (H) represent a fairly typical breakdown of vegetation and soil types, factors which certainly influence the distribution of these animals but about which more information is required. Habitat Qualifiers (F) deal with certain types of habitat and are only used when relevant. Further information on numbers, etc., can be put in when available, as can any points of importance not easily classified (53–59).

There are certainly a number of shortcomings in the card as used at present; most obviously it represents a compromise between the conflicting requirements of terrestrial, freshwater and maritime species and between the needs for specific and general habitat information. It must of course be modified in the light of experience.

Use of the scheme

All the sections on the card will be seen to be coded numerically and each card will be translated into an 80-column punched card for each recorded species. Mechanical analysis of the data will then be carried out for mapping purposes and could also be used to obtain habitat data. It will be possible to use the data on either a habitat or species basis to gain an insight into a number of features of the ecology of these animals.

Table I shows the type of sorting that will become possible. It represents a hand sorting of only a very small number of cards but even with this a few general trends can be seen. One species was used, *Cylindroiulus punctatus* (Leach), and examination of the habitats Arable, Garden, Grassland, Scrubland and Woodland shows its habitat preferences in general terms. Obviously these results are hardly significant because of the small size of the sample but it shows the type of sorting that will be possible using much greater numbers of cards.

The cards will form a continually increasing store of data derived not only from distribution studies but from a variety of sources, hence increasing its value. Further, the need for publications of locality lists as in the past will be reduced with the incorporation of the authority

TABLE I

*Partial analysis of habitat information for **Cylindroiulus punctatus** based on a small number of habitat record cards*

Habitat	Total record cards for habitat	Percentage of cards with *C. punctatus*
		%
All	181	35·9
Coastal		33·8
Inland		37·0
Aquatic	4	0
Dune	3	0
Acid Heath/Moor	5	0
Marsh	3	66·7
Arable	30	0
Garden	15	13·3
Grassland	28	35·7
Scrubland	7	42·8
Woodland	86	56·0
Woodland, dense		57·8
Woodland, open, scrub		64·8
Woodland, open, grass and herbs		45·8

in the stored data to give credit where due. The cards also act as a useful guide to observation and train workers to look for the main features of the animal's environment.

It would seem possible that with appropriate modification of the species list, using new code numbers for those types not already listed and with, perhaps, other minor modifications, the scheme could be extended to other parts of Europe. This would not only increase the quantity of information but also give valuable data on comparative ecology. With suitable amendment there seems no reason why this or a similar habitat card scheme could not be extended to other groups of terrestrial invertebrates.

ACKNOWLEDGEMENTS

We wish to acknowledge with thanks the help and encouragement given to the Myriapod Survey by those workers engaged in the Isopod

Survey Scheme who were the originators of this recording system and in conjunction with whom the scheme has been developed; and also by Mr. J. Heath and the Biological Records Centre, Monks Wood, who arranged for the production of the record cards.

References

Balfour-Browne, F. (1931). A plea for uniformity in the method of recording insect captures. *Entomologists' mon. Mag.* **67**: 183–193.

Biernaux, J. (1971). Myriapodes Blaniulidae et Iulidae. In *Atlas Provisoire des arthropodes non insectes de Belgique*. Leclerq, J. & Lebrun, P. (eds.). Gembloux: Faculté des Sciences Agronomiques de l'Etat.

Elton, C. S. (1966). *The pattern of animal communities*. London: Methuen.

Heath, J. (1970). *Provisional atlas of the insects of the British Isles.* **I** Lepidoptera Rhopalocera. Huntingdon: Biological Records Centre.

Heath, J. (1971a). The European Invertebrate Survey. *Acta ent. Scand.* **28**: 27–29.

Heath, J. (1971b). *European Invertebrate Survey,* Instructions for recorders. Huntingdon: Biological Records Centre.

Hynes, H. B. N. (1958). A key to the adults and nymphs of British stoneflies (Plecoptera). *Scient. Publs Freshwat. biol. Ass. Br. Emp.* No. 17: 1–86.

Perring, F. H. (1971). The Biological Records Centre—a data centre. *Biol. J. Linn. Soc.* **3**: 237–243.

Perring, F. H. & Walters, S. M. (1962). *Atlas of the British flora.* London: Nelson.

Ragge, D. R. (1965). *Grasshoppers, crickets & cockroaches of the British Isles.* London: Warne.

S.P.N.R. (Society for the Promotion of Nature Reserves) (1969). Biological Sites Recording Scheme. *S.P.N.R. Conservation Liaison Comm. Technical Bulletin* 1. Alfold, Lincs.: S.P.N.R.

Symp. zool. Soc. Lond. (1974) No. 32, 621–628.

SOME ASPECTS OF THE ECONOMIC IMPORTANCE OF MILLIPEDES

A. N. BAKER

Broom's Barn Experimental Station, Higham, Bury St. Edmunds, Suffolk, England

SYNOPSIS

Millipedes, especially *Brachydesmus superus* (Latz.) and *Blaniulus guttulatus* (Bosc.) can stunt and even kill sugar-beet seedlings in the spring by their aggregated feeding on the young roots. Fortunately, seedlings are at risk over only a limited period, up to about mid-June, after which the millipedes retreat deeper into the soil. The intensity of damage from millipedes varies considerably from year to year and is difficult to forecast, being largely dependent both on rate and vigour of seedling growth and on the numbers and activity of the millipedes. Present investigations, both in the laboratory and in the field, aim at a better understanding of the movements of populations in the soil in order to develop means of protecting the germinating seedling until it is established.

INTRODUCTION

The undesirable effect of millipedes on cultivated plants has been recognized for at least 80 years, but the true nature of the relationship is still controversial; they feed not only upon decaying organic material but also, under some circumstances, upon living plant tissue; Ormerod (1890) observed *Blaniulus guttulatus* attacking strawberry fruits and germinating mangold seed in 1885. Millipede feeding habits were investigated by Brade-Birks (1930), and Rolfe (1937–39) listed species and reported millipede damage to a wide range of horticultural and field crops including beans, peas, cucumbers, cabbage, cereals, lettuce, potatoes and sugar-beet; Wilson (1943) commented that, although they are usually secondary pests, millipedes will feed on the skin of soft skinned varieties of potatoes. Baurant (1964) has described the damage to sugar-beet seedlings attributable to two blaniulid species in Belgium. Spring-sown crops are particularly prone to attack because they are at their tenderest stage when soil conditions are conducive to millipede activity. Warm, moist, glasshouse soils ensure large populations and intense activity throughout the year and *Oxidus gracilis* (Koch), a tropical species, is common in glasshouses both in the British Isles and in south and west U.S.A. (Edwards & Gunn, 1961; Henneberry & Taylor, 1961).

This paper deals primarily with the damage and field behaviour of the most common species that occur where sugar-beet is grown in eastern England and is part of the results from a three-year study on some soil-inhabiting pests of sugar-beet.

MILLIPEDES AS PESTS OF SUGAR-BEET SEEDLINGS

Millipedes cause loss of sugar-beet seedlings; this is especially important with current agricultural trends to minimize labour by sowing fewer seeds at wide spacing and by increased herbicide usage which eliminates any other competing seedlings of other species. Monogerm sugar-beet seed is sown in March or April, between three and eight in. apart in rows 20 in. apart. At this time temperatures in the seedbed may range from 5°C to 7°C and up to five weeks may elapse before half the seeds have germinated. If millipedes are active in the seedbed they may feed on the emerging radicle, unfolding cotyledons and hypocotyl below soil level, making lesions which sometimes coalesce and cause complete separation of these parts, often resulting in death of the seedling. More usually, however, millipedes attack later when the cotyledons are above ground; root damage impairs the seedling's ability to extract water and nutrients from the soil and, if severe, may kill the plant. Nutrient deficiency and wilting are common symptoms of millipede damage to roots (Baker, 1971a).

THE IMPORTANT SPECIES

Advisory Entomologists report that blaniulids cause most damage, especially *Blaniulus guttulatus* (see Cloudsley-Thompson, 1950) and sometimes *Boreoiulus tenuis* (Bigler) and *Archeboreoiulus pallidus* (Brade-Birks). The polydesmids are also sometimes implicated, especially *Brachydesmus superus* which is very common in arable fields and gardens.

A consistent and important feature of the behaviour of millipedes, and of other soil-inhabiting pests of sugar-beet seedlings, is their tendency to move into the seed rows after sowing and to aggregate around the seedlings. This is particularly noticeable with the blaniulids and up to 121 per seedling have been recorded. Soft plant materials are the best food source for millipedes and seedlings are susceptible to damage for a limited period; this usually extends from germination to the four-leaf stage—a period of eight weeks at the most.

Damage is usually most severe in May and is rare by July. Soil conditions conducive to rapid germination and growth also encourage movement and feeding of millipedes in the seedbed. During July the millipedes move deeper into the soil and any fresh damage to the taproot is unlikely. *B. superus and B. guttulatus* differ not only morphologically but also in their field behaviour, and their relative importance as pests reflects these behavioural differences.

BRACHYDESMUS SUPERUS: A SOIL SURFACE SPECIES

B. superus, the commonest polydesmid of arable land, is a most active animal and lives in the loose topsoil except when moulting. If undisturbed it can be found throughout the year near the soil surface or under surface vegetation, but ploughing the land early in winter buries most of the millipedes and they are then difficult to find. In late winter this millipede sometimes aggregates around ploughed-in cereal stubble, manure, or sugar-beet crowns which have decayed enough to be attractive. *B. superus* may cause damage to early-sown sugar-beet before the arrival of the blaniulids in the seed bed from the deeper soil.

VERTICAL DISPLACEMENTS OF THE BLANIULIDAE

The depth distributions in sugar-beet fields of the blaniulids, *B. guttulatus* and *A. pallidus*, have been extensively studied in Belgium (Pierrard, Bonte & Baurant, 1963; Biernaux & Baurant, 1964). Successive 10 cm deep soil samples provided some evidence of a vertical displacement, apparently controlled by differences in soil temperature and moisture; Dowdy (1943–1944) and Blower (1970) also mention vertical movements of millipedes in response to temperature. Most of the blaniulids were found in the 30–90 cm deep soil zone in winter but by the end of April, the time when attacks on sugar-beet seedlings first occur, most are in the top 30 cm. This upward movement of blaniulids has been shown to start when the temperature of their surroundings rises above 4–6°C (Pierrard *et al.*, 1963). During the summer months they move downwards from the top soil; Biernaux (1968) demonstrated their avoidance of soil with a moisture content of less than 12%. After a brief reappearance near the soil surface in the autumn, when the soil is remoistened, they retreat to deeper soil layers.

AGE-STRUCTURES, ACTIVITY AND DAMAGE

B. superus is an annual; the adults produce one brood and then die (semelparity). The duration of moulting, the time spent in each stadium, and the longevity of adults are determined by the temperature of the habitat. Because they are surface-inhabiting animals, they will be directly influenced by seasonal temperature fluctuations. Populations have relatively narrow age-structures. In two consecutive years, populations in the autumn have had a preponderance of adults (64% in December at one site) with a sex ratio of up to three females to one male. Mating can be readily observed in September and October on the soil surface, in soil crevices and under leaves. *Brachydesmus* nests, with egg clusters, have been found occasionally on compacted soil at plough depth in January but oviposition occurs most readily in Spring. No hatching occurs at 2–4°C, but takes only about 20 days at 15–20°C. The threshold value for hatching, in the region of 7–11°C, agrees well with the occurrence of stadium I in soil samples taken in April and May when these temperatures exist in the top 10 cm of soil. At 15–20°C larvae attain stadium V within three months, whereas they fail to moult at 2–4°C. The adult and penultimate stadia are the most harmful to plants and there is evidence that severe damage occurs only when seedling germination coincides with a large adult population; thus factors which control the rate of development of *Brachydesmus* can affect its potential as a pest (Baker, 1972a).

B. guttulatus adults may breed in successive years (iteroparity), producing a wide spectrum of ages in the population. This millipede may eventually reach stadia XVI or XVII; assuming three moults per year the calculated life span could be five or six years (Biernaux, 1967b). Field observations suggest both spring and autumn oviposition peaks (Biernaux, 1967b; Kinkel, 1955). Information about the rate of development indicates that larvae from eggs laid at the end of April can reach stadium V by October and then overwinter; the adult stage is reached only in the second year of growth (Biernaux, 1967a). All stadia, except I, of *Blaniulus* have been observed feeding upon sugar-beet seedlings and they may be found in the seedbed from March until late June (Baker, 1971a; Biernaux, 1967a).

WORK IN PROGRESS ON MILLIPEDE BEHAVIOUR

Studies of behaviour in the laboratory may help in understanding what happens in the field. My investigations suggest that to interpret millipede movements solely as a response to variations in temperature and moisture may be an over-simplification of the relationship between millipedes, plants, and their common environment.

Laboratory studies support field evidence that an increase in soil temperature, similar to that in the field in spring, leads to increased activity, an increase in numbers of larvae and a general dispersal.

Some millipedes move upwards from the sub-soil into the seedbed; here, large diurnal temperature fluctuations ensure intense activity. In the laboratory activity is initiated by both increasing and decreasing temperature but constant temperatures have a depressing effect. The millipedes aggregate around the fruit coat as soon as the radicle and embryo has become exposed on germination. When excised parts of seedlings are offered separately they feed preferentially upon young cotyledon tissue; they also aggregate and feed upon 5% agar discs but prefer those incorporating 5% glucose or 5% sucrose solutions (Baker, 1971a).

Conditions which cause millipedes to stop feeding are important because the longer they feed, the greater the damage they cause. In warm soil they tend to moult more frequently, and therefore stop feeding and leave the surface root zone to build moulting chambers deeper in the soil. Also, increasing plant maturity is associated with a dispersal from the seedbed. Laboratory experiments demonstrate a sensitivity both to soil moisture and to low temperatures, the latter resulting in a positive geotaxis and aggregation in the deepest penetrable soil layers; similar effects can be observed in the field.

CONCLUSIONS

For millipedes to cause significant damage to seedlings they must be sufficiently numerous, sufficiently active to aggregate around the germinating seeds and they must find the seedlings an attractive food source on which to continue feeding.

It is difficult to decide what conditions lead to the build-up of large populations. Recent surveys of millipede populations during the autumn suggest that soil type and geographical location, especially in respect of topography, exert an influence. Millipedes, mainly *B. superus* but some *B. guttulatus*, were found most frequently on the heavier soils—clay loams, silts and heavy peats (Baker, 1971b, 1972b). They were particularly common in certain low lying regions of East Anglia near rivers and estuaries. Little is known of the biological regulation of millipede populations in the field, but there is some evidence that the predatory soil mite *Pergamasus quisquiliarum* Canestrini feeds upon stadium I *B. superus*.

Resistence of a seedling to attack depends upon both its genetic make-up and the conditions which govern the rate of growth. Millipede activity appears to be under the control of edaphic factors, mainly temperature and moisture; it is the most difficult aspect of the syndrome to control, since it is largely governed by climate. However, compacting the soil in order to limit the space for movement in the root zone seems a promising, though by no means new, approach to decrease activity.

ACKNOWLEDGEMENTS

The author wishes to thank R. A. Dunning, J. G. Blower and R. Hull for their help in the preparation of the manuscript.

References

Baker, A. N. (1971a). *Rep. Rothamsted exp. Stn* **1970**: 248–250.

Baker, A. N. (1971b). Report on sugar beet investigations on pests and diseases, fertilizers, seed production, agronomy. *S.B.R.E.C. Comm. Pap.* No. 1177: Appendix 1 (c).

Baker, A. N. (1972a). *Rep. Rothamsted. exp. Stn* **1971**: 271–273.

Baker, A. N. (1972b). Report on sugar beet investigations on pests and diseases, fertilizers, seed production, agronomy. *S.B.R.E.C. Comm. Pap.* No. 1252: Appendix 1 (i).

Baurant, R. (1964). Les dégâts d'iules mouchetés sur jeunes betteraves. *Bull. Inst. agron. Stns Rech. Gembloux* **32**: 3–11.

Biernaux, J. (1967a). *La destruction des Iules de la Betterave peut-elle se faire par une seule intervention printanière?* (Communication présentée à la réunion du sous-groupe "Iules" de l'Institut International de Recherches Betteravières (I.I.R.B.) le 22 fevrier 1967). Mimeogr.

Biernaux, J. (1967b). *Biologie des Iules de la Betterave.* (Communication présentée à l'Assemblée mensuelle de l'Association pour les Études et Recherches de Zoologie Appliquée et de Phytopathologie (AERZAP) le 12 avril 1967). Mimeogr.

Biernaux, J. (1968). Influence du taux d'humidité du sol sur la localisation en profondeur de "Iules de la Betterave" au cours de la bonne saison. *Bull. Inst. Rech. agron. Stns Gembloux* N.S. **3**: 234–240.

Biernaux, J. & Baurant, R. (1964). Observations sur l'hibernation de *Archeboreoiulus pallidus* Br.-Bk. *Bull. Inst. agron. Stns Rech. Gembloux* **32**: 290–298.

Blower, J. G. (1970). The millipedes of a Cheshire wood. *J. Zool., Lond.* **160**: 455–496.

Brade-Birks, S. G. (1930). Notes on Myriapoda—XXXIII. The economic status of Diplopoda and Chilopoda and their allies. Part II. *Jl S.-east. agric. Coll. Wye* No. 27: 103–146.

Cloudsley-Thompson, J. L. (1950). The economics of the "Spotted Snakemillipede". *Blaniulus guttulatus* (Bosc.) *Ann. Mag. nat. Hist.* (12) **3**: 1047–1057.

Dowdy, W. W. (1943–1944). The influence of temperature on vertical migration of invertebrates inhabitating different soil types. *Ecology* **25**: 449–460.

Edwards, C. A. & Gunn, E. (1961). Control of the Glasshouse millipede. *Pl. Path.* **10**: 21–24.

Henneberry, T. J. & Taylor, E. A. (1961). Control of millipedes in greenhouse soil. *J. econ. Ent.* **54**: 197–198.

Kinkel, H. (1955). Zur Biologie und Ökologie des getüpfelten Tausendfusses *Blaniulus guttulatus* Gerv. *Z. angew. Ent.* **37**: 401–436.

Ormerod, E. A. (1890). *A manual of injurious insects and methods of prevention.* London: Simpkin, Marshall, Hamilton, Kent & Co. Ltd.

Pierrard, G., Bonte, E. & Baurant, R. (1963). Observations sur l'hibernation de *Blaniulus guttulatus* Bosc. *Bull. Inst. agron. Stns Rech. Gembloux* **31**: 127–141.

Rolfe, S. W. (1937–1939). Notes on Diplopoda IV–VI. The recognition of some millipedes of economic importance. *Jl S.-east agric. Coll. Wye* No. 40: 99–107 (1937): No. 42: 214–215 (1938); No. 44: 180–182 (1939).

Wilson, G. F. (1943). Potato tuber injury due to soil pests. *Jl R. hort. Soc.* **68**: 206–214.

Discussion

Enghoff: Have you ever found *Boreoiulus tenuis* as a pest of sugar beet?

Baker: Yes, it has been identified as the main seedling pest at Shouldham, near Downham Market, Norfolk, and at Bottisham, Cambridgeshire, both shallow, chalky loam soils over chalk. It does exactly the same sort of damage as the other blaniulids—but in general it is less common.

Biernaux: In Belgium, we have *Archeboreoiulus pallidus* and *Boreoiulus tenuis* together with *Blaniulus guttulatus*.

Jeekel: I think it was Cloudsley-Thompson (1950)* who postulated the idea that a period of drought would increase the severity of millipede damage; he suggested the millipedes would attack the crop to find water. Does this conflict with your idea that they go down when it is dry?

Baker: I have shown in the laboratory that millipedes will damage seedlings even in soil with suitable moisture content, when they are not suffering from a water deficit. I find it hard to accept the hypothesis that they attack plants to find water. I think that millipedes become active as the result of increased soil temperatures, after the prolonged winter inactivity. They start moving; an individual comes across a seedling and attacks it. There seems to be some attraction between millipedes; once one has started to feed, others come along and they form a swarm; the sugars, too, may help to maintain their interest in the plant. Of course when a plant, especially a young seedling, is being deprived of water then growth is restricted, and because it can neither mature enough to harden its root tissue against millipede attack, nor compensate by growing replacement water- or nutrient-absorbing tissue, the seedling would therefore be expected to succumb more easily to a given intensity of millipede attack under "dry" soil conditions. My conclusions, as a result of field and laboratory observations, only go as far as proving that millipedes can cause primary damage to seedlings when the seedlings are grown in soil with adequate moisture for growth. I have not been able to compare directly the amount of damage

* See list of references, p. 626.

done to seedlings grown in "dry" versus "wet" soils. But I have done an experiment where 5% agar discs incorporating a sugar were substituted for seedlings and found no significant difference in the amount of material eaten by the millipedes. I therefore find it hard to accept the hypothesis put forward.

BRADE-BIRKS: Do you have any idea how an attack begins? Often a crop remains unattacked one year and then there is a big influx of millipedes the next. Is it because climatic conditions in spring are suitable for the development of eggs or because the introduction of manure and organic waste material from the factories provides the right conditions for development?

BAKER: I think both these factors may be important, but since 1935 the growing of sugar beet in successive years was prohibited following the introduction of a clause in factory contracts; there is usually a three or four year gap between the growing of beet during which time the land is under cereals.

TURNER: Do millipedes transmit viruses?

BAKER: I do not know. I would not have thought they would be ideal vectors because blaniulids, at least, are not very mobile.

HEATH: On what types of soil do you get the most severe attacks?

BAKER: A severe attack usually occurs on land which supports large numbers of millipedes and which usually has a history of millipede trouble—land which is low lying, has a high organic content and a high cereal stubble content. The soils usually tend to be the silty loams, silts and the heavier peats; the soils must be porous to allow the millipedes to escape adverse conditions; these are the features of a good millipede soil!

Symp. zool. Soc. Lond. (1974) No. 32, 629–643.

NOTE À PROPOS DES DIPLOPODES NUISIBLES AUX CULTURES TEMPÉRÉES ET TROPICALES

G. PIERRARD

Institut de Recherches du Coton et des Textiles exotiques, Station de M. Pesoba, Mali

and

J. BIERNAUX

Centre de Zoologie et d'Entomologie Appliquées, Faculté des Sciences Agronomiques de l'État, Gembloux, Belgique

SYNOPSIS

Millipedes usually attack those parts of the plant within or in contact with the soil, such as roots, storage organs and seeds. They confine their attention to seeds, seedlings and the new fruits in course of formation in the soil. It is difficult to determine from the literature whether millipedes were the primary cause of damage or whether they were merely profiting from the attack of another organism.

A list of temperate crops damaged and the species responsible is given. Damage to sugar beet in the seedling stage is the most frequently reported. Blaniulidae, Iulidae and less frequently Polydesmidae are implicated. The majority of cases of primary damage are ascribed to *Blaniulus guttulatus* but along with this are two other species of Blaniulidae not normally distinguished from it. Damage by iulids and polydesmids is usually secondary. For sugar beet, control of the millipedes by chlorinated hydrocarbons, especially Heptachlor, has been most effective since the effect persists up to the 6–8 leaf stage after which the seedlings are immune from attack. In place of the organochlorines now banned in some countries, organo-phosphorous compounds and carbamates are used. These are less dangerous—and less effective!

Examples of diplopod damage to tropical crops, mainly cotton and groundnuts in Central Africa, are given. Cotton seed is attacked both before and after germination and as seedlings up to the 4-leaf stage. Groundnuts are similarly attacked but are also vulnerable during the formation of new seed. The species responsible belong to the Spirostreptidae and, especially, the Odontopygidae. The biology of these millipedes is unknown. Control by a seed-dressing of insecticide is possible for cotton but groundnuts are susceptible to damage during formation of the new generation of seed and this method is not appropriate; a bait of an attractive substance in which toxic materials are incorporated, is used.

INTRODUCTION

Il est bien connu que les diplopodes présentent un intérêt économique positif du fait du rôle non négligeable qu'ils jouent en certains biotopes dans la fragmentation des composants des litières forestières, y compris les bois morts, ainsi que dans l'incorporation des matières organiques dans le sol.

Malheureusement on ne peut ignorer les méfaits dont certains d'entre eux sont responsables et qui sont de deux types:

1. les envahissements des lieux habités par des mille-pattes tels *Schizophyllum sabulosum* (Linné) et *Tachypodoiulus niger* (Leach) dans les pays d'Europe occidentale et par quelques autres espèces en d'autres lieux.
2. Les déprédations par morsures aux graines en germination ou aux jeunes plantules ainsi qu'aux fruits en formation dans le sol ou en contact avec lui.

Ce sont ces dégâts aux plantes cultivées, aussi bien en régions tempérées que tropicales, que nous nous efforcerons d'examiner.

Il y a lieu de faire remarquer, en guise de préambule, qu'en matière de myriapodologie appliquée, nous ne connaissons encore que peu de choses: d'une part, la biologie de la plupart des diplopodes est loin d'être connue et, d'autre part, la nature et l'importance des dégâts, de même que les espèces responsables, demeurent encore ignorées, particulièrement dans les régions où l'agriculture est encore peu développée. Or, la connaissance de la biologie des espèces dommageables est fondamentale en matière de recherches des moyens de protection à préconiser dans la lutte contre les déprédateurs.

GÉNÉRALITÉS

Le régime alimentaire des diplopodes en relation avec les dégâts qu'ils occasionnent

Les diplopodes paraissent bien être des animaux peu spécialisés en matière de régime alimentaire. Il semble en effet que toutes les espèces sont relativement polyphages: parmi celles qui sont connues, aucune ne dépend exclusivement pour son alimentation d'une seule plante-hôte ou des plantes-hôtes d'une même famille comme cela s'observe pour certains insectes.

Lorsqu'on examine les contenus stomacaux et les fèces des diplopodes, on observe toujours des particules minérales en quantitiés importantes. Une autre partie des matières ingérées est composée de débris, en voie de décomposition. Il peut s'agir soit de parties de plantes ligneuses telles que feuilles mortes, brindilles et morceaux de bois tombés sur le sol également de fragments provenant de branches ou de troncs d'arbres morts, debouts ou abattus, en état de décomposition plus ou moins avancé, soit de parties tendres, non ligneuses, de végétaux en décomposition, soit encore de cadavres d'animaux. Enfin on peut dire que certaines espèces, probablement plus nombreuses qu'on se l'imagine, éprouvent le besoin, voire la nécessité d'ingérer, à certaines époques critiques de leur développement ou de la reproduction, des matières végétales vivantes, de préférence des tissus jeunes, méristématiques (Breny, 1964).

Lorsque ces derniers concernent des plantes bien développées, celles-une paraissent pas souffrir des déprédations; par contre, lorsqu'il s'agit de plantes délicates à un stade particulier, telles que des graines en voie de germination ou des plantules, les déprédations causées par les diplopodes peuvent provoquer des retards de croissance, des affaiblissements ou même la mort du végétal, que celle-ci soit directe ou indirecte. S'il s'agit d'une plante cultivée et que les déprédations sont répétées sur de nombreux pieds, on parle alors de dégâts. Dans ces cas, on observe des morsures sur toutes les parties hypogées des plantes telles que les racines, les graines en germination qui peuvent être vidées de leur contenu, les tigelles, les cotylédons ou les fruits en formation dans le sol; plus rarement les dégâts sont observés sur les parties basses des plantes plus ou moins en contact avec le sol.

Dans d'autres cas, certains diplopodes semblent rechercher de préférence les parties aqueuses, souvent riches en sucres, des organes de végétaux se trouvant dans le sol ou en contact avec lui, tels les melons, concombres, pommes de terre, fraises, bulbes divers. Lorsqu'ils en ont l'occasion, ils n'hésitent pas à y pénétrer et même à y séjourner; ils y trouvent à la fois la nourriture ainsi qu'une humidité ambiante élevée qu'ils recherchent en beaucoup de circonstances.

Le problème de la détermination exacte des dégâts

Dans un certain nombre de circonstances, il est malaisé de déterminer si on a affaire à un dégât réel, "primaire", c'est-à-dire occasionné à un végétal intact et en bon état sanitaire, ou si le diplopode a profité d'une ouverture réalisée dans la plante par un autre déprédateur ou s'il a été attiré par une partie de la plante en mauvais état, en voie de décomposition plus ou moins avancée et donc déjà endommagée et souvent sans valeur économique.

Lorsque l'on passe en revue la bibliographie qui se rapporte aux dégâts de diplopodes, on remarque que dans un bon nombre de cas les auteurs signalent avoir observé la présence d'iules sur telle ou telle partie de végétal; ils ne spécifient pas s'il s'agit d'une partie du végétal encore utile. C'est ainsi qu'on ne peut considérer comme dégât réel le cas des tubercules de pommes de terre abandonnés sur le terrain après la culture ou de pommes ou encore de poires tombées de l'arbre et sur lesquels se sont rassemblés des iules parfois en nombres importants. On ne peut pas non plus incriminer des diplopodes qui dévorent des cotylédons de pois ou de haricots (provenant de semence) après la germination et la parfaite levée des plantules.

Il y a donc lieu d'être circonspect lorsqu'on signale des déprédations de diplopodes: il faut s'assurer que le végétal a bien été attaqué par eux et par eux seuls, et qu'il était parfaitement sain avant le début des morsures. Il convient encore d'évaluer la gravité des dégâts en spécifiant dans toute la mesure du possible le pourcentage des plantes attaquées ou la superficie ayant subit les dommages et surtout l'influence de ceux-ci sur la production.

Nous avons certes la conviction que, dans beaucoup de cas, les observations rapportées par les auteurs ne se rapportent pas à des dégâts réellement importants mais constituent en réalité des indications sur les diverses nourritures pouvant être appréciées par les diplopodes.

Le problème de la détermination des espèces de diplopodes nuisibles aux cultures

La détermination des espèces de diplopodes est relativement malaisée pour un non-spécialiste. Il en résulte assez fréquemment que des dégâts de diplopodes soient rapportés sans préciser l'espèce en cause. Il est bien certain aussi que bon nombre des dégâts attribués à telle ou telle espèce sont en réalité le fait d'autres espèces. Nous sommes ainsi persuadés que certaines déprédations aux betteraves attribuées au seul *Blaniulus guttulatus* sont dues en partie à *Archeboreoiulus pallidus* et à *Boreoiulus tenuis*. Nous nous basons pour affirmer cela sur le fait que l'aspect externe des trois espèces est fort semblable et qu'en Europe occidentale toutes les trois sont habituellement présentes autour des graines et des plantules à l'époque des semis. Or, on ne parlait jusqu'à ces dernières années que de dégâts de *Blaniulus guttulatus*.

C'est donc avec certaines réserves que nous énumérons ci-après les diverses espèces de diplopodes citées comme dommageables dans la littérature examinée.

Il est en outre bien connu que pour beaucoup de régions il reste de nombreuses espèces à découvrir et à décrire. Il est certain que parmi elles plusieurs peuvent occasionner des dommages aux cultures. Il y a donc lieu chaque fois qu'une telle espèce est observée, d'expédier des échantillons à un myriapodologiste systématicien.

DIPLOPODES NUISIBLES AUX CULTURES TEMPÉRÉES

Schubart (1942) a publié une liste bibliographique de 282 travaux consacrés aux rapports des myriapodes avec l'agriculture. Après lui,

Remy (1950) fit paraître une liste complémentaire comprenant 63 références de publication traitant de cette question. Depuis lors de très nombreux travaux ont encore été consacrés au même sujet. Le problème est donc vaste et il ne peut être question d'énumérer ici tous les cas observés.

Principales plantes cultivées endommagées par les diplopodes

Etant donné la polyphagie remarquable des diplopodes nuisibles, il est normal que les déprédations causées par ces arthropodes soient assez variées.

A titre indicatif, nous citerons les principales plantes cultivées sur lesquelles des dégâts ont été signalés:

Sur graines en germination

betterave—*Beta vulgaris*
haricot—*Phaseolus vulgaris*
pois—*Pisum sativum*
carotte—*Daucus carota*
luzerne—*Medicago sativa*
avoine—*Avena sativa*
orge—*Hordeum* sp.
blé—*Secale cereale*
maïs—*Zea maïs*

Sur racines

betterave—*Beta vulgaris*
carotte—*Daucus carota*
concombre—*Cucumis sativus*
laitue—*Lactuca sativa*
choux—*Brassica oleracea*

Sur bulbes

oignon—*Alium cepa*
ail—*Alium sativum*
tulipes—*Tulipa* sp.
lys—*Lilium candidum*
jacinthes—*Hyacinta* sp.

Sur tubercules ou racines tuberculées

pomme de terre—*Solanum tuberosum*
rutabaga—*Brassica napobrassica*

Sur fruits tombés sur le sol ou le touchant

citrouille—*Citrilus vulgaris*
concombre—*Cucumis sativus*
courge—*Cucurbita pepo*
haricot—*Phaseolus vulgaris*
fraisier—*Fragaria vesca*

Sur bourgeons et tiges enterrés

houblon—*Humulus lupulus*
vigne—*Vitis vinifera*
asperge—*Asparagus officinalis*

Divers

artichaut—*Cynara scolymus*
champignons
cactus

Si on essaye de classifier les principales plantes endommagées par ordre d'importance économique décroissant, et en ne tenant compte que des dégâts signalés au cours des dernières années, on peut adopter l'ordre approximatif suivant: betterave sucrière, pois, haricot, fraisier, pomme de terre, choux, concombre, champignons.

Principales espèces de diplopodes responsables

La liste disposée ci-dessous énumère les principales espèces de diplopodes dommageables aux cultures avec, entre parenthèses, les plantes le plus souvent citées comme attaquées.

Blaniulidae

Blaniulus guttulatus Bosc (betterave sucrière, fraisier, carotte, courge, concombre, maïs, haricot, pois, houblon, oignon, ail, tulipes, jacinthes, laitue, pomme de terre, champignons, froment, choux, asperge)*.

Boreoiulus tenuis Bigler (betterave sucrière)

Archeboreoiulus pallidus Brade-Birks (betterave sucrière)

Iulidae

Cylindroiulus londinensis Leach—syn.: *C. teutonicus* Pocock—(pomme de terre, melon, carotte)

C. latestriatus Curtis—syn.: *C. frisius* Verhoeff—(pomme de terre, légumes et fleurs ornementales diverses)

C. britannicus Verhoeff (laitue, champignons)

Brachyiulus littoralis Verhoeff—syn.: *Br. pusillus* Leach—(artichaut, houblon, carotte)

Tachypodoiulus niger Leach—syn.: *T. albipes* Koch—(pomme de terre, betterave).

Polydesmidae

Polydesmus coriaceus Porat (blé, fraisier)

P. angustus Latzel—syn.: *P. complanatus*—(betterave sucrière, cactus, choux, haricot, fraisier, oignon, carotte, artichaut, blé, pois, pomme de terre)

* Comme déjà signalé, il est certain que des dégâts attribués à *Blaniulus guttulatus* sont également dus à *Boreoiulus tenuis* et *Archeboreoiulus pallidus*.

Brachydesmus superus Latzel (betterave sucrière)

Strongylosomidae

Orthomorpha gracilis Koch—syn.: *Oxidus gracilis* Koch et *Paradesmus gracilis* Cook—(tomate).

C'est dans les pays d'Europe occidentale que des dégâts de diplopodes aux plantes cultivées ont été le plus souvent décrits. Cela est certes dû au fait que c'est dans ces régions que les techniques culturales sont très poussées et les rendements très élevés. De ce fait, les cultures font l'objet de soins attentifs et les moindres dégâts pouvant provoquer des baisses de rendement sont rapidement décelés et signalés. Quelques déprédations ont aussi été observées en d'autres régions, notamment en Amérique du nord et dans la partie méridionale de l'Amérique du sud.

En Europe, si l'on excepte les quelques rares espèces dont la nuisance a été décrite en l'une ou l'autre circonstance, ce sont les "iules de la betterave" (*Blaniulus guttulatus*, *Boreoiulus tenuis* et *Archeboreoiulus pallidus*) qui préoccupent le plus les cultivateurs (Biernaux, 1966). Les *Polydesmus* peuvent également endommager les plantes cultivées quoique dans une mesure sensiblement moindre.

Il semble bien que *Cylindroiulus londinensis* et *C. latestriatus*, ainsi que *Brachyiulus littoralis*, *Brachydesmus superus* et *Tachypodoiulus niger* ne sont que rarement dommageables aux cultures. Il s'agit plutôt de déprédateurs occasionnels qui, le plus souvent, ne s'attaquent qu'à des parties de plantes en mauvais état sanitaire ou déjà endommagées par d'autres animaux.

Les dégâts d'iules de la betterave peuvent avoir des conséquences considérables sur les récoltes. Ainsi on a pu évaluer que dans un champ moyennement attaqué, la diminution de rendement était de l'ordre de 22% en poids de betterave et de 25% en poids de sucre (Biernaux, 1966).

Moyens de lutte

Les ennemis naturels des iules de la betterave et des autres diplopodes nuisibles ne paraissent pas susceptibles de réduire leurs populations d'une manière efficace. Il semble que ce soient plutôt les facteurs écologiques qui régissent la grandeur des populations et donc le dépassement du seuil de nuisance.

Ce sont certaines conditions de température et d'humidité des couches superficielles de sol au moment du semis et de la germination qui semblent favoriser les déprédations. Il y a donc lieu de traiter les cultures de betterave préventivement, tout au moins dans les régions où la présence des iules a été décelée antérieurement. Lorsque les dégâts sont commis et qu'on s'en aperçoit, il est toujours trop tard pour effectuer un traitement curatif par application d'un produit chimique en surface: en effet il a été démontré que la majorité des iules de la betterave vivent profondément enfouis dans le sol (Pierrard, Bonte & Baurant, 1963; Biernaux & Baurant, 1964); ils viennent commettre leurs dégâts en s'approchant des radicelles par en-dessous; ils ne sont donc pas incommodés par les produits insecticides épandus en surface qui ne peuvent pénétrer à plus de un ou deux centimètres dans le sol. Il est donc conseillé de traiter le sol au moment ou avant le semis et d'incorporer si possible le produit dans les six premiers centimètres de façon à imprégner la terre se trouvant autour des graines et sous elles.

Comme matière active efficace contre les iules de la betterave ce sont les organo-chlorés et plus spécialement l'heptachlore qui ont procuré les meilleurs résultats. Ils possèdent un très bon pouvoir iulicide et présentent également l'avantage de garder leur efficacité dans le sol pendant un temps suffisamment long pour protéger la betterave pendant la période critique de germination et de première croissance, jusqu'au moment où elle atteint le stade de six à huit feuilles; à ce moment elle ne peut plus être gravement endommagée par les iules (Baurant, 1964).

Cependant l'utilisation de ce groupe d'insecticides est dès à présent interdit dans certains pays et il est probable qu'il le sera dans les autres, dans des temps plus ou moins longs. Il a donc été nécessaire de rechercher d'autres produits valables, notamment des insecticides organo-phosphorés ou des carbamates. Les recherches sont en cours depuis un certain temps déjà mais on peut dire que jusqu'à présent on n'a pas trouvé de produits de remplacement présentant des qualités équivalentes à celles de l'heptachlore.

En Belgique, où le problème des iules de la betterave est important, on préconise actuellement l'usage des produits ou mélanges de produits suivants (avec entre parenthèses les doses à utiliser, en kilogrammes de matière active par hectare—Kg m.a./ha) (Biernaux & Seutin, 1969):

Produits organo-chlorés

Heptachlore (3·750 Kg m.a./ha)
Aldrin (3·750 Kg)
Lindane (1·500 Kg)

Produits organo-phosphorés

Fenthion (2·750 Kg)
Mecarbam (2·750 Kg)

Mélange de produits organo-chlorés et organo-phosphorés
Lindane (1·600 Kg) + Diazinon (0·450 Kg)
Lindane (0·700 Kg) + Mecarbam (1·900 Kg)

Ces doses sont valables pour des traitements en plein, c'est-à-dire sur toute la surface. Si on applique le traitement sur les bandes de semis seulement il y a lieu de n'utiliser qu'un tiers des doses préconisées.

En France, où les insecticides organo-chlorés sont en voie de suppression, on propose l'utilisation du phorate et du disyston qui sont des organo-phosphorés et de l'aldicarb qui est un carbamate.

En Angleterre et aux Pays-Bas la protection des cultures contre les iules ne paraît pas aussi préoccupante et on se contente généralement d'appliquer dans le champ de betteraves un traitement préventif au Lindane pour les préserver de l'ensemble des déprédateurs hypogés, qu'il s'agisse de myriapodes ou d'hexapodes.

En Allemagne, il semble bien que, sauf exception, les iules ne constituent pas un problème alarmant pour les planteurs de betteraves.

DIPLOPODES NUISIBLES AUX CULTURES TROPICALES

Des déprédations de diplopodes aux cultures tropicales ont été signalées en Amérique du sud, en Afrique centrale et en Asie.

Pour le premier de ces continents Schubart (1942) a signalé qu'au Costa Rica des graines de castilloa (*Castilloa elastica*), plante à caoutchouc de la famille des moracées, sont rongées par des diplopodes.

Au Brésil, des déprédations de divers diplopodes ont été observées par divers auteurs: *Rhinocricus* sp. (Rhinocricidae) sur bananier (*Musa* sp.) (Fonseca, 1944); *Orthomorpha coarctata* Saussure et *O. gracilis Koch* (Strongylosomidae) sur plantules de caféiers (*Coffea* sp.) ainsi que *Pseudonannolene sp.* (Pseudonannolidae) sur patates douces (Lordello, 1954).

En Asie, plus précisément aux Indes, dans l'état de Mysore, Puttarudriah (1958) a rapporté des attaques de mille-pattes sur patates douces (*Ipomea batatas*), coriandre (*Coriandrum sativum*), piments (*Capsicum frutescens*), cotonniers (*Gossypium* sp.) ainsi que sur diverses plantes potagères, tandis que Ramakrishna Ayyar (1940) a noté que les arachides (*Arachis hypogea*) pouvaient aussi être attaquées. Dans l'état de Kerala, Das, Nair & Jacobs (1966) ont observé des dégâts de *Harpurostreptus* sp. (Harpagophoridae) sur les jeunes racines et sur les bourgeons de plantes de manioc (*Manihot* sp.) et de piments.

Cependant c'est en Afrique, au sud du Sahara, que les relations de nuisibilité des diplopodes aux plantes cultivées sont les plus nombreuses;

elles ont été rapportées dans une publication récente (Pierrard, 1969). Sur ce continent, mis à part les mentions de déprédations de polydesmes sur cotonniers à Madagascar (Delattre, 1958), toutes les espèces nuisibles aux cultures tropicales appartiennent aux familles des Spirostreptidae et Odontopygidae, les espèces faisant partie de cette dernière étant les plus fréquemment citées. Dans la relation qui va suivre on ne considérera que les plantes cultivées sur une grande échelle et pouvant subir des dommages économiquement importants de la part des diplopodes connus du point de vue systématique. C'est dans cette optique qu'on considérera successivement les cotonniers, les arachides et les céréales.

Déprédations sur cotonniers

Dans les cotonneries, les diplopodes rongent les graines, avant et après leur germination, ainsi que la tige des plantules qui n'ont pas atteint le stade de quatre feuilles; lors du fouissage, ils soulèvent la terre des poquets, ce qui peut entraîner le flétrissement des plantules.

En République centrafricaine, les attaques de diplopodes peuvent être importantes mais sont très localisées, se limitant à un champ ou à une partie de champ. Dans le cas des plus fortes infestations, plus de 25% des poquets de graines avaient subi des dommages de myriapodes; ceux-ci s'ajoutant aux autres facteurs de réduction de la levée (pouvoir germinatif des semences souvent inférieur à 75%, microrganismes pathogènes, déprédations d'insectes) concourent à réduire la densité d'occupation des plantes. Les espèces les plus dommageables sont: *Peridontopyge schoutedeni* Attems, *P. demangei* Pierrard et *Tibiomus gossypii* Pierrard; elles font toutes partie de la famille des Odontopygidae. Les attaques sont toujours le fait de plusieurs espèces; en cas d'infestation massive, la dernière espèce citée dominait numériquement.

Au Tchad, dans la région frontalière du Nord Cameroun où des déprédations importantes de diplopodes furent parfois enregistrées, on identifia plusieurs espèces de Spirostreptoïdes (Demange, 1957). Dans la région sud de ce pays, par contre, l'espèce la plus fréquente dans les semis de cotonniers était *Tibiomus bebedjaensis* Pierrard (Pierrard, 1970).

Au Mali, il a été noté il y a plusieurs décades que, dans une zone au nord de Bamako, *Peridontopyge spinosissima* Silvestri et *Syndesmogenus mineuri* Brölemann détruisaient certaines années, jusqu'à plus de 15% des graines semées. Depuis lors, la culture cotonnière s'est considérablement développée et les dégâts de diplopodes sont inexistants ou négligeables, sauf dans la région ouest du pays où on tente d'introduire la culture moderne du coton.

Au Sénégal, dans la zone ouest de l'aire de culture cotonnière, le Sine Saloum, les diplopodes constituent un facteur limitant l'intensification de cette spéculation. Lors d'échantillonnages destinés à estimer la gravité des dégâts, le pourcentage de poquets détruits était de l'ordre 50% ; les espèces capturées étaient *Peridontopyge spinosissima* Silvestri, *P. conani* Brölemann, *Haplothysanus chappelei* Demange et *Tibiomus* sp., soit tous des Odontopygidae.

Déprédations sur arachides

Pour cette légumineuse, les dégâts sur graines et plantules sont analogues à ceux signalés pour le cotonnier. Cependant, la plante peut être une nouvelle fois attaquée lors de la phase de fructification: les myriapodes s'attaquent alors aux fruits en formation dans le sol; ils pratiquent en rongeant une ouverture dans le péricarde de la gousse, s'introduisent par ce trou et dévorent le contenu des graines. Le plus souvent, les cultivateurs et même la plupart des spécialistes ignorent ce type de dégâts qui est d'autant plus grave qu'il affecte le produit de récolte en formation.

En Centrafrique on a observé des attaques de diplopodes sur arachides qui réduisaient de moitié la production escomptée; on a établi à cette occasion que le pourcentage de perte en produit marchand correspondait à peu près à celui des gousses attaquées. Les espèces responsables étaient les mêmes que celles identifiées dans la région sur cotonnier avec, en plus, *Haplothysanus oubangaiensis* Pierrard.

Au Sénégal, les méfaits des diplopodes sur arachide constituent actuellement un problème aigu. En effet, l'économie de ce pays est basée principalement sur l'arachide, denrée tropicale dont les cours sont peu élevés. Afin d'obtenir un produit mieux rémunéré on a lancé, dans une région du Sine Saloum, une opération de culture d'arachide de bouche. Ce produit doit répondre à des conditions très strictes de présentation, notamment l'absence de toute lésion aux coques; or, dans la région où se développe ce type de culture, les dégâts de diplopodes sont considérables, pouvant détruire jusqu'à plus de 60% de la récolte. Outre les espèces nuisibles déjà signalées pour ce pays sur le cotonnier, on a observé *Peridontopyge rubescens* Attems et *P. pervittata* Silvestri. Ces deux espèces seraient particulièrement nombreuses lors de la germination des graines tandis que *Tibiomus* sp. serait le plus abondant sur les gousses en formation.

Il faut noter encore que, dans le Sine Saloum, d'après les résultats d'examen d'échantillons prélevés dans les cultures d'arachide d'huilerie les pertes de récolte en 1970 ont été estimées à environ 20%.

Déprédations sur céréales

Dans la zone centrale de Haute Volta, on a observé des dégâts importants, causés par des diplopodes aux semis de mils et de sorghos. A la station de recherches de l'I.R.A.T.* à Saria, les espèces nuisibles identifiées étaient *Peridontopyge spinosissima* Silvestri et *P. clavigera* Demange (Demange, 1966).

Moyens de lutte

On ne connaît encore que très peu de choses sur la biologie des myriapodes tropicaux; il n'est pas étonnant, dès lors, qu'aucune méthode destinée à préserver les plantes attaquées n'ait été expérimentée jusqu'à ce jour, si ce n'est celle de l'emploi de produits insecticides chimiques; encore faut-il signaler que même en ce domaine les données sont encore très peu nombreuses.

La mise au point de la technique de lutte la mieux appropriée ne pourra être réalisée pour une espèce ou un ensemble d'espèces, que lorsque les études biologiques approfondies préalables auront été menées à bien. Il serait à ce moment peut-être possible de tirer parti de certaines particularités éthologiques des déprédateurs pour réduire ou éliminer leurs dégâts; c'est ainsi par exemple qu'au Sénégal on a observé qu'en fin de saison des pluies, le pied de certains arbres ainsi que les termitières constituaient des refuges où abondaient les diplopodes; dans ces dernières on a trouvé parfois plus d'un millier d'individus appartenant à diverses espèces; la plus commune était *Spirostreptus* sp. qui ne serait toutefois pas nuisible aux cultures.

En attendant que de telles recherches aient pu être réalisées, l'agronome a donc recours à la lutte directe par utilisation de produits insecticides de synthèse. Cette lutte contre les diplopodes nuisibles aux semis serait très rentable si elle pouvait être réalisée grâce à l'enrobage des semences au moyen d'un insecticide efficace; tel est probablement le seul procédé à retenir dans les cas où on ne redoute que des dégâts sporadiques; cette technique présente l'avantage de ne mettre en oeuvre qu'une quantité minimum de produit toxique. La protection des fruits d'arachide en formation dans le sol pose un problème plus délicat à résoudre du fait que la période d'attaque par le déprédateur s'étendrait sur plus d'un mois et qu'on ne peut tolérer la présence d'une certaine teneur en insecticide dans le produit de récolte; dans ce cas des appâts composés à la fois de substances attractives pour diplopodes et de produit toxique constitueraient, pour autant qu'ils soient efficaces, la méthode de lutte idéale.

* Institut de Recherches Agronomiques Tropicales et des cultures vivrières.

Nous énumérons ci-après les données les plus récentes en matières de lutte chimique.

En Haute Volta, l'I.R.A.T. utilise, comme moyen de protection de ses parcelles d'essai de céréales, des appâts à base de paraffine additionée de "Warfarin"! il n'est pas précisé si l'appât est seulement destiné à détourner les arthropodes des semences de céréales ou si on escompte l'intoxication des diplopodes.

Au Sénégal, l'I.R.H.O.* a obtenu dans des parcelles d'essais d'arachides une mortalité considerable des déprédateurs en épandant sur le sol un appât empoisonné composé de 50 Kg de son de mil, de 150 l d'eau et de 400 g de lindane à 90% de matière active; ces quantités sont valables pour 1 ha de culture.

Au Sénégal encore, l'I.R.C.T.† a comparé l'efficacité de divers traitements des graines de cotonnier par trempage dans une bouillie insecticide. Après 8 jours, le pourcentage de poquets ayant germés par rapport aux poquets semés était de 61·0% pour le traitement au gardona, 55·2% pour la phosalone et 37·2% dans le cas de poquets non traités; il faut noter que le pourcentage de levée obtenu avec des graines traitées à sec au moyen de 4% d'heptachlore à 25% de matière active n'était pas supérieur au témoin.

En Centrafrique, dans des essais de lutte avec infestations contrôlées des odontopyges nuisibles au cotonnier, l'heptachlore et le dialdrine utilisés en traitment des graines à sec avaient donné une levée de l'ordre de 20% supérieure à celle obtenue dans le témoin non traité (Pierrard, 1969).

La différence d'efficacité de l'heptachlore vis-à-vis de Spirostreptoïdes constatée dans les deux dernières régions (Sénégal et République Centrafricaine) pourrait être due à une toxicité spécifique différente du poison; une constation analogue ressortait déjà de bioessais conduits en Afrique du Sud (Fiedler, 1965).

Bibliographie

Baurant, R. (1964). Les dégâts d'iules mouchetés sur jeunes betteraves. *Bull. Inst. agron. Stns Rech. Gembloux* **32**: 3–11.

Biernaux, J. (1966). Incidence économique des Iules en culture betteravière. *Meded. Rijksfac. Landbete. Gent* **31**: 717–729.

Biernaux, J. & Baurant, R. (1964). Observations sur l'hibernation de *Archiboreoiulus pallidus* Br.-Bk. *Bull. Inst. agron. Stns Rech. Gembloux* **32**: 290–298.

* Institut de Recherches pour les Huiles et les Oléagineux.

† Institut de Recherches du Coton et des Textiles exotiques.

Biernaux, J. & Seutin, E. (1969). Efficacité de quelques insecticides non encore utilisés dans la pratique contre les "Iules de la betterave". *Bull. Rech. agron. Gembloux* N.S. **4**: 181–188.

Breny, R. (1964). Considérations actuelles sur le problème des iules mouchetés en culture betteravière. *Bull. Inst. agron. Stns Rech. Gembloux* **32**: 12–25.

Das, N. M., Nair, N. M. G. K. & Jacobs, Abraham (1966). On the occurrence and control of *Harpurostreptus* sp., a new millipede pest of cultivated crops in Kerala. *Indian J. Ent.* **28**: 563–565.

Delattre, R. (1958). Les parasites du cotonnier à Madagascar. *Coton Fib. trop.* **13**: 335–352.

Demange, J. M. (1957). Myriapodes Diplopodes du Tchad (A.E.F.) nuisibles au cotonnier. *Bull. Mus. natn. Hist. nat. Paris* (2) **29**: 96–105.

Demange, J. M. (1966). Une nouvelle espèce du genre *Peridontopyge*, nuisible aux plantations tropicales (Myriapode, Diplopode, Spirostreptoïdea) *Bull. Inst. fond. Afr. noire* (A) **28**: 986–988.

Fiedler, O. G. H. (1965). Notes on the susceptibility of Millipedes (Diplopoda) to insecticides *J. ent. Soc. sth. Afr.* **27**: 219–225.

Fonseca, J. P. da (1944). Millipes (*Rhinocricus*) em bananeira. *Biologico, S. Paulo* **10**: 55.

Lordello, L. G. (1954). Observaçoè sobre algum diplopodes de interesse agricola. *Anais Esc. Sup. Agric. "Luiz Queiroz"* **11**: 69–76.

Pierrard, G. (1969). Nocivité des diplopodes aux plantes cultivées dans les pays chauds. *Coton Fibr. trop.* **24**: 429–441.

Pierrard, G. (1970). *Tibiomus bebedjaensis* n. sp. (*Odontopygidae*) myriapode nuisible à la culture du cotonnier au Tchad. *Coton Fibr. trop.* **25**: 355–358.

Pierrard, G., Bonte, E. & Baurant, R. (1963). Observations sur l'hibernation de *Blaniulus guttulatus* Bosc. *Bull. Inst. agron. Stns Rech. Gembloux* **31**: 127–141.

Puttarudriah, M. (1958). Millipedes damage crops. *Mysore agric. J.* **33**: 130–132.

Ramakrishna Ayyar, T. V. (1940). *Handbook of economic entomology for South India.* Madras: Government Press.

Remy, P. A. (1950). Les myriapodes et les plantes cultivées (Bibliographie). *Bull. mens. Soc. linn. Lyon* **19**: 232–334.

Schubart, O (1942). Os Myriapodes e suas relações com a agricultura. *Papéis Dep. Zool. S. Paulo* **2**: 205–234.

Discussion

DEMANGE: I found your contribution very interesting but it caused me great anxiety. The reserve of living millipedes in the area is perhaps large, but it must be limited. It has been said, that after treatment with bait, in the fields of groundnuts there was a carpet of dead millipedes.

PIERRARD: The destruction of these harmful myriapods must have an effect on the whole ecosystem; just what effect, we do not know. We should know. The insecticides used will affect other animals, especially those at the ends of food chains along which the chemicals are concentrated. However, we must also consider the human population relying on the crop, with which the millipedes are in direct competition. We must determine both the long and short term results of the application of these poisons. The carpet of dead millipedes is close to the truth. In the zone of Sine Saloum in Senegal, the number of millipedes is unbelievable. I know the diplopods of Tropical Africa fairly well, but was amazed by the large numbers which walked on the roads, as well as the fields, just at the beginning of the rains. It is the aggregation of this vast population which makes this enormous kill possible.

DEMANGE: What is the nature of surrounding areas? Could some other type of cultivation nearby act as a reservoir for the infestation?

PIERRARD: The cultivated zone is small compared with the area left uncultivated. The cultivated zone appears especially attractive to the millipedes and the populations from the neighbouring uncultivated areas may drain off into the fields.

HERBAUT: Are there as many diplopods per square metre in the cultivated parts as in the uncultivated parts? Presumably cultivated land is not as favourable for their multiplication?

PIERRARD: There is no information; there are no previous studies of diplopod populations here. It appears that the cultivated fields are very attractive to the millipedes as the seeds germinate or when the new nuts are being formed. At these times the population in the cultivated fields appears to be much larger than in the surrounding uncultivated areas.

ENGHOFF: Have you ever found European species in addition to the tropical species you mentioned?

PIERRARD: No, none. Among the harmful species, we have only the Odontopygidae which are restricted to Africa south of the Sahara, and the Spirostreptidae which have a wider distribution in Asia, America and Africa.

HERBAUT: Are there organizations controlling the use of pesticides in the countries where you have worked?

PIERRARD: No. In most places, in the cotton plantations, the principal insecticide used is, or has been, endrin; this insecticide has been banned in Belgium and has never been authorized in France because of its toxicity. DDT is also used in association with endrin. These two are the cheapest of the insecticides which are effective. No doubt the farmers would be prepared to introduce less toxic products if they were as effective and as cheap as endrin and DDT.

Symp. zool. Soc. Lond. (1974) No. 32, 645–655.

SOME EFFECTS OF INSECTICIDES ON MYRIAPOD POPULATIONS

C. A. EDWARDS

Rothamsted Experimental Station, Harpenden, Hertfordshire, England

SYNOPSIS

The effects of aldrin, DDT, chlorfenvinphos, diazinon, disulfoton, parathion and phorate on populations of Pauropoda, Symphyla, Diplopoda and Chilopoda, were studied in four field experiments. Plots 3 m square were treated and the insecticide rotovated into the soil. Samples were taken at monthly or two-monthly intervals. Pauropoda were extremely susceptible to all the insecticides, Symphyla and Chilopoda less so and Diplopoda were little affected by any insecticide.

INTRODUCTION

Knowledge of how insecticides affect populations of Pauropoda, Symphyla, Diplopoda and Chilopoda differs greatly between these groups, and seems to be related to their feeding habits; much more is known about those that are pests.

Pauropoda are often overlooked in soil invertebrate surveys because they are so small. Usually they occur in only small numbers (Wallwork, 1970) but Starling (1944) found many in North American woodland soils, and I have found them to be numerous in English woodland and arable soils. Little is known of their feeding habits, but they are thought to feed mostly on decaying organic matter, fungi and bacteria. There is no evidence that they are plant pests and there are few reports of the effects of insecticides on them (Edwards, Dennis & Empson, 1967; Edwards, Lofty & Thompson, 1967; Long, Anderson & Isa, 1967; Edwards, Thompson & Beynon, 1968).

Symphyla are larger than Pauropoda and much more common, both in woodland and arable soils (Edwards, 1958). They are omnivorous but most species of Scutigerellidae prefer to feed on plant tissues and are serious pests of a wide range of crops, especially in Great Britain, France and the U.S.A. Because symphylids are pests, there are many reports of the effects of insecticides upon them (Edwards & Thompson, 1973) and, in general, they seem to be very difficult to kill with pesticides, probably because they spend much time deep in the subsoil where no pesticides can penetrate.

Millipedes are much larger and sometimes occur in very great numbers. Their food is decaying organic matter, but they also eat tender plant tissues, and cause considerable damage to susceptible crops such as potatoes, sugar beet and young seedlings of many kinds. Much information on how they are affected by pesticides has been obtained from investigations into suitable methods of control (e.g. Edwards & Gunn, 1961) and from more general studies of the effects of insecticides on the soil fauna (Edwards & Thompson, 1973). They differ considerably in their susceptibility to different pesticides, but, in general, are not readily killed, and most insecticides have little or no effect on them.

Centipedes are almost exclusively predatory and because they are never pests, little is known about which insecticides kill them (Edwards & Thompson, 1973).

During the last decade my colleagues and I have studied the effects of most herbicides, fungicides, nematicides, fumigants and insecticides that are normally applied to soil, on populations of soil invertebrates, including myriapods. Many of these studies have been reported elsewhere (Edwards & Dennis, 1960; Edwards, 1965, 1969; Edwards, Dennis *et al.*, 1967; Edwards, Lofty *et al.*, 1967; Edwards, Thompson *et al.*, 1968; Edwards & Lofty, 1971) and an extensive discussion of them is not possible here. Instead, the effects of a few persistent organochlorine and non-persistent organophosphate insecticides will be discussed, because these clearly show the relative susceptibilities of myriapods to insecticides.

METHODS

The same basic technique was used in all experiments. The experimental site was ploughed and rotovated thoroughly, and plots 3 m square with 1 m guard rows between plots marked out. There were four replicates of treated and untreated plots. Insecticides were applied as a surface spray or dust which was rotovated thoroughly into the soil immediately after application. From each plot, four 5 cm diam. × 15 cm soil cores were taken at random before treatment and at one- or two-monthly intervals thereafter. Invertebrates were extracted from the soil in modified Macfadyen high gradient Tullgren funnels (Edwards & Lofty, 1972), and stored in a mixture of 70% alcohol and 5% glycerol until they were identified and counted.

Aldrin and DDT

Arable

Aldrin (4·7 kg/ha) and DDT (12·6 kg/ha) were applied to plots on a field that had previously grown dredge corn, swedes, barley and lucerne in successive years. After treatment, plots were fallowed and soil samples were taken throughout a year at two-monthly intervals.

Pasture

An old pasture was ploughed out and aldrin (4·5 kg/ha) and DDT (6·7 and 67·0 kg/ha) were applied to plots and rotovated in thoroughly. The plots were kept fallow for three years and then sown with grass. Soil samples were taken at two-monthly intervals for five years after treatment.

Organophosphate insecticides

Pasture

In two separate experiments (A and B), old pasture was ploughed and organophosphate insecticides sprayed on to the surface and rotovated into the soil. Soil samples were taken at monthly intervals for one year after treatment. The insecticides tested were: diazinon, chlorfenvinphos, parathion (used in both experiments A and B), phorate and disulfoton (used only in one experiment, B). All the insecticides were applied at 4·2 kg/ha and all plots cropped with wheat.

RESULTS

The more important results concerning myriapod populations are summarized in Figs 1–12. The vertical axes of the histograms represent the total number of animals collected in one year's sampling. The graphs (Figs 8–11) show that the effect of some organophosphate insecticides changed with time.

DISCUSSION

In the first experiment, there were too few millipedes and centipedes to show how populations were affected, but the aldrin almost eradicated Pauropoda from the soil and DDT greatly decreased their numbers (Fig. 1). By contrast, these insecticides had a much smaller effect on the Symphyla (Fig. 1).

In the long-term study on old pasture, the numbers of all animals were greater before treatment, so a more accurate assessment of the effects of the two insecticides on all the myriapods was possible during five consecutive years after treatment. Once again, aldrin and the large dose of DDT decreased numbers of pauropods to a very low level throughout the experiment. The small dose of DDT decreased numbers of pauropods less in the fifth year than in earlier ones (Fig. 2). All of the insecticidal treatments had much less effect on numbers of symphylids than on pauropods, and in the fourth and fifth years both aldrin and the small dose of DDT had very little residual effect on symphylids.

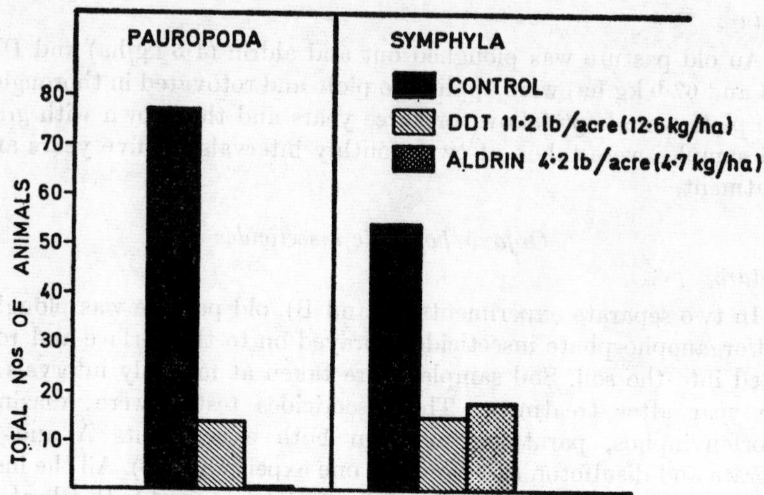

Fig. 1. Effects of DDT and aldrin on myriapods.

(Fig. 3). Numbers of millipedes (Diplopoda) were very small in the first three years when the plots were fallow, but there were enough in the fourth and fifth years to show that neither aldrin nor the low dose of DDT had much effect on them, although the large dose of DDT was

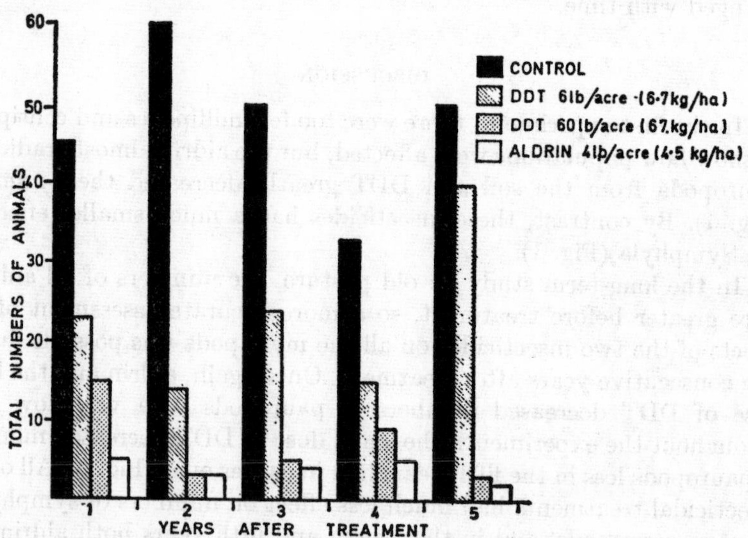

Fig. 2. Effects of DDT and aldrin on Pauropoda.

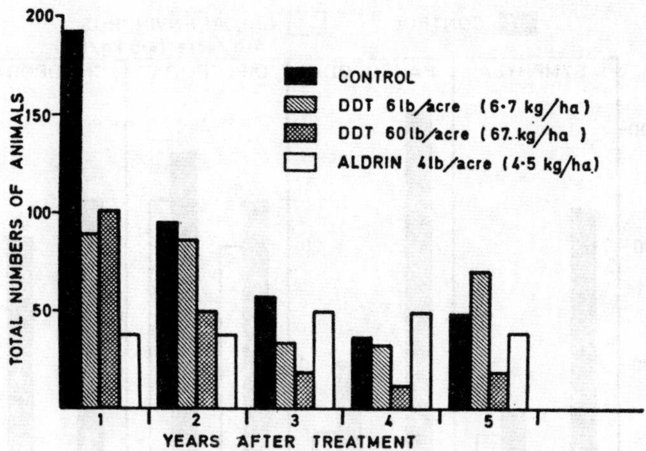

Fig. 3. Effects of DDT and aldrin on Symphyla.

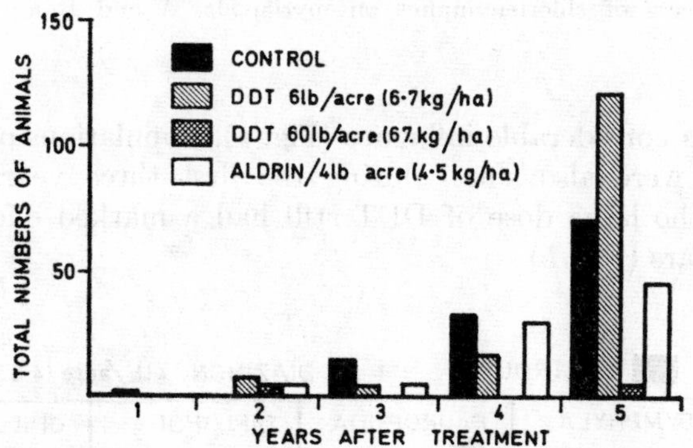

Fig. 4. Effects of DDT and aldrin on Diplopoda.

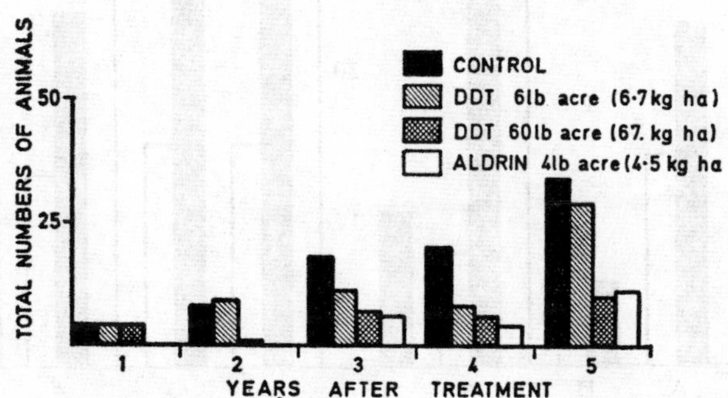

Fig. 5. Effects of DDT and aldrin on Chilopoda.

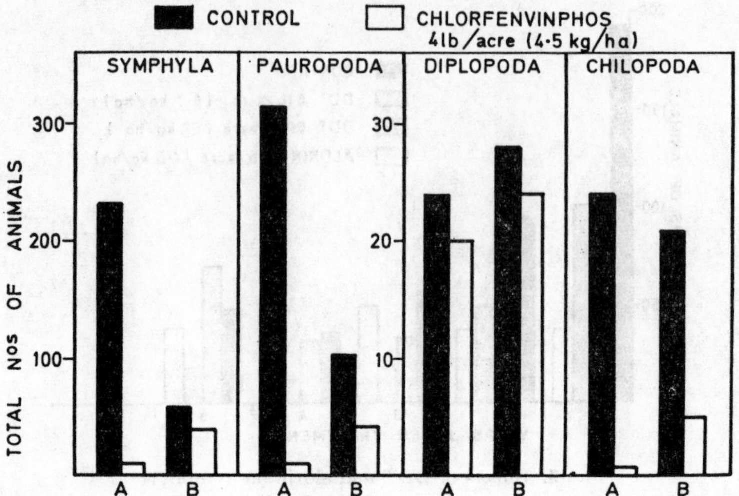

FIG. 6. Effects of chlorfenvinphos on myriapods. A and B are two separate experiments.

still having a considerable influence (Fig. 4). Populations of centipedes (Chilopoda) were also small during the first three years, but both aldrin and the large dose of DDT still had a marked effect on them after five years (Fig. 5).

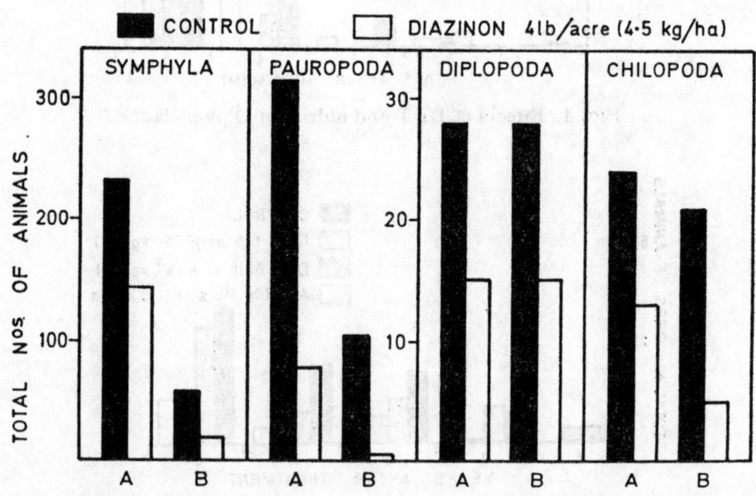

FIG. 7. Effects of diazinon on myriapods. A and B are two separate experiments.

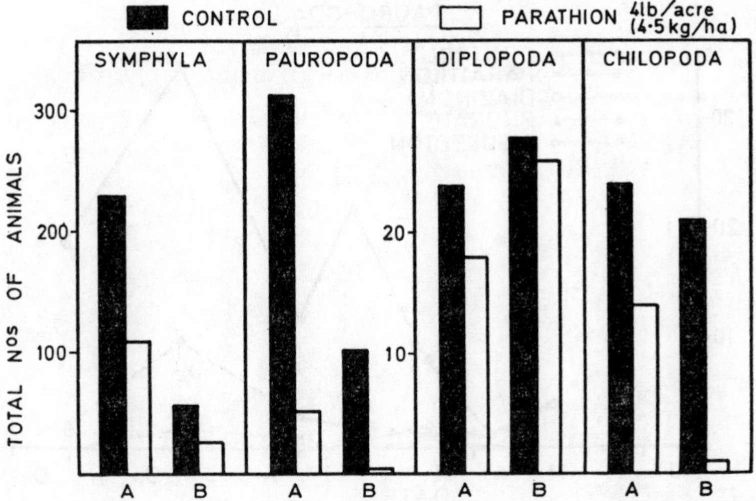

Fig. 8. Effects of parathion on myriapods. A and B are two separate experiments.

The effects of organophosphate insecticides on myriapods were not quite so consistent as those of organochlorines (Figs 6–12). However, many pauropods (Figs 6–10) were killed by all the insecticides tested, phorate being the least toxic to them, with little difference between the effects of the others. Although these chemicals persist for a few weeks at most, populations of pauropods had still not recovered nine months after treatment (Figs 9 & 10). Most of the organophosphates tested

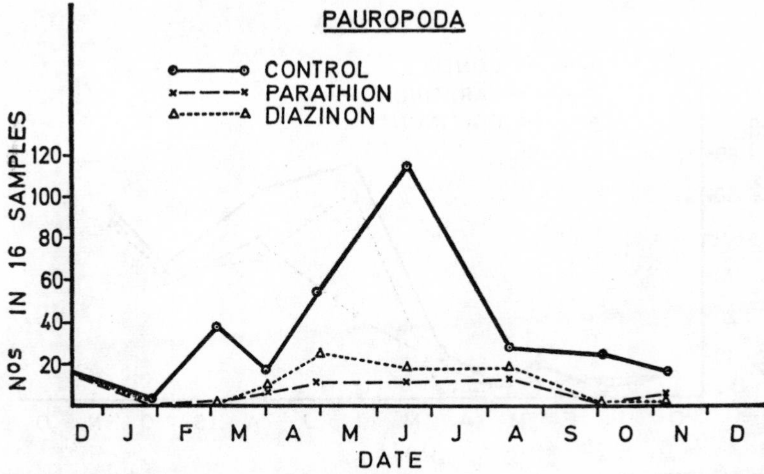

Fig. 9. The effects of organophosphorus insecticides on Pauropoda (Experiment A).

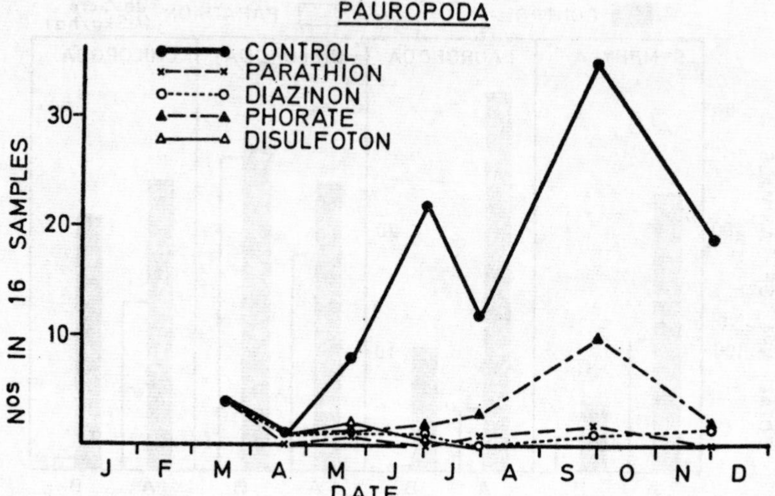

FIG. 10. The effects of organophosphorus insecticides on Pauropoda (Experiment B).

killed about half of the symphylids, with little significant difference between the effects of different compounds. Populations recovered completely within six months of treatment (Figs 11 & 12).

Millipedes were little affected by any of the organophosphates tested (Figs 6–8). These chemicals had very similar effects on centipedes to those on symphylids, the decreases in population being remarkably similar for all the compounds (Figs 6–8).

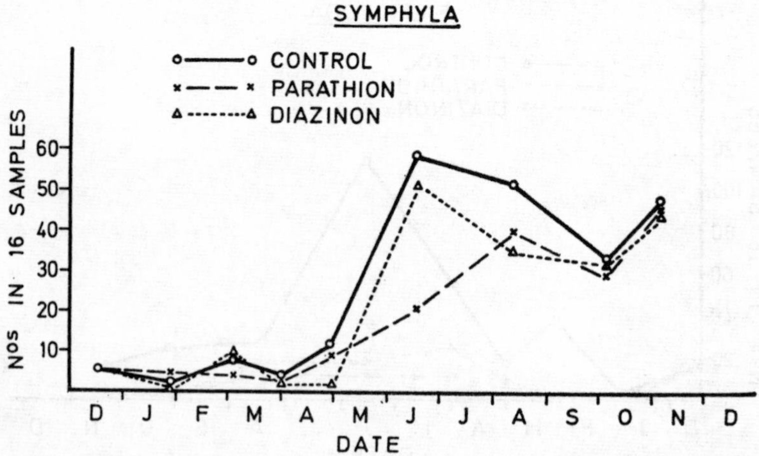

FIG. 11. The effects of organophosphorus insecticides on Symphyla (Experiment A).

From this series of field experiments, it is clear that myriapods differ greatly in their susceptibility to insecticides. The pauropods are much the most sensitive, being almost completely eliminated; unfortunately not enough is known about their habits to speculate on the reason for this susceptibility. The symphylids and centipedes were intermediate in susceptibility; they sometimes escape excessive exposure

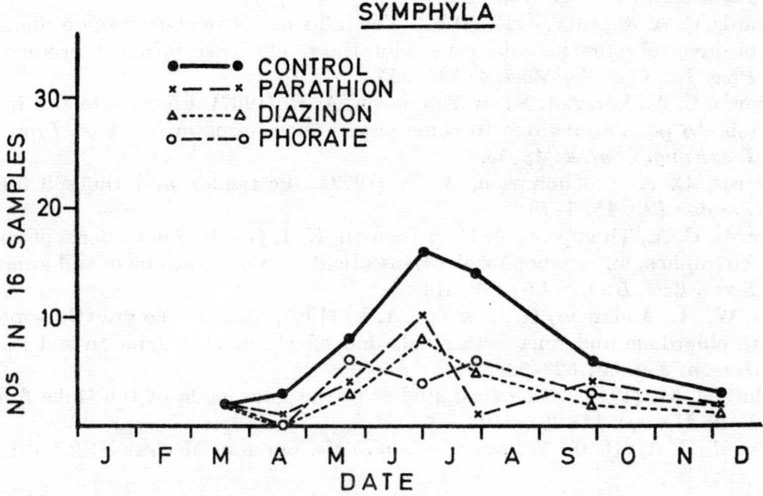

Fig. 12. The effects of organophosphorus insecticides on Symphyla (Experiment B).

to the insecticides due to their habits of penetrating deep into soil or hiding inactive in cracks and crevices in soil, but both symphylids and centipedes are extremely active animals and, in the course of their rapid movements through soil, may pick up a lethal dose of insecticide quicker than sluggish animals such as millipedes, which are the least susceptible to insecticides of all the myriapods. Currently, insufficient attention is paid to the effects of insecticides on the centipedes which are beneficial by preying on pest invertebrates. Millipedes are important in soil formation in woodlands but populations do not seem to be at hazard from any of the insecticides tested.

References

Edwards, C. A. (1958). The ecology of Symphyla. Pt. I. Populations. *Entomologia exp. appl.* **1**: 308–319.

Edwards, C. A. (1965). Effects of pesticide residues on soil invertebrates and plants. *Symp. Br. ecol. Soc.* **5**: 239–261.

Edwards, C. A. (1969). Soil pollutants and soil animals. *Scient. Am.* **220** (4): 88–99.

Edwards, C. A. & Dennis, E. B. (1960). Some effects of aldrin and DDT on the soil fauna of arable land. *Nature, Lond.* **188**: 767–768.

Edwards, C. A., Dennis, E. B. & Empson, D. W. (1967). Pesticides and the soil fauna: effects of aldrin and DDT in an arable field. *Ann. appl. Biol.* **60**: 11–22.

Edwards, C. A. & Gunn, E. B. (1961). Control of the glasshouse millipede. *Pl. Path.* **10**: 21–24.

Edwards, C. A. & Lofty, J. R. (1971). Nematicides and the soil fauna. *Proc. Br. Ins. Fungi Conf.* **61**: 158–166.

Edwards, C. A. & Lofty, J. R. (1972). The influence of temperature on numbers of invertebrates in soil, especially those affecting primary production. *Proc. Int. Coll. Soil Zool.* **4**: 545–555.

Edwards, C. A., Lofty, J. R., & Thompson, A. R. (1967). Changes in soil invertebrate populations due to some organophosphorus insecticides. *Proc. Br. Ins. Fungi Conf.* **4**: 48–55.

Edwards, C. A. & Thompson, A. R. (1973). Pesticides and the soil fauna. *Residue Rev.* **45**: 1–79.

Edwards, C. A., Thompson, A. R. & Beynon, K. I. (1968). Some effects of chlorfenvinphos, an organophosphate insecticide, on populations of soil animals. *Revue Ecol. Biol. Sol* **5**: 199–214.

Long, W. H., Anderson, H. L. & Isa, A. L. (1967). Sugar cane growth response to chlordane and microarthropods and effects of chlordane on soil fauna. *J. econ. Ent.* **60**: 623–629.

Starling, J. H. (1944). Ecological studies of the Pauropoda of the Duke forest. *Ecol. Monogr.* **14**: 291–310.

Wallwork, J. A. (1970). *Ecology of soil animals.* London: McGraw-Hill.

Discussion

Brade-Birks: Are all your experiments done on the same soil series and is the insecticide incorporated at the same depth?

Edwards: The experiments were conducted in several places, with different soils, but the insecticide was always incorporated to the same depth.

Brade-Birks: Do you mean depth or horizon?

Edwards: Most of the agricultural soils we have used do not have distinct horizons for the first nine inches or so.

Brade-Birks: I see, it is in fact an artificial horizon, but a real horizon nevertheless.

Edwards: I might add that I have also used radiation to sterilize soil, both as soil cores (5 cm diam. and 15 cm deep) in aluminium cans, and in the field. Centipedes were extremely sensitive to radiation, but symphylids, pauropods and millipedes were much less so. In general the more active the animal, the more susceptible it is to radiation.*

*Edwards, C. A. (1967). Effects of gamma irradiation on populations of soil invertebrates. *2nd Symp. Radioecology*: 68–77. Ann Arbor, Michigan, 1967, pub. U.S.A.E.C.

PIERRARD: What was the object of sterilizing the soil by irradiation?

EDWARDS: We were interested in whether soil arthropods contributed to soil fertility. The idea was to kill the animals and then follow the growth of a crop. We hoped we might be able to kill the animals with a dose lower than that which we knew from experience would kill the microflora. Our hopes were justified and we established a dose which kills all the animals but does not destroy the microfloral activity for more than two or three weeks. These experiments are running at the moment.

BLOWER: To what depth does the radiation penetrate?

EDWARDS: This is difficult to determine. We suspect that in the field we do not get deeper than a foot and so we have not killed everything below and there may be recolonization.

BLOWER: What is the source you use in the field?

EDWARDS: A cobalt source, carefully shielded of course. There was a possibility of using the method in horticulture but it is too dangerous for general use, I think.

BLOWER: Have you been able to follow the depth distribution of the kill?

EDWARDS: No, not yet.

BLOWER: Was there any residual radiation?

EDWARDS: No, it does not create any radiation.

Symp. zool. Soc. Lond. (1974) No. 32, 657.

CENTRE INTERNATIONAL DE MYRIAPODOLOGIE

Le C.I.M. a été créé pendant le Ier Congrès international de Paris, avril 1968. Ses buts étaient de constituer un annuaire des myriapodologistes, une liste de leurs spécialités et de leurs projets de recherches, actuels et futurs, et de diffuser, chaque année, une liste des travaux publiés et sous presse. L'initiative de MM. J.-M. Demange et J.-P. Mauriès et du Professeur O. Kraus a été applaudie par le Congrès et il leur a été demandé de constituer le secrétariat permanent.

Les membres du Ier Congrès ont reçu la première liste des publications (jusqu'à 1967). Au début de chacune des années suivant le Congrès une liste des publications a été diffusée parmi les membres. Vers la fin de chaque année les membres reçoivent un questionnaire émanant du C.I.M. et leur demandant de communiquer les titres de leurs publications, leurs projets de recherches, etc. Il y a une contribution volontaire d'un minimum de 20 francs français payable à M. J.-M. Demange, CCP Paris 86–11 ou Banque "Société Générale" (Agence A. L. Gobelins).

De plus amples informations peuvent être obtenues au secrétariat permanent du C.I.M.:

Messieurs J.-M. Demange et J.-P. Mauriès,
Laboratoire de Zoologie (Arthropodes),
Muséum National d'Histoire Naturelle,
61 rue de Buffon,
Paris V^e, France.

Professor O. Kraus,
Zoologisches Institut und Zoologisches Museum,
Universität Hamburg,
2 Hamburg, 13,
Papendamm 3, Germany.

Symp. zool. Soc. Lond. (1974) No. 32, 659.

English Translation

CENTRE INTERNATIONAL DE MYRIAPODOLOGIE

The C.I.M. was created during the 1st International Congress of Myriapodology in Paris, April 1968. Its aims were to keep a directory of myriapodologists, listing their specialities and their research projects, both actual and planned, and to circulate, each year, a list of papers published and in the press. The initiative of MM. J.-M. Demange and J.-P. Mauriès and Professor O. Kraus was applauded by Congress and they were asked to serve as the permanent secretariat.

Members of the first Congress received the first list of publications (up to 1967). At the beginning of each year following that of the first Congress, a list of publications has been circulated to members. Towards the end of each year, members received a questionnaire from C.I.M. asking them to give titles of papers, research projects, etc. There is a voluntary subscription of a minimum of 20F payable to M. J.-M. Demange, CCP Paris 86–11 or Banque, "Société Générale" (Agence A. L. Gobelins). Further information may be obtained from the permanent secretariat:

Messieurs J.-M. Demange et J.-P. Mauriès,
Laboratoire de Zoologie (Arthropodes),
Muséum National d'Histoire Naturelle,
61 rue de Buffon,
Paris V^{e}, France.

Professor O. Kraus,
Zoologisches Institut und Zoologisches Museum,
Universität Hamburg,
2 Hamburg, 13,
Papendamm 3, Germany.

Symp. zool. Soc. Lond. (1974) No. 32, 661–662.

RAPPORT SUR LES ACTIVITÉS DU C.I.M.

J.-P. MAURIÈS

Muséum national d'Histoire naturelle, Paris, France

Le Tableau montre:

1. Une augmentation du nombre des myriapodologistes (*sensu lato*).
2. Une diminution du nombre des réponses aux questionnaires et des contributions volontaires.
3. Une diminution du nombre des titres de travaux pour 1972.

Ces chiffres sont inquiétants dans leur nudité, mais ils doivent être accompagnés de deux remarques:

1. La diminution du pourcentage du nombre des réponses et du nombre de contributions volontaires s'explique par le fait que, entre 1968 et 1972, il y a un double mouvement:

(a) un mouvement centrifuge de la part des myriapodologistes stricts (moins nombreux), ce qui est très rassurant.
(b) un mouvement centripète de la part des myriapodologistes occasionnels (bien plus nombreux), ce qui est normal.

Le secrétariat du C.I.M. renouvelle néanmoins son appel de 1968 pour *qu'un nombre croissant de nos collègues se sentent concernés par le fonctionnement du centre.**

2. La diminution du nombre des travaux pour 1972 semble correspondre plutôt à une période de renouvellement (afflux de nombreux jeunes collègues ayant peu publié) qu'à une baisse d'activité générale.

Malgré les avertissements que nous donnent les chiffres du Tableau, nous pensons pouvoir être optimistes sur l'avenir du C.I.M. Le C.I.M. n'est pas concurrent d'autres organismes de documentation (*Zool. Record; Biol. Abstr.; Entom. Abstr.*), mais peut aider ces organismes. Son but essentiel, rappelons le, n'est pas seulement d'informer rapidement, mais de faciliter les échanges entre myriapodologistes, et de coordonner de manière non autoritaire, leurs travaux sur le plan international.

* Cet appel semble avoir été entendu s'il on en juge par le nombre de réponses et le nombre de contributions sans précédent qui nous sont parvenus en ce début 1973.

Tableau illustrant la participation des myriapodologistes aux activités du C.I.M. (Centre International de Myriapodologie) depuis sa création en 1968 (1° Congrès International, Paris)

	1968	1969	1970	1971	1972
Myriapodologistes recensés (y compris les occasionnels)	140	168	167	163	175
Réponses aux questionnaires	72	65	70	60	—
%	51	38	41	36	
Contributions financières	34	43	39	31	34*
%	24	25	23	19	
Nombre de titres de travaux	258	255	288	223	—

* au 15 mars 1972.

Symp. zool. Soc. Lond. (1974) No. 32, 663.

English Translation

REPORT ON THE ACTIVITIES OF THE C.I.M.

J.-P. MAURIÈS

Muséum national d'Histoire naturelle, Paris, France

The Table (see Report in French, p. 661) shows:

1. An increase in the number of myriapodologists (*sensu lato*).
2. A decrease in the number of replies to the questionnaire and of the voluntary contributions.
3. A decrease in the number of titles of papers for 1972.

The figures by themselves are disturbing but they must be qualified by two remarks:

1. The percentage decrease in the replies and in the voluntary contributions can be explained by the fact that between 1968 and 1972 there has been a double movement:

(a) an expansion on the part of strict myriapodologists (less numerous) which is very reassuring.
(b) a contraction on the part of occasional myriapodologists (more numerous) which is normal.

The secretariat of the C.I.M. nevertheless renew their appeal of 1968,* *that a growing number of our colleagues should become involved with the working of the centre.*

2. The decrease in the number of papers for 1972 appears to correspond rather to a period of renewal (arrival of numerous young colleagues having published little) than to a lowering of general activity.

Notwithstanding the prominence we give to the figures in the Table, we feel able to be optimistic about the future of the C.I.M.

The C.I.M. is not competing with the other organizations for documentation (*Zool. Record*; *Biol. Abstr.*; *Entom. Abstr.*), but is able to help these bodies. The essential aim, let us remember, is not only to inform quickly, but to facilitate exchanges between myriapodologists and to co-ordinate their work informally, at international level.

* This appeal appears to have been heeded, judging from the unprecedented number of replies and voluntary contributions received by the beginning of 1973.

Symp. zool. Soc. Lond. (1974) No. 32, 665.

PLAN POUR UN RECENSEMENT DES COLLECTIONS INTERNATIONALES DE MYRIAPODES

J.-M. DEMANGE

Muséum national d'Histoire naturelle, Paris, France

Qui d'entre-nous, taxonomiste ou non, n'a été dans l'obligation d'étudier certains specimens décrits par un prédécesseur, soit que la description ait été publiée sans illustration, soit, qu'elle paraisse incomplète vis à vis des exigences modernes, soit tout simplement, qu'un nouvel examen s'impose pour contrôler tel ou tel caractère à des fins de révision par exemple?

Le spécialiste est confronté bien souvent, dans de telles conditions, à des difficultés parfois insurmontables et contraint à effectuer des recherches fort longues (perte de temps) lorsqu'il s'agit de déterminer le Musée détenteur de la précieuse collection. Encore n'est-il pas assuré de trouver cette collection dans le Musée en question sur la seule foi que l'auteur y a fait carrière ou d'y retrouver la totalité du matériel étudié; la collection peut être renvoyée à son Musée d'origine (collecteur) ou fragmentée par distribution à plusieurs spécialistes.

Il en est ainsi pour les collections déterminées, mais qui n'a eu à effectuer également de longues et fastidieuses recherches dans les Musées du monde pour rassembler des collections de myriapodes non identifiés appartenant, soit à une région géographique particulière soit, plus spécialement, à un groupe systématique choisi?

L'ignorance où l'on est actuellement de l'état réel du matériel scientifique rassemblé et conservé depuis des générations dans le monde entier (matériel mis à la disposition des chercheurs), et par conséquent de la méconnaissance ou d'une connaissance incomplète d'un matériel de référence, base fondamentale de toute recherche, n'est plus digne de notre époque moderne dont l'information scientifique en est l'un des fondements.

On a donc pensé que le C.I.M., dont les adhérents représentent plus des trois quarts des myriapodologistes, devrait intervenir dans ce domaine et précisément avec l'aide des spécialistes, tous directement concernés, et prendre, s'il le faut, quelques initiatives.

Le projet de base proposé se situe à deux niveaux:

1. établissement d'un inventaire des collections des myriapodes, déterminés et non déterminés, conservés dans les Etablissements

Scientifiques (Musées, Universités, Fondations, etc. .) avec indication, naturellement, des types *sensu lato*.

2. accords internationaux sur les échanges éventuels de ce matériel scientifique (consultation, prêt, études originales, etc.).

La réalisation de ce programme est, en fait, entre les mains des spécialistes, c'est-à-dire nous-mêmes. Il est indispensable que les bonnes volontés se fassent jour, notamment, pour débuter, parmi ceux de nos collègues qui ont déjà une place assise dans l'un de ces établissements et ensuite ou simultanément que s'effectue, *dans chaque pays*, un regroupement des spécialistes dans le but d' entreprendre les recherches nécessaires à la réalisation du plan au niveau de *leur propre nation*.

On aura ainsi réalisé rapidement l'inventaire des collections; au moins des principaux pays du monde car si l'on veut bien consulter le Bulletin du C.I.M., au chapitre de la répartition des chercheurs dans le monde, on s'aperçoit que tous ont, parmi leur cadres scientifiques, des myriapodologistes.

Il a été commencé a Paris l'inventaire des collections du Muséum, parmi les plus riches (collections Brölemann, Ribaut, Gervais, Chalande, Saussure et Zehntner (Madagascar) etc . . .). On peut espérer la publier en un fascicule spécial du C.I.M. avant le prochain Congrès de Myriapodologie. En outre, grâce à l'extrême amabilité du Professeur E. Tremblay, Directeur de la Fondation F. Silvestri (Portici), on travaille actuellement à la remise en ordre de l'immense collection de F. Silvestri et à en inventorier les éléments. On peut dans ce domaine, s'attendre à des surprises (Cf. J.M. Demange, 1971, *Découverte de spécimens cotypes de la collection du Musée de Dresde, entièrement détruite avec le Musée pendant la guerre*).

Il n'est pas dans nos intentions de proposer ici une structure particulière pour la réalisation de ce projet mais il paraît indispensable de constituer, dès maintenant si nos collègues le désirent, un "Comité International". Ce Comité aurait pour tout premier rôle de susciter des bonnes volontés et mettre au point les détails techniques de réalisation, notamment la manière dont seront présentés ces inventaires avant que d'en assurer l'exécution.

Si l'on insiste sur la création, au moins dans son principe, et la constitution d'un "Comité International", c'est pour donner à notre entreprise une forme officielle afin d'acquérir plus de poids et passer de l'abstrait au concret. La prise de position des spécialistes, réunis en Congrès International, marquant leur accord sur l'utilité d'une telle oeuvre d'une part, et le commencement de réalisation confirmant leur volonté d'aboutir d'autre part, ne peut que cautionner auprès d' organismes officiels nationaux et internationaux (U.N.E.S.C.O. par exemple) le sérieux nécessaire à une demande motivée, d'appuis financiers par exemple.

En conclusion, nous avons l'intention de demander à nos collègues réunis en Congrès au cours de la Séance réservée au C.I.M.:

1. s'ils approuvent le projet, au moins dans son principe;

2. si quelques uns d'entre eux souhaitent avoir une part active dans cette oeuvre collective et de longue haleine. Dans ce cas il semble indispensable *qu'ils veuillent bien se manifester dès la fin de la séance ou bien par lettre au C.I.M.*

Symp. zool. Soc. Lond. (1974) No. 32, 669–671.

English Translation

PLANS FOR A CENSUS OF THE INTERNATIONAL COLLECTIONS OF MYRIAPODS (IDENTIFIED AND NON-IDENTIFIED) AND ITS PUBLICATION

J.-M. DEMANGE

Who amongst us, be he a taxonomist or not, has not had to study certain specimens described by a predecessor, either because the description was published without illustrations, or was incomplete according to modern standards, or quite simply because a new examination is required to observe a certain character for revisionary reasons?

In such a situation the specialist is very often confronted by difficulties which are sometimes insurmountable and he is forced to undertake extremely lengthy research (wasting time) when it comes to discovering the museum which houses the collection he needs. He cannot be certain even then of finding the collection in a museum just because the author used information found there, or of uncovering the whole of the material he is seeking; the collection may have been sent to another museum or divided up for distribution among several specialists. Such is the case of the known collections, but who has not had to undertake lengthy and fastidious research in the museums of the world to gather together collections of non-identified myriapods, belonging either to a specific geographic region or, more particularly, to a chosen systematic group?

The ignorance in which we find ourselves today concerning the actual state of scientific material collected and conserved for generations throughout the world (material available for people doing research) and consequently the lack or partial lack of reference material, which is the basis of all research, cannot be tolerated in this modern age in which scientific knowledge is so important.

It is therefore proposed that the C.I.M., whose adherents represent more than three quarters of myriapodologists, should intervene and, if necessary, with the help of the specialists who are all directly concerned, take some initiative.

The proposed project basically consists of two stages:

1. The establishment of an inventory of the collections of myriapods (identified and non-identified) which are kept in scientific establishments

(museums, universities, foundations, etc.). The *sensu lato* types would naturally be indicated.

2. International agreement on eventual exchanges of this scientific material (consultation, lending, original studies, etc.).

The realization of this programme is really in the hands of the specialists, in other words, ourselves. It is vital that a willingness is shown, especially at the beginning, among those of our colleagues who already hold a position in this kind of establishment, and afterwards or simultaneously that *in each country* specialists form groups for the purpose of undertaking the necessary research for the realization of the plan *in their own particular country.*

In this way the inventories of the collections in the principal countries of the world, at least, can be quickly drawn up. If one consults the Bulletin of the C.I.M. and looks at the chapter on the world distribution of research workers, one sees that all these countries have among their scientists some myriapodologists.

The inventory of the museum collections (including some very rich ones—the Brölemann, Ribaut, Gervais, Chalande, Saussure and Zehntner (Madagascar) collections) was begun in Paris. It is hoped to draw up this list in a special publication of the C.I.M. before the next Myriapod Congress. In addition, thanks to the extreme kindness of Professor E. Tremblay, Director of the F. Silvestri Foundation (Portici), the vast collection of F. Silvestri is now being put in order and its elements are being categorized. Some surprises can be expected in this field (*cf.* J.-M. Demange, 1971, *Discovery of cotype specimens of the collection in Dresden Museum completely destroyed with the Museum during the War*).

It is not our intention here to propose a specific plan for the realization of this project, but it seems indispensable to form here and now an "International Committee". The Committee's first function would be to promote goodwill and to formulate the technicalities involved in the realization of the project, notably the method of presentation of the inventories, before assuring their execution.

If we emphasize the need for the creation and formation of an "International Committee", it is to give our enterprise an official form in order to carry more weight. An indication of the agreement of the specialists gathered at the Congress concerning the usefulness of such an undertaking on the one hand, and the beginning of the realization confirming their willingness to complete it on the other hand, are evidence of our serious intent when it comes to making requests, for example for financial support, from national and international bodies, such as U.N.E.S.C.O.

In conclusion we intend asking our colleagues gathered at the Congress during the course of the meeting set aside for the C.I.M.

1. whether they approve of the project at least in principle and,
2. if any among them wish to take an active part in this collective long-term enterprise. In this case it is essential that *they come forward at the meeting itself or indicate their willingness by letter to the C.I.M.*

Symp. zool. Soc. Lond. (1974) No. 32, 673–675.

RÉUNION PLÉNIÈRE DU CONGRÈS

Lundi 10 avril 1972, 16 heures

PRÉSIDENT: PROFESSEUR O. KRAUS

Un petit comité de membres du Congrès a été réuni par le Président dans la soirée précédente pour discuter de sujets soulevés par les rapports et propositions du C.I.M., présentés au Congrès par J.-M. Demange et J.-P. Mauriès le samedi 8 avril (voir pp. 657) et également pour discuter de l'organisation, du lieu et de la date du prochain Congrès. Le président demande au Professeur O. Kraus de prendre la présidence de cette réunion plénière et de porter les propositions devant les membres pour en délibérer et prendre les décisions que s'imposent.

Le secrétariat permanent du C.I.M.

Le comité propose que M. J.-M. Demange, M. J.-P. Mauriès et le professeur O. Kraus continuent à représenter le secrétariat permanent du C.I.M. et poursuivent l'excellent travail commencé à la création du C.I.M., quatre ans auparavant.

Catalogue mondial des collections de myriapodes

Le comité considère, nonobstant le soutien unanime du Congrès à la proposition de M. J.-M. Demange, qu'un catalogue mondial des collections soit élaboré; la proposition sera ultérieurement discutée en détail dans le but de donner pleins pouvoirs et autorité au C.I.M. pour lancer et réaliser le projet. Le Dr. Crabill se demande si la liste doit se rapporter aux espèces, taxa ou aux types. Le président émet l'opinion que les collections principales aient une priorité. M. Demange est d'accord et suggère que la liste des types soit réalisée plus tard. Le président fait allusion au prochain Congrès international de taxonomie et de biologie évolutive aux U.S.A. et suggère que le C.I.M. cherche une coopération des Musées par l'influence de ce plus grand corps. M. Demange propose qu'en plus du catalogue des collections le C.I.M. puisse utilement faciliter les échanges de matériel et demande l'approbation du Congrès pour cette extension de son activité. Les membres estiment que le C.I.M. n'a nullement besoin du consentement du Congrès et qu'une si utile activité est bien en rapport avec ses options originales.

Il a été *résolu* que:

Le second Congrès International de Myriapodologie souhaiterait que les myriapodologistes du monde entier aident le C.I.M. à recueillir toutes les informations nécessaires à l'établissement d'une liste mondiale des collections. Il souhaiterait également que toutes les institutions et les responsables de ces collections contribuent à cette oeuvre dans la mesure de leurs possibilités.

Correspondants régionaux du C.I.M.

Se référant au rapport de M. Mauriès révélant que le nombre de réponses aux questionnaires et les contributions financières du C.I.M. sont en baisse, le comité propose que certains membres, représentant un pays ou une région, soient sollicités pour agir comme intermédiaire entre les membres de son pays ou de sa région et le secrétariat du C.I.M. dans le but de hâter le regroupement des réponses et des contributions volontaires et d'économiser le coût de la distribution de la liste des publications. M. Demange rapporte que le tranfers des fonds par la poste peut occasionner des difficultés à quelques membres; il ouvrira un compte dans une banque à laquelle pourront être envoyées les contributions. Le président pense qu'un autre rôle du correspondant serait de recruter de nouveaux membres et de faire de la publicité pour le C.I.M. dans sa zone d'influence.

Il a été *résolu* que:

Le second Congrès International de Myriapodologie autorise le secrétariat du C.I.M. de s'efforcer à trouver et de désigner des myriapodologistes qui voudraient bien servir de "correspondants" dans leur pays ou dans leur région.

Comité permanent des Congrès

Le président estime que le comité d'organisation du présent Congrès a eu la chance d'avoir été aidé et appuyé non seulement par la Société zoologique de Londres, ce dont chacun lui est très reconnaissant, mais aussi par le secrétariat permanent du C.I.M. Le comité discute enfin de la façon dont sera menée l'organisation du prochain Congrès et il est considéré qu'un comité permanent sera chargé des responsabilités du prochain Congrès et portera assistance et donnera avis aux organisateurs locaux de la même manière que les présents organisateurs les reçurent du C.I.M. Après un accord de principe à la création d'un tel comité une discussion s'engage pour savoir si la composition du comité sera décidée avant la détermination du lieu du prochain Congrès.

Il est finalement décidé de former le Comité en premier car il est considéré comme plus important d'avoir à sa disposition un corps responsable pour assurer la continuité avant que de s'engager dans l'étude des détails du prochain Congrès.

Il a été *résolu* que:

> Le second Congrès International de Myriapodologie crée un comité permanent durant la période comprise entre ce présent Congrès et le prochain, dans le but de conseiller et d'aider les organisateurs locaux du prochain Congrès en ce qui concerne les questions importantes. Le comité sera composé du secrétariat du C.I.M., M. J. G. Blower et de l'organisateur du prochain Congrès.

Le troisième Congrès international de Myriapodologie

Après discussion relative aux différents mérites d'une périodicité de trois ans, intention primitive, ou de quatre ans laps de temps présentement écoulé, il a été décidé que trois ans est le meilleur intervalle. Le président annonce que deux invitations ont été faites avant la séance par des membres, une pour la Belgique (Gembloux) et une pour l'Allemagne (Hambourg). Le comité considérant que les deux premiers Congrès ont été tenus dans des pays de langue française et de langue anglaise, accepte l'invitation de l'Allemagne pour 1975 et retient l'invitation de la Belgique pour 1978. Les membres du Congrès, à l'exception de ceux des pays retenus, votent en de l'Allemagne et de Hambourg ou se tiendra le 3e Congrès en 1975.

Résumés

Le professeur Sahli suggère que les détails des travaux, devant être présentés au Congrès soient distribués avant la réunion afin que les membres possèdent à l'avance des informations détaillées sur le sujet qui sera exposé. Le professeur Kraus le remercie pour cette suggestion et espère que cela sera possible.

La réunion est terminée à 17.05 h.

Symp. zool. Soc. Lond. (1974) No. 32, 677–679.

English Translation

PLENARY MEETING OF CONGRESS

Monday, 10th April 1972, 1600 h

CHAIRMAN: PROFESSOR OTTO KRAUS

A small committee of members of Congress had been called together by the President the previous evening to discuss matters arising from the report and proposals of the Centre International de Myriapodologie presented to Congress by J.-M. Demange and J.-P. Mauriès on Saturday, 8th April (see pp. 669) and also to discuss the organization, venue and date of the next Congress. The President asked Professor Kraus to take the chair at this plenary meeting and to place the committee's proposals before members for their consideration and action.

The permanent secretariat of C.I.M.

The committee proposed that Monsieur J.-M. Demange, Monsieur J.-P. Mauriès and Professor O. Kraus should continue to serve as the permanent secretariat of C.I.M. and to continue the excellent work which they began at the inception of C.I.M. just four years ago.

World list of collections of Myriapoda

The committee considered that, notwithstanding the unanimous support of Congress for the earlier proposal of M. Demange that a world list of collections be prepared, the proposal should be further discussed and itemized in order to give full support and authority to C.I.M. to launch and to realize the project. Dr Crabill wondered whether the list was to refer to species, taxa or types. The chairman was of the opinion that the major collections should be listed first. M. Demange agreed but suggested that perhaps the listing of types might be attempted later. The chairman referred to the forthcoming International Congress on Taxonomy and Evolutionary Biology in the United States and suggested that the C.I.M. might seek to gain the co-operation of Museums through the influence of this larger body. M. Demange proposed that in addition to compiling a list of preserved material, the C.I.M. might usefully facilitate the exchanges of material and requested the approval of Congress for this extension of their activity. It was thought by members that the C.I.M. did not need the permission of Congress and that such activity was well within their original terms of reference.

It was RESOLVED that

> The Second International Congress of Myriapodology recommends that myriapodologists throughout the world help the C.I.M. to bring together the information needed for a world-list of collections in this field. It further recommends institutions and curators in charge of such collections to contribute as far as possible.

Regional correspondents of C.I.M.

Referring to the report of M. Mauriès that the number of replies to the questionnaires and the financial contributions to C.I.M. had been declining, the committee proposed that certain members representing a country or region be requested to act as intermediaries between members of their country or region and the secretariat of C.I.M. in order to expedite the collection of replies and financial contributions and to economize in the cost of distributing the List of Publications. M. Demange reported that transfer of funds to a post centre might prove difficult for some members and that he would open a bank account to which contributions could be sent. The chairman thought that another useful function of the correspondent would be to recruit new members and to publicize C.I.M. in his own area.

It was RESOLVED that

> The Second International Congress of Myriapodology authorizes the secretariat of the C.I.M. to try to find and to appoint myriapodologists who are willing to serve as "correspondents" for special countries or regions.

Standing Congress Committee

The chairman said that the organizing committee of the present Congress had been fortunate to have had the help and support, not only of the Zoological Society of London, to whom all were very grateful, but also of the permanent secretariat of the C.I.M. The committee discussed at length how the organization of the next Congress should be handled and it was considered that there should be a standing committee which should be charged with the responsibility for the next Congress and to render such assistance and advice to the local organizers as the present organizers had received from C.I.M. Having agreed in principle to the setting-up of such a committee there was discussion as to whether its composition should be decided before considering where to hold the next congress. It was finally decided to form the committee first, since it was considered more important to have a body responsible for continuity before dealing with the details of further congresses.

It was RESOLVED that

> The Second International Congress of Myriapodology appoints a standing committee for the period between this and the forthcoming Congress in order to give advice to the local organizers of the next Congress on questions of principal importance. This commmittee shall consist of the secretariat of the C.I.M., Mr. J. G. Blower, and the organiser of the next congress.

The Third International Congress of Myriapodology

After some discussion of the relative merits of a period of three years, which had been the original intention, or of four years, which had actually elapsed, it was decided that three years was the better interval. The chairman said that two invitations to the Third Congress were before members, one from Belgium (Gembloux) and one from Germany (Hamburg). The committee had considered that since the first two Congresses had been held in French-speaking and English-speaking countries, the invitation from Germany should be accepted for 1975 and the invitation from Belgium be considered for 1978. Congress members, except those from the interested countries, voted in favour of accepting the invitation from Germany to hold the Third Congress in Hamburg in 1975.

Abstracts

Professor Sahli suggested that details of papers to be presented at the Congress should be circulated previous to the meeting to enable members to have more detailed advance information of the areas to be covered. Professor Kraus thanked him for the suggestion and hoped that this would be possible.

The meeting ended at 17.05 h.

AUTHOR INDEX

Numbers in italics refer to pages which include a full reference to an original paper. Numbers within brackets refer to pages which include contributions to the discussions following the papers. Authorities of systematic categories and persons mentioned in the acknowledgements are not included in this index.

E

F

G

H

R

S

T

SUBJECT INDEX

Page numbers in italics refer to a figure or table; bold type indicates the first page of an extensive treatment.

SYSTEMATIC INDEX

Names of families, sub-families and tribes are printed in roman type, genera and species in italic. Non-myriapod species are indicated by an asterisk. Page numbers against the generic name include a non-specific reference to the genus. Page numbers in italics refer to a figure or a table, numbers in bold type indicate the first page of an extensive treatment.

CLASSIFICATION OF THE MYRIAPODA

The purpose of this scheme is to enable the non-specialist to place, in a general context, the various genera and families mentioned throughout this volume. It is *not* offered as a definitive classification. The diplopod scheme follows closely that of Jeekel (1970)* but some well-used older names are added, ostensibly to give clarity, though perhaps at the expense of systematic respectability.

All major groups are included but only the families and genera referred to in the text. Groups receiving major systematic treatment in this volume are indicated by appropriate page numbers—to which please refer for further details.

Class PAUROPODA

Pauropidae *Pauropus, Allopauropus*, (s.g. *Decapauropus*), *Cauvetauropus, Stylopauropus, Polypauropus*

Class DIPLOPODA

Sub-class PENICILLATA (PSELAPHOGNATHA, SCHIZOCEPHALA) Bristly millipedes, without calcified cuticle

Order POLYXENIDA

Polyxenidae *Polyxenus*

Sub-class CHILOGNATHA with calcified cuticle

Super-order PENTAZONIA (OPISTHANDRIA) Free pleurites and sternites, posterior gonopods.

ONISCOMORPHA

Pill millipedes

Order GLOMERIDA (PLESIOCERATA) Mainly Palaearctic pills, 11 or 12 tergites

Glomeridae *Glomeris, Loboglomeris, Spelaeoglomeris, Haploglomeris, Trachysphaera*

Glomeridellidae *Glomeridella, Protoglomeris*

Order SPHAEROTHERIIDA (CHORIZOCERATA) Giant pills, 13 tergites, see pp. 41–5

Sphaerotheriidae *Sphaerotherium*

Sphaeropoeidae

* Jeekel, C. A. W. *Nomenclator generum et familiarum Diplopodorum*: A List of the genus and family-group names in the Class Diplopoda from the 10th edition of Linnaeus, 1758, to the end of 1957. Amsterdam 1970.

LIMACOMORPHA

Pentazonian, but not pills, 22 tergites

Order GLOMERIDESMIDA
- Glomeridesmidae
- Zephroniodesmidae

Super-order HELMINTHOMORPHA (PROTERANDIA) Pleurites fused to tergites, anterior gonopods.

TRIZONIA (NEMATOPHORA)

Free sternites, silk-spinning millipedes.

Order CRASPEDOSOMATIDA

Sub-order CRASPEDOSOMATIDEA (CHORDEUMIDA, ASCOSPERMOPHORA)

Short-bodied, often with tergal keels, quadrangular (flat-backs), 26, 28, 30 or 32 rings.

Craspedosomidae	*Craspedosoma, Polymicrodon*
Brachychaeteumidae	*Brachychaeteuma*
Chordeumidae	*Chordeuma, Microchordeuma*
Haplobainosomidae	*Cantabrosoma, Turdulisoma, Aragosoma, Pyreneosoma*
Attemsiidae	
Anthogonidae	

Sub-order LYSIOPETALIDEA (CALLIPODOIDEA)

long-bodied, 'Juliform', Mediterranean, 40 + rings

Order STEMMIULIDA Long-bodied, Juliform, tropical, 39–60 rings

Stemmiulidae *Diopsiulus*

MONOZONIA

Sternites, pleurites and tergites fused together into a complete cylindrical sclerite.

Order POLYDESMIDA (PROTEROSPERMOPHORA)

Short-bodied quadrangular (flat-backed) monozonia, 19, 20, 21 or 22 rings, usually 20

Sub-order PARADOXOSOMATIDEA (STRONGYLOSOMOIDEA)

Paradoxosomatidae	(=Strongylosomidae) *Oxidus, Orthomorpha, Habrodesmus, Xanthodesmus, Entothalassinum*
Pratinidae	*Jonespeltis*

Sub-order POLYDESMIDAE

Polydesmidae	*Polydesmus, Brachydesmus, Ophiodesmus, Macrosternodesmus*
Pterodesmidae	*Aporodesmus*

Sub-order SPHAERIODESMIDEA

Platyrrhacidae	*Platyrrhacus, Cantabrodesmus?*
Euryuridae	
Oxydesmidae	*Coromus*
Gomphodesmidae	*Ulodesmus, Tymbodesmus, Sphenodesmus*

JULIFORMIA*

long-bodied cylindrical monozonia

Order JULIDA* (SYMPHYOGNATHA, OPISTHOSPERMOPHORA)

Sub-order PARAIULIDEA (BLANIULIDEA) mainly holarctic

Paraiulidae, see pp. 23–39

Mongoliulidae

Blaniulidae (Nemasomidae) *Blaniulus, Proteroiulus, Archeboreoiulus, Boreoiulus, Nopoiulus, Isobates.*

Sub-order JULIDAE* mainly palaearctic

Julidae* *Julus,* Ophyiulus, Ommatoiulus* (=*Schizophyllum*), *Cylindroiulus, Leptoiulus, Leptophyllum, Brachyiulus, Tachypodoiulus*

Order SPIROBOLIDA mainly pan-tropical

Spirobolidae *Spirobolus, Narceus*

Rhinocricidae *Chersastus, Rhinocricus*

Pachybolidae *Aulacobolus, Pachybolus*

Trigoniulidae

Spiromimidae

Order SPIROSTREPTIDA mainly pan-tropical, especially African

Sub-order SPIROSTREPTIDAE

Spirostreptidae *Graphidostreptus, Alloporus*

Odontopygidae *Odontopyge, Peridontopyge, Tibiomus, Haplothysanus*

Harpagophoridae *Harpurostreptus, Ktenostreptus, Poratophilus*

Sub-order CAMBALIDAE

Pseudonannolenidae

POLYZONIA

Various unions of tergites, pleurites and sternites into a more-or-less complete ring.

Super-order COLOBOGNATHA

Order POLYZONIIDA

Class CHILOPODA

EPIMORPHA hatching with full number of segments

Order SCOLOPENDROMORPHA usually cursorial epimorphs with 21 or 23 pairs of legs.

Scolopendridae *Scolopendra, Ethmostigmus, Rhysida, Asanada*

Cryptopidae *Cryptops*

Order GEOPHILOMORPHA Burrowing epimorphs, 35 or more pairs of legs.

Himantariidae *Haplophilus, Orya, Himantarium*

Geophilidae *Geophilus, Strigamia*

* The spelling of *Julus*, Julidae, Julidea and Julida is correct according to Jeekel (1970) *loc. cit.*, but the alternative 'I' spelling remains in places throughout the volume.

ANAMORPHA hatching with fewer than adult numbers of segments cursorial, with 15 pairs legs.

Order LITHOBIOMORPHA see pp. 65–73 and 75–87

Lithobiidae

Gosibiidae

Henicopidae

Order SCUTIGEROMORPHA see pp. 89–98

Scutigeridae

Family of uncertain affinity:

Craterostigmidae see pp. 182–187

Class SYMPHYLA see pp. 168–173

Scutigerellidae *Scutigerella, Hanseniella*

Scolopendrellidae *Symphylella*